Coulson and Richardson's Chemical Engineering

T0295379

Coulson and Richardson's Chemical Engineering

Volume 2B: Separation Processes

Sixth Edition

Edited by

Ajay Kumar Ray

Chemical and Biochemical Engineering Department,
Western University, London, ON, Canada

Butterworth-Heinemann
An imprint of Elsevier

Butterworth-Heinemann is an imprint of Elsevier
The Boulevard, Langford Lane, Kidlington, Oxford OX5 1GB, United Kingdom
50 Hampshire Street, 5th Floor, Cambridge, MA 02139, United States

Notices
Knowledge and best practice in this field are constantly changing. As new research and experience broaden our understanding, changes in research methods, professional practices, or medical treatment may become necessary.

Practitioners and researchers must always rely on their own experience and knowledge in evaluating and using any information, methods, compounds, or experiments described herein. In using such information or methods they should be mindful of their own safety and the safety of others, including parties for whom they have a professional responsibility.

To the fullest extent of the law, neither the Publisher nor the authors, contributors, or editors, assume any liability for any injury and/or damage to persons or property as a matter of products liability, negligence or otherwise, or from any use or operation of any methods, products, instructions, or ideas contained in the material herein.

ISBN: 978-0-08-101097-6

For information on all Butterworth-Heinemann publications
visit our website at https://www.elsevier.com/books-and-journals

Publisher: Susan Dennis
Acquisitions Editor: Anita Koch
Editorial Project Manager: Kyle Gravel
Production Project Manager: Bharatwaj Varatharajan
Cover Designer: Victoria Pearson

Typeset by STRAIVE, India

Working together
to grow libraries in
developing countries

www.elsevier.com • www.bookaid.org

Contents

Contributors

Raj Barchha
Department of Chemical and Biochemical Engineering, University of Western Ontario, London, ON, Canada

Amarjeet S. Bassi
Chemical and Biochemical Engineering, Western University, London, ON, Canada

Arvind Rajendran
Department of Chemical and Materials Engineering, University of Alberta, Edmonton, AB, Canada

Ajay Kumar Ray
Department of Chemical and Biochemical Engineering, University of Western Ontario, London, ON, Canada

Yan Zhang
Process Engineering, Memorial University of Newfoundland, St John's, NL, Canada

About Professor Coulson

John Coulson, who died on 6 January 1990 at the age of 79, came from a family with close involvement with education. Both he and his twin brother Charles (a renowned physicist and mathematician), who predeceased him, became professors. John completed his undergraduate studies at Cambridge and then moved to Imperial College where he took the postgraduate course in chemical engineering—the normal way to qualify at that time—and then carried out research on the flow of fluids through packed beds. He then became an assistant lecturer at Imperial College and, after wartime service in the Royal Ordnance Factories, returned as lecturer and was subsequently promoted to a readership. At Imperial College, he initially had to run the final year of the undergraduate course almost single-handed, a very demanding assignment. During this period, he collaborated with Sir Frederick (Ned) Warner to write a model design exercise for the Institution of Chemical Engineers (IChemE) home paper on 'The Manufacture of Nitrotoluene'. He published research papers on heat transfer and evaporation, distillation, and liquid extraction and co-authored this textbook of chemical engineering. He carried out valiant work for IChemE that awarded him its Davis medal in 1973 and was also a member of the advisory board for what was then a new Pergamon journal, *Chemical Engineering Science*.

In 1954, he was appointed to the newly established chair at Newcastle upon Tyne, where chemical engineering became a separate department and independent of mechanical engineering of which it was formerly a part and remained there until his retirement in 1975. He took a period of secondment with Heriot Watt University where, following the splitting of the joint Department of Chemical Engineering with Edinburgh, he acted as adviser and de facto head of department. The Scottish university awarded him an honorary DSc in 1973.

John's first wife Dora sadly died in 1961; they had two sons, Anthony and Simon. He remarried in 1965 and is survived by Christine.

John F. Richardson

About Professor Richardson

Professor **John Francis Richardson**, Jack to all who knew him, was born at Palmers Green, North London, on 29 July 1920 and attended the Dame Alice Owen's School in Islington. Subsequently, after studying chemical engineering at Imperial College, he embarked on research into the suppression of burning liquids and fires. This early work contributed much to our understanding of the extinguishing properties of foams, carbon dioxide, and halogenated hydrocarbons, and he spent much time during the war years on large-scale fire control experiments in Manchester and at the Llandarcy Refinery in South Wales. At the end of the war, Jack returned to Imperial College as a lecturer where he focussed on research in the broad area of multiphase fluid mechanics, especially sedimentation and fluidisation, two-phase flow of a gas and a liquid in pipes. This laid the foundation for the design of industrial processes like catalytic crackers and led to a long-lasting collaboration with the Nuclear Research Laboratories at Harwell. This work also led to the publication of the famous paper, now common knowledge, the so-called Richardson-Zaki equation, which was selected as the Week's citation classic (*Current Contents*, 12 February 1979).

After a brief spell with Boake Roberts in East London, where he worked on the development of novel processes for flavours and fragrances, he was appointed professor of chemical engineering at the then University College of Swansea (now University of Swansea) in 1960. He remained there until his retirement in 1987 and thereafter continued as an emeritus professor until his death on 4 January 2011.

Throughout his career, his major thrust was on the well-being of the discipline of chemical engineering. In the early years of his teaching duties at Imperial College, he and his colleague John Coulson recognised the lack of satisfactory textbooks available in the field of chemical engineering. They set about rectifying the situation, and this is how the now well-known Coulson-Richardson series of books on chemical engineering was born. The fact that this series of books (six volumes) is as relevant today as it was at the time of their first appearance is a testimony to the foresight of John Coulson and Jack Richardson.

Throughout his entire career spanning almost 40 years, Jack contributed significantly to all facets of professional life, teaching, research in multiphase fluid mechanics, and service to the Institution of Chemical Engineers (IChemE, the United Kingdom). His professional work and long-standing public service was well recognised. Jack was the president of IChemE during the period 1975–76; he was named a fellow of the Royal Academy of Engineering in 1978 and was awarded an OBE in 1981.

In his spare time, Jack and his wife Joan were keen dancers, having been founder members of the Society of International Folk Dancing, and they also shared a love of hill walking.

Raj Chhabra

Introduction

Welcome to the next generation of the Coulson–Richardson series of books on chemical engineering. I would like to convey my feelings about this series of books that have evolved over the past 30 years and are based on numerous conversations with Jack Richardson himself (1981 onwards until his death in 2011) and with some of the other contributors to previous editions, including Tony Wardle, Ray Sinnott, Bill Wilkinson, and John Smith. So what follows here is the essence of these interactions combined with what the independent (solicited and unsolicited) reviewers had to say about this series of books on several occasions.

The Coulson–Richardson series of books has served the academia, students, and working professionals extremely well since their first publication more than 50 years ago. This is a testimony to their robustness and, to some extent, their timelessness. I have often heard much praise, from different parts of the world, for these volumes both for their informal and user-friendly yet authoritative style and for their extensive coverage. Therefore, there is a strong case for continuing with its present style and pedagogical approach.

On the other hand, advances in our discipline in terms of new applications (energy, bio, microfluidics, nanoscale engineering, smart materials, new control strategies, and reactor configurations, for instance) are occurring so rapidly and in such a significant manner that it will be naive, even detrimental, to ignore them. Therefore, although we have tried to retain the basic structure of this series, the contents have been thoroughly revised. Wherever the need was felt, the material has been updated, revised, and expanded as deemed appropriate. Therefore, the reader, whether a student, a researcher, or a working professional, should feel confident that what is in the book is the most up-to-date, accurate, and reliable piece of information on the topic he/she is interested in.

Evidently, this is a massive undertaking, which cannot be managed by a single individual. Therefore, we now have a team of volume editors responsible for each volume, with individual chapters being written by experts in some cases. I am most grateful to all of them for having joined us in this endeavour. Furthermore, based on extensive deliberations and feedback from a large number of individuals, some structural changes were deemed appropriate as detailed here. Due to their size, each volume has been split into two sub-volumes as follows:

Volume 1A: Fluid Flow
Volume 1B: Heat and Mass Transfer
Volume 2A: Particulate Technology and Processing
Volume 2B: Separation Processes
Volume 3A: Chemical Reactors
Volume 3B: Process Control

Undoubtedly, the success of a project of such a vast scope and magnitude hinges on the cooperation and assistance of many individuals. In this regard, we have been extremely fortunate in working with some of the outstanding individuals at

Butterworth-Heinemann, a few of whom deserve to be singled out: Jonathan Simpson, Anita Koch, Fiona Geraghty, Maria Convey, Ashlie Jackman, Joshua Bayliss, Afzal Ali, and Kyle Gravel, who have taken personal interest in this project and have come to our rescue whenever needed, going much beyond the call of duty.

This series has had a glorious past, but I sincerely hope that its future will be even brighter by presenting the best possible books to the global chemical engineering community for the next 50 years, if not for longer. I sincerely hope that the new edition of this series will meet (if not exceed) your expectations! Finally, a request to the readers, please continue to do the good work by letting me know if, no not if, but when you spot a mistake so that it can be corrected at the first opportunity.

Raj Chhabra
Editor-in-Chief
Dadri, Greater Noida, September 2022

Introduction on separation processes

Separation of mixtures represents an important genre of operation in the chemical process industry. Some mixtures are responsive to separation entirely by mechanical means. Typical examples are separation of mixtures of solid particles of different sizes by screening; separation of small suspended particles from a gas by a cyclone or filter bag; separation of suspended solids from a liquid by filtration, settling, or centrifugation; or separation of two immiscible liquids by phase separation followed by decantation. However, there are many other mixtures, like gas or liquid mixtures (solutions), that cannot be separated by any of the preceding mechanical methods. The approach for separation of such mixtures is based on the use of an externally supplied means. Typical examples are use of thermal energy in distillation for separation of more volatile components from a solution; use of water to separate ammonia from a mixture of another gas by absorption; or the separation of an organic vapour from a mixture with another gas by adsorption. The separation processes of this kind are based on the principles of mass transfer and have traditionally been called mass transfer operations.

Mass transfer is the transport of a species from one point to another in a single phase or from one phase to another generally in the presence of a difference in concentration (or partial pressure), called driving force. There are numerous examples of this phenomenon in daily life. Aquatic life uses oxygen dissolved in water for survival and the supply of oxygen mostly comes from air. The concentration of oxygen in natural water is less than what it should be at saturation or at equilibrium. As a result, oxygen is absorbed in the water of lakes, rivers, and oceans. The phenomenon of transport of oxygen from air to water is a mass transfer process since it is caused by a concentration driving force. The driving force gives a measure of how far a system is away from equilibrium. The larger the departure of a system from equilibrium, the greater is the driving force and the higher is the rate of transport. Water exposed to air will absorb oxygen faster if it has a low concentration of oxygen.

Mass transfer may be diffusional or convective. A purely diffusional xmass transfer phenomena occurs in the absence of any bulk motion in the medium. The mitigation of moisture within a grain during drying is purely diffusional. The transport of a reactant or a product through the pores of a catalyst pellet occurs by diffusion. When mass transfer occurs in a fluid medium which is in some sort of motion, the rate of mass transfer increases greatly. This is convective mass transfer, although molecular diffusion has an inherent role to play. The stronger the flow field and mixing and turbulence in a medium, the higher is the rate of mass transfer.

Mass transfer is a transport process. It is one of the three major transport processes (along with momentum and heat transfer) central to chemical engineering operations. One notable difference between heat and mass transfer operations is that in most practical heat transfer operations transport occurs through indirect contact.

The two phases exchanging heat are separated by a thermally conductive medium. On the other hand, most mass transfer operations occur through direct contact. The phases exchanging mass through direct contact are either immiscible or partially miscible. If the phases are miscible, a selectively permeating barrier is placed between them to effect mass transfer without physical mixing of the phases. Membrane gas separation is an example of selective transport through a permeating barrier and is an indirect contact process. Another important difference between heat and mass transfer is the fact that the two phases in thermal equilibrium have the same temperature, but the two phases in equilibrium in mass transfer do not necessarily have the same concentration of the solute. In fact, in most cases, they do not have the same concentration. The concentrations of the species in the two phases have a thermodynamic relation called the *equilibrium relation*. The simplest equilibrium relation is a linear relation of Henry's law type. Moreover, the driving force in mass transfer is expressed in many different ways depending upon convenience, in contrast to heat transfer, in which the driving force is always expressed in terms of temperature difference. Also, mass transfer is likely to affect the flow rates of the phases as well as the molecular motion near the phase boundary. All these factors make the principles of mass transfer and their applications seemingly a bit more involved than heat transfer counterparts.

The principles of mass transfer, both diffusional and convective, form the basis of the separation processes (a technique that transforms a mixture of substances into two or more products differing in composition) used in the chemical process industries. In this textbook, only concentration-driven separation processes are considered and not those involving mechanical separation processes. To understand the individual concentration-driven separation processes, the principles of mass transfer, essentially the principles of diffusion and convective mass transport, must be understood. The principles of mass transfer from one phase to another and these how principles are applied to a number of separation processes that are described in various chapters of the book. Many separation processes or mass transfer operations are carried out in a stage-wise fashion in chemical process industries. Separation of mixtures accounts for about 40% to 70% of both capital and operating costs of a chemical process industry. The cost of separation of high-value products from a dilute solution (as in the case of recovery and concentration of bio-products, for example active pharmaceutical ingredients) may entail up to 90% of the operating cost of a plant. Separation processes are necessary in almost every stage, from purification of raw materials to product separation and treatment of effluent streams. The core separation processes in the chemical industry are gas absorption and stripping, distillation, liquid–liquid and solid–liquid extraction, drying of a wet solid, adsorption, crystallisation, membrane separation, and separation of multi-component mixtures. All these separation processes involve mass transfer from one phase to another. Such two-phase systems may be categorised as gas (or vapour)–liquid, liquid–liquid, gas–solid, and liquid–solid.

Separation of a soluble species from a gas mixture by using a solvent is called gas absorption. This is a gas–liquid contacting operation. A typical application is the separation of CO_2 from the ammonia synthesis gas using a solvent such as aqueous ethanolamine or carbonate-bicarbonate buffer. The absorbed gas is recovered by 'stripping' when the solvent also is regenerated and is reused. The most common method of separation of a liquid mixture is fractional distillation. This technique relies upon the difference in volatility of a component over another in the mixture. Heat is supplied to vaporise a part of the liquid which flows up through a properly designed column and enters into intimate contact with a down-flowing liquid when the exchange of the component occurs. The more volatile components migrate into the vapour phase and the less volatile ones are transferred into the liquid phase. The concentration of the more volatile ones increases up the column and that of the less volatiles increases down the column. The top product, which comes up as a vapour, is condensed. The less volatile part is taken out as the bottom product. If some or all the components have low volatility and are thermally unstable in a liquid mixture or solution, it becomes convenient to introduce a suitable liquid, called a solvent, into the separation device to extract the target compounds. This is liquid–liquid extraction. The desired species is recovered from the extract by using another separation technique, such as distillation or crystallisation. For example, penicillin is extracted from the fermentation broth by using an ester such as butyl acetate and the final product recovery is done by crystallisation.

One type of solid–liquid contact operation aims at solubilising or extracting a target substance from a solid matrix. This is solid–liquid extraction, or leaching. Some typical examples are acid-leaching of ores, or extraction of oil from flaked oil seeds or of fragrance from flowers. Another type of solid–liquid operation involves simultaneous separation and purification of a substance from a supersaturated solution by crystallisation. A dissolved solute, particularly at a low concentration, can be conveniently separated by using a solid agent that adsorbs the target substance. The solid adsorbent loaded with the solute is now further treated to recover the product. Here, the adsorbent plays the same role as a solvent does in gas absorption. Another important case of solid–liquid contact is the separation of an ionic species from a solution by a solid ion exchange resin.

Separation of a solute from a gas mixture can sometimes be done by adsorption in a solid material. The adsorbed solute is frequently recovered by thermal stripping. Another important application of gas–solid contacting is for drying of moist solids using a drying gas. The hot gas is brought in intimate contact with the solid when the moisture leaves the solid and migrates into the gas as a vapour. A common application is the separation of organic vapours present in low concentration in a gaseous emission by using a bed of active carbon. Gas–solid contact also occurs in a less common separation like gas chromatography.

From all of the preceding descriptions, it appears that a separation process needs a 'separating agent' to split a mixture into more than two streams of different concentrations. For example, a gas mixture is separated when it is in contact with a solvent that selectively or preferentially absorbs one or more components. Here the solvent is the separating agent. The separating agent is left out after a job is completed. Thermal energy is the separating agent in distillation. It is supplied for vaporisation of a liquid mixture and is removed from the top vapour when it is condensed. In liquid–liquid extraction, the externally added solvent is the separating agent. It is removed from the extract to recover the product. A hot gas is the common separating agent in drying and an adsorbent is the agent for separation by adsorption.

A good understanding of the construction and operating principles of separation equipment is essential for an overall grasp of the subject area. Considerable space has been devoted in respective chapters to address these aspects. Recent developments in equipment have been included as far as possible. The equipment selection criteria from several available alternatives for a particular separation job have also been discussed. The procedure of equipment design and sizing has been illustrated by simple examples. Since gas–solid and vapour–liquid contacting, particularly distillation, is more frequent in process industries, an entire chapter has been devoted to the description of the basic features of the relevant equipment, including recent modifications.

Separation using a membrane as the separating agent or separating barrier is now an established strategy. Spectacular developments have taken place in membrane casting and module design to exploit the separation capability of membranes for a variety of applications. An overview has been given of different applications and aspects of membrane separation.

Humidification and water cooling are necessary in every process industry. This operation cannot be considered a separation process since the basic objective is not the separation of a stream. However, it is a very useful practical example of simultaneous heat and mass transfer. During the last few decades major developments have taken place in several new areas such as materials, biotechnology, pharmaceuticals, and alternative fuels. These developments are accompanied by challenging separation problems, often beyond the scope of traditional mass transfer operations. Improvements in selectivity of separation by using better separating agents, concentration of high-value products from dilute solutions, increasing efficiency and reducing equipment size, and reducing energy requirements are a few of these challenges. In addition, growing concern for the environment and rising energy costs are two major factors having significant influence on reshaping separation processes. From a broader perspective, the two important present-day criteria for evaluation of a process, whether new or conventional, are whether it is 'sustainable' and 'green'. Alongside development and adoption of green technologies has come the idea of 'green separation' processes. Some of their attributes have been mentioned throughout several chapters in this book. So far as the chemical process industries are concerned, green technologies and green separation processes are complementary.

Distillation

Raj Barchha and Ajay Kumar Ray

Department of Chemical and Biochemical Engineering, University of Western Ontario, London,
ON, Canada

Nomenclature

		Units in SI system	Dimensions in M, N, L, T, θ
A	component **A**	—	—
A	cross-sectional area of column	m^2	$\mathbf{L}^2$
A_{12}, A_{21}	Margules and Van Laar constants	—	—
A_D	Margules third constant	—	—
A	interfacial surface per unit volume of column	m^2/m^3	$\mathbf{L}^{-1}$
B	component **B**	—	—
B'	constant in Eq. (1.170)	—	—
B''	constant in Eq. (1.171)	—	—
b_m, b_n	factors in Eq. (1.113)	—	—
C	concentration in moles/unit volume	$kmol/m^3$	$\mathbf{NL}^{-3}$
C	total annual cost	£/year	
c_a	annual cost of column per unit area of plate	£/m²year	
c_b	annual cost of heat exchanger equipment per unit area including depreciation and interest	£/m²year	
c_c	annual cost of column per kmol of distillate	£/kmol year	
c_d	annual cost of steam and cooling water	£/year	
c_h	annual cost of reboiler and condenser	£/year	

Continued

Coulson and Richardson's Chemical Engineering. https://doi.org/10.1016/B978-0-08-101097-6.00001-8

		Units in SI system	Dimensions in M, N, L, T, θ
c_w	annual variable cost	£/year	
c_1	function in Eq. (1.11)	K	θ
c_2	function in Eq. (1.11)	—	—
c_5	function in Eq. (1.12)	—	—
D	moles (or mass) of product per unit time	kmol/s or (kg/s)	NT^{-1} (MT^{-1})
D_b	moles of product in batch distillation	kmol	N
D_L	diffusivity in the liquid phase	m²/s	$L^2 T^{-1}$
D_v	diffusivity in the vapour phase	m²/s	$L^2 T^{-1}$
D	bubble diameter	m	L
d_c	column diameter	m	L
d_r	diameter of ring	m	L
E	average overall plate efficiency, n/n_p	—	—
E_a	plate efficiency allowing for entrainment e	—	—
E_m	local Murphree plate efficiency	—	—
E_M	average Murphree plate efficiency	—	—
E	fractional voidage of packing	—	—
e'	entrainment (moles per unit time and unit cross-section)	kmol/m²s	$NL^{-2} T^{-1}$
F	cost factor in Eq. (1.67)	£/year	
F	molar or mass feed per unit time	kmol/s, kg/s	NT^{-1}, MT^{-1}
F_{lv}	parameter $L'/G'\sqrt{(\rho_v/\rho_L)}$	—	—
$\bar{F}$	Parameter $u\sqrt{\rho_v}$ used in Eq. (1.137)	kg$^{1/2}$/m$^{1/2}$s	$M^{1/2}L^{-1/2}T^{-1}$
G'	molar flow rate of vapour per unit time and unit cross-section	kmol/m²s	$NL^{-2} T^{-1}$
G	acceleration due to gravity	m/s²	LT^{-2}
H	enthalpy per mole or unit mass	J/kmol, J/kg	$MN^{-1}L^2T^{-2}$, L^2T^{-2}
H^L	enthalpy per mole or unit mass of liquid	J/kmol, J/kg	$MN^{-1}L^2T^{-2}$, L^2T^{-2}
H^V	enthalpy per mole or unit mass of vapour	J/kmol, J/kg	$MN^{-1}L^2T^{-2}$, L^2T^{-2}
H'_d	$H^L_d+(Q_c/D)$	J/kmol, J/kg	$MN^{-1}L^2T^{-2}$, L^2T^{-2}
H'_w	$H^L_w-(Q_B/W)$	J/kmol, J/kg	$MN^{-1}L^2T^{-2}$, L^2T^{-2}
H	height of transfer unit	m	L
$\mathbf{H}_G$	height of transfer unit—gas film	m	L
$\mathbf{H}_{OG}$	height of transfer unit—overall (gas concentrations)	m	L

Continued

		Units in SI system	Dimensions in M, N, L, T, θ
H_L	height of transfer unit—liquid film	m	**L**
H_{OL}	height of transfer unit—overall (liquid concentration)	m	**L**
H'	Henry's constant (P_A/x_A)	N/m^2	$\mathbf{ML^{-1}T^{-2}}$
J	rate of change in composition with height for unit driving force	m^{-1}	$\mathbf{L^{-1}}$
j_d	j-factor for mass transfer	—	—
K	equilibrium constant (y/x) or coefficient	—	—
K_g	overall mass transfer coefficient	m/s	$\mathbf{LT^{-1}}$
K_l	overall mass transfer coefficient (mol/unit time-unit area-unit mole fraction driving force)	kmol/m^{2}s	$\mathbf{NL^{-2}T^{-1}}$
K_l	overall mass transfer coefficient	m/s	$\mathbf{LT^{-1}}$
K'_l	overall mass transfer coefficient	kmol/m^{2}s	$\mathbf{NL^{-2}T^{-1}}$
k_g, k'_g	film coefficients corresponding to K_g, K'_g	m/s, kmol/m^{2}s	$\mathbf{LT^{-1}}, \mathbf{NL^{-2}T^{-1}}$
k_l, k'_l	film coefficients corresponding to K_l, K'_l	m/s, kmol/m^{2}s	$\mathbf{LT^{-1}}, \mathbf{NL^{-2}T^{-1}}$
k_1	constant in Eq. (1.6)	—	—
k_2	constant in Eq. (1.6)	K	θ
k_3	constant in Eq. (1.6)	K	θ
k_4	constant in Eq. (1.7)	—	—
k_5	constant in Eq. (1.7)	K	θ
k_6	constant in Eq. (1.7)	—	—
k_7	constant in Eq. (1.7)	K^{-6}	θ^{-6}
k_8	constant in Eq. (1.8)	—	—
k_9	constant in Eq. (1.8)	K	θ
k_{10}	constant in Eq. (1.11)	—	—
k_{11}	constant in Eq. (1.11)	—	—
k_{12}	constant in Eq. (1.11)	K^{-6}	θ^{-6}
L	liquid flow rate in mass or moles/unit time	kg/s, kmol/s	$\mathbf{MT^{-1}}, \mathbf{NT^{-1}}$
L'	liquid flow rate, in moles/unit time-unit area	kmol/m^{2}s	$\mathbf{NL^{-2}T^{-1}}$
L_b	mole of liquid in batch distillation	kmol	**N**
L_p	liquid rate per unit periphery	m^3/ms	$\mathbf{L^2T^{-1}}$
l_m	metal thickness	m	**L**
M	Molecular weight	kg/kmol	$\mathbf{MN^{-1}}$
	or difference stream below feed plate	kmol/s	$\mathbf{NT^{-1}}$

Continued

		Units in SI system	Dimensions in M, N, L, T, θ
M_m	mean molecular weight	kg/kmol	$\mathbf{MN^{-1}}$
M	mass of material	kg	$\mathbf{M}$
	or gradient of equilibrium curve	—	—
N	molar rate of transfer per unit area of interface	kmol/m^{2}s	$\mathbf{NL^{-2}T^{-1}}$
	or difference stream above feed plate	kmol/s	$\mathbf{NT^{-1}}$
N''	vapour handling capacity of boiler and condenser	kmol/m^{2}s	$\mathbf{NL^{-2}T^{-1}}$
$\mathbf{N}$	number of transfer units	—	—
$\mathbf{N}_G$, $\mathbf{N}_{OG}$	number of gas film and overall gas transfer units	—	—
$\mathbf{N}_L$, $\mathbf{N}_{OL}$	number of liquid film and overall liquid transfer units	—	—
N	number of (theoretical) plates	—	—
n_m	number of plates at total reflux	—	—
n_p	number of actual plates	—	—
P	total pressure	N/m^2	$\mathbf{ML^{-1}T^{-2}}$
P_A, P_B	partial pressure of **A, B**	N/m^2	$\mathbf{ML^{-1}T^{-2}}$
P_A°, P_B°	vapour pressure of **A, B**	N/m^2	$\mathbf{ML^{-1}T^{-2}}$
$-\Delta P_{dry}$	pressure drop over dry tray expressed as head	m	$\mathbf{L}$
$-\Delta P_T$	total pressure drop expressed as head	m	$\mathbf{L}$
P	constant in Eq. (1.159)	—	—
P_c	critical pressure	N/m^2	$\mathbf{ML^{-1}T^{-2}}$
P_r	reduced pressure	—	—
Q_B	heat per unit time supplied to boiler for bottom product W	W	$\mathbf{ML^2T^{-3}}$
Q_C	heat per unit time removed in condenser for product D	W	$\mathbf{ML^2T^{-3}}$
Q_R	heat to still in batch distillation to provide reflux	J	$\mathbf{ML^2T^{-2}}$
Q'	liquid rate (volumetric)	m^3/s	$\mathbf{L^3T^{-1}}$
Q	heat to vaporise 1 mol of feed divided by molar latent heat		—
q_c	Q_c/D	W/kmol	$\mathbf{MN^{-1}L^2T^{-3}}$
R	reflux ratio	—	—
R_m	minimum reflux ratio	—	—
R'	resistance force per unit area, as used in Chapter 4	N/m^2	$\mathbf{ML^{-1}T^{-2}}$
$\mathbf{R}$	universal gas constant	J/kmolK	$\mathbf{MN^{-1}L^2T^{-2}\theta^{-1}}$

Continued

		Units in SI system	Dimensions in M, N, L, T, θ
R	radius of an orifice	m	**L**
r_m, r_n, r_f	ratio of concentrations of key components on plates m, n and in the feed	–	–
S	molar quantity of material in the still	kmol	**N**
S', S'', S'''	molar quantity or mass of material per unit time in sidestreams 1, 2, 3	kmol/s	**NT**$^{-1}$
T	absolute temperature	K	θ
T_B	boiling point	K	θ
T	time	s	**T**
T_c	critical temperature	K	θ
T_r	reduced temperature	–	–
t_a	annual period of operation	h/year	–
U	vapour velocity	m/s	**LT**$^{-1}$
u'	allowable vapour velocity	m/s	**LT**$^{-1}$
V	vapour flow in moles or mass per unit time	kmol/s, kg/s	**NT**$^{-1}$, **MT**$^{-1}$
Y	volumetric vapour flow rate	m^3/s	**L**^{3}**T**$^{-1}$
w	bottom product in moles or mass per unit time	kmol/s kg/s	**NT**$^{-1}$, **MT**$^{-1}$
X	mole or mass fraction of a component in the liquid phase	–	–
Y	mole fraction of a component in the gas phase	–	–
Z	height of packed column	m	**L**
Z_p	plate spacing	m	**L**
Z_t	height of equivalent theoretical plate	m	**L**
Z	distance	m	**L**
A	relative volatility or volatility relative to heavy key	–	–
α_{AB}	volatility of **A** relative to **B**	–	–
Γ	activity coefficient	–	–
E	$m(G'/L')$	–	–
Λ	latent heat per mole	J/kmol	**MN**$^{-1}$**L**^{2}**T**$^{-2}$
μ_L	viscosity of liquid	Ns/m^2	**ML**$^{-1}$**T**$^{-1}$
μ_v	viscosity of vapour	Ns/m^2	**ML**$^{-1}$**T**$^{-1}$
ρ_L	density of liquid	kg/m^3	**ML**$^{-3}$
ρ_v	density of vapour	kg/m^3	**ML**$^{-3}$
ρ_m	density of metal	kg/m^3	**ML**$^{-3}$

Continued

		Units in SI system	Dimensions in M, N, L, T, θ
Σ	moles per unit time of steam for steam distillation	kmol/s	NT^{-1}
Σ	surface tension	J/m^2 (or N/m)	MT^{-2}
Θ	root of Eq. (1.115)	—	—
ϕ	intercept of operating line on Y-axis $[x_d/(R+1)]$	—	—
Ψ	r_m/r_n at minimum reflux or fractional entrainment	—	—
Ω	fraction of packing wetted	—	—
Re	Reynolds number	—	—
Sc	Schmidt number	—	—
Suffixes			
1, 2	inlet, outlet		
A, B, C, D	materials **A, B, C, D**		
B	bottom		
C	intersection of equilibrium and operating lines at minimum reflux		
D	top product		
E	equilibrium		
F	feed		
Fs	feed at its boiling point		
G	azeotrope		
H	component heavier than heavy key		
I	interface		
L, l	liquid		
m, n	plates m, n below and above feed plate respectively		
Q	intersection of operating lines		
S', S'', S'''	sidestreams, 1, 2, 3		
S	still		
T	top		
V	vapour		
W	bottom product		
Av	average		

1.1 **Introduction**

The separation of liquid mixtures into their various components is one of the major operations in the process industries, and distillation, the most widely used method of achieving this end, is the key operation in any oil refinery. In processing, the demand for purer products, coupled with the need for greater efficiency, has promoted continued research into the techniques of distillation. In engineering terms, distillation columns have to be designed with a larger range in capacity than any other types of processing equipment, with single columns 0.3 – 10 m in diameter and 3 – 75 m in height. Designers are required to achieve the desired product quality at minimum cost and also to provide constant purity of product even though there may be variations in feed composition. A distillation unit should be considered together with its associated control system, and it is often operated in association with several other separate units.

An important requirement of a distillation unit is the provision of intimate contact between the vapour and liquid streams so that equilibrium is approached. The vertical cylindrical column provides, in a compact form and with the minimum of ground requirements, a large number of separate stages of vaporisation and condensation. In this chapter, the basic problems of design are considered and it may be seen that not only the physical and chemical properties, but also the fluid dynamics inside the unit, determine the number of stages required and the overall layout of the unit.

The separation of benzene from a mixture with toluene, for example, requires only a simple single unit as shown in Fig. 1.1, and virtually pure products may be obtained. A more complex arrangement is shown in Fig. 1.2 where the columns for the purification of crude styrene formed by the dehydrogenation of ethyl benzene are shown. It may be seen that, in this case, several columns are required and that it is necessary to recycle some of the streams to the reactor.

In this chapter, consideration is given to the theory of the process, methods of distillation and calculation of the number of stages required for both binary and multicomponent systems, and discussion on design methods is included for plate and packed columns incorporating a variety of column internals.

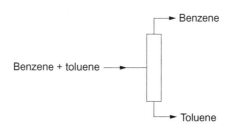

FIG. 1.1

Separation of a binary mixture.

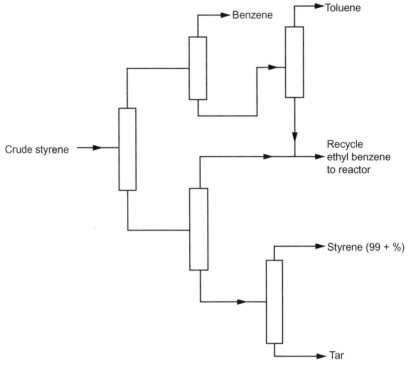

FIG. 1.2

Multicomponent separation.

1.2 Vapour – liquid equilibrium

The composition of the vapour in equilibrium with a liquid of given composition is determined experimentally using an equilibrium still. The results are conveniently shown on a temperature – composition diagram as shown in Fig. 1.3. In the normal case shown in Fig. 1.3A, the curve ABC shows the composition of the liquid which boils at any given temperature, and the curve ADC the corresponding composition of the vapour at that temperature. Thus, a liquid of composition x_1 will boil at temperature T_1, and the vapour in equilibrium is indicated by point D of composition y_1. It is seen that for any liquid composition x, the vapour formed will be richer in the more volatile component, where x is the mole fraction of the more volatile component in the liquid, and y in the vapour. Examples of mixtures giving this type of curve are benzene – toluene, n-heptane – toluene, and carbon disulphide – carbon tetrachloride.

In Figs 1.3B and C, there is a critical composition x_g where the vapour has the same composition as the liquid, so that no change occurs on boiling. Such critical mixtures are called azeotropes. Special methods which are necessary to effect

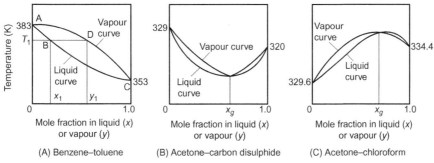

FIG. 1.3

Temperature composition diagrams.

separation of these are discussed in Section 1.8. For compositions other than x_g, the vapour formed has a different composition from that of the liquid. It is important to note that these diagrams are for constant pressure conditions, and that the composition of the vapour in equilibrium with a given liquid will change with pressure.

For distillation purposes, it is more convenient to plot y against x at a constant pressure since the majority of industrial distillations take place at substantially constant pressure. This is shown in Fig. 1.4 where it should be noted that the temperature varies along each of the curves.

1.2.1 Partial vaporisation and partial condensation

If a mixture of benzene and toluene is heated in a vessel, closed in such a way that the pressure remains atmospheric and no material can escape and the mole fraction of the more volatile component in the liquid, that is benzene, is plotted as abscissa, and the temperature at which the mixture boils as ordinate, then the boiling curve is obtained as shown by ABCJ in Fig. 1.5. The corresponding dew-point curve ADEJ shows the temperature at which a vapour of composition y starts to condense.

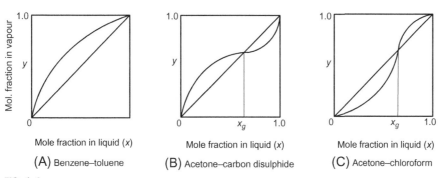

FIG. 1.4

Vapour composition as a function of liquid composition at constant pressure.

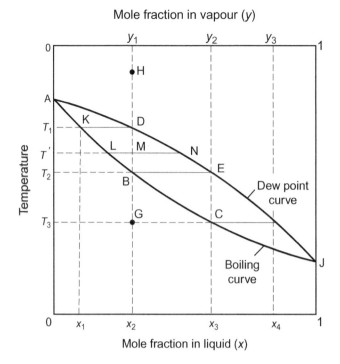

FIG. 1.5

Effect of partial vaporisation and condensation at the boiling point.

If a liquid mixture of composition x_2 is at a temperature T_3 below its boiling point, T_2, as shown by point G on the diagram, then on heating at constant pressure the following changes will occur:

(a) When the temperature reaches T_2, the liquid will start to boil, as shown by point B, and the initial vapour of composition y_2, shown by point E, is formed.

(b) On further heating, the composition of the liquid will change because of the loss of the more volatile component to the vapour and the boiling point will therefore rise to some temperature T'. At this temperature, the liquid will have a composition represented by point L, and the vapour a composition represented by point N. Since no material is lost from the system, there will be a change in the proportion of liquid to vapour, where the ratio is:

$$\frac{\text{Liquid}}{\text{Vapour}} = \frac{\text{MN}}{\text{ML}}$$

(c) On further heating to a temperature T_1, all of the liquid is vaporised to give vapour D of the same composition y_1 as the original liquid composition x_2.

It may be seen that partial vaporisation of the liquid gives a vapour richer in the more volatile component than the liquid. If the vapour initially formed, as for instance at point E, is at once removed by condensation, then a liquid of composition x_3 is obtained, represented by point C. The step BEC may be regarded as representing an ideal stage, since the liquid passes from composition x_2 to a liquid of composition x_3, which represents a greater enrichment in the more volatile component than can be obtained by any other single stage of vaporisation.

Starting with superheated vapour represented by point H, on cooling to D condensation commences, and the first drop of liquid has a composition K. Further cooling to T' gives liquid L and vapour N. Thus, partial condensation brings about enrichment of the vapour in the more volatile component in the same manner as partial vaporisation. The industrial distillation column is, in essence, a series of units in which these two processes of partial vaporisation and partial condensation are affected simultaneously.

1.2.2 Partial pressures, and Dalton's, Raoult's, and Henry's laws

The partial pressure P_A of component **A** in a mixture of vapours is the pressure that would be exerted by component **A** at the same temperature, if present in the same volumetric concentration as in the mixture.

By Dalton's law of partial pressures, $P = \Sigma P_A$, that is the total pressure is equal to the summation of the partial pressures of each constituent **A**, **B**, **C**, … of the vapour mixture P_A, P_B, P_C,.. Since in an ideal gas or vapour, the partial pressure is proportional to the mole fraction of the constituent, y_A, then:

$$P_A = y_A P \tag{1.1}$$

For an *ideal mixture*, the partial pressure is related to the concentration in the liquid phase by Raoult's law which may be written as:

$$P_A = P_A^o x_A \tag{1.2}$$

where P_A^o is the vapour pressure of pure **A** at the same temperature and x_A is the mole fraction in the liquid phase. This relation is usually found to be true only for high values of x_A, or correspondingly low values of x_B, although mixtures of organic isomers and some hydrocarbons follow the law closely.

For dilute solutions, where values of x_A are low, a linear relation between P_A and x_A again exists. In this case, the proportionality factor is Henry's constant H', and not the vapour pressure P_A^o of the pure material.

For a partial pressure developed by a liquid solute **A** in a solvent liquid **B**, Henry's law takes the form:

$$P_A = H' x_A \tag{1.3}$$

If the mixture follows Raoult's law, then the vapour pressure of a mixture may be obtained graphically from a knowledge of the vapour pressure of the two components. Thus, in Fig. 1.6, OA represents the partial pressure P_A of **A** in a mixture,

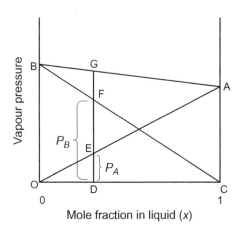

FIG. 1.6

Partial pressures of ideal mixtures.

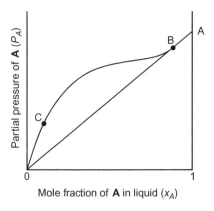

FIG. 1.7

Partial pressures of non-ideal mixtures.

and CB the partial pressure of **B**, with the total pressure being shown by the line BA. In a mixture of composition D, the partial pressure P_A is given by DE, P_B by DF, and the total pressure P by DG, from the geometry of Fig. 1.6.

Fig. 1.7 shows the partial pressure of one component **A** plotted against the mole fraction for a mixture that is not ideal. It is found that over the range OC the mixture follows Henry's law, and over BA it follows Raoult's law. Although most mixtures show wide divergences from ideality, one of the laws is usually followed at very high and very low concentrations.

If the mixture follows Raoult's law, then the values of y_A for various values of x_A may be calculated from a knowledge of the vapour pressures of the two components at various temperatures.

$$\text{Thus}: P_A = P_A^o x_A$$

$$\text{and}: P_A = P y_A$$

$$\text{so that}: y_A = \frac{P_A^o x_A}{P}, \text{ and } y_B = \frac{P_B^o x_B}{P} \tag{1.4}$$

$$\text{But}: y_A + y_B = 1$$

$$\frac{P_A^o x_A}{P} + \frac{P_B^o (1 - x_A)}{P} = 1$$

giving:

$$x_A = \frac{P - P_B^o}{P_A^o - P_B^o} \tag{1.5}$$

Example 1.1

The vapour pressures of n-heptane (**A**) and toluene (**B**) at 373 K are 106 and 73.7 kN/m2 respectively. What are the mole fractions of n-heptane in the vapour and in the liquid phase at 373 K if the total pressure is 101.3 kN/m2?

Solution

$$\text{At } 373\,\text{K}, P_A^o = 106\,\text{kN/m}^2 \text{ and } P_B^o = 73.7\,\text{kN/m}^2$$

Thus, in Eq. (1.5):

$$x_A = (P - P_B^o)/(P_A^o - P_B^o) = \frac{(101.3 - 73.7)}{(106 - 73.7)} = \underline{0.854}$$

and, in Eq. (1.4):

$$y_A = P_A^o x_A / P = \frac{(106 \times 0.854)}{101.3} = \underline{0.894}$$

Equilibrium data usually have to be determined by tedious laboratory methods. Proposals have been made which enable the complete diagram to be deduced with reasonable accuracy from a relatively small number of experimental values. Some of these methods are discussed by Robinson and Gilliand [1] and by Thornton and Garner [2].

One of the most widely used correlations of saturated vapour pressure is that proposed by ANTOINE [3]. This takes the form:

$$\ln P^o = k_1 - k_2/(T + k_3) \tag{1.6}$$

where the constants, k_1, k_2 and k_3 must be determined experimentally [4,5,6] although many values of these constants are available in the literature [6,7,8,9,10]. Eq. (1.6) is valid only over limited ranges of both temperature and pressure, although the correlation interval may be extended by using the equation proposed by Riedel [11]. This takes the form:

$$\ln P° = k_4 - k_5/T + k_6 \ln T + k_7 T^6 \tag{1.7}$$

If only two values of the vapour pressure at temperatures T_1 and T_2 are known, then the Clapeyron equation may be used:

$$\ln P° = k_8 - k_9/T \tag{1.8}$$

where:

$$k_8 = \ln P_1^o + k_9/T_1 \tag{1.9}$$

and:

$$k_9 = \ln(P_2/P_1)/[(1/T_1) - (1/T_2)] \tag{1.10}$$

Equation (1.8) may be used for the evaluation of vapour pressure over a small range of temperature, although large errors may be introduced over large temperature intervals. If the critical values of temperature and pressure are available along with one other vapour pressure point such as, for example, the normal boiling point, then a reduced form of the Riedel equation may be used; this takes the form:

$$\ln P_r^o = k_9 - k_{10}/T_r + k_{11} \ln T_r + k_{12}T_r^6 \tag{1.11}$$

where: P_r^o = reduced vapour pressure $= (P°/P_c)$, T_r = reduced temperature $= (T/T_c)$, $k_9 = 35c_1$, $k_{10} = -36c_1$, $k_{11} = 42c_1 + c_2$, $k_{12} = -c_1$ and $c_1 = 0.0838$ $(3.758 - c_2)$. c_2 is determined by inserting the other known vapour pressure point into Eq. (1.11) and solving for c_2. This gives:

$$c_2 = [(0.315c_5 - \ln P_{r1}^o)/(0.0938c_5 - \ln T_{r1})] \tag{1.12}$$

where: $c_5 = -35 + 36/T_{r1} + 42 \ln T_{r1} - T_{r1}^6$

Example 1.2

The following data have been reported for acetone by Ambrose et al. [12]: $P_c = 4700\,\text{kN/m2}$, $T_c = 508.1\,\text{K}$, $P_1^o = 100.666\,\text{kN/m2}$ when $T_1 = 329.026\,\text{K}$. What is $P°$ when $T = 350.874\,\text{K}$?

Solution

$T_{r1} = (329.026/508.1) = 0.64756$, $\quad P_{r1} = (100.666/4700.0) = 0.021418$ and hence, in Eq. (1.12):

$$\text{and}: c_5 = -35 + (36/0.64756) + 42 \ln 0.64756 - (0.64756)^6 = 2.2687$$

$$c_2 = [((0.315 \times 2.2687) - \ln 0.021418)/((0.0838 \times 2.2687) - \ln 0.64756)]$$
$$= 7.2970$$

$$c_1 = 0.0838(3.758 - 7.2970 = -0.29657.$$

$$k_9 = -35(-029657) = 10.380$$

$$k_{10} = -36(-0.29657) = 10.677$$

$$k_{11} = 42(-0.29657) + 7.2970 = -5.1589$$

$$k_{12} = 1(0.29657) = 10.380$$

Substituting these values into Eq. (1.11) together with a value of $T_r = (350.874/508.1) = 0.69056$, then:

$$\ln P_r^o = 10.380 - (10.677/0.69056) - 5.1589 \ln 0.69056 + 0.29657(0.69056)^6$$
$$= -3.1391$$

From which: $P_r^o = 0.043322$

$$\text{and}: P^o = (0.043322 \times 4700.0) = \underline{\underline{203.61 \text{ kN/m}^2}}$$

This may be compared with an experimental value of 201. 571 kN/m2.

Example 1.3

The constants in the Antoine equation, Eq. (1.6), are:

For benzene:	$k_1 = 6.90565$	$k_2 = 1211.033$	$k_3 = 220.79$
For toluene:	$k_1 = 6.95334$	$k_2 = 1343.943$	$k_3 = 219.377$

where P^o is in mm Hg, T is in °C and $\log_{10}$ is used instead of $\log_e$.

Determine the vapour phase composition of a mixture in equilibrium with a liquid mixture of 0.5 mol fraction benzene and 0.5 mol fraction of toluene at 338 K. Will the liquid vaporise at a pressure of 101.3 kN/m2?

Solution

The saturation vapour pressure of benzene at 338 K $= 65$°C is given by:

$$\log_{10} P_B^o = 6.90565 - [1211.033/(65 + 220.79)] = 2.668157$$

from which: $P_B^o = 465.75$ mm Hg or 62.10 kN/m^2

Similarly for toluene at $338\,\text{K} = 65°\text{C}$:

$$\log_{10} P_T^o = 6.95334 - [1343.943/(65 + 219.377)] = 2.22742$$

$$\text{and}: P_T^o = 168.82 \text{ mm Hg or } 22.5 \text{ kN/m}^2$$

The partial pressures in the mixture are:

$$P_B = (0.50 \times 62.10) = 31.05\,\text{kN/m2}$$

$$\text{and}: P_T = (0.50 \times 22.51) = 11.255\,\text{kN/m2} - \text{a total pressure of } 42.305\,\text{kN/m2}$$

Using Eq. (1.1), the composition of the vapour phase is:

$$y_B = (31.05/42.305) = \underline{0.734}$$

$$\text{and}: y_T = (11.255/42.305) = \underline{0.266}$$

Since the total pressure is only 42.305 kN/m2, then with a total pressure of 101.3 kN/m2, the liquid will not vaporise unless the pressure is decreased.

Example 1.4
What is the boiling point of an equimolar mixture of benzene and toluene at 101.3 kN/m2?

Solution
The saturation vapour pressures are calculated as a function of temperature using the Antoine equation, Eq. (1.6), and the constants given in Example 1.3, and then, from Raoult's Law, Eq. (1.1), the actual vapour pressures are given by:

$$P_B = x_B P_B^o \text{ and } P_T = x_T P_T^o$$

It then remains, by a process of trial and error, to determine at which temperature: $(P_B + P_T) = 101.3\,\text{kN/m2}$. The data, with pressures in kN/m2, are:

T (K)	P_B^o	P_T^o	P_B	P_T	$(P_B + P_T)$
373	180.006	74.152	90.003	37.076	127.079
353	100.988	38.815	50.494	19.408	69.902
363	136.087	54.213	68.044	27.106	95.150
365	144.125	57.810	72.062	28.905	100.967
365.1	144.534	57.996	72.267	28.998	101.265

101.265 kN/m2 is essentially 101.3 kN/m2 and hence, at this pressure, the boiling or the bubble point of the equimolar mixture is $\underline{365.1}$ K which lies between the boiling points of pure benzene, 353.3 K, and pure toluene, 383.8 K.

Example 1.5

What is the dew point of an equimolar mixture of benzene and toluene at $101.3\,kN/m^2$?

Solution

From Raoult's Law, Eqs (1.1) and (1.2):

$$P_B = x_B P_B^o = y_B P$$

$$\text{and}: P_T = x_T P_T^o = y_T P$$

Since the total pressure is $101.3\,kN/m^2$, $P_B = P_T = 50.65\,kN/m^2$ and hence:

$$x_B = P_B/P_B^o = 50.65/P_B^o \text{ and } x_T = 50.65/P_T^o$$

It now remains to estimate the saturation vapour pressures as a function of temperature, using the data of Example 1.3, and then determine, by a process of trial and error, when $(x_B + x_T) = 1.0$. The data, with pressures in kN/m^2 are:

$T(K)$	P_B^o	x_B	P_T^o	x_T	$(x_B + x_T)$
373.2	180.984	0.2799	74.603	0.6789	0.9588
371.2	171.390	0.2955	70.189	0.7216	1.0171
371.7	173.751	0.2915	71.273	0.7107	1.0022
371.9	174.702	0.2899	71.710	0.7063	0.9962
371.8	174.226	0.2907	71.491	0.7085	0.9992

As 0.9992 is near enough to 1.000, the dew point may be taken as 371.8 K.

1.2.3 Relative volatility

The relationship between the composition of the vapour y_A and of the liquid x_A in equilibrium may also be expressed in a way, which is particularly useful in distillation calculations. If the ratio of the partial pressure to the mole fraction in the liquid is defined as the volatility, then:

$$\text{Volatility of } \mathbf{A} = \frac{P_A}{x_A} \text{ and volatility of } \mathbf{B} = \frac{P_B}{x_B}$$

The ratio of these two volatilities is known as the relative volatility α given by:

$$\alpha = \frac{P_A x_B}{x_A P_B}$$

Substituting $P y_A$ for P_A, and $P y_B$ for P_B:

$$\alpha = \frac{y_A x_B}{y_B x_A} \tag{1.13}$$

or:

$$\frac{y_A}{y_B} = \alpha \frac{x_A}{x_B} \tag{1.14}$$

This gives a relation between the ratio of **A** and **B** in the vapour to that in the liquid. Since with a binary mixture $y_B = 1 - y_A$, and $x_B = 1 - x_A$ then:

$$\alpha = \left(\frac{y_A}{1 - y_A}\right)\left(\frac{1 - x_A}{x_A}\right)$$

or:

$$y_A = \frac{\alpha x_A}{1 + (\alpha - 1)x_A} \tag{1.15}$$

and:

$$y_A = \frac{y_A}{a - (\alpha - 1)y_A} \tag{1.16}$$

This relation enables the composition of the vapour to be calculated for any desired value of x, if α is known. For separation to be achieved, α must not equal 1 and, considering the more volatile component, as α increases above unity, y increases and the separation becomes much easier. Eq. (1.14) is useful in the calculation of plate enrichment and finds wide application in multicomponent distillation.

From the definition of the volatility of a component, it is seen that for an ideal system the volatility is numerically equal to the vapour pressure of the pure component. Thus, the relative volatility α may be expressed as:

$$\alpha = \frac{P_A^o}{P_B^o} \tag{1.17}$$

This also follows by applying Eq. (1.1) from which $P_A/P_B = y_A/y_B$, and Eq. (1.2) from which $P_A/P_B = P_A^o x_A / P_B^o x_B$ so that:

$$\alpha = \frac{P_A x_B}{P_B x_A} = \frac{P_A^o x_A x_B}{P_B^o x_B x_A} = \frac{P_A^o}{P_B^o}$$

Whilst α does vary somewhat with temperature, it remains remarkably steady for many systems, and a few values to illustrate this point are given in Table 1.1.

It may be seen that α increases as the temperature falls, so that it is sometimes worthwhile reducing the boiling point by operating at reduced pressure. When Eq. (1.16) is used to construct the equilibrium curve, an average value of α must be taken over the whole column. As Frank [13] points out, this is valid if the relative volatilities at the top and bottom of the column differ by less than 15%. If they differ by more than this amount, the equilibrium curve must be constructed incrementally by calculating the relative volatility at several points along the column.

Another frequently used relationship for vapour – liquid equilibrium is the simple equation:

$$yA = KxA \tag{1.18}$$

Table 1.1 Relative volatility of mixtures of benzene and toluene.

Temperature (K)	353	363	373	383
α (−)	2.62	2.44	2.40	2.39

For many systems, K is constant over an appreciable temperature range and Eq. (1.11) may be used to determine the vapour composition at any stage. The method is particularly suited to multicomponent systems, discussed further in Section 1.7.1.

1.2.4 Non-ideal systems

Equation (1.4) relates x_A, y_A, P_A^o and P. For a *non-ideal* system the term γ, the activity coefficient, is introduced to give:

$$y_A = \frac{\gamma_1 P_A^o x_A}{P} \quad \text{and} \quad y_B = \frac{\gamma_2 P_B^o x_B}{P} \tag{1.19}$$

or in Eq. (1.18):

$$y_A = K\gamma_1 x_A \quad \text{and} \quad y_B = K\gamma_2 x_B \tag{1.20}$$

The liquid phase activity coefficients γ_1 and γ_2 depend upon temperature, pressure, and concentration. Typical values taken from Perry's Chemical Engineers' Handbook [14] are shown in Fig. 1.8 for the systems n-propanol–water and acetone–chloroform. In the former, the activity coefficients are considered positive, that is greater than unity, whilst in the latter, they are fractional so that the logarithms of the values are negative. In both cases, γ approaches unity as the liquid concentration approaches unity and the highest values of γ occur as the concentration approaches zero.

The fundamental thermodynamic equation relating activity coefficients and composition is the *Gibbs–Duhem* relation which may be expressed as:

$$x_1 \left(\frac{\partial \ln \gamma_1}{\partial x_1} \right)_{T,P} - x_2 \left(\frac{\partial \ln \gamma_2}{\partial x_2} \right)_{T,P} = 0 \tag{1.21}$$

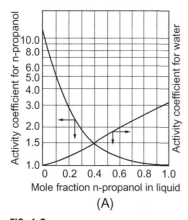
(A)

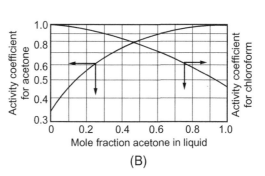
(B)

FIG. 1.8

Activity coefficient data.

This equation relates the slopes of the curves in Fig. 1.8 and provides a means of testing experimental data. It is more convenient, however, to utilise integrated forms of these relations. A large number of different solutions to the basic Gibbs–Duhem equation are available, each of which gives a different functional relationship between log γ and x. Most binary systems may be characterised, however, by either the three- or four-suffix equations of Margules, or by the two-suffix van Laar equations, given as follows in the manner of Wohl [15,16]. The three-suffix Margules binary equations are:

$$\log \gamma_1 = x_2^2[A_{12} + 2x_1(A_{21} - A_{12})] \tag{1.22}$$

$$\log \gamma_2 = x_1^2[A_{21} + 2x_2(A_{12} - A_{21})] \tag{1.23}$$

Constants A_{12} and A_{21} are the limiting values of log γ as the composition of the component considered approaches zero. For example, in Eq. (1.22), $A_{12} = \log \gamma_1$ when $x_1 = 0$.

The four-suffix Margules binary equations are:

$$\log \gamma_1 = x_2^2\left[A_{12} + 2x_1(A_{21} - A_{12} - A_D) + 3A_Dx_1^2\right] \tag{1.24}$$

$$\log \gamma_2 = x_1^2\left[A_{21} + 2x_2(A_{12} - A_{21} - A_D) + 3A_Dx_2^2\right] \tag{1.25}$$

A_{12} and A_{21} have the same significance as before and A_D is a third constant. Equations (1.24) and (1.25) are more complex than Eqs (1.22) and (1.23) though, because they contain an additional constant A_D, they are more flexible. When A_D becomes zero in Eqs (1.24) and (1.25), they become identical to the three-suffix equations.

The two-suffix van Laar binary equations are:

$$\log \gamma_1 = \frac{A_{12}}{[1 + (A_{12}x_1/A_{21}x_2)]^2} \tag{1.26}$$

$$\log \gamma_2 = \frac{A_{21}}{[1 + (A_{21}x_2/A_{12}x_1)]^2} \tag{1.27}$$

These equations become identical to the three-suffix Margules equations when $A_{12} = A_{21}$, and the functional form of these two types of equations is not greatly different unless the constants A_{12} and A_{21} differ by more than about 50%.

The Margules and van Laar equations apply only at *constant temperature and pressure*, as they were derived from Eq. (1.21), which also has this restriction. The effect of pressure upon γ values and the constants A_{12} and A_{21} is usually negligible, especially at pressures far removed from the critical. Correlation procedures for activity coefficients have been developed by Balzhiser et al. [17], Frendenslund et al. [18], Praunsitz et al. [19], Reid et al. [20], van Ness and Abbott [21], and Walas [22] and actual experimental data may be obtained from the PPDS system of the National Engineering Laboratory, UK [23]. When the liquid and vapour compositions are the same, that is, $x_A = y_A$, point x_g in Figs 1.3 and 1.4, the system is said to form an azeotrope, a condition which is discussed in Section 1.8.

1.3 Methods of distillation—Two component mixtures

From curve a of Fig. 1.4, it is seen that, for a binary mixture with a normal $y - x$ curve, the vapour is always richer in the more volatile component than the liquid from which it is formed. There are three main methods used in distillation practice which all rely on this basic fact. These are:

(a) Differential distillation.
(b) Flash or equilibrium distillation, and
(c) Rectification.

Of these, rectification is much the most important, and it differs from the other two methods in that part of the vapour is condensed and returned as liquid to the still, whereas, in the other methods, all the vapour is either removed as such, or is condensed as product.

1.3.1 Differential distillation

The simplest example of batch distillation is a single stage, differential distillation, starting with a still pot, initially full, heated at a constant rate. In this process, the vapour formed on boiling the liquid is removed at once from the system. Since this vapour is richer in the more volatile component than the liquid, it follows that the liquid remaining becomes steadily weaker in this component, with the result that the composition of the product progressively alters. Thus, whilst the vapour formed over a short period is in equilibrium with the liquid, the total vapour formed is not in equilibrium with the residual liquid. At the end of the process, the liquid which has not been vaporised is removed as the bottom product. The analysis of this process was first proposed by Rayleigh [24].

If S is the number of moles of material in the still, x is the mole fraction of component **A** and an amount dS, containing a mole fraction y of **A**, is vaporised, then a material balance on component **A** gives:
    and:

$$y\,dS = d(Sx)$$
$$= S\,dx + x\,dS$$
$$\int_{S_0}^{S} \frac{dS}{S} = \int_{x_0}^{x} \left(\frac{dx}{y - x} \right) \tag{1.28}$$
$$\ln \frac{S}{S_0} = \int_{x_0}^{x} \left(\frac{dx}{y - x} \right)$$

The integral on the right-hand side of this equation may be solved graphically if the equilibrium relationship between y and x is available. In some cases, a direct integration is possible. Thus, if over the range concerned the equilibrium relationship is a straight line of the form $y = mx + c$, then:

$$\ln \frac{S}{S_0} = \left(\frac{1}{m-1}\right) \ln \left[\frac{(m-1)x + c}{(m-1)x_0 + c}\right]$$

or :

$$\frac{S}{S_0} = \left(\frac{y - x}{y_0 - x_0}\right)^{1/(m-1)}$$

and :

$$\left(\frac{y - x}{y_0 - x_0}\right) = \left(\frac{S}{S_0}\right)^{m-1} \qquad (1.29)$$

From this equation, the amount of liquid to be distilled in order to obtain a liquid of given concentration in the still may be calculated, and from this, the average composition of the distillate may be found by a mass balance.

Alternatively, if the relative volatility is assumed constant over the range concerned, then $y = \alpha x/(1 + (\alpha - 1)x)$, Eq. (1.15) may be substituted in Eq. (1.28). This leads to the solution:

$$\ln \frac{S}{S_0} = \left(\frac{1}{\alpha - 1}\right) \ln \left[\frac{x(1 - x_0)}{x_0(1 - x)}\right] + \ln \left[\frac{1 - x_0}{1 - x}\right] \qquad (1.30)$$

As this process consists of only a single stage, a complete separation is impossible unless the relative volatility is infinite. Application is restricted to conditions where a preliminary separation is to be followed by a more rigorous distillation, where high purities are not required, or where the mixture is very easily separated.

1.3.2 Flash or equilibrium distillation

Flash or equilibrium distillation, frequently carried out as a continuous process, consists of vaporising a definite fraction of the liquid feed in such a way that the vapour evolved is in equilibrium with the residual liquid. The feed is usually pumped through a fired heater and enters the still through a valve where the pressure is reduced. The still is essentially a separator in which the liquid and vapour produced by the reduction in pressure have sufficient time to reach equilibrium. The vapour is removed from the top of the separator and is then usually condensed, whilst the liquid leaves from the bottom.

In a typical pipe still where, for example, a crude oil might enter at 440 K and at about 900 kN/m2, and leave at 520 K and 400 kN/m2, some 15% may be vaporised in the process. The vapour and liquid streams may contain many components in such an application, although the process may be analysed simply for a binary mixture of **A** and **B** as follows:

If F = moles per unit time of feed of mole fraction x_f of **A**,

V = moles per unit time of vapour formed with y the mole fraction of **A**, and.

S = moles per unit time of liquid with x the mole fraction of **A**, then an overall mass balance gives:

$F = V + S$.

so $S = F - V$.

and for the more volatile component:

$$Fx_f = V_y + Sx$$

$$\text{Thus}: \frac{V}{F} = \left(\frac{x_f - x}{y - x}\right)$$

$$\text{or}: y = \frac{F}{V}x_f - x\left(\frac{F}{V} - 1\right) \tag{1.31}$$

Equation (1.31) represents a straight line of slope:

$$-\left(\frac{F - V}{V}\right) = \frac{-S}{V}$$

passing through the point (x_f, x_f). The values of x and y required must satisfy, not only the equation, but also the appropriate equilibrium data. Thus these values may be determined graphically using an $x - y$ diagram as shown in Fig. 1.9.

In practice, the quantity vaporised is not fixed directly but it depends upon the enthalpy of the hot incoming feed and the enthalpies of the vapour and liquid leaving the separator. For a given feed condition, the fraction vaporised may be increased by lowering the pressure in the separator.

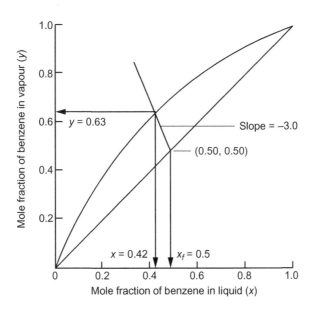

FIG. 1.9

Equilibrium data for benzene–toluene for Example 1.6.

Example 1.6

An equimolar mixture of benzene and toluene is subjected to flash distillation at 100 kN/m2 in the separator. Using the equilibrium data given in Fig. 1.9, determine the composition of the liquid and vapour leaving the separator when the feed is 25% vaporised. For this condition, the boiling point diagram in Fig. 1.10 may be used to determine the temperature of the exit liquid stream.

Solution

The fractional vaporisation $= V/F = f$ (say).
The slope of Eq. (1.31) is:

$$-\left(\frac{F-V}{V}\right) = -\left(\frac{1-f}{f}\right)$$

When $f = 0.25$, the slope of Eq. (1.31) is therefore:
$-(1-0.25)/0.25 = -3.0$.
and the construction is made as shown in Fig. 1.9 to give $x = 0.42$ and $y = 0.63$.
From the boiling point diagram, in Fig. 1.10, the liquid temperature when $x = 0.42$ is seen to be 366.5 K.

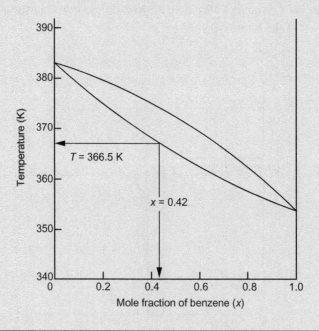

FIG. 1.10

Boiling point diagram for benzene–toluene for Example 1.6.

1.3.3 **Rectification**

In the two processes considered, the vapour leaving the still at any time is in equilibrium with the liquid remaining, and normally there will be only a small increase in concentration of the more volatile component. The essential merit of rectification is that it enables a vapour to be obtained that is substantially richer in the more volatile component than is the liquid left in the still. This is achieved by an arrangement known as a fractionating column which enables successive vaporisation and condensation to be accomplished in one unit. Detailed consideration of this process is given in Section 1.4.

1.3.4 **Batch distillation**

In batch distillation, which is considered in detail in Section 1.6, the more volatile component is evaporated from the still which therefore becomes progressively richer in the less volatile constituent. Distillation is continued, either until the residue of the still contains a material with an acceptably low content of the volatile material, or until the distillate is no longer sufficiently pure in respect of the volatile content.

1.4 **The fractionating column**
1.4.1 **The fractionating process**

The operation of a typical fractionating column may be followed by reference to Fig. 1.11. The column consists of a cylindrical structure divided into sections by a series of perforated trays which permit the upward flow of vapour. The liquid reflux flows across each tray, over a weir and down a downcomer to the tray below. The vapour rising from the top tray passes to a condenser and then through an accumulator or reflux drum and a reflux divider, where part is withdrawn as the overhead product D, and the remainder is returned to the top tray as reflux R.

The liquid in the base of the column is frequently heated, either by condensing steam or by a hot oil stream, and the vapour rises through the perforations to the bottom tray. A more commonly used arrangement with an external reboiler is shown in Fig. 1.11 where the liquid from the still passes into the reboiler where it flows over the tubes and weir and leaves as the bottom product by way of a bottoms cooler, which preheats the incoming feed. The vapour generated in the reboiler is returned to the bottom of the column with a composition y_s, and enters the bottom tray where it is partially condensed and then revaporised to give vapour of composition y_1. This operation of partial condensation of the rising vapour and partial vaporisation of the reflux liquid is repeated on each tray. Vapour of composition y_t from the top tray is condensed to give the top product D and the reflux R, both of the same composition y_t. The feed stream is introduced on some intermediate tray where the liquid has approximately the same composition as the feed. The part of the column above the feed point is known as the rectifying section and the lower portion is known as the stripping section.

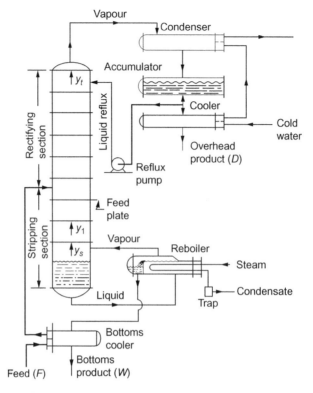

FIG. 1.11

Continuous fractionating column with rectifying and stripping sections.

The vapour rising from an ideal tray will be in equilibrium with the liquid leaving, although in practice a smaller degree of enrichment will occur.

In analysing the operation on each tray, it is important to note that the vapour rising to it, and the reflux flowing down to it, are not in equilibrium, and adequate rates of mass and heat transfer are essential for the proper functioning of the tray.

The tray as described is known as a sieve tray and it has perforations of up to about 12 mm diameter, although there are several alternative arrangements for promoting mass transfer on the tray, such as valve units, bubble caps and other devices described in Section 1.10.1. In all cases the aim is to promote good mixing of vapour and liquid with a low drop in pressure across the tray.

On each tray, the system tends to reach equilibrium because:

(a) Some of the less volatile component condenses from the rising vapour into the liquid thus increasing the concentration of the more volatile component (MVC) in the vapour.
(b) Some of the MVC is vaporised from the liquid on the tray thus decreasing the concentration of the MVC in the liquid.

The number of molecules passing in each direction from vapour to liquid and in reverse is approximately the same since the heat given out by 1 mol of the vapour on condensing is approximately equal to the heat required to vaporise 1 mol of the liquid. The problem is thus one of equimolecular counterdiffusion, described in Volume 1, Chapter 10. If the molar heats of vaporisation are approximately constant, the flows of liquid and vapour in each part of the column will not vary from tray to tray. This is the concept of constant molar overflow which is discussed under the heat balance heading in Section 1.4.2. Conditions of varying molar overflow, arising from unequal molar latent heats of the components, are discussed in Section 1.5.

In the arrangement discussed, the feed is introduced continuously to the column and two product streams are obtained, one at the top much richer than the feed in the MVC and the second from the base of the column weaker in the MVC. For the separation of small quantities of mixtures, a batch still may be used. Here, the column rises directly from a large drum which acts as the still and reboiler and holds the charge of feed. The trays in the column form a rectifying column and distillation is continued until it is no longer possible to obtain the desired product quality from the column. The concentration of the MVC steadily falls in the liquid remaining in the still so that enrichment to the desired level of the MVC is not possible. This problem is discussed in more detail in Section 1.6.

A complete unit will normally consist of a feed tank, a feed heater, a column with boiler, a condenser, an arrangement for returning part of the condensed liquid as reflux, and coolers to cool the two products before passing them to storage. The reflux liquor may be allowed to flow back by gravity to the top plate of the column or, as in larger units, it is run back to a drum from which it is pumped to the top of the column. The control of the reflux on very small units is conveniently affected by hand-operated valves, and with the larger units by adjusting the delivery from a pump. In many cases the reflux is divided by means of an electromagnetically operated device which diverts the top product either to the product line or to the reflux line for controlled time intervals.

1.4.2 Number of plates required in a distillation column

In order to develop a method for the design of distillation units to give the desired fractionation, it is necessary, in the first instance, to develop an analytical approach which enables the necessary number of trays to be calculated. First the heat and material flows over the trays, the condenser, and the reboiler must be established. Thermodynamic data are required to establish how much mass transfer is needed to establish equilibrium between the streams leaving each tray. The required diameter of the column will be dictated by the necessity to accommodate the desired flow rates, to operate within the available drop in pressure, whilst at the same time affecting the desired degree of mixing of the streams on each tray.

Four streams are involved in the transfer of heat and material across a plate, as shown in Fig. 1.12 in which plate n receives liquid L_{n+1} from plate $n+1$ above, and

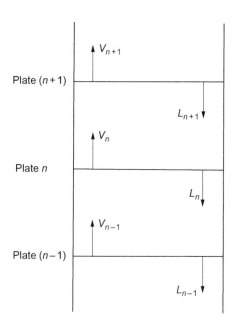

FIG. 1.12

Material balance over a plate.

vapour V_{n-1} from plate $n-1$ below. Plate n supplies liquid L_n to plate $n-1$, and vapour V_n to plate $n+1$.

The action of the plate is to bring about mixing so that the vapour V_n, of composition y_n, approaches equilibrium with the liquid L_n, of composition x_n. The streams L_{n+1} and V_{n-1} cannot be in equilibrium and, during the interchange process on the plate, some of the more volatile component is vaporised from the liquid L_{n+1}, decreasing its concentration to x_n, and some of the less volatile component is condensed from V_{n-1}, increasing the vapour concentration to y_n. The heat required to vaporise the more volatile component from the liquid is supplied by partial condensation of the vapour V_{n-1}. Thus the resulting effect is that the more volatile component is passed from the liquid running down the column to the vapour rising up, whilst the less volatile component is transferred in the opposite direction.

Heat balance over a plate

A heat balance across plate n may be written as:

$$L_{n+1}H_{n+1}^L + V_{n-1}H_{n-1}^V = V_nH_n^V + L_nH_n^L + \text{losses} + \text{heat of mixing} \qquad (1.32)$$

where: H_n^L is the enthalpy per mole of the liquid on plate n, and
H_n^V is the enthalpy per mole of the vapour rising from plate n.

This equation is difficult to handle for the majority of mixtures, and some simplifying assumptions are usually made. Thus, with good column insulation, the heat losses will be small and may be neglected, and for an ideal system the heat of mixing is zero. For such mixtures, the molar heat of vaporisation may be taken as constant and independent of the composition. Thus, one mole of vapour V_{n-1} on condensing releases sufficient heat to liberate one mole of vapour V_n. It follows that $V_n = V_{n-1}$, so that the molar vapour flow is constant up the column unless material enters or is withdrawn from the section. The temperature change from one plate to the next will be small, and H_n^L may be taken as equal to H_{n+1}^L. Applying these simplifications to Eq. (1.32), it is seen that $L_n = L_{n+1}$, so that the moles of liquid reflux are also constant in this section of the column. Thus, V_n and L_n are constant over the rectifying section, and V_m and L_m are constant over the stripping section.

For these conditions, there are two basic methods for determining the number of plates required. The first is due to Sorel [25] and later modified by Lewis [26], and the second is due to McCabe and Thiele [27]. The Lewis method is used here for binary systems, and also in Section 1.7.4 for calculations involving multicomponent mixtures. This method is also the basis of modern computerised methods. The McCabe–Thiele method is particularly important since it introduces the idea of the operating line which is an important common concept in multistage operations. The best assessment of these methods and their various applications is given by Underwood [28].

When the molar heat of vaporisation varies appreciably and the heat of mixing is no longer negligible, these methods have to be modified, and alternative techniques are discussed in Section 1.5.

Calculation of number of plates using the Lewis–Sorel method
If a unit is operating as shown in Fig. 1.13, so that a binary feed F is distilled to give a top product D and a bottom product W, with x_f, x_d, and x_w as the corresponding mole fractions of the more volatile component, and the vapour V_t rising from the top plate

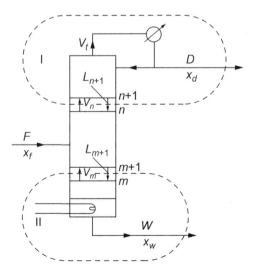

FIG. 1.13

Material balances at top and bottom of column.

is condensed, and part is run back as liquid at its boiling point to the column as reflux, the remainder being withdrawn as product, then a material balance above plate n, indicated by the loop I in Fig. 1.13 gives:

$$V_n = L_{n+1} + D \tag{1.33}$$

Expressing this balance for the more volatile component gives:
    Thus:

$$y_n V_n = L_{n+1} x_{n+1} + D x_d$$
$$y_n = \frac{L_{n+1}}{V_n} x_{n+1} + \frac{D}{V_n} x_d \tag{1.34}$$

This equation relates the composition of the vapour rising to the plate to the composition of the liquid on any plate above the feed plate. Since the molar liquid overflow is constant, L_n may be replaced by L_{n+1} and:

$$y_n = \frac{L_n}{V_n} x_{n+1} + \frac{D}{V_n} x_d \tag{1.35}$$

Similarly, taking a material balance for the total streams and for the more volatile component from the bottom to above plate m, as indicated by the loop II in Fig. 1.13, and noting that $L_m = L_{m+1}$ gives:

$$L_m = V_m + W \tag{1.36}$$

$$\text{and}: y_m V_m = L_m x_{m+1} - W x_w$$

Thus:

$$y_m = \frac{L_m}{V_m} x_{m+1} - \frac{W}{V_m} x_w \tag{1.37}$$

This equation, which is similar to Eq. (1.35), gives the corresponding relation between the compositions of the vapour rising to a plate and the liquid on the plate, for the section below the feed plate. These two equations are the equations of the operating lines.

    In order to calculate the change in composition from one plate to the next, the equilibrium data are used to find the composition of the vapour above the liquid, and the enrichment line to calculate the composition of the liquid on the next plate. This method may then be repeated up the column, using Eq. (1.37) for sections below the feed point, and Eq. (1.35) for sections above the feed point.

Example 1.7

A mixture of benzene and toluene containing 40 mol% benzene is to be separated to give a product containing 90 mol% benzene at the top, and a bottom product containing not more than 10 mol% benzene. The feed enters the column at its boiling point, and the vapour leaving the column which is condensed but not cooled, provides reflux and product. It is proposed to operate the unit with a reflux ratio of 3 kmol/kmol product. It is required to find the number of theoretical plates needed and the position of entry for the feed. The equilibrium diagram at $100\,kN/m2$ is shown in Fig. 1.14.

Solution

For 100 kmol of feed, an overall mass balance gives:

$100 = D + W$.

A balance on the MVC, benzene, gives:

$(100 \times 0.4) = 0.9\,D + 0.1\,W$.

Thus: $40 = 0.9(100 - W) + 0.1\,W$

and: $W = 62.5$ and $D = 37.5$ kmol

Using the notation of Fig. 1.13 then:

$L_n = 3D = 112.5$ kmol.

and: $V_n = L_n + D = 150$ kmol

Thus, the top operating line from Eq. (1.35) is:

$$\text{or:} \quad y_n = \left(\frac{112.5}{150}\right)x_{n+1} + \frac{(37.5 \times 0.9)}{150} \quad \text{(i)}$$

$$y_n = 0.75x_{n+1} + 0.225$$

Since the feed is all liquid at its boiling point, this will all run down as increased reflux to the plate below.

Thus: $L_m = L_n + F$

$= (112.5 + 100) = 212.5$ kmol

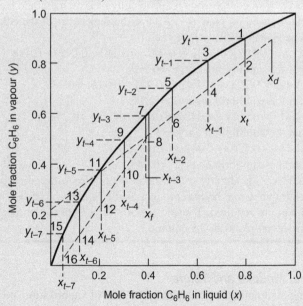

FIG. 1.14

Calculation of the number of plates by the Lewis–Sorel method for Example 1.7.

Also: $V_m = L_m - W$
$= 212.5 - 62.5 = 150 = V_n$

$$\text{Thus}: y_m = \left(\frac{212.5}{150}\right)x_{m+1} - \left(\frac{62.5}{150}\right) \times 0.1 \qquad (1.37)$$

or: $y_m = 1.417x_{m+1} - 0.042$ (ii)

With the two Equations (i) and (ii) and the equilibrium curve, the composition on the various plates may be calculated by working either from the still up to the condenser, or in the reverse direction. Since all the vapour from the column is condensed, the composition of the vapour y_t from the top plate must equal that of the product x_d, and that of the liquid returned as reflux x_r. The composition x_t of the liquid on the top plate is found from the equilibrium curve and, since it is in equilibrium with vapour of composition, $y_t = 0.90, x_t = 0.79$.

The value of y_{t-1} is obtained from Equation (i) as:
$y_{t-1} = (0.75 \times 0.79) + 0.225 = (0.593 + 0.225) = 0.818$
x_{t-1} is obtained from the equilibrium curve as 0.644
$y_{t-2} = (0.75 \times 0.644) + 0.225 = (0.483 + 0.225) = 0.708$
x_{t-2} from equilibrium curve $= 0.492$
$y_{t-3} = (0.75 \times 0.492) + 0.225 = (0.369 + 0.225) = 0.594$
x_{t-3} from the equilibrium curve $= 0.382$

This last value of composition is sufficiently near to that of the feed for the feed to be introduced on plate $(t - 3)$. For the lower part of the column, the operating line Equation (ii) will be used.

Thus: $y_{t-4} = (1.415 \times 0.382) - 0.042 = (0.540 - 0.042) = 0.498$
x_{t-4} from the equilibrium curve $= 0.298$.
$y_{t-5} = (1.415 \times 0.298) - 0.042 = (0.421 - 0.042) = 0.379$
x_{t-5} from the equilibrium curve $= 0.208$
$y_{t-6} = (1.415 \times 0.208) - 0.042 = (0.294 - 0.042) = 0.252$
x_{t-6} from the equilibrium curve $= 0.120$
$y_{t-7} = (1.415 \times 0.120) - 0.042 = (0.169 - 0.042) = 0.127$
x_{t-7} from the equilibrium curve $= 0.048$

This liquid x_{t-7} is slightly weaker than the minimum required 0.1 and it may be withdrawn as the bottom product as the liquid x_{t-6} at 0.12 does not meet the minimum required. Thus, x_{t-7} will correspond to the reboiler, and there will be seven plates in the column.

The method of McCabe and Thiele

The simplifying assumptions of constant molar heat of vaporisation, no heat losses, and no heat of mixing, lead to a constant molar vapour flow and a constant molar reflux flow in any section of the column, that is $V_n = V_{n+1}$, $L_n = L_{n+1}$, and so on. Using these simplifications, the two enrichment equations are obtained:

$$y_n = \frac{L_n}{V_n}x_{n+1} + \frac{D}{V_n}x_d \qquad (1.35)$$

for the upper operating line which is often referred to as enrichment operating line or rectifying line.

and:

$$y_m = \frac{L_m}{V_m}x_{m+1} - \frac{W}{V_m}x_w \qquad (1.37)$$

for the lower operating line which is often referred to as stripping operating line.

These equations are used in the Lewis–Sorel method to calculate the relation between the composition of the liquid on a plate and the composition of the vapour rising to that plate. McCabe and Thiele [27] pointed out that, since these equations represent straight lines connecting y_n with x_{n+1} and y_m with x_{m+1}, they can be drawn on the same diagram as the equilibrium curve to give a simple graphical solution for the number of stages required. Thus, the line of Eq. (1.35), the enrichment operating line, will pass through the points 2, 4, 6, and 8 shown in Fig. 1.14, and similarly the line of Eq. (1.37), the stripping operating line, will pass through points 8, 10, 12, and 14.

If $x_{n+1}=x_d$ in Eq. (1.35), then:

$$y_n = \frac{L_n}{V_n}x_d + \frac{D}{V_n}x_d = x_d \qquad (1.38)$$

and this equation represents a line passing through the point $y_n =x_{n+1} =x_d$. If x_{n+1} is put equal to zero, then $y_n =Dx_d/V_n$, giving a second easily determined point. The enrichment operating line is therefore drawn through two points of coordinates (x_d, x_d) and $(0, (Dx_d/V_n))$.

For the stripping operating line, Eq. (1.30), if $x_{m+1}=x_w$, then:

$$y_m = \frac{L_m}{V_m}x_w - \frac{W}{V_m}x_w \qquad (1.39)$$

Since $V_m =L_m -W$, it follows that $y_m =x_w$. Thus the stripping operating line, passes through the point C, that is (x_w, x_w), and has a slope L_m/V_m. When the two operating lines have been drawn in, the number of stages required may be found by drawing steps between the operating line and the equilibrium curve starting from point A as shown in Fig. 1.15.

This method is one of the most important concepts in chemical engineering and is an invaluable tool for the solution of distillation problems. The assumption of constant molar overflow is not limiting since in very few systems do the molar heats of vaporisation differ by more than 10%. The method does have limitations, however, and should not be employed when the relative volatility is less than 1.3 or greater than 5, when the reflux ratio is less than 1.1 times the minimum, or when more than 25 theoretical trays are required [13]. In these circumstances, the Ponchon–Savarit method described in Section 1.5 should be used.

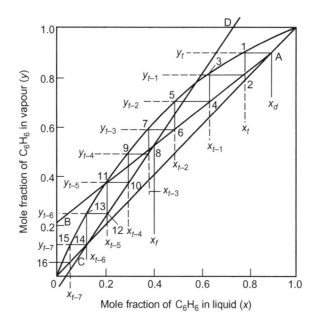

FIG. 1.15

Determination of number of plates by the McCabe–Thiele method (Example 1.8).

Example 1.8

Example 1.7 is now worked using this method. Thus, with a feed composition, $x_f = 0.4$, the top composition, x_d is to have a value of 0.9 and the bottom composition, x_w is to be 0.10. The reflux ratio, $L_n/D = 3$.

Solution

(a) From a material balance for a feed of 100 kmol:
   $V_n = V_m = 150\,\text{kmol}$; $L_n = 112.5\,\text{kmol}$; $L_m = 212.5\,\text{kmol}$; $D = 37.5\,\text{kmol}$ and $W = 62.5\,\text{kmol}$.

(b) The equilibrium curve and the diagonal line are drawn in as shown in Fig. 1.15.

(c) The equation of the enrichment operating line, is:
   $y_n = 0.75x_{n+1} + 0.225$ (i)
   Thus, the line AB is drawn through the two points A (0.9, 0.9) and B (0, 0.225).

(d) The equation of the stripping operating line is:
   $y_m = 1.415x_{m+1} - 0.042$ (ii)
   This equation is represented by the line CD drawn through C (0.1, 0.1) at a slope of 1.415.

(e) Starting at point A, the horizontal line is drawn to cut the equilibrium line at point 1. The vertical line is dropped through 1 to the operating line at point 2 and this procedure is repeated to obtain points 3 to 6.

(f) A horizontal line is drawn through point 6 to cut the equilibrium line at point 7 and a vertical line is drawn through point 7 to the stripping operating line at point 8. This procedure is repeated in order to obtain points 9 to 16.
(g) The number of stages is then counted, that is points 2, 4, 6, 8, 10, 12, and 14 which gives the <u>number of plates required as 7</u>.

Enrichment in still and condenser

Point 16 in Fig. 1.15 represents the concentration of the liquor in the still (or reboiler). The concentration of the vapour is represented by point 15, so that the enrichment represented by the increment $16-15$ is achieved in the reboiler or still body. Again, the concentration on the top plate is given by point 2, but the vapour from this plate has a concentration given by point 1, and the condenser by completely condensing the vapour gives a product of equal concentration, represented by point A. The still and condenser together, therefore, provide enrichment $(16-15)+(1-A)$, which is equivalent to one ideal stage. Thus, the actual number of theoretical plates required in the column is one less than the number of stages shown on the diagram. From a liquid in the reboiler (or still), point 16 to the product, point A, there are eight steps, so the column need only contain seven theoretical plates.

The intersection of the operating lines

It is seen from the example shown in Fig. 1.15 in which the feed enters as liquid at its boiling point that the two operating lines intersect at a point having an X-coordinate of x_f. The locus of the point of intersection of the operating lines is of considerable importance since, as will be seen, it is dependent on the temperature and physical condition of the feed.

If the two operating lines intersect at a point with coordinates (x_q, y_q), then from Eqs (1.35) and (1.37):

$$V_n y_q = L_n x_q + D x_d \tag{1.40}$$

$$\text{and}: V_m y_q = L_m x_q - W x_w \tag{1.41}$$

$$\text{or}: y_q(V_m - V_n) = (L_m - L_n)x_q - (D x_d + W x_w) \tag{1.42}$$

A material balance over the feed plate in Fig. 1.13 gives:

$$\text{or}: \quad \frac{F + L_n + V_m = L_m + V_n}{V_m - V_n = L_m - L_n - F} \tag{1.43}$$

To obtain a relation between L_n and L_m, it is necessary to make an enthalpy balance over the feed plate, and to consider what happens when the feed enters the column. If the feed is all in the form of liquid at its boiling point, the reflux L_m overflowing to the plate below will be $L_n + F$. If however the feed is a liquid at a temperature T_f, that is less than the boiling point, some vapour rising from the plate below will condense to provide sufficient heat to bring the feed liquor to the boiling point.

If H_f is the enthalpy per mole of feed, and H_{fs} is the enthalpy of 1 mol of feed at its boiling point, then the heat to be supplied to bring feed to the boiling point is $F(H_{fs} - H_f)$, and the number of moles of vapour to be condensed to provide this heat is $F(H_{fs} - H_f)/\lambda$, where λ is the molar latent heat of the vapour.

The reflux liquor is then:

$$L_m = L_n + F + \frac{F(H_{fs} - H_f)}{\lambda}$$

where :

$$= L_n + F\left(\frac{\lambda + H_{fs} - H_f}{\lambda}\right) \tag{1.44}$$

$$= L_n + qF$$

$$q = \frac{\text{heat to vaporise 1 mole of feed}}{\text{molar latent heat of the feed}}$$

Thus, from Eq. (1.43):

$$V_m - V_n = qF - F = F(q - 1) \tag{1.45}$$

A material balance of the more volatile component over the whole column gives:

$$Fx_f = Dx_d + Wx_w$$

Thus, from Eqs (1.42), (1.44), and (1.45):

$$F(q - 1)y_q = qFx_q - Fx_f$$

or :

$$y_q = \left(\frac{q}{q - 1}\right)x_q - \left(\frac{x_f}{q - 1}\right) \tag{1.46}$$

This equation is commonly known as the equation of the q-line. If $x_q = x_f$, then $y_q = x_f$. Thus, the point of intersection of the two operating lines lies on the straight line of slope $q/(q - 1)$ passing through the point (x_f, x_f). When $y_q = 0$, $x_q = x_f/q$. The line may thus be drawn through two easily determined points. From the definition of q, it follows that the slope of the q-line is governed by the nature of the feed as follows.

(a) Cold feed as liquor $q > 1$ q line slope / (positive slope)
(b) Feed at boiling point $q = 1$ q line slope | (vertical)
(c) Feed partly vapour $0 < q < 1$ q line slope \ (negative slope)
(d) Feed saturated vapour $q = 0$ q line slope — (horizontal)
(e) Feed superheated vapour $q < 0$ q line slope /(positive slope)

These various conditions are indicated in Fig. 1.16.

Altering the slope of the q-line will alter the liquid concentration at which the two operating lines cut each other for a given reflux ratio. This will mean a slight alteration in the number of plates required for the given separation. Whilst the change in the number of plates is usually rather small, if the feed is cold, there will be an increase in reflux flow below the feed plate, and hence an increased heat consumption from the reboiler per mole of distillate.

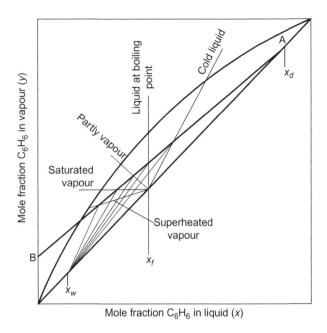

Mole fraction C_6H_6 in vapour (y)

Mole fraction C_6H_6 in liquid (x)

FIG. 1.16

Effect of the condition of the feed on the intersection of the operating lines for a fixed reflux ratio.

1.4.3 **The importance of the reflux ratio**

Influence on the number of plates required

The ratio L_n/D, that is the ratio of the top overflow to the quantity of product, is denoted by R, and this enables the equation of the operating line to be expressed in another way, which is often more convenient. Since, $V_n = L_n + D$ Eq. (1.35) becomes,

$$y_n = \left(\frac{L_n}{L_n + D}\right)x_{n+1} + \left(\frac{D}{L_n + D}\right)$$ (1.47)

Thus dividing numerators and denominators by D, and introducing $R = L_n/D$ in Eq. (1.47) gives:

$$y_n = \left(\frac{R}{R + 1}\right)x_{n+1} + \left(\frac{x_d}{R + 1}\right)$$ (1.48)

Any change in the reflux ratio R will therefore modify the slope of the enrichment line and, as may be seen from Fig. 1.15, this will alter the number of plates required for a given separation. If R is known, the top line is most easily drawn by joining point A (x_d, x_d) to B (0, $x_d/(R+1)$) as shown in Fig. 1.17. This method avoids the calculation of the actual flow rates L_n and V_n, when the number of plates only is to be estimated.

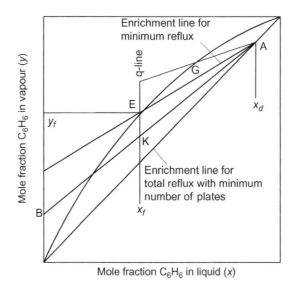

FIG. 1.17

Influence of reflux ratio on the number of plates required for a given separation.

If no product is withdrawn from the still, that is $D = 0$, then the column is said to operate under conditions of total reflux and, as seen from Eq. (1.47), the enrichment line has its maximum slope of unity, and coincides with the line $x = y$. If the reflux ratio is reduced, the slope of the enrichment line is reduced and more stages are required to pass from x_f to x_d, as shown by the line AK in Fig. 1.17. Further reduction in R will eventually bring the enrichment line to AE, where an infinite number of stages is needed to pass from x_d to x_f. This arises from the fact that under these conditions the steps become very close together at liquid compositions near to x_f, and no enrichment occurs from the feed plate to the plate above. These conditions are known as *minimum reflux*, and the reflux ratio is denoted by R_m. Any small increase in R beyond R_m will give a workable system, although a large number of plates will be required. It is important to note that any line such as AG, which is equivalent to a smaller value of R than R_m, represents an impossible condition, since it is impossible to pass beyond point G towards x_f as the q-line and the line AG meet outside the operating envelope between the equilibrium curve and the line $x = y$. Two important deductions may be made. Firstly that the minimum number of plates is required for a given separation at conditions of total reflux, and secondly that there is a minimum reflux ratio below which it is impossible to obtain the desired enrichment, however many plates are used.

Calculation of the minimum reflux ratio

Fig. 1.17 represents conditions where the q-line is vertical, and the point E lies on the equilibrium curve and has co-ordinates (x_f, y_f). The slope of the line AE is then given by.

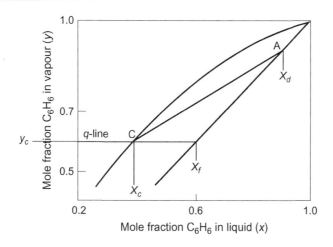

FIG. 1.18

Minimum reflux ratio with feed as saturated vapour.

$(y_d - y_f)/(x_d - x_f)$. Since the point A is on line $x = y$ hence $x_d = y_d$,

$$\left(\frac{R_m}{R_m + 1}\right) = \left(\frac{x_d - y_f}{x_d - x_f}\right)$$

or :

$$R_m = \left(\frac{x_d - y_f}{y_f - x_f}\right)$$

(1.49)

If the q-line is horizontal as shown in Fig. 1.18, the enrichment line for minimum reflux is given by AC, where C has coordinates (x_c, y_c). Thus:
or, since $y_c = x_f$:

$$\left(\frac{R_m}{R_m + 1}\right) = \left(\frac{x_d - y_c}{x_d - x_c}\right)$$

$$R_m = \left(\frac{x_d - y_c}{y_c - x_c}\right) = \left(\frac{x_d - x_f}{x_f - x_c}\right)$$

(1.50)

Underwood and Fenske equations

For ideal mixtures, or where over the concentration range concerned the relative volatility may be taken as constant, R_m may be obtained analytically from the physical properties of the system as discussed by Underwood [28]. Thus, if x_{nA} and x_{nB} are the mole fractions of two components **A** and **B** in the liquid on any plate n, then a material balance over the top portion of the column above plate n gives:

$$V_n y_{nA} = L_n x_{(n+1)A} + D x_{dA}$$

(1.51)

$$\text{and} : V_n y_{nB} = L_n x_{(n+1)B} + D x_{dB}$$

(1.52)

Under conditions of minimum reflux, a column has to have an infinite number of plates, or alternatively the composition on plate n is equal to that on plate $n+1$. Dividing Eq. (1.51) by Eq. (1.52) and using the relations $x_{(n+1)A} = x_{nA}$ and $x_{(n+1)B} = x_{nB}$, then:

Thus:

$$\frac{\alpha x_{nA}}{x_{nB}} = \frac{y_{nA}}{y_{nB}} = \frac{L_n x_{nA} + D x_{dA}}{L_n x_{nB} + D x_{dB}}$$

$$R_m = \left(\frac{L_n}{D}\right)_{min} = \frac{1}{\alpha - 1}\left[\frac{x_{dA}}{x_{nA}} - \alpha\left(\frac{x_{dB}}{x_{nB}}\right)\right] \tag{1.53}$$

In this analysis, α is taken as the volatility of **A** relative to **B**. There is, in general, therefore a different value of R_m for each plate. In order to produce any separation of the feed, the minimum relevant value of R_m is that for the feed plate, so that the minimum reflux ratio for the desired separation is given by:

$$R_m = \frac{1}{(\alpha - 1)}\left[\frac{x_{dA}}{x_{fA}} - \alpha\frac{x_{dB}}{x_{fB}}\right] \tag{1.54}$$

For a binary system, this becomes:

$$R_m = \frac{1}{(\alpha - 1)}\left[\frac{x_{dA}}{x_{fA}} - \alpha\frac{(1 - x_{dA})}{(1 - x_{fA})}\right] \tag{1.55}$$

This relation may be obtained by putting $y = \alpha x/[1 + (\alpha - 1)x]$ from Eq. (1.15), in Eq. (1.49) to give:

$$R_m = \frac{x_d - \left(\frac{\alpha x_f}{1 + (\alpha-1)x_f}\right)}{\left(\frac{\alpha x_f}{1 + (\alpha-1)x_f}\right) - x_f} = \frac{1}{(\alpha - 1)}\left[\frac{x_d}{x_f} - \frac{\alpha(1 - x_d)}{(1 - x_f)}\right] \tag{1.56}$$

The number of plates at total reflux. Fenske's method

For conditions in which the relative volatility is constant, Fenske [29] derived an equation for calculating the required number of plates for a desired separation. Since no product is withdrawn from the still, the equations of the two operating lines become:

$$y_n = x_{n+1} \quad \text{and} \quad y_m = x_{m+1} \tag{1.57}$$

If for two components **A** and **B**, the concentrations in the still are x_{sA} and x_{sB}, then the composition on the first plate is given by:

$$\left(\frac{x_A}{x_B}\right)_1 = \left(\frac{y_A}{y_B}\right)_s = \alpha_s\left(\frac{x_A}{x_B}\right)_s$$

where the subscript outside the bracket indicates the plate, and s the still or the reboiler.

For plate 2: $\left(\frac{x_A}{x_B}\right)_2 = \left(\frac{x_A}{x_B}\right)_1 = \left(\frac{y_A}{y_B}\right)_s = \alpha_s\left(\frac{x_A}{x_B}\right)_s$

and for plate n:

$$\left(\frac{x_A}{x_B}\right)_n = \left(\frac{y_A}{y_B}\right)_{n-1} = \alpha_1\alpha_2\alpha_3\ldots\alpha_{n-1}\alpha_s\left(\frac{x_A}{x_B}\right)_s$$

If an average value of α is used, then:

$$\left(\frac{x_A}{x_B}\right)_n = \alpha_{av}^n\left(\frac{x_A}{x_B}\right)_s$$

In most cases total condensation occurs in the condenser, so that:

$$\left(\frac{x_A}{x_B}\right)_d = \left(\frac{y_A}{y_B}\right)_n = \alpha_n\left(\frac{x_A}{x_B}\right)_n = \alpha_{av}^{n+1}\left(\frac{x_A}{x_B}\right)_s$$

$$n+1 = \frac{\log\left[\left(\frac{x_A}{x_B}\right)_d\left(\frac{x_B}{x_A}\right)_s\right]}{\log\alpha_{av}}$$

(1.58)

and n is the required number of theoretical plates in the column.

It is important to note that, in this derivation, only the relative volatilities of two components have been used. The same relation may be applied to two components of a multicomponent mixture, as is seen in Section 1.7.6.

Example 1.9

For the separation of a mixture of benzene and toluene, considered in Example 1.7, $x_d = 0.9$, $x_w = 0.1$, and $x_f = 0.4$. If the mean volatility of benzene relative to toluene is 2.4, what is the number of plates required at total reflux?

Solution

The number of plates at total reflux is given by:

$$n+1 = \frac{\log\left[\left(\frac{0.9}{0.1}\right)\left(\frac{0.9}{0.1}\right)\right]}{\log 2.4} = 5.0$$

(1.58)

Thus the number of theoretical plates, n, in the column is $\underline{4}$, a value which is independent of the feed composition.

If the feed is liquid at its boiling point, then the minimum reflux ratio R_m is given by:

$$R_m = \frac{1}{\alpha-1}\left[\frac{x_d}{x_f} - \alpha\frac{(1-x_d)}{(1-x_f)}\right]$$

$$= \frac{1}{2.4-1}\left[\frac{0.9}{0.4} - \frac{(2.4 \times 0.1)}{0.6}\right]$$

(1.56)

$$= \underline{1.32}$$

Using the graphical construction shown in Fig. 1.18, with $y_f = 0.61$, the value of R_m using Eq. (1.5) is:

$$R_m = \frac{x_d - y_f}{y_f - x_f} = \frac{(0.9 - 0.61)}{(0.61 - 0.4)} = \underline{\underline{1.38}}$$

Selection of economic reflux ratio

The cost of a distillation unit includes the capital cost of the column, determined largely by the number and diameter of the plates, and the operating costs, determined by the steam and cooling water requirements. The depreciation charges may be taken as a percentage of the capital cost, and the two together taken as the overall charges. The steam required will be proportional to V_m, which may be taken as V_n where the feed is liquid at its boiling point. From a material balance over the top portion of the column, $V_n = D(R + 1)$, and hence the steam required per mole of product is proportional to $(R + 1)$. This will be a minimum when R equals R_m, and will steadily rise as R is increased. The relationship between the number of plates n and the reflux ratio R, as derived by Gilliland [30], is discussed in Section 1.7.7.

The reduction in the required number of plates as R is increased beyond R_m will tend to reduce the cost of the column. For a column separating a benzene–toluene mixture, for example, where $x_f = 0.79$, $x_d = 0.99$ and $x_w = 0.01$, the numbers of theoretical plates as given by the McCabe–Thiele method for various values of R are given as follows. The minimum reflux ratio for this case is 0.81.

Reflux ratio R	0.81	0.9	1.0	1.1	1.2
Number of plates	∞	25	22	19	18

Thus, an increase in R, at values near R_m, gives a marked reduction in the number of plates, although at higher values of R, further increases have little effect on the number of plates. Increasing the reflux ratio from R_m therefore affects the capital and operating costs of a column as follows:

(a) The operating costs rise and are approximately proportional to $(R + 1)$.
(b) The capital cost initially falls since the number of plates falls off rapidly at this stage.
(c) The capital cost rises at high values of R, since there is then only a very small reduction in the number of plates, although the diameter, and hence the area, continually increases because the vapour load becomes greater. The associated condenser and reboiler will also be larger and hence more expensive.

The total charges may be obtained by adding the fixed and operating charges as shown in Fig. 1.19, where curve A shows the steam costs and B the fixed costs. The final total is shown by curve C which has a minimum value corresponding to the economic reflux ratio. There is no simple relation between R_m and the optimum value, although practical values are generally 1.1–1.5 times the minimum, with much higher values being employed, particularly in the case of vacuum distillation. It may be noted that, for a fixed degree of enrichment from the feed to the top product, the number of trays required increases rapidly as the difficulty of separation increases, that is as the relative volatility approaches unity. A demand for a higher purity of product necessitates a very considerable increase in the number of trays, particularly

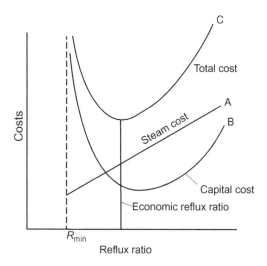

FIG. 1.19

Influence of reflux ratio on capital and operating costs of a still.

when α is near unity. In these circumstances, only a limited improvement in product purity may be obtained by increasing the reflux ratio. The designer must be careful to consider the increase in cost of plant resulting from specification of a higher degree of purity of production and at the same time assess the highest degree of purity that may be obtained with the proposed plant.

In general, the greater the reflux ratio, the lower is the number of plates or transfer units required although the requirements of steam in the reboiler cooling water in the condenser and power requirements for the reflux pump are increased and a column of larger diameter is required in order to achieve acceptable vapour velocities. In addition, larger condenser, reboiler, and reflux pumps are required. An optimum value of the reflux ratio may be obtained by using the following argument which is based on the work of Colburn [31].

The annual capital cost of a distillation column, c_c per mole of distillate, including depreciation, interest and overheads, may be written as:

$$c_c = c_a An/(Et_a D) \tag{1.59}$$

where c_a is the annual cost of the column per unit area of plate, A is the cross-sectional area of the column, n is the number of theoretical plates, E is the plate efficiency, t_a is the annual period of operation and D is the molar flow rate of distillate. The cross-sectional area of the column is given by:

$$A = V/u' \tag{1.60}$$

where V is the molar flow of vapour and u' is the allowable molar vapour velocity per unit area. Since $V = D(R+1)$, where R is the reflux ratio, then the cost of the column is:

$$c_c = c_a n(R+1)/(Et_a u') \tag{1.61}$$

The annual cost of the reboiler and the condenser, c_h per mole of distillate may be written as:

$$c_h = c_b A_h / (t_a D) \tag{1.62}$$

where c_b is the annual cost of the heat exchange equipment per unit area including depreciation and interest and A_h is the area for heat transfer. $A_h = V/N''$ where N'' is the vapour handling capacity of the boiler and condenser in terms of molar flow per unit area. Thus, $A_h = D(R+1)/N''$ and the cost of the reboiler and the condenser is:

$$c_h = c_d (R+1) / (t_a N'') \tag{1.63}$$

As far as operating costs are concerned, the important annual variable costs are that of the steam in the reboiler and that of the cooling water in the condenser. These may be written as:

$$c_w = c_d V / D = c_3 (R+1) \tag{1.64}$$

where c_d is the annual cost of the steam and the cooling water. Pump power and cooling water costs can be assumed to be small relative to steam costs. The total annual cost, c per mole of distillate, is the cost of the steam and the cooling water plus the costs of the column, reboiler, and condenser, or:

$$c = (R+1)[(c_a n / Et_a u') + (c_b / t_a N'') + c_d] \tag{1.65}$$

As the number of plates, n, is a function of R, Eq. (1.65) may be differentiated with respect to R to give:

$$dc/dR = c_a n / (Et_a V') + [(c_a / (Et_a u'))(R+1)dn/dR] + c_b / (t_a N'') + c_d \tag{1.66}$$

Equating to zero for minimum cost, the optimum value of the reflux ratio is:

$$R_{opt} + 1 = (n_{opt} + F)/(-dn/dR) \tag{1.67}$$

where n_{opt} is the optimum number of theoretical plates corresponding to R_{opt} and the cost factor, F, is:

$$F = [c_d + c_b / (t_a N'')][(Et_a u')/c_a] \tag{1.68}$$

Because there is no simple equation relating n and dn/dR, it is not possible to obtain an expression for R_{opt} although a method of solution is given in the Example 1.18 which is based on the work of Harker [32].

In practice, values of 110% to 150% of the minimum reflux ratio are used although higher values are sometimes employed particularly in vacuum distillation. Where a high-purity product is required, only limited improvements can be obtained by increasing the reflux ratio and since there is a very large increase in the number of trays required, an arrangement by which the minimum acceptable purity is achieved in the product is usually adopted.

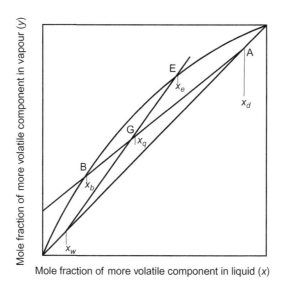

FIG. 1.20

Location of feed point.

1.4.4 Location of feed point in a continuous still

From Fig. 1.20, it may be seen that, when stepping off plates down the top operating line AB, the bottom operating line CE cannot be used until the value of x_n on any plate is less than x_e. Again it is essential to pass to the lower line CE by the time $x_n = x_b$. The best conditions are those where the minimum number of plates is used. From the geometry of the figure, the largest steps in the enriching section occur down to the point of intersection of the operating lines at $x = x_q$. Below this value of x, the steps are larger on the lower operating line. Thus, although the column will operate for a feed composition between x_e and x_b, the minimum number of plates will be required if $x_f = x_q$. For a binary mixture at its boiling point, this is equivalent to making x_f equal to the composition of the liquid on the feed plate.

1.4.5 Multiple feeds and sidestreams

In general, a sidestream is defined as any product stream other than the overhead product and the residue such as the streams S', S'', and S''' in Fig. 1.21. In a similar way, F_1 and F_2 are separate feed streams to the column. Sidestreams are most often removed with multicomponent systems, although they may be used with binary mixtures.

A binary system is now considered, with one sidestream, as shown in Fig. 1.22. S' represents the rate of removal of the sidestream and x_s' its composition.

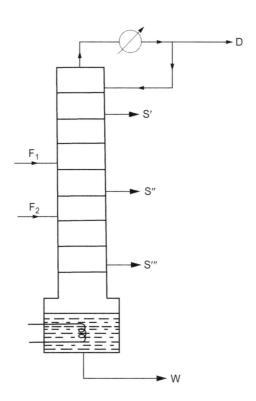

FIG. 1.21

Column with multiple feeds and sidestreams.

Assuming constant molar overflow, then for the part of the column above the sidestream the operating line is given by:

$$y_n = \frac{L_n}{V_n}x_{n+1} + \frac{Dx_d}{V_n} \tag{1.35}$$

as before. Balances for the part of the tower above a plate between the feed plate and the sidestream give:

$$V_s = L_s + S' + D \tag{1.69}$$

$$\text{and}: V_s y_n = L_s x_{n+1} + S'x_{s'} + Dx_d \tag{1.70}$$

$$\text{Thus}: y_n = \frac{L_s}{V_s}x_{n+1} + \frac{S'x_{s'} + Dx_d}{V_s} \tag{1.71}$$

Since the sidestream is normally removed as a liquid, $L_s = (L_n - S')$ and $V_s = V_n$.

The line represented by Eq. (1.35) has a slope L_n/V_n and passes through the point (x_d, x_d). Equation (1.71) represents a line of slope L_s/V_s, which passes through the point $y = x = (S'x_{s'} + Dx_d)/(S' + D)$, which is the mean molar composition of the

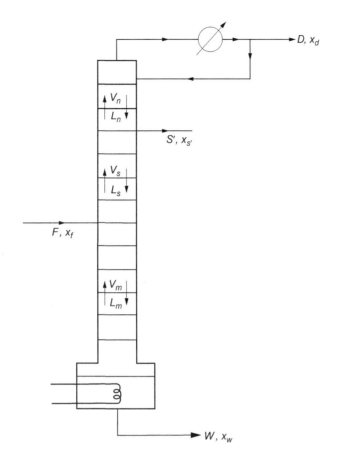

FIG. 1.22

Column with a sidestream.

overhead product and sidestream. Since $x_{s'} < x_d$, and $L_s < L_n$, this additional operating line cuts the $y = x$ line at a lower value than the upper operating line though it has a smaller slope, as shown in Fig. 1.23. The two lines intersect at $x = x_{s'}$. Plates are stepped off as before between the appropriate operating line and the equilibrium curve. It may be seen that the removal of a sidestream increases the number of plates required, due to the decrease in liquid rate below the sidestream.

The effect of any additional sidestream or feed is to introduce an additional operating line for each stream. In all other respects the method of calculation is identical with that used for the straight separation of a binary mixture.

The Ponchon–Savarit method, using an enthalpy–composition diagram, may also be used to handle sidestreams and multiple feeds, though only for binary systems. This is dealt with in Section 1.5.

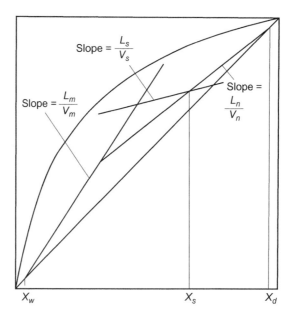

FIG. 1.23

Effect of a sidestream.

1.5 Conditions for varying overflow in non-ideal binary systems

1.5.1 The heat balance

In previous sections, the case of constant molar latent heat has been considered with no heat of mixing, and hence a constant molar rate of reflux in the column. These simplifying assumptions are extremely useful in that they enable a simple geometrical method to be used for finding the change in concentration on the plates and, whilst they are rarely entirely true in industrial conditions, they often provide a convenient start for design purposes. For a non-ideal system, where the molar latent heat is no longer constant and where there is a substantial heat of mixing, the calculations become much more tedious. For binary mixtures of this kind a graphical model has been developed by Ruhemann [33], Ponchon [34], and Savarit [35], based on the use of an enthalpy–composition chart. A typical enthalpy–composition or $H - x$ chart is shown in Fig. 1.24, where the upper curve V is the dew-point curve, and the lower curve L is the boiling-point curve. The use of this diagram is based on the geometrical properties, as illustrated in Fig. 1.25. A quantity of mixture in any physical state is known as a 'phase' and is denoted by mass, composition and enthalpy. The phase is shown upon the diagram by a point which shows enthalpy and composition, though it

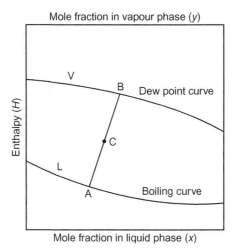

FIG. 1.24

Enthalpy–composition diagram, showing the enthalpies of liquid and vapour.

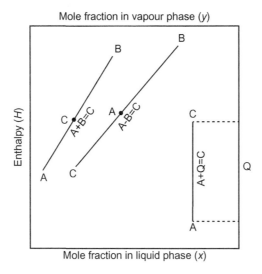

FIG. 1.25

Combination and separation of a mixture on an enthalpy–composition diagram.

does not show the mass. If m is the mass, x the composition and H the enthalpy per unit mass, then the addition of two phases A and B to give phase C is governed by:

$$m_A + m_B = m_C \tag{1.72}$$

$$m_A x_A + m_B x_B = m_C x_C \tag{1.73}$$

$$\text{and}: m_A H_A + m_B H_B = m_C H_C \tag{1.74}$$

Similarly, if an amount Q of heat is added to a mass m_A of a phase, the increase in enthalpy from H_A to H_C will be given by:

$$H_A + \frac{Q}{m_A} = H_C \qquad (1.75)$$

Thus, the addition of two phases A and B is shown on the diagram by point C on the straight line joining the two phases, whilst the difference $(A - B)$ is found by a point C on the extension of the line AB. If, as shown in Fig. 1.24, a phase represented by C in the region between the dew-point and boiling-point curves is considered, then this phase will divide into two phases A and B at the ends of a tie line through the point C, so that:

$$\frac{m_A}{m_B} = \frac{CB}{CA} \qquad (1.76)$$

The $H - x$ chart, therefore, enables the effect of adding two phases, with or without the addition of heat, to be determined geometrically. The diagram may be drawn for unit mass or for 1 mol of material, although as a constant molar reflux does not now apply, it is more convenient to use unit mass as the basis. Thus, working with unit mass of product, the mass of the individual streams as proportions of the product are calculated.

Fig. 1.26 represents a continuous distillation unit operating with a feed F of composition x_f, and giving a top product D of composition x_d and a bottom product W of composition x_w. In this analysis, the quantities in the streams V of rising vapour and L of reflux are given in mass units, such as kg/s, and the composition of the streams as mass fractions, x referring to the liquid and y to the vapour streams as usual.

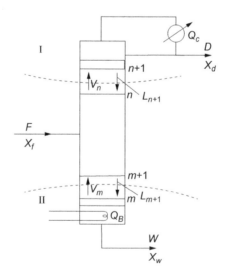

FIG. 1.26

Continuous distillation column.

The plates are numbered from the bottom upwards, subscript n indicating the rectifying (or enrichment) and m the stripping section.

H^V and H^L represent the enthalpy per unit mass of a vapour and liquid stream respectively.

Q_C is the heat removed in the condenser. In this case no cooling of product is considered.

Q_B is the heat added in the reboiler.

The following relationships are then obtained by taking material and heat balances:

$$\text{or}: \quad \begin{aligned} V_n &= L_{n+1} + D \\ V_n - L_{n+1} &= D \end{aligned} \tag{1.77}$$

$$\text{or}: \quad \begin{aligned} V_n y_n &= L_{n+1} x_{n+1} + D x_d \\ V_n y_n - L_{n+1} x_{n+1} &= D x_d \end{aligned} \tag{1.78}$$

$$\text{or}: \quad \begin{aligned} V_n H_n^V &= L_{n+1} H_{n+1}^L + D H_d^L + Q_c \\ V_n H_n^V - L_{n+1} H_{n+1}^L &= D H_d^L + Q_c \end{aligned} \tag{1.79}$$

Putting $H_d' = H_d^L + Q_C/D$ then Eq. (1.79) may be written as:

$$\text{or}: \quad \begin{aligned} V_n H_n^V &= L_{n+1} H_{n+1}^L + D H_d' \\ V_n H_n^V - L_{n+1} H_{n+1}^L &= D H_d' \end{aligned} \tag{1.80}$$

From Eqs (1.77) and (1.78):

$$\frac{L_{n+1}}{D} = \frac{x_d - y_n}{y_n - x_{n+1}} \tag{1.81}$$

and from Eqs (1.77) and (1.80):

$$\frac{L_{n+1}}{D} = \frac{H_d' - H_n^V}{H_n^V - H_{n+1}^L} \tag{1.82}$$

$$\text{or}: \quad \frac{H_d' - H_n^V}{H_n^V - H_{n+1}^L} = \frac{x_d - y_n}{y_n - x_{n+1}} \tag{1.83}$$

$$\text{and}: \quad y_n = \left[\frac{H_d' - H_n^V}{H_d' - H_{n+1}^L}\right] x_{n+1} + \left[\frac{H_n^V - H_{n+1}^L}{H_d' - H_{n+1}^L}\right] x_d \tag{1.84}$$

Equation (1.84) represents any operating line relating the composition of the vapour y_n rising from a plate to the composition of the liquid reflux entering the plate, or alternatively it represents the relation between the composition of the vapour and liquid streams between any two plates. From Eq. (1.83), it may be seen that all such operating lines pass through a common pole N of coordinates x_d and H_d'.

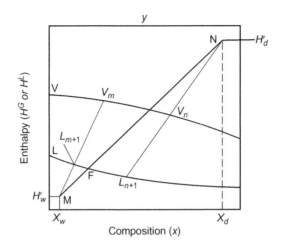

FIG. 1.27

Enthalpy–composition diagram.

Alternatively, noting that the right-hand side of Eqs (1.77), (1.78), and (1.79) are independent of conditions below the feed plate, a stream N may be defined with mass equal to the difference between the vapour and liquid streams between two plates, of composition x_d and of enthalpy H'_d. The three quantities V_n, L_{n+1}, and N are then on a straight line passing through N, as shown in Fig. 1.27.

Below the feed plate a similar series of equations for material and heat balances may be written as:

or:

$$V_m + W = L_{m+1}$$
$$-V_m + L_{m+1} = W$$

(1.85)

or :

$$V_m y_m + W x_w = L_{m+1} x_{m+1}$$
$$-V_m y_m + L_{m+1} x_{m+1} = W x_w$$

(1.86)

or :

$$V_m H_m^V + W H_w^L = L_{m+1} H_{m+1}^L + Q_B$$
$$-V_m H_m^V + L_{m+1} H_{m+1}^L = W H_w^L - Q_B$$

(1.87)

Putting:

$$H'_w = H_w^L - \frac{Q_B}{W}$$

(1.88)

then:

$$-V H_m^V + L_{m+1} H_{m+1}^L = W H'_w$$

(1.89)

Then :

$$\frac{L_{m+1}}{W} = \frac{-x_w + y_m}{y_m - x_{m+1}}$$

(1.90)

$$\text{and}: \frac{L_{m+1}}{W} = \frac{-H'_w + H^V_m}{H^V_m - H^L_{m+1}} \tag{1.91}$$

$$\text{Thus}: \frac{-H'_w + H^V_m}{H^V_m - H^L_{m+1}} = \frac{-x_w + y_m}{y_m - x_{m+1}} \tag{1.92}$$

Equation (1.92) represents any operating line below the feed plate, and it shows that all such lines pass through a common pole M of coordinates x_w and H'_w. As with the rectifying section, a stream M may be defined by mass $L_{m+1} - V_m$, composition x_w and enthalpy H'_w, Thus:

$$F = M + N \tag{1.93}$$

$$\text{and}: Fx_f = Mx_w + Nx_d \tag{1.94}$$

It therefore follows that phases F, M, and N are on a straight line on the $H - x$ chart, as shown in Figs 1.27 and 1.28.

1.5.2 Determination of the number of plates on the $H - x$ diagram

The determination of the number of plates necessary for a desired separation is shown in Fig. 1.28. The position of the feed (F, x_f) is shown at F on the boiling line and the pole N is located as (x_d, H'_d), where:

$$H'_d = H^L_d + \frac{Q_C}{D} \tag{1.95}$$

Pole M is located as on the extension of NF cutting the ordinate at x_w in M.

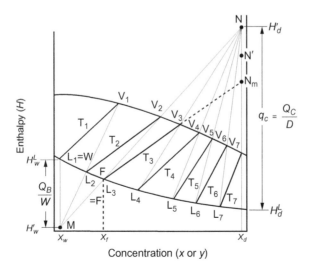

FIG. 1.28

Determination of the number of plates using the enthalpy–composition diagram.

The condition of the vapour leaving the top plate is shown at V_7 on the dew-point curve with abscissa x_d. The condition of the liquid on the top plate is then found by drawing the tie line T_7 from V_7 to L_7 on the boiling curve. The condition V_6 of the vapour on the second plate is found, from Eq. (1.77), by drawing L_7N to cut the dew-point curve on V_6. L_6 is then found on the tie line T_6. The conditions of vapour and liquid V_5, V_4, V_3 and L_5, L_4 are found in the same way. Tie line T_3 gives L_3, which has the same composition as the feed. V_2 is then found using the line MFV_2, as this represents the vapour on the top plate of the stripping section. L_2, L_1 and V_1 are then found by a similar construction. L_1 has the required composition of the bottoms, x_w.

Alternatively, calculations may start with the feed condition and proceed up and down the column.

1.5.3 Minimum reflux ratio

The pole N has coordinates $[x_d, H_d^L + Q_C/D]$. Q_C/D is the heat removed in the condenser per unit mass of product, as liquid at its boiling point and is represented as shown in Fig. 1.28. The number of plates in the rectifying section is determined, for a given feed x_f and product x_d, by the height of this pole N. As N is lowered to say N' the heat q_c falls, although the number of plates required increases. When N lies at N_m on the isothermal through F, q_c is a minimum although the number of plates required becomes infinite. Since the tie lines have different slopes, it follows that there is a minimum reflux for each plate, and the tie line cutting the vertical axis at the highest value of H will give the minimum practical reflux. This will frequently correspond to the tie line through F.

From Eqs (1.83) and (1.95) and writing $Q_C/D = q_c$, then:

$$\frac{H_d^L + q_c - H_n^V}{H_n^V - H_{n+1}^L} = \frac{x_d - y_n}{y_n - x_{n+1}} \tag{1.96}$$

$$\text{or}: q_c = \left(H_n^V - H_{n+1}^L\right)\left(\frac{x_d - y_n}{y_n - x_{n+1}}\right) + H_n^V - H_d^L \tag{1.97}$$

$$\text{and}: (q_c)_{min} = \left(H_f^V - H_{f+1}^L\right)\left(\frac{x_d - y_f}{y_f - x_{f+1}}\right) + H_f^V - H_d^L \tag{1.98}$$

The advantage of the $H - x$ chart lies in the fact that the heat quantities required for the distillation are clearly indicated. Thus, the higher the reflux ratio the more heat must be removed per mole of product, and point N rises. This immediately shows that both q_c and Q_B are increased. The use of this method is illustrated by considering the separation of ammonia from an ammonia–water mixture, as occurs in the ammonia absorption unit for refrigeration.

Example 1.10

It is required to separate 1 kg/s (3.6 t/h) of a solution of ammonia in water, containing 30% by mass of ammonia, to give a top product of 99.5% purity and a weak solution containing 10% by mass of ammonia.

Calculate the heat required in the boiler and the heat to be rejected in the condenser, assuming a reflux 8% in excess of the minimum and a column pressure of $1000\,kN/m^2$. The plates may be assumed to have an ideal efficiency of 60%.

Solution

Taking a material balance for the whole throughput and for the ammonia gives:

$D+W=1.0$.

$0.995D+0.1\,w=(1.0\times0.3)$

Thus: $D=0.22\,kg/s$

and: $W=0.78\,kg/s$

The enthalpy–composition chart for this system is shown in Fig. 1.29. It is assumed that the feed F and the bottom product W are both liquids at their boiling points.

Location of the poles N and M

N_m for minimum reflux is found by drawing a tie-line through F, representing the feed, to cut the line $x=0.995$ at N_m.

$$\text{The minimum reflux ratio,}\ \ R_m=\frac{\text{length}\,N_m A}{\text{length}\,AL}$$

$$=\frac{(1952-1547)}{(1547-295)}=0.323$$

Since the actual reflux is 8% above the minimum, then:

$NA=1.08\,N_mA$.

$=(1.08\times405)=437$

Point N therefore has an ordinate of $(437+1547)=1984$ and an abscissa of 0.995.

Point M is found by drawing NF to cut the line $x=0.10$, through W, at M.

The number of theoretical plates is found, as on the diagram, to be 5+.

The number of plates to be provided $=(5/0.6)=8.33$, say 9. Always rounded up to next integer as we can't have partial plates in a tray column.

The feed is introduced just below the third ideal plate from the top, or just below the fifth actual plate.

The heat input at the boiler per unit mass of bottom product is:

$\frac{Q_B}{W}=582-(-209)=791\,kJ/kg$

Since W is 0.78 kg/s,

Heat input to boiler $=(791\times0.78)=\underline{617\,kW}$.

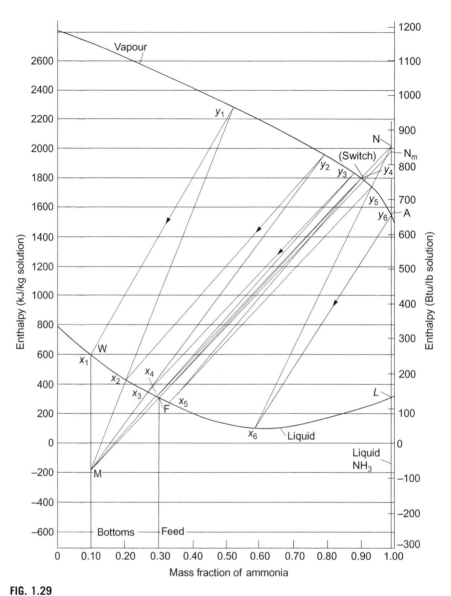

FIG. 1.29

Enthalpy–composition diagram for ammonia–water at 1.0 MN/m2 pressure (Example 1.10).

Condenser duty $=$ length $NL \times D$.
where, $NL = (1984 - 296) = 1688$ kJ/kg,
Since D is 0.22 kg/s.
Condenser duty $= 1688 \times 0.22 = \underline{372\,kW.}$

1.5.4 **Multiple feeds and sidestreams**

The enthalpy–composition approach may also be used for multiple feeds and side-streams for binary systems. For the condition of constant molar overflow, each additional sidestream or feed adds a further operating line and pole point to the system.

Taking the same system as used in Fig. 1.22, with one sidestream only, the procedure is as shown in Fig. 1.30.

The upper pole point N is located as before. The effect of removing a sidestream S' from the system is to produce an effective feed F', where $F' = F - S'$ and where $F'S'/F'F = F/S'$. Thus, once S' and F' have been located in the diagram, the position of F' may also be determined. The position of the lower pole point M, which must lie on the intersection of $x = x_w$ and the straight line drawn through NF', may then be found. N relates to the section of the column above the sidestream and M to that part below the feed plate. A third pole point must be defined to handle that part of the column between the feed and the sidestream.

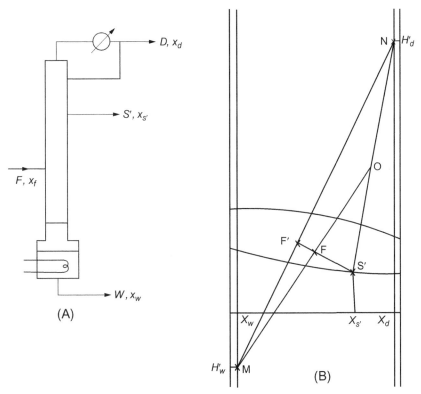

FIG. 1.30

Enthalpy-composition diagram for a system with one sidestream.

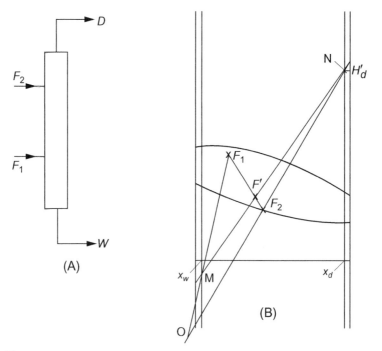

FIG. 1.31

Enthalpy–composition diagram for a system with two feeds.

The pole point for the intermediate section must be on the limiting operating line for the upper part of the column, that is NS'. This must also lie on the limiting operating line for the lower part of the column, that is MF or its extension. Thus the intersection of NS' and MF extended gives the position of the intermediate pole point O.

The number of stages required is determined in the same manner as before, using the upper pole point N for that part of the column between the sidestream and the top, the intermediate pole point O between the feed and the sidestream, and the lower pole point M between the feed and the bottom.

For the case of multiple feeds, the procedure is similar and may be followed by reference to Fig. 1.31.

Example 1.11

A mixture containing equal parts by mass of carbon tetrachloride and toluene is to be fractionated to give an overhead product containing 95 mass per cent carbon tetrachloride, a bottom product of 5 mass per cent carbon tetrachloride, and a sidestream containing 80 mass per cent carbon tetrachloride. Both the feed and sidestream may be regarded as liquids at their boiling points.

The rate of withdrawal of the sidestream is 10% of the column feed rate and the external reflux ratio is 2.5. Using the enthalpy composition method, determine the number of theoretical stages required, and the amounts of bottom product and distillate as percentages of the feed rate.

It may be assumed that the enthalpies of liquid and vapour are linear functions of composition. Enthalpy and equilibrium data are provided.

Solution

Basis: 100 kg feed.

An overall material balance gives:

$F = D + W + S'$.

or: $100 = D + W + 10$

$Fx_f = Dx_d + Wx_w + S'x_{s'}$.

and: $50 = 0.95D + 0.05W + 8$

Thus: $\underline{D = 41.7\%}$; $\underline{W = 48.3\%}$

From the enthalpy data and the reflux ratio, the upper pole point M may be located as shown in Fig. 1.32. Points F and S' are located on the liquid line, and the position of the effective feed, such that $F'S'/F'F = 10$. NF' is joined and extended to cut $x = x_w$ at M, the lower pole point.

MF is Joined and extended to cut NS' at O, the immediate pole point. The number of stages required is then obtained from the figure and in this case 13 theoretical stages are required.

1.6 Batch distillation

1.6.1 The process

Distillation emerged as a process to produce ethanol in early days when simple batch stills were used. Crude ethanol was placed in a still and heated, and the vapour drawn from the still was condensed for consumption. Lamp oil was later produced using the same method, with crude oil heated in batch stills.

In the previous sections conditions have been considered in which there has been a continuous feed to the still and a continuous withdrawal of products from the top and bottom. In many instances processes are carried out in batches, and it is more convenient to distil each batch separately. In these cases the whole of a batch is run into the boiler of the still and, on heating, the vapour is passed into a fractionation column, as shown in Fig. 1.33. As with continuous distillation, the composition of the top product depends on the still composition, the number of plates in the column and on the reflux ratio used. When the still is operating, since the top product will be relatively rich in the more volatile component, the liquid remaining in the still will become steadily weaker in this component.

As a result, the purity of the top product will steadily fall. Thus, the still may be charged with S_1 moles of a mixture containing a mole fraction x_{s1} of the more volatile component. Initially, with a reflux ratio R_1, the top product has a composition x_{d1}. If

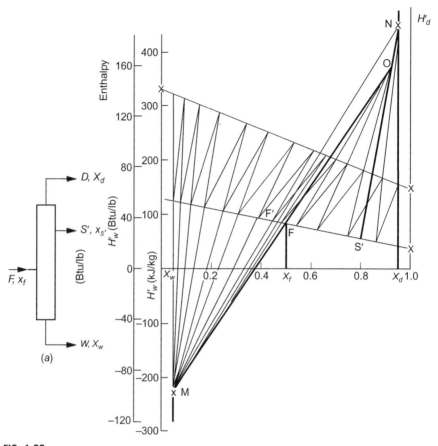

FIG. 1.32

Enthalpy–composition diagram for carbon tetrachloride–toluene separation with one sidestream—Example 1.11.

after a certain interval of time the composition of the top product starts to fall, then, if the reflux ratio is increased to a new value R_2, it will be possible to obtain the same composition at the top as before, although the composition in the still is weakened to x_{s2}. This method of operating a batch still requires a continuous increase in the reflux ratio to maintain a constant quality of the top product.

An alternative method of operation is to work with a constant reflux ratio and allow the composition of the top product to fall. For example, if a product of composition 0.9 with respect to the more volatile component is required, the composition initially obtained may be 0.95, and distillation is allowed to continue until the composition has fallen to some value below 0.9, say 0.82. The total product obtained will then have the required composition, provided the amounts of a given purity are correctly chosen.

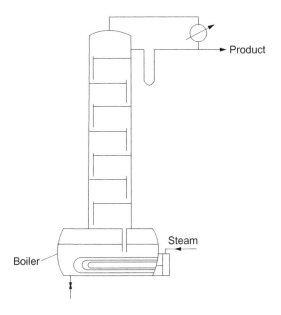

FIG. 1.33

Column for batch distillation.

One of the added merits of batch distillation is that more than one product may be obtained. Thus, a binary mixture of alcohol and water may be distilled to obtain initially a high quality alcohol. As the composition in the still weakens with respect to alcohol, a second product may be removed from the top with a reduced concentration of alcohol. In this way, it is possible to obtain not only two different quality products, but also to reduce the alcohol in the still to a minimum value. This method of operation is particularly useful for handling small quantities of multi-component organic mixtures, since it is possible to obtain the different components at reasonable degrees of purity, in turn. To obtain the maximum recovery of a valuable component, the charge remaining in the still after the first distillation may be added to the next batch.

1.6.2 Operation at constant product composition

The case of a column with four ideal plates used to separate a mixture of ethyl alcohol and water may be considered. Initially, there are S_1 moles of liquor of mole fraction x_{s1} with respect to the more volatile component, alcohol, in the still. The top product is to contain a mole fraction x_d, and this necessitates a reflux ratio R_1. If the distillation is to be continued until there are S_2 moles in the still, of mole fraction x_{s2}, then, for the same number of plates the reflux ratio will have been increased to R_2. If the amount of product obtained is D_b moles, then a material balance gives:

$$S_1 x_{s1} - S_2 x_{s2} = D_b x_d \qquad (1.96)$$

$$\text{and}: S_1 - S_2 = D_b \qquad (1.97)$$

Thus: $S_1 x_{s1} - (S_1 - D_b)x_{s2} = D_b x_d$

and: $S_1 x_{s1} - S_1 x_{s2} = D_b x_d - D_b x_{s2}$

$$\text{and}: D_b = S_1 \left[\frac{x_{s1} - x_{s2}}{x_d - x_{s2}} \right] = \left(\frac{a}{b} \right) S_1 \tag{1.98}$$

where a and b are as shown in Fig. 1.34. If ϕ is the intercept on the Y-axis for any operating line, Eq. (1.48), then:

$$\frac{x_d}{R+1} = \varphi, \quad or \quad R = \frac{x_d}{\varphi} - 1 \tag{1.99}$$

These equations enable the final reflux ratio to be determined for any desired end concentration in the still, and they also give the total quantity of distillate obtained. What is important, in comparing the operation at constant reflux ratio with that at constant product composition, is the difference in the total amount of steam used in the distillation, for a given quantity of product, D_b.

If the reflux ratio R is assumed to be adjusted continuously to keep the top product at constant quality, then at any moment the reflux ratio is given by $R = dL_b/dD_b$. During the course of the distillation, the total reflux liquor flowing down the column is given by:

$$\int_0^{L_b} dL_b = \int_{R=R_1}^{R=R_2} R \, d \, D_b \tag{1.100}$$

To provide the reflux dL_b the removal of a quantity of heat equal to $\lambda \, dL_b$ in the condenser is required, where λ is the latent heat per mole. Thus, the heat to be supplied in the boiler to provide this reflux during the total distillation Q_R is given by:

$$Q_R = \lambda \int_0^{L_b} dL_b = \lambda \int_{R=R_1}^{R=R_2} R dD_b \tag{1.101}$$

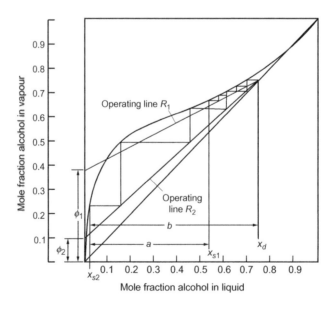

FIG. 1.34

Graphical representation of batch distillation.

This equation may be integrated graphically if the relation between R and D_b is known. For any desired value of R, x_s may be obtained by drawing the operating line, and marking off the steps corresponding to the given number of stages. The amount of product D_b is then obtained from Eq. (1.88) and, if the corresponding values of R and D_b are plotted, graphical integration will give the value of $\int R \, dD_b$.

The minimum reflux ratio R_m may be found for any given still concentration x_s from Eq. (1.56).

1.6.3 Operation at constant reflux ratio

If the same column is operated at a constant reflux ratio R, the concentration of the more volatile component in the top product will continuously fall. Over a small interval of time dt, the top-product composition with respect to the more volatile component will change from x_d to $x_d + dx_d$, where dx_d is negative for the more volatile component. If in this time the amount of product obtained is dD_b, then a material balance on the more volatile component gives:

More volatile component removed in product $= dD_b\left[x_d + \frac{dx_d}{2}\right]$.
which, neglecting second-order terms, gives:

$$= x_d \, dD_b \tag{1.102}$$

and: $x_d \, dD_b = -d(Sx_s)$
But $dD_b = -dS$,
and hence: $-x_d \, dS = -S \, dx_s - x_s \, dS$
and: $S \, dx_s = dS(x_d - x_s)$

$$\text{Thus}: \quad \int_{S_1}^{S_2} \frac{dS}{S} = \int_{x_{s1}}^{x_{s2}} \frac{dx_s}{x_d - x_s}$$
$$\ln \frac{S_1}{S_2} = \int_{x_{s2}}^{x_{s1}} \frac{dx_s}{x_d - x_s} \tag{1.103}$$

The right-hand side of this equation may be integrated by plotting $1/(x_d - x_s)$ against x_s. This enables the ratio of the initial to final quantity in the still to be found for any desired change in x_s, and hence the amount of distillate D_b. The heat to be supplied to provide the reflux is $Q_R = \lambda RD_b$ and hence the reboil heat required per mole of product may be compared with that from the first method.

Example 1.12

A mixture of ethyl alcohol and water with 0.55 mol fraction of alcohol is distilled to give a top product of 0.75 mol fraction of alcohol. The column has four ideal plates and the distillation is stopped when the reflux ratio has to be increased beyond 4.0. Assume the final still product falls to 0.05 mol fraction alcohol.

What is the amount of distillate obtained, and the heat required per kmol of product?

Solution

For various values of R the corresponding values of the intercept ϕ and the concentration in the still x_s are calculated. Values of x_s are found as shown in Fig. 1.35 for the two values of R of 0.85 and 4. The amount of product is then found from Eq. (1.98). Thus, for $R = 4$:

$$D_b = 100 \left[\frac{0.55 - 0.05}{0.75 - 0.05} \right] = 100 \left(\frac{0.5}{0.7} \right) = \underline{\underline{71.4 \, \text{kmol}}}$$

Values of D_b found in this way are:

R	ϕ	x_s	D_b
0.85	0.405	0.55	0
1.0	0.375	0.50	20.0
1.5	0.3	0.37	47.4
2.0	0.25	0.20	63.8
3.0	0.187	0.075	70.5
4.0	0.15	0.05	71.4

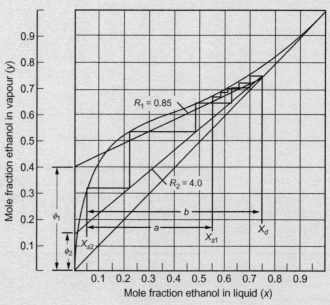

FIG. 1.35

Batch distillation–constant product composition.

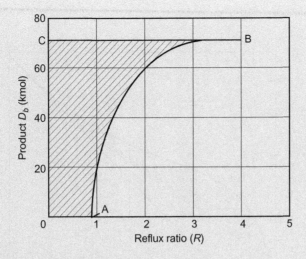

FIG. 1.36

Graphical integration for Example 1.12.

The relation between D_b and R is shown in Fig. 1.36 and the $\int_{R=0.85}^{R=4.0} R dD_b$ is given by area OABC as <u>96 kmol</u>.

Assuming an average latent heat for the alcohol–water mixtures of 4000 kJ/kmol, the heat to be supplied to provide the reflux, Q_R is $(96 \times 4000)/1000$ or approximately 380 MJ.

The heat to be supplied to provide the reflux per kmol of product is then $(380/71.4) = 5.32$ MJ and the total heat is $(5.32 + 4.0) = \underline{9.32\,MJ/kmol\ product}$.

Example 1.13

If the same batch as in Example 1.12 is distilled with a constant reflux ratio of $R = 2.1$, what will be the heat required and the average composition of the distillate if the distillation is stopped when the composition in the still has fallen to 0.105 mol fraction of ethanol?

Solution
Since we are seeking a top product of close to 0.75 mol fraction, use a range close this specification. So the initial composition of the top product will be 0.78, as shown in Fig. 1.37, and the final composition will be 0.74. Values of $x_d, x_s, x_d - x_s$ and of $1/(x_d - x_s)$ for various values of x_s and a constant reflux ratio are:

x_s	x_d	$x_d - x_s$	$1/(x_d - x_s)$
0.550	0.780	0.230	4.35
0.500	0.775	0.275	3.65
0.425	0.770	0.345	2.90
0.310	0.760	0.450	2.22
0.225	0.750	0.525	1.91
0.105	0.740	0.635	1.58

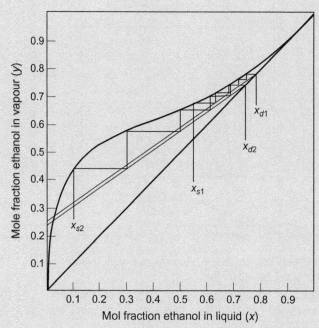

FIG. 1.37

Batch distillation–constant reflux ratio (Example 1.13).

Values of x_s and $1/(x_d - x_s)$ are plotted in Fig. 1.38 from which $\int_{0.105}^{0.55} (dx_s/(x_d - x_s)) = 1.1$.

From Eq. (1.103): $\ln(S_1/S_2) = 1.1$ and $(S_1/S_2) = 3.0$.

Product obtained, $D_b = S_1 - S_2 = (100 - 100/3) = 66.7$ kmol.

Amount of ethanol in product $= x_1 S_1 - x_2 S_2$.

$$= (0.55 \times 100) - (0.105 \times 33.3) = 51.5 \text{ kmol.}$$

Thus: average composition of product $= (51.5/66.7) = \underline{0.77 \text{ mol fraction ethanol}}$.

The heat required to provide the reflux $= (4000 \times 2.1 \times 66.7) = 560,380$ kJ.

Heat required to provide reflux per kmol of product $= (560,380/66.7) = 8400$ kJ.

Thus in Example 1.12 the total heat required per kmol of product is $(5320 + 4000) = 9320$ kJ and at constant reflux ratio (Example 1.13) it is $(8400 + 4000) = 12,400$ kJ, although the average quality of product is 0.77 for the second case and only 0.75 for the first.

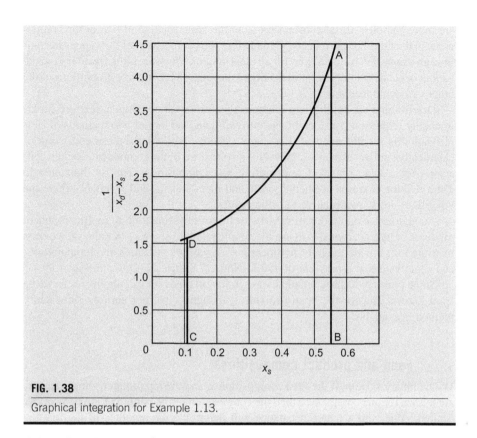

FIG. 1.38

Graphical integration for Example 1.13.

1.6.4 Batch or continuous distillation

A discussion on the relative merits of batch and continuous distillation is given by Ellis [36], who shows that when a large number of plates is used and the reflux ratio approaches the minimum value, then continuous distillation has the lowest reflux requirement and hence operating costs. If a smaller number of plates are used and high-purity product is not required, then batch distillation is probably more attractive.

1.7 Multicomponent mixtures

1.7.1 Equilibrium data

For a binary mixture under constant pressure conditions, the vapour–liquid equilibrium curve for either component is unique so that, if the concentration of either component is known in the liquid phase, the compositions of the liquid and of the vapour are fixed. It is on the basis of this single equilibrium curve that the McCabe–Thiele method was developed for the rapid determination of the number of theoretical plates required for a given separation. With a ternary system the conditions of equilibrium are more complex, for at constant pressure the mole fraction of two of the components in the liquid phase must be given before the composition of the vapour in equilibrium can be determined, even for an ideal system. Thus, the mole fraction y_A in the vapour depends not only on x_A in the liquid, but also on the relative proportions of the other two components.

Determining the equilibrium relationships for a multicomponent mixture experimentally requires a considerable quantity of data, and one of two methods of simplification is usually adopted. For many systems, particularly those consisting of chemically similar substances, the relative volatilities of the components remain constant over a wide range of temperature and composition. This is illustrated in Table 1.2 for mixtures of phenol, ortho and meta-cresols, and xylenols, where the volatilities are shown relative to orthocresol.

An alternative method, particularly useful for the separation of multicomponent mixtures of hydrocarbons, is to use the simple relation $y_A = Kx_A$. K values have been measured for a wide range of hydrocarbons at various pressures and temperatures, and some values at a pressure of 790.3 kN/m2 (7.8 atm) are shown in Fig. 1.39.

Some progress has been made in presenting methods for calculating ternary data from known data for the binary mixtures, though as yet not entirely satisfactory method is available.

1.7.2 **Feed and product compositions**

With a binary system, if the feed composition x_f and the top product composition x_d are known for one component, then the composition of the bottoms x_w may have any desired value, and a material balance will determine the amounts of the top and bottom products D and W. This freedom of selecting the compositions does not apply for mixtures with three or more components. Gilliland and Reed [37] have determined the number of degrees of freedom for the continuous distillation of a multicomponent mixture. For the common case in which the feed composition, nature of the feed, and operating pressure are given, there remain only four variables that may be selected. If the reflux ratio R is fixed and the number of plates above and below the feed plate are chosen to give the best use of the plates, then only two variables remain. The complete composition of neither the top nor bottom product can then be fixed at will. This means that some degree of trial and error is unavoidable in

Table 1.2 Volatilities relative to o-cresol

	Temperature (K)		
	353	393	453
Phenol	1.25	1.25	1.25
o-cresol	1	1	1
m-cresol	0.57	0.62	0.70
Xylenols	0.30	0.38	0.42

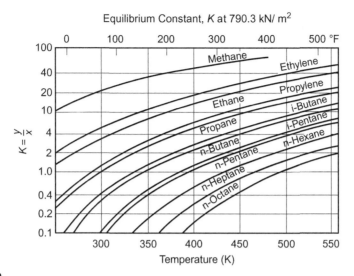

Equilibrium Constant, K at 790.3 kN/ m^2

FIG. 1.39

Vapour–liquid equilibrium data for hydrocarbons.

calculating the number of plates required for any desired separation. Thus, if a trial composition is taken, and it is found that for the given bottom composition the desired top composition is not obtained with the selected reflux ratio, then an adjustment must be made in the bottom composition. An exact fit in a calculation of this kind is not essential since the equilibrium data and the plate efficiency are known with only limited accuracy. This problem is frequently simplified if a sharp cut is to be made between the components, so that all of the more volatile components appear in the top and all of the less volatile in the bottom product.

1.7.3 Light and heavy key components

In the fractionation of multicomponent mixtures, the essential requirement is often the separation of two components. Such components are called the key components and by concentrating attention on these it is possible to simplify the handling of complex mixtures. If a four-component mixture **A–B–C–D**, in which **A** is the most volatile and **D** the least volatile, is to be separated as shown in Table 1.3, then **B** is the lightest component appearing in the bottoms and is termed the light key component. **C** is the heaviest component appearing in the distillate and is called the heavy key component. The main purpose of the fractionation is the separation of **B** from **C**.

Table 1.3 Separation of multicomponent mixture

Feed	Top product	Bottoms
A	A	
B	B	B
C	C	C
D		D

1.7.4 The calculation of the number of plates required for a given separation

One of the most successful methods for calculating the number of plates necessary for a given separation is due to Lewis and Matheson [38]. This is based on the Lewis–Sorel method, described previously for binary mixtures. If the composition of the liquid on any plate is known, then the composition of the vapour in equilibrium is calculated from a knowledge of the vapour pressures or relative volatilities of the individual components. The composition of the liquid on the plate above is then found by using an operating equation, as for binary mixtures, although in this case there will be a separate equation for each component.

If a mixture of components **A, B, C, D,** and so on has mole fractions x_A, x_B, x_C, x_D, and so on in the liquid and y_A, y_B, y_C, y_D, and so on in the vapour, then:

$$y_A + y_B + y_C + y_D + \cdots = 1 \tag{1.104}$$

$$\text{and}: \frac{y_A}{y_B} + \frac{y_B}{y_B} + \frac{y_C}{y_B} + \frac{y_D}{y_B} + \ldots = \frac{1}{y_B}$$

$$\alpha_{AB}\frac{x_A}{x_B} + \alpha_{BB}\frac{x_B}{x_B} + \alpha_{CB}\frac{x_C}{x_B} + \alpha_{DB}\frac{x_D}{x_B} + \ldots = \frac{1}{y_B}$$

$$\Sigma(\alpha_{AB}x_A) = \frac{x_B}{y_B} \tag{1.105}$$

$$y_B = \frac{x_B}{\Sigma(\alpha_{AB}x_A)} \tag{1.106}$$

and, similarly:

$$y_A = \frac{x_A\alpha_{AB}}{\Sigma(\alpha_{AB}x_A)}; \quad y_C = \frac{x_C\alpha_{CB}}{\Sigma(\alpha_{AB}x_A)}; \quad y_D = \frac{x_D\alpha_{DB}}{\Sigma(\alpha_{AB}x_A)} \tag{1.107}$$

Thus, the composition of the vapour is conveniently found from that of the liquid by use of the relative volatilities of the components. Examples 11.14 – 11.17 which follow illustrate typical calculations using multicomponent systems. Such solutions are now computerised, as discussed further in Volume 6.

Example 1.14

A mixture of *ortho*, *meta*, and *para*-mononitrotoluenes containing 60, 4, and 36 mol% respectively of the three isomers is to be continuously distilled to give a top product of 98 mol% ortho, and the bottom is to contain 12.5 mol% ortho. The mixture is to be distilled at a temperature of 410 K requiring a pressure in the boiler of about 6.0 kN/m2. If a reflux ratio of 5 is used, how many ideal plates will be required and what will be the approximate compositions of the product streams? The volatility of ortho relative to the para isomer may be taken as 1.70 and of the meta as 1.16 over the temperature range of 380 to 415 K.

Solution

As a first estimate, it is supposed that the distillate contains 0.6 mol% meta and 1.4 mol% para. A material balance then gives the composition of the bottoms.

For 100 kmol of feed with D and W kmol of product and bottoms, respectively and x_{do} and x_{wo} the mole fraction of the ortho in the distillate and bottoms, then an overall material balance gives:

$$100 = D + W.$$

An ortho balance gives:

$$60 = Dx_{do} + Wx_{wo}.$$

and:

$$60 = (100 - W)0.98 + 0.125W$$

from which:

$$D = 55.56 \text{ kmol} \quad \text{and} \quad W = 44.44 \text{ kmol}$$

The compositions and amounts of the streams are then be obtained as:

Component	Feed (kmol)	Feed (mole per cent)	Distillate (kmol)	Distillate (mole per cent)	Bottoms (kmo l)	Bottoms (mole per cent)
Ortho o	60	60	54.44	98.0	5.56	12.5
Meta m	4	4	.33	0.6	3.67	8.3
Para p	36	36	0.79	1.4	35.21	79.2
	100	100	55.56	100	44.44	100

Equations of operating lines.
The liquid and vapour streams in the column are obtained as follows:
Above the feed-point:
Liquid downflow, $L_n = 5D = 277.8$ kmol
Vapour up, $V_n = 6D = 333.4$ kmol
Below the feed-point, assuming the feed is liquid at its boiling point then:
Liquid downflow, $L_m = L_n + F = (277.8 + 100) = 377.8$ kmol

Vapour up, $V_m = L_m - W = (377.8 - 44.44) = 333.4$ kmol

The equations for the operating lines may then be written as:

below the feed plate:

$$y_m = \frac{L_m}{V_m}x_{m+1} - \frac{W}{V_m}x_w \qquad (1.37)$$

ortho :

$$y_{mo} = \left(\frac{377.8}{333.4}\right)x_{m+1} - \left(\frac{44.44}{333.4}\right)x_w$$

$$
\left.
\begin{aligned}
&= 1.133x_{m+1} - 0.0166 \\
\text{meta :} \quad y_{mm} &= 1.133x_{m+1} - 0.011 \\
\text{para :} \quad y_{mp} &= 1.133x_{m+1} - 0.105
\end{aligned}
\right\} \qquad (i)
$$

and above the feed plate:

$$y_n = \frac{L_n}{V_n}x_{n+1} - \frac{D}{V_n}x_d \qquad (1.35)$$

ortho :

$$y_{no} = \left(\frac{277.8}{333.4}\right)x_{n+1} + \left(\frac{55.56}{333.4}\right)0.98$$

$$
\left.
\begin{aligned}
&= 0.833x_{n+1} + 0.163 \\
\text{meta :} \quad y_{nm} &= 0.833x_{x+1} + 0.001 \\
\text{para :} \quad y_{np} &= 0.833x_{n+1} + 0.002
\end{aligned}
\right\} \qquad (ii)
$$

Composition of liquid on first plate.

The temperature of distillation is fixed by safety considerations at 410 K and, from a knowledge of the vapour pressures of the three components, the pressure in the still is found to be about 6 kN/m2. The composition of the vapour in the still is found from the relation $y_{so} = \alpha_o x_{so}/\Sigma\alpha x_s$.

The liquid composition on the first plate is then found from Equation (i) and for ortho:

$0.191 = (1.133x_1 - 0.0166)$

$x_1 = 0.183.$

and:

The values of the compositions as found in this way are shown in the following table. The liquid on plate 7 has a composition with the ratio of the concentrations of ortho and para about that in the feed, and the feed will therefore be introduced on this plate. Above this plate the same method is used but the operating equations are Equation (ii). The vapour from the sixteenth plate has the required concentration of the ortho isomer, and the values the meta and para are sufficiently near to take this as showing that 16 ideal plates will be required.

Using the relation $v_{mo} = \alpha_o x_{mo}/\Sigma\alpha x_m$:

Plate compositions below the feed plate

Component	x_7	αx_7	y_7	x_8	αx_8	y_8	x_9
o	0.125	0.211	0.191	0.183	0.308	0.270	0.253
m	0.083	0.096	0.088	0.088	0.102	0.090	0.089
p	0.792	0.792	0.721	0.729	0.729	0.640	0.658
	1	1.099	1	1	1.139	1	1

	αx_2	y_2	x_3	αx_3	y_3	x_4	αx_4
o	0.430	0.357	0.330	0.561	0.450	0.411	0.698
m	0.103	0.086	0.086	0.100	0.080	0.080	0.093
p	0.658	0.557	0.584	0.584	0.470	0.509	0.509
	1.191	1	1	1.245	1	1	1.300

Component	y_4	x_5	αx_5	y_5	x_6	αx_6	y_6
o	0.537	0.488	0.830	0.613	0.556	0.944	0.674
m	0.071	0.072	0.083	0.061	0.063	0.073	0.052
p	0.392	0.440	0.440	0.326	0.381	0.381	0.274
	1	1	1.353	1	1	1.398	1

	$x-$
o	0.609
m	0.055
p	0.336
	1

Plate compositions above the feed plate

Component	x_7	αx_7	y_7	x_8	αx_8	y_8	x_9
o	0.609	1.035	0.721	0.669	1.136	0.770	0.728
m	0.055	0.064	0.044	0.051	0.059	0.040	0.047
p	0.336	0.336	0.235	0.280	0.280	0.190	0.225
	1	1.435	1	1	1.475	1	1

	αx_9	y_9	x_{10}	αx_{10}	y_{10}	x_{11}	αx_{11}
o	1.238	0.816	0.782	1.330	0.856	0.832	1.415
m	0.054	0.035	0.041	0.047	0.030	0.035	0.040
p	0.225	0.149	0.177	0.177	0.144	0.133	0.133
	1.517	1	1	1.554	1	1	1.588

	y_{11}	x_{12}	αx_{12}	y_{12}	x_{13}	αx_{13}	y_{13}
o	0.891	0.874	1.485	0.920	0.907	1.542	0.940
m	0.025	0.029	0.033	0.020	0.023	0.027	0.017
p	0.084	0.097	0.097	0.060	0.070	0.070	0.043
	1	1	1.615	1	1	1.639	1

	x_{14}	αx_{14}	y_{14}	x_{15}	αx_{15}	y_{15}	x_{16}
o	0.932	1.585	0.957	0.953	1.620	0.970	0.968
m	0.019	0.022	0.013	0.014	0.016	0.010	0.010
p	0.049	0.049	0.030	0.033	0.033	0.020	0.022
	1	1.656	1	1	1.669	1	1

	αx_{16}	y_{16}
o	1.632	0.980
m	0.012	0.007
p	0.022	0.013
	1.666	1

1.7.5 **Minimum reflux ratio**

In the distillation of binary mixtures, the minimum reflux ratio is given by the operating line which joins the product-composition to the point where the q-line cuts the equilibrium curve. Thus, if Fig. 1.40 represents the McCabe–Thiele diagram for a binary mixture of the two key components of a multicomponent mixture, then DF and WF give the minimum reflux ratios for the rectifying and stripping sections respectively. Moving along these operating lines, the change in composition on adjacent plates becomes less and less until it becomes negligible at the feed plate. At F the compositions are said to be 'pinched'.

In a multicomponent mixture, the pinch does not necessarily occur at the position corresponding to the feed plate. In general, there are relatively few plates above the feed plate in which the concentrations of components heavier than the heavy key are reduced to negligible proportions, and then a true pinched condition occurs with only the heavy key and more volatile components present. Similarly, there is a region in which the concentrations of materials lighter than the light key are reduced to very low values, and thus there is a second pinch below the feed plate. In multicomponent distillation there are thus two pinched-in-regions. In locating these pinched-in regions, it may be noted that:

(a) If there are no components lighter than the light key, then all of the components appear in the bottoms and the pinch in the stripping section will be near the feed plate.

(b) If there are no components heavier than the heavy key, then all of the components will appear in the top and the upper pinch is also at the feed plate.

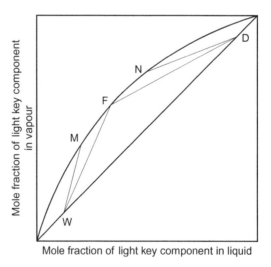

FIG. 1.40

McCabe–Thiele diagram for two key components.

If both of these conditions are true, then the two pinches coincide at the feed plate, as for a binary system. For the general case, a number of proposals have been made for locating the pinched regions and hence the minimum reflux ratio R_m. Of these, the methods of Colburn and Underwood are considered. It may be noted that between the feed plate and the enriching pinch, the concentrations of components heavier than the heavy key fall off rapidly, so that the upper pinch may be regarded as containing the heaviest key and lighter components. Similarly the lower pinch has the lightest key and heavier components.

Colburn's method for minimum reflux

If **A** and **B** are the light and heavy key components of a multicomponent mixture, then applying the method given earlier for binary mixtures, Eq. (1.53), the minimum reflux ratio R_m is obtained from:

$$R_m = \frac{1}{(\alpha_{AB} - 1)} \left[\frac{x_{dA}}{x_{nA}} - \alpha_{AB} \frac{x_{dB}}{x_{nB}} \right] \tag{1.108}$$

where: x_{dA} and x_{nA} are the top and pinch compositions of the light key component,

x_{dB} and x_{nB} are the top and pinch compositions of the heavy key component, and

α_{AB} is the volatility of the light key relative to the heavy key component.

The difficulty in using this equation is that the values of x_{nA} and x_{nB} are known only in the special case where the pinch coincides with the feed composition. Colburn [39] has suggested that an approximate value for x_{nA} is given by:

$$x_{nA} \approx \frac{r_f}{(1 + r_f)(1 + \Sigma \alpha x_{fh})} \tag{1.109}$$

and:

$$x_{nB} \approx \frac{x_{nA}}{r_f} \tag{1.110}$$

where: r_f is the estimated ratio of the key components on the feed plate. For an all liquid feed at its boiling point, r_f equals the ratio of the key components in the feed. Otherwise r_f is the ratio of the key components in the liquid part of the feed.

x_{fh} is the mole fraction of each component in the liquid portion of feed heavier than the heavy key, and

α is the volatility of the component relative to the heavy key.

Using this approximate value for R_m, Eq. (1.109) may be rearranged to give the concentrations of all the light components in the upper pinch as:

$$x_n = \frac{x_d}{(\alpha - 1)R_m + \alpha(x_{dB}/x_{nB})} \tag{1.111}$$

The concentration of the heavy key in the upper pinch is then obtained by difference, after obtaining the values for all the light components. The second term in the denominator is usually negligible, as the concentration of the heavy key in the top product is small.

A similar condition occurs in the stripping section, and the concentration of all components heavier than the light key is given by:

$$x_m = \frac{\alpha_{AB}x_w}{(\alpha_{AB} - \alpha)(L_m/W) + \alpha(x_{wA}/x_{mA})} \tag{1.112}$$

where: x_m and x_w are the compositions of a given heavy component at the pinch and in the bottoms,

x_{mA} and x_{wA} are the compositions of the light key component at the pinch and in the bottoms,

L_m/W is the molar ratio of the liquid in the stripping section to the bottom product,

α_{AB} is the volatility of the light key relative to the heavy key, and

α is the volatility of the component relative to the heavy key.

Again, the second term in the denominator may usually be neglected.

The essence of Colburn's method is that an empirical relation between the compositions at the two pinches for the condition of minimum reflux is provided. This enables the assumed value of R_m to be checked. This relation may be written as:

$$\frac{r_m}{r_n} = \psi = \frac{1}{(1 - \Sigma b_m \alpha x_m)(1 - \Sigma b_n x_n)}. \tag{1.113}$$

where: r_m is the ratio of the light key to the heavy key in the stripping pinch,

r_n is the ratio of the light key to the heavy key in the upper pinch,

$\Sigma b_m \alpha x_m$ is the summation of $b_m \alpha x_m$ for all components heavier than the heavy key in the lower pinch,

$\Sigma b_n x_n$ is the summation of $b_n x_n$ for all components lighter than the light key in the upper pinch, and

b_m, b_n are the factors shown in Fig. 1.41.

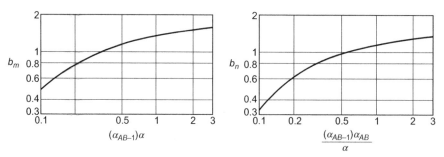

FIG. 1.41

Factors in Colburn's solution.

Example 1.15

A mixture of n-C_4 to n-C_7 hydrocarbons is to be distilled to give top and bottom products as follows. The distillation is effected at $800\,kN/m2$ and the feed is at $372\,K$. The equilibrium values, K, are shown in Fig. 1.39. It is required to find the minimum reflux ratio. No cooling occurs in the condenser.

Component	Feed F	Distillate D	x_d	Bottoms W	x_w
n-C_4 (light key)	40	39	0.975	1	0.017
C_5 (heavy key)	23	1	0.025	22	0.367
C_6	17			17	0.283
C_7	20			20	0.333
Totals:	F = 100	D = 40	1.000	W = 60	1.000

Solution

1. Estimation of top temperature T_d

By a dew-point calculation, $\Sigma x_d = \Sigma(x_d/K)$.

Component	x_d	$T_d = 344\,K$ K	x_d/K	$T_d = 343\,K$ K	x_d/K
n-C_4	0.975	1.05	0.929	1.04	0.938
C_5	0.025	0.41	0.061	0.405	0.062
	1		0.990		1.000

Hence the top temperature $T_d = 343\,K$.

2. Estimation of still temperature T_s

$\Sigma x_w = \Sigma K x_w$.

Component	x_w	$T_s = 419\,K$ K	Kx_w	$T_s = 416\,K$ K	Kx_w
n-C_4	0.017	3.05	0.052	2.93	0.050
C_5	0.367	1.6	0.586	1.54	0.565
C_6	0.283	0.87	0.246	0.82	0.232
C_7	0.333	0.49	0.163	0.46	0.153
	1		1.047		1.000

Hence the still temperature $T_s = 416\,K$.

3. Calculation of feed condition

To determine the nature of the feed, its boiling point T_B must be found, that is where $\Sigma K x_f = 1$.

Component	x_f	$T_B = 377\,K$		$T_B = 376\,K$	
		K	Kx_f	K	Kx_f
$n\text{-}C_4$	0.40	1.80	0.720	1.78	0.712
C_5	0.23	0.81	0.186	0.79	0.182
C_6	0.17	0.39	0.066	0.38	0.065
C_7	0.20	0.19	0.038	0.185	0.037
	1		1.010		0.996

Hence T_B is approximately $376\,K$, and since the feed is at $372\,K$ it may be assumed to be all liquid at its boiling point.

4. Calculation of pinch temperatures

The temperatures of the pinches are taken in the first place at one third and two thirds of the difference between the still and top temperatures.

Thus, the upper pinch temperature, $T_n = 343 + 0.33(416 - 343) = 367\,K$ and the lower pinch temperature, $T_m = 343 + 0.67(416 - 343) = 391\,K$

5. Calculation of approximate minimum reflux ratio

The calculations may be laid out as:

Component	$T_n = 367\,K$	$T_m = 391\,K$	x'_{th}	αx_{th}
	α	α		
$n\text{-}C_4$	2.38	2.00		
C_5	1.00	1.00		
C_6	0.455	0.464	0.17	0.077
C_7	0.220	0.254	0.20	0.044
				0.121

Thus :
$$r_f = \frac{x_{f4}}{x_{f5}} = \left(\frac{0.40}{0.23}\right) = 1.740$$

From Eq. (1.109):
$$x_{n4} = \frac{1.740}{(1 + 1.740)(1 + 0.121)} = 0.565$$

and :
$$x_{n5} = \left(\frac{0.565}{1.740}\right) = 0.325$$

From Eq.(1.108) :
$$R_m = \left(\frac{1}{2.38 - 1}\right)\left(\frac{0.975}{0.563}\right) - 2.38\left(\frac{0.025}{0.325}\right)$$
$$= 1.12$$

6. The streams in the column

$L_n = DR_m = (40 \times 1.12) = 44.8$ kmol

$V_n = L_n + D = (44.8 + 40) = 84.8$ kmol

$L_m = L_n + F = (44.8 + 100) = 144.8$ kmol

$V_m = L_m - W = (144.8 - 60) = 84.4$ kmol

$L_m/W = (144.8/60) = 2.41$

7. Check on minimum reflux ratio

$x_n = \frac{x_d}{(\alpha-1)R_m}$ for components lighter than heavy key.

For n-C_4:

$$x_n = \frac{0.975}{(2.38 - 1)1.12} = 0.630$$

and for n-C_5:

$$x_n = (1 - 0.630) = 0.370$$

Temperature check for upper pinch. $\Sigma K x_n = 1.0$

$$\Sigma K x_n = (1.62 \times 0.630) + (0.68 \times 0.370) = 1.273$$

Thus an upper pinch temperature of 367 K is incorrect. A value of $T_n = 355$ K will be tried.

Component	K_{355}	α	x_n	Kx_n
n-C_4	1.35	2.55	0.562	0.759
C_5	0.53	1	0.438	0.233
			1.000	0.992

This is sufficiently near, and the upper pinch temperature will be taken as 355 K.

Thus : $r_n = (0.562/0.438) = 1.282$

Since there is no component lighter than light key $\Sigma b_n x_n = 0$. In the lower pinch, (from Eq. 1.112):

$$x_m = \frac{\alpha_{AB} x_w}{(\alpha_{AB} - \alpha)L_m/W}$$

Component	K_{391}	α	$\alpha_{AB} - \alpha$	x_m	Kx_m
$n-C_4^*$	2.2	2		0.384	0.845
C_5	1.1	1	1.000	0.305	0.335
C_6	0.51	0.464	1.536	0.153	0.078
C_7	0.28	0.254	1.746	0.158	0.044
				1.000	1.302

*x_m by difference.

Component	K_{372}	α	$\dfrac{\alpha_{AB}}{-\alpha}$	x_m	Kx_m	b_m	$b_m\alpha_m x_m$
$n-C_4^*$	1.70	2.30	–	0.428	0.729		
C_5	0.74	1	1.30	0.270	0.200		
C_6	0.35	0.47	1.83	0.148	0.052	1.22	0.085
C_7	0.17	0.23	2.07	0.154	0.026	0.91	0.032
				1.000	1.007		0.117

x_m by difference.

As the temperatures do not check T_m is not 391 K and a value of 372 K will be tried.

This is sufficiently near, and $T_m = 372$ K.

Thus: $r_m = (0.428/0.270) = 1.586$

$\alpha r_m/r_n = (1.586/1.282) = 1.235$.

$\psi = 1/(1.\,0.117) = (1/0.883)$.

$= 1.132$.

Hence R_m is not quite equal to 1.12.

8. Second approximation to reflux ratio

A value of $R_m = 1.08$ will be tried.

$L_n = (40 \times 1.08) = 43.2$.

$V_n = (40 + 43.2) = 83.2 = V_m$.

$L_m = 143.2$.

Taking $T_n = 355$ K as before, then:

Thus :
$$r_n = (0.582/0.418) = 1.393$$

Component	K_{355}	α	x_n	Kx_n	
$n\text{-}C_4$	1.35	2.55	0.582	0.785	
C_5	0.53	1	0.418	0.221	
			1.000	1.006	Checks

Component	K_{372}	α	$\alpha_{AB} - \alpha$	x_m	Kx_m	b_m	$b_m\alpha_m x_m$
$n-C_4^*$	1.70	2.30	–	0.424	0.721		
C_5	0.74	1	1.30	0.272	0.201		
C_6	0.35	0.47	1.83	0.149	0.052	1.22	0.085
C_7	0.17	0.23	2.07	0.155	0.026	0.91	0.034
				1.000	1.000		0.119

x_m by difference.

Taking $T_m = 372\,\text{K}$ as before, then:

Thus : $r_m = (0.424/0.272) = 1.558$ $\alpha r_m/r_n = (1.558/1.393) = 1.12$

and : $\psi = 1/(1 - 0.119) = 1.13$

Thus : $\underline{R_m = 1.08}$ is near enough.

Since there are no components lighter than the light key, the lower pinch is expected to be near the feed plate as it is, although the general method of taking the pinch temperature as one-third and two-thirds up the column was used above.

The small change in R from 1.12 to 1.08 gives a change in r_n though very little change in r_m. It is seen that the first estimation for R_m of 1.12 based on Eq. (1.108) for locating the upper pinch composition is nearly correct but that it gives the wrong pinch composition.

Minimum reflux ratio, using Underwood's method

For conditions where the relative volatilities remain constant, Underwood [40] developed the following two equations from which R_m may be calculated:

$$\frac{\alpha_A x_{fA}}{\alpha_A - \theta} + \frac{\alpha_B x_{fB}}{\alpha_B - \theta} + \frac{\alpha_C x_{fC}}{\alpha_C - \theta} + \cdots = 1 - q \tag{1.114}$$

and :
$$\frac{\alpha_A x_{dA}}{\alpha_A - \theta} + \frac{\alpha_B x_{dB}}{\alpha_B - \theta} + \frac{\alpha_C x_{dC}}{\alpha_c - \theta} + \cdots = R_m + 1 \tag{1.115}$$

where: $x_{fA}, x_{fB}, x_{fC}, x_{dA}, x_{dB}, x_{dC}$, etc., are the mole fractions of components **A, B, C,** etc., in the feed and distillate, **A** being the light and **B** the heavy key,

q is the ratio of the heat required to vaporise 1 mol of the feed to the molar latent heat of the feed, as in Eq. (1.44),

$\alpha_A, \alpha_B, \alpha_C$, etc., are the volatilities with respect to the least volatile component, and

θ is the root of Eq. (1.114), which lies between the values of α_A and α_B.

If one component in the system has a relative volatility falling between those of the light and heavy keys, it is necessary to solve for two values of θ.

Example 1.16

A mixture of hexane, heptane, and octane is to be separated to give the following products. What will be the value of the minimum reflux ratio, if the feed is liquid at its boiling point?

Component	Feed F (kmol)	x_f	Product D (kmol)	x_d	Bottoms W (kmol)	x_w	Relative volatility
Hexane	40	0.40	40	0.534	0	0	2.70
Heptane	35	0.35	34	0.453	1	0.04	2.22
Octane	25	0.25	1	0.013	24	0.96	1.0

Solution

From Eq. (1.114), the light key (**A**) is heptane and the heavy key (**B**) is octane. With $q = 1$ then:

$$\left(\frac{2.70 \times 0.40}{2.70 - \theta}\right) + \left(\frac{2.22 \times 0.35}{2.22 - \theta}\right) + \left(\frac{1 \times 0.25}{1 - \theta}\right) = 0$$

The required value of θ must satisfy the relation $\alpha_B < \theta < \alpha_A$, that is $1.0 < \theta < 2.22$. Assuming $\theta = 1.15$, then:

$$\Sigma \frac{\alpha x_f}{\alpha - \theta} = -0.243$$

Assuming $\theta = 1.17$, then : $\Sigma \dfrac{\alpha x_f}{\alpha - \theta} = -0.024$

This is near enough and from Eq. (1.115):

$$\left(\frac{2.70 \times 0.534}{2.70 - 1.17}\right) + \left(\frac{2.22 \times 0.453}{2.22 - 1.17}\right) + \left(\frac{1.00 \times 0.013}{1.00 - 1.17}\right) = 1.827$$

Thus : $R_m = \underline{0.827}$

1.7.6 Number of plates at total reflux

The number of plates required for a desired separation under conditions of total reflux can be found by applying Fenske's equation, Eq. (1.59), to the two key components.

Thus:

$$n + 1 = \frac{\log\left[\left(\frac{x_A}{x_B}\right)_d \left(\frac{x_B}{x_A}\right)_s\right]}{\log\left(\alpha_{AB}\right)_{av}} \tag{1.116}$$

Example 1.17

For the separation of hexane, heptane, and octane as in Example 1.16, determine the number of theoretical plates required.

Solution

$$n+1 = \frac{-\log \left[\dfrac{0.0453}{0.013} \times \dfrac{0.960}{0.040} \right]}{-\log 2.22} = \frac{(\log 835)}{(\log 2.22)}$$

$$= 8.5$$

Thus, the minimum number of plates $= 7.5$, say 8.

1.7.7 Relation between reflux ratio and number of plates

Gilliland [30] has given an empirical relation between the reflux ratio R and the number of plates n, in which only the minimum reflux ratio R_m and the number of plates at total reflux n_m are required. This is shown in Fig. 1.42, where the group $[(n+1) - (n_m+1)]/(n+2)$ is plotted against $(R - R_m)/(R+1)$.

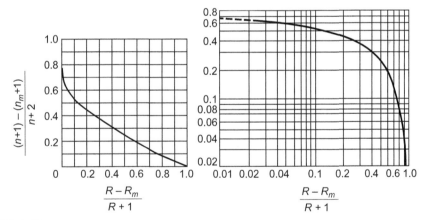

FIG. 1.42

Relation between reflux ratio and number of plates.

Example 1.18

Using the data of Example 1.17, investigate the change in n with R using Fig. 1.42. It may be assumed that $R_m = 0.83$ and $(n_m + 1) = 8.5$.

Solution

R	$\frac{R-R_m}{R+1}$	$\frac{(n+1)-(n_m+1)}{n+2}$	n
1	$\frac{0.17}{2} = 0.085$	0.55	19
2	$\frac{1.17}{3} = 0.390$	0.32	11.8
5	$\frac{4.17}{6} = 0.695$	0.15	9.1
10	$\frac{9.17}{11} = 0.833$	0.08	8.2

This Example shows that the number of plates falls off rapidly at first, though more slowly later, and a value of $R \approx 2$ is probably the most economic.

For calculations of the type illustrated by Example 1.18 a convenient nomograph has recently been produced by Zanker [41] which relates R, R_m, n, and n_m so that any variable may be quickly found if the other three are known.

Example 1.19

In the separation of a mixture of 100 kmol of hexane, heptane, and octane, the flows and concentrations are:

	hexane	heptane	octane
Relative volatility	2.70	2.22	1.0
Feed:			
kmol	40	35	25
mole fraction	0.40	0.35	0.25
Overheads:			
kmol	40	34	1
mole fraction	0.534	0.453	0.013
Bottom product:			
kmol	0	1	24
mole fraction	0	0.04	0.96

Assuming operation for 8000 h/year, a plate efficiency of 0.95, allowable vapour velocities of $u' = 2 \times 10^{-3}$ and $N'' = 1.35 \times 10^{-3}$ kmol/m^2 s respectively and the following incremental costs:

$c_a = £400/m^2$ year plate.

$c_b = £25/m^2$ year, both of which allow 50%/year for depreciation, interest and maintenance.

and $c_d = £0.05$/kmol based of an overall coefficient of 0.5 kW/m^2 K and a temperature difference of 15 K in both the condenser and the reboiler.

Estimate the optimum reflux ratio.

Solution

The minimum reflux ratio, R_m is calculated using Underwood's method (Example 1.16) as 0.83 and, using Fenske's method, Example 1.17, the number of plates at total reflux is $n_m = 8$. The following data have been taken from Fig. 1.42, attributable to Gilliland [30]:

$(R - R_m)/(R + 1)$	$[(n + 1) - (n_m + 1)]/(n + 2)$
0	0.75
0.02	0.62
0.04	0.60
0.06	0.57
0.08	0.55
0.1	0.52
0.2	0.45
0.4	0.30
0.6	0.18
0.8	0.09
1.0	0

Substituting $R_m = 0.83$ and $n_m = 8$, the following data are obtained, together with values of $-\mathrm{d}n/\mathrm{d}R$ obtained from a plot of R against n:

R	n	$-\mathrm{d}n/\mathrm{d}R$
0.92	28.6	110.0
1.08	22.8	34.9
1.25	16.9	9.8
1.75	13.5	3.8
2.5	11.7	1.7
3.5	10.5	0.6
5.0	9.8	0.4
7.0	9.2	0.2
9.0	8.95	0.05

On the basis of the cost data, $F = 7.72$ and hence from Eq. (1.67):

$$R_{opt} + 1 = (n_{opt} + 7.72)/(-\mathrm{d}n/\mathrm{d}R)$$

Values of R are now selected and substituted in this equation with the corresponding values of n and $-\mathrm{d}n/\mathrm{d}R$ taken from the previous table until a balance is attained. This occurs when $R = 1.25$ or approximately 150% of the minimum reflux condition.

1.7.8 Multiple column systems

In considering multicomponent systems, a feed of say **A, B,** and **C** has been fed to a single column and the top product, mainly pure **A,** is obtained with bottoms of mainly **B** and **C**. If each component is required pure, then a two-column system is required in which the bottom stream is fed to the second column to separate **B** and **C**. With more components in the feed, more columns will be required and their arrangement becomes more complex. This is a typical problem in the petrochemical industry and a paper by Eliceche and Sargent [42] offers particularly helpful advice.

1.8 Azeotropic and extractive distillation

In the systems considered previously, the vapour becomes steadily richer in the more volatile component up the successive plates. There are two types of mixture where this steady increase in the concentration of the more volatile component, either does not take place, or it takes place so slowly that an uneconomic number of plates is required.

If, for example, a mixture of ethanol and water is distilled, the concentration of the alcohol steadily increases until it reaches 96% by mass, when the composition of the vapour equals that of the liquid, and no further enrichment occurs. This mixture is called an azeotrope, and it cannot be separated by straightforward distillation. Such a condition is shown in the $y - x$ curves of Fig. 1.43 where it is seen that the equilibrium curve crosses the diagonal, indicating the existence of an azeotrope. A large number of azeotropic mixtures have been found, some of which are of great industrial importance, such as water-nitric acid, water-hydrochloric acid, and water-alcohols. The problem of non-ideality is discussed in Section 1.2.4 where the determination of the equilibrium data is considered. When the activity coefficient is greater than unity, giving a positive deviation from Raoult's law, the molecules of the components in the system repel each other and exert a higher partial pressure than if their behaviour were ideal. This leads to the formation of a 'minimum boiling' azeotrope shown in Fig. 1.43A. For values of the activity coefficient less than unity, negative deviation from Raoult's law results in a lower partial pressure and the formation of a 'maximum boiling' azeotrope, as shown in Fig. 1.43B.

The second type of problem occurs where the relative volatility of a binary mixture is very low, in which case continuous distillation of the mixture to give nearly pure products will require high reflux ratios with correspondingly high heat requirements. In addition, it will necessitate a tower of large cross-section containing many trays. An example of the second type of problem is the separation of n-heptane from methyl cyclohexane in which the relative volatility is only 1.08 and a large number of plates are required to achieve separation.

The principle of azeotropic and of extraction distillation lies in the addition of a new substance to the mixture so as to increase the relative volatility of the two key components, and thus make separation relatively easy. Benedict and Rubin [43] have

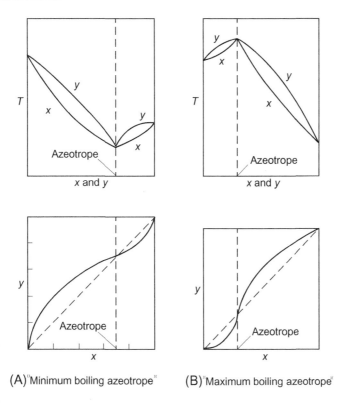

(A) "Minimum boiling azeotrope"   (B) "Maximum boiling azeotrope"

FIG. 1.43

Types of azeotropic behaviour.

defined these two processes as follows. In azeotropic distillation, the substance added forms an azeotrope with one or more of the components in the mixture, and as a result is present on most of the plates of the column in appreciable concentrations. With extractive distillation, the substance added is relatively non-volatile compared with the components to be separated, and it is therefore fed continuously near the top of the column. This extractive agent runs down the column as reflux and is present in appreciable concentrations on all the plates. The third component added to the binary mixture is sometimes known as the *entrainer* or the *solvent*.

1.8.1 Azeotropic distillation

Young [44], found in 1902, that if benzene is added to the ethanol–water azeotrope, then a ternary azeotrope is formed with a boiling point of 338.0 K, that is less than that of the binary azeotrope, 351.3 K. The industrial production of ethanol from the azeotrope, using this principle, has been described by Guinot and Clark [45] and the

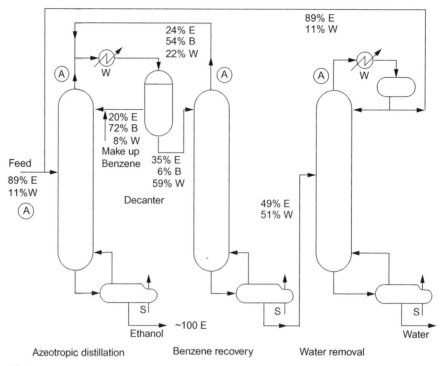

FIG. 1.44

Azeotropic distillation for the separation of ethanol from water using benzene as entrainer. Compositions are given in mole per cent. E = Ethanol, B = Benzene, W = Water, S = Steam.

general arrangement of the plant is as shown in Fig. 1.44. This requires the use of three atmospheric pressure fractionating columns, and a continuous two-phase liquid separator or decanter.

The azeotrope in the ethanol–water binary system has a composition of 89 mol% of ethanol [14]. Starting with a mixture containing a lower proportion of ethanol, it is not possible to obtain a product richer in ethanol than this by normal binary distillation. Near azeotropic conditions exist at points marked Ⓐ in Fig. 1.44. The addition of the relatively non-polar benzene entrainer serves to volatilise water, a highly polar molecule, to a greater extent than ethanol, a moderately polar molecule, and a virtually pure ethanol product may be obtained. Equilibrium conditions for this system have been discussed by Norman [46] who shows how the number of plates required may be determined.

The first tower in Fig. 1.44 gives the ternary azeotrope as an overhead vapour, and nearly pure ethanol as bottom product. The ternary azeotrope is condensed and splits into two liquid phases in the decanter. The benzene-rich phase from the decanter serves as reflux, whilst the water–ethanol-rich phase passes to two towers, one for benzene recovery and the other for water removal. The azeotropic overheads from these successive towers are returned to appropriate points in the primary tower.

Fig. 1.45 shows a composition profile for the azeotropic distillation column in the process shown in Fig. 1.44. This is taken from a solution presented by Robinson and Gilliland [1].

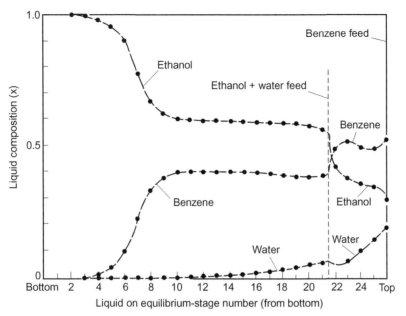

FIG. 1.45

Composition profile for the azeotropic distillation of ethanol and water, with benzene as entrainer.

1.8.2 Extractive distillation

Extractive distillation is a method of rectification similar in purpose to azeotropic distillation. To a binary mixture which is difficult or impossible to separate by ordinary means, a third component, termed a *solvent*, is added which alters the relative volatility of the original constituents, thus permitting the separation. The added solvent is, however, of low volatility and is itself not appreciably vaporised in the fractionator.

For a non-ideal binary mixture the partial pressure may be expressed as:

$$P_A = \gamma_A P_A^\circ x_A \tag{1.117}$$

$$P_B = \gamma_B P_B^\circ x_B \tag{1.118}$$

where γ_A and γ_B are the activity coefficients for the two components. The relative volatility α may thus be written as:

$$\begin{aligned} \alpha &= \frac{P_A \, x_B}{P_B \, x_A} \\ &= \frac{\gamma_A}{\gamma B} \frac{P_A^\circ}{P_B^\circ} \end{aligned} \tag{1.119}$$

The solvent added to the mixture in extractive distillation differentially affects the activities of the two components, and hence the relative volatility, α.

Such a process depends upon the difference in departure from ideality between the solvent and the components of the binary mixture to be separated. In the following example, both toluene and isooctane separately form non-ideal liquid solutions with phenol, although the extent of the non-ideality with isooctane is greater than that with toluene. When all three substances are present, therefore, the toluene and iso-octane themselves behave as a non-ideal mixture, and their relative volatility becomes high.

An example of extractive distillation given by Treybal [47] is the separation of toluene, boiling point 384 K, from paraffin hydrocarbons of approximately the same molecular weight. This is either very difficult or impossible, owing to low relative volatility or azeotrope formation, yet such a separation is necessary in the recovery of toluene from certain petroleum hydrocarbon mixtures. Using isooctane of boiling point 372.5 K, as an example of a paraffin hydrocarbon, Fig. 1.46A shows that iso-octane in this mixture is the more volatile, although the separation is obviously difficult. In the presence of phenol, boiling point 454.6 K, however, the relative volatility of isooctane increases, so that, with as much as 83 mol% phenol in the liquid, the separation from toluene is relatively easy. A flowsheet of a process for accomplishing this is shown in Fig. 1.46B, where the binary mixture is introduced more or less centrally into the extractive distillation tower (1), and phenol as the solvent is introduced near the top so as to be present in high concentration upon most of the trays in the tower. Under these conditions isooctane is readily distilled as an overhead product, whilst toluene and phenol are removed as a residue. Although phenol is relatively high-boiling, its vapour pressure is nevertheless sufficient for some to appear in the overhead product. The solvent-recovery section of the tower, which

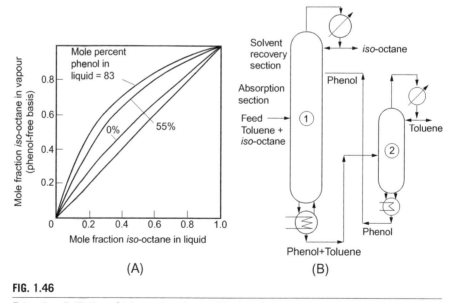

(A)   (B)

FIG. 1.46

Extractive distillation of toluene-isooctane with phenol.

may be relatively short, serves to separate the phenol from the isooctane. The residue from the tower must be rectified in the auxiliary tower (2) to separate toluene from the phenol which is recycled, but this is a relatively easy separation. In practice, the paraffin hydrocarbon is a mixture rather than pure isooctane, although the principle of the operation remains the same.

The solvent to be used is selected on the basis of selectivity, volatility, ease of separation from the top and bottom products, and the cost. The selectivity is most easily assessed by determining the effect on the relative volatility of the two key components of addition of the solvent. The more volatile the solvent, the greater the percentage of solvent in the vapour, and the poorer the separation for a given heat consumption in the boiler. It is important to note that the solvent must not form an azeotrope with any of the components. Some of the problems of selecting the solvent are discussed by Scheibel [48] who points out that use may be made of the fact that, when two compounds show deviations from Raoult's law, then one of these compounds shows the same type of deviation with any member of the homologous series of the other component. Thus the azeotropic mixture acetone (b.p. 329.6 K)–methanol (b.p. 337.9 K) has 20 mol% acetone and boils at 328.9 K, that is less than the boiling point of either component. Thus any member of the series ethanol (b.p. 357.5 K), propanol (b.p. 370.4 K), water (b.p. 373.2 K), butanol (b.p. 391.0 K) may be used as an extractive agent, or in the series of ketones, methyl n-propyl ketone (b.p. 375 K), and methyl isobutyl ketone (b.p. 389.2 K). The advantage of using a solvent from the alcohol series is that the more volatile acetone will be taken overhead, though water would have the advantage of cheapness. Pratt [49] has given details of a method of calculation for extractive distillation, using the system acetonitrile–trichloroethylene–water as an example.

Extractive distillation is usually more desirable than azeotropic distillation since no large quantities of solvent have to be vaporised. In addition, a greater choice of added component is possible since the process is not dependent upon the accident of azeotrope formation. It cannot, however, be conveniently carried out in batch operations.

Azeotropic and extractive-distillation equipment may be designed using the general methods for multicomponent distillation, and detailed discussion is available elsewhere [1,42] and presented by Hoffman [50] and Smith [51].

1.9 Steam distillation

Where a material to be distilled has a high boiling point, and particularly where decomposition might occur if direct distillation is employed, the process of steam distillation may be used. Steam is passed directly into the liquid in the still and the solubility of the steam in the liquid must be very low. Steam distillation is perhaps the most common example of differential distillation.

Two cases may be considered. The steam may be superheated and so provide sufficient heat to vaporise the material concerned, without itself condensing. Alternatively, some of the steam may condense, producing a liquid water phase. In either case, assuming the ideal gas laws to apply, the composition of the vapour produced may be obtained from:

$$\left(\frac{m_A}{M_A}\right) / \left(\frac{m_B}{M_B}\right) = \frac{P_A}{P_B} = \frac{y_A}{y_B} = \frac{P_A}{P - P_A} \qquad (1.120)$$

where the subscript A refers to the component being recovered, and B to steam, and:

$m =$ mass,

$M =$ molecular weight,

$P_A, P_B =$ partial pressure of **A, B**, and

$P =$ total pressure.

If there is no liquid phase present, then from the phase rule there will be two degrees of freedom, both the total pressure and the operating temperature can be fixed independently, and $P_B = P - P_A$, which must not exceed the vapour pressure of pure water, if no liquid phase is to appear.

When a liquid water phase is present, there will be only one degree of freedom, and selecting the temperature or pressure fixes the system, with the water and the other component each exerting a partial pressure equal to its vapour pressure at the boiling point of the mixture. In this case, the distillation temperature will always be less than that of boiling water at the total pressure in question. Consequently, a high boiling organic material may be steam-distilled at temperatures below 373 K at atmospheric pressure. By using reduced operating pressures, the distillation temperature may be reduced still further, with a consequent economy of steam.

A convenient method of calculating the temperature and composition of the vapour, for the case where the liquid water phase is present, is by using the diagram shown in Fig. 1.47 which is due to Hausbrand [52], where the parameter $(P - P_B)$ is plotted for total pressures of 101.3, 40 and 9.3 kN/m2, and the vapour pressures of a number of other materials are plotted directly against temperature. The intersection of the two appropriate curves gives the temperature of distillation, and the molar ratio of water to organic material is given by $(P - P_A)/P_A$. Thus, if nitrobenzene (molecular weight $= 123$) is distilled at atmospheric pressure (101.3 kN/m2) with live saturated steam (molecular weight $= 18$), the boiling point will be about 372 K (with nitrobenzene vapour pressure of 2.7 kN/m2) and the mass-ratio of water to nitrobenzene in the vapour will be:

$$\frac{101.3 - 2.7}{2.7}\left(\frac{18}{123}\right) = 5.34$$

Where there is no liquid water phase present, the steam consumption will be high unless the steam is very highly superheated. With a water phase present, the boiling point of the mixture will be low, and consequently P_A will have a low value. Thus, on a molar basis, the steam consumption will again be high, although due to the relatively low molecular weight of steam, the consumption may not be excessive. Steam economy may be effected by using indirect heating of the still, having no liquid water phase present, or by operating under reduced pressure.

In an operation of this kind, illustrated in Fig. 1.48, it is essential that the separation of the material being distilled from the water should be a relatively simple operation.

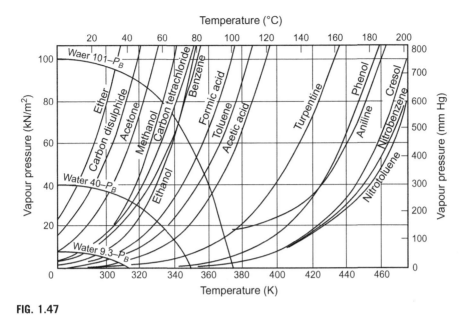

FIG. 1.47

Vapour pressure curves for steam distillation calculations.

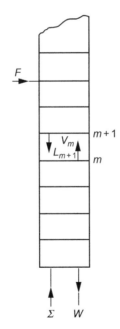

FIG. 1.48

Steam distillation.

In determining the number of stages required to efect a steam distillation, the steam flow must be included in the operating-line equation for the lower part of the column. Using indirect heating and assuming constant molar overflow, the lower operating line for the organic material is given by:

$$y_m = \frac{L_m}{V_m} x_{m+1} - \frac{W}{V_m} x_w \qquad (1.37)$$

This has a slope L_m/V_m and cuts the $y=x$ line at $x=x_w$.

If Σ kmol of live steam is used as shown in Fig. 1.48, then considering that part of the column below the plate $(m+1)$:

$$V_m = L_{m+1} + \Sigma - W \qquad (1.121)$$

and organic material component balance:

$$V_m y_m = L_{m+1} x_{m+1} - W x_w \qquad (1.122)$$

Assuming constant molar overflow, $L_m = L_{m+1} = W$ and $V_m = \Sigma$, and:

$$y_m = \frac{L_m}{V_m} x_{m+1} - \frac{W}{V_m} x_w \qquad (1.123)$$

This equation also has a slope of L_m/V_m although it cuts the $y=x$ line at $x=Wx_w/(W-\Sigma)$ as shown in Fig. 1.49. When $x=x_w$, $y=0$, corresponding to the composition of the vapour, steam, rising to the bottom plate. For a given external reflux ratio and feed condition, L_m/V_m will be the same whether direct or indirect steam is used and the lower operating line must cut the $y=x$ line at the same value in each case, though, of course, x_m for the indirect steam will be higher than x_w for direct steam.

When stepping off the theoretical stages, the bottom step should start at $y=0$, $x=x_w$. In this way, the use of direct steam, although eliminating the still, dilutes the bottom material, and so increases the number of stages required in the lower part of the column.

1.10 Plate columns

Distillation may be carried out in plate columns in which each plate constitutes a single stage, or in packed columns where mass transfer is between a vapour and liquid in continuous countercurrent flow. Plate columns are now considered, and packed columns are discussed in Section 1.11.

The number of theoretical stages required to effect a required separation, and the corresponding rates for the liquid and vapour phases, may be determined by the procedures described previously. In order to translate these quantities into an actual design, the following factors should be considered:

(a) The type of plate or tray.
(b) The vapour velocity, which is the major factor in determining the diameter of the column.

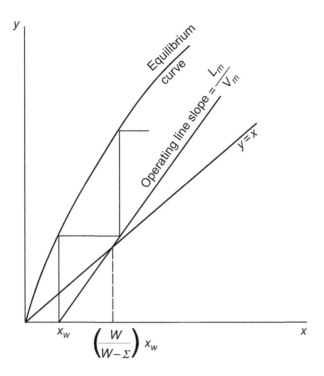

FIG. 1.49

Operating lines for steam distillation.

(c) The plate spacing, which is the major factor fixing the height of the column when the number of stages is known.

Design methods are discussed later, although it is first necessary to consider the range of trays available and some of their important features.

1.10.1 Types of trays

The main requirement of a tray is that it should provide intimate mixing between the liquid and vapour streams, that it should be suitable for handling the desired rates of vapour and liquid without excessive entrainment or flooding, that it should be stable in operation, and that it should be reasonably easy to instal and maintain. In many cases, particularly with vacuum distillation, it is essential that the drop in pressure over the tray should be a minimum.

The arrangements for the liquid flow over the tray depend largely on the ratio of liquid to vapour flow. Three layouts are shown in Fig. 1.50, of which the cross-flow arrangement is much the most frequently used. Considering these in turn:

(a) *Cross-flow*. Normal, with a good length of liquid path giving a good opportunity for mass transfer.

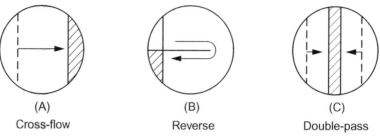

(A)   (B)   (C)
Cross-flow   Reverse   Double-pass

FIG. 1.50

Arrangements for liquid flow over a tray.

(b) *Reverse.* Downcomers are much reduced in area, and there is a very long liquid path. This design is suitable for low liquid–vapour ratios.

(c) *Double-pass.* As the liquid flow splits into two directions, this system will handle high liquid–vapour ratios.

(d) *Radial flow tray:* The liquid enters in the middle of the tray through circular downcomer and flows radially over the tray to a circumferential weir near the tower wall. They are suitable for high liquid loadings and enable higher separation efficiencies but cost more to instal.

The liquid reflux flows across each tray and enters the downcomer by way of a weir, the height of which largely determines the amount of liquid on the tray. The downcomer extends beneath the liquid surface on the tray below, thus forming a vapour seal. The vapour flows upwards through risers into caps, or through simple perforations in the tray.

The bubble-cap tray. This is the most widely used tray because of its range of operation, although it is being superseded by newer types, such as the valve tray discussed later. The general construction is shown in Fig. 1.51. The individual caps are mounted on risers and have rectangular or triangular slots cut around their sides. The caps are held in position by some form of spider, and the areas of the riser and the annular space around the riser should be about equal. With small trays, the reflux

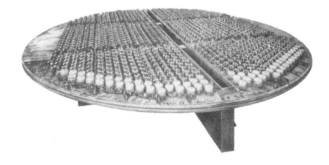

FIG. 1.51

A bubble-cap tray.

passes to the tray below over two or three circular weirs, and with the larger trays through segmental downcomers.

Sieve or perforated trays. These are much simpler in construction, with small holes in the tray. The liquid flows across the tray and down the segmental downcomer. Fig. 1.52 indicates the general form of tray layout.

Valve trays. These may be regarded as a cross between a bubble-cap and a sieve tray. The construction is similar to that of cap types, although there are no risers and no slots. It may be noted that with most types of valve tray the opening may be varied by the vapour flow, so that the trays can operate over a wide range of flow rates. Because of their flexibility and price, valve trays are tending to replace bubble-cap trays. Fig. 1.53 shows a typical tray.

FIG. 1.52

A sieve or perforated plate tray.

FIG. 1.53

A valve tray.

Radial flow trays: These trays are not used as extensively because of higher costs but are suitable for high liquid loadings and enable higher separation efficiencies and are often considered for debottleneck retrofits.

These four types of trays have a common feature in that they all have separate downcomers for the passage of liquid from each tray to the one below. There is another class of tray which has no separate downcomers and yet it still employs a tray type of construction giving a hydrodynamic performance between that of a packed and a plate column. Two examples of this type of device are the Kittel plate and a Turbogrid tray [53]. Design data for these trays are sparse in the literature and the manufacturer's recommendations should be sought.

1.10.2 Factors determining column performance

The performance of a column may be judged in relation to two separate but related criteria. Firstly, if the vapour and liquid leaving a tray are in equilibrium this constitutes a theoretical tray and provides a standard of performance. Secondly, the relative performance of, say, two columns of the same diameter must be considered in relation to their capacity for liquid and vapour flow. The main features are:

(a) Liquid and vapour velocities.
(b) Physical properties of the liquid and vapour.
(c) Extent of entrainment of liquid by rising vapour streams.
(d) The hydraulics of the flow of liquid and vapour across and through the tray.

It has been found by Carey [54], Carey et al. [55], and Souders and Brown [56] that the vapour velocity is a prime factor in determining the diameter of a column. Kirschbaum [57] using an equimolar mixture of ethanol and water on a 400 mm diameter plate containing 15 bubble caps, obtained the following results:

(a) For all plate spacings the efficiency E_{Mv}, which is defined in Eq. (1.124) as the ratio of the actual change in liquid composition on a plate to that which would be obtained if the liquid left in equilibrium with the vapour, decreases as the velocity is increased due mainly to the reduction in contact time between the phases.
(b) The decrease in efficiency is much less with high plate spacings.
(c) The capacity is limited by the ability of the downcomers to carry the reflux, rather than that of the caps to handle the vapour.

The effects of liquid viscosity on tray efficiency have been studied by Drickamer and Bradford [58] and O'Connell [59] and these are discussed in Section 1.10.5. Surface tension influences operation with sieve trays, in relation both to foaming and to the stability of bubbles.

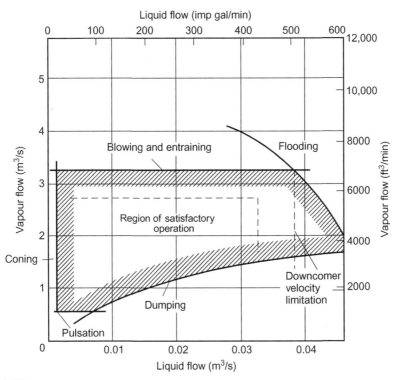

FIG. 1.54

Capacity graph for a typical bubble-cap tray.

1.10.3 Operating ranges for trays

For a given tray layout there are certain limits for the flows of vapour and liquid within which stable operation is obtained. The range is shown in Fig. 1.54, which relates to a bubble-cap plate. The region of satisfactory operation is bounded by areas where undesirable phenomena occur. Coning occurs at low liquid rates, where the vapour forces the liquid back from the slots and passes out as a continuous stream, with a consequent loss in efficiency. Low vapour rates result in pulsating vapour flow or dumping. With low liquid rates, vapour passes through the slots intermittently, though with higher liquid rates some slots dump liquid rather than passing vapour. Both pulsating vapour flow and dumping, which may be referred to jointly as weeping, result in poor efficiency. At very high vapour rates, the vapour bubbles carry liquid as spray or droplets to the plate above, giving excessive entrainment. With high liquid rates, a point is reached where the drop in pressure across the plate equals the liquid head in the downcomer. Beyond this point, the liquid builds up and floods the tray.

The extent of entrainment of the liquid by the vapour rising over a plate has been studied by many workers. The entrainment has been found to vary with the vapour velocity in the slot or perforation, and the spacing used. Strang [60], using an air–water system, found that entrainment was small until a critical vapour velocity was reached, above which it increased rapidly. Similar results from Peavy and Baker [61] and Colburn [62] have shown the effect on tray efficiency, which is not seriously affected until the entrainment exceeds 0.1 kmol of liquid per kmol of vapour. The entrainment on sieve trays is discussed in Section 1.10.4.

The design of the tray fittings and the downcomers influences the column performance. It is convenient to consider this separately for bubble-cap trays and sieve trays. In design, the important factors are the diameter of the tower, the tray spacing and the detailed design of the tray.

1.10.4 General design methods

In designing a column for a given separation, the number of stages required and the flow rates of the liquid and vapour streams must first be determined using the general methods outlined previously. In the mechanical design of the column, tower diameter, tray spacing, and the detailed layout of each tray is considered. Initially, a diameter is established, based on the criterion of absence from liquid entrainment in the vapour stream, and then the weirs and the downcomers are designed to handle the required liquid flow. It is then possible to consider the tray geometry in more detail, and, finally, to examine the general operating conditions for the tray and to establish its optimum range of operation. This approach to design is covered in detail in Volume 6 where the different methods applicable to bubble caps, sieve trays and valve trays are discussed in detail.

Bubble-cap trays

Bubble-cap trays are rarely used for new installations on account of their high cost and their high pressure drop. In addition, difficulties arise in large columns because of the large hydraulic gradients which are set up across the trays. Bubble-cap trays are capable of dealing with very low liquid rates and are therefore useful for operation at low reflux ratios. There are still many bubble-cap columns in use and the design considerations presented in Volume 6 are given to enable, in particular, existing equipment to be assessed for new applications and duties.

Sieve trays

Sieve trays offer several advantages over bubble-cap trays, and their simpler and cheaper construction has led to their increasing use. The general form of the flow on a sieve tray is typical of a cross-flow system with perforations in the tray taking the place of the more complex bubble caps. The hydraulic flow conditions for such a tray are discussed in Volume 6 in the same manner as for the bubble-cap tray, by considering entrainment, flooding, pressure loss, and so on. The key differences in operation between these two types of tray should be noted. With the sieve tray the vapour passes vertically through the holes into the liquid on the tray, whereas with the bubble cap the vapour issues in an approximately horizontal direction from the slots. With the sieve plate the vapour velocity through the perforations must be greater than a certain minimum value in order to prevent the weeping of the liquid stream down through the holes. At the other extreme, a very high vapour velocity leads to excessive entrainment and loss of tray efficiency. The capacity graph for a sieve tray is similar to that shown in Fig. 1.54 for bubble-cap trays.

Valve trays

The valve tray, which may be regarded as intermediate between the bubble-cap and the sieve tray, offers advantages over both. The important feature of the tray is that liftable caps act as variable orifices which adjust themselves to changes in vapour flow. The valves are either metal discs of up to about 38 mm diameter, or metal strips which are raised above the openings in the tray deck as vapour passes through the trays. The caps are restrained by legs or spiders which limit the vertical movement and some types are capable of forming a total liquid seal when the vapour flow is insufficient to lift the cap.

Advantages claimed for valve trays include:

(a) Operation at the same capacity and efficiency as sieve trays.
(b) A low pressure drop which is fairly constant over a large portion of the operating range.
(c) A high turndown ratio, that is, it can be operated at a small fraction of design capacity.
(d) A relatively simple construction which leads to a cost of only 20% higher than that of a comparable sieve tray.

Valve trays, because of their proprietary nature, are usually designed by manufacturers, although it is possible to obtain an estimate of design and performance from published literature [63] and from the methods summarised in Volume 6.

1.10.5 **Plate efficiency**

The number of ideal stages required for a desired separation may be calculated by one of the methods discussed previously, although in practice more trays are required than ideal stages. The ratio n/n_p of the number of ideal stages n to the number of actual trays n_p represents the overall efficiency E of the column, which may be 30 to 100% [4]. The main reason for loss in efficiency is that the kinetics for the rate of approach to equilibrium, and the flow pattern on the plate, may not permit an equilibrium between the vapour and liquid to be attained. Some empirical equations have been developed from which values of efficiency may be calculated, and this approach is of considerable value in giving a general picture of the problem. The proportion of liquid and vapour, and the physical properties of the mixtures on the trays, will vary up the column, and conditions on individual trays must be examined, as suggested by Murphree [64].

For a single ideal tray, the vapour leaving is in equilibrium with the liquid leaving, and the ratio of the actual change in composition achieved to that which would occur if equilibrium between y_n and x_n were attained is known as the Murphree plate efficiency E_M. Using the notation shown in Fig. 1.55, the plate efficiency expressed in vapour terms is given by:

$$E_{Mv} = \frac{y_n - y_{n-1}}{y_e - y_n - 1} \tag{1.124}$$

where y_e is the composition of the vapour that would be in equilibrium with the liquid of composition x_n actually leaving the plate. This equation gives the efficiency in vapour terms, although if the concentrations in the liquid streams are used then the plate efficiency E_{MI} is given by:

$$E_{MI} = \frac{x_n + 1 - x_n}{x_{n+1} - x_e} \tag{1.125}$$

where x_e is the composition of the liquid that would be in equilibrium with the composition y_n of the vapour actually leaving the plate.

The ratio E_{Mv} is shown graphically in Fig. 1.56 where for any operating line AB the enrichment that would be achieved by an ideal plate is BC, and that achieved with an actual plate is BD. The ratio BD/BC then represents the plate efficiency. The efficiency may vary from point to point on a tray. Local values of the Murphree efficiency are designated E_{mv} and E_{ml}.

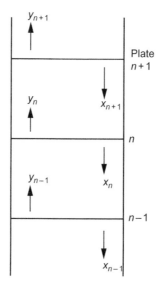

FIG. 1.55

Compositions of liquid and vapour streams from plates.

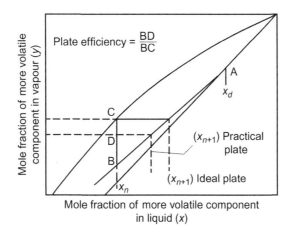

FIG. 1.56

Graphical representation of plate efficiency *EMr*.

Empirical expressions for plate efficiency

The efficiency of the individual plates is expected to depend on the physical properties of the mixture, the geometrical arrangements of the trays, and the flow rates of the two phases. A simple empirical relationship for the overall efficiency, E, of columns handling petroleum hydrocarbons is given by Drickamer and Bradford [58] who relate efficiency of the column to the average viscosity of the feed by:

$$E = 0.17 - 0.616 \log_{10} \Sigma \left[x_f (\mu_L / \mu_w) \right] \tag{1.126}$$

where: x_f is the mole fraction of the component in the feed,

μ_L is the viscosity at the mean tower temperature, and

μ_w is the viscosity of water at 293 K (approximately $1 \, \text{mNs/m}^2$).

Further work, mainly with larger towers 3 m in diameter, suggested that higher efficiencies were obtained with larger diameters because of the longer liquid path. Thus, compared with a 0.9 m diameter tray, one of 3 m diameter might give up to 25% greater efficiency.

Example 1.20

Using Eq. (1.126), determine the plate efficiency for the following data on the separation of a stream of C_3 to C_6 hydrocarbons.

Component	Mole fraction in feed x_f	μ_L (mNs / m²)	$x_f(\mu_L/\mu_w)$
C_3	0.2	0.048	0.0096
C_4	0.3	0.11	0.0336
C_5	0.2	0.145	0.0290
C_6	0.3	0.188	0.0564
			0.1286

Solution

From Eq. (1.126):

$$E = (0.17 - 0616 \log 0.1286)$$
$$= \underline{0.72}.$$

O'Connell [59] found that a rather better relation may be obtained by plotting the overall efficiency as a function of the product of the viscosity and the relative volatility of the key components. This relation has also been presented by Lockhart and Leggett [65], as shown in Fig. 1.57. Thus, in Example 1.20, taking C_3 and C_5 as key components, the relative volatility α is about 1.76 and the mean viscosity about 0.15 mNs/m^2, giving a product of 0.26. From Fig. 1.57, E is then found as 70%.

Chu et al. [66] has given a more complex correlation for overall efficiency E by including the relative flow rates L and V of the phases and the effective submergence of the liquid h_L. This takes the form:

$$\log_{10}E = 1.67 + 0.30 \log_{10}(L/V) - 0.25 \log_{10}(\mu_L\alpha) + 0.30h_L \qquad (1.127)$$

where: L, V are the liquid and vapour flow rates (kmol/s),
μ_L is the viscosity of the liquid feed (mNs/m^2),

Product of viscosity and relative volatility

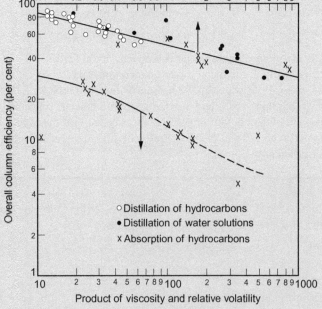

FIG. 1.57

Overall column efficiency E as function of viscosity–relative volatility product $(\mu lxf)(65)$.
α is the relative volatility of the key components, and
h_L is the effective submergence (m), taken as the distance from the top of the lot to the weir lip plus half the slot height.

Expressions for plate efficiency related to mass transfer

By assuming that the vapour issuing from slots is in the form of spherical bubbles, Chu [66], Geddes [67], and Bakowski [68] derived methods for expressing the efficiency E in terms of transfer coefficients, k_g, k_l and tray parameters such as the slot dimensions. These methods have proved very difficult to use because of the unreliability of data for calculating transfer coefficients, and the greater problem of calculating the interfacial areas. Probably the most successful analysis for determining efficiency in terms of mass transfer functions is that obtained by the American Institute of Chemical Engineers Research Project [69], and an outline of this work is given here. Whilst this is a complex analysis containing parameters which are known only approximately, the method does outline some of the important factors involved and shows that some parameters are of little importance.

Plate efficiency in terms of transfer units

The process of mass transfer across a phase boundary is discussed in Volume 1, Chapter 10. A resistance to mass transfer exists within the fluid on each side of the interface, and the overall transfer rate of a component in a mixture depends on the sum of these resistances and the total driving force.

The concept of a transfer unit for a countercurrent mass transfer process, introduced in Volume 1, is developed further for distillation in packed columns in Section 1.11. The number of transfer units is defined as the integrated value of the ratio of the change in composition to the driving force. Thus, considering the vapour phase, the number of overall gas transfer, units N_{OG} is given by:

$$N_{OG} = \int \frac{dy}{y_e - y} \qquad (1.128)$$

For the liquid phase, the corresponding number of overall liquid transfer units N_{OL} is given by:

$$N_{OL} = \int \frac{dx}{x - x_e} \qquad (1.129)$$

Equations (1.128) and (1.129) are derived in the same way as Eqs (1.157) and (1.158).

Noting Eq. (1.164), the relation between N_{OG} and N_{OL} is given by:

$$\frac{N_{OL}}{N_{OG}} = \frac{mG'}{L'} = \varepsilon \qquad (1.130)$$

where: m is the slope of the vapour–liquid equilibrium line (y_e vs x), and

G' and L' are the molar rates of flow of vapour and liquid, respectively, per unit cross-section of column.

The equation for transfer units may be applied to the mass transfer over a tray, and thus relate the local Murphree efficiency E_{mv} to the overall transfer units N_{OG}. With the notation in Fig. 1.55, the vapour y_{n-1} rises from plate $n - 1$, crosses the liquid on plate n and leaves with composition y_n. The liquid flowing from plate $n + 1$ through the downcomer crosses tray n and leaves with composition x_n. It is supposed for this argument that there is no change in the composition of the liquid in a vertical plane through the liquid. Applying the mass transfer equation for the flow of vapour on a vertical path and over a small element of plate area gives:

$$N_{OG} = \int \frac{dy}{y_e - y} = -\ln\left(\frac{y_e - y_n}{y_e - y_{n-1}}\right) \tag{1.131}$$

or :

$$\exp\left(-\mathbf{N}_{OG}\right) = \left(\frac{y_e - y_n}{y_e - y_{n-1}}\right)$$

and :

$$1 - \exp\left(-\mathbf{N}_{OG}\right) = \left(\frac{y_e - y_n}{y_e - y_{n-1}}\right) = E_{mv} \tag{1.132}$$

This analysis refers to a small area for vertical flow, and E_{mv} is therefore the *point or local* Murphree efficiency. The relation between this point efficiency and the tray efficiency depends on the nature of the liquid mixing on the tray. If there is complete mixing of the liquid, $x = x_n$ for the liquid, and y_e and y will also be constant over a horizontal plane. The tray efficiency $E_{Mv} = E_{mv}$. With no mixing of the liquid, the liquid may be considered to be in plug flow. If $y_e = mx + b$ and E_{mv} is taken as constant over tray, it may be shown [69] that:

$$E_{mv} = \frac{1}{\varepsilon}\left[\exp\left(\varepsilon E_{Mv}\right) - 1\right] \tag{1.133}$$

$$\text{where} : \varepsilon = \frac{mG'}{L'}$$

Intermediate cases where partial mixing of liquid occurs are dealt with in the A.I.Ch. E. Manual [69].

Plate efficiency in terms of liquid concentrations

With the same concept for tray layout as in Fig. 1.55, relations for E_{ml} and E_{Ml} may be derived. Assuming that the vapour concentration does not change in a horizontal plane, a similar analysis to that above gives:

$$E_{ml} = 1 - \exp\left(-\mathbf{N}_{OL}\right) \tag{1.134}$$

The efficiencies E_{mv} and E_{ml} may be related by using the relation between $\mathbf{N}_{OG}$ and $\mathbf{N}_{OL}$ given in Eq. (1.130) to give:

$$\ln(1 - E_{ml}) = \varepsilon \ \ln(1 - E_{mv}) \tag{1.135}$$

Effect of entrainment on efficiency

For conditions where the entrainment may be assumed constant across a tray, Colburn [62] has suggested that the following expression gives, for entrainment e'

(moles/unit time. Unit area), a correction to E_{Mv}, so that the new value of efficiency E_a is given by:

$$E_a = \frac{E_{Mv}}{1 + (e'E_{Mv})/L'} \qquad (1.136)$$

Experimental work from A.I.Ch.E. programme

Having noted the way in which the tray efficiencies may be related to the values of N_{OG} and N_{OL}, experimentally determined results are now required for expressing the mass transfer in terms of degree of mixing, entrainment, geometrical arrangements on the trays and the operating conditions including mass flow rates. These are provided from experimental work, which gives expressions for the number of film transfer units N_G and N_L, outlined in Section 1.11.3.

Gas phase transfer

The value of N_G is expressed in terms of weir height h_w, gas flow expressed as $\overline{F}$, liquid flow L_p and the Schmidt number Sc_v for the vapour phase. The two key relations are:

$$N_G = \left[0.776 + 0.0046h_w - 0.24\overline{F} + 105L_p\right]Sc_v^{-0.5} \qquad (1.137)$$

$$\text{and}: N_G = -\ln(1 - E_{mv}) \qquad (1.138)$$

Equation (1.138) gives the point efficiency for cases where all the resistance occurs in the gas phase. In these equations:

h_w is the exit weir height (mm),

$\overline{F} = u\sqrt{\rho_v}$, where u is the vapour rate (m/s) based on the bubbling area, and ρ_v is vapour density (kg/m³),

L_p is the liquid flow (m³/s per m liquid flow path),

μ_v is the vapour viscosity (Ns/m²),

D_v is the vapour diffusivity (m²/s), and.

Sc_v, is the Schmidt number $\mu_v/\rho_v D_v$.

Liquid phase transfer

The value of N_L is expressed in terms of the $\overline{F}$-factor for vapour flow, the time of contact $t_L(s)$, and the liquid diffusivity $D_L(m^2/s)$. Experimental work gives:

$$N_L = \left[4.13 \times 10^8 D_L\right]^{0.5}\left[0.21\overline{F} + 0.15\right]t_L \qquad (1.139)$$

The residence time t_L in seconds is expressed by:

$$t_L = Z_c Z_L/L_p \qquad (1.140)$$

where Z_c is the hold-up of liquid on the tray in m³/m² of effective cross-section, given by:

$$Z_c = 0.043 + 1.91 \times 10^{-4}h_w - 0.013\overline{F} + 2.5L_p \qquad (1.141)$$

and Z_L is the distance between the weirs in metres.

Relationships for N_G and N_L

From a knowledge of $\mathbf{N}_G$ and $\mathbf{N}_L$, the value of $\mathbf{N}_{OG}$ is obtained from Eq. (1.142) which is derived in the same way as Eq. (1.163).

$$\text{Thus}: \frac{1}{\mathbf{N}_{OG}} = \frac{1}{\mathbf{N}_G} + \frac{mG'}{L'}\frac{1}{\mathbf{N}_L} \tag{1.142}$$

The point efficiency E_{mv} is then obtained from:

$$E_{mv} = 1 - \exp\left(-\mathbf{N}_{OG}\right) \tag{1.132}$$

Whilst these expressions are difficult to use and involve some inconsistent assumptions about the liquid and vapour flow, they do bring out some useful features in relation to the tray efficiency. Thus $\mathbf{N}_G$ varies linearly with h_w, $\overline{F}$, and L_p, although the important relation between $\mathbf{N}_G$ and E_{mv} is complex. The A.I.Ch.E. Manual [69] gives guideline figures.

1.11 Packed columns for distillation

In bubble-cap and perforated plate columns, a large interfacial area between the rising vapour and the reflux is obtained by causing the vapour to bubble through the liquid. An alternative arrangement, which also provides the necessary large interfacial area for diffusion, is the packed column, in which the cylindrical shell of the column is filled with some form of packing. A common arrangement for distillation is as indicated in Fig. 1.58, where the packing may consist of rings, saddles, or other shaped particles, all of which are designed to provide a high interfacial area for transfer. These are referred to in Chapter 4. In packed columns the vapour flows steadily up and the reflux steadily down the column, giving a true countercurrent system in contrast to the conditions in bubble-cap columns, where the process of enrichment is stage wise.

1.11.1 Packings

The selection of a suitable packing material is based on the same arguments as for absorption towers considered in Chapter 12, although for industrial units the most usual packings are rings, and the material of construction is determined by the corrosive nature of the fluids, or otherwise. It is important to note that in a distillation system operating at high reflux ratios the mass of reflux is approximately equal to the mass of vapour, although at low reflux ratios the flow of the liquid is only a small fraction of that of the vapour. Since the vapour has a much lower density than the liquid, the process is really one in which a small quantity of liquid passes through the vapour, and the establishment of good distribution of the liquid is more difficult than in absorption towers, where the two streams are more nearly balanced.

In Chapter 4, the characteristics of packings and their influence on column hydraulics are considered, and in Chapter 12, the mass transfer aspects are covered.

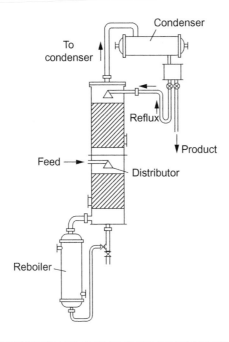

FIG. 1.58

A packed column for distillation.

The packing for a particular application may be selected using this information, although it may be noted that, in the case of vacuum distillation, for example, pressure-drop considerations may be of overriding importance, and there may be problems associated with the wetting of the packing because of the low liquid loadings. For distillation in packed towers, it is normal practice to increase the calculated height of packing by 40% to allow for liquid mal-distribution and wetting problems due to channelling.

1.11.2 **Calculation of enrichment in packed columns**

With plate columns, the vapour leaving an ideal plate is richer in the more volatile component than the vapour entering the plate, by one equilibrium step. Peters [70] suggests that this same enrichment of the vapour will occur in a certain height of packing, which is termed the *height equivalent of a theoretical plate* (HETP). As all sections of the packing are physically the same, it is assumed that one equilibrium stage is represented by a given height of packing. Thus the required height of packing for any desired separation is given by HETP × (No. of ideal stages required).

This is a simple method of representation which has been widely used as a method of design. Despite this fact, there have been few developments in the theory. Murch [71] gives the following relationships for the HETP from an analysis of the results of a number of workers. Columns 50 to 750 mm diameter and packed over heights of 0.9 to 3.0 m with rings, saddles, and other packings have been considered. Most of the results were for conditions of total reflux, with a vapour rate of 0.18 to 2.5 kg/m²s which corresponded to 25 to 80% of flooding. The relationship is:

$$\text{HETP} = C_1 G'^{c2} d_c^{c3} Z^{1/3} \left(\frac{\alpha \mu L}{\rho L} \right) \tag{1.143}$$

where the values of C_1, C_2, C_3 varied with packings as given in Table 1.4.

It may also be noted that the mixtures considered were mainly hydrocarbons with values of relative volatilities only slightly in excess of 3. In Eq. (1.143):

G' = mass velocity of vapour (kg/m² s of tower area),
d_c = column diameter (m),
Z = packed height (m),
α = relative volatility,
μ_L = liquid viscosity (N s/m²), and.
ρ_L = liquid density (kg/m³).

Ellis [36] presented the following dimensionally consistent equation for the HETP (Z_t) of packed columns using 25 and 50 mm Raschig rings:

$$Z_t = 18 d_r + 12m \left[\frac{G'}{L'} - 1 \right] \tag{1.144}$$

where: d_r is the diameter of the rings.
m is the average slope of equilibrium curve,
G'' is the vapour flow rate, and
L'' is the liquor flow rate.

In practice, the HETP concept is used to convert empirically the number of theoretical stages to packing height. As most data in the literature have been derived from small-scale operations, these do not provide a good guide to the values which

Table 1.4 Constants for use in Eq. (1.143).

Type of packing	Size (mm)	C_1 (×10⁻⁵)	C_2	C_3
Rings	6			1.24
	9	0.77	−0.37	1.24
	12.5	7.43	−0.24	1.24
	25	1.26	−0.10	1.24
	50	1.80	0	1.24
Saddles	12.5	0.7	−0.4	1.11
	25	0.80	−0.14	1.11
Raschig rings of protruded metal	6	0.28	0.2	0.30
	9	0.29	0.0	0.30
	19	0.4	0.30	0.30
	25	0.92	0.12	0.30

Table 1.5 Values of HETP [14] for full-scale plant.

Type of packing, application	HETP (m)
25mm diam. Packing	0.46
38mm diam. Packing	0.66
50mm diam. Packing	0.9
Absorption duty	1.5 – 1.8
Small diameter columns (<0.6m diam.)	column diameter
Vacuum columns	values as above +0.1m

will be obtained on full-scale plant. The values given in Table 1.5 may, however, be used as a guide.

For a particular type of packing, the ratio, HETP/pressure drop, is fairly constant for all sizes so that there is no advantage in attempting to improve the HETP by using a smaller packing, since the disadvantages of the higher pressure drop will offset the savings made by reduction of packed height.

Further data on HETP for packings smaller than 38mm are presented in Fig. 1.59, where it may be seen that some of the newer packings, such as Pall rings and Mini rings, give a relatively constant value of HETP over a wide range of vapour rates.

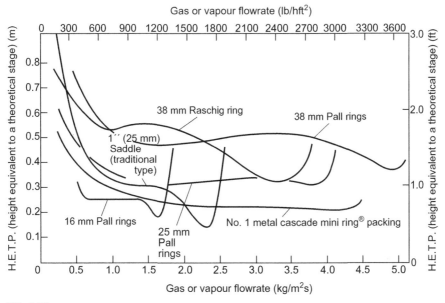

FIG. 1.59

Height equivalent to a theoretical stage for common packings.

1.11.3 The method of transfer units

The proposals of Peters [70] are really the application of the stage-wise mechanism for the plate column to the packed tower, where the process is one of continuous countercurrent mass transfer. The degree of separation is represented by the rate of change of composition of a component with height of packing, that is, dy/dZ. This rate of change of composition is dependent upon the equipment, the operating conditions, and on the diffusional potential across the two films. Over the vapour film this driving force is measured by $y_i - y$ or $(\Delta y)_f$ where y_i is the mole fraction of the diffusing component at the interface and y the value for the vapour.

The performance of the column may therefore be represented by J, the change in composition with height for unit driving force where:

$$J = \frac{dy/dZ}{(\Delta y)_f} \tag{1.145}$$

A relation between dy/dZ and $(\Delta y)_f$ may be obtained on the basis of the two-film theory of mass transfer. For the vapour film, Fick's law, Volume 1, Chapter 10, gives:

$$N_A = -D_v \frac{dC}{dz} \tag{1.146}$$

where: N_A is the molar rate of transfer per unit area of interface of component **A**,
  D_v is the vapour diffusion coefficient,
  C_A is the concentration of **A** in moles per unit volume, and
  z is the distance in direction of diffusion.
  For an ideal gas, this gives:

$$N_A = \frac{D_v}{RT} \frac{dp_A}{dz} = -\frac{D_v P}{RT} \frac{dy}{dz} \tag{1.147}$$

where y is the mole fraction of **A**.

The negative sign occurs because z is taken in the direction of diffusion from the interface, and y decreases in this direction.

For equimolecular counterdiffusion, integration gives:

$$N_A = -\frac{D_v P}{RT_z}(\Delta y)f \tag{1.148}$$

or :
$$N_A = k'_g (\Delta y)f \tag{1.149}$$

where :
$$k'_g = -\frac{D_v P}{RT_z} \tag{1.150}$$

Considering the column shown in Fig. 1.60, in which the concentration of the more volatile component increases from y_b to y_t, the rate of transfer over a small height of the column dZ may be written as:

$$k'_g a A(y_i - y)dZ \tag{1.151}$$

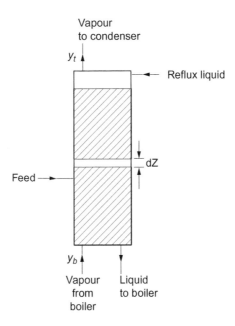

FIG. 1.60

Arrangement for a packed column.

where: a is the active interfacial area for transfer per unit volume of the column and,
   A is the cross-sectional area of the column.
   The moles of component **A** diffusing = (total moles of vapour × change in mole fraction).

$$= G'A \, dy$$

   where: G' is the vapour rate in the column in moles/unit time and unit cross-section,
   and k'_g is known as the gas film transfer coefficient, and is measured as moles/unit time unit area-unit mole fraction difference, and hence:

$$G'A \, dy = k'_g aA(y_i - y)dZ \qquad (1.152)$$

or :
$$J = \frac{dy/dZ}{y_i - y} = \frac{k'_g a}{G'} \qquad \text{(from 1.145)}$$

and :
$$\int_{yb}^{yt} \frac{dy}{y_i - y} = \frac{k'_g a}{G'} Z \qquad (1.153)$$

where $k'_g a$ is taken as constant over the column.

   The group on the left-hand side of this equation represents the integrated ratio of the change in composition to the driving force tending to bring this change about. This group has been defined by Chilton and Colburn [72] as the number of transfer units N_G. The quantity $G'/(k'_g a)$, which is the reciprocal of the efficiency J and has the dimensions of length, is defined as the height of a transfer unit, H_G. Equation (1.153) may be written as:

$$Z = H_G N_G \tag{1.154}$$

The concentrations y_i, y refers to conditions on either side of the gas film, and hence N_G is the number of gas film transfer units, and H_G the height of a gas film transfer unit.

For packed columns $(k'_g a)/G'$ represents a useful value for the efficiency, and the performance of a packed column is commonly represented by the simple term H_G, where a low value of H_G corresponds to an efficient column. If H_G is known, the necessary height of a column is found from Eq. (1.154), since N_G is determined from the change in concentration required and the shape of the equilibrium curve. Application of this technique is discussed in Chapter 12 and reference may also be made to Volume 1, Chapter 10.

The same number of moles passes through the liquid film and a similar series of equations may be obtained in terms of concentrations across the liquid film, that is:

$$G'\, dy = L'\, dx \tag{1.155}$$

where: L' is the molar flow rate of liquid/unit area, and

x is the mole fraction of the more volatile component in the liquid.

If k'_l is the mass transfer coefficient for the liquid phase in moles/unit time, unit area, unit mole fraction driving force, then:

$$AL'\, dx = k'_l a A (x - x_i)\, dZ$$

Thus :

$$\int_{xb}^{xt} \frac{dx}{x - x_i} = \frac{k'_l a}{L'} Z \tag{1.156}$$

or :

$$N_L = \frac{1}{H_L} Z$$

where: N_L is the number of liquid film transfer units and

H_L is the height of a liquid film transfer unit, which for distillation applications is presented in Table 1.6, taken from the work of Gilliland and Sherwood [73], as a function of type and size of packing.

Overall transfer coefficients and transfer units.

The driving force over the gas film is taken as $(y_i - y)$ and over the liquid film as $(x - x_i)$. If y_e is the concentration in the gas phase in equilibrium with concentration x in the liquid phase, then $(y_e - y)$ is taken as the overall driving force expressed in terms of y. Similarly $(x - x_e)$ is taken as the overall driving force in terms of x, where x_e is the concentration in the liquid in equilibrium with a concentration y in the vapour.

The overall driving forces $(\Delta y)_o$ and $(\Delta x)_o$ may then be written as:

$$(\Delta y)_o = y_e - y = (y_e - y_i) + (y_i - y)$$
$$(\Delta x)_o = x - x_e = (x - x_i) + (x_i - x_e)$$

Table 1.6 Values of H_L for distillation [73].

Packing size (mm)		12	18	2	40	0
(in.)		0.	0.7	1.0	1.	2.0
Raschig type	m	0.073	0.092	0.104	0.143	0.177
	ft	0.24	0.30	0.34	0.47	0. 8
Intalox	m	0.061	0.079	0.089	0.122	–
	ft	0.20	0.26	0.29	0.40	–
Pall rings	m	–	–	–	0.122	0.150
	ft	–	–	–	0.40	0.49

By analogy with the derivation for film coefficients, a series of overall transfer coefficients and overall transfer units based on these overall driving forces may be defined.

Thus, the number of overall gas transfer units is given by:

$$\mathbf{N}_{OG} = \int \frac{dy}{y_e - y} \tag{1.157}$$

and the number of overall liquid transfer units is given by:

$$\mathbf{N}_{OL} = \int \frac{dx}{x - x_e} \tag{1.158}$$

The heights of the overall transfer units are:

$$\mathbf{H}_{OG} = \frac{G'}{K'_g a} \tag{1.159}$$

and :

$$\mathbf{H}_{OL} = \frac{L'}{K'_l a} \tag{1.160}$$

where $K'_g a$ and $K'_l a$ are overall transfer coefficients, based on gas or liquid concentrations in moles/unit time-unit volume-unit mole fraction driving force.

Relation between overall and film transfer units
From these definitions, the following equation may be written:

$$\frac{dZ}{\mathbf{H}_{OG}} = \frac{dy}{y_e - y} \quad \frac{dZ}{\mathbf{H}_G} = \frac{dy}{y_i - y}$$
$$\frac{dZ}{\mathbf{H}_{OL}} = \frac{dx}{x - x_e} \quad \frac{dZ}{\mathbf{H}_L} = \frac{dx}{x - x_i}$$

Thus :

$$\frac{\mathbf{H}_{OG}}{y_e - y} = \frac{\mathbf{H}_G}{y_i - y}$$

and :

$$\mathbf{H}_{OG} = \mathbf{H}_G \left[\frac{y_e - y_i + y_i - y}{y_i - y} \right] = \mathbf{H}_G \left[1 + \frac{y_e - y_i}{y_i - y} \right] \tag{1.161}$$

If the equilibrium curve is linear over the range $y = y_e$ to $y = y_i$, then assuming equilibrium at the interface:

$$y_e - y_i = m(x - x_i)$$

and :

$$\mathbf{H}_{OG} = \mathbf{H}_G + m\left(\frac{x - x_i}{y_i - y}\right)\mathbf{H}_G$$

and :

$$\frac{x - x_i}{y_i - y} = \frac{dx\,\mathbf{H}_L}{dy\,\mathbf{H}_G} = \frac{G'\,\mathbf{H}_L}{L'\,\mathbf{H}_G}$$

since $G'\,dy = L'\,dx$.

Thus:

$$\mathbf{H}_{OG} = \mathbf{H}_G + \frac{mG'}{L'}\mathbf{H}_L \tag{1.162}$$

Similarly:

$$\mathbf{H}_{OL} = \mathbf{H}_L + \frac{L'}{mG'}\mathbf{H}_G \tag{1.163}$$

Dividing Eq. (1.162) by Eq. (1.163) gives:

$$\frac{\mathbf{H}_{OG}}{\mathbf{H}_{OL}} = \frac{mG'}{L'} = \frac{\mathbf{N}_{OL}}{\mathbf{N}_{OG}} \tag{1.164}$$

This form of relationship may be written in terms of transfer coefficients, as discussed in Chapter 12, to give:

$$\frac{1}{K'_g a} = \frac{1}{k'_g a} + \frac{m}{k'_l a} \tag{1.165}$$

Relation of HTU to HETP

In a theoretical plate, the mole fraction of the more volatile component in the vapour will increase from y to y_e, so that the total mass transfer is $G'(y_e - y)$. If the equilibrium curve may be considered linear over the height of column equivalent to a theoretical plate Z_t, the logarithmic mean driving force may be used. Thus, referring to Fig. 1.61:

$$G'(y_e - y) = k'_g a Z_t (y_e - y)\frac{(mG'/L') - 1}{\ln(mG'/L')}$$

Thus:

$$
\begin{aligned}
Z_t &= \frac{G'}{K'_g a}\frac{\ln(mG'/L')}{(mG'/L') - 1} = \mathbf{H}_{OG}\frac{\ln(mG'/L')}{(mG'/L') - 1} \\
&= \mathbf{H}_{OG}\frac{\ln\{1 - [1 - (mG'/L')]\}}{(mG'/L') - 1} \\
&= \mathbf{H}_{OG}\left\{1 + \frac{1}{2}\left[1 - (mG'/L') + \frac{1}{3}[1 - (mG'/L')]^2 + \cdots\right]\right\}
\end{aligned}
\tag{1.166}
$$

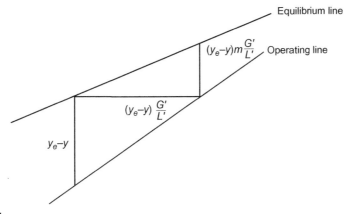

Equilibrium line

$(y_e-y)m\dfrac{G'}{L'}$ Operating line

$(y_e-y)\dfrac{G'}{L'}$

y_e-y

FIG. 1.61

Height equivalent of a theoretical plate.

If the operating and equilibrium lines are parallel, $mG'/L' = 1$, and:

$$Z_t = \mathbf{H}_{OG} \tag{1.167}$$

Thus, the ratio of the height equivalent to a theoretical plate to the height of the transfer unit ($Z_t/\mathbf{H}_{OG}$) may be greater or less than unity, according to whether the slope of the operating line is greater or less than that of the equilibrium curve.

Experimental determination of transfer units

There have been a number of reports presented by Furnas and Taylor [74], Duncan et al. [75], and Sawistowski and Smith [76] on the influence of flow parameters and physical properties on the value of the height of a transfer unit. Most of the work has been carried out in small laboratory columns and great care must be exercised if these data are applied to large diameter units. Some general indication of the values of $\mathbf{H}_{OG}$ are given in Table 1.7, which gives values obtained by Furnas and Taylor [74] for experiments with ethanol–water mixtures at atmospheric pressure in a column 305 mm diameter operating at total reflux.

Table 1.7 Values of the height of the transfer unit $\mathbf{H}_{OG}$ [74].

Packing	Depth of packing (m)	Liquid rate (kg/m²s)	$\mathbf{H}_{OG}$ (m)
50 mm Raschig rings	3.0	1.06	0.670
25 mm Raschig rings	3.0	1.02	0.366
25 mm Berl saddles	2.75	0.195	0.427
25 mm Berl saddles	2.75	1.25	0.335
12.5 mm Berl saddles	3.0	0.25	0.457
12.5 mm Berl saddles	3.0	1.196	0.274
9.5 mm Raschig rings	2.44	0.416	0.396
9.5 mm Raschig rings	2.44	0.780	0.305

Values of HTU in terms of flow rates and physical properties
Wetted-wall columns

The proposals made for calculating transfer coefficients from physical data of the system and the liquid and vapour rates are all related to conditions existing in a simpler unit in the form of a wetted-wall column. In the wetted-wall column, discussed in Chapter 12, vapour rising from the boiler passes up the column which is lagged to prevent heat loss. The liquid flows down the walls, and it thus provides the simplest form of equipment giving countercurrent flow. The mass transfer in the unit may be expressed by means of the j-factor of Chilton and Colburn which is discussed in Volume 1, Chapter 10. Thus:

$$j_d = \frac{k'_g}{G'}\left(\frac{\mu}{\rho D}\right)^{2/3}_v = 0.023\left(\frac{d_c u \rho}{\mu}\right)^{-0.17}_v \tag{1.168}$$

where the flow rate and physical properties refer to the vapour. This type of unit has been studied by Gilliland and Sherwood [73], Chari and Storrow [77], Surowiec and Furnas [78], and others.

For a wetted-wall column:

$$\frac{\text{area of interface}}{\text{volume of column}} = (4/d_c) = a$$

Thus:

$$\mathbf{H}_G = \frac{G'}{k'_g a} = 10.9 d_c \, Re_v^{0.17} Sc_v^{2/3} \tag{1.169}$$

where the linear characteristic length is taken as the diameter of the column d_c. Re_v and Sc_v are the Reynolds and Schmidt numbers with respect to the vapour. Surowiec and Furnas were able to express their results, obtained with alcohol and water, in this form. For transfer through the liquid film, an expression was derived based on the analysis of heat transfer from a tube to a liquid flowing under viscous conditions down the inside of the tube.

The equation was presented as:

$$\mathbf{H}_L = B'Z\left(\frac{M}{M_m}\right) Re_l^{8/9} Sc_l^{10/9}\left(\frac{D_L^2}{gZ^3}\right)^{2/9} \tag{1.170}$$

where: B' is a constant,
M_m is the mean molecular weight of the liquid,
M is the point value of the molecular weight,
Z is the height of the tube, and
Re_l and Sc_l are the Reynolds and Schmidt numbers with respect to the liquid.

It has been suggested, however, that for mass transfer, the transfer in the liquid phase is from a vapour–liquid interface where the liquid velocity is a maximum to the wall where it is zero. With a liquid flowing inside the tube the heat transfer is from a layer of zero velocity at the wall to the fluid all the way to the centre of the tube where it is moving with a maximum velocity. Hatta [79] based his analysis on the more closely related process of diffusion of a gas into a liquid, and obtained the expression:

$$\mathbf{H}_L = B''Z\left(\frac{M}{M_m}\right)Re_l^{2/3}Sc_l^{5/6}\left(\frac{D_L^2}{gZ^3}\right)^{1/6} \tag{1.171}$$

It may be seen that, despite the difference in the arguments, the two equations are really of a similar nature.

Packed columns

The application of the ideas for wetted-wall columns to the more complex case of packed columns requires the assumptions: (a) that the mechanism is unchanged and (b) that the expressions are valid over the much wider ranges of flow rates used in packed columns. This has been attempted by Sawistowski and Smith [76] and Pratt [80]. Pratt started from the basic equation:

$$j_d = \frac{k'_g}{G'}\left(\frac{\mu}{\rho D}\right)_v^{2/3} = \text{const } Re^{-0.2} \tag{1.172}$$

and suggested, from the examination of the available data, that the importance of the degree of wetting may be taken into account by writing this as:

$$\frac{k_g}{G'}e\left(\frac{\mu}{\rho D}\right)_v^{2/3} = p\omega\left(\frac{d_e G'}{\mu e}\right)^{-0.25} \tag{1.173}$$

where: G' is the mass velocity (mass rate per unit area),
$\quad d_e$ is the hydraulic mean diameter for the packing,
$\quad p$ is a constant,
$\quad \omega$ is the fraction of the packing wetted, and
$\quad e$ is the fractional voidage of the packing.

Pratt [80] gives several plots of $p\omega$ against L_p, as shown in Fig. 1.62, where L_p is the liquid rate based on the periphery of the packing, in $m^3/s\,m$. The periphery is taken as $a^{-1}(m^3/m^2)$, although this is only correct for geometrical systems such as stacked rings. The problems of the wetting of packings are discussed in Chapter 4 where other methods are presented.

1.12 **Catlytic/reactive distillation** [81]

Catalytic Distillation (CD) is a recently adapted reactor technology that utilises the dynamics of simultaneous reaction and separation in a single process unit to achieve a more compact, economical, efficient, and optimised process design when compared to the traditional multi-unit designs. The process combines a packed catalyst tower which doubles as a reactor/distillation tower. The catalyst bed emulates the packed distillation tower/a reactor bed. It came in vogue in the late seventies when

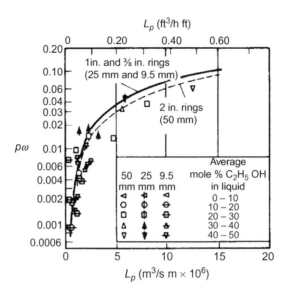

FIG. 1.62

Effect of liquid rate on the degree of wetting of packing [80].

demands for benzene reduction in gasoline and addition of oxygenates for clean gasoline were mandated in various countries. One of the first applications involved alkylation of aromatics in gasoline to their corresponding cycloparafins. Production of methyl tertiary butyl ether, MTBE, utilising reaction of methanol and isobutylene followed soon along with conversion of benzene to ethyl benzene, a prestep in production of styrene monomer. Today it has now grown to include many other processes including hydrogenation.

1.13 Aspen HYSIS process design and simulator [82]

HYSYS was first conceived and created by the Canadian company Hyprotech, founded by researchers from the University of Calgary. Now known as **Aspen HYSYS,** it is a chemical process simulator used to mathematically model chemical processes, from unit operations to full chemical plants and refineries. HYSYS is able to perform many of the core calculations of chemical engineering, including those concerned with mass balance, energy balance, vapour–liquid equilibrium, heat transfer, mass transfer, chemical kinetics, fractionation, and pressure drop. HYSYS is used extensively in industry and academia for steady-state and dynamic simulation, process design, performance modelling, and optimisation.

References

[1] C.S. Robinson, E.R. Gilliland, Elements of Fractional Distillation, fourth ed., McGraw-Hill, New York, 1950.

[2] J.D. Thornton, F.H. Garner, Vapour – liquid equilibria in hydrocarbon-non-hydrocarbon systems. 1: the system benzene – cyclohexane – furfuraldehyde, J. Appl. Chem. Suppl. 1 (1951) 61.

[3] C. Antoine, Tensions des vapeurs: nouvelle relation entre les tensions et les températures, Compt. Rendus 107 (1888). 681, 836.

[4] T. Boublik, V. Fried, E. Hala, The Vapour Pressure of Pure Substances, Elsevier, New York, 1973.

[5] T.E. Jordan, Vapour Pressure of Organic Compounds, Interscience, New York, 1954.

[6] D.R. Stull, Vapour pressures of pure substances – organic compounds, Ind. Eng. Chem. 39 (1947) 517–540. *Ind. Eng. Chem.* 39 (1947) 540 – 550. Vapour pressures of pure substances – inorganic compounds.

[7] S. Ohe, Computer Aided Data Book of Vapour Pressure, Data Book Publishing Co., Tokyo, 1976.

[8] R.C. Reid, J.M. Prausnitz, T.K. Sherwood, The Properties of Liquids and Gases, third ed., McGraw-Hill, New York, 1977.

[9] B.J. Zwolinski, Selected Values of Properties of Hydrocarbons and Related Compounds (API Research Project 44, College Station, Texas).

[10] B.J. Zwolinski, Selected Values of Properties of Chemical Compounds (Thermodynamics Research Center, Texas A & M University, College Station, Texas).

[11] L. Riedel, Liquid density in the saturated state. Extension of the theory of corresponding states, Chem. Ing. Tech. 26 (1954) 259–264.

[12] D. Ambrose, C.H.S. Sprake, R. Townsend, Thermodynamic properties of organic oxygen compounds XXXIII. The vapour pressure of acetone, J. Chem. Thermodyn. 6 (1974) 693–700.

[13] O. Frank, Distillation design, Chem. Eng. Albany 84 (1977) 111.

[14] R.H. Perry, D.W. Green, J.O. Maloney (Eds.), Perry's Chemical Engineers' Handbook, seventh ed., McGraw-Hill Book Company, New York, 1997.

[15] K. Wohl, Thermodynamic evaluation of binary and ternary liquid systems, Trans. Am. Inst. Chem. Eng. 42 (1946) 215.

[16] K. Wohl, Thermodynamic evaluation of binary and ternary liquid systems, Chem. Eng. Prog. 49 (1953) 218.

[17] R.E. Balzhiser, M.R. Samuels, J.D. Eliassen, Chemical Engineering Thermodynamics: The Study of Energy, Entropy, and Equilibrium, Prentice-Hall, Englewood Cliffs, NJ, 1972.

[18] A. Fredenslund, J. Bmehling, P. Rasmussen, Vapor-Liquid Equilibria Using UNIFAC: A Group-Contribution Method, Elsevier, Amsterdam, 1977.

[19] J.M. Prausntiz, T.F. Anderson, E.A. Grens, C.A. Eckert, R. Hsieh, J.P. O'Connell, Computer Calculations for Multicomponent Vapor – Liquid and Liquid – Liquid Equilibria, Prentice-Hall, Englewood Cliffs, NJ, 1980.

[20] R.C. Reid, J.M. Prausnitz, B.E. Poling, The Properties of Gases and Liquids, fourth ed., McGraw-Hill, New York, 1987.

[21] H.C. van Ness, M.M. Abbott, Classical Thermodynamics of Non-Electrolyte Solutions, McGraw-Hill, New York, 1982.

[22] S.M. Walas, Phase Equilibria in Chemical Engineering, Butterworth, Boston, 1985.

[23] The National Engineering Laboratory, UK.

[24] L. Rayleigh, On the distillation of binary mixtures, Phil. Mag. 4 (23) (1902) 521 (vi).

[25] E. Sorel, Distillation et Rectification Industrielle, G. Carré et C, Naud, 1899.

[26] W.K. Lewis, The theory of fractional distillation, Ind. Eng. Chem. 1 (1909) 522.

[27] W.L. McCabe, E.W. Thiele, Graphical design of fractionating columns, Ind. Eng. Chem. 17 (1925) 605.

[28] A.J.V. Underwood, The theory and practice of testing stills, Trans. Inst. Chem. Eng. 10 (1932) 112.

[29] M.R. Fenske, Fractionation of straight-run Pennsylvania gasoline, Ind. Eng. Chem. 24 (1932) 482.

[30] E.R. Gilliland, Multicomponent rectification. Estimation of the number of theoretical plates as a function of the reflux ratio, Ind. Eng. Chem. 32 (1940) 1220.

[31] A.P. Colburn, Division of Chemical Engineering Lecture Notes, University of Delaware, Newark, U.S.A., 1943.

[32] J.H. Harker, Processing, Economic Balance in Distillation, April 1979, p. 39.

[33] M. Ruhemann, The ammonia absorption machine. Ibid. 152. A study of the generator and rectifier of an ammonia absorption machine, Trans. Inst. Chem. Eng. 25 (1947) 143.

[34] M. Ponchon, Etude graphique de la distillation fractionnée industrielle, Technique Moderne 13 (1921). 20 and 55.

[35] P. Savarit, Eléments de distillation, Arts et Métiers 75 (1922) 65.

[36] S.R.M. Ellis, H.E.T.P. values in ring packed columns, Birmingham Univ. Chem. Eng. 5 (1) (1953) 21.

[37] E.R. Gilliland, C.E. Reed, Degrees of freedom in multicomponent absorption and rectification columns, Ind. Eng. Chem. 34 (1942) 551.

[38] W.K. Lewis, G.L. Matheson, Studies in distillation. Design of rectifying columns for natural and refining gasoline, Ind. Eng. Chem. 24 (1932) 494.

[39] A.P. Colburn, The calculation of minimum reflux ratio in the distillation of multicomponent mixtures, Trans. Am. Inst. Chem. Eng. 37 (1941) 805.

[40] A.J.V. Underwood, Fractional distillation of multi-component mixtures—calculation of minimum reflux ratio, J. Inst. Petrol. 32 (1946) 614.

[41] A. Zanker, Nomograph replaces Gilliland plot, Hydrocarb. Process. 56 (5) (1977) 263.

[42] A.M. Eliceche, R.W.H. Sargent, Synthesis and design of distillation systems, in: Cost Savings in Distillation. I. Chem. E Symposium Series No. 61, 1981, p. 1.

[43] M. Benedict, L.C. Rubin, Extractive and azeotropic distillation, Trans. Am. Inst. Chem. Eng. 41 (1945) 353.

[44] S. Young, Fractional Distillation, Macmillan, London, 1902.

[45] H. Guinot, F.W. Clark, Azeotropic distillation in industry, Trans. Inst. Chem. Eng. 16 (1938) 189.

[46] W.S. Norman, The dehydration of ethanol by azeotropic distillation. *Ibid.* 89. Design calculations for azeotropic dehydration columns, Trans. Inst. Chem. Eng. 23 (1945) 66.

[47] R.E. Treybal, Mass Transfer Operations, second ed., McGraw-Hill Book Co., New York, 1968.

[48] E.G. Scheibel, Principles of extractive distillation, Chem. Eng. Prog. 44 (1948) 927.

[49] H.R.C. Pratt, Continuous purification and azeotropic dehydration of acetonitrile produced by the catalytic acetic acid – ammonia reaction, Trans. Inst. Chem. Eng. 25 (1947) 43.

[50] E.J. Hoffman, Azeotropic and Extractive Distillation, Interscience Publishers Inc., New York, 1964.

[51] B.D. Smith, Design of Equilibrium Stage Processes, McGraw-Hill, New York, 1963.

[52] E. Hausbrand, Principles and Practice of Industrial Distillation, sixth ed., Wiley, 1926. translated by Tripp, E. H.

[53] Engineering Staff, Shell Development Company, Emeryville, California, Turbogrid distillation trays, Chem. Eng. Prog. 50 (1954) 57.

[54] J.S. Carey, Plate-type distillation columns, Chem. Met. Eng. 46 (1939) 314.

[55] J.S. Carey, J. Griswold, W.K. Lewis, W.H. McAdams, Plate efficiencies in rectification of binary mixtures, Trans. Am. Inst. Chem. Eng. 30 (1934) 504.

[56] M. Souders, G.G. Brown, Design of fractionating columns, Ind. Eng. Chem. 26 (1934) 98.

[57] E. Kirschbaum, Distillation and Rectification, Chemical Publishing Co., 1948.

[58] H.G. Drickamer, J.R. Bradford, Overall plate efficiency of commercial hydrocarbon fractionating columns as a function of viscosity, Trans. Am. Inst. Chem. Eng. 39 (1943) 319.

[59] H.E. O'Connell, Plate efficiency of fractionating columns and absorbers, Trans. Am. Inst. Chem. Eng. 42 (1946) 741.

[60] L.C. Strang, Entrainment in a bubble-cap fractionating column, Trans. Inst. Chem. Eng. 12 (1934) 169.

[61] C.C. Peavy, E.M. Baker, Efficiency and capacity of a bubble-plate fractionating column, Ind. Eng. Chem. 29 (1937) 1056.

[62] A.P. Colburn, Effect of entertainment on plate efficiency in distillation, Ind. Eng. Chem. 28 (1936) 526.

[63] Ballast Tray Manual, Bulletin No. 4900 (revised), Fritz Glitsch and Sons Inc, Dallas, Texas, 1970.

[64] E.V. Murphree, Rectifying column calculations—with particular reference to N component mixtures, Ind. Eng. Chem. 17 (1925) 747.

[65] F.J. Lockhart, C.W. Leggett, Chapter 6, New fractionating-tray designs, in: J.A. Kobe, J. J. McKeytta (Eds.), Advances in Petroleum Chemistry and Refining, vol. 1, Interscience, 1958.

[66] J.C. Chu, J.R. Donovan, B.C. Bosewell, L.C. Furmeister, Plate efficiency correlation in distilling columns and gas absorbers, J. Appl. Chem. 1 (1951) 529.

[67] R.L. Geddes, Local efficiencies of bubble plate fractionators, Trans. Am. Inst. Chem. Eng. 42 (1946) 79.

[68] S. Bakowski, A new method for predicting the plate efficiency of bubble-cap columns, Chem. Eng. Sci. 1 (1951/1952) 266.

[69] Bubble Tray Design Manual, American Institute of Chemical Engineers, New York, 1958.

[70] W.A. Peters, The efficiency and capacity of fractionating columns, Ind. Eng. Chem. 14 (1922) 476.

[71] D.P. Murch, Height of equivalent theoretical plate in packed fractionation columns, Ind. Eng. Chem. 45 (1953) 2616.

[72] T.H. Chilton, A.P. Colburn, Distillation and absorption in packed columns, Ind. Eng. Chem. 27 (1935) 255. 904.

[73] E.R. Gilliland, T.K. Sherwood, Diffusion of vapours into air streams, Ind. Eng. Chem. 26 (1934) 516.

[74] C.C. Furnas, M.L. Taylor, Distillation in packed columns, Trans. Am. Inst. Chem. Eng. 36 (1940) 135.

[75] D.W. Duncan, J.H. Koffolt, J.R. Withrow, The effect of operating variables on the performance of a packed column still, Trans. Am. Inst. Chem. Eng. 38 (1942) 259.

[76] H. Sawistowski, W. Smith, Performance of packed distillation columns, Ind. Eng. Chem. 51 (1959) 915.

[77] K.S. Chari, J.A. Storrow, Film resistances in rectification, J. Appl. Chem. 1 (1951) 45.

[78] A.J. Surowiec, C.C. Furnas, Distillation in a wetted-wall tower, Trans. Am. Inst. Chem. Eng. 38 (1942) 53.

[79] S. Hatta, On the theory of absorption of gases by liquids flowing as a thin layer, J. Soc. Chem. Ind. Japan 37 (1934) 275.

[80] H.R.C. Pratt, The performance of packed absorption and distillation columns with particular reference to wetting, Trans. Inst. Chem. Eng. 29 (1951) 195.

[81] McDermott International through its merger with Chicago Bridge & Iron Company owns the technology formerly developed by **CDTECH** is a partnership between ABB **Lummus** Global and Chemical Research & Licensing Company. They develops and license advanced refining and petrochemical processes based on **CDTECH's** proprietary catalytic distillation technology.

[82] AspenTech is the owner of Aspen HYSYS a chemical process simulator extensively used in chemical, refining and petrochemical industries to mathematically model chemical processes, from unit operations to full chemical plants and refineries.

Further reading

J.R. Backhurst, J.H. Harker, Process Plant Design, Heinemann Educational Books, London, 1973.

R. Billet, Distillation Engineering, Heyden and Sons Ltd, 1979.

P.S. Buckley, W.L. Luyben, J.P. Shunta, Design of Distillation Column Control Systems, Edward Arnold, New York, 1985.

E.J. Hoffman, Azeotropic and Extractive Distillation, Interscience Publishers, Inc, New York, 1964.

C.D. Holland, Fundamentals of Multicomponent Distillation, McGraw-Hill Book Co., New York, 1981.

C.J. King, Separation Processes, second ed., McGraw-Hill Book Co., New York, 1981.

M.J. Lockett, Distillation Fundamentals, Cambridge University Press, 1986.

M.J. Lockett, Distillation Tray Fundamentals, Cambridge University Press, 1986.

W.L. McCabe, J.C. Smith, P. Harriott, Unit Operations of Chemical Engineering, fourth ed., McGraw-Hill Book Co., New York, 1985.

H. Sawistowski, W. Smith, Mass Transfer Process Calculations, Wiley, Chichester, 1963.

T.K. Sherwood, R.L. Pigford, C.R. Wilke, Mass Transfer, McGraw-Hill Book Co., New York, 1974.

B.D. Smith, Design of Equilibrium Stage Processes, McGraw-Hill Book Co., New York, 1963.

R.E. Treybal, Mass Transfer Operations, third ed., McGraw-Hill Book Co., New York, 1980.

P.C. Wankat, Equilibrium Staged Separations: Separations for Chemical Engineers, Elsevier, New York, 1988.

Absorption and stripping of gases

Raj Barchha and Ajay Kumar Ray

Department of Chemical and Biochemical Engineering, University of Western Ontario, London, ON, Canada

Nomenclature

		Units in SI system	Dimensions in M, L, T θ
A	Cross-sectional area of column	m^2	$\mathbf{L}^2$
A	Absorption factor	–	–
a	Surface area of interface per unit volume of column	m^2/m^3	$\mathbf{L}^{-1}$
$a_1, a_2, \ldots$	Constants in Eq. (2.41)	–	–
a'	Specific surface area (Eq. 2.100)	m^{-1}	$\mathbf{L}^{-1}$
B	A constant in Eq. (2.24)	–	–
B'	A constant in Eq. (2.25)	–	–
C	Molar concentration	kmol/m^3	$\mathbf{N}\mathbf{L}^{-3}$
C_A, C_B	Molar concentrations of **A**, **B**	kmol/m^3	$\mathbf{N}\mathbf{L}^{-3}$
C_{AL}, C_{BL}	Molar concentrations of **A**, **B** in bulk of liquid phase	kmol/m^3	$\mathbf{N}\mathbf{L}^{-3}$
C_{Ae}	Molar concentration of **A** in liquid phase in equilibrium with partial pressure P_{AG} in gas phase	kmol/m^3	$\mathbf{N}\mathbf{L}^{-3}$
C_{Ai}	Molar concentration of **A** at interface	kmol/m^3	$\mathbf{N}\mathbf{L}^{-3}$
C_{AL}	Molar concentration of **A** in bulk of liquid	kmol/m^3	$\mathbf{N}\mathbf{L}^{-3}$
C_T	Total molar concentration	kmol/m^3	$\mathbf{N}\mathbf{L}^{-3}$
c	Constant term in equation of equilibrium line	–	–
c_G	Gas mixture constant ($\rho_r/\mu_r)^{0.25}/(D_{Vr})^{0.5}$ in cgs units	[(cm^2/s)$^{-3/4}$]	$\mathbf{L}^{-3/2}\,\mathbf{T}^{3/4}$

Continued

Coulson and Richardson's Chemical Engineering. https://doi.org/10.1016/B978-0-08-101097-6.00002-X

		Units in SI system	Dimensions in M, L, T θ
D_L	Liquid phase diffusivity	m^2/s	$L^2 T^{-1}$
D_V	Vapour phase diffusivity	m^2/s	$L^2 T^{-1}$
d	Column diameter	m	L
d_i	Impeller diameter	m	L
d_0	Sauter mean diameter	m	L
d_p	Packing size	m	L
d_t	Tank diameter	m	L
e	Voidage	–	–
F'	Fractional conversion (Eq. 2.41)	–	–
f	Fraction of surface renewed per unit time	s^{-1}	T^{-1}
G'_m	Molar rate of flow of inert gas per unit cross-section	$kmol/m\ s$	$NL^{-2} T^{-1}$
G'	Gas flow rate (mass) per unit cross-section	kg/m^2s	$ML^{-2} T^{-1}$
$\mathbf{H}$	Height of transfer unit	m	L
h	Heat transfer coefficient	W/m^2K	$MT^{-3} \theta^{-1}$
h_D	Mass transfer coefficient (D_V/z_G)	m/s	LT^{-1}
h_p	Height of packing	m	L
$\mathcal{H}$	Henry's constant	$(N/m^2)/(kmol/m^3)$	$MN^{-1} L^2 T^{-2}$
i	Number of mole of **B** reacting with 1 mol of **A**	–	–
j_d	j-factor for mass transfer	–	–
K_G	Overall gas-phase transfer coefficient	s/m	$L^{-1} T$
K_L	Overall liquid-phase transfer coefficient	m/s	LT^{-1}
K''_G	Overall gas-phase transfer coefficient in terms of mole fractions	$kmol/m^2s$	$NL^{-2} T^{-1}$
K''_L	Overall liquid-phase transfer coefficient in terms of mole fractions	$kmol/m^2s$	$NL^{-2} T^{-1}$
k	Thermal conductivity	$W/m\ K$	$MLT^{-3} \theta^{-1}$
k_G	Gas-film transfer coefficient $(D_V P/RTz_G P_{Bm})$	s/m	$L^{-1} T$
k'_G	Gas-film transfer coefficient (D_V/RTz_G)	s/m	$L^{-1} T$
k''_G	Gas-film transfer coefficient in terms of mole fractions	$kmol/m^2s$	$NL^{-2} T^{-1}$
k_L	Liquid-film transfer coefficient	m/s	LT^{-1}
k''_L	Liquid-film transfer coefficient in terms of mole fractions	$kmol/m^2s$	$NL^{-2} T^{-1}$
k_2	Reaction rate constant for second-order reaction	$m^3/kmols$	$N^{-1} L^3 T^{-1}$

Continued

		Units in SI system	Dimensions in M, L, T θ
L'_m	Molar rate of flow of solute-free liquor per unit cross-section	kmol/s m^2	$\mathbf{NL^{-2}\,T^{-1}}$
L_v	Volumetric liquid rate	m^3/s	$\mathbf{L^3\,T^{-1}}$
L'	Liquid flow rate (mass) per unit cross-section	kg/s m^2	$\mathbf{ML^{-2}T^{-1}}$
m	Slope of equilibrium line	–	–
N_A, N_B	Molar rate of diffusion of **A**, **B** per unit area	kmol/s m^2	$\mathbf{NL^{-2}T^{-1}}$
N'_A, N'_B	Molar rate of absorption of **A**, **B** per unit area .	kmol/s m^2	$\mathbf{NL^{-2}\,T^{-1}}$
N''_A	Molar rate of absorption of **A** per unit area with chemical reaction	kmol/s m^2	$\mathbf{NL^{-2}\,T^{-1}}$
N	Number of transfer units	–	–
N'	Impeller speed	s^{-1} (Hz)	$\mathbf{T^{-1}}$
n	Number of plates from bottom	–	–
P	Total pressure	N/m^2	$\mathbf{ML^{-1}\,T^{-2}}$
P_A, P_B	Partial pressures of **A** and **B**	N/m^2	$\mathbf{ML^{-1}\,T^{-2}}$
P_{Bm}	Logarithmic mean value of P_B	N/m^2	$\mathbf{ML^{-1}\,T^{-2}}$
P_{Ae}	Partial pressure of **A** in equilibrium with concentration C_{AL} in liquid phase	N/m^2	$\mathbf{ML^{-1}\,T^{-2}}$
P_{AG}	Partial pressure of **A** in bulk of gas phase	N/m^2	$\mathbf{ML^{-1}\,T^{-2}}$
P_{Ai}	Partial pressure of **A** at interface	N/m^2	$\mathbf{ML^{-1}\,T^{-2}}$
ΔP_{Alm}	Log mean driving force for **A**	N/m^2	$\mathbf{ML^{-1}\,T^{-2}}$
P$_V$	Power input per unit volume	W/m^3	$\mathbf{ML^{-1}\,T^{-3}}$
R	Universal gas constant	J/kmol K	$\mathbf{NM^{-1}\,L^2\,T^{-2}\,\theta^{-1}}$
r	Ratio of effective film thickness for absorption without and with chemical reaction	–	–
S	Specific surface of packing	m^{-1}	$\mathbf{L^{-1}}$
s	Total number of plates in column	–	–
T	Absolute temperature	K	θ
t	Time	s	$\mathbf{T}$
u	Gas velocity	m/s	$\mathbf{LT^{-1}}$
u_0	Terminal rise velocity	m/s	$\mathbf{LT^{-1}}$
u_s	Superficial gas velocity (based on inlet conditions)	m/s	$\mathbf{LT^{-1}}$
V	Volume of packed section of column	m^3	$\mathbf{L^3}$
X	Moles of solute gas **A** per mole of solvent in liquid phase	–	–
x	Mole fraction of **A** in liquid phase	–	–

Continued

		Units in SI system	Dimensions in M, L, T θ
Y	Molar ratio of solute gas **A** to inert gas **B** in gas phase	–	–
y	Mole fraction of **A** in gas phase	–	–
Z	Height of packed column	m	**L**
z	Distance of direction of mass transfer	m	**L**
z_G	Thickness of gas film	m	**L**
z_L	Thickness of liquid film	m	**L**
α	A coefficient in Eq. (2.30)	$s^{1.8}/kg^{0.8}m^{0.4}$	$\mathbf{M^{-0.8}\,L^{-0.4}\,T^{1.8}}$
β	A coefficient	$1/m^{1.25}$	$\mathbf{L^{-5/4}}$
β'	A coefficient (Eq. 2.35)	–	–
μ	Viscosity of gas	Ns/m^2	$\mathbf{ML^{-1}\,T^{-1}}$
μ_L	Viscosity of liquid	Ns/m^2	$\mathbf{ML^{-1}\,T^{-1}}$
ρ	Density of gas	kg/m^3	$\mathbf{ML^{-3}}$
ρ_L	Density of liquid	kg/m^3	$\mathbf{ML^{-3}}$
σ	Surface tension	J/m^2	$\mathbf{MT^{-2}}$
ϕ	Correction factor for concentrated solutions	–	
Ga	Galileo number	–	–
Pr	Prandtl number	–	–
Re	Reynolds number	–	–
Sc	Schmidt number	–	–
Sh	Sherwood number	–	–
Suffixes			
1	denotes conditions at bottom of packed column, or at plane 1		
2	denotes conditions at top of packed column, or at plane 2		
A	denotes soluble gas		
B	denotes insoluble gas		
e	denotes equilibrium value		
f	denotes film value		
i	denotes value at interface		
G	denotes gas phase		
L	denotes liquid phase		
lm	denotes logarithmic mean value		
n	denotes values on plate n		
r	denotes reference state (293 K, 101.3 kN/m^2)		

LG, OG, L, OL refer to gas film, overall gas, liquid film, and overall liquid transfer units

2.1 **Introduction**

The removal of one or more selected components or contaminants, e.g. ammonia or benzene from a mixture of process gases, hydrogen sulphide from desulphurisation process off gases or carbon dioxide from steam-hydrocarbon reformer product gases by absorption into a suitable liquid is the second major operation of chemical engineering that is based on interphase mass transfer controlled largely by rates of diffusion. Thus, acetone can be recovered from an acetone–air mixture by passing the gas stream into water in which the acetone dissolves while the air passes out. Similarly, ammonia may be removed from an ammonia–air mixture by absorption in water. In each of these examples the process of absorption of the gas in the liquid may be treated as a physical process, the chemical reaction having no appreciable effect. When oxides of nitrogen are absorbed in water to give nitric acid, however, or when carbon dioxide is absorbed in a solution of sodium hydroxide, a chemical reaction occurs, the nature of which influences the actual rate of absorption. Absorption processes are therefore conveniently divided into two groups, those in which the process is solely physical and those where a chemical reaction is occurring. In today's regulatory environment requiring minimisation of waste disposal and containment of gaseous components or contaminants from being emitted to atmosphere, reversible physical and chemical processes dominate absorption. Irreversible chemical reactions only apply where the resulting product has a value-added outlet. Some examples of reversible physical absorption processes include absorption of benzene or lighter hydrocarbons using oil as a solvent and the reversible chemical reaction adsorption processes include ones such as hydrogen sulphide or carbon dioxide absorbed in ethanolamine solutions. Reversible processes include a stripping unit in tandem with the absorber so that the absorbed component is stripped either with an inert gas or steam to be recovered as a useful product, avoiding its emission to atmosphere and returning the stripped or lean solvent back to the absorber. Even though most of the research and earlier work was focused on absorption, both absorption and stripping processes involve same mass transfer theories and associated equations with diffusion directions reversed. In considering the design of equipment to achieve gas absorption and stripping, the main requirement is that the gas should be brought into intimate contact with the liquid, and the effectiveness of the equipment will largely be determined by the success with which it promotes contact between the two phases. The general form of equipment is similar to that described for distillation in Chapter 11, and packed and plate towers are generally used for large installations. The method of operation, as will be seen later, is not the same. In absorption, the feed is a selected components rich or contaminant rich gas introduced at the bottom of the column, and the lean solvent from the stripper is fed to the top, as a liquid absorbent; the absorbed gas and contaminant rich solvent leave at the bottom, and the selected components or contaminants depleted treated gas leaves from the top. In stripping, the stripping gas or the steam are introduced at the bottom of the column, and the rich solvent from the absorber is fed to the top; the contaminant rich gas leaves at the top of the column and the stripped solvent leaves at the bottom to be cooled if necessary and returned to the absorber as the lean solvent. The essential difference between distillation and absorption is that in the former the vapour has to be produced in each stage by partial vaporisation of the liquid which is therefore at its boiling point, whereas in absorption, the liquid is well below its boiling point. Typically, absorption is carried out at close to isothermal conditions and where the process is exothermic, side coolers are often used at various tower heights. In distillation, there is a diffusion of molecules in both directions, so that for an ideal system, equimolecular counterdiffusion takes place; however, in absorption gas, molecules are diffusing into the liquid, with negligible transfer in the reverse direction, as discussed in Volume 1, Chapter 10. In the stripper, the molecules are stripped or diffused from the liquid. In general, the ratio of the liquid to the gas flow rate is considerably greater in absorption than in distillation with the result that layout of the trays is different in the two cases. Furthermore, with the higher liquid rates in absorption, packed columns are much more commonly used.

2.2 Conditions of equilibrium between liquid and gas

When two phases are brought into contact, they eventually reach equilibrium. Thus, water in contact with air evaporates until the air is saturated with water vapour, and the air is absorbed by the water until it becomes saturated with the individual gases. In any mixture of gases, the degree to which each gas is absorbed is determined by its partial pressure. At a given temperature and concentration, each dissolved gas exerts a definite partial pressure. Three types of gases may be considered from this aspect— a very soluble one, such as ammonia, a moderately soluble one, such as sulphur dioxide, and a slightly soluble one, such as oxygen. The values in Table 2.1 show the concentrations in kilograms per 1000 kg of water that are required to develop a partial pressure of 1.3, 6.7, 13.3, 26.7, and 66.7 kN/m^2 at 303 K. It may be seen that a slightly soluble gas requires a much higher partial pressure of the gas in contact with the liquid to give a solution of a given concentration. Conversely, with a very soluble gas a

Table 2.1 Partial pressures and concentrations of aqueous solutions of gases at 303 K.

Partial pressure of solute in gas phase (kN/m^2)	Concentration of solute in water kg/1000 kg water		
	Ammonia	Sulphur dioxide	Oxygen
1.3	11	1.9	–
6.7	50	6.8	–
13.3	93	12	0.008
26.7	160	24.4	0.013
66.7	315	56	0.033

given concentration in the liquid phase is obtained with a lower partial pressure in the vapour phase. At 293 K a solution of 4 kg of sulphur dioxide per 1000 kg of water exerts a partial pressure of 2.7 kN/m². If a gas is in contact with this solution with a partial pressure of SO_2 greater than 2.7 kN/m², sulphur dioxide will be absorbed. The most concentrated solution that can be obtained is that in which the partial pressure of the solute gas is equal to its partial pressure in the gas phase. These equilibrium conditions fix the limits of operation of an absorption unit. Thus, in an ammonia–air mixture containing 13.1% of ammonia, the partial pressure of the ammonia is 13.3 kN/m² and the maximum concentration of the ammonia in the water at 303 K is 93 kg per 1000 kg of water.

While the solubility of a gas is not substantially affected by the total pressure in the system for pressures up to about 500 kN/m², it is important to note that the solubility falls with a rise of temperature. As a result, correspondingly higher partial pressure is needed to obtain the same concentration in the absorbing liquid. Thus, for a concentration of 25% by mass of ammonia in water, the equilibrium partial pressure of the ammonia is 30.3 kN/m² at 293 K and 46.9 kN/m² at 303 K.

In many instances, the absorption is accompanied by the evolution of heat, and it is therefore necessary to fit coolers to the equipment to keep the temperature sufficiently low for an adequate degree of absorption to be obtained.

For dilute concentrations of most gases, and over a wide range for some gases, the equilibrium relationship is given by Henry's law. This law, as used in Chapter 11, can be written as:

$$P_A = \mathcal{H} C_A \tag{2.1}$$

where P_A is the partial pressure of the component **A** in the gas phase, C_A is the concentration of the component in the liquid, typically expressed in mol fraction and $\mathcal{H}$ is Henry's constant.

2.3 **The mechanism of absorption**

2.3.1 **The two-film theory**

The most useful concept of the process of absorption is given by the two-film theory due to Whitman [1], and this is explained fully in Volume 1, Chapter 10. According to this theory, material is transferred in the bulk of the phases by convection currents, and concentration differences are regarded as negligible except in the vicinity of the interface between the phases. On either side of this interface, it is supposed that the currents die out and that there exists a thin film of fluid through which the transfer is affected solely by molecular diffusion. This film will be slightly thicker than the laminar sub-layer, because it offers a resistance equivalent to that of the whole boundary layer. According to Fick's law (Volume 1, eq. 10.1), the rate of transfer by diffusion is proportional to the concentration gradient and to the area of interface over which the diffusion is occurring. Fick's law is limited to cases where the concentration of the absorbed component is low. At high concentrations, bulk flow occurs and the mass transfer rate, which is increased by a factor C_T/C_B, is governed by Stefan's

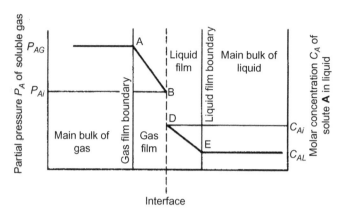

FIG. 2.1

Concentration profile for absorbed component **A**.

law, Eq. (2.2). Under these circumstances, the concentration gradient is no longer constant throughout the film and the lines AB and DE are curved. This question has been discussed in Chapter 10 of Volume 1, but some of the important features will be given here.

The direction of transfer of material across the interface is not dependent solely on the concentration difference, but also on the equilibrium relationship. Thus, for a mixture of ammonia or hydrogen chloride and air which is in equilibrium with an aqueous solution, the concentration in the water is many times greater than that in the air. There is, therefore, a very large concentration gradient across the interface, although this is not the controlling factor in the mass transfer, as it is generally assumed that there is no resistance at the interface itself, where equilibrium conditions will exist. The controlling factor will be the rate of diffusion through the two films where all the resistance is considered to lie. The change in concentration of a component through the gas and liquid phases is illustrated in Fig. 2.1. P_{AG} represents the partial pressure in the bulk of the gas phase and P_{Ai} the partial pressure at the interface. C_{AL} is the concentration in the bulk of the liquid phase and C_{Ai} the concentration at the interface. Thus, according to this theory, the concentrations at the interface are in equilibrium, and the resistance to transfer is centred in the thin films on either side. This type of problem is encountered in heat transfer across a tube, where the main resistance to transfer is shown to lie in the thin films on either side of the wall; here the transfer is by conduction.

2.3.2 Application of mass transfer theories

The preceding analysis of the process of absorption is based on the two-film theory of Whitman [1]. It is supposed that the two films have negligible capacity, but offer all the resistance to mass transfer. Any turbulence disappears at the interface or free surface, and the flow is thus considered to be laminar and parallel to the surface.

An alternative theory described in detail in Volume 1, Chapter 10, has been put forward by Higbie [2], and later extended by Danckwerts [3] and Danckwerts and Kennedy [4] in which the liquid surface is considered to be composed of a large number of small elements each of which is exposed to the gas phase for an interval of time, after which they are replaced by fresh elements arising from the bulk of the liquid.

All three of these proposals give the mass transfer rate N'_A directly proportional to the concentration difference $(C_{Ai} - C_{AL})$ so that they do not directly enable a decision to be made between the theories. However, in the Higbie–Danckwerts theory $N'_A \propto \sqrt{D_L}$ whereas $N'_A \propto D_L$ in the two-film theory. Danckwerts [3] applied this theory to the problem of absorption coupled with chemical reaction but, although in this case the three proposals give somewhat different results, it has not been possible to distinguish between them.

The application of the penetration theory to the interpretation of experimental results obtained in wetted-wall columns has been studied by Lynn et al. [5]. They absorbed pure sulphur dioxide in water and various aqueous solutions of salts and found that, in the presence of a trace of Teepol which suppressed ripple formation, the rate of absorption was closely predicted by the theory. In very short columns, however, the rate was overestimated because of the formation of a region in which the surface was stagnant over the bottom 1 cm length of column. The studies were extended to columns containing spheres and again the penetration theory was found to hold, there being very little mixing of the surface layers with the bulk of the fluid as it flowed from one layer of spheres to the next.

Absorption experiments in columns packed with spheres, 37.8 mm diameter, were also carried out by Davidson et al. [6] who absorbed pure carbon dioxide into water. When a small amount of surface active agent was present in the water no appreciable mixing was found between the layers of spheres. With pure water, however, the liquid was almost completely mixed in this region.

Davidson [7] built up theoretical models of the surfaces existing in a packed bed, and assumed that the liquid ran down each surface in laminar flow and was then fully mixed before it commenced to run down the next surface. The angles of inclination of the surfaces were taken as random. In the first theory, it was assumed that all the surfaces were of equal length, and in the second that there was a random distribution of surface lengths up to a maximum. Thus, the assumptions regarding age distribution of the liquid surfaces were similar to those of Higbie [2] and Danckwerts [3]. Experimental results were in good agreement with the second theory. All random packings of a given size appeared to be equivalent to a series of sloping surfaces, and therefore the most effective packing would be that which gave the largest interfacial area.

In an attempt to test the surface renewal theory of gas absorption, Danckwerts and Kennedy [8] measured the transient rate of absorption of carbon dioxide into various solutions by means of a rotating drum which carried a film of liquid through the gas. Results so obtained were compared with those for absorption in a packed column and it was shown that exposure times of at least 1 s were required to give a strict comparison; this was longer than could be obtained with the rotating drum. Roberts and Danckwerts [9] therefore used a wetted-wall column to extend the times of contact up to 1.3 s. The column was carefully designed to eliminate entry and exit effects and the formation of ripples. The experimental results and conclusions are reported by Danckwerts et al. [10] who showed that they could be used, on the basis of the penetration theory model, to predict the performance of a packed column to within about 10%.

There have been many recent studies of the mechanism of mass transfer in a gas absorption system. Many of these have been directed toward investigating whether there is a significant resistance to mass transfer at the interface itself. In order to obtain results which can readily be interpreted, it is essential to operate with a system of simple geometry. For that reason, a laminar jet has been used by a number of workers.

Cullen and Davidson [11] studied the absorption of carbon dioxide into a laminar jet of water. When the water issued with a uniform velocity over the cross-section, the measured rate of absorption corresponded closely with the theoretical value. When the velocity profile in the water was parabolic, the measured rate was lower than the calculated value; this was attributed to a hydrodynamic entry effect.

The possible existence of an interface resistance in mass transfer has been examined by Raimondi and Toor [12] who absorbed carbon dioxide into a laminar jet of water with a flat velocity profile, using contact times down to 1 ms. They found that the rate of absorption was not more than 4% less than that predicted on the assumption of instantaneous saturation of the surface layers of liquid. Thus, the effects of interfacial resistance could not have been significant. When the jet was formed at the outlet of a long capillary tube so that a parabolic velocity profile was established, absorption rates were lower than predicted because of the reduced surface velocity. The presence of surface-active agents appeared to cause an interfacial resistance, although this effect is probably attributable to a modification of the hydrodynamic pattern.

Sternling and Scriven [13] have examined interfacial phenomena in gas absorption and have explained the interfacial turbulence which has been noted by a number of workers in terms of the Marangoni effect which gives rise to movement at the interface due to local variations in interfacial tension. Some systems have been shown to give rise to stable interfaces when the solute is transferred in one direction, although instabilities develop during transfer in the reverse direction.

Goodridge and Robb [14] used a laminar jet to study the rate of absorption of carbon dioxide into sodium carbonate solutions containing a number of additives including glycerol, sucrose, glucose, and arsenites. For the short times of exposure used, absorption rates into sodium carbonate solution or aqueous glycerol corresponded to those predicted on the basis of pure physical absorption. In the presence of the additives, however, the process was accelerated as the result of chemical reaction.

Absorption of gases and vapour by drops has been studied by Garner and Kendrick [15] and Garner and Lane [16] who developed a vertical wind tunnel in which drops could be suspended for considerable periods of time in the rising gas stream. During the formation of each drop, the rate of mass transfer was very high because of the high initial turbulence. After the initial turbulence had subsided, the mass transfer rate approached the rate for molecular diffusion provided that the circulation had stopped completely. In a drop with stable natural circulation, the rate was found to approach 2.5 times the rate for molecular diffusion.

2.3.3 Diffusion through a stagnant gas

The process of absorption may be regarded as the diffusion of a soluble gas **A** into a liquid. The molecules of **A** have to diffuse through a stagnant gas film and then through a stagnant liquid film before entering the main bulk of liquid. The absorption of a gas consisting of a soluble component **A** and an insoluble component **B** is a problem of mass transfer through a stationary gas to which Stefan's law (Volume 1, Chapter 10) applies:

$$N'_A = -D_V \frac{C_T}{C_B} \frac{dC_A}{dz} \tag{2.2}$$

where N'_A is the overall rate of mass transfer (moles/unit area and unit time), D_V is the gas-phase diffusivity, z is distance in the direction of mass transfer, and C_A, C_B, and C_T are the molar concentrations of **A**, **B**, and total gas, respectively.

Integrating over the whole thickness z_G of the film, and representing concentrations at each side of the interface by suffixes 1 and 2:

$$N'_A = \frac{D_V C_T}{z_G} \ln \frac{C_{B2}}{C_{B1}} \tag{2.3}$$

Since $C_T = P/\mathbf{R}T$, where $\mathbf{R}$ is the gas constant, T the absolute temperature, and P the total pressure. For an ideal gas, then:

$$N'_A = \frac{D_V P}{\mathbf{R}T z_G} \ln \frac{P_{B2}}{P_{B1}} \tag{2.4}$$

Writing P_{Bm} as the log mean of the partial pressures P_{B1} and P_{B2}, then:

$$P_{Bm} = \frac{P_{B2} - P_{B1}}{\ln(P_{B2}/P_{B1})} \tag{2.5}$$

$$N'_A = \frac{D_V P}{\mathbf{R}T z_G} \frac{P_{B2} - P_{B1}}{P_{Bm}}$$

$$= \frac{D_V P}{\mathbf{R}T z_G} \left[\frac{P_{A1} - P_{A2}}{P_{Bm}} \right] \tag{2.6}$$

Hence the rate of absorption of **A** per unit time per unit area is given by:

$$N'_A = k'_G P \left[\frac{P_{A1} - P_{B2}}{P_{Bm}} \right] \tag{2.7}$$

or

$$N'_A = k_G (P_{A1} - P_{A2}) \tag{2.8}$$

where

$$k'_G = \frac{D_V}{RT z_G} \quad \text{and} \quad k_G = \frac{D_V P}{RT z_G P_{Bm}} = \frac{k'_G P}{P_{Bm}} \tag{2.9}$$

In the great majority of industrial processes, the film thickness is not known, so that the rate equation of immediate use is Eq. (2.8) using k_G. k_G is known as the gas-film transfer coefficient for absorption and is a direct measure of the rate of absorption per unit area of interface with a driving force of unit partial pressure difference.

2.3.4 Diffusion in the liquid phase

The rate of diffusion in liquids is much slower than in gases, and mixtures of liquids may take a long time to reach equilibrium unless agitated. This is partly due to the much closer spacing of the molecules, as a result of which the molecular attractions are more important.

While there is at present no theoretical basis for the rate of diffusion in liquids comparable with the kinetic theory for gases, the basic equation is taken as similar to that for gases, or for dilute concentrations:

$$N'_A = -D_L \frac{dC_A}{dz} \tag{2.10}$$

On integration:

$$N'_A = -D_L \left[\frac{C_{A2} - C_{A1}}{z_L} \right] \tag{2.11}$$

where C_A, C_B are the molar concentrations of **A** and **B**, z_L is the thickness of liquid film through which diffusion occurs, and D_L is the diffusivity in the liquid phase.

Since the film thickness is rarely known, Eq. (2.11) is usually rewritten as:

$$N'_A = k_L (C_{A1} - C_{A2}) \tag{2.12}$$

which is similar to Eq. (2.8) for gases.

In Eq. (2.12), k_L is the liquid-film transfer coefficient, which is usually expressed in kmol/s m^2 (kmol/m^3) = m/s. For dilute concentrations:

$$k_L = \frac{D_L}{z_L}$$

2.3.5 Rate of absorption

In a steady-state process of absorption, the rate of transfer of material through the gas film will be the same as that through the liquid film, and the general equation for mass transfer of a component **A** may be written as:

$$N'_A = k_G (P_{AG} - P_{Ai}) = k_L (C_{Ai} - C_{AL}) \tag{2.13}$$

where P_{AG} is the partial pressure in the bulk of the gas, C_{AL} is the concentration in the bulk of the liquid, and P_{Ai} and C_{Ai} are the values of concentration at the interface where equilibrium conditions are assumed to exist. Therefore:

$$\frac{k_G}{k_L} = \frac{C_{Ai} - C_{AL}}{P_{AG} - P_{Ai}} \qquad (2.14)$$

These conditions may be illustrated graphically as in Fig. 2.2, where ABF is the equilibrium curve for the soluble component **A**.

Point D (C_{AL}, P_{AG}) represents conditions in the bulk of the gas and liquid.

P_{AG} is the partial pressure of **A** in the main bulk of the gas stream.

C_{AL} is the average concentration of **A** in the main bulk of the liquid stream.

Point A (C_{Ae}, P_{AG}) represents a concentration of C_{Ae} in the liquid in equilibrium with P_{AG} in the gas.

Point B (C_{Ai}, P_{Ai}) represents the concentration of C_{Ai} in the liquid in equilibrium with P_{Ai} in the gas, and gives conditions at the interface.

Point F (C_{AL}, P_{Ae}) represents a partial pressure P_{Ae} in the gas phase in equilibrium with C_{AL} in the liquid.

Then, the driving force causing transfer in the gas phase is:

$$(P_{AG} - P_{Ai}) \equiv DE$$

and the driving force causing transfer in the liquid phase is:

$$(C_{Ai} - C_{AL}) \equiv BE$$

Then

$$\frac{P_{AG} - P_{Ai}}{C_{Ai} - C_{AL}} = \frac{k_L}{k_G}$$

and the concentrations at the interface (point B) are found by drawing a line through D of slope $-k_L/k_G$ to cut the equilibrium curve in B.

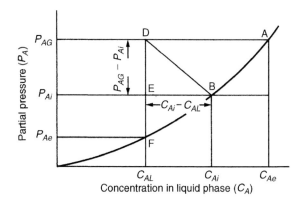

FIG. 2.2

Driving forces in the gas and liquid phases.

Overall coefficients

In order to obtain a direct measurement of the values of k_L and k_G, the measurement of the concentration at the interface would be necessary. These values can only be obtained in very special circumstances, and it has been found of considerable value to use two overall coefficients K_G and K_L defined by:

$$N'_A = K_G(P_{AG} - P_{Ae}) = K_L(C_{Ae} - C_{AL}) \tag{2.15}$$

where K_G and K_L are known as the overall gas and liquid phase coefficients, respectively.

Relation between film and overall coefficients

The rate of transfer of **A** may now be written as:

$$N'_A = k_G[P_{AG} - P_{Ai}] = k_L[C_{Ai} - C_{AL}] = K_G[P_{AG} - P_{Ae}] = K_L[C_{Ae} - C_{AL}]$$

Thus:

$$\frac{1}{K_G} = \frac{1}{k_G}\left[\frac{P_{AG} - P_{Ae}}{P_{AG} - P_{Ai}}\right]$$

$$= \frac{1}{k_G}\left[\frac{P_{AG} - P_{Ai}}{P_{AG} - P_{Ai}}\right] + \frac{1}{k_G}\left[\frac{P_{Ai} - P_{Ae}}{P_{AG} - P_{Ai}}\right] \tag{2.16}$$

From the previous discussion:

$$\frac{1}{k_G} = \frac{1}{k_L}\left[\frac{P_{AG} - P_{Ai}}{C_{Ai} - C_{AL}}\right]$$

Thus

$$\frac{1}{K_G} = \frac{1}{k_G} + \frac{1}{K_L}\left[\frac{P_{AG} - P_{Ai}}{C_{Ai} - C_{AL}}\right]\left[\frac{P_{Ai} - P_{Ae}}{P_{AG} - P_{Ai}}\right]$$

$$= \frac{1}{k_G} + \frac{1}{K_L}\left[\frac{P_{Ai} - P_{Ae}}{C_{Ai} - C_{AL}}\right]$$

$(P_{Ai} - P_{Ae})/(C_{Ai} - C_{AL})$ is the average slope of the equilibrium curve and, when the solution obeys Henry's law, $\mathcal{H} = dP_A/dC_A \approx (P_{Ai} - P_{Ae})/(C_{Ai} - C_{AL})$.
   Therefore:

$$\frac{1}{k_G} = \frac{1}{k_G} + \frac{H}{k_L} \tag{2.17}$$

Similarly:

$$\frac{1}{k_L} = \frac{1}{k_L} + \frac{1}{Hk_G} \tag{2.18}$$

and

$$\frac{1}{k_G} = \frac{H}{k_L} \tag{2.19}$$

À more detailed discussion of the relationship between film and overall coefficients is given in Volume 1, Chapter 10.

The validity of using Eqs (2.17) and (2.18) in order to obtain an overall transfer coefficient has been examined in detail by King [17]. He has pointed out that the equilibrium constant $\mathcal{H}$ must be constant, there must be no significant interfacial resistance, and there must be no interdependence of the values of the two film coefficients.

Rates of absorption in terms of mole fractions

The mass transfer equations can be written as:

$$N'_A = k''_G(y_A - y_{Ai}) = K''_G(y_A - y_{Ae})$$

(2.20)

and

$$N'_A = k''_L(x_{Ai} - x_A) = K''_L(x_{Ae} - x_A)$$

(2.21)

where x_A, y_A are the mole fractions of the soluble component **A** in the liquid and gas phases, respectively.

$k''_G, k''_L, K''_G, K''_L$ are transfer coefficients defined in terms of mole fractions by Eqs (2.20) and (2.21).

If m is the slope of the equilibrium curve [approximately $(y_{Ai} - y_{Ae})/(x_{Ai} - x_A)$], it can then be shown that:

$$\frac{1}{K''_G} = \frac{1}{k''_G} + \frac{m}{k''_L}$$

(2.22)

which is similar to eq. (11.151) used for distillation.

Factors influencing the transfer coefficient

The influence of the solubility of the gas on the shape of the equilibrium curve, and the effect on the film and overall coefficients, may be seen by considering three cases in turn—very soluble, almost insoluble, and moderately soluble gases.

(a) *Very soluble gas.* Here the equilibrium curve lies close to the concentration-axis and the points E and F are very close to one another as shown in Fig. 2.2. The driving force over the gas film (DE) is then approximately equal to the overall driving force (DF), so that k_G is approximately equal to K_G.

(b) *Almost insoluble gas.* Here the equilibrium curve rises very steeply so that the driving force $(C_{Ai} - C_{AL})$ (EB) in the liquid film becomes approximately equal to the overall driving force $(C_{Ae} - C_{AL})$ (AD). In this case, k_L will be approximately equal to K_L.

(c) *Moderately soluble gas.* Here both films offer an appreciable resistance, and the point B at the interface must be located by drawing a line through D of slope $-(k_L/k_G) = -(P_{AG} - P_{Ai})/(C_{Ai} - C_{AL})$.

In most experimental work, the concentration at the interface cannot be measured directly, and only the overall coefficients are therefore found. To obtain values for the film coefficients, the relations between k_G, k_L, and K_G are utilised as discussed previously.

2.4 Determination of transfer coefficients

In the design of an absorption tower, the most important single factor is the value of the transfer coefficient or the height of the transfer unit. While the total flow rates of the gas and liquid streams are fixed by the process, it is necessary to determine the most suitable flow per unit area through the column. The gas flow is limited by the fact that the flooding rate must not be exceeded and there will be a serious drop in performance if the liquid rate is very low. It is convenient to examine the effects of flow rates of the gas and liquid on the transfer coefficients, and also to investigate the influence of variables such as temperature, pressure, and diffusivity.

In the laboratory, wetted-wall columns have been used by a number of workers and they have proved valuable in determining the importance of the various factors, and have served as a basis from which correlations have been developed for packed towers.

2.4.1 Wetted-wall columns

In many early studies, the rate of vaporisation of liquids into an air stream was measured in a wetted-wall column, similar to that shown in Fig. 2.3. Logarithmic plots of d/z_G and $Re = du\rho/\mu$ gave a series of approximately straight lines and d/z_G was proportional to $Re^{0.83}$ where d is the diameter of tube, z_G is the thickness of gas film, u is the gas velocity, ρ is the gas density, μ is the gas viscosity, and B is a constant.

The unknown film thickness z_G may be eliminated as follows:

$$k_G = \frac{D_V P}{\mathbf{R} T z_G P_{Bm}}$$

(2.23)

Thus

$$\frac{k_G \mathbf{R} T P_{Bm}}{D_V P} = \frac{1}{z_G} = \frac{B}{d} Re^{0.83}$$

or

$$\frac{h_D d P_{Bm}}{D_V P} = B Re^{0.83}$$

(2.24)

where $h_D = k_G \mathbf{R} T$ is the mass transfer coefficient with the driving force expressed as a molar concentration difference.

Gilliland and Sherwood's data [18], expressed by Eq. (2.24), are shown in Fig. 2.4 for a number of systems. To allow for the variation in the physical properties, the Schmidt Group Sc is introduced, and the general equation for mass transfer in a wetted-wall column is then given by:

$$\frac{h_D d}{D_V} \frac{P_{Bm}}{P} = B' Re^{0.83} Sc^{0.44}$$

(2.25)

Values of B' 0.021–0.027 have been reported and a mean value of 0.023 may be taken, which means that Eq. (2.25) very similar to the general heat transfer

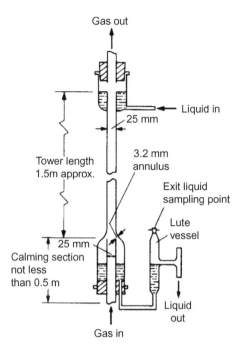

Gas out

Liquid in

25 mm

3.2 mm
annulus

Tower length
1.5m approx.

Exit liquid
sampling point

Lute
vessel

25 mm

Calming section
not less
than 0.5 m

Liquid
out

Gas in

FIG. 2.3

Diagram of a typical laboratory wetted-wall column.

equation for forced convection in tubes (Volume 1, Chapter 9). The data shown in Fig. 2.4 are replotted as $(h_D d/D_V)(P_{Bm}/P)Sc^{-0.44}$ in Fig. 2.5 and, in this way, they may be correlated by means of a single line.

In comparing the results of various workers, it is important to ensure that the inlet arrangements for the air are similar. Modifications of the inlet give rise to various values for the index on the Reynolds number, as found by Hollings and Silver [19]. A good calming length is necessary before the inlet to the measuring section, if the results are to be reproducible.

Eq. (2.25) is frequently rearranged as:

$$\frac{h_D d}{D_V}\frac{P_{Bm}}{P}\frac{\mu}{du\rho}\left[\frac{\mu}{\rho D_V}\right]^{0.56} = B'\, Re^{-0.17}\left(\frac{\mu}{\rho D_V}\right)$$

or

$$\frac{h_D}{u}\frac{P_{Bm}}{P}\left[\frac{\mu}{\rho D_V}\right]^{0.56} = B'\, Re^{-0.17} = j_d \tag{2.26}$$

where j_d is the j-factor for mass transfer as introduced by Chilton and Colburn [20] and discussed in Volume 1, Chapter 10. The main feature of this type of work is that $h_D \propto G'^{0.8}$, $D_V^{0.56}$ and P/P_{Bm}. This form of relation is the basis for correlating data on packed towers.

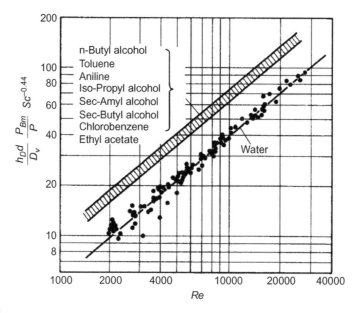

FIG. 2.4

Vaporisation of liquids in a wetted-wall column.

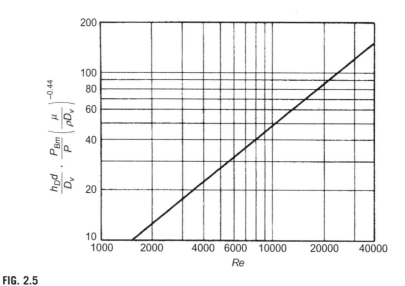

FIG. 2.5

Correlation of data on the vaporisation of liquids in wetted-wall columns.

Example 2.1

The overall liquid transfer coefficient, K_La, for the absorption of SO_2 in water in a column is 0.003 kmol/s m^3 (kmol/m^3). By assuming an expression for the absorption of NH_3 in water at the same liquor rate and varying gas rates, derive an expression for the overall liquid-film coefficient K_La for absorption of NH_3 in water in this equipment at the same water rate though with varying gas rates. The diffusivities of SO_2 and NH_3 in air at 273 K are 0.103 and 0.170 cm^2/s. SO_2 dissolves in water, and Henry's constant is equal to 50 (kN/m^2)/(kmol/m^3). All data are expressed for the same temperature.

Solution
From Eq. (2.18):

$$\frac{1}{K_La} = \frac{1}{k_La} + \frac{1}{Hk_Ga} = \frac{1}{0.003} = 333.3$$

For the absorption of a moderately soluble gas, it is reasonable to assume that the liquid and gas phase resistances are of the same order of magnitude, assuming them to be equal.

$$\frac{1}{k_La} = \frac{1}{Hk_Ga} = \left(\frac{333}{2}\right) = 166.7$$

or

$$k_La = Hk_Ga = 0.006 \ \text{kmol/s m}^3 \ (\text{kmol/m}^3)$$

Thus, for SO_2: $k_Ga = 0.006/\mathcal{H} = 0.006/50 = 0.00012$ kmol/s m^3(kN/m^2). From Eq. (2.26): k_Ga is proportional to (diffusivity)$^{0.56}$.
Hence for NH_3: $k_Ga = 0.00012(0.17/0.103)^{0.56} = 0.00016$ kmol/s m^3 (kN/m^2).
For a very soluble gas such as NH_3, $k_Ga \sim K_Ga$.
For NH_3 the liquid-film resistance will be small, and

$$k_Ga = K_Ga = 0.00016 \ \text{kmol/s m}^3 (\text{kN/m}^2)$$

In early work on wetted-wall columns, Morris and Jackson [21] represented the experimental data for the mass transfer coefficient for the gas film h_D in a form similar to Eq. (2.26), though with slightly different indices, to give:

$$\frac{h_D}{u} = 0.04\left[\frac{ud\rho}{\mu}\right]^{-0.25}\left[\frac{\mu}{\rho D_V}\right]^{-0.5}\left[\frac{P}{P_{Bm}}\right]$$

The velocity u of the gas is strictly the velocity relative to the surface of the falling liquid film, though little error is introduced if it is taken as the superficial velocity in the column.

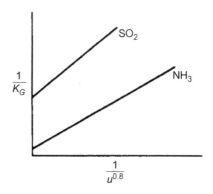

FIG. 2.6

Plot of $1/K_G$ vs $1/u0.8$ for ammonia and for sulphur dioxide.

Compounding of film coefficients

Assuming k_G is approximately proportional to $G'^{0.8}$, Eq. (2.17) may be rearranged to give:

$$\frac{1}{K_G} = \frac{1}{k_G} + \frac{H}{k_L} = \frac{1}{\psi u^{0.8}} + \frac{H}{k_L} \qquad (2.28)$$

If k_L is assumed to be independent of the gas velocity, then a plot of $1/K_G$ against $1/u^{0.8}$ will give a straight line with a positive intercept on the vertical axis representing the liquid-film resistance $\mathcal{H}/k_L$, as shown for ammonia and for sulphur dioxide in Fig. 2.6. It may be seen that in each case a straight line is obtained. The lines for ammonia pass almost through the origin showing that the liquid-film resistance is very small, although the line for sulphur dioxide gives a large intercept on the vertical axis, indicating a high value of the liquid-film resistance.

For a constant value of Re, the film thickness z_G should be independent of temperature, since $\mu/\rho D_V$ is almost independent of temperature. k_G will then vary as $\sqrt{T}$, because $D_V \propto T^{3/2}$ and $k_G \propto D_V/T$. This is somewhat difficult to test accurately since the diffusivity in the liquid phase also depends on temperature. Thus, the data for sulphur dioxide, shown in Fig. 2.6, qualitatively support the theory for different temperatures, although the increase in value of k_L masks the influence of temperature on k_G.

Example 2.2

A wetted-wall column is used for absorbing sulphur dioxide from air by means of a caustic soda solution. At an air flow of $2\,kg/m^2 s$, corresponding to a Reynolds number of 5160, the friction factor $R/\rho u^2$ is 0.0200.

Calculate the mass transfer coefficient in kg $SO_2/s\,m^2$ (kN/m^2) under these conditions if the tower is at atmospheric pressure. At the temperature of absorption, the following values may be used:

The diffusion coefficient for $SO_2 = 0.116 \times 10^{-4} \, m^2/s$, the viscosity of gas $= 0.018 \, mNs/m^2$, and the density of gas stream $= 1.154 \, kg/m^3$.

Solution

For wetted-wall columns, the data are correlated by:

$$\left(\frac{h_d}{u}\right)\left(\frac{P_{Bm}}{P}\right)\left(\frac{\mu}{\rho D}\right)^{0.56} = B' \, Re^{-0.17} = j_d \qquad (2.29)$$

From Volume 1, Chapter 10: $j_d \simeq R/\rho u^2$

In this problem: $G' = 2.0 \, kg/m^2 s$, $Re = 5160$, and $R/\rho u^2 = 0.020$

$$D = 0.116 \times 10^{-4} m^2/s, \quad \mu = 1.8 \times 10^{-5} Ns/m^2, \quad \text{and} \quad \rho = 1.154 \, kg/m^3$$

Substituting these values gives:

$$\left(\frac{\mu}{\rho D}\right)^{0.56} = \left(\frac{1.8 \times 10^{-5}}{1.154 \times 0.116 \times 10^{-4}}\right)^{0.56} = 1.18$$

Thus

$$\left(\frac{h_d}{u}\right)\left(\frac{P_{Bm}}{P}\right) = (0.020/1.18) = 0.0169$$

$$G' = \rho u = 2.0 \, kg/m^2 s$$

and

$$u = (2.0/1.154) = 1.73 \, m/s$$

Thus

$$h_d(P_{Bm}/P) = (0.0169 \times 1.73) = 0.0293$$

d may be obtained from $d = Re\mu/\rho u = 0.046 \, m$ (46 mm), which is the same order of size of wetted-wall column as that which was originally used in the research work.

$$k_G = \left(\frac{h_d}{RT}\right)\left(\frac{P_{Bm}}{P}\right)$$

where **R** is the gas constant, $= 8314 \, m^3 \, (N/m^2)/K \, kmol$ and T will be taken as 298 K, and hence:

$$k_G = [0.0293/(8314 \times 298)]$$

$$= 1.18 \times 10^{-8} \, kmol/m^2 s(N/m^2)$$

$$= 7.56 \times 10^{-4} \, kg \, SO_2/m^2 s(kN/m^2)$$

2.4.2 Coefficients in packed towers

The majority of published data on transfer coefficients in packed towers are for rather small laboratory units, and there is still some uncertainty in extending the data for use in industrial units. One of the great difficulties in correlating the performance of packed towers is the problem of assessing the effective wetted area for interphase transfer. It is convenient to consider separately conditions where the gas-film is controlling, and then where the liquid film is controlling. The general method of expressing results is based on that used for wetted-wall columns.

Gas-film controlled processes

The absorption of ammonia in water has been extensively studied by a number of workers. Kowalke et al. [22] used a tower of 0.4 m internal diameter with a packing 1.2 m deep, and expressed their results as:

$$K_G a = \alpha G'^{0.8} \tag{2.30}$$

where K_G is expressed in kmol/s m^2 (kN/m^2) and a is the interfacial surface per unit volume of tower (m^2/m^3). Thus $K_G a$ is a transfer coefficient based on unit volume of tower. G' is in kg/s m^2, and varies with the nature of the packing and the liquid rate. It was noted that α increased with L' for values up to 1.1 kg/s m^2, after which further increase gave no significant increase in $K_G a$. It was thought that the initial increase in the coefficient was occasioned by a more effective wetting of the packing. On increasing the liquid rate so that the column approached flooding conditions, it was found that $K_G a$ decreased. Other measurements by Borden and Squires [23] and Norman [24] confirm the applicability of Eq. (2.30).

Fellinger [25] used a 450 mm diameter column with downcomers and risers in an attempt to avoid the problem of determining any entrance or exit effects. Some of the results for H_{OG} are shown in Table 2.2, taken from Perry's Chemical Engineers' Handbook [26]. Further discussion on the use of transfer units is included in Section 2.8.8 and in Chapter 11.

Molstad et al. [27] also measured the absorption of ammonia in water using a tower of 384 mm side packed with wood grids, or with rings or saddles, and obtained

Table 2.2 Height of the transfer unit H_{OG} in metres.

Raschig rings size (mm)	G' (kg/m²s)	H$_{OG}$ (L' = 0.65 kg/m²s)	H$_{OG}$ (L' = 1.95 kg/m²s)
9.5	0.26	0.37	0.23
	0.78	0.60	0.32
25	0.26	0.40	0.22
	0.78	0.64	0.34
50	0.26	0.60	0.34
	0.78	1.04	0.58

K_Ga by direct experiment. The value of k_Ga was then calculated from the following relation based on Eq. (2.17):

$$\frac{1}{K_Ga} = \frac{1}{k_Ga} + \frac{H}{k_La} \tag{2.31}$$

The simplest method of representing data for gas-film coefficients is to relate the Sherwood number $[(h_Dd/D_V)(P_{Bm}/P)]$ to the Reynolds number (Re) and the Schmidt number $(\mu/\rho D_V)$. The indices used vary between investigators though van Krevelen and Hoftijzer [28] have given the following expression, which is claimed to be valid over a wide range of Reynolds numbers:

$$\frac{h_Dd}{D_V}\frac{P_{Bm}}{P} = 0.2\,Re^{0.8}\left(\frac{\mu}{\rho D_V}\right)^{0.33} \tag{2.32}$$

Later work suggests that 0.11 is a more realistic value for the coefficient.

Semmelbauer [29] has recommended the following correlation for $100 < (Re)_G < 10,000$ and $0.01\,m < d_p < 0.05\,m$:

$$(Sh)_G = \beta(Re)_G^{0.59}(Sc)_G^{0.33} \tag{2.33}$$

where $\beta = 0.69$ for Raschig rings and 0.86 for Berl saddles,
$$(Sh)_G = h_Dd_p/D_G, \quad (Re)_G = G'd_p/\mu_G,$$
$$(Sc)_G = \mu_G/\rho_GD_G, \quad \text{and} \quad d_P = \text{packing size.}$$

Processes controlled by liquid-film resistance

The absorption of carbon dioxide, oxygen, and hydrogen in water are three examples in which most, if not all, of the resistance to transfer lies in the liquid phase. Sherwood and Holloway [30] measured values of k_La for these systems using a tower of 500 mm diameter packed with 37 mm rings. The results were expressed in the form:

$$\frac{k_La}{D_L} = \beta\left[\frac{L'}{\mu_L}\right]^{0.75}\left[\frac{\mu_L}{\rho_LD_L}\right]^{0.50} \tag{2.34}$$

It may be noted that this equation has no term for characteristic length on the right-hand side and therefore it is not a dimensionally consistent equation. If values of k_La are plotted against value L' on logarithmic scales as shown in Fig. 2.7, a slope of about 0.75 is obtained for values of L' 0.5–20 kg/s m^2. Beyond this value of L', it was found that k_La tended to fall because the loading point for the column was reached. These values of k_La were found to be affected by the gas rate. Subsequently, Cooper et al. [31] established that, at the high liquid rates and low gas rates used in practice, the transfer rates were much lower than given by Eq. (2.34). This was believed to be due to maldistribution at gas velocities as low as 0.03 m/s. The results of Cooper et al. [31] and Sherwood and Holloway [30] are compared in Fig. 2.8, where the height of the transfer unit H_{OL} is plotted against the liquid rate for various gas velocities.

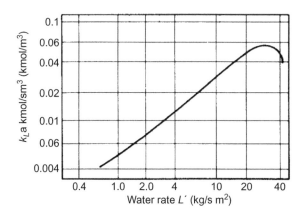

FIG. 2.7

Variation of liquid-film coefficient with liquid flow for the absorption of oxygen in water.

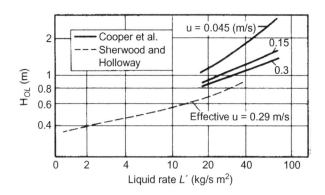

FIG. 2.8

Effect of liquid rate on height of transfer unit HOL. Comparison of the results of Sherwood and Holloway [30], and Cooper et al. [31].

In an equation similar to Eq. (2.33), Semmelbauer [29] produced the following correlation for the liquid-film mass transfer coefficient k_L for $3 < Re_L < 3000$ and $0.01\,\text{m} < d_p < 0.05\,\text{m}$:

$$(Sh)_L = \beta'(Re)_L^{0.59}(Sc)_L^{0.5}\left(d_p^3 g \rho_L^2 / \mu_L^2\right)^{0.17} \tag{2.35}$$

where $\beta' = 0.32$ and 0.25 for Raschig rings and Berl saddles, respectively
$(Sh)_L = k_L d_p / D_L$, $(Re)_L = L' d_p / \mu_L$, and $(Sc)_L = \mu_L / \rho_L D_L$.

Nonhebel [32] emphasises that values of the individual film mass transfer coefficients obtained from this equation must be used with caution when designing large-scale towers and appropriately large safety factors should be incorporated.

2.4.3 **Coefficients in spray towers**

It is difficult to compare the performance of various spray towers since the type of spray distributor used influences the results. Data from Hixson and Scott [33] and others show that $K_G a$ varies as $G'^{0.8}$, and is also affected by the liquid rate. More reliable data with spray columns might be expected if the liquid were introduced in the form of individual drops through a single jet into a tube full of gas. Unfortunately, the drops tend to alter in size and shape and it is not possible to get the true interfacial area very accurately. This has been investigated by Whitman et al. [34], who found that k_G for the absorption of ammonia in water was about 0.035 kmol/s m^2 (N/m^2), compared with 0.00025 for the absorption of carbon dioxide in water.

Some values obtained by Pigford and Pyle [35] for the height of a transfer unit $\mathbf{H}_L$ for the stripping of oxygen from water are shown in Fig. 2.9. For short heights, the efficiency of the spray chamber approximates closely to that of a packed tower although, for heights greater than 1.2 m, the efficiency of the spray tower drops off rather rapidly. While it might be possible to obtain a very large active interface by producing small drops, in practice it is impossible to prevent these coalescing, and hence the effective interfacial surface falls off with height, and spray towers are not used extensively.

2.5 **Absorption associated with chemical reaction**

In the instances so far considered, the process of absorption of the gas in the liquid has been entirely a physical one. There are, however, a number of cases in which the gas, on absorption, reacts chemically with a component of the liquid phase [36]. The topic of mass transfer accompanied by chemical reaction is treated in detail in Volume 1, Chapter 10.

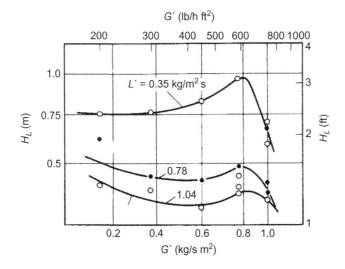

FIG. 2.9

Height of the transfer unit $\mathbf{H}_L$ for stripping of oxygen from water in a spray tower.

In the absorption of carbon dioxide by caustic soda, the carbon dioxide reacts directly with the caustic soda and the process of mass transfer is thus made much more complicated. Again, when carbon dioxide is absorbed in an ethanolamine solution, there is direct chemical reaction between the amine and the gas. In such processes, the conditions in the gas phase are similar to those already discussed, though in the liquid phase, there is a liquid film followed by a reaction zone. The process of diffusion and chemical reaction may still be represented by an extension of the film theory by a method due to Hatta [37]. In the case considered, the chemical reaction is irreversible and of the type in which a solute gas **A** is absorbed from a mixture by a substance **B** in the liquid phase, which combines with **A** according to the equation $\mathbf{A + B \rightarrow AB}$. As the gas approaches the liquid interface, it dissolves and reacts at once with **B**. The new product **AB**, thus formed, diffuses toward the main body of the liquid. The concentration of **B** at the interface falls; this results in diffusion of **B** from the bulk of the liquid phase to the interface. Since the chemical reaction is rapid, **B** is removed very quickly, so that it is necessary for the gas **A** to diffuse through part of the liquid film before meeting **B**. There is thus a zone of reaction between **A** and **B** which moves away from the gas–liquid interface, taking up some position towards the bulk of the liquid. The final position of this reaction zone will be such that the rate of diffusion of **A** from the gas–liquid interface is equal to the rate of diffusion of **B** from the main body of the liquid. When this condition has been reached, the concentrations of **A**, **B**, and **AB** may be indicated as shown in Fig. 2.10, where the concentrations are shown as ordinates and the positions of a plane relative to the interface as abscissae. In this figure, the plane of the interface between gas and liquid is shown by U, the reaction zone by R, and the outer boundary of liquid film by S. Then **A** diffuses through the gas film as a result of the driving force $(P_{AG} - P_{Ai})$ and diffuses to the reaction zone as a result of the driving force C_{Ai} in the liquid phase. The component **B** diffuses from the main body of the liquid to the reaction

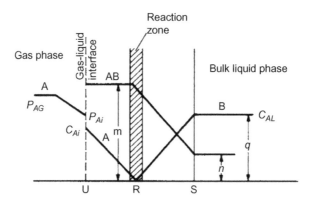

FIG. 2.10

Concentration profile for absorption with chemical reaction.

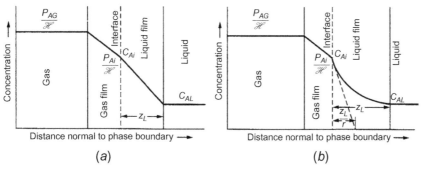

FIG. 2.11

Concentration profiles for absorption (A) without chemical reaction and (B) with chemical reaction. The scales for concentration in the two phases are not the same and are chosen so that PAi/H in the gas phase and CAi for the liquid phase are at the same position in the diagrams.

zone under a driving force q, and the non-volatile product **AB** diffuses back to the main bulk of the liquid under a driving force $(m - n)$.

The difference between a physical absorption, and one in which a chemical reaction occurs, can also be shown by Fig. 2.11A and B, taken from a paper by van Krevelen and Hoftijzer [28]. Fig. 2.11A shows the normal concentration profile for physical absorption while Fig. 2.11B shows the profile modified by the chemical reaction.

For transfer in the gas phase:

$$N'_A = k_G(P_{AG} - P_{Ai}) \tag{2.36}$$

and in the liquid phase:

$$N'_A = k_L(C_{Ai} - C_{AL}) \tag{2.37}$$

The effect of the chemical reaction is to accelerate the removal of **A** from the interface, and supposing that it is now r times as great then:

$$N''_A = rk_L(C_{Ai} - C_{AL}) \tag{2.38}$$

In Fig. 2.11A, the concentration profile through the liquid film of thickness z_L is represented by a straight line such that $k_L = D_L/z_L$. In b, component **A** is removed by chemical reaction, so that the concentration profile is curved. The dotted line gives the concentration profile if, for the same rate of absorption, **A** were removed only by diffusion. The effective diffusion path is $1/r$ times the total film thickness z_L. Thus:

$$N''_A = \frac{rD_L}{z_L}(C_{Ai} - C_{AL}) = rk_L(C_{Ai} - C_{AL}) \tag{2.39}$$

van Krevelen and Hoftyzer [28] showed that the factor r may be related to C_{Ai}, D_L, k_L, to the concentration of **B** in the bulk liquid C_{BL}, and to the second-order reaction rate constant k_2 for the absorption of CO_2 in alkaline solutions. Their relationship is shown in Fig. 2.12, in which r, that is $N''_A/k_L C_{Ai}$, is plotted against $(k_2 D_L C_{BL})^{1/2}/k_L$ for various values of C_{BL}/iC_{Ai}, where i is the number of kmol of **B** combining with 1 kmol of **A**.

Fig. 2.2 illustrates three conditions:

(a) If k_2 is very small, $r \simeq 1$, and conditions are those of physical absorption.
(b) If k_2 is very large, $r \simeq C_{BL}/iC_{Ai}$, and the rate of the process is determined by the transport of **B** toward the phase boundary.
(c) At moderate values of k_2, $r \simeq (jD_L C_{BL})^{1/2}/k_L$, and the rate of the process is determined by the rate of the chemical reaction.

Thus, from Eq. (2.39):

$$N''_A = k_L(C_{Ai} - C_{AL})\frac{(k_2 D_L C_{BL})^{1/2}}{k_L} = (C_{Ai} - C_{AL})(k_2 D_L C_{BL})^{1/2} \tag{2.40}$$

and the controlling parameter is now k_2.

The results of this work have been confirmed by Nijsing et al. [38].

As an illustration of combined absorption and chemical reaction, the results of Tepe and Dodge [39] on the absorption of carbon dioxide by sodium hydroxide

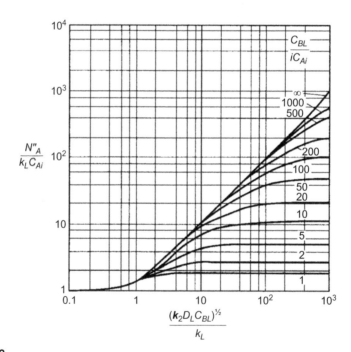

FIG. 2.12

$N'A/kLCAi$ vs $(k2DLCBL)1/2/kL$ for various values of $CBL/iCAi$.

solution may be considered. A 150 mm diameter tower filled to a depth of 915 mm with 12.5 mm carbon Raschig rings was used. Some of the results are indicated in Fig. 2.13. K_Ga increases rapidly with increasing sodium hydroxide concentration up to a value of about $2\,kmol/m^3$. Changes in the gas rate were found to have negligible effect on K_Ga, indicating that the major resistance to absorption was in the liquid phase. The influence of the liquid rate was rather low, and was proportional to $L'^{0.28}$. It may be assumed that, in this case, the final rate of the process is controlled by the resistance to diffusion in the liquid, by the rate of the chemical reaction, or by both together.

Cryder and Maloney [40] presented data on the absorption of carbon dioxide in diethanolamine solution, using a 200 mm tower filled with 20 mm rings, and some of their data are shown in Fig. 2.14. The coefficient K_Ga is found to be independent of the gas rate but to increase with the liquid rate, as expected in a process controlled by the resistance in the liquid phase.

It is difficult to deduce the size of tower required for an absorption combined with a chemical reaction, and a laboratory scale experiment should be carried out in all cases. Stephens and Morris [41] have used a small disc-type tower illustrated in Fig. 2.15 for preliminary experiments of this kind. It was found that a simple wetted-wall column was unsatisfactory where chemical reactions took place. In this unit a series of discs, supported by means of a wire, was arranged one on top of the other as shown.

The absorption of carbon dioxide into aqueous amine solutions has been investigated by Danckwerts and McNeil [42] using a stirred cell. It was found that the reaction proceeded in two stages: first a fast reaction to give amine carbamate,

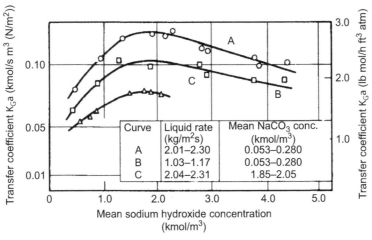

FIG. 2.13

Absorption of carbon dioxide in sodium hydroxide solution $G' = 0.24$–$0.25\,kg/m^2 s$, temperature $= 298\,K$.

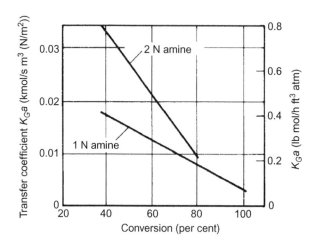

FIG. 2.14

Absorption of carbon dioxide in diethanolamine solutions. Liquid rate $= 1.85\,kg/m^2s$.

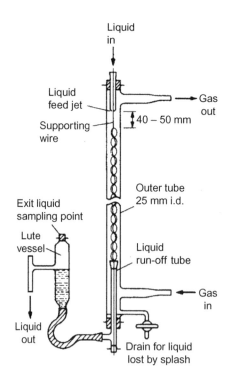

FIG. 2.15

Small disc-tower for absorption tests.

and secondly a slow reaction in the bulk of the liquid in which the carbamate was partially hydrolysed to bicarbonate. The use of sodium arsenite as catalyst considerably accelerated this second reaction, showing that the overall capacity of an absorber could be substantially increased by a suitable catalyst.

A comprehensive review of work on the absorption of carbon dioxide by alkaline solutions has been carried out by Danckwerts and Sharma [43] who applied results of research to the design of industrial scale equipment. Subsequently, Sahay and Sharma [44] showed that the mass transfer coefficient may be correlated with the gas and liquid rates and the gas and liquid compositions by:

$$K_G a = \text{const. } L'^{a1} G'^{a2} \exp (a_3 F' + a_4 y) \tag{2.41}$$

where a_1, a_2, a_3, a_4 are experimentally determined constants, $F' = $ fractional conversion of the liquid, and $y = $ mole fraction of CO_2 in the gas.

Eckert [45], by using the same reaction, determined the mass transfer performance of packings in terms of $K_G a$ as:

$$K_G a = \frac{N}{V(\Delta P_A)_{1m}} \tag{2.42}$$

where $N = $ number of moles of CO_2 absorbed, $V = $ packed volume, and $(\Delta P_A)_{1m} = $ log mean driving force.

Data obtained from this work are limited by the conditions under which they were obtained. It is both difficult and dangerous to extrapolate over the entire range of conditions encountered on a full-scale plant.

2.6 Absorption accompanied by the liberation of heat

In some absorption processes, especially where a chemical reaction occurs, there is a liberation of heat. This generally gives rise to an increase in the temperature of the liquid, with the result that the position of the equilibrium curve is adversely affected.

In the case of plate columns, a heat balance may be performed over each plate and the resulting temperature determined. For adiabatic operation, where no heat is removed from the system, the temperature of the streams leaving the absorber will be higher than those entering, due to the heat of solution. This rise in temperature lowers the solubility of the solute gas so that a large value of L_m/G_m and a larger number of trays will be required than for isothermal operation.

For packed columns, the temperature rise will affect the equilibrium curve, and differential equations for heat and mass transfer, together with heat and mass balances, must be integrated numerically. An example of this procedure is given in Volume 1, Chapter 13, for the case of water cooling. For gas absorption under non-isothermal conditions, reference may be made to specialist texts [46,47] for a detailed description of the methods available. As an approximation, it is sometimes assumed that all the heat evolved is taken up by the liquid, and that temperature rise of the gas may be neglected. This method gives an overestimate of the rise in temperature of the liquid and results in the design of a tower which is taller than necessary. Fig. 2.16 shows the effect of the temperature rise on the equilibrium curve for an adiabatic absorption process of ammonia in water. If the amount of heat liberated is very large, it may be necessary to cool the liquid. This is most conveniently done in a plate column, either with heat exchangers connected between consecutive plates, or with cooling coils on the plate, as shown in Fig. 2.17.

The overall heat transfer coefficient between the gas–liquid dispersion on the tray and the cooling medium in the tubes is dependent upon the gas velocity, as pointed out by Poll and Smith [48], but is usually in the range 500–2000 W/m² K.

With packed towers it is considerably more difficult to arrange for cooling, and it is usually necessary to remove the liquid stream at intervals down the column and to cool externally. Coggan and Bourne [49] have presented a computer programme to

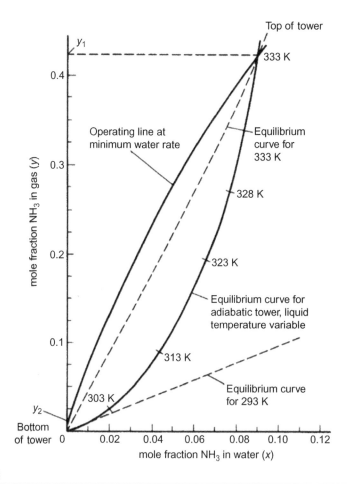

FIG. 2.16

Equilibrium curve modified to allow for the heat of solution of the solute [46].

FIG. 2.17

Glitsch 'truss type' bubble-tray in stainless steel for a 1.9 m absorption column.

enable the economic decision to be made between an adiabatic absorption tower, or a smaller isothermal column with interstage cooling.

2.7 **Packed towers for gas absorption**

From the analysis given already of the diffusional nature of absorption, one of the outstanding requirements is to provide as large an interfacial area of contact as possible between the phases. For this purpose, columns similar to those used for distillation are suitable. However, whereas distillation columns are usually tall and thin absorption columns are more likely to be short and fat. In addition, equipment may be used in which gas is passed into a liquid which is agitated by a stirrer. A few special forms of units have also been used, although it is the packed column which is most frequently used for gas absorption applications.

2.7.1 **Construction**

The essential features of a packed column, as discussed in Chapter 4, are the shell, the arrangements for the gas and liquid inlets and outlets and the packing with its necessary supporting and redistributing systems. Reference may be made to Chapter 4 and to Volume 6 for details of these aspects, while this section is largely concerned with the determination of the height of packing for a particular duty. In installations where the gas is fed from a previous stage of a process where it is under pressure, there is no need to use a blower for the transfer of the gas through the column. When this is not the case, a simple blower is commonly used, and such blowers have been described in Volume 1, Chapter 8. The pressure drop across the column may be calculated by the methods presented in Chapter 4 of this volume and the blower sized accordingly. A pressure drop exceeding 30 mm of water per metre of packing is said to improve gas distribution though process conditions may not permit a figure as high as this. The packed height should not normally exceed 6 m in any section of the tower and for some packings a much lower height must be used.

In the design of an absorption tower, it is necessary to take into account the characteristics of the packing elements and the flow behaviour discussed in Chapter 4, together with the considerations given in the following sections concerning the performance of columns under operating conditions.

2.7.2 Mass transfer coefficients and specific area in packed towers

Traditional methods of assessing the capacity of tower packings, which involve the use of the specific surface area S and the voidage e, developed from the fact that these properties could be readily defined and measured for a packed bed of granular material such as granite, limestone, and coke which were some of the earliest forms of tower packings. The values of S and e enabled a reasonable prediction of hydraulic performance to be made. With the introduction of Raschig rings and other specially shaped packings, it was necessary to introduce a basis for comparing their relative efficiencies. Although the commonly published values of specific surface area S provide a reasonable basis of comparison, papers such as that by Shulman et al. [50] showed that the total area offered by Raschig rings was not used, and varied considerably with hydraulic loading.

Further evidence of the importance of the wetted fraction of the total area came with the introduction of the Pall type ring. A Pall ring having the same surface area as a Raschig ring is up to 60% more efficient, though many still argue the relative merits of packings purely on the basis of surface area.

The selection of a tower packing is based on its hydraulic capacity, which determines the required cross-sectional area of the tower, and the efficiency, $K_G a$ typically, which governs the packing height. Here a is the area of surface per unit volume of column and is therefore equal to $S(1 - e)$. Table 2.3 [51] shows the capacity of the commonly available tower packings relative to 25 mm Raschig rings, for which a considerable amount of information is published in the literature. The table lists the packings in order of relative efficiency, $K_G a$, evaluated at the same approach to the hydraulic capacity limit determined by flooding in each case.

2.7.3 Capacity of packed towers

The drop in pressure for the flow of gas and liquid over packings is discussed in Chapter 4. It is important to note that, during operation, the tower does not reach flooding conditions. In addition, every effort should be made to have as high a liquid rate as possible, in order to attain satisfactory wetting of the packing.

With low liquid rates, the whole of the surface of the packing is not completely wetted. This may be seen very readily by allowing a coloured liquid to flow over packing contained in a glass tube. From the flow patterns, it is obvious how little

Table 2.3 Capacity of commonly available packings relative to 25 mm Raschig rings [51].

Relative $K_G a$	Raschig rings	Traditional saddles	Pall rings	Ceramic pall rings	Ceramic cascade mini ring[c]	Super Intalox saddles	Hypak[a]	Tellerettes[b]	Cascade mini-ring[c]
Materials available for this relative $K_G a$	Ceramic	Ceramic plastic (P)	Metal (M)	Ceramic	Ceramic	Ceramic plastic	Metal	Plastic	Metal (M) plastic (P)
0.6–0.7	75 mm								
0.7–0.8	50 mm								
0.8–0.9	37 mm								
0.9–1.0	25 mm								
1.0–1.1	12 mm	75 mm	87 mm		No. 5	No. 3		Size L	
1.1–1.2		50 mm	50 mm				No. 3		
1.2–1.3		37 mm	50 mm		No. 3	No. 2	No. 2		
1.3–1.4		25 mm	37 mm	50 mm					No. 4 (M)
1.4–1.5				37 mm					No. 3 (P)
1.5–1.6			25 mm	25 mm					No. 3 (M)
1.6–1.7			25 mm		No. 2	No. 1	No. 1	Size S	No. 2 (P)
1.7–1.8			16 mm						
1.8–1.9									No. 2 (M)
1.9–2.0									No. 1 (P)
2.0–2.1									
2.1–2.2									No. 1 (M)

Gas capacity before hydraulic limit (flooding) relative to 25 mm Raschig rings (also approx. The reciprocal of tower cross-sectional area relative to 25 mm Raschig rings for the same pressure drop throughout loading range). All relative capacity figures are valid for the same liquid to gas mass rate ratio:

Note: Relative $K_G a$ for all systems controlled by mass transfer coefficient (K_G) and wetted area (a) per unit volume of column. Some variation should be expected when liquid reaction rate is controlling (not liquid diffusion rate). In these cases liquid hold-up becomes more important. In general a packing having high liquid hold-up which is clearly greater than that in the falling film has poor capacity.

[a]Trade Mark of Norton Company, USA (Hydronyl UK).
[b]Trade Mark of Celicote Company.
[c]Trade Mark of Mass Transfer Ltd. (& Inc.).

of the surface is wetted until the rate is quite high. This difficulty of wetting can sometimes be overcome by having considerable recirculation of the liquid over the tower, although in other cases, such as vacuum distillation, poor wetting will have to be accepted because of the low volume of liquid available. In selecting a packing, it is desirable to choose the form which will give as near complete wetting as possible. The minimum liquid rate below which the packing will no longer perform satisfactorily is known as the minimum wetting rate, discussed in Chapter 4.

The following treatment is a particular application of the more general approach adopted in Volume 1, Chapter 10.

Fig. 2.18 illustrates the conditions that occur during the steady operation of a countercurrent gas–liquid absorption tower. It is convenient to express the concentration of the streams in terms of moles of solute gas per mole of inert gas in the gas phase, and as moles of solute gas per mole of solute free liquid in the liquid phase. The actual area of interface between the two phases is not known, and the term a is introduced as the interfacial area per unit volume of the column. On this basis the general Eq. (2.13), for mass transfer can be written as:

$$N'_A A\mathrm{d}Za = k_G a(P_{AG} - P_{Ai})A\mathrm{d}Z$$
$$= k_L a(C_{Ai} - C_{AL})A\mathrm{d}Z \tag{2.43}$$

where $N'_A =$ kmol of solute absorbed per unit time and unit interfacial area, $a=$ surface area of interface per unit volume of column, $A=$ cross-sectional area of column, and $Z=$ height of packed section.

The interfacial area for transfer $= a\,\mathrm{d}V = aA\,\mathrm{d}Z.$ $\qquad$ (2.44)

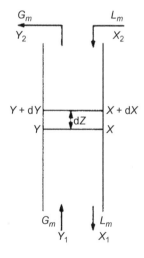

FIG. 2.18

Countercurrent absorption tower.

2.7.4 **Height of column based on conditions in the gas film**

If G_m = moles of inert gas/(unit time) (unit cross-section of tower), L_m = moles of solute-free liquor/(unit time) (unit cross-section of tower), Y = moles of solute gas **A**/mole of inert gas **B** in gas phase, and X = moles of solute **A**/mole of inert solvent in liquid phase and at any plane at which the molar ratios of the diffusing material in the gas and liquid phases are Y and X, then over a small height dZ, the moles of gas leaving the gas phase will equal the moles taken up by the liquid.

Thus

$$AG_m dY = AL_m \ dX \tag{2.45}$$

But

$$G_m A \, dY = N_A'(a \, dV) = k_G a (P_{Ai} - P_{AG}) A \, dZ \tag{2.46}$$

It may be noted that, in a gas absorption process, gas and liquid concentrations will decrease in the upward direction and both dX and dY will be negative.

Since:

$$P_{AG} = \frac{Y}{1+Y} P$$

$$G_m dY = k_G a P \left[\frac{Y_i}{1+Y_i} - \frac{Y}{1+Y} \right] dZ$$

$$= k_G a P \left[\frac{Y_i - Y}{(1+Y)(1+Y_i)} \right] dZ$$

Hence the height of column Z required to achieve a change in Y from Y_1 at the bottom to Y_2 at the top of the column is given by:

$$\int_0^Z dZ = Z = \frac{G_m}{k_G a P} \int_{Y_1}^{Y_2} \frac{(1+Y)(1+Y_i) dY}{Y_i - Y} \tag{2.47}$$

which for dilute mixtures may be written as:

$$Z = \frac{G_m}{k_G a P} \int_{y_1}^{y_2} \frac{dY}{Y_i - Y} \tag{2.48}$$

In this analysis, it has been assumed that k_G is a constant throughout the column, and provided the concentration changes are not too large this will be reasonably true.

2.7.5 **Height of column based on conditions in liquid film**

A similar analysis may be made in terms of the liquid film. Thus from Eqs (2.43) and (2.44):

$$A L_m \ dX = k_L a (C_{Ai} - C_{AL}) A \, dZ \tag{2.49}$$

where the concentrations C are in terms of moles of solute per unit volume of liquor. If C_T = (moles of solute + solvent) (volume of liquid), then:

$$\frac{C_A}{C_T - C_A} = \frac{\text{moles of solute}}{\text{moles of solvent}} = X$$

whence

$$C_A = \frac{X}{1 + X} C_T \tag{2.50}$$

The transfer Eq. (2.49) may now be written as:

$$L_m dX = k_L a C_T \left[\frac{X}{1 + X} - \frac{X_i}{1 + X_i} \right] dZ$$

$$= k_L a C_T \left[\frac{X - X_i}{(1 + X_i)(1 + X)} \right] dZ$$

Thus

$$\int_0^Z dZ = Z = \frac{L_m}{k_L a C_T} \int_{X_1}^{X_2} \frac{(1 + X_i)(1 + X) dX}{X - X_i} \tag{2.51}$$

and for dilute concentrations, this gives:

$$Z = \frac{L_m}{k_L a C_T} \int_{X_1}^{X_2} \frac{dX}{X - X_i} \tag{2.52}$$

where C_T and k_L have been taken as constant over the column.

2.7.6 Height based on overall coefficients

If the driving force based on the gas concentration is written as $(Y - Y_e)$ and the overall gas transfer coefficient as K_G, then the height of the tower for dilute concentrations becomes:

$$Z = \frac{G_m}{k_G a P} \int_{Y_1}^{Y_2} \frac{dY}{Y_e - Y} \tag{2.53}$$

or in terms of the liquor concentration as:

$$Z = \frac{L_m}{k_L a C_T} \int_{X_1}^{X_2} \frac{dX}{X - X_e} \tag{2.54}$$

Equations for dilute concentrations

As the mole fraction is approximately equal to the molar ratio at dilute concentrations then considering the gas film:

$$Z = \frac{G_m}{k_G a P} \int_{Y_1}^{Y_2} \frac{dX}{Y_e - Y} = \frac{G_m}{K_G a P} \int_{y_1}^{y_2} \frac{dy}{y_e - y} \tag{2.55}$$

and considering the liquid film:

$$Z = \frac{L_m}{k_L a C_T} \int_{X_1}^{X_2} \frac{dX}{X - X_e} = \frac{L_m}{K_L a C_T} \int_{x_1}^{x_2} \frac{dx}{x - x_e}$$ (2.56)

2.7.7 The operating line and graphical integration for the height of a column

Taking a material balance on the solute from the bottom of the column to any plane where the mole ratios are Y and X gives for unit area of cross-section:

$$G_m(Y_1 - Y) = L_m(X_1 - X)$$ (2.57)

or

$$Y_1 - Y = \left(\frac{L_m}{G_m}\right)(X_1 - X)$$ (2.58)

This is the equation of a straight line of slope L_m/G_m, which passes through the point (X_1, Y_1). It may be seen by making a material balance over the whole column that the same line passes through the point (X_2, Y_2). This line, known as the operating line, represents the conditions at any point in the column. It is similar to the operating line used in Chapter 11. Fig. 2.19 illustrates typical conditions for the case of moist air

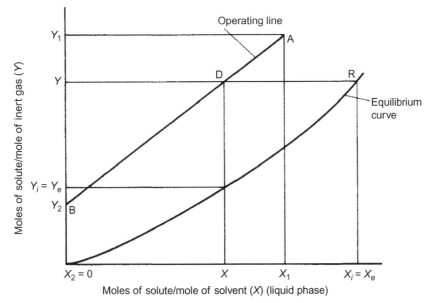

FIG. 2.19

Driving force in gas and liquid-film controlled processes. The figure shows the operating line BDA and the equilibrium curve FR.

and sulphuric acid or caustic soda solution, where the main resistance lies in the gas phase.

The equilibrium curve is represented by the line FR, and the operating line is given by AB, A corresponding to the concentrations at the bottom of the column and B to those at the top of the column. D represents the condition of the bulk of the liquid and gas at any point in the column, and has coordinates X and Y. Then, if the gas film is controlling the process, Y_i equals Y_e, and is given by a point F on the equilibrium curve, with coordinates X and Y_i. The driving force causing transfer is then given by the distance DF. It is therefore possible to evaluate the expression:

$$\int_{Y_1}^{Y_2} \frac{dY}{Y_i - Y}$$

by selecting values of Y, reading off from the figure the corresponding values of Y_i, and thus calculating $1/(Y_i - Y)$. It may be noted that, for gas absorption, $Y > Y_i$ and $Y_i - Y$ and dY in the integral are both negative.

If the liquid film controls the process, X_i equals X_e and the driving force $X_i - X$ is given in Fig. 2.19 by the line DR. The evaluation of the integral:

$$\int_{X_1}^{X_2} \frac{dX}{X - X_i}$$

may be effected in the same way as for the gas film.

Special case when equilibrium curve is a straight line

If over the range of concentrations considered the equilibrium curve is a straight line, it is permissible to use a mean value of the driving force over the column. For dilute concentrations, over a small height dZ of column, the absorption is given by:

$$N'_A Aa\,dZ = G_m A\,dy = K_G aAP(y_e - y)dZ \tag{2.59}$$

If

$$y_e = mx + c \tag{2.60}$$

then

$$y_{e2} = mx_2 + c$$

and

$$y_{e1} = mx_1 + c$$

so that

$$m = \frac{y_{e1} - y_{e2}}{x_1 - x_2} \tag{2.61}$$

Further, taking a material balance over the lower portion of the columns gives:

$$L_m(x_1 - x) = G_m(y_1 - y)$$

and

$$x = x_1 - \frac{G_m}{L_m}(y_1 - y) \tag{2.62}$$

From Eq. (2.59):

$$\int_0^Z \frac{K_G a P}{G_m} dZ = \int_{y_1}^{y_2} \frac{dy}{ye - y} \tag{2.63}$$

$$= \int_{y_1}^{y_2} \frac{dy}{m[x_1 + (G_m/L_m)(y - y_1)] + c - y}$$

(from Eqs 2.58 and 2.60)

$$= \frac{1}{1 - (mG_m/L_m)} \ln \frac{mx_1 + c - y_1}{y_2 - m[x_1 + (G_m/L_m)(y_2 - y_1)] - c}$$

$$= \frac{1}{1 - \dfrac{y_{e1} - y_{e2}}{x_1 - x_2} \cdot \dfrac{x_1 - x_2}{y_1 - y_2}} \ln \frac{y_{e1} - y_1}{y_{e1} - y_{e2} - \left(\dfrac{y_{e1} - y_{e2}}{x_1 - x_2} \dfrac{x_1 - x_2}{y_1 - y_2} y_1 - y_2 \right)}$$

(from Eqs 2.58–2.60)

$$= \frac{y_1 - y_2}{(y - y_e)_1 - (y - y_e)_2} \ln \frac{(y - y_e)_1}{(y - y_e)_2}$$

$$= \frac{y_1 - y_2}{(y - y_e)_{1m}}$$

where $(y - y_e)_{lm}$ is the logarithmic mean value of $y - y_e$.
  Substituting in Eq. (2.63):

$$\frac{K_G a P}{G_m} Z = \frac{y_1 - y_2}{(y - y_e)_{1m}}$$

Thus:

$$a A Z N_A' = G_m(y_1 - y_2)A = K_G a A P(y - y_e)_{1m} Z \tag{2.64}$$

and in terms of mole ratios:

$$a A Z N_A' = G_m(Y_1 - Y_2)A = K_G a A P(Y - Y_e)_{1m} Z \tag{2.65}$$

Thus, the logarithmic mean of the driving forces at the top and the bottom of the
column may be used.
  For concentrated solutions:

$$a A Z N_A' = G_m(Y_1 - Y_2)A = K_G a A \phi P(Y - Y_e)_{1m} Z \tag{2.66}$$

It is necessary to introduce the factor ϕ since Y is not directly proportional to P. The
value of ϕ may be found from the relation:

$$\varphi Y = \frac{Y}{1 + Y} \tag{2.67}$$

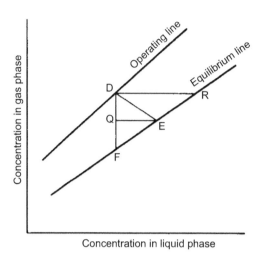

FIG. 2.20

Driving force when equilibrium curve is a straight line.

from which $\phi = 1/(1 + Y)$. Although the value of ϕ will change slightly over the column, a mean value will generally be acceptable.

It is of interest to note from Fig. 2.20, that, as long as the ratio k_L/k_G remains constant (that is, if the slope of DE is constant), then the ratio of DQ, the driving force through the gas phase, divided by DF, the driving force assuming all the resistance to be in the gas phase, will be a constant. Thus, the use of the driving force DF is satisfactory even if the resistance does not lie wholly in the gas phase. The coefficient k_G on this basis is not an accurate value for the gas-film coefficient, although is proportional to it. It follows that, if the equilibrium curve is straight, either the gas-film or the liquid-film coefficient may be used. This simplification is of considerable value.

2.7.8 Capacity of tower in terms of partial pressures for high concentrations

A material balance taken between the bottom of the column and some plane where the partial pressure in the gas phase is P_{AG} and the concentration in the liquid is X gives:

$$L_m(X_1 - X) = G_m \left[\frac{P_{AG1}}{P - P_{AG1}} - \frac{P_{AG}}{P - P_{AG}} \right] \qquad (2.68)$$

Over a small height of the column dZ, therefore:

$$-L_m dX = \frac{-G_m P}{(P - P_{AG})^2} dP_{AG} = k_G a(P_{AG} - P_{Ai})dZ$$

$$= \frac{k_G P_{Bm}}{P} aP \frac{(P_{AG} - P_{Ai})}{P_{Bm}} dZ \tag{2.69}$$

$$= k'_G aP \frac{(P_{AG} - P_{Ai})}{P_{Bm}} dZ$$

Thus:

$$\int_0^Z dZ = G_m \int_{P_{AG_1}}^{P_{AG_2}} \frac{-P_{Bm} dP_{AG}}{k'_G a(P - P_{AG})^2 (P_{AG} - P_{Ai})} \tag{2.70}$$

The advantage of using k'_G instead of k_G is that k'_G is independent of concentration, although this equation is almost unmanageable in practice. If a substantial amount of the gas is absorbed from a concentrated mixture, k'_G will still change as a result of a reduced gas velocity, although it is independent of concentration.

2.7.9 The transfer unit

The group $\int (dy/y_e - y)$, which is used in Chapter 11, has been defined by Chilton and Colburn [52] as the number of overall gas transfer units $\mathbf{N}_{OG}$. The concept of the transfer unit is also introduced in Volume 1, Chapter 10. The application of this group to the countercurrent conditions in the absorption tower is now considered.

Over a small height dZ, the partial pressure of the diffusing component **A** will change by an amount dP_{AG}. Then the moles of **A** transferred are given by:

(change in mole fraction) × (total moles of gas)

Therefore:

$$K_G a(P_{AG} - P_{Ae}) dZ = \frac{-dP_{AG}}{P} G'_m \tag{2.71}$$

(for dilute concentrations)

Thus

$$\int_{P_{AG_2}}^{P_{AG_1}} \frac{dP_{AG}}{P_{Ae} - P_{AG}} = \int_0^Z \frac{K_G a P}{G'_m} dZ \tag{2.72}$$

or in terms of mole fractions:

$$\mathbf{N}_{OG} = \int_{y_1}^{y_2} \frac{dy}{y_e - y} = \int_0^Z K_G a \frac{P}{G'_m} dZ = K_G a \frac{P}{G'_m} Z \tag{2.73}$$

The number of overall gas transfer units $\mathbf{N}_{OG}$ is an integrated value of the change in composition per unit driving force, and therefore represents the difficulty of the separation.

In many cases in gas absorption, $(y - y_e)$ is very small at the top of the column, and consequently $1/(y - y_e)$ is very much greater at the top than at the bottom of the column. Thus, Eq. (2.81) may lead to the use of an integral which is difficult to evaluate graphically because of the very steep slope of the curve.

Now:

$$N_{OG} = \int_{y_1}^{y_2} \frac{dy}{y_e - y} = \int_{y_1}^{y_2} \frac{y d(\ln y)}{y_e - y} \tag{2.74}$$

In these circumstances, the new form of the integral is much more readily evaluated, as pointed out by Rackett [53].

As in Chapter 11 (eq. 11.140), Eq. (2.81) may be written as:

$$N_{OG} = \frac{\text{Height of column}}{\text{Height of transfer unit}} = \frac{Z}{H_{OG}} \tag{2.75}$$

The height of the overall gas transfer unit is then $H_{OG} = \dfrac{G'_m}{PK_G a}$. $\quad$ (2.76)

If the driving force is taken over the gas-film only, the height of a gas-film transfer unit $H_G = G'_m/Pk_G a$ is obtained. Similarly for the liquid film, the height of the overall liquid-phase transfer unit H_{OL} is given by:

$$H_{OL} = \frac{L'_m}{K_L a C_T} \tag{2.77}$$

The height of the liquid-film transfer unit is given by:

$$H_L = \frac{L'_m}{K_L a C_T} \tag{2.78}$$

where C_T is the mean molar density of the liquid.

In this analysis, it is assumed that the total number of moles of gas and liquid remain the same. This is true in absorption only when a small change in concentration takes place. With distillation, the total number of moles of gas and liquid does remain more nearly constant so that no difficulty then arises. In Chapter 11, the following relationships between individual and overall heights of transfer units are obtained and methods of obtaining the values of H_G and H_L are discussed:

$$H_{OG} = H_G + \frac{mG'_m}{L'_m} H_L \tag{2.79}$$

$$H_{OL} = H_L + \frac{L'_m}{mG'_m} H_G \tag{2.80}$$

For absorption duties, Semmelbauer [29] presented the following equations to evaluate H_G and H_L for Raschig rings and Berl saddles:

$$H_G = \beta \left[\frac{G'^{0.41} \mu_G^{0.26} \mu_L^{0.46} \sigma^{0.5}}{L'^{0.46} \rho_G^{0.67} \rho_L^{0.5} D_G^{0.67} d_P^{0.05}} \right] \tag{2.81}$$

Table 2.4 Range of application of Eqs (2.81) and (2.82).

L'	0.1–10	kg/m²s	μ_L	0.2–2	mN s/m²
G'	0.1–1.0	kg/m²s	μ_G	0.005–0.03	mN s/m²
d_p	0.006–0.06	m	Σ	$(20-200)\times 10^{-3}$	J/m²
ρ_L	600–1400	kg/m³	T	273–373	K
ρ_G	0.4–4	kg/m³	d/d_p	2.5–25	–
D_L	$(3–30\times 10^{-10})$	m²/s	h_p/d_p	10–100	–
D_G	$(3–90\times 10^{-6})$	m²/s			

$$\mathbf{H}_L = \beta \left[\frac{\mu_L^{0.88}\sigma^{0.5}}{L'^{0.05}\rho_L^{1.33}D_L^{0.5}d_P^{0.55}} \right] \qquad (2.82)$$

where $\beta = 30$ for Raschig rings and $\beta = 21$ for Berl saddles respectively, and L' and G' are *mass* flow rates per unit area.

The limits of validity and the units for the terms in Eqs (2.81) and (2.82) are given in Table 2.4.

For a range of packings, Morris and Jackson [21] have presented values of the heights of the individual film transfer-units as shown in Table 2.5. For Pall rings

Table 2.5 Height of a transfer unit for various packings [54].

Material	Size (mm)			Height of a transfer unit (m)	
Grids	Pitch	Height	Thickness	H_G	H_L
Plain grids					
Metal	25	25	1.6	1	0.5
	25	50	1.6	1.2	0.6
Wood	25	25	6.4	0.9	0.5
	25	50	6.4	1.2	0.6
Serrated grids					
Wood	100	100	13	6.8	0.7
	50	50	9.5	1.8	0.6
	38	38	4.8	1.6	0.6
Solid material	Nominal size				
Coke	75			0.7	0.9
	38			0.25	0.8
	25			0.2	0.7

Continued

Table 2.5 Height of a transfer unit for various packings [54]—*cont'd*

Material	Size (mm)			Height of a transfer unit (m)	
Grids	**Pitch**	**Height**	**Thickness**	**H$_G$**	**H$_L$**
Quartz	50			0.5	0.8
	25			0.16	0.8
	Diameter	Height	Thickness		
Stacked Raschig rings stoneware	100	100	9.5	1.8	0.7
	75	75	9.5	1.1	0.6
	75	75	6.4	1.4	0.6
	50	50	6.4	0.7	0.6
	50	50	4.8	0.8	0.6
Random Raschig rings metal	50	50	1.6	0.5	0.6
	25	25	1.6	0.2	0.5
	13	13	0.8	0.1	0.5
Stoneware	75	75	9.5	0.8	0.7
	50	50	6.4	0.5	0.6
	50	50	4.8	0.5	0.6
	38	38	4.8	0.3	0.6
	25	25	2.5	0.2	0.5
	19	19	2.5	0.15	0.5
	13	13	1.6	0.1	0.5
Carbon	50	50	6.4	0.5	0.6
	25	25	4.8	0.2	0.5
	13	13	3.2	0.1	0.5

and Intalox saddles, the nomographs in Figs. 2.21 and 2.22 [54] may be used though Fig. 2.22 must not be used to estimate H_L for distillation applications. Table 11.6 gives the value as a function of size and type of packing. It is, however, satisfactory for absorption and stripping duties.

Concentrated solutions

With concentrated solutions, allowance must be made for the change in the total number of moles flowing, because the molar flow will decrease up the column if the amount of absorption is large.

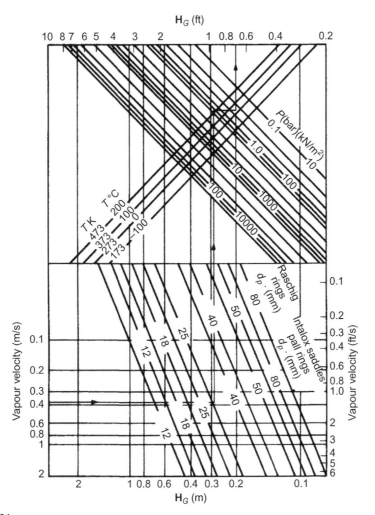

FIG. 2.21

Nomograph for the estimation of the height of a gas-phase transfer unit [54].

Colburn [55] has shown that, under these conditions, the number of transfer units is given by:

$$N_{OG} \int_{y_1}^{y_2} \frac{dy}{y_e - y} \frac{(1-y)_{lm}}{1-y} \qquad (2.83)$$

where $(1-y)_{lm}$ is the logarithmic mean of $(1-y)$ and $(1-y_i)$.

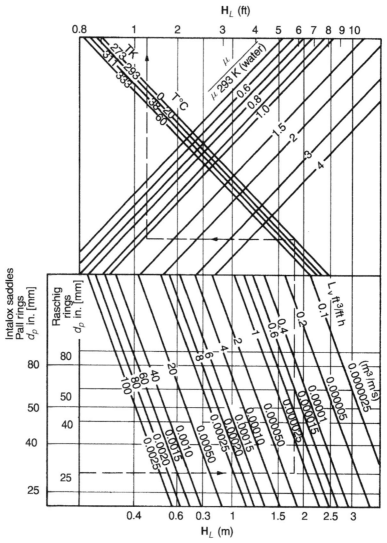

FIG. 2.22

Nomograph for the estimation of the height of a liquid-phase transfer unit [54].

Example 2.3

An acetone–air mixture containing 0.015 mol fraction of acetone has the mole fraction reduced to 1% of this value by countercurrent absorption with water ($x_2=0$, pure water) in a packed tower. The gas flow rate G' is $1\,kg/m^2s$ of air and the water ($x_2=0$) enters at $1.6\,kg/m^2s$. For this system, Henry's law holds and $y_e = 1.75x$, where y_e is the mole fraction of acetone in the vapour in equilibrium with a mole fraction x in the liquid. How many overall transfer units are required?

Solution

As the system is dilute, mole fractions are approximately equal to mole ratios.

At the bottom of the tower: $y_1=0.015$, $G'=1.0\,kg/m^2s$, x_1 is unknown.

At the top of the tower: $y_2=0.00015$, $x_2=0$ and $L'=1.6\,kg/m^2s$

Thus

$$L_m = (1.6/18) = 0.0889 \text{ kmol/m}^2\text{s}$$

and

$$G_m = (1.0/29) = 0.0345 \text{ kmol/m}^2\text{s}$$

(On solute free basis, molar mass of water is $18\,kg/kg$ mole and that of air is $29\,kg/kg$ mole)

An overall mass balance gives:

$$G_m(y_1 - y_2) = L_m(x_1 - x_2)$$

or

$$0.0345(0.015 - 0.00015) = 0.0889(x_1 - x_2), \text{ hence } x_1 = 0.00576 \text{ as } x_2 = 0$$

Thus:

$$y_{e_1} = (1.75 \times 0.00576) = 0.0101$$

The number of overall transfer units is defined by:

$$N_{OG} = \int_{y_2}^{y_1} \frac{dy}{y - y_e} = \frac{y_1 - y_2}{(y - y_e)_{lm}}$$

(from Eqs 2.60 and 2.70)

Top driving force, $(y_2 - y_{e2}) = 0.00015$ since $x_2 = 0$.

Bottom driving force, $= (y_1 - y_{e1}) = (0.015 - 0.0101) = 0.0049$.

Thus: $(y - y_e)_{lm} = (0.0049 - 0.00015)/\ln(0.0049/0.00015) = 0.00136$

and $N_{OG} = (0.015 - 0.00015)/0.00136 = 10.92$.

Also: $N_{OL} = N_{OG} \times mG_m/L_m$, where m is the slope of the equilibrium line, 1.75.

$$= (10.92 \times 1.75 \times 0.0345)/0.0889 = \underline{7.42}.$$

2.7.10 The importance of liquid and gas flow rates and the slope of the equilibrium curve

For a packed tower operating with dilute concentrations, since $x \simeq X_1$ and $y \simeq Y_1$, then:

$$G'_m(y_1 - y_2) = L'_m(x_1 - x_2) \tag{2.84}$$

where, as before, x and y are the mole fractions of solute in the liquid and gas phases, and G'_m and L'_m are the gas and liquid molar flow rates per unit area on a solute free basis.

A material balance between the top and some plane where the mole fractions are x, y gives:

$$G'_m(y - y_2) = L'_m(x - x_2) \tag{2.85}$$

If the entering solvent is free from solute, then $x_2 = 0$ and:

$$x = \frac{G'_m}{L'_m}(y - y_2) \tag{2.86}$$

But the number of overall transfer units is given by:

$$N_{OG} = \int_{y_1}^{y_2} \frac{dy}{y_e - y}$$

For dilute concentrations, Henry's law holds and $y_e = mx$. Therefore:
Thus:

$$N_{OG} = \int_{y_1}^{y_2} \frac{dy}{\frac{mG'_m}{L'_m}(y - y_2) - y}$$

$$= \int_{y_1}^{y_2} \frac{dy}{\left[\frac{mG'_m}{L'_m} - 1\right]y - \frac{mG_m}{L_m}y_2}$$

and

$$N_{OG} = \frac{1}{1 - \frac{mG'_m}{L'_m}} \ln\left[\left(1 - \frac{mG'_m}{L'_m}\right)\frac{y_1}{y_2} + \frac{mG'_m}{L'_m}\right] \tag{2.87}$$

Colburn [55] has shown that this equation may usefully be plotted as shown in Fig. 2.23 which is taken from his paper. In this plot the number of transfer units N_{OG} is shown for values of y_1/y_2 using mG'_m/L'_m as a parameter and it may be seen that the greater mG_m/L_m, the greater is the value of N_{OG} for a given ratio of y_1/y_2. From Eq. (2.86):
Thus:

$$\frac{L'_m}{G'_m} = \frac{y_1 - y_2}{x_1} = \frac{y_1 - y_2}{y_e/m}$$

$$\frac{mG'_m}{L'_m} = \frac{y_{e1}}{y_1 - y_2}$$

where y_{e1} is the value of y in equilibrium with x_1.

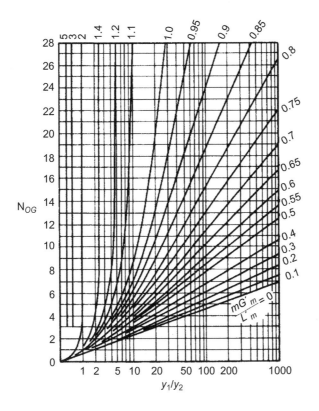

FIG. 2.23

Number of transfer units **N**OG as a function of $y1$, $y2$, with $mG'm/L'm$ as parameter.

On this basis, the lower the value of mG'_m/L'_m, the lower will be y_{e1}, and hence the weaker the exit liquid. Colburn has suggested that the economic range for mG'_m/L'_m is 0.7–0.8. If the value of $\mathbf{H}_{OG}$ is known, the quickest way of obtaining a good indication of the required height of the column is by using Fig. 2.23.

Example 2.4

Gas, from a petroleum distillation column, has a concentration of H_2S reduced from 0.03 (kmol H_2S/kmol of inert hydrocarbon gas) to 1% of this value by scrubbing with a triethanolamine–water solvent in a countercurrent tower, operating at 300 K and atmospheric pressure. The equilibrium relation for the solution may be taken as $Y_e = 2X$.

The solvent enters the tower free of H_2S and leaves containing 0.013 kmol of H_2S/kmol of solvent. If the flow of inert gas is 0.015 kmol/s m^2 of tower cross-section, calculate:

a) the height of the absorber necessary and

Continued

Example 2.4—Cont'd

b) the number of transfer units N_{OG} required.

The overall coefficient for absorption K''_Ga may be taken as $0.04 \, \text{kmol/s m}^3$ (unit mole fraction driving force).

Solution

Driving force at top of column $= (Y_2 - Y_{2e}) = 0.0003$, since $X = 0.0$ so $Y_{2e} = 2 \times 0.0$.

Driving force at bottom of column $= (Y_1 - Y_{1e}) = (0.03{-}0.026) = 0.004$, where $Y_{1e} = 2 \times 0.013$.

$$\text{Logarithmic mean driving force} = \frac{(0.004 - 0.0003)}{\ln\left(\frac{0.004}{0.0003}\right)}$$

$$= 0.00143$$

From Eq. (2.64): $G'_m(Y_1 - Y_2)S = K_Ga \, P \, (Y - Y_e)_{lm} \, SZ$

That is: $G'_m(Y_1 - Y_2) = K''_Ga \, (Y - Y_e)_{lm}Z$

Thus: $0.015(0.03{-}0.0003) = 0.04 \times 0.00143 \, Z$ and

$$Z = \frac{0.000446}{0.0000572} = 7.79 = 7.8 \, \text{m (say)}$$

Height of transfer unit $\mathbf{H}_{OG} = \frac{G'_m}{K''_Ga}$

$$= \frac{0.015}{0.04} = 0.375 \ \text{m}$$

Number of transfer units $\mathbf{N}_{OG} = \frac{7.79}{0.375} = 20.7 = \underline{\underline{21 \, (\text{say})}}$

Example 2.5

Ammonia is to be removed from a 10% ammonia–air mixture by countercurrent scrubbing with water in a packed tower at 293 K so that 99% of the ammonia is removed when working at a total pressure of $101.3 \, \text{kN/m}^2$.

If the gas rate is $0.95 \, \text{kg/m}^2\text{s}$ of tower cross-section and the liquid rate is $0.65 \, \text{kg/m}^2\text{s}$, find the necessary height of the tower if the absorption coefficient $K_Ga = 0.001 \, \text{kmol/m}^3\text{s} \, (\text{kN/m}^2)$ partial pressure difference. The equilibrium data are:

kmol NH₃/kmol water							
Partial pressure NH$_3$	0.021	0.031	0.042	0.053	0.079	0.106	0.159
(mm Hg)	12.0	18.2	24.9	31.7	50.0	69.6	114.0
(kN/m^2)	1.6	2.4	3.3	4.2	6.7	9.3	15.2

Solution

If the compositions of the gas are given as per cent by volume, at the bottom of the tower are $y_1=0.10$ and $Y_1=0.10/(1-0.10)=0.111$.

At the top of the tower: $y_2=0.001\simeq Y_2$.

Mass flow rate of gas $=0.95\,kg/m^2s$.

Mass per cent of air $=[0.9\times 29/(0.1\times 17+0.9\times 29)]\times 100=93.8$.

Thus: mass flow rate of air $=(0.938\times 0.95)=0.891\,kg/m^2s$ and

$$G'm = (0.891/29) = 0.0307\ kmol/m^2s$$

$$L'm = 0.65/18 = 0.036\ kmol/m^2s$$

A mass balance between a plane in the tower where the compositions are X and Y and the top of the tower gives:

But:

$$G'_m(Y - Y_2) = L'_m(X - X_2)$$
$$X_2 = 0$$

Thus: $0.0307(Y - 0.001)=0.036x$ or $Y= 1.173x+0.001$.

This is the equation of the operating line in terms of mole ratios.

The given equilibrium data may be converted to the same basis since and

$$P_G = yP = \frac{YP}{1 + Y}$$
$$Y = \frac{P_G}{P - P_G}$$

using these equations and the given equilibrium data, the following data are obtained:

kmol NH$_3$/kmol H$_2$O	0.021	0.031	0.042	0.053	0.079	0.106	0.159
Partial pressure P_G (mm)	12	18.2	24.9	31.7	50.0	69.6	114.0
$P - P_G = 760 - P_G$ (mm)	748	741.8	735.1	728.3	710	690.4	646
$Y=P_G/(P - P_G)$	0.016	0.0245	0.0339	0.0435	0.0704	0.101	0.176

These data are plotted in Fig. 2.24.

From a mass balance over the column, the height Z is given by:

$$Z = \frac{G'_m}{k_G a P} \int_{Y_2}^{Y_1} \frac{(1 + Y)(1 + Y_i)}{(Y - Y_i)} dY \qquad (2.88)$$

Continued

Example 2.5—Cont'd

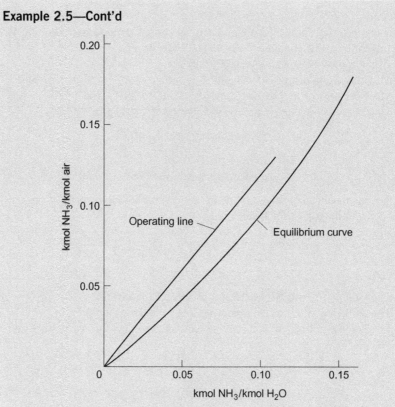

FIG. 2.24

Operating and equilibrium lines for Example 2.5.

Fig. 2.25 may be used to evaluate the integral as follows:

Y	Y_i	$(1+Y)(1+Y_i)$	$\frac{(1+Y)(1+Y_i)}{(Y-Y_i)}$
0.111	0.089	1.21	55.0
0.10	0.078	1.185	53.8
0.08	0.059	1.14	54.3
0.06	0.042	1.11	61.4
0.04	0.027	1.067	82.0
0.02	0.013	1.035	148
0.01	0.006	1.016	254
0.005	0.0026	1.010	421
0.001	0	1.0	1000

The area under the curve in Fig. 2.25 is $12.6\,\text{m}^2/\text{m}^3$. For a very soluble gas $k_Ga \simeq k_Ga$ so that:

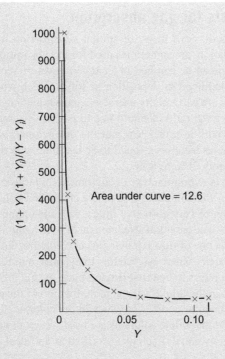

FIG. 2.25

Determination of column height for Example 2.5.

$$Z = \frac{0.0307}{(0.001 \times 101.3)} \times 12.6 = \underline{\underline{3.82}} \text{ m}$$

If the equilibrium line is assumed to be straight, then:

$$G'_m(Y_2 - Y_1) = K_G a Z \Delta P_{lm}$$

Top driving force $= \Delta Y_2 = 0.022$. Bottom driving force $= \Delta Y_1 = 0.001$. Thus:

$$\Delta Y_{lm} = 0.0068, \quad \Delta P_{lm} = 0.688 \text{ kN/m}^2$$

and

$$Z = \frac{(0.0307 \times 0.11)}{(0.001 \times 0.688)} = \underline{\underline{4.91}} \text{ m.}$$

2.8 Plate towers for gas absorption

Bubble-cap columns or sieve trays, of similar construction to those described in Chapter 11 on distillation, are sometimes used for gas absorption, particularly when the load is more than can be handled in a packed tower of about 1 m diameter and when there is any likelihood of deposition of solids which would quickly choke a packing. Plate towers are particularly useful when the liquid rate is sufficient to flood a packed tower. Since the ratio of liquid rate to gas rate is greater than with distillation, the slot area will be rather less and the downcomers rather larger. On the whole, plate efficiencies have been found to be less than with the distillation equipment, and to range from 20% to 80%.

The plate column is a common type of equipment for large installations, although when the diameter of the column is less than 2 m, packed columns are more often used. For the handling of very corrosive fluids, packed columns are frequently preferred for larger units. The essential arrangement of such a unit is shown in Fig. 2.26, where L'_m is the molar rate of flow per unit area of solute free liquid, G'_m is the molar rate of flow per unit area of inert gas, n refers to the plate numbered from the bottom upward (and suffix n refers to material leaving plate n), x is the mole fraction of the absorbed component in the liquid, y is the mole fraction of the absorbed component in the gas, and s is the total number of plates in the column.

It may be assumed that dilute solutions are used so that mole fractions and mole ratios are approximately equal. Each plate is taken as an 'ideal' unit, so that the gas

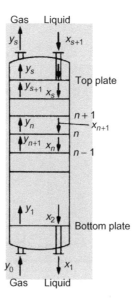

FIG. 2.26

Plate tower—nomenclature for fluid streams.

leaving of composition y_n is in equilibrium with the liquid of composition x_n leaving the plate n.

A material balance for the absorbed component from the bottom to a plane above plate n gives:

$$G'_m y_n + L'_m x_1 = G'_m y_0 + L'_m x_{n+1} \qquad (2.89)$$

or

$$y_n = \frac{L'_m}{G'_m} x_{n+1} + y_0 - \frac{L'_m}{G'_m} x_1 \qquad (2.90)$$

This is the equation of a straight line of slope L'_m/G'_m, relating the composition of the gas entering a plate to the liquid leaving the plate, and is known as the *operating line*. As shown in Fig. 2.27, such a line passes through two points $B(x_{s+1}, y_s)$ and $A(x_1, y_0)$, representing the terminal concentrations in the column. The equilibrium curve is shown in this figure as PQR.

Point A represents conditions at the bottom of the tower. The gas rising from the bottom plate is in equilibrium with a liquid of concentration x_1 and is shown as point 3 on the operating line. Then point 4 indicates the concentration of the liquid on the second plate from the bottom. In this way steps may be drawn to point B, giving the gas y_s rising from the top plate and the liquid x_{s+1} entering the top of the absorber.

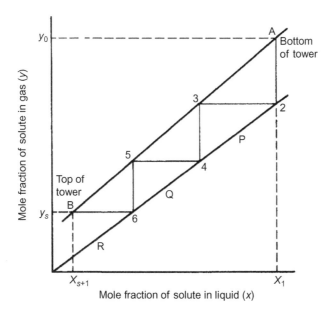

FIG. 2.27

Diagrammatic representation of changes in a plate column.

2.8.1 **Number of plates by use of absorption factor**

If the equilibrium curve can be represented by the relation $y_e = mx$, then the number of plates required for a given degree of absorption can conveniently be found by a method due to Kremser [56] and Souders and Brown [57]. The same treatment is applicable for concentrated solutions provided concentrations are expressed as mole ratios, and if the equilibrium curve can be represented approximately by $Y_e = mX$.

A material balance over plate n gives:

$$L'_m(x_n - x_{n+1}) = G'_m(y_{n-1} - y_n) \tag{2.91}$$

For an ideal plate, $y_n = mx_n$; and

$$\frac{L'_m}{mG'_m}(y_n - y_{n+1}) = y_{n-1} - y_n \tag{2.92}$$

This group L'_m/mG'_m, which will be taken as constant, is called the *absorption factor A*.

Thus:

$$y_n = \frac{y_{n-1} + Ay_{n+1}}{1 + A} \tag{2.93}$$

Applying this relation to the bottom plate and taking y_0 as the mole fraction of absorbed component in the gas entering the column, then:

$$y_1 = \frac{y_0 + Ay_2}{1 + A}$$

And for the second plate from the bottom:

$$y_2 = \frac{y_1 + Ay_3}{1 + A}$$

$$= \frac{A(1 + A)y_3 + Ay_2 + y_0}{(1 + A)^2}$$

Simplifying:

$$y_2 = \frac{y_0(1 + A) + A^2 y_3}{A^2 + A + 1}$$

And for the third plate from the bottom:

$$y_3 = \frac{y_0(1 + A + A^2) + A^3 y_4}{A^3 + A^2 + A + 1}$$

which may be written as:

$$y_3 = \frac{[(A^3 - 1)/(A - 1)]y_0 + A^3 y_4}{(A^4 - 1)/(A - 1)}$$

$$= \frac{(A^3 - 1)y_0 + A^3(A - 1)y_4}{A^4 - 1}$$

Proceeding thus until plate n is reached:

$$y_n = \frac{(A^n - 1)y_0 + A^n(A-1)y_{n+1}}{A^{n+1} - 1}$$

$$y_0 = \frac{(A^{n+1} - 1)y_n - A^n(A-1)y_{n+1}}{A^n - 1}$$

Thus:

$$y_0 - y_n = \frac{(A^{n+1} - A^n)y_n - A^n(A-1)y_{n+1}}{A^n - 1}$$

and

$$y_0 - y_{n+1} = \frac{(A^{n+1} - 1)y_n - (A^{n+1} - 1)y_{n+1}}{A^n - 1}$$

Dividing:

$$\frac{y_0 - y_n}{y_0 - y_{n+1}} = \frac{(A^{n+1} - A^n)y_n - A^n(A-1)y_{n+1}}{(A^{n+1} - 1)y_n - (A^{n+1} - 1)y_{n+1}}$$

$$= \frac{A^n(A-1)(y_n - y_{n+1})}{(A^{n+1} - 1)(y_n - y_{n+1})}$$

or

$$\frac{y_0 - y_n}{y_0 - y_{n+1}} = \frac{A^{n+1} - A}{A^{n+1} - 1}$$

Applying this relation over the whole column and putting $n = s$ gives $(y_0 - y_s) =$ actual change in composition of gas, and $(y_0 - y_{s+1}) =$ maximum possible change in composition of gas, that is if the gas leaving the absorber is in equilibrium with the entering liquid (or $y_s = mx_{s+1}$).

Then:

$$\frac{y_0 - y_s}{y_0 - mx_{s+1}} = \frac{(L'_m/mG'_m)^{s+1} - (L'_m/mG'_m)}{(L'_m/mG'_m)^{s+1} - 1} \tag{2.94}$$

This equation is conveniently represented, as suggested by Souders and Brown [57], by Figs. 2.28 and 2.29, and it is easy to use such a diagram to determine the number of plates required.

A high degree of absorption can be obtained, either by using a large number of plates, or by using a high absorption factor L'_m/mG'_m. Since m is fixed by the system, this means that L'_m/G'_m must be large if a high degree of absorption is to be obtained, although this will result in a low value of x for the liquid leaving at the bottom. This problem is to some extent met by recirculating the liquid over the tower, although the advantages of a countercurrent flow system are then lost. A value of mG'_m/L'_m of about 0.7–0.8 is probably the most economic, that is L'_m/mG'_m 1.3.

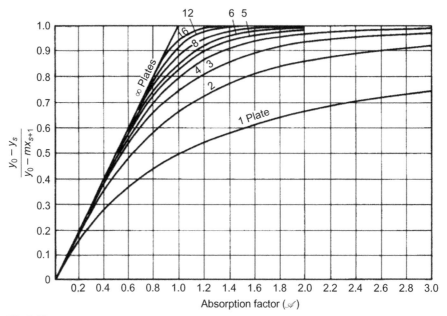

FIG. 2.28

Graphical representation of the effect of the absorption factor and the number of plates on the degree of absorption.

It is important to note that, if L'_m/mG'_m is less than 1, then a very large number of plates are required to achieve a high recovery, and even an infinite number will not give complete recovery. L'_m/mG'_m is the ratio of the slope of the operating line L_m/G_m to the slope of the equilibrium curve m, so that if $L'_m/G'_m < m$, or $L'_m/mG'_m < 1$, then the operating line will never cut the equilibrium curve and the gas leaving the top of the column will not therefore reach equilibrium with the entering liquid.

Absorption and stripping—Complementary processes

As was mentioned in Section 1, in today's regulatory environment requiring minimisation of waste disposal and containment of gaseous components or contaminants from being emitted to atmosphere, reversible physical and chemical processes dominate absorption. Reversible processes include a stripping unit in tandem with the absorber so that the absorbed component is stripped either with an inert gas or steam to be recovered as a useful product, avoiding its emission to atmosphere and returning the stripped or lean solvent back to the absorber.

- The operating line equations for both are the same, the only difference being that in the stripper, the solute is transferred from liquid to gas.

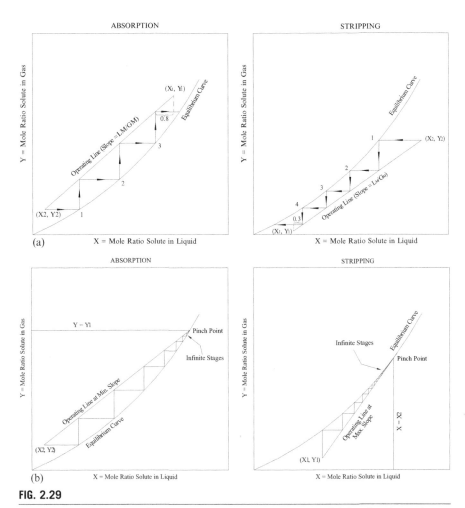

FIG. 2.29

(A) Construction of stages for absorption and stripping columns. (B) Pinch point and determining minimum liquid flow slope for an absorber and minimum gas flow for a stripper.

The method of McCabe and Thiele

In section 11.4.2, under Chapter 11, distillation, the graphical method by McCabe and Thiele introduced the idea of operating line concept in multistage operations. This concept also lends itself for use in absorption and stripping operations. The method is a graphical solution, using a $X–Y$ chart with equilibrium curve. In the chart an operating line is constructed, the slope of which must be constant.

- For absorption the operating line is above the equilibrium line, because solute is being transferred from the gas to the liquid.
- For stripping the opposite is the case. The operating line is below the equilibrium curve, because solute is transferred from liquid to gas.

The slope of the operating line is, L_m/G_m which must be constant and will be the case with the following assumptions:

1. The heat of absorption is negligible
2. The operation is isothermal
3. Solvent is non-volatile, i.e., L_m in all stages is constant
4. Carrier gas is insoluble, i.e., G_m in all stages is constant

Referring to Fig. 2.18, in Section 7.3, the operating line for the column is derived by overall column component mass balance.

$$G_m Y_1 + L_m X_2 = G_m Y_2 + L_m X_1$$

Rearranging the terms gives the operating line equation as:

$$L_m/G_m = (Y_1 - Y_2)/(X_1 - X_2)$$

Stage calculations

Using the equilibrium diagram relating X and Y of the component in question.

1. Construct the operating line from the mass balance equation

$$L_m/G_m = (Y_1 - Y_2)/(X_1 - X_2)$$

Note:

- (X_1, Y_1) represent conditions at the bottom of the tower, and (X_2, Y_2) top of the tower.
- X_2 and Y_1 are the input conditions, X_1 and Y_2 output conditions, which are usually set before the design. When a pure absorption solvent is used, $X_2 = 0$.
- The slope L_m/G_m gives the required ratio of liquid flow to gas flow for the tower to operate in order to meet those conditions.
- Operating line is always *above* the equilibrium line in an absorber and below the equilibrium line in a stripper.

2. Construct stages stepping off between operating line and equilibrium line.

Number of stages in graph = number of theoretical plates (NTP)
     Note:

- stages are counted horizontally, not vertically as in distillation
- fractional stage applies, as given by fractional length of the horizontal line.

Minimum and maximum slope of the operating line

- Slope of operating line = L_m/G_m. Minimum slope means minimum liquid flow rate, maximum slope means minimum gas flow rate. The former is a key reference setting for absorption operations, the latter is for stripping operations.

- At minimum or maximum slope the operating line meets (touches as a tangent, or crosses, depending on the shape of the equilibrium line) the equilibrium line. X, Y will be in equilibrium in such condition.
- In either case infinite number of stages are required.
- The minimum or maximum slope can be obtained graphically.
- In absorption operations the purity of the solvent feed (X_2), the required cleanliness of the treated gas (Y_2), and the concentration of the solute in the feed gas (Y_1) are known. Minimum L_m/G_m can be determined by drawing a straight line from (X_2, Y_2) to the intersection of $Y = Y_1$ and the equilibrium curve.
- In stripping operations, the purity of the stripping gas feed (Y_1), the required cleanliness of the treated liquid (X_1), and the concentration of the solute in the feed liquid (X_2) known. Maximum L_m/G_m can be determined by drawing a straight line from (X_1, Y_1) to the intersection of $X = X_2$ and the equilibrium curve.
- For column design, it is common to use an optimum slope $= 1.2$–1.5 of the minimum or maximum.

2.8.2 Tray types for absorption

It has already been noted that trays which are suitable for distillation may be used for absorption duties though in general lower efficiencies will be obtained. In Chapter 11, the design of trays for common contacting devices is considered and the methods presented in that chapter are generally applicable. The most commonly used tray types are shown in fig. 11.50A with the crossflow tray being the most popular.

At high liquid flow rates, the liquid gradient on the tray can become excessive and lead to poor vapour distribution across the plate. This problem may be overcome by the shortening of the liquid flow-path as in the case of the double-pass and cascade trays. The whole design process is discussed in Volume 6.

Example 2.6
A bubble-cap column with 30 plates is to be used to remove n-pentane from a solvent oil by means of steam stripping. The inlet oil contains 6 kmol of n-pentane per 100 kmol of pure oil and it is desired to reduce the solute content to 0.1 kmol per 100 kmol of solvent. Assuming isothermal operation and an overall plate efficiency of 30%, find the specific steam consumption, that is the kmol of steam required per kmol of solvent oil treated, and the ratio of the specific and minimum steam consumptions. How many plates would be required if this ratio were 2.0?

The equilibrium relation for the system may be taken as $Y_e = 3.0X$, where Y_e and X are expressed in mole ratios of pentane in the gas and liquid phases respectively.

Continued

Example 2.6—Cont'd

Solution

Number of theoretical plates $= (30 \times 0.3) = 9$.

At the bottom of the tower:

Flow rate of steam $= G'_m$ (kmol/m^{2}s).

Mole ratio of pentane in steam $= Y_1$, and

Mole ratio of pentane in oil $= X_1 = 0.001$.

At the top of the tower:

exit steam composition $= Y_2$, inlet oil composition $= X_2 = 0.06$,

flow rate of oil $= L'_m$ (kmol/m^{2}s).

The minimum steam consumption occurs when the exit steam stream is in equilibrium with the inlet oil, that is when:

$$Y_{e2} = (0.06 \times 3) = 0.18$$
$$L'_{min}(X_2 - X_1) = G'_{min}(Y_2 - Y_1)$$

If $Y_1 = 0$, that is the inlet steam is pentane-free, then:

and:

$$L'_{min}(0.06 - 0.001) = \left(G'_{min} \times 0.18\right)$$
$$(G'/L')_{min} = (0.06 - 0.001)0.18 = 0.328$$

The operating line may be fixed by trial and error as it passes through the point $(0.001, 0)$, and 9 theoretical plates are required for the separation. Thus, it is a matter of selecting the operating line which, with 9 steps, will give $X_2 = 0.001$ when $X_1 = 0.06$. This is tedious but possible, and the problem may be better solved analytically since the equilibrium line is straight.

Use may be made of the absorption factor method where

$$\frac{Y_1 - Y_2}{Y_1 - mX_2} = \frac{A^{N+1} - A}{A^{N+1} - 1} \qquad (2.95)$$

where A is the absorption factor $= L'_m/mG'_m$ and N is the number of theoretical plates.

The corresponding expression for a stripping operation is:

$$\frac{X_2 - X_1}{X_2 - Y_1/m} = \frac{(1/A)^{N+1} - (1/A)}{(1/A)^{N+1} - 1}$$

In this problem, $N = 9$, $X_2 = 0.06$, $X_1 = 0.001$, and $Y_1 = 0$

Thus:

$$\frac{(0.06 - 0.001)}{0.06} = 0.983 = \frac{(1/A)^{10} - (1/A)}{(1/A)^{10} - 1} \qquad \text{from which } (1/A) = 1.37$$

Thus:

$$\frac{mG'_m}{L'_m} = 1.37, \quad \frac{G'_m}{L'_m} = \frac{1.37}{3} = 0.457$$

and

$$\frac{\text{actual } G'_m/L'_m}{\text{minimum } G'_m/L'_m} = \frac{0.457}{0.328} = \underline{\underline{1.39}}$$

If (actual G'_m/L'_m)/(min G'_m/L'_m) = 2, actual $G'_m/L'_m = 0.656$.
Thus:

$$1/A = mG'_m/L'_m = 1.968$$

and

$$0983 = \frac{(1.968)^{N+1} - 1.968}{(1.968)^{N+1} - 1} \quad \text{from which } N = 4.9$$

The actual number of plates = (4.9/0.3) = 16.3 $\underline{\underline{(\text{say } 17)}}$.

2.9 Other equipment for gas absorption

2.9.1 The use of vessels with agitators

A gas may be dissolved in a liquid by dispersing it through holes in a pipe immersed in the liquid which is stirred with some form of agitator, as shown in 2.30. Although this type of equipment will give only one theoretical stage per unit, but it often provides a useful method of saturating a liquid with a gas. Cooper et al. [58] have studied the absorption of oxygen from air in an aqueous solution of sodium sulphite using simple vessels of 0.15–2.44 m diameter fitted with four simple baffles. Air was just below the agitator which was a vaned-disc or flat-paddle. It was found that the absorption coefficient $K_G a$ varied almost directly with $\mathbf{P}_V$, the power input per unit volume. For constant values of $\mathbf{P}_V$, the following relation was obtained:

$$K_G a \propto u_s^{0.67} \qquad (2.96)$$

where u_s is the superficial gas velocity based on the volume of gas at inlet and the cross-section of tank. A general correlation was obtained by plotting $K_G a/u_s^{0.67}$ against the power input per unit volume $\mathbf{P}_V$, as shown in Fig. 2.31 taken from this investigation. Ayerst and Herbert [59] have given some data on the use of this type of unit for the absorption of carbon dioxide into ammoniacal solutions.

The interfacial area, a, was the subject of an investigation by Westerterp et al. [60] though the correlations proposed are complex. Maximum values of a are about 1000 m^2/m^3. Further work on the interfacial area in agitated vessels has been

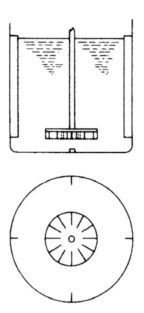

FIG. 2.30

Vessel fitted with vaned-disc agitator.

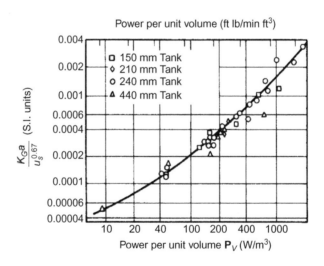

FIG. 2.31

General correlation of data for a vessel (height=diameter) with vaned-disc agitator.

reviewed and summarised by Sridar and Potter [61] who found that the correlation of Calderbank [62] was applicable for most situations. Calderbank proposed that, for pure liquids, the specific interfacial area, that is the surface area per unit volume of aerated suspension, $a(m^2/m^3)$ is given by:

$$a = 24,200 \ (\mathbf{P}_V)^{0.4} \ \left(\frac{\rho_L^{0.2}}{\sigma^{0.6}}\right) \left(\frac{u_s}{u_0}\right)^{0.5} \tag{2.97}$$

where surface aeration is negligible, that is when:

$$\left(\frac{N'd_t^2\rho_L}{\mu_L}\right)^{0.7} \left(\frac{N'd_i}{u_s}\right)^{0.3} < 25,000 \tag{2.98}$$

When the surface aeration is significant, then the interfacial area is:

$$\frac{a'}{a} = 10^{-4} \left\{ \left[\left(\frac{N'd_t^2\rho_L}{\mu_L}\right)^{0.7} \left(\frac{N'd_i}{u_s}\right)^{0.2} \right] - 25,000 \right\} \tag{2.99}$$

In these equations, a' is the specific interfacial area for a significant degree of surface aeration (m^2/m^3), $\mathbf{P}_V$ is the agitator power per unit volume of vessel (W/m^3), ρ_L is the liquid density, σ is the surface tension (N/m), u_s is the superficial gas velocity (m/s), u_0 is the terminal bubble-rise velocity (m/s), N' is the impeller speed (Hz), d_i is the impeller diameter (m), d_t is the tank diameter (m), μ_L is the liquid viscosity (Ns/m^2) and d_0 is the Sauter mean bubble diameter defined in Chapter 1, section 1.2.4.

The effects of gas hold-up and bubble diameter have also been studied by Sridhar and Potter and, again, the correlations obtained by Calderbank are recommended.

The liquid-phase mass transfer coefficient, k_L, in agitated vessels has been measured and data correlated by several workers. Sideman et al. [63] and Valentin [64] have presented reviews of the early work and more recent work has been published by Yagi and Yoshida [65], Zlokarnik [66], van't Riet [67], and Hoker et al. [68]. For small bubbles (<2.5 mm diameter) produced in well-agitated vessels, Calderbank [62] suggests the following correlation for bubbles in agitated electrolytes:

$$K_L = 0.31 \left(\frac{\Delta\rho\mu_L g}{\rho_L^2}\right)^{1/3} (Sc)^{-2/3} \tag{2.100}$$

where $\Delta\rho$ = density difference between gas and liquid, ρ_L, μ_L = density and viscosity of the liquid, and Sc = Schmidt number for transport in the liquid.

Joshi and Sharma [69] and Fukada et al. [70] have investigated the performance of vessels with multiple impellers on horizontal shafts.

Several investigations have been carried out into the power requirements for agitation of aerated liquids including those of Yung et al. [71] and Luong and Volesky [72] and it is generally concluded that the power required is less for an aerated system than for a non-aerated system.

Although, as described by Bjerle et al. [73], liquid jet-type absorbers are also used, one relatively recent application of mass transfer in agitated tanks with chemical reaction is the absorption of pollutants from flue gases and, in particular, the scrubbing of sulphur dioxide by a slurry containing fine limestone particles. In this case, the concentration of sulphur dioxide is usually very low and the mechanism of the absorption is complicated due to the presence of solids in the liquid phase where the rate of solid dissolution may significantly affect the absorption rate.

Studies on the dissolution of solids in the liquid phase include that of Hixson and Baum [74] whose correlation of data in terms of Reynolds, Sherwood, and Schmidt numbers, discussed in detail in section 10.2 in connection with mass transfer during leaching, is one of the most frequently used methods for calculating the mass transfer coefficient for the solid dissolution.

Further work on the absorption of sulphur dioxide by Uchida et al. [75] has shown that the absorption rate changes with the surface area of the limestone particles which in turn varies with the size and the number of particles, and that the rate of dissolution plays a very important role on the absorption. It was further found the absorption rate does not vary significantly with temperature and that the reactions involved may be considered as being instantaneous.

2.9.2 The centrifugal absorber

In an attempt to obtain the benefits of repeated spray formations, a centrifugal type absorber has been developed from the ideas of Piazza for a still head. The principle of the unit is shown in Fig. 2.32. A set of stationary concentric rings intermeshes with a second set of rings attached to a rotating plate. Liquid fed to the centre of the plate is carried up the first ring, splashes over to the baffle and falls into the through between the rings. It then runs up the second ring and in a similar way passes from ring to ring through the unit. The gas stream can be introduced at the top to give cocurrent flow, or at the bottom if countercurrent flow is desired. Some of the features of this unit are discussed by Ahmed [76] who found that the depth of the ring was not very important and that most of the transfer took place as the gas mixed with the liquid spray leaving the top of the rings. Chambers and Wall [77] have given some particulars of the performance of the 510 mm diameter unit shown in Fig. 2.33, for the absorption of carbon dioxide from air containing 10%–15% of carbon dioxide, using mono-ethanolamine solution. Some values of absorption rates are given in Table 2.6.

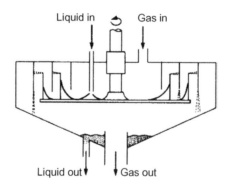

FIG. 2.32

The centrifugal absorber.

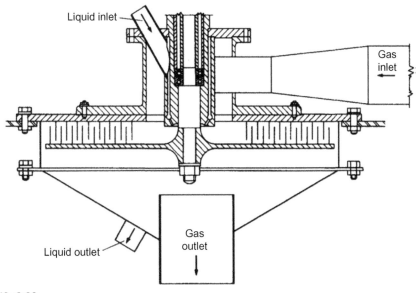

FIG. 2.33

Details of a 510 mm diameter centrifugal absorber.

Table 2.6 Results for absorption in a 510 mm diameter absorber.

Gas flow (m³/s)	Liquid flow (m³/s)	Per cent CO₂ in gas		Absorption rate (kg/s)
		In	Out	
0.016	1.07×10^{-4}	16.3	2.3	0.0044
0.024	1.07×10^{-4}	15.8	4.5	0.0055
0.031	1.07×10^{-4}	14.3	6.6	0.0051
0.039	1.07×10^{-4}	16.3	8.7	0.0065

2.9.3 Spray towers

In the spray tower, the gas enters at the bottom and the liquid is introduced through a series of sprays at the top. The performance of these units is generally rather poor, because the droplets tend to coalesce after they have fallen through a few metres, and the interfacial surface is thereby seriously reduced. Although there is considerable turbulence in the gas phase, there is little circulation of the liquid within the drops, and the resistance of the equivalent liquid film tends to be high. Spray towers are therefore useful only where the main resistance to mass transfer lies within the gas phase, and have consequently been used with moderate success for the absorption of ammonia in water. They are also used as air humidifiers, in which case the whole of the resistance lies within the gas phase.

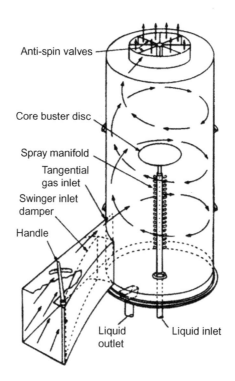

FIG. 2.34

Centrifugal spray tower [78].

Centrifugal spray tower

Fig. 2.34, taken from the work of Kleinschmidt and Anthony [78], illustrates a spray tower in which the gas stream enters tangentially, so that the liquid drops are subjected to centrifugal force before they are taken out of the gas stream at the top.

References

[1] W.G. Whitman, The two-film theory of absorption, Chem. Metall. Eng. 29 (1923) 147.
[2] R. Higbie, The rate of absorption of pure gas into a still liquid during short periods of exposure, Trans. Am. Inst. Chem. Eng. 31 (1935) 365.
[3] P.V. Danckwerts, Significance of liquid-film coefficients in gas absorption, Ind. Eng. Chem. 43 (1951) 1460.
[4] P.V. Danckwerts, A.M. Kennedy, Kinetics of liquid-film processes in gas absorption, Trans. Inst. Chem. Eng. 32 (1954) S49.
[5] S. Lynn, J.R. Straatemeier, H. Kramers, Absorption studies in the light of the penetration theory. I. Long wetted-wall columns. II. Absorption by short wetted-wall columns. III. Absorption by wetted-spheres, singly and in columns, Chem. Eng. Sci. 4 (1955) 49. 58, 63.

[6] J.F. Davidson, E.J. Cullen, D. Hanson, D. Roberts, The hold-up and liquid film coefficient of packed towers. Part I. Behaviour of a string of spheres, Trans. Inst. Chem. Eng. 37 (1959) 122.

[7] J.F. Davidson, The hold-up and liquid film coefficient of packed towers. Part II: statistical models of the random packing, Trans. Inst. Chem. Eng. 37 (1959) 131.

[8] P.V. Danckwerts, A.M. Kennedy, The kinetics of absorption of carbon dioxide into neutral and alkaline solutions, Chem. Eng. Sci. 8 (1958) 201.

[9] D. Roberts, P.V. Danckwerts, Kinetics of CO_2 absorption in alkaline solutions. I. Transient absorption rates and catalysis by arsenite, Chem. Eng. Sci. 17 (1962) 961.

[10] P.V. Danckwerts, A.M. Kennedy, D. Roberts, Kinetics of CO_2 absorption in alkaline solutions. II. Absorption in a packed column and tests of surface renewal models, Chem. Eng. Sci. 18 (1963) 63.

[11] E.J. Cullen, J.F. Davidson, Absorption of gases in liquid jets, Trans. Faraday Soc. 53 (1957) 113.

[12] P. Raimondi, H.L. Toor, Interfacial resistance in gas absorption, AIChE J. 5 (1959) 86.

[13] C.V. Sternling, L.E. Scriven, Interfacial turbulence: hydrodynamic instability and the Marangoni effect, AIChE J. 5 (1959) 514.

[14] F. Goodridge, I.D. Robb, Mechanism of interfacial resistance, Ind. Eng. Chem. Fundam. 4 (1965) 49.

[15] F.H. Garner, P. Kendrick, Mass transfer to drops of liquid suspended in a gas stream. Part I—a wind tunnel for the study of individual liquid drops, Trans. Inst. Chem. Eng. 37 (1959) 155.

[16] F.H. Garner, J.J. Lane, Mass transfer to drops of liquid suspended in a gas stream. Part II: experimental work and results, Trans. Inst. Chem. Eng. 37 (1959) 162.

[17] C.J. King, The additivity of individual phase resistances in mass transfer operations, AIChE J. 10 (1964) 671.

[18] E.R. Gilliand, T.K. Sherwood, Diffusion of vapours into air streams, Ind. Eng. Chem. 26 (1934) 516.

[19] H. Hollings, L. Silver, The washing of gas, Trans. Inst. Chem. Eng. 12 (1934) 49.

[20] T.H. Chilton, A.P. Colburn, Mass transfer (absorption) coefficients—prediction from data on heat transfer and fluid friction, Ind. Eng. Chem. 26 (1934) 1183.

[21] G.A. Morris, J. Jackson, Absorption Towers, Butterworths, London, 1953.

[22] O.L. Kowalke, O.A. Hougen, K.M. Watson, Transfer Coefficients of Ammonia in Absorption Towers, Bull. Univ. Wisconsin Eng. Sta. Ser. No. 68, 1925.

[23] H.M. Borden, W. Squires, Absorption of Ammonia in a Ring-Packed Tower, Massachusetts Institute of Technology, 1937. SM thesis. (cited in Reference 30).

[24] W.S. Norman, The performance of grid-packed towers, Trans. Inst. Chem. Eng. 29 (1951) 226.

[25] L.L. Fellinger, Absorption of Ammonia by Water and Acids in Various Standard Packings, Massachusetts Institute of Technology, 1941. DSc thesis.

[26] R.H. Perry, D.W. Green, J.O. Maloney (Eds.), Perry's Chemical Engineers' Handbook, seventh ed., McGraw-Hill Book Company, New York, 1997.

[27] M.C. Molstad, J.F. McKinney, R.G. Abbey, Performance of drip-point grid tower packings, III. Gas-film mass transfer coefficients: additional liquid-film mass transfer coefficients, Trans. Am. Inst. Chem. Eng. 39 (1943) 605.

[28] D.W. Van Krevelen, P.J. Hoftijzer, Kinetics of gas–liquid reactions. Part I. General theory, Recl. Trav. Chim. Pays-Bas 67 (1948) 563.

[29] R. Semmelbauer, Die Berechnung der Schütthöhe bei absorptions-vorgängen in Füllkörperkolonnen. (Calculation of the height of packing in packed towers.), Chem. Eng. Sci. 22 (1967) 1237.

[30] T.K. Sherwood, F.A.L. Holloway, Performance of packed towers—experimental studies of absorption and desorption, 39, 181—liquid film data for several packings, Trans. Am. Inst. Chem. Eng. 36 (1940) 21.

[31] C.M. Cooper, R.J. Christl, L.C. Peery, Packed tower performance at high liquor rates—the effect of gas and liquor rates upon performance in a tower packed with two-inch rings, Trans. Am. Inst. Chem. Eng. 37 (1941) 979.

[32] G. Nonhebel, Gas Purification Processes for Air Pollution Control, second ed., Newnes–Butterworth, London, 1972.

[33] A.W. Hixson, C.E. Scott, Absorption of gases in spray towers, Ind. Eng. Chem. 27 (1935) 307.

[34] W.G. Whitman, L. Long, H.Y. Wang, Absorption of gases by a liquid drop, Ind. Eng. Chem. 18 (1926) 363.

[35] R.L. Pigford, C. Pyle, Performance characteristics of spray-type absorption equipment, Ind. Eng. Chem. 43 (1951) 1649.

[36] W.S. Norman, Absorption, Distillation and Cooling Towers, Longmans, London, 1961.

[37] S. Hatta, On the absorption velocity of gases by liquids. II. Theoretical considerations of gas absorption due to chemical reaction, in: Tech. Repts. Tohoku Imp. Univ, vol. 10, 1932, p. 119.

[38] R.A.T.O. Nijsing, R.H. Hendriksz, H. Kramers, Absorption of CO_2 in jets and falling films of electrolyte solutions, with and without chemical reaction, Chem. Eng. Sci. 10 (1959) 88.

[39] J.B. Tepe, B.F. Dodge, Absorption of carbon dioxide by sodium hydroxide solutions in a packed column, Trans. Am. Inst. Chem. Eng. 39 (1943) 255.

[40] D.S. Cryder, J.O. Maloney, The rate of absorption of carbon dioxide in diethanolamine solutions, Trans. Am. Inst. Chem. Eng. 37 (1941) 827.

[41] E.J. Stephens, G.A. Morris, Determination of liquid-film absorption coefficients. A new type of column and its application to problems of absorption in presence of chemical reaction, Chem. Eng. Prog. 47 (1951) 232.

[42] P.V. Danckwerts, K.M. McNeil, The absorption of carbon dioxide into aqueous amine solutions and the effects of catalysis, Trans. Inst. Chem. Eng. 45 (1967) 32.

[43] P.V. Danckwerts, M.M. Sharma, The Absorption of Carbon Dioxide into Solutions of Alkalis and Amines (With Some Notes on Hydrogen Sulphide and Carbonyl Sulphide), Chem. Engr, London, Oct. 1966, p. CE244. No. 202.

[44] B.N. Sahay, M.M. Sharma, Effective interfacial areas and liquid and gas side mass transfer coefficients in a packed column, Chem. Eng. Sci. 28 (1973) 41.

[45] J.S. Eckert, How tower packings behave, Chem. Eng. 82 (1975) 70.

[46] T.K. Sherwood, R.L. Pigford, C.R. Wilke, Mass Transfer, McGraw-Hill Book Company, New York, 1980.

[47] R.E. Treybal, Mass Transfer Operations, third ed., McGraw-Hill Book Co, New York, 1980.

[48] A. Poll, W. Smith, Froth contact heat exchanger, Chem. Eng. 71 (1964) 111.

[49] C.G. Coggan, J.R. Bourne, The design of gas absorbers with heat effects, Trans. I. Chem. E. 47 (1969) T96. T160.

[50] H.L. Shulman, C.F. Ullrich, A.Z. Proulx, J.O. Zimmerman, Interfacial areas—gas and liquid phase mass transfer rates, AIChE J. 1 (1955) 2. 253.

[51] Eastham, I.E.: Private communication (1977).

[52] T.H. Chilton, A.P. Colburn, Distillation and absorption in packed columns, Ind. Eng. Chem. 27 (1935) 255.

[53] H.G. Rackett, Modified graphical integration for determining transfer units, Chem. Eng. Albany 71 (1964) 108.

[54] Norton Chemical Process Products Div., Box 350, Akron, Ohio; Hydronyl Ltd., King St., Fenton, Stokeon-Trent, UK.

[55] A.P. Colburn, The simplified calculation of diffusional processes. General consideration of two-film resistances, Trans. Am. Inst. Chem. Eng. 35 (1939) 211.

[56] A. Kremser, Theoretical analysis of absorption processes, Natl. Pet. News 22 (1930) 43.

[57] M. Souders, G.G. Brown, Fundamental design of high pressure equipment involving paraffin hydrocarbons. IV. Fundamental design of absorbing and stripping columns for complex vapours, Ind. Eng. Chem. 24 (1932) 519.

[58] C.M. Cooper, G.A. Fernstrom, S.A. Millers, Performance of agitated gas–liquid contactors, Ind. Eng. Chem. 36 (1944) 504.

[59] R.R. Ayerst, L.S. Herbert, A study of the absorption of carbon dioxide in ammonia solutions in agitated vessels, Trans. Inst. Chem. Eng. 32 (1954) S68.

[60] K.R. Westerterp, L.L. van Dierendonck, J.R. de Kraa, Interfacial areas in agitated gas–liquid contactors, Chem. Eng. Sci. 18 (1963) 157.

[61] T. Sridhar, O.E. Potter, Interfacial areas in gas–liquid stirred vessels, Chem. Eng. Sci. 35 (1980) 683.

[62] P.H. Calderbank, Gas Absorption from Bubbles, Chem. Eng., 1967, p. CE209. No. 212.

[63] S.O. Sideman, O. Hortacsu, J.W. Fulton, Mass transfer in gas–liquid contacting systems, Ind. Eng. Chem. 58 (1966) 32.

[64] F.H.H. Valentin, Mass transfer in agitated tanks, Br. Chem. Eng. 12 (1967) 1213.

[65] H. Yagi, F. Yoshida, Gas absorption by Newtonian and non-Newtonian fluids in sparged agitation vessels, Ind. Eng. Chem. Proc. Des. Dev. 14 (1975) 488.

[66] M. Zlokarnik, Sorption characteristics for gas–liquid contacting in mixing vessels, Adv. Biochem. Eng. 8 (1978) 133.

[67] K. van't Riet, Review of measuring methods and results in nonviscous gas–liquid mass transfer in stirred tanks, Ind. Eng. Chem. Proc. Des. Dev. 18 (1979) 357.

[68] H. Höcker, G. Langer, W. Udo, Mass transfer in aerated Newtonian and non-Newtonian liquids, Ger. Chem. Eng. 4 (1981) 51.

[69] J.B. Joshi, M.M. Sharma, Mass transfer characteristics of horizontal agitated contactors, Can. J. Chem. Eng. 54 (1976) 460.

[70] H. Fukuda, K. Idogawa, K. Ikeda, K. Endoh, Volumetric gas-phase mass transfer coefficients in baffled horizontal stirred tanks, J. Chem. Eng. Jpn. 13 (1980) 298.

[71] C.H. Yung, C.W. Wong, C.L. Chang, Gas holdup and aerated power consumption in mechanically stirred tanks, Can. J. Chem. Eng. 59 (1979) 672.

[72] H.T. Luong, B. Volesky, Mechanical power requirements of gas–liquid agitated systems, AIChE J. 25 (1970) 893.

[73] I. Bjerle, S. Bengtsson, K. Färnkvist, Absorption of SO_2 in $CaCO_3$-slurry in a laminar jet absorber, Chem. Eng. Sci. 27 (1972) 1853.

[74] A.W. Hixson, S.J. Baum, Agitation: heat and mass transfer coefficients in liquid-solid systems, Ind. Eng. Chem. 33 (1941) 478.

[75] S. Uchida, H. Moriguchi, H. Maejima, K. Koide, S. Kageyama, Absorption of sulphur dioxide into limestone slurry in a stirred tank reactor, Can. J. Chem. Eng. 56 (1978) 690.

[76] N. Ahmed, Design of gas scrubber based upon thin films and sprays, University of London, 1949. PhD thesis.

[77] H.H. Chambers, R.C. Wall, Some factors affecting the design of centrifugal gas absorbers, Trans. Inst. Chem. Eng. 32 (1954) S96.

[78] R.V. Kleinschmidt, A.W. Anthony, Recent development of Pease–Anthony gas scrubber, Trans. Am. Soc. Mech. Eng. 63 (1941) 349.

Further reading

T. Hobler, Mass Transfer and Absorbers, Pergamon Press, Oxford, 1966.

W.L. McCabe, J.C. Smith, P. Harriott, Unit Operations of Chemical Engineering, seventh ed., McGraw-Hill, New York, 2005.

W.S. Norman, Absorption, Distillation and Cooling Towers, Longmans, London, 1961.

T.K. Sherwood, R.L. Pigford, Absorption and Extraction, McGraw-Hill Book Co, New York, 1952.

T.K. Sherwood, R.L. Pigford, C.R. Wilke, Mass Transfer, McGraw-Hill Book Co, New York, 1975.

B.D. Smith, Design of Equilibrium Stage Processes, McGraw-Hill Book Co, New York, 1963.

R.E. Treybal, Mass Transfer Operations, third ed., McGraw-Hill Book Co, New York, 1980.

P.C. Wankat, Equilibrium Staged Separations: Separations for Chemical Engineers, Elsevier, New York, 1988.

R. Zarzycki, A. Chacuk, Absorption. Fundamentals and Applications, Pergamon Press, Oxford, 1993.

F.A. Zenz, Design of gas absorption towers, in: P.A. Schweitzer (Ed.), Handbook of Separation Techniques for Chemical Engineers, second ed., McGraw Hill, New York, 1988. Chapter 21.

Applications in humidification and water cooling

Ajay Kumar Ray

Department of Chemical and Biochemical Engineering, University of Western Ontario, London,
ON, Canada

Nomenclature

		Units in SI system	Dimensions in M, N, L, T, θ
A	interfacial area	m^2	L^2
A_b	base area of hyperbolic tower	m^2	L^2
a	interfacial area per unit volume of column	m^2/m^3	L^{-1}
b	psychrometric ratio ($h/h_D\rho_A s$)	–	$L^2 T^{-2}\theta^{-1}$
C_a	specific heat of gas at constant pressure	J/kgK	
C_L	specific heat of liquid	J/kgK	$L^2 T^{-2}\theta^{-1}$
C_p	specific heat of gas and vapour mixture at constant pressure	J/kgK	$L^2 T^{-2}\theta^{-1}$
C_s	specific heat of solid	J/kgK	$L^2 T^{-2}\theta^{-1}$
C_t	performance coefficient or efficiency factor	–	–
C_w	specific heat of vapour at constant pressure	J/kgK	$L^2 T^{-2}\theta^{-1}$
c	mass concentration of vapour	kg/m^3	ML^{-3}
C_0	mass concentration of vapour in saturated gas	kg/m^3	ML^{-3}
D_t	duty coefficient of tower (Eq. 3.31)	–	–
f	correction factor for mean driving force	–	–

Continued

Coulson and Richardson's Chemical Engineering. https://doi.org/10.1016/B978-0-08-101097-6.00003-1

		Units in SI system	Dimensions in M, N, L, T, θ
G'	mass rate of flow of gas per unit area	kg/m² s	$\mathbf{ML^{-2}T^{-1}}$
H	enthalpy of humid gas per unit mass of dry gas	J/kg	$\mathbf{L^2T^{-2}}$
H_a	enthalpy per unit mass, of dry gas	J/kg	$\mathbf{L^2T^{-2}}$
H_w	enthalpy per unit mass, of vapour	J/kg	$\mathbf{L^2T^{-2}}$
H_1	enthalpy of stream of gas, per unit mass of dry gas	J/kg	$\mathbf{L^2T^{-2}}$
H_2	enthalpy of another stream of gas, per unit mass of dry gas	J/kg	$\mathbf{L^2T^{-2}}$
H_3	enthalpy per unit mass of liquid or vapour	J/kg	$\mathbf{L^2T^{-2}}$
H'	modified enthalpy of humid gas defined by (5.69)	J/kg	$\mathbf{L^2T^{-2}}$
ΔH	enthalpy driving force ($H_f - H_G$)	J/kg	$\mathbf{L^2T^{-2}}$
$\Delta H'$	change in air enthalpy on passing through tower	J/kg	$\mathbf{L^2T^{-2}}$
h	heat-transfer coefficient	W/m² K	$\mathbf{MT^{-3}\theta^{-1}}$
h_D	mass-transfer coefficient	kmol/(kmol/m³) m²s	$\mathbf{LT^{-1}}$
h_G	heat-transfer coefficient for gas phase	W/m² K	$\mathbf{MT^{-3}\theta^{-1}}$
h_L	heat-transfer coefficient for liquid phase	W/m² K	$\mathbf{MT^{-3}\theta^{-1}}$
H	humidity	kg/kg	–
H_s	humidity of gas saturated at the adiabatic saturation temperature	kg/kg	–
H_w	humidity of gas saturated at the wet-bulb temperature	kg/kg	–
H_0	humidity of saturated gas	kg/kg	–
H_1	humidity of a gas stream	kg/kg	–
H_2	humidity of second gas stream	kg/kg	–
L'	mass rate of flow of liquid per unit area	kg/m² s	$\mathbf{ML^{-2}T^{-1}}$
M_A	molecular weight of gas	kg/kmol	$\mathbf{MN^{-1}}$
M_w	molecular weight of vapour	kg/kmol	$\mathbf{MN^{-1}}$
m, m_1, m_2	masses of dry gas	kg	$\mathbf{M}$
m_3	mass of liquid or vapour	kg	$\mathbf{M}$
P	total pressure	N/m²	$\mathbf{ML^{-1}T^{-2}}$
P_A	mean partial pressure of gas	N/m²	$\mathbf{ML^{-1}T^{-2}}$
P_w	partial pressure of vapour	N/m²	$\mathbf{ML^{-1}T^{-2}}$

Continued

		Units in SI system	Dimensions in M, N, L, T, θ
P_{w0}	partial pressure of vapour in saturated gas	N/m^2	$\mathbf{ML^{-1}T^{-2}}$
Q	rate of transfer of heat to liquid surface	W	$\mathbf{ML^{-1}T^{-2}}$
$\mathbf{R}$	universal gas constant	8314 J/kmol K	$\mathbf{MN^{-1}L^{-2}T^{-2}\theta^{-1}}$
S	humid heat of gas	J/kgK	$\mathbf{L^2T^{-2}\theta^{-1}}$
T	absolute temperature	K	$\boldsymbol{\theta}$
ΔT	change in water temperature in passing through the tower	K	$\boldsymbol{\theta}$
$\Delta T'$	(temperature of air leaving packing—ambient dry-bulb temperature)	K	$\boldsymbol{\theta}$
V	active volume per plan area of column	m^3/m^2	$\mathbf{L}$
W_L	water loading on tower	kg/s	$\mathbf{MT^{-1}}$
w	rate of evaporation	kg/s	$\mathbf{MT^{-1}}$
Z	percentage humidity	–	–
z	height from bottom of tower	m	$\mathbf{L}$
Z_t	height of cooling tower	m	$\mathbf{L}$
θ	temperature of gas stream	K	$\boldsymbol{\theta}$
θ_0	reference temperature, taken as the melting point of the material	K	$\boldsymbol{\theta}$
θ_s	adiabatic saturation temperature	K	$\boldsymbol{\theta}$
θ_w	wet-bulb temperature	K	$\boldsymbol{\theta}$
13.8.1.1.1. λ	latent heat of vaporisation per unit mass, at datum temperature	J/kg	$\mathbf{L^2T^{-2}}$
λ_f	latent heat of freezing per unit mass, at datum temperature	J/kg	$\mathbf{L^2T^{-2}}$
λ'	modified latent heat of vaporisation per unit mass defined by (5.68)	J/kg	$\mathbf{L^2T^{-2}}$
ρ	mean density of gas and vapour	kg/m^3	$\mathbf{ML^{-3}}$
ρ_A	mean density of gas at partial pressure P_A	kg/m^3	$\mathbf{ML^{-3}}$
Pr	Prandtl number	–	–
Sc	Schmidt number	–	–

Suffixes 1, 2, f, L, G denote conditions at the bottom of the tower, the top of the tower, the interface, the liquid, and the gas, respectively.
Suffix m refers to the mean water temperature.

3.1 Introduction

In the processing of materials, it is often necessary either to increase the amount of vapour present in a gas stream, an operation known as *humidification*, or to reduce the vapour present, a process referred to as *dehumidification*. In humidification, the vapour content may be increased by passing the gas over a liquid that then evaporates into the gas stream. This transfer into the main stream takes place by diffusion, and at the interface, simultaneous heat and mass transfer take place according to the relations considered in previous chapters. In the reverse operation, that is, dehumidification, partial condensation must be effected and the condensed vapour removed.

The most widespread application of humidification and dehumidification involves the air–water system, and a discussion of this system forms the greater part of the present chapter. Although the drying of wet solids is an example of a humidification operation, the reduction of the moisture content of the solids is the main objective, and the humidification of the airstream is a secondary effect. Much of the present chapter is, however, of vital significance in any drying operation. Air-conditioning and gas drying also involve humidification and dehumidification operations. For example, moisture must be removed from wet chlorine so that the gas can be handled in steel equipment that otherwise would be severely corroded. Similarly, the gases used in the manufacture of sulphuric acid must be dried or dehumidified before entering the converters, and this is achieved by passing the gas through a dehydrating agent such as sulphuric acid, in essence an absorption operation, or by an alternative dehumidification process discussed later.

In order that hot condenser water may be reused in a plant, it is normally cooled by contact with an airstream. The equipment usually takes the form of a tower in which the hot water is run in at the top and allowed to flow downwards over a packing against a countercurrent flow of air that enters at the bottom of the cooling tower. The design of such towers forms an important part of the present chapter, though at the outset it is necessary to consider basic definitions of the various quantities involved in humidification, in particular *wet-bulb* and *adiabatic saturation temperatures*, and the way in which humidity data are presented on charts and graphs. Whilst the present discussion is devoted to the very important air–water system, which is in some ways unique, the same principles may be applied to other liquids and gases, and this topic is covered in a final section.

3.2 Humidification terms

3.2.1 Definitions

The more important terms used in relation to humidification are defined as follows:

Humidity (*H*)	Mass of vapour associated with unit mass of dry gas
Humidity of saturated gas (*H*₀)	Humidity of the gas when it is saturated with vapour at a given temperature
Percentage humidity	100 (*H*/*H*₀)
Humid heat (*s*)	Heat required to raise unit mass of dry gas and its associated vapour through unit temperature difference at constant pressure or $s = C_a + HC_w$ where C_a and C_w are the specific heat capacities of the gas and the vapour, respectively. (For the air–water system, the humid heat is approximately $s = 1.00 + 1.9H$ kJ/Kg K)
Humid volume	Volume occupied by unit mass of dry gas and its associated vapour
Saturated volume	Humid volume of saturated gas
Dew point	Temperature at which the gas is saturated with vapour. As a gas is cooled, the dew point is the temperature at which condensation will first occur
Percentage relative humidity	$\left(\dfrac{\text{Partial pressure of vapour in gas}}{\text{Partial pressure of vapour in saturated gas}}\right) \times 100$

The above nomenclature conforms with the recommendations of BS1339 [1], although there are some ambiguities in the standard.

The relationship between the partial pressure of the vapour and the humidity of a gas may be derived as follows. In unit volume of gas,

$$\text{Mass of vapour} = \frac{P_w M_w}{\mathbf{R}T}$$

and

$$\text{Mass of noncondensable gas} = \frac{(P - P_w)M_A}{\mathbf{R}T}$$

The humidity is therefore given by

$$H = \frac{P_w}{P - P_w}\left(\frac{M_w}{M_A}\right) \tag{3.1}$$

and the humidity of the saturated gas is

$$H_0 = \frac{P_{w0}}{P - P_{w0}}\left(\frac{M_w}{M_A}\right) \tag{3.2}$$

where P_w is the partial pressure of vapour in the gas, P_{w0} the partial pressure of vapour in the saturated gas at the same temperature, M_A the mean molecular weight of the dry gas, M_w the molecular mass of the vapour, P the total pressure, $\mathbf{R}$ the gas constant (8314 J/kmol K in SI units), and T the absolute temperature.

For the air–water system, P_w is frequently small compared with P, and hence substituting for the molecular masses,

$$H = \frac{18}{29}\left(\frac{P_w}{P}\right)$$

The relationship between the percentage humidity of a gas and the percentage relative humidity may be derived as follows:

The percentage humidity, by definition $= 100H/H0$.

Substituting from Eqs (3.1), (3.2) and simplifying,

$$\text{Percentage humidity} = \left(\frac{P - P_{w0}}{P - P_w}\right) \cdot \left(\frac{P_w}{P_{w0}}\right) \times 100$$

$$= \frac{(P - P_{w0})}{(P - P_w)} \times (\text{percentage relative humidity})$$

(3.3)

When $(P - P_{w0})/(P - P_w) \approx 1$, the percentage relative humidity and the percentage humidity are equal. This condition is approached when the partial pressure of the vapour is only a small proportion of the total pressure or when the gas is almost saturated, that is, as $P_w \rightarrow P_{w0}$.

Example 3.1

In a process in which it is used as a solvent, benzene is evaporated into dry nitrogen. At 297 K and 101.3 kN/m^2, the resulting mixture has a percentage relative humidity of 60. It is required to recover 80% of the benzene present by cooling to 283 K and compressing to a suitable pressure. What should this pressure be? The vapour pressure of benzene is 12.2 kN/m^2 at 297 K and 6.0 kN/m^2 at 283 K.

Solution

From the definition of percentage relative humidity (RH),

$$P_w = P_{w0}\left(\frac{RH}{100}\right)$$

$$\text{At } 297\text{K}, P_w = (12.2 \times 1000) \times \left(\frac{60}{100}\right) = 7320\,\text{N/m}^2$$

In the benzene–nitrogen mixture,

$$\text{Mass of benzene} = \frac{P_w M_w}{RT} = \frac{(7320 \times 78)}{8314 \times 297} = 0.231\,\text{kg}$$

$$\text{Mass of nitrogen} = \frac{(P - P_w)M_A}{RT} = \frac{[(101.3 - 732) \times 1000 \times 28]}{(8314 \times 297)} = 1.066\,\text{kg}$$

Hence, the humidity is

$$H = \left(\frac{0.231}{1.066}\right) = 0.217\,\text{kg/kg}$$

In order to recover 80% of the benzene, the humidity must be reduced to 20% of the initial value. As the vapour will be in contact with liquid benzene, the nitrogen will be saturated with benzene vapour, and hence at 283 K,

$$H_0 = \frac{(0.217 \times 20)}{100} = 0.0433\,\text{kg/kg}$$

Thus, in Eq. (3.2),

$$0.0433 = \left(\frac{6000}{P - 6000}\right)\left(\frac{78}{28}\right)$$

from which

$$P = 3.92 \times 10^5\,\text{N/m}^2 = 392\text{kN/m}^2$$

Example 3.2

In a vessel at 101.3 kN/m^2 and 300 K, the percentage relative humidity of the water vapour in the air is 25. If the partial pressure of water vapour when air is saturated with vapour at 300 K is 3.6 kN/m^2, calculate.
(1) the partial pressure of the water vapour in the vessel,
(2) the specific volumes of the air and water vapour,
(3) the humidity of the air and humid volume,
(4) the percentage humidity.

Solution

(a) From the definition of percentage relative humidity,

$$P_w = P_{w0}\frac{RH}{100} = 3600 \times \left(\frac{25}{100}\right) = 900\text{N/m}^2 = 0.9\text{KN/m}^2$$

(b) In 1 m^3 of air,

$$\text{Mass of water vapour} = \frac{(900 \times 18)}{8314 \times 300} = 0.0065\,\text{kg}$$

$$\text{Mass of air} = \frac{[(101.3 - 0.9) \times 1000 \times 29]}{(8314 \times 300)} = 1.167\text{kg}$$

Hence, specific volume of water vapour at 0.9kN/m^2 = $\left(\frac{1}{0.0065}\right)$ = 154m^3/kg.

specific volume of air at 100.4kN/m^2 = $\left(\frac{1}{1.167}\right)$ = 0.857 m^3/kg.

(c) Humidity,

$$H = \left(\frac{0.0065}{1.1673}\right) = 0.0056\text{kg/kg}$$

(Using the approximate relationship,

$$H = \frac{(18 \times 900)}{(29 \times 101.3 \times 1000)} = 0.0055\,\text{kg/kg})$$

∴ humid volume, volume of 1 kg air + associated vapour = specific volume of air at 100.4 kN/m²

$$= 0.857\,\text{m}^3/\text{kg}$$

(d) From Eq. (3.3),

$$\text{Percentage humidity} = \frac{[(101.3 - 3.6) \times 1000]}{[(101.3 - 0.9) \times 1000]} \times 25$$
$$= 24.3$$

3.2.2 Wet-bulb temperature

When a stream of unsaturated gas is passed over the surface of a liquid, the humidity of the gas is increased due to evaporation of the liquid. The temperature of the liquid falls below that of the gas, and heat is transferred from the gas to the liquid. At equilibrium, the rate of heat transfer from the gas just balances that required to vaporise the liquid, and the liquid is said to be at the *wet-bulb temperature*. The rate at which this temperature is reached depends on the initial temperatures and the rate of flow of gas past the liquid surface. With a small area of contact between the gas and the liquid and a high gas flow rate, the temperature and the humidity of the gas stream remain virtually unchanged.

The rate of transfer of heat from the gas to the liquid can be written as

$$Q = hA(\theta - \theta_w) \tag{3.4}$$

where Q is the heat flow, h the coefficient of heat transfer, A the area for transfer, and θ and θ_w are the temperatures of the gas and liquid phases.

The liquid evaporating into the gas is transferred by diffusion from the interface to the gas stream as a result of a concentration difference $(c_0 - c)$, where c_0 is the concentration of the vapour at the surface (mass per unit volume) and c is the concentration in the gas stream. The rate of evaporation is then given by

$$W = h_D A(c_0 - c) = h_D A \frac{M_w}{\mathbf{R}T}(P_{w0} - P_w) \tag{3.5}$$

where h_D is the coefficient of mass transfer.

The partial pressures of the vapour, P_w and P_{w0}, may be expressed in terms of the corresponding humidities H and H_w by Eqs (3.1), (3.2).

If P_w and P_{w0} are small compared with P, $(P - P_w)$ and $(P - P_{w0})$ may be replaced by a mean partial pressure of the gas P_A and

$$W = h_{DA}A \frac{(H_w - H)M_w}{RT} \cdot \left(P_A \frac{M_A}{M_w} \right)$$

$$= h_D A \rho_A (H_w - H) \tag{3.6}$$

where ρ_A is the density of the gas at the partial pressure P_A.

The heat transfer required to maintain this rate of evaporation is

$$Q = h_D A \rho_A (H_w - H) \lambda \tag{3.7}$$

where λ is the latent heat of vaporisation of the liquid.

Thus, equating Eqs (3.4), (3.7),

$$(H - H_w) = -\frac{h}{h_v \rho_A \lambda}(\theta - \theta_w) \tag{3.8}$$

Both h and h_D are dependent on the equivalent gas-film thickness, and thus, any decrease in the thickness, as a result of increasing the gas velocity, for example, increases both h and h_D. At normal temperatures, (h/h_D) is virtually independent of the gas velocity provided this is greater than about 5 m/s. Under these conditions, heat transfer by convection from the gas stream is large compared with that from the surroundings by radiation and conduction.

The wet-bulb temperature θ_w depends only on the temperature and the humidity of the gas, and values normally quoted are determined for comparatively high gas velocities, such that the condition of the gas does not change appreciably as a result of being brought into contact with the liquid and the ratio (h/h_D) has reached a constant value. For the air–water system, the ratio $(h/h_D\rho_A)$ is about 1.0 kJ/kg K and varies from 1.5 to 2.0 kJ/kg K for organic liquids.

Example 3.3

Moist air at 310 K has a wet-bulb temperature of 300 K. If the latent heat of vaporisation of water at 300 K is 2440 kJ/kg, estimate the humidity of the air and the percentage relative humidity. The total pressure is 105 kN/m², and the vapour pressure of water vapour at 300 K is 3.60 and 6.33 kN/m² at 310 K.

Solution

The humidity of air saturated at the wet-bulb temperature is given by.

$$H_w = \frac{P_{w0}}{P0P_{w0}} \frac{M_w}{M_A} \text{ (Eq.13.2)}$$

$$= \left(\frac{3.6}{105.0 - 3.6} \right) \left(\frac{18}{29} \right) = 0.0220 \text{kg/kg}$$

Therefore, taking $(h/h_D\rho_A)$ as 1.0 kJ/kg K, in Eq. (3.8),

$$(0.0220 - H) = \left(\frac{1.0}{2440} \right)(310 - 300)$$

$$H = \underline{0.018 \text{kg/kg}}$$

or

$$\text{at } 310K, \ P_{w0} = 6.33\text{kN/m}^2$$

In Eq. (3.2),

$$0.0780 = \frac{18P_w}{(105.0 - P_w)29}$$

$$\therefore P_w = 2.959\,\text{kN/m}^2$$

and the percentage relative humidity

$$= \frac{(100 \times 2.959)}{6.33} = \underline{\underline{46.7\%}}$$

3.2.3 Adiabatic saturation temperature

In the system just considered, neither the humidity nor the temperature of the gas is appreciably changed. If the gas is passed over the liquid at such a rate that the time of contact is sufficient for equilibrium to be established, the gas will become saturated, and both phases will be brought to the same temperature. In a thermally insulated system, the total sensible heat falls by an amount equal to the latent heat of the liquid evaporated. As a result of continued passage of the gas, the temperature of the liquid gradually approaches an equilibrium value that is known as the *adiabatic saturation temperature*.

These conditions are achieved in an infinitely tall thermally insulated humidification column through which gas of a given initial temperature and humidity flows countercurrently to the liquid under conditions where the gas is completely saturated at the top of the column. If the liquid is continuously circulated around the column and if any fresh liquid that is added is at the same temperature as the circulating liquid, the temperature of the liquid at the top and bottom of the column and of the gas at the top approaches the adiabatic saturation temperature. Temperature and humidity differences are a maximum at the bottom and zero at the top, and therefore, the rates of transfer of heat and mass decrease progressively from the bottom to the top of the tower. This is illustrated in Fig. 3.1.

Making a heat balance over the column, it is seen that the heat of vaporisation of the liquid must come from the sensible heat in the gas. The temperature of the gas falls from θ to the adiabatic saturation temperature θ_s, and its humidity increases from H to H_s (the saturation value at θ_s). Then, working on the basis of unit mass of dry gas, or

$$(\theta - \theta_s)s = (H_s - H)\lambda$$

$$(H - H_s) = -\frac{s}{\lambda}(\theta - \theta_s) \tag{3.9}$$

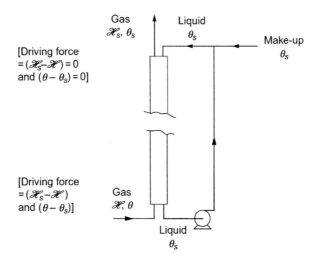

Gas
$\mathscr{H}_s, \theta_s$

Liquid
θ_s

Make-up
θ_s

[Driving force
$= (\mathscr{H}_s - \mathscr{H}) = 0$
and $(\theta - \theta_s) = 0$]

[Driving force
$= (\mathscr{H}_s - \mathscr{H})$
and $(\theta - \theta_s)$]

Gas
$\mathscr{H}, \theta$

Liquid
θ_s

FIG. 3.1

Adiabatic saturation temperature θ_s.

where s is the humid heat of the gas and λ the latent heat of vaporisation at θ_s. s is almost constant for small changes in H.

Eq. (3.9) indicates an approximately linear relationship between humidity and temperature for all mixtures of gas and vapour having the same adiabatic saturation temperature θ_s. A curve of humidity versus temperature for gases with a given adiabatic saturation temperature is known as an *adiabatic cooling line*. For a range of adiabatic saturation temperatures, a family of curves, approximating to straight lines of slopes equal to $-(s/\lambda)$, is obtained. These lines are not exactly straight and parallel because of variations in λ and s.

Comparing Eqs (3.8), (3.9), it is seen that the adiabatic saturation temperature is equal to the wet-bulb temperature when $s = h/h_D \rho_A$. This is the case for most water vapour systems and accurately so when $H = 0.047$. The ratio $(h/h_D \rho_A s) = b$ is sometimes known as the *psychrometric ratio*, and as indicated, b is approximately unity for the air–water system. For most systems involving air and an organic liquid, $b = 1.3–2.5$ and the wet-bulb temperature is higher than the adiabatic saturation temperature. This was confirmed in 1932 by Sherwood and Comings [2] who worked with water, ethanol, n-propanol, n-butanol, benzene, toluene, carbon tetrachloride, and n-propyl acetate and found that the wet-bulb temperature was always higher than the adiabatic saturation temperature except in the case of water.

In Chapter 4, it is shown that when the Schmidt and Prandtl numbers for a mixture of gas and vapour are approximately equal to unity, the *Lewis relation* applies or

$$h_D = \frac{h}{C_p \rho} \quad \text{(Eq. 4.105)}$$

where C_p and ρ are the mean specific heat and density of the vapour phase.
Therefore,

$$\frac{h}{h_D \rho_A} = \frac{C_p \rho}{\rho_A} \tag{3.10}$$

Where the humidity is relatively low, $C_p \approx s$ and $\rho \approx \rho_A$, and hence,

$$s \approx \frac{h}{h_D \rho_A} \tag{3.11}$$

For systems containing vapour other than that of water, s is only approximately equal to $h/h_D \rho_A$, and the difference between the two quantities may be as high as 50%.

If an unsaturated gas is brought into contact with a liquid that is at the adiabatic saturation temperature of the gas, a simultaneous transfer of heat and mass takes place. The temperature of the gas falls, and its humidity increases (Fig. 3.2). The temperature of the liquid at any instant tends to change and approaches the wet-bulb temperature corresponding to the particular condition of the gas at that moment. For a liquid other than water, the adiabatic saturation temperature is less than the wet-bulb temperature, and therefore in the initial stages, the temperature of the liquid rises. As the gas becomes humidified, however, its wet-bulb temperature falls and consequently the temperature to which the liquid is tending decreases as evaporation takes place. In due course, therefore, a point is reached where the liquid actually reaches the wet-bulb temperature of the gas in contact with it. It does not remain at this temperature, however, because the gas is not then completely saturated, and further humidification is accompanied by a continued lowering of the wet-bulb temperature. The temperature of the liquid therefore starts to fall and continues to fall until the gas is completely saturated. The liquid and gas are then both at the adiabatic saturation temperature.

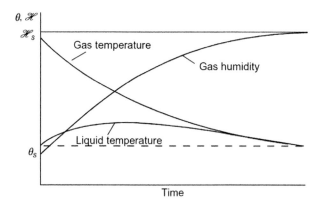

FIG. 3.2

Saturation of gas with liquid other than water at the adiabatic saturation temperature.

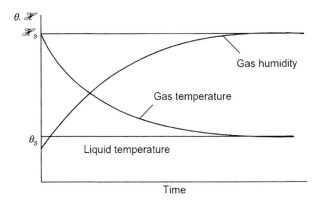

FIG. 3.3

Saturation of air with water at adiabatic saturation temperature.

The air–water system is unique, however, in that the Lewis relation holds quite accurately, so that the adiabatic saturation temperature is the same as the wet-bulb temperature. If, therefore, an unsaturated gas is brought into contact with water at the adiabatic saturation temperature of the gas, there is no tendency for the temperature of the water to change, and it remains in a condition of dynamic equilibrium through the whole of the humidification process (Fig. 3.3). In this case, the adiabatic cooling line represents the conditions of gases of constant wet-bulb temperatures and constant adiabatic saturation temperatures. The change in the condition of a gas as it is humidified with water vapour is therefore represented by the adiabatic cooling line, and the intermediate conditions of the gas during the process are readily obtained. This is particularly useful because only partial humidification is normally obtained in practice.

3.3 **Humidity data for the air–water system**

To facilitate calculations, various properties of the air–water system are plotted on a *psychrometric* or *humidity chart*. Such a chart is based on either the temperature or the enthalpy of the gas. The temperature–humidity chart is the more commonly used though the enthalpy–humidity chart is particularly useful for determining the effect of mixing two gases or of mixing a gas and a liquid. Each chart refers to a particular total pressure of the system.

A humidity–temperature chart for the air–water system at atmospheric pressure, based on the original chart by Grosvenor [3], is given in Fig. 3.4, and the corresponding humidity–enthalpy chart is given in Fig. 3.5.

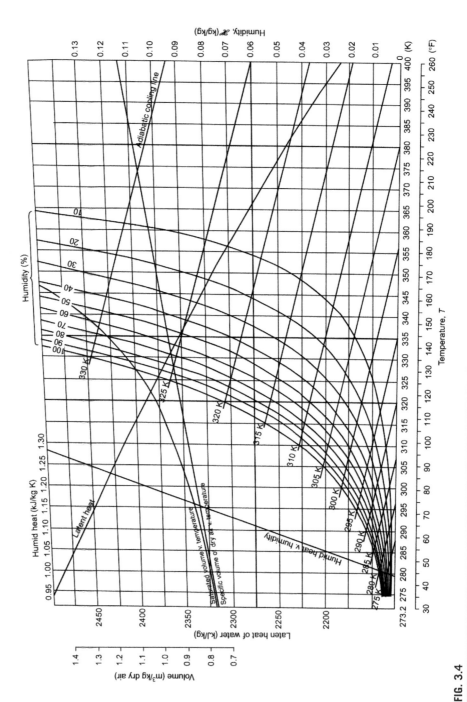

FIG. 3.4

Humidity–temperature chart (see also the Appendix).

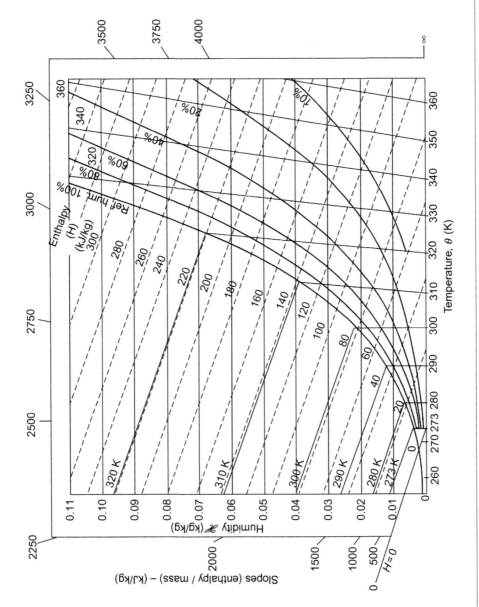

FIG. 3.5

Humidity–enthalpy diagram for air–water vapour system at atmospheric pressure.

3.3.1 Temperature–humidity chart

In Fig. 3.4, it will be seen that the following quantities are plotted against temperature:

(i) The *humidity H* for various values of the percentage humidity. For saturated gas,

$$H_0 = \frac{P_{w0}}{P - P_{w0}} \left(\frac{M_w}{M_A}\right) \tag{13.2}$$

From Eq. (3.1) for a gas with a humidity less than the saturation value,

$$H = \frac{P_w}{P - P_w} \left(\frac{M_w}{M_A}\right) = H_0 \frac{P_w}{P_{w0}} \frac{P_w - P_{w0}}{P - P_w} \tag{3.12}$$

(ii) *The specific volume of dry gas.* This is a linear function of temperature.
(iii) *The saturated volume.* This increases more rapidly with temperature than the specific volume of dry gas because both the quantity and the specific volume of vapour increase with temperature. At a given temperature, the humid volume varies linearly with humidity, and hence, the humid volume of unsaturated gas can be found by interpolation.
(iv) The latent heat of vaporisation.

In addition, the *humid heat* is plotted as the abscissa in Fig. 3.4 with the humidity as the ordinate.

Adiabatic cooling lines are included in the diagram, and as already discussed, these have a slope of $-(s/\lambda)$, and they are slightly curved since s is a function of H. On the chart, they appear as straight lines, however, since the inclination of the axis has been correspondingly adjusted. Each adiabatic cooling line represents the composition of all gases whose adiabatic saturation temperature is given by its point of intersection with the 100% humidity curve. For the air–water system, the adiabatic cooling lines represent conditions of constant wet-bulb temperature as well and, as previously mentioned, enable the change in composition of a gas to be followed as it is humidified by contact with water at the adiabatic saturation temperature of the gas.

Example 3.4

Air containing 0.005 kg water vapour per kg of dry air is heated to 325 K in a dryer and passed to the lower shelves. It leaves these shelves at 60% humidity and is reheated to 325 K and passed over another set of shelves, again leaving at 60% humidity. This is again repeated for the third and fourth sets of shelves, after which the air leaves the dryer. On the assumption that the material on each shelf has reached the wet-bulb temperature and that heat losses from the dryer may be neglected, determine.

(a) the temperature of the material on each tray,
(b) the amount of water removed in kg/s if 5 m³/s moist air leaves the dryer,
(c) the temperature to which the inlet air would have to be raised to carry out the drying in a single stage.

Solution

For each of the four sets of shelves, the condition of the air is changed to 60% humidity along an adiabatic cooling line.

Initial condition of air: $\theta = 325$ K, $H = 0.005$ kg/kg.
On humidifying to 60% humidity,

$$\theta = 301 \text{ K}, \quad H = 0.015 \text{ kg/kg and } \theta_w = 296 \text{ K}$$

At the end of the second pass,

$$\theta = 308 \text{ K}, \quad H = 0.022 \text{ kg/kg and } \theta_w = 301 \text{ K}$$

At the end of the third pass,

$$\theta = 312 \text{ K}, \quad H = 0.027 \text{ kg/kg and } \theta_w = 305 \text{ K}$$

At the end of the fourth pass,

$$\theta = 315 \text{ K}, \quad H = 0.032 \text{ kg/kg and } \theta_w = 307 \text{ K}$$

Thus, the temperatures of the material on each of the trays are

$$\underline{296 \text{ K}, \quad 301 \text{ K}, \quad 305 \text{ K}, \quad \text{and} \quad 307 \text{ K}}$$

Total increase in humidity

$$= (0.032 - 0.005) = 0.027 \text{ kg/kg}$$

The air leaving the system is at 315 K and 60% humidity.
From Fig. 3.4, specific volume of dry air $= 0.893$ m³/kg.
Specific volume of saturated air *(saturated volume)* $= 0.968$ m³/kg.
Therefore, by interpolation, the humid volume of air of 60% humidity $= 0.937$ m³/kg.

$$\text{Mass of air passing through the dryer} = \left(\frac{5}{0.937}\right) = 5.34 \text{ kg/s.}$$
$$\text{Mass of water evaporated} = (5.34 \times 0.027) = \underline{0.144 \text{ kg/s}}$$

If the material is to be dried by air in a single pass, the air must be heated before entering the dryer such that its wet-bulb temperature is 307 K.

For air with a humidity of 0.005 kg/kg, this corresponds to a dry-bulb temperature of $\underline{370 \text{ K}}$. The various steps in this calculation are shown in Fig. 3.6.

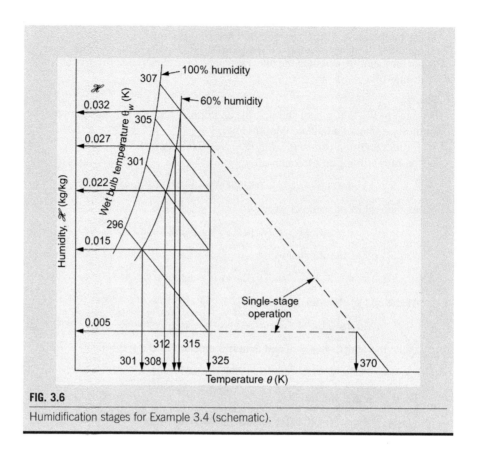

FIG. 3.6

Humidification stages for Example 3.4 (schematic).

3.3.2 Enthalpy–humidity chart

In the calculation of enthalpies, it is necessary to define some standard reference state at which the enthalpy is taken as zero. It is most convenient to take the melting point of the material constituting the vapour as the reference temperature and the liquid state of the material as its standard state.

If H is the enthalpy of the humid gas per unit mass of dry gas, H_a the enthalpy of the dry gas per unit mass, H_w the enthalpy of the vapour per unit mass, C_a the specific heat of the gas at constant pressure, C_w the specific heat of the vapour at constant pressure, θ the temperature of the humid gas, θ_0 the reference temperature, λ the latent heat of vaporisation of the liquid at θ_0, and H the humidity of the gas, then, for an unsaturated gas,

$$H = H_a + H_w H \qquad (3.13)$$

where

$$H_a = C_a(\theta - \theta_0) \qquad (3.14)$$

and

$$H_w = C_w(\theta - \theta_0) + \lambda \qquad (3.15)$$

Thus, in Eq. (3.13),

$$H = (C_a + HC_w)(\theta - \theta_0) + H\lambda$$
$$= (\theta - \theta_0)(s + H\lambda) \tag{3.16}$$

If the gas contains more liquid or vapour than is required to saturate it at the temperature in question, either the gas will be supersaturated or the excess material will be present in the form of liquid or solid according to whether the temperature θ is greater or less than the reference temperature θ_0. The supersaturated condition is unstable and will not be considered further.

If the temperature θ is greater than θ_0 and if the humidity H is greater than the humidity H_0 of saturated gas, the enthalpy H per unit mass of dry gas is given by

$$H = C_a(\theta - \theta_0) + H_0[C_w(\theta - \theta_0) + \lambda] + C_L(H - H_0)(\theta - \theta_0) \tag{3.17}$$

where C_L is the specific heat of the liquid.

If the temperature θ is less than θ_0, the corresponding enthalpy H is given by

$$H = C_a(\theta - \theta_0) + H_0[C_w(\theta - \theta_0) + \lambda] + (H - H_0)[C_s(\theta - \theta_0) + \lambda_f] \tag{3.18}$$

where C_s is the specific heat of the solid and λ_f is the latent heat of freezing of the liquid, a negative quantity.

Eqs (3.16)–(3.18) give the enthalpy in terms of the temperature and humidity of the humid gas for the three conditions: $\theta = \theta_0$, $\theta > \theta_0$, and $\theta < \theta_0$, respectively. Thus, given the percentage humidity and the temperature, the humidity may be obtained from Fig. 3.4, the enthalpy calculated from Eqs (3.16)–(3.18) and plotted against the humidity, usually with enthalpy as the abscissa. Such a plot is shown in Fig. 3.7 for the air–water system, which includes the curves for 100% humidity and for some lower value, say $Z\%$.

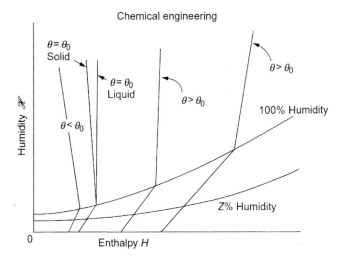

FIG. 3.7

Humidity–enthalpy diagram for air–water system–rectangular axes.

Considering the nature of the isothermals for the three conditions dealt with previously, at constant temperature θ, the relation between enthalpy and humidity for an unsaturated gas is

$$H = \text{constant} + [C_w(\theta - \theta_0) + \lambda]H \qquad (3.19)$$

Thus, the isothermal is a straight line of slope $(C_w(\theta - \theta_0) + \lambda)$ with respect to the humidity axis. At the reference temperature θ_0, the slope is λ; at higher temperatures, the slope is greater than λ; and at lower temperatures, it is less than λ. Because the latent heat is normally large compared with the sensible heat, the slope of the isothermals remains positive down to very low temperatures. Since the humidity is plotted as the ordinate, the slope of the isothermal relative to the X-axis decreases with increase in temperature. When $\theta > \theta_0$ and $H > H_0$ the saturation humidity, the vapour phase consists of a saturated gas with liquid droplets in suspension. The relation between enthalpy and humidity at constant temperature θ is.

$$H = \text{constant} + C_L(\theta - \theta_0)H \qquad (3.20)$$

The isothermal is therefore a straight line of slope $C_L(\theta - \theta_0)$. At the reference temperature θ_0, the slope is zero, and the isothermal is parallel to the humidity axis. At higher temperatures, the slope has a small positive value. When $\theta < \theta_0$ and $H > H_0$, solid particles are formed, and the equation of the isothermal is

$$H = \text{constant} + \left[C_s(\theta - \theta_0) + \lambda_f \right]H \qquad (3.21)$$

This represents a straight line of slope $(C_s(\theta - \theta_0) + \lambda_f)$. Both $C_s(\theta - \theta_0)$ and λ_f are negative, and therefore, the slopes of all these isothermals are negative. When $\theta = \theta_0$, the slope is λ_f. In the supersaturated region, therefore, there are two distinct isothermals at temperature θ_0: one corresponds to the condition where the excess vapour is present in the form of liquid droplets and the other to the condition where it is present as solid particles. The region between these isothermals represents conditions where a mixture of liquid and solid is present in the saturated gas at the temperature θ_0.

The shape of the humidity–enthalpy line for saturated air is such that the proportion of the total area of the diagram representing saturated, as opposed to supersaturated, air is small when rectangular axes are used. In order to enable greater accuracy to be obtained in the use of the diagram, oblique axes are normally used, as in Fig. 3.5, so that the isothermal for unsaturated gas at the reference temperature θ_0 is parallel to the humidity axis.

It should be noted that the curves of humidity plotted against either temperature or enthalpy have a discontinuity at the point corresponding to the freezing point of the humidifying material. Above the temperature θ_0, the lines are determined by the vapour–liquid equilibrium and below it by the vapour–solid equilibrium.

Two cases may be considered to illustrate the use of enthalpy–humidity charts. These are the mixing of two streams of humid gas and the addition of liquid or vapour to a gas.

Mixing of two streams of humid gas

Consider the mixing of two gases of humidities H_1 and H_2, at temperatures θ_1 and θ_2, and with enthalpies H_1 and H_2 to give a mixed gas of temperature θ, enthalpy H, and humidity H. If the masses of dry gas concerned are m_1, m_2, and m, respectively, then taking a balance on the dry gas, vapour, and enthalpy,

$$m_1 + m_2 = m \tag{3.22}$$

$$m_1 H_1 + m_2 H_2 = mH \tag{3.23}$$

and

$$m_1 H_1 + m_2 H_2 = mH \tag{3.24}$$

Elimination of m gives and

$$m_1(H - H_1) = m_2(H_2 - H) \tag{3.25}$$

Dividing these two equations,

$$\frac{(H - H_1)}{H - H_1} = \frac{(H - H_2)}{H - H_2} \tag{3.26}$$

The condition of the resultant gas is therefore represented by a point on the straight line joining (H_1 and H_1) and (H_2 and H_2). The humidity H is given, from Eq. (3.25), by

$$\frac{(H - H_1)}{(H_2 - H)} = \frac{m_2}{m_1} \tag{3.27}$$

The gas formed by mixing two unsaturated gases may be either unsaturated, saturated, or supersaturated. The possibility of producing supersaturated gas arises because the 100% humidity line on the humidity–enthalpy diagram is concave towards the humidity axis.

Example 3.5

In an air-conditioning system, 1 kg/s air at 350 K and 10% humidity is mixed with 5 kg/s air at 300 K and 30% humidity. What is the enthalpy, humidity, and temperature of the resultant stream?

Solution

From Fig. 3.4,
   at $\theta_1 = 350$ K and humidity $= 10\%$; $H_1 = 0.043$ kg/kg.
   at $\theta_2 = 300$ K and humidity $= 30\%$; $H_2 = 0.0065$ kg/kg.
   Thus, in Eq. (3.23),

$$(1 \times 0.043) + (5 \times 0.0065) = (1 + 5)H$$

and

$$\underline{H = 0.0125 \text{kg/kg}}$$

From Fig. 3.5,

at $\theta_1 = 350\text{K}$ and $H_1 = 0.043\text{kg/kg}$; $H_1 = 192\text{kJ/kg}$.

at $\theta_2 = 300\text{K}$ and $H_2 = 0.0065\text{kg/kg}$; $H_2 = 42\text{kJ/kg}$.

Thus, in Eq. (3.25),

$$1(H - 192) = 5(42 - H)$$

and

$$\underline{\underline{H = 67\text{kJ/kg}}}$$

From Fig. 3.5,

at $H = 67\text{kJ/kg}$ and $H = 0.0125\text{kg/kg}$

$$\underline{\underline{\theta = 309\text{K}}}$$

The data used in this example are shown in Fig. 3.8.

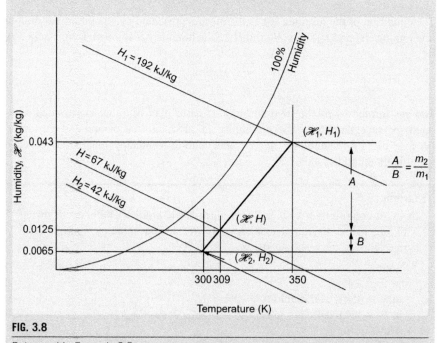

FIG. 3.8

Data used in Example 3.5.

Addition of liquid or vapour to a gas

If a mass m_3 of liquid or vapour of enthalpy H_3 is added to a gas of humidity H_1 and enthalpy H_1 and containing a mass m_1 of dry gas, then

$$m_1(H - H_1) = m_3 \tag{3.28}$$

$$m_1(H - H_1) = m_3 H_3 \tag{3.29}$$

Thus,

$$\frac{H - H_1}{H - H_1} = H_3 \tag{3.30}$$

where H and H are the humidity and enthalpy of the gas produced on mixing.

The composition and properties of the mixed stream are therefore represented by a point on the straight line of slope H_3, relative to the humidity axis, which passes through the point (H_1 and H_1). In Fig. 3.5, the edges of the plot are marked with points that, when joined to the origin, give a straight line of the slope indicated. Thus, in using the chart, a line of slope H_3 is drawn through the origin and a parallel line drawn through the point (H_1 and H_1). The point representing the final gas stream is then given from Eq. (3.28):

$$(H - H_1) = \frac{m_3}{m_1}$$

It can be seen from Fig. 3.5 that for the air–water system, a straight line, of slope equal to the enthalpy of dry saturated steam (2675 kJ/kg), is almost parallel to the isothermals, so that the addition of live steam has only a small effect on the temperature of the gas. The addition of water spray, even if the water is considerably above the temperature of the gas, results in a lowering of the temperature after the water has evaporated. This arises because the latent heat of vaporisation of the liquid constitutes the major part of the enthalpy of the vapour. Thus, when steam is added, it gives up a small amount of sensible heat to the gas, whereas when hot liquid is added, a small amount of sensible heat is given up, and a very much larger amount of latent heat is absorbed from the gas.

Example 3.6

0.15 kg/s steam at atmospheric pressure and superheated to 400 K is bled into an airstream at 320 K and 20% relative humidity. What is the temperature, enthalpy, and relative humidity of the mixed stream if the air is flowing at 5 kg/s? How much steam would be required to provide an exit temperature of 330 K and what would be the humidity of this mixture?

Solution

Steam at atmospheric pressure is saturated at 373 K at which the latent heat

$$= 2258 \text{kJ/kg}$$

Taking the specific heat of superheated steam as 2.0 kJ/kg K,

Enthalpy of the steam : $H_3 = 4.18(373 - 273) + 2258 + 2.0(400 - 273)$
$$= 2730 \text{ kJ/kg}$$

From Fig. 3.5,
at $\theta_1 = 320\,\text{K}$ and 20% relative humidity, $H_1 = 0.013\,\text{kg/kg}$ and $H_1 = 83\,\text{kJ/kg}$.
The line joining the axis and slope $H_3 = 2730\,\text{kJ/kg}$ at the edge of the chart
is now drawn in, and a parallel line is drawn through (H_1, H_1).
Thus,

$$(H - H_1) = \frac{m_3}{m_1} = \left(\frac{0.15}{5}\right) = 0.03 \text{ kg/kg}$$

and

$$H = (0.03 + 0.013) = 0.043 \text{ kg/kg}$$

At the intersection of $H = 0.043\,\text{kg/kg}$ and the line through $(H_1$ and $H_1)$,

$$\underline{H = 165 \text{ kJ/kg and } \theta = 324 \text{ K}}$$

When $\theta = 330\,\text{K}$, the intersection of this isotherm and the line through
H gives an outlet stream in which $H = 0.094\,\text{kg/kg}$ (83% relative humidity)
and $H = 300\,\text{kJ/kg}$.
Thus, in Eq. (3.28),

$$m_3 = 5(0.094 - 0.013) = \underline{\underline{0.41 \text{ kg/s}}}$$

The data used in this example are shown in Fig. 3.9.

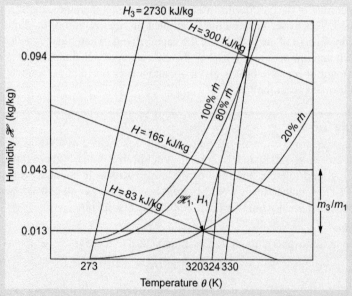

FIG. 3.9

Data used in Example 3.6 (schematic).

3.4 **Determination of humidity**

The most important methods for determining humidity are as follows:

(1) *Chemical methods.* A known volume of the gas is passed over a suitable absorbent, the increase in mass of which is measured. The efficiency of the process can be checked by arranging a number of vessels containing absorbent in series and ascertaining that the increase in mass in the last of these is negligible. The method is very accurate but is laborious. Satisfactory absorbents for water vapour are phosphorus pentoxide dispersed in pumice and concentrated sulphuric acid.

(2) *Determination of the wet-bulb temperature.* Eq. (3.8) gives the humidity of a gas in terms of its temperature, its wet-bulb temperature, and various physical properties of the gas and vapour. The wet-bulb temperature is normally determined as the temperature attained by the bulb of a thermometer, which is covered with a piece of material that is maintained saturated with the liquid. The gas should be passed over the surface of the wet bulb at a high enough velocity (>5 m/s) (a) for the condition of the gas stream not to be affected appreciably by the evaporation of liquid, (b) for the heat transfer by convection to be large compared with that by radiation and conduction from the surroundings, and (c) for the ratio of the coefficients of heat and mass transfer to have reached a constant value. The gas should be passed long enough for equilibrium to be attained, and for accurate work, the liquid should be cooled nearly to the wet-bulb temperature before it is applied to the material.

   The stream of gas over the liquid surface may be produced by a small fan or other similar means (Fig. 3.10A). The crude forms of wet-bulb thermometer, which make no provision for the rapid passage of gas, cannot be used for accurate determinations of humidity.

(3) *Determination of the dew point.* The dew point is determined by cooling a highly polished surface in the gas and observing the highest temperature at which condensation takes place (Fig. 3.10B). The humidity of the gas is equal to the humidity of saturated gas at the dew point. The instrument illustrated in Fig. 3.10C incorporates a polished gold mirror that is cooled using a thermoelectric module that utilises the *Peltier effect.*

(4) *Measurement of the change in length of a hair or fibre.* The length of a hair or fibre is influenced by the humidity of the surrounding atmosphere. Many forms of apparatus for automatic recording of humidity depend on this property. The method has the disadvantage that the apparatus needs frequent calibration because the zero tends to shift. This difficulty is most serious when the instrument is used over a wide range of humidities. A typical hair hygrometer is shown in Fig. 3.10D.

(5) *Measurement of conductivity of a fibre.* If a fibre is impregnated with an electrolyte, such as lithium chloride, its electrical resistance will be governed by its moisture content, which in turn depends on the humidity of the atmosphere in which it is situated.

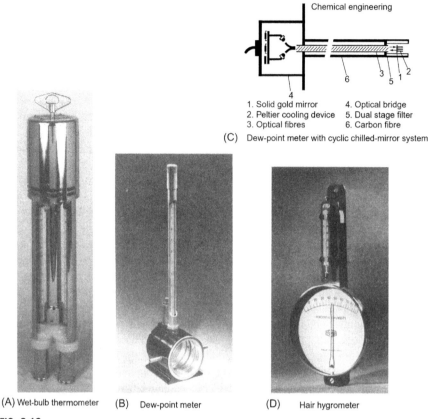

Chemical engineering

1. Solid gold mirror
2. Peltier cooling device
3. Optical fibres
4. Optical bridge
5. Dual stage filter
6. Carbon fibre

(C) Dew-point meter with cyclic chilled-mirror system

(A) Wet-bulb thermometer   (B)   Dew-point meter   (D)   Hair hygrometer

FIG. 3.10

Hygrometers.

In a lithium chloride cell, a skein of very fine fibres is wound on a plastic frame carrying the electrodes, and the current flowing at a constant applied voltage gives a direct measure of the relative humidity.

(6) *Measurement of heat of absorption on to a surface.*

(7) *Electrolytic hygrometry* in which the quantity of electricity required to electrolyse water absorbed from the atmosphere on to a thin film of desiccant is measured.

(8) *Piezoelectric hygrometry* employing a quartz crystal with a hygroscopic coating in which moisture is alternately absorbed from a wet gas and desorbed in a dry gas stream, the dynamics is a function of the gas humidity.

(9) *Capacitance meters* in which the electric capacitance is a function of the degree of deposition of moisture from the atmosphere.

(10) *Observation of colour changes* in active ingredients, such as cobaltous chloride.

Further details of instruments for the measurement of humidity are given in Volume 3. Reference should also be made to standard works on psychrometry [4–6].

3.5 **Humidification and dehumidification**
3.5.1 **Methods of increasing humidity**
The following methods may be used for increasing the humidity of a gas:

1. Live steam may be added directly in the required quantity. It has been shown that this produces only a slight increase in the temperature, but the method is not generally favoured because any impurities that are present in the steam may be added at the same time.
2. Water may be sprayed into the gas at such a rate that, on complete vaporisation, it gives the required humidity. In this case, the temperature of the gas will fall as the latent heat of vaporisation must be supplied from the sensible heat of the gas and liquid.
3. The gas may be mixed with a stream of gas of higher humidity. This method is frequently used in laboratory work when the humidity of a gas supplied to an apparatus is controlled by varying the proportions in which two gas streams are mixed.
4. The gas may be brought into contact with water in such a way that only part of the liquid is evaporated. This is perhaps the most common method and will now be considered in more detail.

In order to obtain a high rate of humidification, the area of contact between the air and the water is made as large as possible by supplying the water in the form of a fine spray; alternatively, the interfacial area is increased by using a packed column. Evaporation occurs if the humidity at the surface is greater than that in the bulk of the air, that is, if the temperature of the water is above the dew point of the air.

When humidification is carried out in a packed column, the water that is not evaporated can be recirculated so as to reduce the requirements of fresh water. As a result of continued recirculation, the temperature of the water will approach the adiabatic saturation temperature of the air, and the air leaving the column will be cooled—in some cases to within 1 K of the temperature of the water. If the temperature of the air is to be maintained constant, or raised, the water must be heated.

Two methods of changing the humidity and temperature of a gas from A (θ_1 and H_1) to B (θ_2 and H_2) may be traced on the humidity chart as shown in Fig. 3.11. The first method consists of saturating the air by water artificially maintained at the dew point of air of humidity H_2 (line AC) and then heating at constant humidity to θ_2 (line CB). In the second method, the air is heated (line AD) so that its adiabatic saturation temperature corresponds with the dew point of air of humidity H_2. It is then saturated by water at the adiabatic saturation temperature (line DC) and heated at constant humidity to θ_2 (line CB). In this second method, an additional operation—the

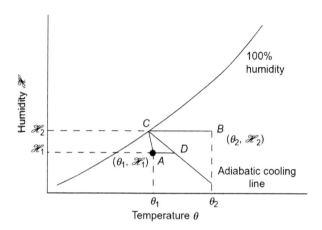

FIG. 3.11

Two methods of changing conditions of gas from (θ_1 and H_1) to (θ_2 and H_2).

preliminary heating—is carried out on the air, but the water temperature automatically adjusts itself to the required value.

Since complete humidification is not always attained, an allowance must be made when designing air humidification cycles. For example, if only 95% saturation is attained, the adiabatic cooling line should be followed only to the point corresponding to that degree of saturation, and therefore, the gas must be heated to a slightly higher temperature before adiabatic cooling is commenced.

Example 3.7

Air at 300 K and 20% humidity is to be heated in two stages with intermediate saturation with water to 90% humidity so that the final stream is at 320 K and 20% humidity. What is the humidity of the exit stream and the conditions at the end of each stage?

Solution

At $\theta_1 = 300$ K and 20% humidity: $H_1 = 0.0045$ kg/kg, from Fig. 13.4, and

at $\theta_2 = 320$ K and 20% humidity: $H_2 = \underline{0.0140}$ kg/kg

When $H_2 = 0.0140$ kg/kg, air is saturated at 292 K and has a humidity of 90% at 293 K.

The adiabatic cooling line corresponding to 293 K intersects with $H = 0.0045$ kg/kg at a temperature $\theta = 318$ K.

Thus, the stages are.

(i) heat the air at $H = 0.0045$ kg/kg from 300 to 318 K,

(ii) saturate with water at an adiabatic saturation temperature of 293 K until 90% humidity is attained. At the end of this stage,

$$H = 0.0140\,\text{kg/kg and}\,\theta = 294.5\,\text{K}$$

(iii) heat the saturated air at $\underline{H = 0.0140\,\text{kg/kg from 294.5 to 320K}}$

3.5.2 Dehumidification

Dehumidification of air can be effected by bringing it into contact with a cold surface, either liquid or solid. If the temperature of the surface is lower than the dew point of the gas, condensation takes place, and the temperature of the gas falls. The temperature of the surface tends to rise because of the transfer of latent and sensible heat from the air. It would be expected that the air would cool at constant humidity until the dew point was reached, and that subsequent cooling would be accompanied by condensation. It is found in practice that this occurs only when the air is well mixed. Normally, the temperature and humidity are reduced simultaneously throughout the whole of the process. The air in contact with the surface is cooled below its dew point, and condensation of vapour therefore occurs before the more distant air has time to cool. Where the gas stream is cooled by cold water, countercurrent flow should be employed because the temperature of the water and air are changing in opposite directions.

The humidity can be reduced by compressing air, allowing it to cool again to its original temperature, and draining off the water that has condensed. During compression, the partial pressure of the vapour is increased, and condensation takes place as soon as it reaches the saturation value. Thus, if air is compressed to a high pressure, it becomes saturated with vapour, but the partial pressure is a small proportion of the total pressure. Compressed air from a cylinder therefore has a low humidity. Gas is frequently compressed before it is circulated so as to prevent condensation in the mains.

Many large air-conditioning plants incorporate automatic control of the humidity and temperature of the issuing air. Temperature control is affected with the aid of a thermocouple or resistance thermometer and humidity control by means of a thermocouple recording the difference between the wet- and dry-bulb temperatures.

3.6 Water cooling
3.6.1 Cooling towers

Cooling of water can be carried out on a small scale either by allowing it to stand in an open pond or by the spray pond technique in which it is dispersed in spray form and then collected in a large, open pond. Cooling takes place both by the transference of sensible heat and by evaporative cooling as a result of which sensible heat in the water provides the latent heat of vaporisation. On the large scale, air and water are brought into countercurrent contact in a cooling tower that may employ either natural draught or mechanical draught. The water flows down over a series of wooden slats that give a large interfacial area and promote turbulence in the liquid. The air is humidified and heated as it rises, whilst the water is cooled mainly by evaporation.

The natural-draught cooling tower depends on the chimney effect produced by the presence in the tower of air and vapour of higher temperature and therefore of lower density than the surrounding atmosphere. Thus, atmospheric conditions and the temperature and quantity of the water will exert a very important effect on the operation of the tower. Not only will these factors influence the quantity of air drawn through the tower, but also they will affect the velocities and flow patterns and hence the transfer coefficients between gas and liquid. One of the prime considerations in design therefore is to construct a tower in such a way that the resistance to airflow is low. Hence, the packings and distributors must be arranged in open formation. The draught of a cooling tower at full load is usually only about $50\,\text{N/m}^2$ [7] and the air velocity in the region of 1.2–1.5 m/s, so that under the atmospheric conditions prevailing in the United Kingdom, the air usually leaves the tower in a saturated condition. The density of the airstream at outlet is therefore determined by its temperature. Calculation of conditions within the tower is carried out in the manner described in the following pages. It is, however, necessary to work with a number of assumed airflow rates and to select the one which fits both the transfer conditions and the relationship between air rate and pressure difference in the tower.

The *natural-draught cooling tower* consists of an empty shell, constructed either of timber or ferroconcrete, where the upper portion is empty and merely serves to increase the draught. The lower portion, amounting to about 10%–12% of the total height, is usually fitted with grids on to which the water is fed by means of distributors or sprays as shown in Fig. 3.12. The shells of cooling towers are now generally constructed in ferroconcrete in a shape corresponding approximately to a hyperboloid of revolution. The shape is chosen mainly for constructional reasons, but it does take account of the fact that the entering air will have a radial velocity component; the increase in cross-section towards the top causes a reduction in the outlet velocity, and there is a small recovery of kinetic energy into pressure energy.

The *mechanical-draught cooling tower* may employ forced draught with the fan at the bottom, or induced draught with the fan driving the moist air out at the top. The air velocity can be increased appreciably above that in the natural-draught tower, and a greater depth of packing can be used. The tower will extend only to the top of the packing unless atmospheric conditions are such that a chimney must be provided in order to prevent recirculation of the moist air. The danger of recirculation is considerably less with the induced-draught type because the air is expelled with a higher velocity. Mechanical-draught towers are generally confined to small installations and to conditions where the water must be cooled to as low a temperature as possible. In some cases, it is possible to cool the water to within 1 K of the wet-bulb temperature of the air. Although the initial cost of the tower is less, maintenance and operating costs are of course higher than in natural-draught towers that are now used for

FIG. 3.12

Water-cooling tower. View of spray distribution system.

all large installations. A typical steel-framed mechanical-draught cooling tower is shown in Fig. 3.13.

The operation of the conventional water-cooling tower is often characterised by the discharge of a plume consisting of a suspension of minute droplets of water in air. This is formed when the hot humid air issuing from the top of the tower mixes with the ambient atmosphere, and precipitation takes place as described earlier (Section 3.3.2). In the *hybrid* (or wet/dry) cooling tower [8], mist formation is avoided by cooling *part* of the water in a finned-tube exchanger bundle that thus generates a supply of warm dry air that is then blended with the air issuing from the evaporative section. By adjusting the proportion of the water fed to the heat exchanger, the plume can be completely eliminated.

In the cooling tower, the temperature of the liquid falls and the temperature and humidity of the air rise, and its action is thus similar to that of an air humidifier. The limiting temperature to which the water can be cooled is the wet-bulb temperature corresponding to the condition of the air at inlet. The enthalpy of the airstream does not remain constant since the temperature of the liquid changes rapidly in the upper portion of the tower. Towards the bottom, however, the temperature of the liquid changes less rapidly because the temperature differences are smaller. At the top of the tower, the temperature falls from the bulk of the liquid to the interface and then again from the interface to the bulk of the gas. Thus, the liquid is cooled by

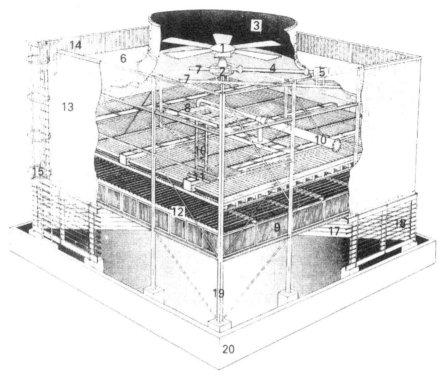

FIG. 3.13

Visco 2000 series steel-framed, mechanical-draught, water-cooling tower. (1) fan assembly; (2) gearbox; (3) fan stack; (4) drive shaft assembly; (5) motor; (6) fan deck; (7) mechanical equipment supports; (8) drift eliminators (PVC or timber–timber shown); (9) cooling tower packing (plastic plate or wooden lath); (10) inlet water distribution pipe; (11) open-type distribution system; (12) timber laths for even water distribution; (13) cladding; (14) cladding extended to form handrail; (15) access ladder; (16) internal access ladder to distribution system and drift eliminators; (17) diagonal wind baffles; (18) air inlet louvres; (19) steel structures with horizontal and diagonal ties; (20) cold water sump. *Some structural members have been omitted for clarity.*

transfer of sensible heat and by evaporation at the surface. At the bottom of a tall tower, however, the temperature gradient in the liquid is in the same direction, though smaller, but the temperature gradient in the gas is in the opposite direction. Transfer of sensible heat to the interface therefore takes place from the bulk of the liquid and from the bulk of the gas, and all the cooling is caused by the evaporation at the interface. In most cases, about 80% of the heat loss from the water is accounted for by evaporative cooling.

3.6.2 Design of natural-draught towers

The airflow through a natural-draught or hyperbolic-type tower (Fig. 3.14) is due largely to the difference in density between the warm air in the tower and the external ambient air; thus, a draught is created in the stack by a chimney effect that eliminates the need for mechanical fans. It has been noted by McKelvey and Brooke [9] that natural-draught towers commonly operate at a pressure difference of some 50 N/m^2 under full load, and above the packing, the mean air velocity is typically 1–2 m/s. The performance of a natural-draught tower differs from that of a mechanical-draught installation in that the cooling achieved depends upon the relative humidity and the wet-bulb temperature. It is important therefore, at the design stage, to determine correctly and to specify the density of the inlet and exit airstreams in addition to the usual tower design conditions of water temperature range, how closely the water temperature should approach the wet-bulb temperature of the air, and the quantity of water to be handled. Because the performance depends to a large extent on atmospheric humidity, the outlet water temperature is difficult to control with natural-draught towers.

In the design of natural-draught towers, a ratio of height-to-base diameter of 3:2 is normally used, and a design method has been proposed by Chilton [10]. Chilton has shown that the duty coefficient D of a tower is approximately constant over the normal range of operation and is related to tower size by an efficiency factor or performance coefficient C_t given by

$$D_t = \frac{19.50 A_b z_t^{0.5}}{C_t^{1.5}} \tag{3.31}$$

where for water loadings in excess of $1 \, \text{kg/m}^2 \, \text{s}$, C_t is usually about 5.2 though lower values are obtained with new packings that are being developed.

FIG. 3.14

Natural-draught water-cooling towers.

The duty coefficient is given by the following equation (in which SI units must be used as it is not dimensionally consistent):

$$\frac{W_L}{D_t} = 0.00369 \frac{\Delta H'}{\Delta T} (\Delta T' + 0.0752 \Delta H')^{0.5} \tag{3.32}$$

where W_L (kg/s) is the water load in the tower, $\Delta H'$ (kJ/kg) the change in enthalpy of the air passing through the tower, $\Delta T'$ (°K) the change in water temperature in passing through the tower, and $\Delta T'$ (°K) the difference between the temperature of the air leaving the packing and the dry-bulb temperature of the inlet air. The air leaving the packing inside the tower is assumed to be saturated at the mean of the inlet and outlet water temperatures. Any divergence between theory and practice of a few degrees in this respect does not significantly affect the results as the draught component depends on the ratio of the change of density to change in enthalpy and not on change in temperature alone [11]. The use of Eqs (3.31), (3.32) is illustrated in the following example.

Example 3.8
What are the diameter and height of a hyperbolic natural-draught cooling tower handling 4810 kg/s of water with the following temperature conditions:

$$\text{Water entering the tower} = 301\text{K}$$
$$\text{Water leaving the tower} = 294\text{K}$$
$$\text{Air : dry bulb} = 287\text{K}$$
$$\text{Wet bulb} = 284\text{K}$$

Solution
Temperature range for the water, $\Delta T = (301-294) = 7\,\text{K}$.
At a mean water temperature of $0.5(301+294) = 297.5\,\text{K}$, the enthalpy $= 92.6\,\text{kJ/kg}$.
At a dry-bulb temperature of 287 K, the enthalpy $= 49.5\,\text{kJ/kg}$.

$$\therefore \Delta T' = (297.5 - 287) = 10.5°\text{K}$$

and

$$\Delta H' = (92.6 - 49.5) = 43.1\text{kJ/kg}$$

In Eq. (3.32),

$$\frac{4810}{D_t} = 0.00369 \left(\frac{43.1}{7}\right) [10.5 + (0.0752 \times 43.1)]^{0.5}$$

and

$$D_t = 57,110$$

Taking C_t as 5.0 and assuming as a first estimate a tower height of 100 m, then in Eq. (3.31),

$$57,110 = 19.50 A_b \frac{100^{0.5}}{5.0^{1.5}}$$

and

$$A_b = 3274 \text{m}^2$$

Thus, the internal diameter of the column at sill level $= \left(\frac{3274 \times 4}{\pi}\right)^{0.5}$

$$= \underline{64.6 \text{m}}$$

Since this gives a height/diameter ratio of $(100{:}64.6) \approx 3{:}2$, the design is acceptable.

3.6.3 Height of packing for both natural and mechanical draught towers

The height of a water-cooling tower can be determined [12] by setting up a material balance on the water, an enthalpy balance, and rate equations for the transfer of heat in the liquid and gas and for mass transfer in the gas phase. There is no concentration gradient in the liquid, and therefore, there is no resistance to mass transfer in the liquid phase.

Considering the countercurrent flow of water and air in a tower of height z (Fig. 3.15), the mass rate of flow of air per unit cross-section G' is constant throughout the whole height of the tower, and because only a small proportion of the total supply of water is normally evaporated (1%–5%), the liquid rate per unit area L' can

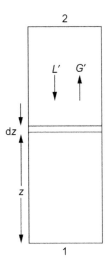

FIG. 3.15

Flow in water-cooling tower.

be taken as constant. The temperature, enthalpy, and humidity will be denoted by the symbols θ, H, and H, respectively; suffixes G, L, 1, 2, and f are being used to denote conditions in the gas and liquid, at the bottom and top of the column, and of the air in contact with the water.

The five basic equations for an incremental height of column, dz., are the following:

(1) Water balance:

$$dL' = G'dH \tag{3.33}$$

(2) Enthalpy balance:

$$G'dH_G = L'dH_L \tag{3.34}$$

since only a small proportion of the liquid is evaporated.

Now,

$$H_G = s(\theta_G - \theta_0) + \lambda H \tag{3.35}$$

and

$$H_L = C_L(\theta_L - \theta_0) \tag{3.36}$$

Thus,

$$G'dH_G = L'C_L d\theta_L \tag{3.37}$$

and

$$dH_G = sd\theta_G + \lambda dH \tag{3.38}$$

Integration of this expression over the whole height of the column, on the assumption that the physical properties of the materials do not change appreciably, gives

$$G'(H_{G2} - H_{G1}) = L'C_L(\theta_{L2} - \theta_{L1}) \tag{3.39}$$

(3) Heat transfer from the body of the liquid to the interface:

$$h_L adz(\theta_L - \theta_f) = L'C_L d\theta_L \tag{3.40}$$

where h_L is the heat-transfer coefficient in the liquid phase and a is the interfacial area per unit volume of column. It will be assumed that the area for heat transfer is equal to that available for mass transfer, though it may be somewhat greater if the packing is not completely wetted.

Rearranging Eq. (3.40):

$$\frac{d\theta_L}{(\theta_L - \theta_f)} = \frac{h_L a}{L'C_L} dz \tag{3.41}$$

(4) Heat transfer from the interface to the bulk of the gas:

$$h_G adz(\theta_f - \theta_G) = G'sd\theta_G \tag{3.42}$$

where h_G is the heat-transfer coefficient in the gas phase.

Rearranging,

$$\frac{d\theta_G}{(\theta_f - \theta_G)} = \frac{h_G a}{G' s} dz \tag{3.43}$$

(5) Mass transfer from the interface to the gas:

$$h_D \rho a dz (H_f - H) = G' dH \tag{3.44}$$

where h_D is the mass-transfer coefficient for the gas and ρ is the mean density of the air (see Eq. 3.6).

Rearranging,

$$\frac{dH}{H_f - H} = \frac{h_D a \rho}{G'} dz \tag{3.45}$$

These equations cannot be integrated directly since the conditions at the interface are not necessarily constant nor can they be expressed directly in terms of the corresponding property in the bulk of the gas or liquid.

If the Lewis relation (Eq. 3.11) is applied, it is possible to obtain workable equations in terms of enthalpy instead of temperature and humidity. Thus, writing h_G as $h_D \rho s$, from Eq. (3.42),

$$G' s d\theta_G = h_D \rho a dz (s\theta_f - s\theta_G) \tag{3.46}$$

and from Eq. (3.44),

$$G' \lambda dH = h_D \rho a dz (\lambda H_f - \lambda H) \tag{3.47}$$

Adding these two equations gives

$$G'(s d\theta_G + \lambda dH) = h_D \rho a dz [(s\theta_f + \lambda H_f) - (s\theta_G + \lambda H)]$$
$$G' dH_G = h_D \rho a dz (H_f - H_G) \quad \text{(from Eq.13.35)} \tag{3.48}$$

or

$$\frac{dH_G}{(H_f - H_G)} = \frac{h_D a \rho}{G'} dz \tag{3.49}$$

The use of an enthalpy driving force, as in Eq. (3.48), was first suggested by Merkel [13], and the following development of the treatment was proposed by Mickley [12].

Combining Eqs (3.37), (3.40), and (3.48) gives

$$\frac{(H_G - H_f)}{(\theta_L - \theta_f)} = -\frac{h_L}{h_D \rho} \tag{3.50}$$

From Eqs (3.46), (3.48),

$$\frac{(H_G - H_f)}{(\theta_G - \theta_f)} = \frac{dH_G}{d\theta_G} \tag{3.51}$$

and from Eqs (3.46), (3.44),

$$\frac{(H - H_f)}{\theta_G - \theta_f} = \frac{dH}{d\theta_G} \tag{3.52}$$

These equations are now employed in the determination of the required height of a cooling tower for a given duty. The method consists of the graphical evaluation of the relation between the enthalpy of the body of gas and the enthalpy of the gas at the interface with the liquid. The required height of the tower is then obtained by integration of Eq. (3.49).

It is supposed that water is to be cooled at a mass rate L' per unit area from a temperature θ_{L2} to θ_{L1}. The air will be assumed to have a temperature θ_{G1}, a humidity H_1, and an enthalpy H_{G1} (which can be calculated from the temperature and humidity), at the inlet point at the bottom of the tower, and its mass flow per unit area will be taken as G'. The change in the condition of the liquid and gas phases will now be followed on an enthalpy–temperature diagram (Fig. 3.16). The enthalpy–temperature curve PQ for saturated air is plotted either using calculated data or from the humidity chart (Fig. 3.4). The region below this line relates to unsaturated air and the region above it to supersaturated air. If it is assumed that the air in contact with the liquid surface is saturated with water vapour, this curve represents the relation between air enthalpy H_f and temperature θ_f at the interface.

The curve connecting air enthalpy and water temperature is now drawn using Eq. (3.39). This is known as the operating line and is a straight line of slope $(L'C_L/G')$, passing through the points $A(\theta_{L1}$ and $H_{G1})$ and $B(\theta_{L2}$ and $H_{G2})$. Since $(\theta_{L1}$ and $H_{G1})$ are specified, the procedure is to draw a line through $(\theta_{L1}$ and $H_{G1})$ of slope $(L'C_L/G')$ and to produce it to a point whose abscissa is equal to θ_{L2}. This point B then corresponds to conditions at the top of the tower, and the ordinate gives the enthalpy of the air leaving the column.

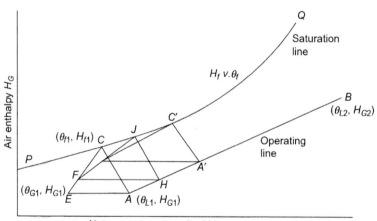

FIG. 3.16

Construction for determining the height of water-cooling tower.

Eq. (3.50) gives the relation between liquid temperature, air enthalpy, and conditions at the interface, for any position in the tower, and is represented by a family of straight lines of slope $-(h_L/h_D\rho)$. The line for the bottom of the column passes through the point $A(\theta_{L1}, H_{G1})$ and cuts the enthalpy–temperature curve for saturated air at the point C, representing conditions at the interface. The difference in ordinates of points A and C is the difference in the enthalpy of the air at the interface and that of the bulk air at the bottom of the column.

Similarly, line $A'C'$, parallel to AC, enables the difference in the enthalpies of the bulk air and the air at the interface to be determined at some other point in the column. The procedure can be repeated for a number of points and the value of (H_f-H_G) obtained as a function of H_G for the whole tower.

Now,

$$\frac{dH_G}{(H_f - H_G)} = \frac{h_D a\rho}{G'}\,dz \qquad (13.49)$$

On integration,

$$z = \int_1^2 dz = \frac{G'}{h_D a\rho}\int_1^2 \frac{dH_G}{(H_f - H_G)} \qquad (3.53)$$

assuming h_D to remain approximately constant.

Since $(H_f - H_G)$ is now known as a function of H_G, $1/(H_f-H_G)$ can be plotted against H_G and the integral evaluated between the required limits. The height of the tower is thus determined.

The integral in Eq. (3.53) cannot be evaluated by taking a logarithmic mean driving force because the saturation line PQ is far from linear. Carey and Williamson [14] have given a useful approximate method of evaluating the integral. They assume that the enthalpy difference $(H_f-H_G)=\Delta H$ varies in a parabolic manner. The three fixed points taken to define the parabola are at the bottom and top of the column (ΔH_1 and ΔH_2, respectively) and ΔH_m, the value at the mean water temperature in the column. The effective mean driving force is $f\,\Delta H_m$, where f is a factor for converting the driving force at the mean water temperature to the effective value. In Fig. 3.17, $(\Delta H_m/\Delta H_1)$ is plotted against $(\Delta H_m/\Delta H_2)$, and contours representing constant values of f are included.

Using the mean driving force, integration of Eq. (3.53) gives

$$\frac{(H_{G2} - H_{G1})}{f\Delta H_m} = \frac{h_D a\rho}{G'}\,z \qquad (3.54)$$

or

$$z = \frac{G'}{h_D a\rho}\frac{(H_{G2} - H_{G1})}{f\Delta H_m}$$

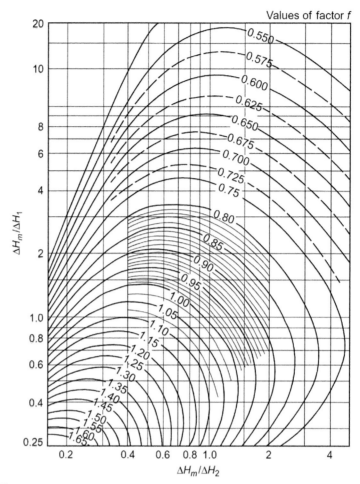

FIG. 3.17

Correction factor f for obtaining the mean effective driving force in column.

3.6.4 Change in air condition

The change in the humidity and temperature of the air is now obtained. The enthalpy and temperature of the air are known only at the bottom of the tower, where fresh air is admitted. Here, the condition of the air may be represented by a point E with coordinates (H_{G1} and θG_1). Thus, the line AE (Fig. 3.16) is parallel to the temperature axis.

Since

$$\frac{H_G - H_f}{\theta_G - \theta_f} = \frac{dH_G}{d\theta_G} \qquad (3.51)$$

the slope of the line EC is $(\mathrm{d}H_G/\mathrm{d}\theta_G)$ and represents the rate of change of air enthalpy with air temperature at the bottom of the column. If the gradient $(\mathrm{d}H_G/\mathrm{d}\theta_G)$ is taken as constant over a small section, the point F, on EC, will represent the condition of the gas at a small distance from the bottom. The corresponding liquid temperature is found by drawing through F a line parallel to the temperature axis. This cuts the operating line at some point H, which indicates the liquid temperature. The corresponding value of the temperature and enthalpy of the gas at the interface is then obtained by drawing a line through H, parallel to AC. This line then cuts the curve for saturated air at a point J, which represents the conditions of the gas at the interface. The rate of change of enthalpy with temperature for the gas is then given by the slope of the line FJ. Again, this slope can be considered to remain constant over a small height of the column, and the condition of the gas is thus determined for the next point in the tower. The procedure is then repeated until the curve representing the condition of the gas has been extended to a point whose ordinate is equal to the enthalpy of the gas at the top of the column. This point is obtained by drawing a straight line through B, parallel to the temperature axis. The final point on the line then represents the condition of the air that leaves the top of the water-cooling tower.

The size of the individual increments of height that are considered must be decided for the particular problem under consideration and will depend, primarily, on the rate of change of the gradient $(\mathrm{d}H_G/\mathrm{d}\theta_G)$. It should be noted that, for the gas to remain unsaturated throughout the whole of the tower, the line representing the condition of the gas must be below the curve for saturated gas. If at any point in the column, the air has become saturated, it is liable to become supersaturated as it passes further up the column and comes into contact with hotter liquid. It is difficult to define precisely what happens beyond this point as partial condensation may occur, giving rise to a mist. Under these conditions, the preceding equations will no longer be applicable. However, an approximate solution is obtained by assuming that once the airstream becomes saturated, it remains so during its subsequent contact with the water through the column.

3.6.5 Temperature and humidity gradients in a water cooling tower

In a water-cooling tower, the temperature profiles depend on whether the air is cooler or hotter than the surface of the water. Near the top, hot water makes contact with the exit air that is at a lower temperature, and sensible heat is therefore transferred both from the water to the interface and from the interface to the air. The air in contact with the water is saturated at the interface temperature, and humidity therefore falls from the interface to the air. Evaporation followed by mass transfer of water vapour therefore takes place, and latent heat is carried away from the interface in the vapour. The sensible heat removed from the water is then equal to the sum of the latent and sensible heats transferred to the air. Temperature and humidity gradients are then as shown in Fig. 3.18A.

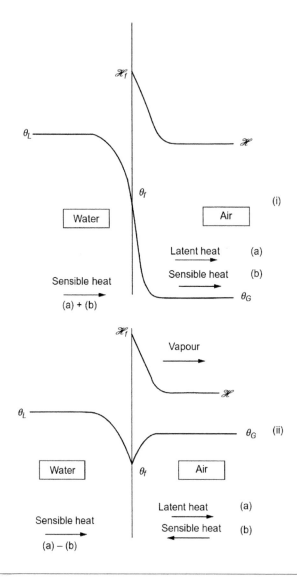

FIG. 3.18

Temperature and humidity gradients in a water-cooling tower, (i) upper sections of tower and (ii) bottom of tower.

If the tower is sufficiently tall, the interface temperature can fall below the dry-bulb temperature of the air (but not below its wet-bulb temperature), and sensible heat will then be transferred from both the air and the water to the interface. The corresponding temperature and humidity profiles are given in Fig. 3.18B. In this part of the tower, therefore, the sensible heat removed from the water will be that transferred as latent heat *less* the sensible heat transferred from the air.

3.6.6 Evaluation of heat and mass transfer coefficients

In general, coefficients of heat and mass transfer in the gas phase and the heat-transfer coefficient for the liquid phase are not known. They may be determined, however, by carrying out tests in the laboratory or pilot scale using the same packing. If, for the air–water system, a small column is operated at steady water and air rates and the temperature of the water at the top and bottom and the initial and final temperatures and humidities of the airstream are noted, the operating line for the system is obtained. Assuming a value of the ratio $-(h_L/h_D\rho)$, for the slope of the tie-line AC, the graphical construction is carried out similar to the construction discussed in Fig. 3.16, starting with the conditions at the bottom of the tower. The condition of the gas at the top of the tower is thus calculated and compared with the measured value. If the difference is significant, another value of $-(h_L/h_D\rho)$ is assumed and the procedure repeated. Now that the slope of the tie line is known, the value of the integral of $dH_G/(H_f - H_G)$ over the whole column can be calculated. Since the height of the column is known, the product $h_D a$ is found by solution of Eq. (3.49). $h_G a$ may then be calculated using the Lewis relation. The values of the three transfer coefficients are therefore obtained at any given flow rates from a single experimental run. The effect of liquid and gas rate may be found by means of a series of similar experiments.

Several workers have measured heat- and mass-transfer coefficients in water-cooling towers and in humidifying towers. Thomas and Houston [15], using a tower 2 m high and 0.3 m square in cross-section, fitted with wooden slats, give the following equations for heat- and mass-transfer coefficients for packed heights >75 mm:

$$h_G a = 3.0 L'^{0.26} G'^{0.72} \tag{3.55}$$

$$h_L a = 1.04 \times 10^4 L'^{0.51} G'^{1.00} \tag{3.56}$$

$$h_D a = 2.95 L'^{0.26} G'^{0.72} \tag{3.57}$$

In these equations, L' and G' are expressed in kg/m^2, s in J/kg K, $h_G a$ and $h_L a$ in W/m^3 K, and $h_D a$ in s^{-1}. A comparison of the gas- and liquid-film coefficients may then be made for a number of gas and liquid rates. Taking the humid heat s as 1.17×10^3 J/kgK,

	$L' = G' = 0.5$ kg/m^2 s	$L' = G' = 1.0$ kg/m^2 s	$L' = G' = 2.0$ kg/m^2 s
hGa	1780	3510	6915
hLa	3650	10, 400	29, 600
hLa/hGa	2.05	2.96	4.28

Cribb [16] quotes values of the ratio h_L/h_G ranging from 2.4 to 8.5.

It is seen that the liquid-film coefficient is generally considerably higher than the gas-film coefficient but that it is not always safe to ignore the resistance to transfer in the liquid phase.

Lowe and Christie [17] used a 1.3 m square experimental column fitted with a number of different types of packing and measured heat- and mass-transfer coefficients and pressure drops.

They showed that in most cases,

$$h_D a \alpha L'^{1-n} G'^n \tag{3.58}$$

The index n was found to vary from about 0.4 to 0.8 according to the type of packing. It will be noted that when $n \approx 0.75$, there is close agreement with the results given by Eq. (3.57).

The heat-transfer coefficient for the liquid is often large compared with that for the gas phase. As a first approximation, therefore, it can be assumed that the whole of the resistance to heat transfer lies within the gas phase and that the temperature at the water–air interface is equal to the temperature of the bulk of the liquid. Thus, everywhere in the tower, $\theta_f - \theta_L$. This simplifies the calculations, since the lines AC, HJ, and so on have a slope of $-\infty$, that is, they become parallel to the enthalpy axis.

Some workers have attempted to base the design of humidifiers on the overall heat-transfer coefficient between the liquid and gas phases. This treatment is not satisfactory since the quantities of heat transferred through the liquid and through the gas are not the same, as some of the heat is utilised in effecting evaporation at the interface. In fact, at the bottom of a tall tower, the transfer of heat in both the liquid and the gas phases may be towards the interface, as already indicated. A further objection to the use of overall coefficients is that the Lewis relation may be applied only to the heat- and mass-transfer coefficients in the gas phase.

In the design of commercial units, nomographs [18,19] are available, which give a performance characteristic (KaV/L'), where K is a mass-transfer coefficient (kg water/m^2 s) and V is the active cooling volume (m^3/m^2 plan area), as a function of θ, θ_W, and (L'/G'). For a given duty, (KaV/L') is calculated from

$$\frac{KaV}{C_L L'} = \int_{\theta_1}^{\theta_2} \frac{d\theta}{(H_f - H_G)} \tag{3.59}$$

and then, a suitable tower with this value of (KaV/L') is sought from performance curves [20,21]. In normal applications, the performance characteristic varies between 0.5 and 2.5.

Example 3.9

Water is to be cooled from 328 to 293 K by means of a countercurrent airstream entering at 293 K with a relative humidity of 20%. The flow of air is 0.68 m^3/ m^2 s, and the water throughput is 0.26 kg/m^2 s. The whole of the resistance to heat and mass transfer may be assumed to be in the gas phase, and the product, $(h_D a)$, may be taken as 0.2 (m/s)(m^2/m^3), that is, 0.2 s^{-1}.

What is the required height of packing and the condition of the exit airstream?

Solution

Assuming the latent heat of water at 273 K = 2495 kJ/kg,
specific heat of air = 1.003 kJ/kg K.
and specific heat of water vapour = 2.006 kJ/kg K.
The enthalpy of the inlet airstream:

$$H_{G1} = 1.003(293 - 273) + H[2495 + 2.006(293 - 273)]$$

From Fig. 3.4,

at θ = 293K and 20% RH, H = 0.003 kg/kg, and hence

$$H_{G1} = (1.003 \times 20) + 0.003[2495 + (2.006 \times 20)]$$
$$= 27.67 \text{ kJ/kg}$$

In the inlet air, water vapour = 0.003 kg/kg dry air or

$$\frac{(0.003/18)}{(1/29)} = 0.005 \text{ kmol/kmol dry air}$$

Thus, flow of dry air = $(1 - 0.005)0.68 = 0.677 \text{m}^3/\text{m}^2\text{s}$

$$\text{Density of air at 293K} = \left(\frac{29}{22.4}\right)\left(\frac{273}{293}\right) = 1.206 \text{kg/m}^3$$

and

Mass flow of dry air = $(1.206 \times 0.677) = 0.817 \text{ kg/m}^2\text{s}$

Slope of operating line:

$$(L'C_L/G') = \frac{(0.26 \times 4.18)}{0.817} = 1.33$$

The coordinates of the bottom of the operating line are

$$\theta_{L1} = 293 \text{K}, H_{G1} = 27.67 \text{kJ/kg}$$

Hence, on an enthalpy–temperature diagram, the operating line of slope 1.33 is drawn through the point (293 and 27.67) = (θ_{L1} and H_{G1}).

The top point of the operating line is given by θ_{L2} = 328 K, and H_{G2} is found to be 76.5 kJ/kg (Fig. 3.19).

From Figs 3.4 and 3.5, the curve representing the enthalpy of saturated air as a function of temperature is obtained and drawn in. Alternatively, this plot may be calculated from

$$H_F = C_a(\theta_f - 273) + H_0[C_w(\theta_f - 273) + \lambda] \text{ kJ/kg}$$

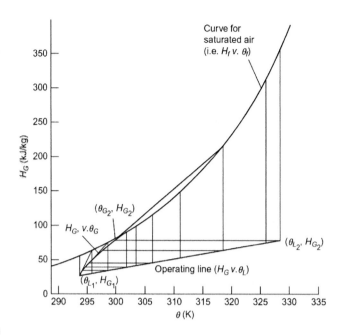

FIG. 3.19

Calculation of the height of a water-cooling tower.

The curve represents the relation between enthalpy and temperature at the interface, that is, H_f as a function of θ_f.

It now remains to evaluate the integral $\int dH_G/(H_f - H_G)$ between the limits, $H_{G1} = 27.7 \, \text{kJ/kg}$ and $H_{G2} = 76.5 \, \text{kJ/kg}$. Various values of H_G between these limits are selected and the value of 9 obtained from the operating line. At this value of θ_1, now θ_f, the corresponding value of H_f is obtained from the curve for saturated air. The working is as follows:

H_G	$\theta = \theta_f$	H_f	$(H_f - H_G)$	$1/(H_f - H_G)$
27.7	293	57.7	30	0.0330
30	294.5	65	35	0.0285
40	302	98	58	0.0172
50	309	137	87	0.0114
60	316	190	130	0.0076
70	323	265	195	0.0051
76.5	328	355	279	0.0035

A plot of $1/(H_f - H_G)$ and H_G is now made as shown in Fig. 3.20 from which the area under the curve $= 0.65$. This value may be checked using the approximate solution of Carey and Williamson [14].

At the bottom of the column,

$$H_{G1} = 27.7 \, \text{kJ/kg,} \quad H_{f1} = 57.7 \, \text{kJ/kg} \quad \therefore \Delta H_1 = 30 \, \text{kJ/kg}$$

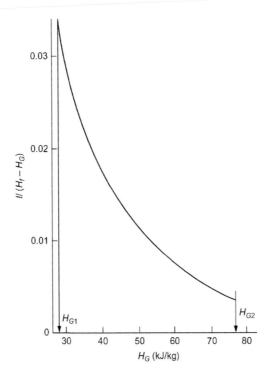

FIG. 3.20

Evaluation of the integral of $dH_G/(H_f - H_G)$.

At the top of the column,

$$H_{G2} = 76.5\,\text{kJ/kg}, \; H_{f2} = 355\,\text{kJ/kg} \; \therefore \Delta H_2 = 279\,\text{kJ/kg}$$

At the mean water temperature of $0.5(328 + 293) - 310.5\,\text{K}$,

$$H_{Gm} = 52\,\text{kJ/kg}, \; H_f = 145\,\text{kJ/kg} \; \therefore \Delta H_m = 93\,\text{kJ/kg}$$

$$\frac{\Delta H_m}{\Delta H_1} = 3.10, \; \frac{\Delta H_m}{\Delta H_2} = 0.333,$$

and from Fig. 3.16, $f = 0.79$.

Thus,

$$\frac{(H_{G2} - H_{G1})}{f \Delta H_m} = \frac{(76.5 - 27.7)}{(0.79 \times 93)} = 0.66$$

which agrees well with the value (0.65) obtained by graphical integration.

Thus, in Eq. (3.53),

$$\text{Height of packing, } z = \int_{H_{G1}}^{H_{G2}} \frac{dH_G}{(H_f - H_G)} \frac{G'}{h_D a \rho}$$

$$= \frac{(0.65 \times 0.817)}{(0.2 \times 1.206)}$$

$$= 2.20\,\underline{\underline{\text{m}}}$$

Assuming that the resistance to mass transfer lies entirely within the gas phase, the lines connecting θ_L and θ_f are parallel with the enthalpy axis.

In Fig. 3.18, a plot of H_G and θ_G is obtained using the construction given in Section 3.6.4 and shown in Fig. 3.15. From this curve, the value of θ_{G2} corresponding to $H_{G2}=76.5\,$kJ/kg is 300 K. From Fig. 3.5, under these conditions, the exit air has a humidity of 0.019 kg/kg that from Fig. 3.4 corresponds to a relative humidity of 83%.

3.6.7 Humidifying towers

If the main function of the tower is to produce a stream of humidified air, the final temperature of the liquid will not be specified, and the humidity of the gas leaving the top of the tower will be given instead. It is therefore not possible to fix any point on the operating line, though its slope can be calculated from the liquid and gas rates. In designing a humidifier, therefore, it is necessary to calculate the temperature and enthalpy and hence the humidity of the gas leaving the tower for a number of assumed water-outlet temperatures and thereby determine the outlet water temperature resulting in the air leaving the tower with the required humidity. The operating line for this water-outlet temperature is then used in the calculation of the height of the tower required to affect this degree of humidification. The calculation of the dimensions of a humidifier is therefore rather more tedious than that for the water-cooling tower.

In a humidifier in which the make-up liquid is only a small proportion of the total liquid circulating, its temperature approaches the adiabatic saturation temperature θ_S, and remains constant, so that there is no temperature gradient in the liquid. The gas in contact with the liquid surface is approximately saturated and has a humidity H_s.

Thus,

$$d\theta_L = 0$$

and

$$\theta_{L1} = \theta_{L2} = \theta_L = \theta_f = \theta_s$$

Hence,

$$-G' s d\,\theta_G = h_G a dz(\theta_G - \theta_s) \text{ (from Eq. 13.42)}$$

and

$$-G'dH = h_D \rho a dz(H - H_s) \text{ (from Eq. 13.44)}$$

Integration of these equations gives

$$\ln\frac{(\theta_{G1} - \theta_s)}{(\theta_{G2} - \theta_s)} = \frac{h_G a}{G' s} z \qquad (3.60)$$

and

$$\ln\frac{(H_s - H_1)}{(H_s - H_2)} = \frac{h_D a \rho}{G'} z \qquad (3.61)$$

assuming h_G, h_D, and s remain approximately constant.

From these equations, the temperature θ_{G2} and the humidity H_2 of the gas leaving the humidifier may be calculated in terms of the height of the tower. Rearrangement of Eq. (3.61) gives

$$\ln\left(1 + \frac{H_1 + H_2}{H_s - H_1}\right) = -\frac{h_D a\rho}{G'}z$$

or

$$\frac{(H_2 - H_1)}{(H_s - H_1)} = 1 - e^{-h_D a\rho z/G'} \tag{3.62}$$

Thus, the ratio of the actual increase in humidity produced in the saturator to the maximum possible increase in humidity (i.e. the production of saturated gas) is equal to $(1 - e^{-h_D a\rho z/G'})$, and complete saturation of the gas is reached exponentially. A similar relation exists for the change in the temperature of the gas stream:

$$\frac{(\theta_{G1} - \theta_{G2})}{(\theta_{G2} - \theta_s)} = 1 - e^{-h_G az/G's} \tag{3.63}$$

Further, the relation between the temperature and the humidity of the gas at any stage in the adiabatic humidifier is given by.

$$\frac{dH}{d\theta_G} = \frac{(H - H_s)}{(\theta_G - \theta_s)} \text{ (from Eq. 13.52)}$$

On integration,

$$\ln\frac{(H_s - H_2)}{(H_s - H_1)} = \ln\frac{(\theta_{G2} - \theta_s)}{(\theta_{G1} - \theta_s)} \tag{3.64}$$

or

$$\frac{(H_s - H_2)}{H_s - H_1} = \frac{(\theta_{G2} - \theta_s)}{(\theta_{G1} - \theta_s)} \tag{3.65}$$

3.7 Systems other than air–water

Calculations involving to systems where the Lewis relation is not applicable are very much more complicated because the adiabatic saturation temperature and the wet-bulb temperature do not coincide. Thus, the significance of the adiabatic cooling lines on the psychrometric chart is very much restricted. They no longer represent the changes that take place in a gas as it is humidified by contact with liquid initially at the adiabatic saturation temperature of the gas but simply give the compositions of all gases with the same adiabatic saturation temperature.

Calculation of the change in the condition of the liquid and the gas in a humidification tower is rendered more difficult since Eq. (3.49), which was derived for the air–water system, is no longer applicable. Lewis and White [22] have developed a method of integration of these equations that gives calculation based on the use of a *modified enthalpy* in place of the true enthalpy of the system.

For the air–water system, from Eq. (3.11),

$$h_G = h_G \rho s \tag{3.66}$$

This relationship applies quite closely for the conditions normally encountered in practice. For other systems, the relation between the heat- and mass-transfer coefficients in the gas phase is given by

$$h_G = b h_D \rho s \tag{3.67}$$

where b is approximately constant and generally has a value greater than unity.

For these systems, Eq. (3.46) becomes

$$G' s \, d\theta_G = d h_D \rho a \, dz \left(s\theta_f - s\theta_G \right) \tag{3.68}$$

Adding Eqs (3.68), (3.47) to obtain the relationship corresponding to Eq. (3.48) gives

$$G' (s \, d\theta_G + \lambda \, dH) = h_D \rho a \, dz \left[(b s\theta_f + \lambda H_f) - (d s\theta_G + \lambda H) \right] \tag{3.69}$$

Lewis and White use a *modified latent heat of vaporisation* λ' defined by

$$b = \frac{\lambda}{\lambda'} \tag{3.70}$$

and a *modified enthalpy* per unit mass of dry gas defined by

$$H'_G = s(\theta_G - \theta_0) + \lambda' H \tag{3.71}$$

Substituting in Eq. (3.67), from Eqs (3.38), (3.70), and (3.71),

$$G' \, dH_G = b h_D \rho a \, dz \left(H'_f - H'_G \right) \tag{3.72}$$

and

$$\frac{dH_G}{\left(H'_f - H'_G \right)} = \frac{b h_D \rho a}{G'} \, dz \tag{3.73}$$

Combining Eqs (3.37), (3.40), and (3.72),

$$\frac{\left(H'_G - H'_f \right)}{(\theta_L - \theta_f)} = -\frac{h_L}{h_D \rho b} \quad \text{(cf.Eq.13.50)} \tag{3.74}$$

From Eqs (3.66), (3.72),

$$\frac{\left(H'_G - H'_f \right)}{(\theta_G - \theta_f)} = \frac{dH_G}{d\theta_G} \quad \text{(cf.Eq.13.51)} \tag{3.75}$$

From Eqs (3.44), (3.67),

$$\frac{(H - H_f)}{(\theta_G - \theta_f)} = b \frac{dH}{d\theta_G} \quad \text{(cf.Eq.13.52)} \tag{3.76}$$

The calculation of conditions within a countercurrent column operating with a system other than air–water is carried out in a similar manner to that already described by applying Eqs (3.73), (3.74), and (3.75) in conjunction with Eq. (3.39) of Vol. 1 A:

$$G'(H_{G2} - H_{G1}) = L'C_L(\theta_{L2} - \theta_{L1}) \text{ (Eq.13.39)}$$

On an enthalpy–temperature diagram (Fig. 3.20), the enthalpy of saturated gas is plotted against its temperature. If equilibrium between the liquid and gas exists at the interface, this curve PQ represents the relation between gas enthalpy and temperature at the interface (H_f vs θ_f). The modified enthalpy of saturated gas is then plotted against temperature (curve RS) to give the relation between H_f' and θ_f. Since b is greater than unity, RS will lie below PQ. By combining Eqs (3.35), (3.70), and (3.72), H_G' is obtained in terms of H_G:

$$H_G' = \frac{1}{b}[H_G + (b - 1)s(\theta_G - \theta_0)] \qquad (3.77)$$

H_G' may be conveniently plotted against H_G for a number of constant temperatures. If b and s are constant, a series of straight lines is obtained. The operating line AB given by Eq. (3.39) is drawn in Fig. 3.21. Point A has coordinates (θ_{L1} and H_{G1}) corresponding to the bottom of the column. Point a has coordinates (θ_{L1} and H_{G1}'), H_{G1}' being obtained from Eq. (3.77).

From Eq. (3.72), a line through a, of slope $-(h_L/h_D\rho b)$, will intersect curve RS at c, (θ_{f1} and H_{f1}') to give the interface conditions at the bottom of the column. The corresponding air enthalpy is given by C, (θ_{f1} and H_{f1}). The difference between the ordinates of c and a then gives the driving force in terms of modified enthalpy at the bottom of the column ($H_{f1}' - H_{G1}'$). A similar construction at other points, such as

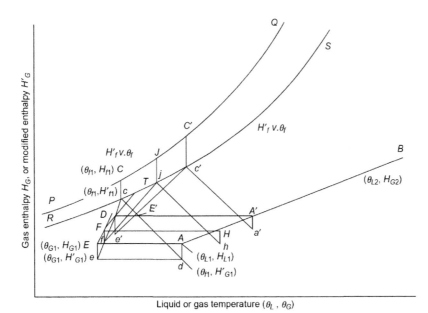

FIG. 3.21

Construction for height of a column for vapour other than water.

A', enables the driving force to be calculated at any other point. Hence, $(H'_{f}-H'_{G1})$ is obtained as a function of H_G throughout the column. The height of column corresponding to a given change in air enthalpy can be obtained from Eq. (3.71) since the left-hand side can now be evaluated.

Thus,

$$\int_{H_{G1}}^{H_{G2}} \frac{\mathrm{d}H_G}{\left(H'_f - H'_G\right)} = \frac{bh_D a\rho}{G'} z \qquad (3.78)$$

The change in the condition of the gas stream is obtained as follows: E, with coordinates (θ_{G1}, H_{G1}), represents the condition of the inlet gas. The modified enthalpy of this gas is given by $e(\theta_{G1}$ and $H'_{G1})$. From Eq. (3.75), it is seen that ec gives the rate of change of gas enthalpy with temperature $(\mathrm{d}H_G/\mathrm{d}\theta_G)$ at the bottom of the column. Thus, ED, parallel to ec, describes the way in which gas enthalpy changes at the bottom of the column. At some arbitrary small distance from the bottom, F represents the condition of the gas, and H gives the corresponding liquid temperature. In exactly the same way, the next small change is obtained by drawing a line hj through h parallel to ac. The slope of fj gives the new value of $(\mathrm{d}H_G/\mathrm{d}\theta_G)$, and therefore, the gas condition at a higher point in the column is obtained by drawing FT parallel to fj. In this way, the change in the condition of the gas through the column can be followed by continuing the procedure until the gas enthalpy reaches the value H_{G2} corresponding to the top of the column.

A detailed description of the method of construction of psychrometric charts is given by Shallcross and Low [23], who illustrate their method by producing charts for three systems: air–water, air–benzene, and air–toluene at pressures of 1 and 2 bar.

Example 3.10

In a countercurrent packed column, n-butanol flows down at a rate of $0.25\,\mathrm{kg/m^2}\,\mathrm{s}$ and is cooled from 330 to 295 K. Air at 290 K, initially free of n-butanol vapour, is passed up the column at the rate of $0.7\,\mathrm{m^3/m^2}$ s. Calculate the required height of tower and the condition of the exit air.

Data:

Mass-transfer coefficient per unit volume, $h_D a = 0.1\,\mathrm{s^{-1}}$.

Psychrometric ratio, $h_G/(h_D\rho_A s) = b = 2.34$.

Heat-transfer coefficients, $h_L = 3h_G$.

Latent heat of vaporisation of n-butanol, $\lambda = 590\,\mathrm{kJ/kg}$.

Specific heat of liquid n-butanol, $C_L = 2.5\,\mathrm{kJ/kg\,K}$.

Humid heat of gas, $s = 1.05\,\mathrm{kJ/kg\,K}$.

Temperature (K)	Vapour pressure of butanol (kN/m²)
295	0.59
300	0.86
305	1.27

310	1.75
315	2.48
320	3.32
325	4.49
330	5.99
335	7.89
340	10.36
345	14.97
350	17.50

Solution

The first stage is to calculate the enthalpy of the saturated gas by way of the saturated humidity, H_0 given by

$$H_0 = \frac{P_{w0}}{P - P_{w0}} \frac{M_w}{M_A} = \frac{P_{w0}}{(101.3 - P_{w0})} \left(\frac{74}{29}\right)$$

The enthalpy is then

$$H_f = \frac{1}{(1 + H_0)} \times 1.001 \left(\theta_f - 273\right) + H_0 \left[2.5 \left(\theta_f - 273\right) + 590\right] \text{ kJ/kg}$$

where 1.001 kJ/kg K is the specific heat of dry air.

Thus,

$$H_f = \frac{1.001\theta_f - 273.27}{(1 + H_0)} + H_0 \left(2.5\theta_f - 92.5\right) \text{kJ/kg moist air}$$

The results of this calculation are presented in the following table, and H_f is plotted against θ_f in Fig. 3.21.

The modified enthalpy at saturation H_f' is given by

$$H_f' = \frac{(1.001\theta_f - 273.27)}{(1 + H_0)} + H_0 \left[2.5 \left(\theta_f - 273\right) + \lambda'\right]$$

where from Eq. (3.70), $\lambda' = \lambda/b = (590/2.34)$ or 252 kJ/kg.

$$\therefore H_f' = \frac{(1.001\theta_f - 273.27)}{(1 + H_0)} + H_0 \left(2.5\theta_f - 430.5\right) \text{ kJ/kg moist air}$$

These results are also given in the following table and plotted as H_f' against θ_f in Fig. 3.21:

θ_f (K)	P_{w0} (kN/m²)	H_0 (kg/ kg)	$(1.001\theta_f-$ 273.27)/ $(1+H_0)$ (kJ/kg)	H_0 (2.5θ_f −92.5) (kJ/kg)	H_f (kJ/kg)	H_0(2.5θ_f −430.5) (kJ/kg)	H_f' (kJ/kg)
295	0.59	0.0149	21.70	9.61	31.31	4.57	26.28
300	0.86	0.0218	24.45	14.33	40.78	6.97	33.42
305	1.27	0.0324	31.03	21.71	52.74	10.76	41.79
310	1.75	0.0448	35.45	30.58	66.03	15.43	50.88
315	2.48	0.0640	39.52	44.48	84.00	22.85	62.37
320	3.32	0.0864	43.31	61.13	104.44	31.92	75.23
325	4.49	0.1183	46.55	85.18	131.73	45.19	91.74
330	5.99	0.1603	49.18	117.42	166.60	63.23	112.41
335	7.89	0.2154	51.07	160.47	211.54	87.67	138.73
340	10.36	0.2905	51.97	220.05	272.02	121.87	173.83
345	14.97	0.4422	49.98	340.49	390.47	191.03	241.01
350	17.50	0.5325	50.30	416.68	466.98	236.70	287.00

The bottom of the operating line (point a) has coordinates $\theta_{L1}=295$ K and H_{G1}, where

$$H_{G1} = 1.05(290 - 273) = 17.9 \text{ kJ/kg}.$$

At a mean temperature of, say, 310 K, the density of air is

$$\left(\frac{29}{22.4}\right)\left(\frac{273}{310}\right) = 1.140 \text{ kg/m}^3$$

and

$$G' = (0.70 \times 1.140) = 0.798 \text{ kg/m}^2\text{s}$$

Thus, the slope of the operating line becomes

$$\frac{L'C_L}{G'} = \frac{(0.25 \times 2.5)}{0.798} = 0.783 \text{ kJ/kg K}$$

and this is drawn in as AB in Fig. 3.22, and at $\theta_{L2}=330$ K, $H_{G2}=46$ kJ/kg. From Eq. (3.77), $H'_G = (H_G + (b-1)s(\theta_G-\theta_0))/b$

$$\therefore H'_{G1} = \frac{17.9 + (2.34 - 1)1.05(290 - 273)}{2.34} = 17.87 \text{ kJ/kg}$$

Point a coincides with the bottom of the column.

$$-\frac{h_L}{h_D\rho b} = -\left(\frac{3h_G}{h_D\rho b}\right)\cdot\left(\frac{h_D\rho s}{h_G}\right).$$

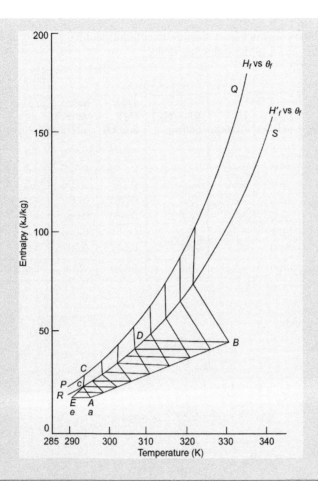

FIG. 3.22

Graphical construction for Example 3.10.

A line is drawn through *a* of slope

$$= -3s = -3.15\,\text{kJ/kg K}$$

This line meets curve *RS* at *c* (θ_{f1} and H'_{f1}) to give the interface conditions at the bottom of the column. The corresponding air enthalpy is given by point *C* whose coordinates are

$$\theta_{f1} = 293\text{K}\; H_{f1} = 29.0 \text{ kJ/kg}$$

The difference between the ordinates of *c* and *a* gives the driving force in terms of the modified enthalpy at the bottom of the column or

$$\left(H'_{f1} - H'_{G1}\right) = (23.9 - 17.9) = 6.0\,\text{kJ/kg}$$

A similar construction is made at other points along the operating line with the results shown in the following table.

θ_f (K)	H_G (kJ/kg)	H_G' (kJ/kg)	H_f' (kJ/kg)	$(H_f' - H_G')$ (kJ/kg)	$1/(H_f' - H_G')$ (kg/kJ)	Mean value in interval	Interval	Value of integral over interval
295	17.9	17.9	23.9	6.0	0.167	0.155	4.1	0.636
300	22.0	22.0	29.0	7.0	0.143	0.126	4.0	0.504
305	26.0	26.0	35.3	9.3	0.108	0.096	4.0	0.384
310	30.0	30.0	42.1	12.1	0.083	0.073	4.0	0.292
315	34.0	34.0	50.0	16.0	0.063	0.057	4.1	0.234
320	38.1	38.1	57.9	19.8	0.051	0.046	3.9	0.179
325	42.0	42.0	66.7	24.7	0.041	0.0375	4.0	0.150
330	46.0	46.0	75.8	29.8	0.034	Value of integral $=2.379$		

from which

$$\int_{H_{G1}}^{H_{G2}} \frac{dH_G}{H_f' - H_G'} = 2.379$$

Substituting in Eq. (3.78),

$$\frac{bh_D\rho az}{G'} = 2.379$$

and

$$z = \frac{(2.379 \times 0.798)}{(2.34 \times 0.1)} = \underline{8.1\,m}$$

It remains to evaluate the change in gas conditions.

Point e, ($\theta_{G1} = 290$K and $H_{G1} = 17.9$kJ/kg) represents the condition of the inlet gas. ec is now drawn in, and from Eq. (3.75), this represents $dH_G/d\theta_G$. As for the air–water system, this construction is continued until the gas enthalpy reaches H_{G2}. The final point is given by D at which $\theta_{G2} = 308$k.

It is fortuitous that, in this problem, $H_G' = H_G$. This is not always the case, and reference should be made to Section 3.7 for elaboration of this point.

References

[1] BS 1339:1965, British Standard 1339 Definitions, Formulae and Constants Relating to the Humidity of the Air, British Standards Institution, London, 1981.

[2] T.K. Sherwood, E.W. Comings, An experimental study of the wet bulb hygrometer, Trans. Am. Inst. Chem. Eng. 28 (1932) 88.

[3] M.M. Grosvenor, Calculations for dryer design, Trans. Am. Inst. Chem. Eng. 1 (1908) 184.

[4] A. Wexler, in: R.E. Ruskin (Ed.), Humidity and Moisture. Measurements and Control in Science and Industry. Principles and Methods of Humidity Measurement in Gases, vol. 1, Reinhold, New York, 1965.

[5] M.J. Hickman, Measurement of Humidity. 4th ed. National Physical Laboratory. Notes on Applied Science No. 4, HMSO, London, 1970.

[6] D.B. Meadowcroft, Chapter 6., Chemical analysis—moisture measurement, in: B.E. Noltingk (Ed.), Instrumentation Reference Book, Butterworth, London, 1988.

[7] B. Wood, P. Betts, A contribution to the theory of natural draught cooling towers, Proc. Inst. Mech. Eng. 163 (1950) 54.

[8] R. Clark, Cutting the fog, Chem. Eng. (London) 529 (1992) 22.

[9] K.K. McKelvcy, M. Brooke, The Industrial Cooling Tower, Elsevier, New York, 1959.

[10] C.H. Chilton, Performance of natural-draught cooling towers, Proc. Inst. Electr. Eng. 99 (1952) 440.

[11] R.H. Perry, D.W. Green (Eds.), Perry's Chemical Engineers' Handbook, sixth ed., McGraw-Hill, New York, 1984.

[12] H.S. Mickley, Design of forced draught air conditioning equipment, Chem. Eng. Prog. 45 (1949) 739.

[13] F. Merkel, Verdunstungs-Kühlung, Ver. Deut. Ing. Forschungsarb (1925) 275.

[14] W.F. Carey, G.J. Williamson, Gas cooling and humidification: design of packed towers from small scale tests, Proc. Inst. Mech. Eng. 163 (1950) 41.

[15] W.J. Thomas, P. Houston, Simultaneous heat and mass transfer in cooling towers, Br. Chem. Eng. 4 (160) (1959) 217.

[16] G. Cribb, Liquid phase resistance in water cooling, Br. Chem. Eng. 4 (1959) 264.

[17] H.J. Lowe, D.G. Christie, Heat transfer and pressure drop data on cooling tower packings, and model studies of the resistance of natural-draught towers to airflow, Inst. Mech. Eng. Symp. Heat Tran. (1962) 933. Paper 113.

[18] B. Wood, P. Betts, A total heat–temperature diagram for cooling tower calculations, Engineer 189 (4912) (1950) 337. (4913) 349.

[19] S.M. Zivi, B.B. Brand, Analysis of cross-flow cooling towers, Refrig. Eng. 64 (8) (1956) 31. 90.

[20] Pritchard (J.F.) & Co, Counter-flow cooling tower performance, J. F. Pritchard & Co, Kansas City, 1957.

[21] Cooling Tower Institute, Performance curves, Cooling Tower Institute, 1967. Houston.

[22] J.G. Lewis, R.R. White, Simplified humidification calculations, Ind. Eng. Chem. 45 (1953) 486.

[23] D.C. Shallcross, S.L. Low, Construction of psychrometric charts for systems other than water vapour in air, Chem. Eng. Res. Des. 72 (1994) 763. Errata: Chem Eng Res Design 1995;73:865.

Further reading

J.R. Backhurst, J.H. Harker, J.E. Porter, Problems in Heat and Mass Transfer, Edward Arnold, London, 1974.

R. Burger, Cooling tower drift elimination, Chem. Eng. Prog. 71 (7) (1975) 73.

J.R. DeMonbrun, Factors to consider in selecting a cooling tower, Chem. Eng. 75 (19) (1968) 106.

J.M. Donohue, C.C. Nathan, Unusual problems in cooling water treatment, Chem. Eng. Prog. 71 (7) (1975) 88.

E.R.G. Eckert, R.M. Drake, Analysis of Heat and Mass Transfer, McGraw-Hill, New York, 1972.

A.W. Elgawhary, Spray cooling system design, Chem. Eng. Prog. 71 (7) (1975) 83.

F. Friar, Cooling-tower basin design, Chem. Eng. 81 (15) (1974) 122.

K.M. Guthrie, Capital cost estimating, Chem. Eng. 76 (6) (1969) 114.

W.A. Hall, Cooling tower plume abatement, Chem. Eng. Prog. 67 (7) (1971) 52.

E.P. Hansen, J.J. Parker, Status of big cooling towers, Power Eng. 71 (5) (1967) 38.

R. Holzhauer, Industrial cooling towers, Plant Eng. 29 (15) (1975) 60.

F.P. Incropera, D.P. De Witt, Fundamentals of Heat and Mass Transfer, fourth ed., Wiley, New York, 1996.

Industrial Water Society, Guide to Mechanical Draught Evaporative Cooling Towers: Selection, Operation and Maintenance, Industrial Water Society, London, 1987.

J. Jackson, Cooling Towers, Butterworths, London, 1951.

D.R. Jordan, M.D. Bearden, W.F. McIlhenny, Blowdown concentration by electrodialysis, Chem. Eng. Prog. 71 (1975) 77.

J.F. Juong, How to estimate cooling tower costs, Hydrocarb. Process. 48 (7) (1969) 200.

G.M. Kelly, Cooling tower design and evaluation parameters, in: ASME Paper, 75-IPWR-9, 1975.

T.D. Kolflat, Cooling tower practices, Power Eng. 78 (1) (1974) 32.

R.W. Maze, Practical tips on cooling tower sizing, Hydrocarb. Process. 46 (2) (1967) 123.

R.W. Maze, Air cooler or water tower—which for heat disposal? Chem. Eng. 83 (1) (1975) 106.

W.L. McCabe, J.C. Smith, P. Harriott, Unit Operations of Chemical Engineering, McGraw-Hill, New York, 1985.

K.K. McKelvey, M. Brooke, The Industrial Cooling Tower, Elsevier, New York, 1959.

J. Meytsar, Estimate cooling tower requirements easily, Hydrocarb. Process. 57 (11) (1978) 238.

W.L. Nelson, What is cost of cooling towers? Oil Gas J. 65 (47) (1967) 182.

W.S. Norman, Absorption, Distillation and Cooling Towers, Longmans, London, 1961.

F.C. Olds, Cooling towers, Power Eng. 76 (12) (1972) 30.

P.M. Paige, Costlier cooling towers require a new approach to water-systems design, Chem. Eng. 74 (14) (1967) 93.

J.E. Park, J.M. Vance, Computer model of crossflow towers, Chem. Eng. Prog. 67 (1971) 55.

M. Picciotti, Design quench water towers, Hydrocarb. Process. 56 (6) (1977) 163.

M. Picciotti, Optimize quench water systems, Hydrocarb. Process. 56 (9) (1977) 179.

A. Rabb, Are dry cooling towers economical? Hydrocarb. Process. 47 (2) (1968) 122.

T. Uchiyama, Cooling tower estimates made easy, Hydrocarb. Process. 55 (12) (1976) 93.

R. Walker, Water Supply, Treatment and Distribution, Prentice-Hall, New York, 1978.

R.B. Wrinkle, Performance of counterflow cooling tower cells, Chem. Eng. Prog. 67 (1971) 45.

Liquid–liquid extraction

Yan Zhang[a] and Ajay Kumar Ray[b]

[a]*Process Engineering, Memorial University of Newfoundland, St John's, NL, Canada,* [b]*Department of Chemical and Biochemical Engineering, University of Western Ontario, London, ON, Canada*

Nomenclature

		Units in SI system	Dimensions in M, N, L, T
A	mass or mass flow rate of solute **A**	kg (kg/s)	$M\ (MT^{-1})$
a	interfacial area per unit volume of tower	m^2/m^3	L^{-1}
a_p	superficial area of packing per unit volume of tower	m^2/m^3	L^{-1}
B	mass or mass flow rate of carrier **B**	kg (kg/s)	$M\ (MT^{-1})$
b	index	–	–
c	concentration of solute	$kmol/m^3$	NL^{-3}
$c*$	value of concentration in equilibrium with second phase	$kmol/m^3$	NL^{-3}
D_T	diameter of tank	m	L
d	particle size	m	L
d_c	diameter of column	m	L
d_p	nominal size of packing	m	L
d_s	surface mean (Sauter mean) diameter of drop	m	L
d_{crit}	critical packing size	m	L
E	mass or mass flow rate of extract	kg (kg/s)	$M\ (MT^{-1})$
e	voidage of packing	–	–
F	mass or mass flow rate of feed	kg (kg/s)	$M\ (MT^{-1})$
g	acceleration due to gravity	m/s^2	LT^{-2}
H	height of the transfer unit	m	L

Continued

		Units in SI system	Dimensions in M, N, L, T
J	coefficient in Eq. (4.45) for flooding data	–	–
j	fractional hold-up of dispersed phase	–	–
K	overall mass transfer coefficient	m/s	$\mathbf{LT^{-1}}$
K'	constant	–	–
k	mass transfer coefficient	m/s	$\mathbf{LT^{-1}}$
L'	volumetric rate of flow per unit area	m^3/m^2s	$\mathbf{LT^{-1}}$
M	total flow of material through the system	kg/s	$\mathbf{MT^{-1}}$
m	slope of the equilibrium line	–	–
N	revolutions per unit time	s^{-1}	$\mathbf{T^{-1}}$
N'	molar rate of transfer per unit area	$kmol/m^2s$	$\mathbf{NL^{-2}T^{-1}}$
$\mathbf{N}$	number of transfer units	–	–
n	number of stages	–	–
$n_1, n_2, \dots$	numbers of particles or droplets	–	–
P	difference stream	kg/s	$\mathbf{MT^{-1}}$
R	mass or mass flow rate of raffinate	kg (kg/s)	$\mathbf{M\ (MT^{-1})}$
r	ratio u_{df}/u_{cf}	–	–
S	mass or mass flow rate of solvent S	kg (kg/s)	$\mathbf{M\ (MT^{-1})}$
u	volumetric flow rate per unit area	m^3/m^2s	$\mathbf{LT^{-1}}$
u_r	velocity of dispersed phase relative to continuous phase	m/s	$\mathbf{LT^{-1}}$
u_0	terminal falling velocity of droplet	m/s	$\mathbf{LT^{-1}}$
$\bar{u}_0$	velocity of single droplet relative to continuous phase	m/s	$\mathbf{LT^{-1}}$
X_f	mass ratio of solute in feed	kg/kg (kmol/kmol)	–
X_1, X_2	mass ratio of solute in raffinate	kg/kg (kmol/kmol)	–
x_A, x_B	mass (or mole) fraction of **A, B**	kg/kg (kmol/kmol)	–
Y_1, Y_2	mass ratio of solute in extract	kg/kg (kmol/kmol)	–
Z	height of packing	m	$\mathbf{L}$
Z_B	height of dispersion band	m	$\mathbf{L}$
β	selectivity ratio (Eq. 4.2)	–	–
ρ	density	kg/m^3	$\mathbf{ML^{-3}}$
$\Delta\rho$	density difference between phases	kg/m^3	$\mathbf{ML^{-3}}$

Continued

		Units in SI system	Dimensions in M, N, L, T
σ	interfacial tension	J/m^2 or N/m	MT^{-2}
σ_T	interfacial tension, water-air at 288 K (0.073 N/m)	J/m^2 or N/m	MT^{-2}
Suffixes			
c, d	continuous and disperse phases		
E, R	extract and raffinate phases		
O	overall (transfer units)		
f	limiting value at flooding point		
i	value of interface		
lm	logarithmic mean value		
1, 2	values at bottom, top of column		

4.1 Introduction

The separation of the components in a liquid mixture by treatment with a solvent in which one or more of the desired components is preferentially soluble is known as liquid–liquid extraction. In this operation, it is essential that the liquid-mixture feed and solvent be at least partially if not completely immiscible, and three stages are involved:

(a) Bringing the feed mixture and the solvent into intimate contact,
(b) Separation of the resulting two phases, and
(c) Removal and recovery of the solvent from each phase.

It is possible to combine stages (a) and (b) into a single piece of equipment such as a column operated continuously. Such an operation is known as differential contacting. Liquid–liquid extraction is also carried out in stage-wise equipment, the prime example being a mixer–settler unit in which the main features are the mixing of two liquid phases by agitation, followed by settling in a separate vessel by gravity. This mixing of two liquids by agitation is of considerable importance, and the topic is discussed in more detail in Chapter 7 of Volume 1.

Extraction is in many ways complementary to distillation and is preferable in the following cases:

(a) Where distillation would require excessive amounts of heat, for example, when the relative volatility is near unity.
(b) When the formation of azeotropes limits the degree of separation obtainable in distillation.
(c) When heating must be avoided.
(d) When the components to be separated are quite different in nature.

Unlike distillation where separation of compounds is based on differences in volatility, extraction separates components based on differences in solubility. The key to an effective extraction process is a suitable solvent, which should have high selectivity and solubility for the solute in addition to being stable, nontoxic, inexpensive, and readily recoverable. In recent years, it has become possible to use computerised techniques to aid in the choice of a solvent with the required selectivity and to 'design' appropriate molecular structures.

Liquid–liquid extraction is an important separation technology for a wide range of applications in the food, petroleum, and pharmaceutical industries. Important applications of liquid–liquid extraction in the petroleum industry include the separation of aromatics from kerosene-based fuel oils to improve their burning qualities and the separation of aromatics from paraffin and naphthenic compounds to improve the temperature-viscosity characteristics of lubricating oils. It is also used to obtain relatively pure compounds such as benzene, toluene, and xylene from catalytically produced reformates in the petro-chemical industry, in the production of anhydrous acetic acid as well as in the extraction of phenol from coal tar liquors.

Liquid–liquid extraction is very important for *metallurgical processes*. The successful development of methods for the purification of uranium fuel and for the recovery of spent fuel elements in the nuclear power industry by extraction methods, mainly based on packed, including pulsed, columns as discussed in Section 5 has led to their application to other metallurgical processes. Of these, the recovery of copper from acid leach liquors and subsequent electro-winning from these liquors are the most extensive, although further applications to nickel and other metals have been developed. In many of these processes, chemical complex is formed between the solute and the solvent so that the kinetics of the process become important. The extraction operation may be either a physical operation, as discussed previously, or a chemical operation. Chemical operations have been classified by Hanson [1] as follows:

(a) Those involving ion exchanges such as the extraction of metals by carboxylic acids, or extraction of anions involving a metal with amines.

(b) Those involving the formation of an additive compound, for example, extraction with neutral organophosphorus compounds. An important operation of this type is the purification of uranium from the nitrate with tri-*n*-butyl phosphate.

The process of metal purification is of particular interest in that it involves applying principles of both chemistry and chemical engineering and necessitates the cost evaluation of alternatives.

In *biotechnology*, many of the usual organic solvents will degrade a sensitive product, such as a protein; this has led to the use of 'mild' aqueous-based solvents, such as water–polyethyleneglycol–phosphate mixtures, which will partition and concentrate the product in one of the two aqueous layers formed.

In recent years, *sub- and supercritical fluids as well as ionic liquids* have been applied as 'green' solvents for the extraction of various types of products. The use of supercritical fluids and ionic liquids is discussed in Section 6.

4.2 **Extraction processes**

Liquid–liquid extraction usually involves a ternary system consisting of two miscible feed components—the *solute* A and the *carrier* B, plus the *solvent* S. Carrier B and solvent S are at most only partially soluble, but solute A is completely or partially soluble in S. During extraction, mass transfer of A from the feed to the solvent occurs, with less transfer of B to the solvent or S to the feed. In practice, liquid–liquid extraction can be carried out either as a batch or as a continuous process.

In a single-stage batch process illustrated in Fig. 4.1, the solvent and the feed are mixed together and then allowed to separate into the two phases—the *extract* E containing the required solute in the added solvent and the *raffinate* R, rich in carrier with some associated solvent. With this simple arrangement mixing and separation take place in the same vessel.

A continuous two-stage operation is shown in Fig. 4.2, where the mixers and separators are shown as separate vessels. There are three main forms of equipment. First there is the mixer–settler as shown in Fig. 4.2, secondly, there is the column type of design with trays or packing as in distillation and, thirdly, there are a variety of units incorporating rotating devices such as the Scheibel and the Podbielniak extractors.

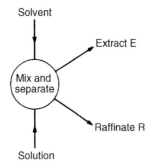

FIG. 4.1

Single-stage batch extraction.

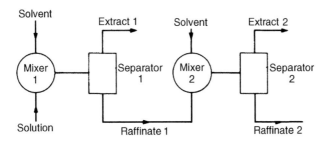

FIG. 4.2

Multiple-contact system with fresh solvent.

In all cases, the extraction units are followed by distillation or a similar operation in order to recover the solvent and the solute. Some indication of the form of these alternative arrangements may be seen by considering two of the processes referred to in Section 1.

One system for separating benzene, toluene, and xylene groups from light feedstock is shown in Fig. 4.3, where n-methylpyrrolidone (NMP) with the addition of glycol is used as the solvent. The feed is passed to a multistage extractor arranged as a tower from which an aromatics-free raffinate is obtained at the top. The extract stream containing the solvent, aromatics, and low boiling non-aromatics is distilled to provide the extractor recycle stream as a top product, and a mixture of aromatics and solvent at the bottom. This stream passes to a stripper from which glycol and the aromatics are recovered. This is a complex system illustrating the need for careful recycling and recovery of solvent.

The concentration of acrylic acid by extraction with ethyl acetate [2] is a rather different illustration of this technique. As shown in Fig. 4.4, the dilute acrylic acid solution of concentration about 20% is fed to the top of the extraction column 1, the ethyl acetate solvent being fed from the bottom of the column. The acetate containing the dissolved acrylic acid and water leaves from the top and is fed to the distillation column 2, where the acetate is removed as an azeotrope with water and the dry acrylic acid is recovered as product from the bottom.

It is clearly seen from these illustrations that successful extraction processes should not be judged simply by the performance of the extraction unit alone, but by assessment of the recovery achieved by the whole plant. This aspect of the process

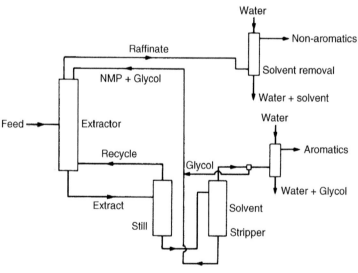

FIG. 4.3

Process for benzene, toluene, and xylene recovery.

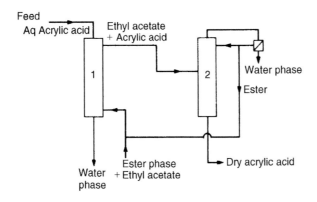

FIG. 4.4

Concentration of acrylic acid by extraction with ethyl acetate [2].

may be complex if chemical reactions are involved. The sections of the plant for mixing and for separation must be considered together when assessing capital cost. The cost of the organic solvents used in the metallurgical processes may also be high.

The mechanism of the transfer of solute from one phase to the second is governed by either molecular or eddy diffusion, and similarly to distillation and absorption processes, the rate of mass transfer depends on phase equilibrium, interfacial area, and surface renewal rate. In liquid–liquid extraction, the rate of mass transfer is affected by mechanical means including pumping and agitation, except in standard packed columns.

In formulating design criteria for extraction equipment, it is necessary to take into account the equilibrium conditions for the distribution of solute between the phases as this determines the maximum degree of separation possible in a single stage. The resistance to diffusion and, in the case of chemical effects, the kinetics are also important in that these determine the residence time required to bring about near equilibrium in a stage-wise unit, or the height of a transfer unit in a differential contactor. The transfer rate is given by the accepted equation:

$$\text{Rate per unit interfacial area} = k\Delta c \qquad (4.1)$$

where k is a mass transfer coefficient and Δc a concentration driving force. A high value of k can be obtained only if turbulent or eddy conditions prevail and, although these may be readily achieved in the continuous phase by some form of agitation, it is very difficult to generate eddies in the drops which constitute the dispersed phase.

4.3 Equilibrium data

The equilibrium condition for the distribution of one solute between two liquid phases is conveniently considered in terms of the distribution law. Thus, at equilibrium, the ratio of the concentrations of the solute in the two phases is given by

$c_E/c_R = K'$, where K' is the distribution coefficient. This relation will apply accurately only if both solvents are immiscible, and if there is no association or dissociation of the solute. If the solute forms molecules of different molecular weights, then the distribution law holds for each molecular species. When the concentrations are quite low, the distribution law usually holds provided no chemical reaction takes place.

The addition of a new solvent to a binary mixture of a solute in a solvent may lead to the formation of several types of mixture:

(a) A homogeneous solution may be formed and the selected solvent is then unsuitable.
(b) The solvent may be completely immiscible with the initial solvent.
(c) The solvent may be partially miscible with the original solvent resulting in the formation of one pair of partially miscible liquids.
(d) The new solvent may lead to the formation of two or three partially miscible liquids.

Of these possibilities, types (b)–(d) all give rise to systems that can be used, although those of types (b) and (c) are the most promising. With conditions of type (b), the equilibrium relation is conveniently shown by a plot of the concentration of solute in one phase against the concentration in the second phase. Conditions given by (c) and (d) are usually represented by triangular diagrams. Equilateral triangles are used, although it is also possible to use right-angled isosceles triangles.

The system, acetone (**A**)–Water (**B**)–methyl isobutyl ketone (**C**), as shown in Fig. 4.5, is of type (c). Here the solute **A** is completely miscible with the carrier **B** and the solvent **C**, although **B** and **C** are only partially miscible with each other. A mixture indicated by point H consists of the three components **A**, **B**, and **C** in the ratio of the perpendiculars HL, HJ, and HK. The distance BN represents the solubility of solvent **C** in **B**, and MC that of **B** in **C**. The area under the curved line NPFQM, the binodal solubility curve, represents a two-phase region which splits

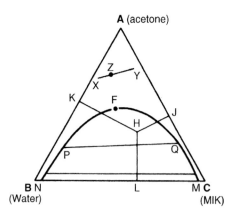

FIG. 4.5

Equilibrium relationship for acetone distributed between water and methyl isobutyl ketone.

up into two layers in equilibrium with each other. These layers have compositions represented by points P and Q, and PQ is known as a 'tie line'. Such lines as shown in the diagram connect the compositions of two phases in equilibrium with each other, and these compositions must be found by experimental measurement. There is one point on the binodal curve at F which represents a single phase that does not split into two phases. F is known as a *plait* point, and this must also be found by experimental measurement. The plait point is fixed if either the temperature or the pressure is fixed. Within the area under the curve, the temperature and composition of one phase will fix the composition of the other. Applying the phase rule to the three-component system at constant temperature and pressure, the number of degrees of freedom is equal to 3 minus the number of phases. In the area where there is only one liquid phase, there are two degrees of freedom and two compositions must be stated. In a system where there are two liquid phases, there is only one degree of freedom.

One of the most useful features of this method of representation is that, if a solution of composition X is mixed with one of composition Y, then the resulting mixture will have a composition shown by Z on a line XY, such that:

$$XZ/ZY = (\text{amount of } Y)/(\text{amount of } X).$$

Similarly, if an extract Y is removed, from a mixture Z the remaining liquor will have composition X.

In Fig. 4.6, two separate two-phase regions are formed for the ternary system of aniline (**A**), water (**B**), and phenol (**C**). Under the conditions shown in Fig. 4.6, **A** and **C** are miscible in all proportions, although **B** and **A**, and **B** and **C** are only partially miscible.

While these diagrams are of considerable use in presenting equilibrium data, Fig. 4.7 is in many ways more useful for determining the selectivity of a solvent, and the number of stages that are likely to be required. In Fig. 4.7, the percentage of solute in one phase is plotted against the percentage in the second phase in equilibrium with it. This is equivalent to plotting the compositions at either end

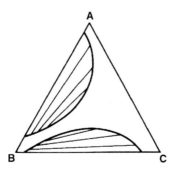

FIG. 4.6

Equilibrium relationships for aniline–water–phenol system.

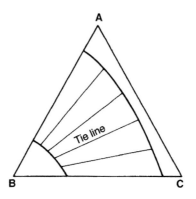

FIG. 4.7

Equilibrium distribution of solute **A** in phases **B** and **C**.

of a tie line. The important factor in assessing the value of a solvent is the ratio of the concentrations of the desired component in the two phases, rather than the actual concentrations. A selectivity ratio is defined in terms of either mass or mole fractions as:

$$\beta = \left[\frac{x_A}{x_B}\right]_E \bigg/ \left[\frac{x_A}{x_B}\right]_R \tag{4.2}$$

where x_A and x_B are the mass or mole fractions of **A** and **B** in the two phases E and R.

For a few systems β tends to be substantially constant, although it usually varies with concentration. The selectivity ratio has the same significance in extraction as relative volatility has in distillation, so that the ease of separation is directly related to the numerical value of β. As β approaches unity, a larger number of stages is necessary for a given degree of separation and the capital and operating costs increase correspondingly. When $\beta = 1$ any separation is impossible.

4.4 Calculation of the number of theoretical stages
4.4.1 Cross-current contact with partially miscible solvents

In calculating the number of ideal stages required for a given degree of separation, the conditions of equilibrium expressed by one of the methods discussed in Section 3 is used. The number of stages where single or multiple contact equipment is involved is considered first, followed by the design of equipment where the concentration change is continuous.

For a general case where the solvents are partially miscible, the feed solution F is brought into contact with the selective solvent S, to give raffinate R_1 and extract E_1. The addition of streams **F** and **S** is shown on the triangular diagram in Fig. 4.8, by the

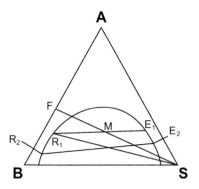

FIG. 4.8

Multiple contact with fresh solvent used at each stage.

point M, where $FM/MS = S/F$. This mixture M breaks down to give extract E_1 and raffinate R_1, at opposite ends of a tie line through M.

If a second stage is used, then the raffinate R_1 is treated with a further quantity of solvent S, and extract E_2 and raffinate R_2 are obtained as shown in the same figure.

The complete process consists in the extraction, and recovering the solvent from the raffinate and extract obtained. Thus, for a single-stage system as shown in Fig. 4.9, the raffinate R is passed into the distillation column where it is separated to give purified raffinate R' and solvent S_R. The extract E is fed to another distillation unit to give extract E' and a solvent stream S_E. It has been assumed in this case that perfect separation is obtained in the stills, so that pure solvent is obtained in the streams S_R and S_E, although the same form of diagram can be used where imperfect separation is obtained. It should be noted that, when ES is a tangent to the binodal curve, then the maximum concentration of solute A in the extract E' is obtained. It also follows that E' then represents the maximum possible concentration of A in the feed. Sufficient solvent S must be used to bring the mixture M within the two-phase area.

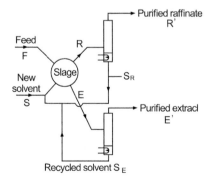

FIG. 4.9

Single-stage process with solvent recovery.

Example 4.1

A two-stage cross-current leaching process is used to extract acetic acid (A) from isopropyl ether (B) using pure water (C) at 20.0 °C as the extracting solvent. The flow rate of feed stream is 150 kg/min, in which 40 wt% is acetic acid and 60 wt% is isopropyl ether. Pure water is introduced cross-currently to each stage with a flow rate of 100.0 kg/min. Find the flow rates and compositions of the raffinate and extract streams for each stage.

The equilibrium data for the system are illustrated in Fig. 4.10.

Solution

Stage 1:

Locate the mixing point M_1 for stage 1, calculating, x_{MA}, x_{MB}, and x_{MC} by mass balance, thus,

$$x_{MA} = Fx_{FA}/(F + S_1) = 150(0.4)/(150 + 100) = 0.24$$

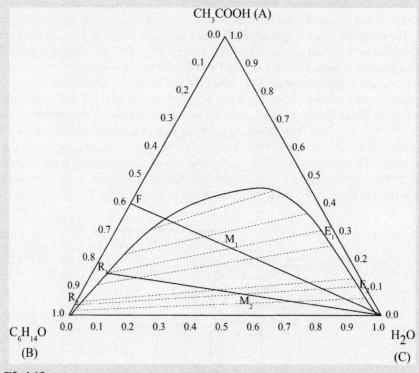

CH$_3$COOH (A)

C$_6$H$_{14}$O (B)   H$_2$O (C)

FIG. 4.10

Equilibrium relationship and mass balance calculation for acetic-acid-isopropyl ether-water system at 20 °C.

$$x_{MB} = Fx_{FB}/(F + S_1) = 150(0.6)/(150 + 100) = 0.36$$

$$x_{MC} = 1 - x_{MA} - x_{MB} = 0.4$$

Find the tie line passing through M_1 and find the compositions of raffinate and extract streams for stage 1.

Raffinate 1: $x_{1A} = 0.15$, $x_{1B} = 0.80$, $x_{1C} = 0.05$.
Extract 1: $y_{1A} = 0.305$, $y_{1B} = 0.04$, $y_{1C} = 0.655$.
The flow rates of raffinate and extract from stage 1 can be calculated by

$$R_1 = (F + S_1)\frac{y_{1A} - x_{MA}}{y_{1A} - x_{1A}} = 250\left(\frac{0.305 - 0.24}{0.305 - 0.15}\right) = 104.84 \, \text{kg/min}$$

$$E_1 = F + S_1 - R_1 = 145.16 \, \text{kg/min}$$

Stage 2:
Locate the mixing point M_2 for stage 2, this time the composition of the mixing point is:

$$x_{MA} = R_1 x_{1A}/(R_1 + S_2) = 104.84(0.15)/(104.84 + 100) = 0.077$$

$$x_{MB} = R_1 x_{1B}/(R_1 + S_2) = 104.84(0.80)/(104.84 + 100) = 0.409$$

$$x_{MC} = 1 - x_{MA} - x_{MB} = 0.514$$

Similarly, we can find the compositions of raffinate and extract streams for stage 2 from the tie line passing through M_2.

Raffinate 2: $x_{2A} = 0.038$, $x_{2B} = 0.946$, $x_{2C} = 0.016$.
Extract 2: $y_{2A} = 0.104$, $y_{2B} = 0.021$, $y_{2C} = 0.875$.
The flow rates of raffinate and extract from stage 2 can be calculated by:

$$R_2 = (R_1 + S_2)\frac{y_{2A} - x_{MA}}{y_{2A} - x_{2A}} = 204.84\left(\frac{0.104 - 0.077}{0.104 - 0.038}\right) = 83.80 \, \text{kg/min}$$

$$E_2 = R_1 + S_2 - R_2 = 121.04 \, \text{kg/min}$$

4.4.2 Cross-current contact with immiscible solvents

In this case, which is illustrated in Fig. 4.11, triangular diagrams are not required. If the initial solution contains a mass B of carrier **B** with a mass ratio X_f of solute **A**, then the selective solvent to be added will be a mass S of solvent **S**. On mixing and separating, a raffinate is obtained containing a mass ratio X_1 of solute, and an extract containing a mass ratio Y_1 of solute. A material balance on solute **A** gives:

$$BX_f = BX_1 + SY_1$$

or

$$\frac{Y_1}{X_1 - X_f} = -\frac{B}{S} \tag{4.3}$$

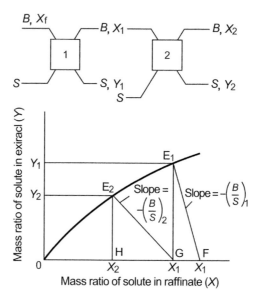

FIG. 4.11

Calculation of number of stages for co-current multiple-contact process, using immiscible solvents.

This process may be illustrated by allowing the point F to represent the feed solution and drawing a line FE_1, of slope $-(B/S)_1$, to cut the equilibrium curve at E_1. This then gives composition Y_1 of the extract and X_1 of the raffinate. If a further stage is then carried out by the addition of solvent S to the stream BX_1, then point E_2 is found on the equilibrium curve by drawing GE_2 of slope $-(B/S)_2$. Point E_2 then gives the compositions X_2 and Y_2 of the final extract and raffinate. This system may be used for any number of stages, with any assumed variation in the proportion of solvent S to raffinate from stage to stage.

If the distribution law is followed, then the equilibrium curve becomes a straight line given by $Y=mX$. The material balance on the solute may then be rewritten as:

$$BX_f = BX_1 + SY_1 = BX_1 + SmX_1 = (B + Sm)X_1$$

and

$$X_1 = \left[\frac{B}{B + Sm}\right] X_f. \tag{4.4}$$

If a further mass S of S is added to raffinate BX_1 to give an extract of composition Y_2 and a raffinate X_2 in a second stage, then:

$$BX_1 = BX_2 + SmX_2 = X_2(B + Sm)$$

and

$$X_2 = \left[\frac{B}{B + Sm}\right] X_1 = \left[\frac{B}{B + Sm}\right]^2 X_f \tag{4.5}$$

For n stages:

$$X_n = \left[\frac{B}{B + Sm}\right]^n X_f \tag{4.6}$$

and the number of stages is given by:

$$n = \frac{\ln\left(X_n/X_f\right)}{\ln\left[\dfrac{B}{B + Sm}\right]} \tag{4.7}$$

4.4.3 Countercurrent contact with immiscible solvents

If a series of mixing and separating vessels is arranged so that the flow is counter-current, then the conditions of flow are represented in Fig. 4.12, where each circle corresponds to a mixer and a separator. The initial solution F of the solute **A** in carrier **B** is fed to the first unit and leaves as raffinate R_1. This stream passes through the units and leaves from the nth unit as stream R_n. The fresh solvent **S** enters the nth unit and passes in the reverse direction through the units, leaving as extract E_1. The following definitions may be made:

$X =$ the ratio of solute to carrier in the raffinate streams and

$Y =$ the ratio of the solute to solvent in the extract streams.

If the two solvents are immiscible, the carrier in the raffinate streams remains as B, and the added solvent in the extract streams as S. The material balances for the solute can be written as.

(a) For the first stage: $BX_f + SY_2 = BX_1 + SY_1$
(b) For the nth stage: $BX_{n-1} + SY_{n+1} = BX_n + SY_n$
(c) For the whole unit: $BX_f + SY_{n+1} = BX_n + SY_1$

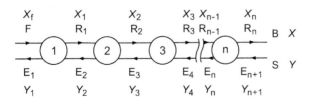

FIG. 4.12

Arrangement for multiple-contact extraction in countercurrent flow.

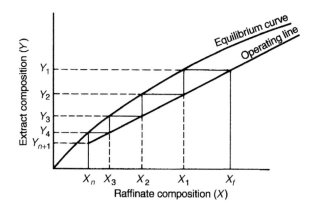

FIG. 4.13

Graphical method for determining the number of stages for the process shown in Fig. 4.12, using immiscible solvents.

or

$$Y_{n+1} = \frac{B}{S}(X_n - X_f) + Y_1 \tag{4.8}$$

This is the equation of a straight line of slope B/S, known as the *operating line*, which passes through the points (X_f, Y_1) and (X_n, Y_{n+1}). In Fig. 4.13, the equilibrium relation, Y_n against X_n, and the operating line are plotted, and the number of stages required to pass from X_f to X_n is found by drawing in steps between the operating line and the equilibrium curve. In this example, four stages are required. It may be noted that the operating line connects the compositions of the raffinate stream leaving and the fresh solvent stream entering a unit, X_n and Y_{n+1}, respectively.

Example 4.2

160 cm³/s of a solvent **S** is used to treat 400 cm³/s of a 10% by mass solution of **A** in **B**, in a three-stage countercurrent multiple-contact liquid–liquid extraction plant. Solvent **S** is immiscible in Carrier **B**. What is the composition of the final raffinate?

Using the same total amount of solvent, evenly distributed between the three stages for a cross-current mode operation, what would be the composition of the final raffinate?

Equilibrium data:

kg **A**/kg **B**	0.05	0.10	0.15
kg **A**/kg **S**	0.069	0.159	0.258
Densities (kg/m³)	$\rho_A = 1200$	$\rho_B = 1000$	$\rho_S = 800$

Solution

a) Countercurrent operation

Considering the solvent **S**, $160\,cm^3/s = 1.6 \times 10^{-4}\,m^3/s$ and

$$\text{mass flow rate of } \mathbf{S} = \left(1.6 \times 10^{-4} \times 800\right) = 0.128\,kg/s.$$

Considering the feed solution, $400\,cm^3/s = 4 \times 10^{-4}\,m^3/s$, containing, say, $a\ m^3/s$ **A** and $(4 \times 10^{-4} - a)m^3/s$ **B**.

Thus, mass flow rate of $\mathbf{A} = 1200a$ kg/s, and

$$\text{mass flow rate of } \mathbf{B} = \left(4 \times 10^{-4} - a\right)1000 = (0.4 - 1000a)\,kg/s.$$

Total mass flow of feed is $(0.4 + 200a)$ kg/s.

The mass fraction of solute in feed is 10%, thus:

$$0.10 = 1200a/(0.4 + 200a),$$

and

$$a = 3.39 \times 10^{-5}\,m^3/s$$

Therefore, mass flow rate of $\mathbf{A} = 0.041$ kg/s, mass flow rate of $\mathbf{B} = 0.366$ kg/s, and

$$X_f = (0.041/0.366) = 0.112.$$

The equilibrium data are plotted in Fig. 4.14 and the value of $X_f = 0.112$ is marked in. The slope of the equilibrium line is:

$$(\text{mass flow rate of } \mathbf{B})/(\text{mass flow rate of } \mathbf{S}) = (0.366/0.128) = 2.86.$$

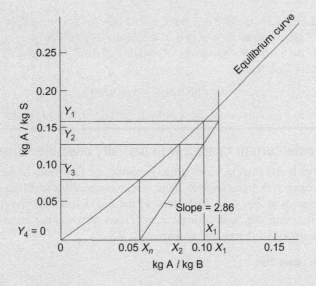

FIG. 4.14

Construction for Example 4.2.

Since pure solvent is added, $Y_{n+1}=0$ and a line of slope 2.86 is drawn in such that stepping off from $X_f=0.112$ to $Y_{n+1}=0$ gives exactly three stages. When $Y_4=0$, $X_n=X_3=0.057\,\text{kg/kg}$,

The composition of final raffinate is 0.057 kg **A**/kg **B**.

b) Cross-current contact

In this case, $(0.128/3)=0.0427\,\text{kg/s}$ of pure solvent **S** is fed to each stage.

Stage 1

$$X_f = (0.041/0.366) = 0.112.$$

From the equilibrium curve, the extract contains 0.18 A/kg S and $(0.18 \times 0.0427)=0.0077\,\text{kg/s}$ **A**.

Thus: raffinate from stage 1 contains $(0.041-0.0077)=0.0333\,\text{kg/s}$ **A** and 0.366 kg/s **B**.

and $X_1=(0.0333/0.366)=0.091$.

Stage 2

$$X_1 = 0.091.$$

From Fig. 4.14 the extract contains 0.14 kg A/kg S, or $(0.14 \times 0.0427)=0.0060\,\text{kg/s}$ **A**.

Thus, the raffinate from stage 2 contains $(0.0333-0.0060)=0.0273\,\text{kg/s}$ **A** and 0.366 kg/s **B**, therefore

$$X_2 = (0.0273/0.366) = 0.075.$$

Stage 3

$$X_2 = 0.075 \text{ kg/kg}.$$

From Fig. 4.14, the extract contains 0.114 kg A/kg S, mass flow of solute A from Stage 3 extract is, $(0.114 \times 0.0427)=0.0049\,\text{kg/s}$ **A**.

Thus, the raffinate from stage 3 contains $(0.0273-0.0049)=0.0224\,\text{kg/s}$ **A** and 0.366 kg/s **B**, and

$$X_3 = (0.0224/0.366) = 0.061.$$

As such, the composition of A in final raffinate $=0.061$ kg **A**/kg **B**.

4.4.4 Countercurrent contact with partially miscible solvents

In case carrier B and solvent S are partially soluble, the flow rates of the extract and raffinate streams vary from stage to stage in a countercurrent multistage extraction process as shown in Fig. 4.12. If the feed F, the final extract E_1, the fresh solvent $S=$ stream E_{n+1} and, the final raffinate R_n are fixed, then making material balances for the total streams entering and leaving each stage:

(a) Over the first unit

$$F + E_2 = R_1 + E_1$$

and

$$F + E_2 = R_1 + E_1 = M \tag{4.9}$$

(b) Over stages $1-n$

$$F + E_{n+1} = R_n + E_1 = M \tag{4.10}$$

and

$$F - E_1 = R_n - E_{n+1} = P \tag{4.11}$$

(c) Over the unit n

$$R_{n-1} + E_{n+1} = R_n + E_n$$

and

$$R_{n-1} - E_n = R_n - E_{n+1} = P \tag{4.12}$$

Thus the difference in quantity between the raffinate leaving a stage R_n, and the extract entering from next stage E_{n+1}, is constant. Similarly, it can be shown that the difference between the amounts of each component in the raffinate and the extract streams is constant. This means that, with the notation of a triangular diagram, lines joining any two points representing R_n and E_{n+1} pass through a common pole. The number of stages required to extract solute A from an initial concentration F to a final raffinate concentration R_n may then be found using a triangular diagram, shown in Fig. 4.15.

If the points F and S representing the compositions of the feed and fresh solvent **S** are joined, then the composition of a mixture of F and **S** is shown by point M where:

$$\frac{\text{MS}}{\text{MF}} = \frac{\text{mass of } F}{\text{mass of } \mathbf{S}}$$

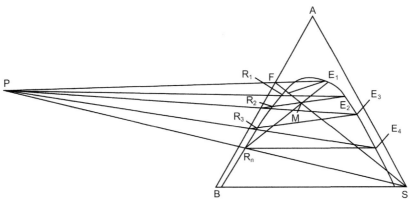

FIG. 4.15

Graphical method for determining the number of stages for the process shown in Fig. 4.12, using partially miscible solvents.

A line is drawn from R_n through M to give E_1 on the binodal curve and E_1F and SR_n to meet at the pole P. It may be noted that P represents an imaginary mixture, as described for the leaching problems discussed in leaching chapter.

In an ideal stage, the extract E_1 leaves in equilibrium with the raffinate R_1, so that the point R_1 is at the end of the tie line through E_1. To determine the extract E_2, PR_1 is drawn to cut the binodal curve at E_2. The points R_2, E_3, R_3, E_4, and so on, may be found in the same way. If the final tie line, say ER_4, does not pass through R_n, then the amount of solvent added is incorrect for the desired change in composition. In general, this does not invalidate the method, since it gives the required number of ideal stages with sufficient accuracy.

Example 4.3

A 50% solution of solute **A** in carrier **B** is extracted with a second solvent **C** in a countercurrent multiple contact extraction unit. The mass of **C** is 25% of that of the feed solution, and the equilibrium data are as given in Fig. 4.16. Determine the number of ideal stages required and the mass and concentration of the first extract if the final raffinate contains 15% of solute **A**.

Solution

The equilibrium data are replotted in Fig. 4.17 and F, representing the feed, is drawn in on AB at $x_{FA} = 0.50$. FC is joined and M located such that FM/MC = 0.25. R_n is located on the equilibrium curve such that $x_{nA} = 0.15$, $x_{nB} = 0.01$, and $x_{nC} = 0.84$. E_1 is located by projecting R_nM on to the curve and the pole P by projecting E_1F and CR_n. R_1 is found by projecting from E_1 along a tie-line and E_2 as the projection of PR_1. The working is continued in this way and it is found that R_5 is below R_n and hence five ideal stages are required.

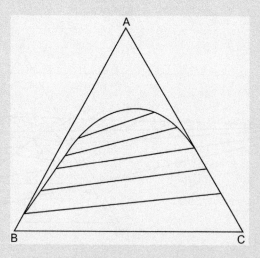

FIG. 4.16

Equilibrium data for Example 4.3.

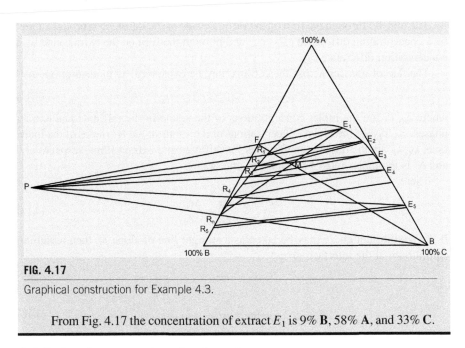

FIG. 4.17

Graphical construction for Example 4.3.

From Fig. 4.17 the concentration of extract E_1 is 9% **B**, 58% **A**, and 33% **C**.

4.4.5 Continuous extraction in columns

As Sherwood and Pigford [3] point out, the use of spray towers, packed towers or mechanical columns enables continuous countercurrent extraction to be obtained in a similar manner to that in gas absorption or distillation. Applying the two-film theory of mass transfer, explained in detail in Volume 1, Chapter 10, the concentration gradients for transfer to a desired solute from a raffinate to an extract phase are as shown in Fig. 4.18, which is similar to Fig. 4.1 for gas absorption.

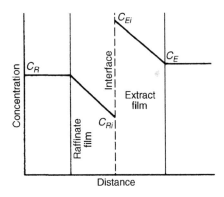

FIG. 4.18

Concentration profile near an interface.

The transfer through the film on the raffinate side of the interface is brought about by a concentration difference $c_R - c_{Ri}$, and through the film on the extract side by a concentration difference $c_{Ei} - c_E$.

The rate of transfer across these films may be expressed, as in absorption, as:

$$N' = k_R(c_R - c_{Ri}) = k_E(c_{Ei} - c_E) \tag{4.13}$$

where c_R, c_E are the molar concentrations of the solute in the raffinate and extract phases, c_{Ri}, c_{Ei} are the molar concentrations of the solute at the two sides of the interface, k_R, k_E are the transfer coefficients for raffinate and extract films, respectively, and N' is the molar rate of transfer per unit area.

Then:

$$\frac{k_R}{k_E} = \frac{c_{Ei} - c_E}{c_R - c_{Ri}} = \frac{\Delta c_E}{\Delta c_R} \tag{4.14}$$

If the equilibrium curve may be taken as a straight line of slope m, then assuming equilibrium at the interface gives:

$$c_{Ei} = mc_{Ri} \tag{4.15}$$

$$c_E = mc_R^* \tag{4.16}$$

and

$$c_E^* = mc_R \tag{4.17}$$

where c_E^* is the concentration in phase E in equilibrium with c_R in phase R, and c_R^* is the concentration in phase R in equilibrium with c_E in phase E.

The relations for mass transfer may also be written in terms of overall transfer coefficients K_R and K_E defined by:

$$N' = K_R(c_R - c_R^*) = K_E(c_E^* - c_E) \tag{4.18}$$

and by a similar reasoning to that used for absorption:

$$\frac{1}{K_R} = \frac{1}{k_R} + \frac{1}{mk_E} \tag{4.19}$$

and

$$\frac{1}{K_E} = \frac{1}{k_E} + \frac{m}{k_R} \tag{4.20}$$

Capacity of a column operating as continuous countercurrent unit

The capacity of a column operating as a countercurrent extractor, as shown in Fig. 4.19, may be derived as follows.

If L'_R, L'_E are the volumetric flow rates of raffinate and extract phases per unit area, a is the interfacial surface per unit volume, and Z is the height of packing, then, over a small height dZ, a material balance gives:

$$L'_R dc_R = L'_E dc_E \tag{4.21}$$

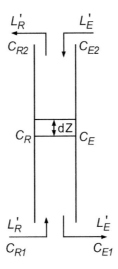

FIG. 4.19

Countercurrent flow in a packed column.

From Eq. (4.13):

$$L'_R dc_R = k_R(c_R - c_{Ri})adZ$$

and

$$\int_{c_{R2}}^{c_{R1}} \frac{dc_R}{c_R - c_{Ri}} = \frac{k_R a}{L'_R} Z \tag{4.22}$$

The integral on the left-hand side of this equation is known as the number of raffinate-film transfer units, $\mathbf{N}_R$, and the height of the raffinate-film transfer unit is:

$$\mathbf{H}_R = \frac{L'_R}{k_R a} \tag{4.23}$$

In a similar manner, and by analogy with absorption:

$$\mathbf{H}_E = \frac{L'_E}{k_E a} = \text{height of extract} - \text{film transfer unit} \tag{4.24}$$

$$\mathbf{H}_{OR} = \frac{L'_R}{K_R a} = \text{height of overall transfer unit based on concentration in raffinate phase}$$
$$\tag{4.25}$$

$$\mathbf{H}_{OE} = \frac{L'_E}{K_E a} = \text{height of overall transfer unit based on concentration in extract phase} \tag{4.26}$$

Since

$$\frac{1}{K_R} = \frac{1}{k_R} + \frac{1}{mk_E}$$

$$\frac{L'_R}{K_R} = \frac{L'_R}{k_R} + \left(\frac{L'_R}{mk_E} \times \frac{L'_E}{L'_E}\right) \tag{4.18}$$

Thus

$$\mathbf{H}_{OR} = \mathbf{H}_R + \frac{L'_R}{mL'_E}\mathbf{H}_E \tag{4.27}$$

and

$$\mathbf{H}_{OE} = \mathbf{H}_E + \frac{mL'_E}{L'_R}\mathbf{H}_R \tag{4.28}$$

These equations are the same form of relation as already obtained for distillation and for absorption, although it is only with dilute solutions that the group mL'_E/L'_R is constant. If Eqs (4.27) and (4.28) are combined, then:

$$\mathbf{H}_{OR} = \frac{L'_R}{mL'_E}\mathbf{H}_{OE} \tag{4.29}$$

The group L'_R/mL'_E is the ratio of the slope of the operating line to that of the equilibrium curve so that, when these two are parallel, it follows that $\mathbf{H}_{OR}$ and $\mathbf{H}_{OE}$ are numerically equal.

In deriving these relationships, it is assumed that L'_R and L'_E are constant throughout the tower. This is not the case if a large part of the solute is transferred from a concentrated solution to the other phase.

It is also assumed that the transfer coefficients are independent of concentration. For dilute solutions and where the equilibrium relation is a straight line, a simple expression may be obtained for determining the required height of a column, by the same method as given in adsorption chapter.

Thus $L'_R\ dc_R = K_R(c_R - c_R^*)a\ dZ$ may be integrated over the height Z and expressed as:

$$L'_R = K_R(\Delta c_R)_{\mathrm{lm}}aZ \tag{4.30}$$

where $(\Delta c_R)_{\mathrm{lm}}$ is the logarithmic mean of $(c_R - c_R^*)_1$ and $(c_R - c_R^*)_2$. This simple relation has been used by workers in the determination of K_R or K_E in small laboratory columns, although care should be taken when applying these results to other conditions.

Eqs (4.27) and (4.28) have been used as a basis of correlating mass transfer measurements in continuous countercurrent contactors. For example, Leibson and Beckmann [4] plotted $\mathbf{H}_{OE}$ against mL'_E/L'_R, as shown in Fig. 4.20, for a variety

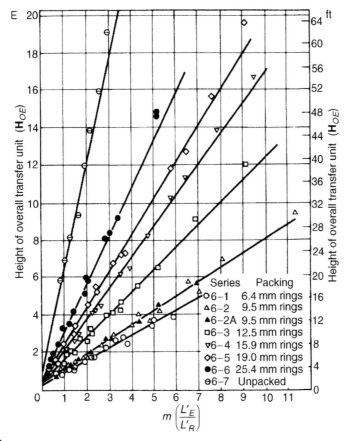

FIG. 4.20

Height of the transfer unit H_{OE} as a function of mL_E/L_R for the transfer of diethylamine from water to dispersed toluene using various types of packing [4].

of column packing materials and obtained good straight lines. Caution must be exercised, however, in drawing conclusions from such plots. Although equation 4.28 suggests that the intercepts and slopes of the lines are numerically equal to H_E and H_R, respectively, this is true only provided that both these quantities are independent of the flow ratio L'_E/L'_R. This is not always the case and, with packed towers, the height of the continuous phase film transfer unit does in fact depend upon the flow ratio, as discussed by Gayler and Pratt [5]. Under these conditions neither Eq. (4.27) nor Eq. (4.28) can be used to apportion the individual resistances to mass transfer between the two phases, and the film coefficients have to be determined by direct measurement.

Example 4.4

In the extraction of acetic acid from an aqueous solution with benzene in a packed column of height 1.4 m and of cross-sectional area 0.0045 m², the concentrations measured at the inlet and outlet of the column are as shown in Fig. 4.21. Determine the overall transfer coefficient and the height of the transfer unit.

Acid concentration in inlet water phase, $c_{W2} = 0.690$ kmol/m³.
Acid concentration in outlet water phase, $c_{W1} = 0.685$ kmol/m³.
Flow rate of benzene phase $= 5.7$ cm³/s or 1.27×10^{-3} m³/m²s.
Inlet benzene phase concentration, $c_{B1} = 0.0040$ kmol/m³.
Outlet benzene phase concentration, $c_{B2} = 0.0115$ kmol/m³.

The equilibrium relationship for this system is: $\frac{c_B^*}{c_W} = 0.0247$.

Solution

The acid transferred to the benzene phase is:

$$5.7 \times 10^{-6}(0.0115 - 0.0040) = 4.275 \times 10^{-8} \text{ kmol/s}$$

From the equilibrium relationship:

$$c_{B_1}^* = (0.0247 \times 0.685) = 0.0169 \text{ kmol/m}^3.$$

and

$$c_{B_2}^* = (0.0247 \times 0.690) = 0.0170 \text{ kmol/m}^3.$$

Thus:
Driving force at bottom, $\Delta c_1 = (0.0169 - 0.0040) = 0.0129$ kmol/m³ and
Driving force at top, $\Delta c_2 = (0.0170 - 0.0115) = 0.0055$ kmol/m³.
Therefore, Log mean driving force, $\Delta c_{lm} = 0.0087$ kmol/m³.

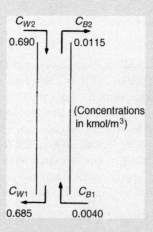

C_{W2} C_{B2}
0.690   0.0115

(Concentrations in kmol/m³)

C_{W1} C_{B1}
0.685   0.0040

FIG. 4.21

Data for Example 4.4.

Thus,

$$K_B a = \frac{\text{moles transferred}}{\text{volume of packing} \times \Delta c_{\text{lm}}} = \frac{(4.275 \times 10^{-8})}{(0.0063 \times 0.0087)}$$

$$= 7.8 \times 10^{-4} \ (1/\text{s})$$

and

$$\mathbf{H}_{OB} = (1.27 \times 10^{-3})/(7.8 \times 10^{-4}) = 1.63 \ \text{m}.$$

4.5 **Extraction equipment**

In most industrial applications, multistage countercurrent contacting is required. The hydrodynamic driving force necessary to induce countercurrent flow and subsequent phase separation may be derived from the differential effects of either gravity or centrifugal force on the two phases of different densities. Essentially there are two types of design by which effective multistage operation may be obtained:

(a) *stage-wise contactors*, in which the equipment includes a series of physical stages where the phases are mixed and separated, and
(b) *differential contactors*, in which the phases are continuously brought into contact with complete phase separation only at the exits from the unit.

The three factors, the inducement of countercurrent flow, stage-wise or differential contacting and the means of effecting phase separation are the basis of a classification of contactors proposed by Hanson [1] which is summarised in Table 4.1.

Typical regions for application of contactors of different types are given in Table 4.2. The choice of a contactor for a particular application requires the consideration of several factors including chemical stability, the value of the products and the rate of phase separation. Occasionally, the extraction system may be chemically unstable and the contact time must then be kept to a minimum by using equipment such as a centrifugal contactor.

4.5.1 **Stage-wise equipment**

The mixer–settler

In the mixer–settler, the solution and solvent are mixed by some form of agitator in the *mixer*, and then transferred to the *settler* where the two phases separate to give an extract and a raffinate. The mixer unit, which is usually a circular or square vessel with a stirrer, may be designed on the principles given in Volume 1, Chapter 7. In the settler, the separation is often gravity-controlled, and the liquid densities and the form of the dispersion are important parameters. It is necessary to establish the principles which determine the size of these units and to have an understanding of the criteria governing their internal construction. While the mixer and settler are first

Table 4.1 Classification of contactors.

Countercurrent flow produced by	Phase interdispersion by	Differential contactors	Stage-wise contactors
Gravity	Gravity	Group A Spray column Packed column	Group B Perforated plate column
	Pulsation	Group C Pulsed packed column Pulsating plate column	Group D Pulsed sieve plate column Controlled cycling column
	Mechanical agitation	Group E Rotating disc contactor Oldshue–Rushton column Zeihl column Graesser contractor	Group F Scheibel column Mixer–settlers
Centrifugal force	Centrifugal force	Group G Podbielniak Quadronic De Laval	Group H Westfalia Robatel

Table 4.2 Typical regions of application of contactor groups listed in Table 4.1 [1].

	System criterion	Modest throughput	High throughput
Small number of stages required	Chemically stable Easy phase separation Low value	A, B or mixer–settler	E or F
	Chemically stable Appreciable value	C or D	E or F (not mixer–settler)
	Chemically unstable Slow phase separation	G or H	G or H
Large number of stages required	Chemically stable Easy phase separation Low value	B, C, D or mixer–settler	E or F
	Chemically stable Appreciable value	C or D	E or F (not mixer–settler)
	Chemically unstable Slow phase separation	G or H	G or H

considered as separate items, it is important to appreciate that they are essential component parts of a single processing unit.

The mixer. As a result of the agitation achieved in a mixer, the two phases are brought to, or near to, equilibrium so that one theoretical stage is frequently obtained in a single mixer where a physical extraction process is taking place. Where a chemical reaction occurs, the kinetics must be established so that the residence time and the hold-up may be calculated. The hold-up is the key parameter in determining size, and scale-up is acceptably reliable, although a reasonably accurate estimate of the power required is important with large units. For a circular vessel, baffles are required to give the optimum degree of agitation and the propeller, which should be about one-third of the diameter of the vessel, and should be mounted just below the interface and operate with a tip speed of 3–15 m/s, depending on the nature of the propeller or turbine. A shroud around the propeller helps to give good initial mixing of the streams, and it also provides some pumping action and hence improves circulation.

As discussed in Volume 1, Chapter 7, the two key parameters determining power are the Reynolds number for the agitator and the power number. The Reynolds number should exceed 10^4 for optimum agitation. This gives a power number of about 6 for a fully baffled tank. It is important to note that the power number and the tip speed cannot both be kept constant in scale-up.

The settler. In this unit, gravitational settling frequently occurs and, in addition, coalescence of droplets must take place. Baffles are fitted at the inlet in order to aid distribution. The rates of sedimentation and coalescence increase with drop size, and therefore, excessive agitation resulting in the formation of very small drops should be avoided. The height of the dispersion band Z_B is influenced by the throughput since a minimum residence time is required for coalescence to occur. This height Z_B is related to the dispersed and continuous phase superficial velocities, u_d and u_c by:

$$Z_B = \text{constant}(u_d + u_c) \tag{4.31}$$

Pilot tests may be necessary to achieve satisfactory design, although the sizing of the settler is a difficult problem when the throughputs are large.

Combined mixer–settler units

Recent work has emphasised the need to consider the combined mixer–settler operation, particularly in metal extraction systems where the throughput may be very large. Thus Warwick and Scuffham [6] give details of a design, shown in Fig. 4.22, in which the two operations are effected in one combined unit. The impeller has swept-back vanes with double shrouds, and the two phases meet in the draught tube. A baffle on top of the agitator reduces air intake and a baffle on the inlet to the settler is important in controlling the flow pattern. This arrangement gives a good performance and is mechanically neat. Raising the impeller above the draught tube increases internal recirculation which in turn improves the stage efficiency, as shown in Fig. 4.23. The effect of agitation on the thickness of the dispersion band is shown in Fig. 4.24. The depth of the dispersion band Z_B varies with the total flow per unit area. While this work was

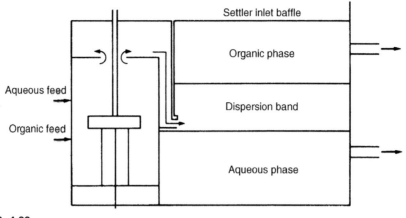

FIG. 4.22

Mixer–settler.

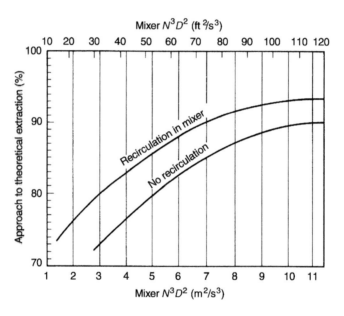

FIG. 4.23

The effect of variation of mixer internal recirculation on extraction efficiency [6].

primarily aimed at a design for copper-extraction processes, it is clear that there is scope for further important applications of these units.

The segmented mixer–settler. Novel features for a combined mixer–settler are incorporated in a unit from Davy International, described by Jackson et al. [7] and illustrated in Fig. 4.25, where specially designed KnitMesh pads are used to

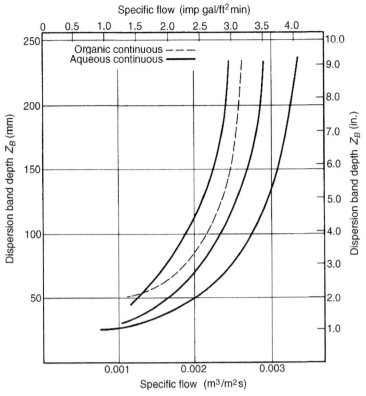

FIG. 4.24

The effect of variation of phase continuity and mixer $N^3 D^2$ on settler dispersion band depth [6].

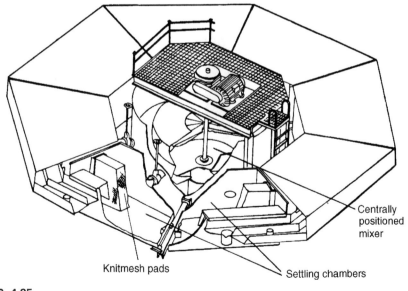

FIG. 4.25

Segmented mixer–settler [7].

speed up the rate of coalescence. The centrally situated mixer is designed to give the required holdup, and the mixture is pumped at the required rate to the settler which is formed in segments around the mixer, each fed by individual pipework. The Knit-Mesh pads which are positioned in each segment are 0.75–1.5 m in depth. One key advantage of this design is that the holdup of the dispersed phase in the settler is reduced to about 20% of that in the mixer, as compared with 50% with simple gravity settlers.

The use of KnitMesh in a coalescer for liquid–liquid separation applications is illustrated in Fig. 4.26 where an oil–water mixture enters the unit and passes through the coalescer element. As it does so, the water droplets coalesce and separation occurs between the oil and the water. After passing through the KnitMesh, the two phases are readily removed from the top and bottom of the unit.

Kühni have recently developed a mixer–settler column which, as its name suggests is a series of mixer–settlers in the form of a column. The unit consists of a number of stages installed one on top of another, each hydraulically separated, and each with a mixing and settling zone as shown in Fig. 4.27. With this design, it is possible to eliminate some of the main disadvantages of conventional mixer–settlers while maintaining stage-wise phase contact. As the mixer turbines do not need to transport the liquids from stage to stage, the speed of rotation can be adjusted so as to achieve optimum droplet sizes. Because it is necessary to settle the liquid phase in every stage, the specific throughput is only $0.003\,m^3/m^2s$ which is considerably lower than with more conventional columns. Residence times are of the order of 900 s and extraction may be controlled by the residence time and pH. It is possible to combine extraction with reaction in such a system.

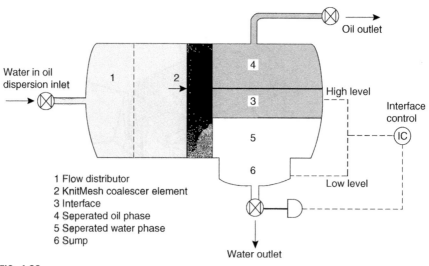

1 Flow distributor
2 KnitMesh coalescer element
3 Interface
4 Seperated oil phase
5 Seperated water phase
6 Sump

FIG. 4.26

Flow in a KnitMesh separator.

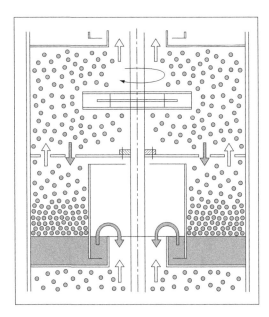

FIG. 4.27

Kühni mixer–settler column.

Baffle-plate columns

These are simple cylindrical columns provided with baffles to direct the flow of the dispersed phase, as shown in Fig. 4.28. The efficiency of each plate is very low, though since the baffles can be positioned very close together at 75–150 mm, it is possible to obtain several theoretical stages in a reasonable height.

The Scheibel column

One of the problems with perforated plate and indeed packed columns is that redispersion of the liquids after each stage is very poor. To overcome this, Scheibel and Karr [8] introduced a unit, shown in Fig. 4.29, in which a series of agitators is mounted on a central rotating shaft. Between the agitators is fitted a wire mesh section which successfully breaks up any emulsions. Some results for a column 292 mm diameter, with 100 mm diameter agitators and with packing sections 230 and 340 mm, are shown in Fig. 4.30. It is found that one theoretical stage is obtained in a height of 0.45–0.75 m. This is a significant improvement on that usually obtained in a packed column. Although there are few data on large units, there should be no fall in efficiency as the diameter is increased.

4.5.2 Differential contact equipment

Spray columns

Two methods of operating spray columns are shown in Fig. 4.31. Either the light or heavy phase may be dispersed. In the former case as illustrated in Fig. 4.31a, the light

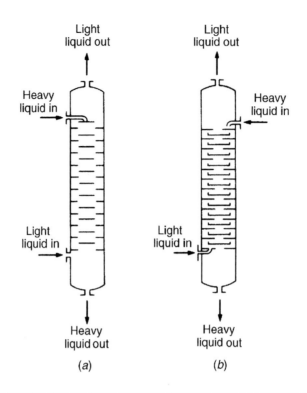

FIG. 4.28

Baffle-plate column.

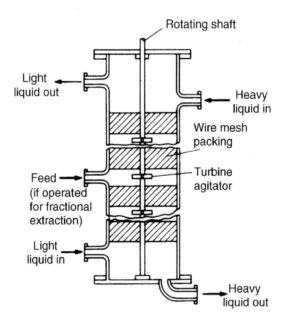

FIG. 4.29

Scheibel column [8].

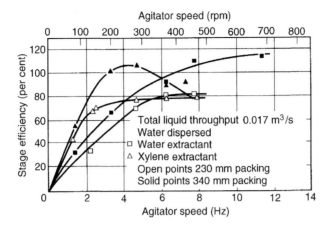

FIG. 4.30

Effect of agitator speed on efficiency for the system acetone–xylene–ester [8].

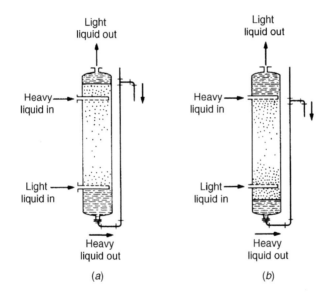

FIG. 4.31

Spray towers.

phase enters from a distributor at the bottom of the column and the droplets rise through the heavier phase, finally coalescing to form a liquid–liquid interface at the top of the tower. Alternatively, the heavier phase may be dispersed, in which case the interface is held at the bottom of the tower as shown in Fig. 4.31b. Although spray towers are simple in construction, they are inefficient because considerable recirculation of the continuous phase takes place. As a result, true countercurrent flow is not maintained and up to 6 m may be required for the height of one theoretical stage. There is very little turbulence in the continuous phase and lack of interface renewal, and appreciable axial mixing results in poor performance.

Because the droplets of dispersed phase rise or fall through the continuous phase under the influence of gravity, it will be apparent that there is a limit to the amount of dispersed phase that can pass through the tower for any given flow rate of continuous phase. Thus referring to Fig. 4.31a, any additional light phase fed to the bottom of the tower, in excess of that which can pass upwards under the influence of gravity, will be rejected from the bottom of the unit and the tower is then said to be flooded, as discussed by Blanding and Elgin [9]. It is therefore important to be able to predict the conditions under which flooding will occur, so that the diameter of tower may be calculated for any required throughput. Although no complete analysis of this problem has, as yet, been achieved, it may be treated approximately in the following way.

Dispersed phase hold-up. Fig. 4.32 represents a section of a spray tower of unit cross-sectional area. The light phase is assumed to be dispersed, and the volumetric flow rates per unit area of the two phases are L'_d and L'_c, respectively. The superficial velocities u_d, u_c of the phases are therefore also equal to L'_d and L'_c. Under steady-state conditions the amount of dispersed phase held up in the tower in the form of droplets is conveniently expressed in terms of the fractional hold-up j, that is the fractional volume of the two-phase dispersion occupied by the dispersed phase. This may also be thought of as the fraction of the cross-sectional area of the tower occupied by the dispersed phase. The velocity of the dispersed phase relative to the tower is therefore L'_d/j. Similarly, the relative velocity of the continuous phase is equal to $L'_c/(1-j)$. If the overall flow is regarded as strictly countercurrent, the sum of these two velocities will be equal to the velocity of the dispersed phase relative to the continuous phase, u_r or:

$$u_r = \frac{L'_d}{j} + \frac{L'_c}{(1-j)} \qquad (4.32)$$

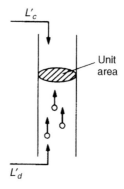

FIG. 4.32

Section of a spray tower.

In the case of spray towers it has been shown by Thornton [10] that u_r is well represented by $\bar{u}_0(1 - j)$ where $\bar{u}_0$ is the velocity of a single droplet relative to the continuous phase, and is termed the droplet characteristic velocity. The term $(1 - j)$ is a correction to $\bar{u}_0$ which takes into account the way in which the characteristic velocity is modified when there is a finite population of droplets present, as opposed to a single droplet. It must be seen therefore that for very dilute dispersions, that is as $j \rightarrow 0$, $\bar{u}_0(1 - j) \rightarrow \bar{u}_0$. On the other hand, as the fractional hold-up increases, the relative velocity of the dispersed phase decreases due to interactions between the droplets. Substituting for u_r, Eq. (4.32) may be written as:

$$\frac{u_d}{j} + \frac{u_c}{1-j} = \bar{u}_0(1 - j) \tag{4.33}$$

This equation relates the hold-up to the flow rates of the phases and column diameter through the characteristic velocity, $\bar{u}_0$. It therefore gives a method of calculating the hold-up for a given set of flow rates if $\bar{u}_0$ is known. Conversely, equation 4.33 may be used to calculate $\bar{u}_0$ from experimental hold-up measurements made at different flow rates. Thus, if hold-up data are plotted with $L'_d + (j/(1 - j))L'_c$ as the ordinate against $j(1 - j)$ as the abscissa, a linear plot is obtained which passes through the origin and which has a slope equal to $\bar{u}_0$ [10].

Flooding-point condition. A plot of equation 4.33 in the form of u_d and u_c against j for a typical value of $\bar{u}_0 = 0.042$ m/s is shown in Fig. 4.33. Although Eq. (4.33) is cubic, only the root which lies between zero and the flooding-point values of u_d and u_c is realisable in practice. These portions of the hold-up curves are shown by full lines in Fig. 4.33. If the flow rate of one of the phases is kept constant, an increase in the flow rate of the other phase will result in an increased value of the hold-up until the flooding-point is reached. The latter corresponds to the maxima in the two curves shown in Fig. 4.33, so that the flooding-point condition is given by $du_d/dj = du_c/dj = 0$. Since those portions of the hold-up curves beyond the flooding-point are unrealistic in practice, the value of j corresponding to the maximum flow rates also represents the limiting hold-up at flooding, although this condition is not obtainable mathematically from Eq. (4.33).

Carrying out the differentiation described previously gives:

$$u_{df} = 2\bar{u}_0 j_f^2 \left(1 - j_f\right) \text{ when } du_c/dj = 0 \tag{4.34}$$

and

$$u_{cf} = \bar{u}_0 \left(1 - j_f\right)^2 \left(1 - 2j_f\right) \text{ when } du_d/dj = 0 \tag{4.35}$$

The value of j_f, the limiting hold-up at the flooding-point, may be obtained by eliminating $\bar{u}_0$ between equations 4.34 and 4.35 and solving for j_f.
Thus:

$$j_f = \frac{(r^2 + 8r)^{0.5} - 3r}{4(1 - r)} \tag{4.36}$$

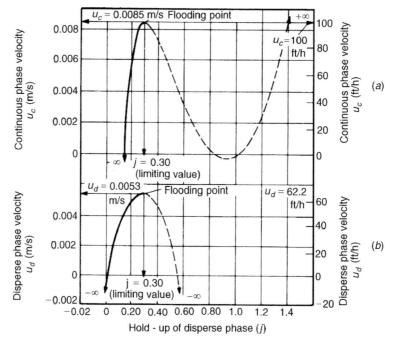

FIG. 4.33

Spray tower (A) Continuous phase velocity (u_c) as function of hold-up of disperse phase (j) ($u_c=0.0085$ m/s, $\bar{u}_0=0.042$ m/s) and (B) Disperse phase velocity (u_d) as a function of hold-up of disperse phase (j) ($u_d=0.0053$ m/s, $\bar{u}_0=0.0042$ m/s).

where

$$r = u_{df}/u_{cf} \qquad (4.37)$$

The derivation of Eqs (4.34) and (4.35) has been carried out assuming that $\bar{u}_0$ is constant and independent of the flow rates, up to and including the flooding-point. This in turn assumes that the droplet size is constant and that no coalescence occurs as the hold-up increases. While this assumption is essentially valid in properly designed spray towers, this is certainly not the case with packed towers. Eqs (4.34) and (4.35) cannot therefore be used to predict the flooding-point in packed towers and a more empirical procedure must be adopted.

A typical form of flooding-point curve for spray towers is shown in Fig. 4.34 where values of u_{df} are plotted against u_{cf}. The limiting values of each flow rate as the other approaches zero may be determined readily from Eqs (4.34) and (4.35). Thus, when $u_{df} \to 0$, $j_f \to 0$ and in the limit $u_{cf}=\bar{u}_0$. Similarly, as $u_{cf} \to 0$, $j_f \to 0.50$ and in the limit $u_{df}=\bar{u}_0/4$. In addition, by combining Eqs (4.34) and (4.35) and differentiating u_{df} with respect to u_{cf}, it is seen that the slope of the flooding-point curve is given by:

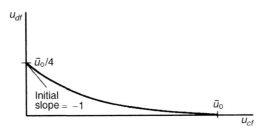

FIG. 4.34

Flooding-point curve for spray tower.

$$\frac{du_{df}}{du_{cf}} = -\frac{j_f}{\left(1 - j_f\right)} \tag{4.38}$$

It is apparent therefore that the curve meets the abscissa tangentially as $u_{df} \to 0$, and meets the ordinate with a slope of -1 as $u_{cf} \to 0$.

Eqs (4.34) and (4.35) may be used for correlating or extending incomplete flooding-point data. Thus, for example, if flooding-point–hold-up data are available for a range of flow rates, a plot of u_{df} as ordinate against $j_f^2(1 - j_f)$ as abscissa will result in a straight line through the origin, of slope $2\bar{u}_0$. Alternatively, if $\bar{u}_0$ is known, either Eqs (4.34) or (4.35) may be used to calculate the tower diameter for a required throughput. The actual area of tower to be used is then taken as twice this area to ensure that the unit operates at below 50% of the flooding-point flow rates.

Droplet characteristic velocity and droplet size. The characteristic velocity may be calculated from hold-up measurements below the flooding-point using Eq. (4.33), or from hold-up measurements at the flooding-point using Eqs (4.34) or (4.35). In many instances, however, such data are not available, and it is then necessary to be able to predict $\bar{u}_0$ from a knowledge of the physical properties of the extraction system. This involves predicting firstly the mean droplet size which is present in the tower, and then the corresponding mean droplet velocity. A full discussion of the problem is beyond the scope of this chapter although for low nozzle velocities, the mean droplet size may be established by means of the correlation proposed by Hayworth and Treybal [11]. The corresponding droplet velocity may then be calculated by the methods of Hu and Kintner [12] or Klee and Treybal [13].

Interfacial area. The interfacial area per unit volume of tower, or specific area, is given by:

$$a = \frac{\text{Total area}}{\text{Volume}} = \frac{\pi d_s^2}{\frac{\pi d_s^3}{6} \frac{1}{j}} = \frac{6j}{d_s} \tag{4.39}$$

where j is the fractional hold-up and d_s is the mean droplet size, defined in Chapter 1, eq. (1.14), as:

$$d_s = \frac{\sum n_1 d_1^3}{\sum n_1 d_1^2} \tag{4.40}$$

where n_1 is the number of droplets of diameter d_1 in a population.

Example 4.5

The number and size of droplets in a given population are:

Diameter (mm)	2	3	4	5	6
Number (–)	30	120	200	80	20

Estimate the surface mean droplet size.

Solution

The mean droplet size is given by Eqs (4.40) and (1.14) as:

$$d_s = \sum n_1 d_1^3 / \sum n_1 d_1^2$$

where n_1 is the number of droplets of diameter d_1 in a population. Based on the calculation results in Table 4.3, finally we have

Table 4.3 Calculation of mean droplet diameter.

n_1	d_1 (mm)	d_1^2	d_1^3	$n_1 d_1^3$	$n_1 d_1^2$
30	2	4	8	240	120
120	3	9	27	3240	1080
200	4	16	64	12,800	3200
80	5	25	25	100,000	2000
200	6	36	216	4320	720
Total				30,600	7120

$$\sum n_1 d_1^3 = 30,600 \quad \text{and} \quad \sum n_1 d_1^2 = 7120$$

and

$$d_s = (30600/7120) = 4.30 \text{ mm}$$

Mass transfer. It is not yet possible to predict the mass transfer coefficient with a high degree of accuracy because the mechanisms of solute transfer are but imperfectly understood as discussed by Light and Conway [14], Coulson and Skinner [15] and Garner and Hale [16]. In addition, the flow in spray towers is not strictly countercurrent due to recirculation of the continuous phase, and consequently the effective overall driving force for mass transfer is not the same as that for true countercurrent flow.

As a first approximation, the dispersed-phase film coefficients may be calculated using the Handlos and Baron [17] model for circulating liquid spheres. The continuous phase film coefficients may be estimated from the correlation of Ruby and Elgin [18], and the overall coefficients then calculated using Eqs (4.19) or (4.20).

Such procedures are, however, only approximate at best, and further work is required before generalised correlations can be developed which take into account the effect of recirculation and the different mass transfer rates at the dispersed phase entry nozzles, during droplet rise and during coalescence at the top of the tower. Typical data [18–21] and standard texts [22–24] for HTU values covering a range of extraction systems are available in the literature. In this respect, scale-up from pilot data raises problems. Reference may also be made to the work of Hanson [1] for further discussion on mass transfer in liquid–liquid systems.

Hitherto no mention has been made of interfacial effects accompanying the mass transfer process. Under certain conditions the presence of an undistributed solute gives rise to the Marangoni effect, discussed by Davies and Rideal [25] and Groothuis and Zuiderweg [26], which results in interfacial turbulence and droplet coalescence. These effects are generally more obvious when solute transfer takes place from an organic solvent droplet into an aqueous continuous phase. In the reverse direction of transfer, interfacial disturbances of this nature are frequently absent. The existence of interfacial phenomena of this type presents yet one more obstacle to a quantitative interpretation of the mass transfer process. In addition, when coalescence is marked, the droplet size is no longer constant up the tower and the derivations of Eqs (4.34) and (4.35), which are based upon constant $\bar{u}_0$ values, are no longer valid. From a practical point of view, these equations may still be used for design purposes, since the effect of droplet coalescence is to enhance the flooding-point beyond the value predicted by Eqs (4.34) and (4.35), so that the tower operates in practice below 50% of the flooding-point.

Packed columns

Although packed columns are similar to those used for distillation and absorption, it must be noted that the flow rates of the phases are very different and two liquid phases are always present. The packing increases the interfacial area, and considerably increases mass transfer rates compared with those obtained with spray columns because of the continuous coalescence and break-up of the drops, though the HTU values are still high. Packed columns are unsuitable for use with dirty liquids, suspensions, or high viscosity liquids. They have proved to be satisfactory in the petroleum industry, though at present they cannot be scaled-up to cope with the very high flows encountered in metallurgical processes. They are economical in the use of ground space.

Minimum packing size. For reproducible results the minimum packing size should be such that the mean void height is not less than the mean droplet diameter. As reported by Gayler et al. [27], this critical packing size is given by:

$$d_{\text{crit}} = 2.42 \left(\frac{\sigma}{\Delta \rho g} \right)^{0.5} \tag{4.41}$$

Dispersed phase hold-up. Three regimes of flow may be distinguished with packings greater than the critical size:

(a) Region of linear hold-up. At low dispersed phase flow rates the droplets move freely within the interstices of the packing and the hold-up increases linearly with dispersed phase flow rate.

(b) Region of rapidly increasing hold-up. Above a hold-up of approximately 10%, corresponding to the lower transition point, the hold-up increases more rapidly with increasing dispersed phase flow. This is due to the onset of hindered movement of the droplets within the packing voids.

(c) Region of constant hold-up. At higher dispersed phase flow rates, the upper transition point is encountered above which droplet coalescence occurs. In this region, the hold-up remains constant as the flow rate is increased until the flooding-point is reached. It is apparent therefore that a plot of u_d against j is of similar form to that shown in Fig. 4.33, except that the point corresponding to the flooding point for spray towers coincides with the upper transition point for packed towers.

Below the upper transition point, hold-up data may be correlated by an equation analogous to that used for spray towers, except as Gayler et al. [27] and Gayler and Pratt [28] point out, a correction must be introduced to take the packing voidage e into account. Thus:

$$\frac{u_d}{j} + \frac{u_c}{(1-j)} = e\bar{u}_0(1-j) \tag{4.42}$$

Droplet size and interfacial area. In the absence of interfacial effects accompanying mass transfer, the droplets break down by impact with elements of packing and finally reach an equilibrium size which is independent of the packing size. Conversely, small droplets gradually coalesce until the equilibrium size is attained. Pratt and his coworkers [5,29] showed that the mean droplet size attained in the tower is well represented by:

$$d_0 = 0.92 \left(\frac{\sigma}{\Delta\rho g}\right)^{0.5} \left(\frac{\bar{u}_0 e j}{u_d}\right) \tag{4.43}$$

Correcting Eq. (4.39) for the packing voidage, the specific area in the tower is given by:

$$a = \frac{6ej}{d_0} \tag{4.44}$$

Droplet characteristic velocity. Gayler et al. [27] proposed a graphical correlation for $\bar{u}_0$ which takes into account the fact that the droplets are periodically halted by collisions with elements of packing and then accelerate to some fraction of their terminal velocity, u_0, before being deflected or stopped again by a further collision. Fig. 4.35 enables u_0 to be determined from a knowledge of the physical properties of the system.

Once $\bar{u}_0$ has been determined, the hold-up for any particular set of flow rates may be calculated from Eq. (4.42), the mean droplet size from Eq. (4.43), and the specific area from Eq. (4.44).

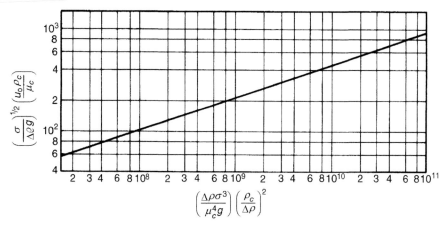

FIG. 4.35

Chart for calculation of droplet terminal velocity, u_0 [27].

Flooding-point. Because the flooding-point is no longer synonymous with that for spray towers, Eqs (4.34) and (4.35) predict only the upper transition point. Dell and Pratt [30] adopted a semi-empirical approach for the flooding-point by consideration of the forces acting on the separate dispersed and continuous phase channels which form when coalescence sets in just below the flooding-point. The following expression correlates data to within $\pm 20\%$:

$$1 + 0.835 \left(\frac{\rho_d}{\rho_c}\right)^{1/4} \left(\frac{u_d}{u_c}\right)^{1/2} = J \left[\frac{u_c^2 a_p}{g e^3} \left(\frac{\rho_c}{\Delta\rho}\right) \left(\frac{\sigma}{\sigma_T}\right)^{1/4}\right]^{-1/4} \tag{4.45}$$

Mass transfer. As in the case of spray columns, it is not yet possible to predict mass transfer rates from first principles. In the absence of any reliable correlations, typical values of overall [20,31] and film [32,33] coefficients can be used. A comprehensive summary is given in Perry's Chemical Engineers' Handbook [22].

Example 4.6

In order to extract acetic acid from dilute aqueous solution with isopropyl ether, the two immiscible phases are passed countercurrently through a packed column 3 m in length and 75 mm in diameter. It is found that, if 0.5 kg/m^{2}s of the pure ether is used to extract 0.25 kg/m^{2}s of 4.0% acid by mass, then the ether phase leaves the column with a concentration of 1.0% acid by mass.

Calculate:

a) the number of overall transfer units based on the raffinate phase and
b) the overall extraction coefficient based on the raffinate phase.

The equilibrium relationship is given by:

$$(\text{kg acid/kg isopropyl ether}) = 0.3 \, (\text{kg acid/kg water}).$$

Solution

Cross-sectional area of packing $=(\pi/4)0.075^2=0.0044\,\mathrm{m}^2$ and
Volume of packing $=(0.0044\times3)=0.0133\,\mathrm{m}^3$.
The concentration of acid in the extract $=1.0\%$ or $0.01\,\mathrm{kg/kg}$. Thus,
Mass flux of acid transferred to the ether $=0.5(0.01-0)=0.005\,\mathrm{kg/m}^2\mathrm{s}$.
Mass flow rate of transferred acid to the ether $=(0.005\times0.0044)=2.2\times10^{-5}\,\mathrm{kg/s}$.
Acid in the aqueous feed $=(0.25\times0.04)=0.01\,\mathrm{kg/m}^2$ s.
Thus: acid in raffinate $=(0.01-0.005)=0.005\,\mathrm{kg/m}^2\mathrm{s}$, and
Concentration of acid in the raffinate $=(0.005/0.25)=0.02\,\mathrm{kg/kg}$.
At the top of the column:

$$c_{E2}=0.01\,\mathrm{kg/kg}\ \text{and}\ c_{R_2}^*=0.01/0.3=0.0333\,\mathrm{kg/kg}.$$

Thus concentration driving force based on raffinate is:

$$\Delta c_2=c_{R_2}-c_{R_2}^*=0.04-0.0333=0.0067\,\mathrm{kg/kg}$$

At the bottom of the column:
$$c_{E1}=0\ \text{and}\ c_{R_1}^*=0.$$
Thus concentration driving force based on raffinate is: $\Delta c_1=(0.02-0)=0.02\,\mathrm{kg/kg}$.
The logarithmic mean driving force based on raffinate stream is:

$$(\Delta c_R)_{\mathrm{lm}}=(0.02-0.0067)/\ln(0.02/0.0067)=0.012\,\mathrm{kg/kg}.$$
$$K_Ra=\text{mass transferred}/[\text{volume of packing}\times(\Delta c_R)_{\mathrm{lm}}]$$
$$=2.2\times10^{-5}/(0.0133\times0.012)=\underline{0.138\,\mathrm{kg/m}^3\mathrm{s}.}$$

From Eq. (4.25), we know the height of an overall transfer unit,

$$\mathbf{H}_{OR}=L'_R/K_Ra=(0.25/0.138)=1.81\,\mathrm{m}$$

and the number of overall transfer units $=(3/1.81)=1.66$.

Rotary annular columns and rotary disc columns

With these columns as described by Thornton and Pratt [34] and Vermijs and Kramers [35], mechanical energy is provided to form the dispersed phase. The equipment is particularly suitable for installations where a moderate number of stages is required, and where the throughput is considerable. A well dispersed system is obtained with this arrangement.

Flooding-point data may be correlated by Eqs (4.34) and (4.35) using the droplet characteristic velocity concept as discussed by Thornton and Pratt [34], since coalescence is absent.

Pulsed columns

In order to prevent coalescence of the dispersed drops, van Dijck [36] and others have devised methods of providing the whole of the continuous phase with a pulsed motion. This may be done, either by some mechanical device, or by the introduction of compressed air.

The pulsation markedly improves performance of packed columns and the HTU is about half that of an unpulsed column. There are advantages in using gauze-type packings since the pulsation operation often breaks ceramic rings. Perforated plates, as used in distillation, may also be used for pulsed extraction. Pulsed packed columns have been used in the nuclear industry though they are limited in size since the pulsation system is difficult to arrange and the pulsation itself demands strengthening of the column.

Flooding-point data have been correlated by equations analogous to 4.34 and 4.35. This procedure is permissible since, as Thornton [37] and Logsdail and Thornton [38] report, pulsed columns can be operated up to the flooding-point with no droplet coalescence.

Centrifugal extractors

If separation is difficult in a mixer–settler unit, a centrifugal extractor may be used in which the mixing and the separation stages are contained in the same unit which operates as a differential contactor.

In the *Podbielniak* contactor, the first of the rotating machines to be developed, the heavy phase is driven outwards by centrifugal force and the light phase is displaced inwards. Referring to Fig. 4.36 which illustrates a unit produced by Baker Perkins, the heavy phases enters at *D*, passes to *J* and is driven out at *B*. The light

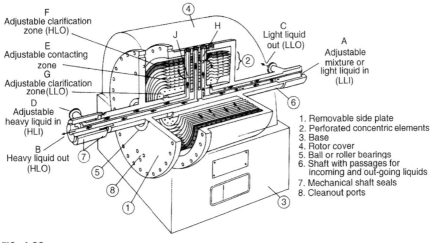

FIG. 4.36

Podbielniak contactor.

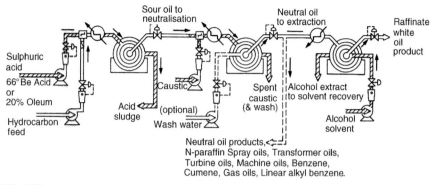

FIG. 4.37

Application of the Podbielniak contactor to the acid treatment of hydrocarbon feeds.

phase enters at *A* and is displaced inwards towards the shaft and leaves at *C*. The two liquids intermix in zone *E* where they are flowing countercurrently through the perforated concentric elements and are separated in the spaces between. In zones *F* and *G*, the perforated elements are surfaces on which the small droplets of entrained liquid can coalesce, the large drops then being driven out by centrifugal force.

The contactor finds extensive use where high performance phase separation and countercurrent extraction or washing in the one unit are required. Particularly important applications are the removal of acid sludges from hydrocarbons, shown in Fig. 4.37, hydrogen peroxide extraction, sulphonate soap and antibiotics extraction, the extraction of rare earths such as uranium and vanadium from leach liquors, and the washing of refined edible oils.

The *Alfa-Laval* contactor shown in Fig. 4.38, has a vertical spindle and the rotor is fitted with concentric cylindrical inserts with helical wings forming a series of spiral passages. The two phases are fed into the bottom, the light phase being led to the periphery from which it flows inwards along the spiral, with the heavy phase flowing countercurrently. High shear forces are thus generated giving high extraction rates.

These units give many ideal stages, run continuously and take up a minimum space. For these reasons they have been adopted in many drug extractions, though they are unsuitable for medium or large throughputs.

4.6 Use of supercritical fluids and ionic liquids

With the extensive use of the liquid–liquid extraction for the separation of complex mixtures into their components, it has been necessary to develop fluids with highly selective characteristics. The metallurgical, nuclear, biotechnolgy and food industries are now the major users of the technique, and many of the recent developments have originated in those fields. The characteristics and properties of two classes of fluids–supercritical fluids and ionic liquids are described in this section.

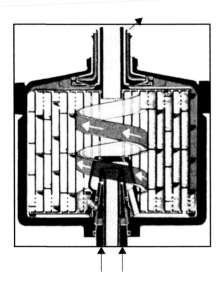

FIG. 4.38

Working principle of Alfa-Laval centrifugal extractor.

Supercritical fluids can be highly selective and their solvent power can be controlled by adjusting the operating pressure. With fluids such as carbon dioxide (CO_2), there is no residual contamination of the product as the solvent evaporates completely at the end of the operation.

Ionic liquids (ILs) have been considered as environment friendly or 'green' solvents for liquid–liquid extraction and have been widely used for the recovery and purification of many different types of solutes. Useful properties of ILs include extremely low volatility, good thermal stability, controllable physicochemical properties and long-term stability. Due to their unique properties, ILs have attracted great attention and are used in the areas of bioseparation, fuel desulphurisation and removal of environmental contaminants.

4.6.1 Supercritical fluids

A supercritical fluid is a substance that is above the highest temperature and pressure at which its vapour and liquid can co-exist at equilibrium. Although some materials decompose at a temperature below what would be their critical temperature, the supercritical regions of many common gases and liquids are easily attained. The two most popular and inexpensive fluids in this respect are CO_2 and water, which are non-toxic and non-flammable and therefore, as pointed out by Brenecke [39], essentially environmentally benign solvents that can be used even for food processing without undue regulatory restrictions. As presented by Reid et al. [40], the critical point of CO_2 (304.3 K and 7.3 MN/m^2) is readily accessible although that of

Table 4.4 Critical properties of various solvents.

	T_c (K)	P_c (MN/m^2)
Ethane	305.4	4.88
Ethylene	282.9	5.04
Ethanol	514.0	6.14
Nitrous oxide	309.8	7.26
Propane	369.9	4.25
Acetone	508.2	4.70
Ammonia	405.7	11.28

water (647.2 K and 22.09 MN/m^2) is more difficult to reach. Data for other supercritical fluids are listed in Table 4.4.

Since the densities of these fluids change drastically with very small variations in temperature or applied pressure, any density-dependent property such as the solubility of a heavy organic solute, can be manipulated, as Friedrich et al. [41] have pointed out, over wide ranges. This feature can be utilised in simple separation schemes in which a compound is extracted at high pressure where its solubility is high, followed by a reduction of the pressure causing the solute to be separated from the solution. Therefore, the supercritical fluid is recycled by re-pressurisation.

The dependence of solubility on pressure can be exploited in the separation of complex mixtures. If the pressure is initially high enough for all the components to be dissolved and is then reduced in stages, precipitation of successive components may be achieved in each stage. Other advantages of supercritical fluids include the fact that their viscosities are less than those of typical liquids and, as the diffusivities of the solutes are closer to those of gases rather than liquids, mass transfer resistances are considerably less than those in normal liquids. Thus, in general, supercritical fluids combine the advantages of having the diffusivities of gases and the solvent power of liquids.

One of the main commercial applications of supercritical fluid extraction (SFE) is in food processing where supercritical carbon dioxide (SC-CO$_2$) is used on a large scale in the de-caffeination of tea and coffee. In addition, McHugh and Krukonis [42] have described the extraction of hops, spices and flavours using SC-CO$_2$. In cosmetic and pharmaceutical industries, SC-CO$_2$ is extensively used for the extraction of a whole range of natural products, including pharmaceutical compounds, health supplements and fragrances. Jennings et al. [43] have described the extraction of taxol, an anti-cancer agent, from a slow-growing variety of the yew tree; Favati et al. [44] have found that gammalinolenic acid, a health aid, can be extracted from evening primrose oil seeds; and Borch-Jensen et al. [45] have shown that the valuable eicosapentaenoic and docosahexaenoic acids in fish oils may be fractionated with SC-CO$_2$. Moyler [46] lists over 80 varieties of seeds, roots, leaves, flowers and fruits which may be extracted with either liquid or SC-CO$_2$, and many SC-CO$_2$ extracts are currently available including celery, ginger, paprika, rosemary, sage and vanilla.

SFE also find applications in the areas of pollution prevention and remediation, and SC-CO_2 is used as a replacement solvent for many hazardous solvents in both extraction and separation processes and also as a reaction medium in materials processing. Although CO_2 is considered as a 'greenhouse gas', there is actually no net increase in the amount of the gas if it is removed from the environment, used as the solvent instead of a hazardous substance, and returned to the environment. In this way, most of the uses of SC-CO_2 may be considered as environmentally friendly. Because the solubility of oils and greases in CO_2 are quite high, it is particularly suited to the cleaning of machinery [47] and, as discussed in the literature [48], it is used as a solvent in textile dyeing operations where it is used to treat any dye-laden wastewater. With the addition of appropriate chelating agents, as described by Saito et al. [49], metals can be effectively extracted from solutions and soils with SC-CO_2 which is also used, as discussed by Yazdi and Beekman [50] to recover uranium from the aqueous solutions produced in the reprocessing of spent nuclear fuels. The main application of supercritical water has been in the oxidation of hazardous organic materials since water is readily miscible with both oxygen and organics and very high degrees of destruction can be achieved with very short residence times. This technology also finds use in the destruction of chemical weapons and stockpiled explosives, as well as in the clean-up of industrial and municipal wastes.

In the fields of bioseparations, Johnston et al. [51] have reported that proteins can be solubilised in reverse micelles formed in CO_2. SC-CO_2 has also been used to isolate or purify amino acids, steroids, enzymes in food and pharmaceutical industries [52]. In general, bioseparations can offset the high cost of attaining the necessary pressures since the products are of high value, they are present in the broth in low concentrations and conventional solvents often lead to recovery problems.

Although perhaps beyond the area of solvent extraction, it may be noted that various processes have been developed in which the solute is dissolved in the supercritical fluid and then the solution is expanded through a nozzle. This gives very high degree of supersaturation and results in the growth and nucleation of very fine particles, as discussed in Chapter 15 Crystallization. Products formed in this way include drug-polymers and materials which promote the gradual release of flavours and fragrances, as discussed by Hutchenson and Forster [53]. The selective precipitation of solutes is used for a wide variety of products including foods, proteins and explosives, as described by Chang and Randolph [54]. Spray drying of a solution in a supercritical liquid has been used for the production of microspheres and microporous fibres [55]. Again, beyond the field of solvent extraction, it may be noted that supercritical fluids have found increasing use in providing benign solvents in which to carry out chemical reactions. A review of this work has been provided by Savage et al. [56], and the processes described include the polymerisation of highly fluorinated acrylic polymers, the production of formic acid, brominations of alkylaromatics and phase-transfer catalysis. A general review of the exploitation of supercritical fluids in reaction chemistry has been provided by Clifford and Bartle [57].

4.6.2 Ionic liquids

ILs are defined as low-melting salts typically with melting points below $100\,°C$. The cations of most ILs are organic-based moieties such as imidazolium, N-alkylpyridinium, tetra-alkylammonium, and tetra-alkylphosphonium ions; whereas their anionic counterparts can be organic or inorganic entities, such as halides, nitrate, acetate, hexafluorophosphate ($[PF_6]$), tetrafluoroborate ($[BF_4]$), and so on [58]. ILs can be hydrophilic and hydrophobic depending on the structures of cations and anions. One of the main advantages of ILs as solvents for liquid–liquid extraction is that they can be designed to create desired properties, which make them very selective in extracting the desired solutes.

ILs have been extensively studied for the extraction and separation of bioactive compounds from diverse origins due to their excellent solvation ability for a wide range of compounds and materials and good stabilising properties for proteins, enzymes, nucleic acids, among others [59]. Typically, the separation and purification of target biocompounds are implemented by the application of hydrophobic ILs or the use of IL-based aqueous biphasic systems (ABS). Hydrophobic ILs ($[PF_6]$-based) were applied by Absalan et al. [60] in the extraction of a plant growth regulator, 3-indole-butyric acid from its aqueous extracts. Prasad and co-workers [61] studied three imidazolium-based ILs and showed that $[C_4C_1im][PF_6]$ was able to extract 65% of the total *trans*-zeatin and 18% of the total indole 3-acetic acid present in the sap. As alternative to imidazolium-based fluids, bistriflimide-based ILs were developed by Larriba et al. [62] for the extraction of tyrosol, a naturally occurring antioxidant from olive mill wastewater. Hydrophobic ILs have also been employed to extract various amino acids and proteins. In 2003, Carda-Broch et al. [63] firstly reported the extraction of amino-acid using hydrophobic ILs. After this pioneer study, a series of research studies [64,65] were conducted for the recovery of L-leucine, L-tryptophan, L-phenylalanine, and L-tyrosine. Shimojo et al. [66] investigated the extraction of heme protein using ILs from an aqueous phase through the addition of dicyclohexano-18-crown-6. Tzeng et al. [67] modified an imidazolium-based IL with a dye and applied this system to the extraction of lysozyme. Although research studies have been found resorting to the use of hydrophobic ILs for liquid–liquid extraction purposes, the number of available hydrophobic water-immiscible ILs are much more limited when compared to water-soluble ones. This imposes severe limitations for a wide range of applications.

IL-based ABS, formed by mixing hydrophilic ILs with inorganic salts in aqueous solution are now widely used for the separation and purification of a variety types of bioactive compounds [68–71], including lipids, amino acids, proteins, polysaccharides and chiral pharmaceutical molecules. Typically, inorganic salts induce the salting-out of ILs in aqueous media and lead to the creation of two-liquid aqueous-rich phases. Due to the large plethora of available water-miscible ILs and the second phase-forming components (salts, carbohydrates and amino acids), it is possible to tailor the polarities of the phases and IL-based ABS have allowed selective and enhanced separation [72]. Many research studies demonstrate that partition of bioactive compounds in ABS is influenced by a large number of parameters such as the cation side alkyl chain length of the ionic liquid, the nature of ionic liquid anions, types of inorganic salts (or carbohydrates, amino acids), ionic strength, pH and temperature. In most cases, conditions for a desired partition have to be determined experimentally, as discussed by Kula et al. [73], although a thermodynamic interpretation has been presented by Brooks et al. [74].

Apart from a wide spectrum applications of ILs in the field of bioseparations, ILs are also found to be extremely effective in fuel desulphurisation processes for removing organosulphur compounds of aliphatic and alicyclic types. Zhang et al. [75] investigated various ILs for simultaneous desulphurisation and denitrogenation process and found that [EMIM][BF4] and [BMIM][PF6] exhibited high selectivity in extracting aromatic sulphurs and nitrogen compounds. Huang et al. [76] prepared CuCl-based imidazolium ILs which demonstrated remarkable ability in removing sulphur from gasoline. Further research indicated that combination of ILs with oxidising agents is able to enhance the sulphur removal efficiency remarkably. Lo et al. [77] reported an oxidative extraction of sulphur compounds in light oils by the use of ILs and H_2O_2. These extractive desulphurisation methods using ILs have shown high energy-efficiency and high effectiveness in getting rid of the low level of sulphur from fuels.

ILs have been investigated for the removal of metal ions and organic contaminants from water as well [78]. The hydrophobic character of some ILs allows them to extract metal cations from aqueous solution by the formation of complexes between metal cations and ILs. The solvation of crown-ether complexes in ILs is more thermodynamically favoured than conventional organic solvents. Representative examples of IL extractions of metal ions include alkali [79], alkaline earth [80], heavy, and radioactive [81,82] metals. Key factors controlling the extraction efficiency include structure of an IL (especially the side chain of its cation), type of extractants (such as crown ethers) and solution pH. The high solubility of the charged organic molecules in ILs has stimulated the development of organic contaminants removed by these 'green' media. The extraction and removal of anionic dyes like methyl orange, eosin yellow and orange G from aqueous phase were achieved with imidazolium-based ILs [83]. Recently, two research articles [84,85] reviewed the applications of ILs for the removal of metal ions and organic pollutants.

References

[1] C. Hanson, The technology of solvent extraction in metallurgical processes, Het. Ingenieursblad. 41 (15–16) (1972) 408–417.

[2] British Patent No. 995472: Acrylic Acid Recovery, Distillers Company Limited, 1964. 29 April.

[3] T.K. Sherwood, R.L. Pigford, Absorption and Extraction, second ed., McGraw-Hill, New York, 1952.

[4] I. Leibson, R.B. Beckmann, The effect of packing size and column diameter on mass transfer in liquid–liquid extraction, Chem. Eng. Prog. 49 (1953) 405.

[5] R. Gayler, H.R.C. Pratt, Liquid–liquid extraction. Part V—further studies of droplet behaviour in packed columns, Trans. Inst. Chem. Eng. 31 (1953) 69.

[6] G.C.I. Warwick, J.B. Scuffham, The design for mixer–settlers for metallurgical duties, Het. Ingenieursblad. 41 (15–16) (1972) 442–449.

[7] I.D. Jackson, G.M. Newrick, G.C.I. Warwick, A recent development in the design of hydrometallurgical mixer–settlers, I. Chem. E. Symp. Ser. 42 (1975) 15.

[8] E.G. Scheibel, A.E. Karr, Semicommercial multistage extraction columns, Ind. Eng. Chem. 42 (1950) 1048.

[9] F.H. Blanding, J.C. Elgin, Limiting flow in liquid–liquid extraction columns, Trans. Am. Inst. Chem. Eng. 38 (1942) 305.

[10] J.D. Thornton, Spray liquid–liquid extraction columns: prediction of limiting holdup and flooding rates, Chem. Eng. Sci. 5 (1956) 201–208.

[11] C.B. Hayworth, R.E. Treybal, Drop formation in two-liquid-phase systems, Ind. Eng. Chem. 42 (1950) 1174.

[12] S. Hu, R.C. Kintner, The fall of single liquid drops through water, AIChE J. 1 (1955) 42.

[13] A.J. Klee, R.E. Treybal, Rate of rise or fall of liquid drops, AIChE J. 2 (1956) 44.

[14] W. Licht, J.B. Conway, Mechanism of solute transfer in spray towers, Ind. Eng. Chem. 42 (1950) 1151.

[15] J.M. Coulson, S.J. Skinner, The mechanism of liquid–liquid extraction across stationary and moving interfaces. Part 1. Mass transfer into single dispersed drops, Chem. Eng. Sci. 1 (1952) 197.

[16] F.H. Garner, A.A. Hale, The effect of surface agents in liquid extraction processes, Chem. Eng. Sci. 2 (1953) 157.

[17] A.E. Handlos, T. Baron, Mass and heat transfer from drops in liquid–liquid extraction, AIChE J. 3 (1957) 127.

[18] C.L. Ruby, J.C. Elgin, Mass transfer between liquid drops and a continuous liquid phase in a countercurrent fluidized system. Liquid–liquid extraction in a spray tower, in: Mass Transfer—Transport Properties. Chem. Eng. Prog. Symp. Series No. 16, vol. 51, 1955, p. 17.

[19] F.J. Appel, J.C. Elgin, Countercurrent extraction of benzoic acid between toluene and water, Ind. Eng. Chem. 29 (1973) 451.

[20] T.K. Sherwood, J.E. Evans, J.V.A. Longcor, Extraction in spray and packed columns, Trans. Am. Inst. Chem. Eng. 35 (1939) 597.

[21] S.B. Row, J.H. Koffolt, J.R. Withrow, Characteristics and performance of a nine-inch liquid–liquid extraction column, Trans. Am. Inst. Chem. Eng. 37 (1941) 559.

[22] R.H. Perry, D.W. Green, J.O. Maloney (Eds.), Perry's Chemical Engineers' Handbook, seventh ed., McGraw-Hill Book Company, New York, 1997.

[23] R.E. Treybal, Liquid Extraction, second ed., McGraw-Hill, New York, 1963.

[24] T.K. Sherwood, R.L. Pigford, C.R. Wilke, Mass Transfer, McGraw-Hill, New York, 1975.

[25] J.T. Davies, E.K. Rideal, Interfacial Phenomena, Academic Press, 1961.

[26] H. Groothuis, F.J. Zuiderweg, Influence of mass transfer on coalescence of drops, Chem. Eng. Sci. 12 (1960) 288.

[27] R. Gayler, N.W. Roberts, H.R.C. Pratt, Liquid–liquid extraction. Part IV. A further study of hold-up in packed columns, Trans. Inst. Chem. Eng. 31 (1953) 57.

[28] R. Gayler, H.R.C. Pratt, Symposium on liquid–liquid extraction. Part II. Hold-up and pressure drop in packed columns, Trans. Inst. Chem. Eng. 29 (1951) 110.

[29] J.B. Lewis, I. Jones, H.R.C. Pratt, Symposium on liquid–liquid extraction. Part III. A study of droplet behaviour in packed columns, Trans. Inst. Chem. Eng. 29 (1951) 126.

[30] F.R. Dell, H.R.C. Pratt, Symposium on liquid–liquid extraction. Part I. Flooding rates for packed columns, Trans. Inst. Chem. Eng. 29 (1951) 89.

[31] H.R.C. Pratt, S.T. Glover, Liquid–liquid extraction: removal of acetone and acetaldehyde from vinyl acetate with water in a packed column, Trans. Inst. Chem. Eng. 24 (1946) 54.

[32] A.P. Colburn, D.G. Welsh, Experimental study of individual transfer resistances in countercurrent liquid–liquid extraction, Trans. Am. Inst. Chem. Eng. 38 (1942) 179.

[33] G.S. Laddha, J.M. Smith, Mass transfer resistances in liquid–liquid extraction, Chem. Eng. Prog. 46 (1950) 195.

[34] J.D. Thornton, H.R.C. Pratt, Liquid–liquid extraction. Part VII. Flooding rates and mass transfer data for rotary annular columns, Trans. Inst. Chem. Eng. 31 (1953) 289.

[35] H.J.A. Vermljs, H. Kramers, Liquid–liquid extraction in a "rotating disc contactor", Chem. Eng. Sci. 3 (1954) 55.

[36] W.J.D. van Dijck, Intimately Contacting Fluids (Immiscible Liquids), U.S. Patent 2,011,186, 1935.

[37] J.D. Thornton, Liquid–liquid extraction. Part XIII. The effect of pulse wave-form and plate geometry on the performance and throughput of a pulsed column, Trans. Inst. Chem. Eng. 35 (1957) 316.

[38] D.H. Logsdail, J.D. Thornton, Liquid–liquid extraction. Part XIV. The effect of column diameter upon the performance and throughput of pulsed plate columns, Trans. Inst. Chem. Eng. 35 (1957) 331.

[39] J.F. Brenecke, New applications of supercritical fluids, Chem. Ind. (1996) 831. 4 November.

[40] R.C. Reid, J.M. Prausnitz, B.E. Poling, The Properties of Liquids and Gases, fourth ed., McGraw-Hill, New York, 1987.

[41] J.P. Friedrich, G.R. List, A.J. Heakin, Petroleum-free extraction of oil from soybeans with supercritical CO_2, J. Am. Oil Chem. Soc. 59 (1982) 288–292.

[42] M.A. McHugh, V.J. Krukonis, Supercritical Fluid Extraction—Principles and Practice, second ed., Butterworth-Heinemann, Oxford, 1994.

[43] D.W. Jennings, F. Chang, V. Bazook, Vapor-liquid equilibria for carbon dioxide plus 1-pentanol, J. Chem. Eng. Data 37 (1992) 337–338.

[44] F. Favati, J.W. King, M. Mazzanti, Supercritical carbon dioxide extraction of evening primrose oil, J. Am. Oil Chem. Soc. 68 (1991) 422–427.

[45] C. Borch-Jensen, A. Staby, J.M. Mollerup, Phase equilibria of urea-fractioned fish oil fatty acid ethyl esters and supercritical carbon dioxide, Ind. Eng. Chem. Res. 33 (1994) 1574–1579.

[46] D.A. Moyler, in: M.B. King, T.R. Bott (Eds.), Extraction of Natural Products Using Near-Critical Solvents, Chapman & Hall, Glasgow, 1993.

[47] Electronic Materials Technology News 9, 1995. No. 3.

[48] R. Steiner, Carbon dioxide's expanding role, Chem. Eng. 100 (3) (March 1993) 114–119.

[49] N. Saito, Y. Ikushima, T. Goto, Liquid-solid extraction of acetylacetone chelates with supercritical carbon dioxide, Bull. Chem. Soc. Jpn. 63 (1990) 1532–1534.

[50] A.V. Yazdi, E.J. Beekman, Design of highly CO_2-soluble chelating agents for carbon dioxide extraction of heavy metals, Mater. Res. 10 (1995) 530–537.

[51] K.P. Johnston, K.L. Harrison, M.J. Clarke, Water in carbon dioxide microemulsions: an environment for hydrophiles including proteins, Science 271 (1996) 624–626.

[52] G.A. Mansoori, K. Schulz, E.E. Martinelli, Bioseparation using supercritical fluid extraction/retrograde condensation, Nat. Biotechnol. 6 (1988) 393–396.

[53] K.W. Hutchenson, N.R. Foster, in: K.W. Hutchenson, N.R. Foster (Eds.), Innovations in Supercritical Fluids. ACS Symposium Series 608, American Chemical Society, Washington, 1995.

[54] C.M.J. Chang, A.D. Randolph, N.E. Croft, Separation of beta-caretone mixtures precipitated from liquid solvents with high pressure CO_2, Biotech. Prog. 7 (1991) 275.

[55] D.J. Dixon, K.P. Johnston, R.A. Bodmeier, Polymeric materials formed by precipitation with a compressed fluid antisolvent, AIChE J. 39 (1993) 127.

[56] P.E. Savage, S. Goplan, T.I. Mizan, C.J. Martino, E.E. Brock, Reactions at supercritical conditions—applications and fundamentals, AIChE J. 41 (1995) 1723–1778.

[57] T. Clifford, K. Bartle, Chemical reactions in supercritical fluids, Chem. Ind. (1996) 449–452.

[58] X. Han, D.W. Armstrong, Ionic liquids in separations, Acc. Chem. Res. 40 (2007) 1079–1086.

[59] S.P.M. Ventrra, F.A. e Silva, M.V. Quental, D. Mondal, M.G. Freire, J.A.P. Coutinho, Ionic-liquid-mediated extraction and separation processes for bioactive compounds: past, present, and future trends, Chem. Rev. 117 (2017) 6984–7052.

[60] G. Absalan, M. Akhond, L. Sheikhian, Extraction and high performance liquid chromatographic determination of 3-indole butyric acid in pea plants by using imidazolium-based ionic liquids as extractant, Talanta 77 (2008) 407–411.

[61] A.K. Das, K. Prasad, Extraction of plant growth regulators present in *Kappaphycus alvarezii* sap using imidazolium based ionic liquids: detection and quantification by using HPLC-DAD technique, Anal. Methods 7 (2015) 9064–9067.

[62] M. Larriba, S. Omar, P. Navarro, J. Garcia, F. Rodriguez, M. Gonzalez-Miquel, Recovery of tyrosol from aqueous streams using hydrophobic ionic liquids: a first step towards developing sustainable processes for olive mill wastewater (OMW) management, RSC Adv. 6 (2016) 18751–18762.

[63] S. Carda-Broch, A. Berthod, D.W. Armstrong, Solvent properties of the 1-butyl-3-methylimidazolium hexafluorophosphate ionic liquid, Anal. Bioanal. Chem. 375 (2003) 191–199.

[64] J. Wang, Y. Pei, Y. Zhao, Z. Hu, Recovery of amino acids by imidazolium based ionic liquids from aqueous media, Green Chem. 7 (2005) 196–202.

[65] L.I.N. Tomé, V.R. Catambas, A.R.R. Teles, M.G. Freire, I.M. Marrucho, J.A.P. Coutinho, Tryptophan extraction using hydrophobic ionic liquids, Sep. Purif. Technol. 72 (2010) 167–173.

[66] K. Shimojo, N. Kamiya, F. Tani, H. Naganawa, Y. Naruta, M. Goto, Extractive solubilization, structural change, and functional conversion of cytochrome C in ionic liquids via crown ether complexation, Anal. Chem. 78 (2006) 7735–7742.

[67] Y.-P. Tzeng, C.-W. Shen, T. Yu, Liquid–liquid extraction of lysozyme using a dye-modified ionic liquid, J. Chromatogr. A 1193 (2008) 1–6.

[68] M.G. Freire, C.L.S. Louros, L.P.N. Rebelo, J.A.P. Coutinho, Aqueous biphasic systems composed of a water-stable ionic liquid + carbohydrates and their applications, Green Chem. 13 (2011) 1536–1545.

[69] M.T. Zafarani-Moattar, S. Hamzehzadeh, Partitioning of amino acids in the aqueous biphasic system containing the water-miscible ionic liquid 1-butyl-3-methylimidazolium bromide and the water-structuring salt potassium citrate, Biotechnol. Prog. 27 (2011) 986–997.

[70] Z. Du, Y.L. Yu, J.H. Wang, Extraction of proteins from biological fluids by use of an ionic liquid/aqueous two-phase system, Chem. Eur. J. 13 (2007) 2130–2137.

[71] M. Zawadzki, F.A. e Silva, U. Domańska, J.A.P. Coutinho, S.P.M. Ventura, Recovery of an antidepressant from pharmaceutical wastes using ionic liquids-based aqueous biphasic systems, Green Chem. 18 (2016) 3527–3536.

[72] M.G. Freire, A.F.M. Claúdio, M.M. Araújo, J.A.P. Coutinho, I.M. Marrucho, J.N. Canongia Lopes, L.P.N. Rebelo, Aqueous biphasic systems: a boost brought about by using ionic liquids, Chem. Soc. Rev. 41 (2012) 4966–4995.

[73] M.-R. Kula, in: L.B. Wingard, E. Katchalski-Katzire, L. Goldstein (Eds.), Applied Biochemistry and Bioengineering. Vol 2: Enzyme Technology, Academic Press, New York, 1979, pp. 71–95.

[74] D.E. Brooks, K.A. Sharp, D. Fisher, in: H. Walter, D.E. Brooks, D. Fisher (Eds.), Partitioning in Aqueous Two Phase Systems, Theory, Methods, Uses and Applications to Biotechnology, Academic Press, New York, 1985, pp. 11–84.

[75] S. Zhang, Q. Zhang, Z.C. Zhang, Extractive desulfurization and denitrogenation of fuels using ionic liquids, Ind. Eng. Chem. Res. 43 (2004) 614–622.

[76] C. Huang, B. Chen, J. Zhang, Z. Liu, Y. Li, Desulfurization of gasoline by extraction with new ionic liquids, Energy Fuel 18 (2004) 1862–1864.

[77] W.-H. Lo, H.-Y. Yang, G.-T. Wei, One-pot desulfurization of light oils by chemical oxidation and solvent extraction with room-temperature ionic liquids, Green Chem. 5 (2003) 639–642.

[78] H. Zhao, S. Xia, P. Ma, Use of ionic liquids as "green" solvents for extractions, J. Chem. Technol. Biotechnol. 80 (2005) 1089–1096.

[79] S. Chun, S.V. Dzyuba, R.A. Bartsch, Influence of structural variation in room-temperature ionic liquids on the selectivity and efficiency of competitive alkali metal salt extraction by a crown ether, Anal. Chem. 73 (2001) 3737–3741.

[80] R.A. Bartsch, S. Chun, S.V. Dzyuba, in: R.D. Rogers, K.R. Seddon (Eds.), Ionic Liquids: Industrial Applications for Green Chemistry, American Chemical Society, Washington, DC, 2002, pp. 58–68.

[81] A.E. Visser, R.P. Swatloski, S.T. Griffin, D.H. Hartman, R.D. Rogers, Liquid/liquid extraction of metal ions in room temperature ionic liquids, Sep. Sci. Technol. 36 (2001) 785–804.

[82] K. Nakashima, F. Kubota, T. Maruyama, M. Goto, Ionic liquids as a novel solvent for lanthanide extraction, Anal. Sci. 19 (2003) 1097–1098.

[83] Y.C. Pei, J.J. Wang, X.P. Xuan, J. Fan, M. Fan, Factors affecting ionic liquids based removal of anionic dyes from water, Environ. Sci. Technol. 41 (2007) 5090–5095.

[84] M. Made, J.-F. Liu, L. Pang, Environmental application, fate, effects and concerns of ionic liquids: a review, Environ. Sci. Technol. 49 (2015) 12611–12627.

[85] J. Ma, X. Hong, Application of ionic liquids in organic pollutants control, J. Environ. Manag. 99 (2012) 104–109.

Further reading

R. Blumberg, Liquid–Liquid Extraction, Harcourt Brace Jovanovich, London, 1988.

A.W. Francis, Handbook for Components in Solvent Extraction, Gordon & Breach, New York, 1972.

J.C. Godfrey, M.J. Slater, Liquid–liquid Extraction Equipment, Wiley, New York, 1994.

A.L. Hines, R.N. Maddox, Mass Transfer Fundamentals and Applications, Prentice-Hall, Englewood Cliffs, 1985.

F.A. Holland, F.S. Chapman, Liquid Mixing and Processing in Stirred Tanks, Reinhold, New York, 1966.

W.D. Jamrack, Base Metal Extraction by Chemical Engineering Techniques, Pergamon Press, Oxford, 1963.

T.C. Lo, M.I. Baird, C. Hanson (Eds.), Handbook of Solvent Extraction, Krieger, 1991.

W.L. McCabe, J.C. Smith, P. Harriott, Unit Operations in Chemical Engineering, seventh ed., McGraw-Hill, New York, 2005.

M.A. McHugh, V.J. Krukonis, Supercritical Fluid Extraction—Principles and Practice, Butterworths, Boston, 1986.

M.E. Paulaitis, S.M.L. Penninger, R.D. Gray, P. Davidson (Eds.), Chemical Engineering at Supercritical Fluid Conditions, Ann Arbor Science, Ann Arbon, 1983.

H.R.C. Pratt, Countercurrent Separation Processes, Elsevier, Amsterdam, 1967.

J.D. Seader, E.J. Henley, D.K. Roper, Separation Process Principles: Chemical and Biochemical Operations, third ed., John Wiley & Sons, Hoboken, 2011.

T.G. Squires, M.E. Paulaitis, Supercritical Fluids—Chemical and Engineering Principles and Applications, ACS Symposium Series 329, American Chemical Society, Washington, 1987.

H. Walter, D.E. Brooks, D. Fisher, Partitioning in Aqueous Two-Phase Systems, Academic Press, New York, 1985.

P.C. Wankat, Separation Process Engineering: Includes Mass Transfer Analysis, third ed., Prentice Hall, New York, 2012.

B.Y. Zaslavsky, Aqueous two-phase partitioning (Marcel Dekker, 1994), in: Solvent Extraction, Proc. Int. Solvent Extraction Conference, The Hague, 1971. Soc. Chem. Ind., London, 1971.

Evaporation

5

Ajay Kumar Ray

Department of Chemical and Biochemical Engineering, University of Western Ontario,
London, ON, Canada

Nomenclature

		Units in SI system	Dimensions in M, L, T, θ
A	heat transfer surface	m^2	$\mathbf{L}^2$
a	constant in Eq. (5.17)	m^{4}K^2/W^2	$\mathbf{M}^{-2}\mathbf{T}^5\theta^2$
b	constant in Eq. (5.17)	m^{4}K^2/W^2	$\mathbf{M}^{-2}\mathbf{T}^6\theta^2$
C_b	variable cost during operation	£/s	$\mathbf{T}^{-1}$
C_c	total cost of a shutdown	£	–
C_p	specific heat of liquid at constant pressure	J/kg K	$\mathbf{L}^2\mathbf{T}^{-2}\theta^{-1}$
C_s	surface factor	–	–
C_T	total cost during period t_P	£	–
D	liquid evaporated or steam condensed per unit time	kg/s	$\mathbf{MT}^{-1}$
d	a characteristic dimension	m	$\mathbf{L}$
d_t	tube diameter	m	$\mathbf{L}$
E	power to compressor	W	$\mathbf{ML}^2\mathbf{T}^{-3}$
E'	net work done on unit mass	J/kg	$\mathbf{L}^2\mathbf{T}^{-2}$
G_F	mass rate of feed	kg/s	$\mathbf{MT}^{-1}$
G_x	mass flow of extra steam dryness fraction	kg/s	$\mathbf{MT}^{-1}$
G_y	mass flow of sea water	kg/s	$\mathbf{MT}^{-1}$
g	acceleration due to gravity	m/s^2	$\mathbf{LT}^{-2}$
H	enthalpy per unit mass of vapour	J/kg	$\mathbf{L}^2\mathbf{T}^{-2}$
h	average value of h_b for a tube bundle	W/m^2 K	$\mathbf{MT}^{-3}\theta^{-1}$
h_b	film heat transfer coefficient for boiling liquid	W/m^2 K	$\mathbf{MT}^{-3}\theta^{-1}$

Continued

Coulson and Richardson's Chemical Engineering. https://doi.org/10.1016/B978-0-08-101097-6.00005-5

		Units in SI system	Dimensions in M, L, T, θ
h_c	film heat transfer coefficient for condensing steam	W/m^2 K	$\mathbf{MT^{-3}\theta^{-1}}$
h_L	liquid-film heat transfer coefficient	W/m^2 K	$\mathbf{MT^{-3}\theta^{-1}}$
h_{tp}	heat transfer coefficient for two phase mixture	W/m^2 K	$\mathbf{MT^{-3}\theta^{-1}}$
k	thermal conductivity of liquid	W/m K	$\mathbf{MLT^{-3}\theta^{-1}}$
m	mass	kg	$\mathbf{M}$
M	mass of cooling water per unit mass of vapour	kg/kg	–
N	number of effects	–	–
P	pressure	N/m^2	$\mathbf{ML^{-1}T^{-2}}$
Q	heat transferred per unit time	W	$\mathbf{ML^2T^{-3}}$
Q_b	total heat transferred during boiling time	J	$\mathrm{ML^2T^{-2}}$
q	heat flux per unit area	W/m^2	$\mathbf{MT^{-3}}$
T	temperature	K	$\boldsymbol{\theta}$
T_b	boiling temperature of liquid	K	$\boldsymbol{\theta}$
T_c	condensing temperature of steam	K	$\boldsymbol{\theta}$
T_f	feed temperature	K	$\boldsymbol{\theta}$
T_w	heater wall temperature	K	$\boldsymbol{\theta}$
ΔT	temperature difference	K	$\boldsymbol{\theta}$
t	time	s	$\mathbf{T}$
t_b	boiling time	s	$\mathbf{T}$
t_c	time for emptying, cleaning, and refilling unit	s	$\mathbf{T}$
t_P	total production time	s	$\mathbf{T}$
U	overall heat transfer coefficient	W/m^2 K	$\mathbf{MT^{-3}\theta^{-1}}$
V	volume	m^3	$\mathbf{L^3}$
G_F	feed rate	kg/s	$\mathbf{MT^{-1}}$
X_{tt}	Lockhart and Martinelli's parameter (Eqs 5.5 and 5.6)	–	–
y	mass fraction of vapour	–	–
Z	hydrostatic head	m	$\mathbf{L}$
γ	ratio of specific heat at constant pressure to specific heat at constant volume	–	–
λ	latent heat of vaporisation per unit mass	J/kg	$\mathbf{L^2T^{-2}}$
η	economy	–	–
η'	efficiency of ejector	–	$\mathbf{-}$
μ_L	viscosity of liquid	Ns/m^2	$\mathbf{ML^{-1}T^{-1}}$
μ_v	viscosity of vapour	Ns/m^2	$\mathbf{ML^{-1}T^{-1}}$
ρ_L	density of liquid	kg/m^3	$\mathbf{ML^{-3}}$
ρ_v	density of vapour	kg/m^3	$\mathbf{ML^{-3}}$
σ	interfacial tension	J/m^2	$\mathbf{MT^{-2}}$

Continued

		Units in SI system	Dimensions in M, L, T, θ
Suffixes			
0	refers to the steam side of the first effect		
1, 2, 3	refer to the first, second and third effects		
av	refers to an average value		
c	refers to the condenser		
i and e	refer to the inlet and exit cooling water		

5.1 Introduction

Evaporation, a widely used method for the concentration of aqueous solutions, involves the removal of water from a solution by boiling the liquor in a suitable vessel, an evaporator, and withdrawing the vapour. If the solution contains dissolved solids, the resulting strong liquor may become saturated so that crystals are deposited. Liquors which are to be evaporated may be classified as follows:

a) Those which can be heated to high temperatures without decomposition, and those that can be heated only to a temperature of about 330 K.
b) Those which yield solids on concentration, in which case crystal size and shape may be important, and those which do not.
c) Those which, at a given pressure, boil at about the same temperature as water, and those which have a much higher boiling point.

Evaporation is achieved by adding heat to the solution to vaporise the solvent. The heat is supplied principally to provide the latent heat of vaporisation, and, by adopting methods for recovery of heat from the vapour, it has been possible to achieve great economy in heat utilisation. While the normal heating medium is generally low-pressure exhaust steam from turbines, special heat transfer fluids or flue gases are also used.

The design of an evaporation unit requires the practical application of data on heat transfer to boiling liquids, together with a realisation of what happens to the liquid during concentration. In addition to the three main features outlined above, liquors which have an inverse solubility curve and which are therefore likely to deposit scale on the heating surface merit special attention.

5.2 Heat transfer in evaporators
5.2.1 Heat transfer coefficients

The rate equation for heat transfer takes the form:

$$Q = UA\Delta T \tag{5.1}$$

where Q is the heat transferred per unit time, U is the overall coefficient of heat transfer, A is the heat transfer surface, and ΔT is the temperature difference between the two streams.

In applying this equation to evaporators, there may be some difficulty in deciding the correct value for the temperature difference because of what is known as the *boiling point rise* (BPR). If water is boiled in an evaporator under a given pressure, then the temperature of the liquor may be determined from steam tables and the temperature difference is readily calculated. At the same pressure, a solution has a boiling point greater than that of water, and the difference between its boiling point and that of water is the BPR. For example, at atmospheric pressure ($101.3\,kN/m^2$), a 25% solution of sodium chloride boils at 381 K and shows a BPR of 8 K. If steam at 389 K were used to concentrate the salt solution, the overall temperature difference would not be $(389–373) = 16\,K$, but $(389–381) = 8\,K$. Such solutions usually require more heat to vaporise unit mass of water, so that the reduction in capacity of a unit may be considerable. The value of the BPR cannot be calculated from physical data of the liquor, though Dühring's rule is often used to find the change in BPR with pressure. If the boiling point of the solution is plotted against that of water at the same pressure, then a straight line is obtained, as shown for sodium chloride in Fig. 5.1. Thus, if the pressure is fixed, the boiling point of water is found from steam tables, and the boiling point of the solution from Fig. 5.1. The boiling point rise is much greater with strong electrolytes, such as salt and caustic soda.

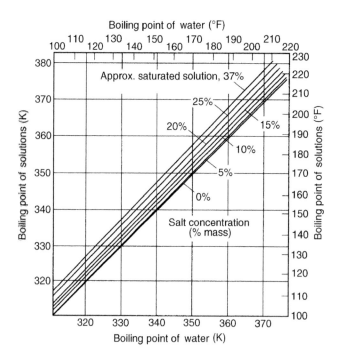

FIG. 5.1

Boiling point of solutions of sodium chloride as a function of the boiling point of water. Dühring lines.

Overall heat transfer coefficients for any form of evaporator depend on the value of the film coefficients on the heating side and for the liquor, together with allowances for scale deposits and the tube wall. For condensing steam, which is a common heating medium, film coefficients are approximately $6\,kW/m^2\,K$. There is no entirely satisfactory general method for calculating transfer coefficients for the boiling film. Design equations of sufficient accuracy are available in the literature, however, although this information should be used with caution.

5.2.2 Boiling at a submerged surface

The heat transfer processes occurring in evaporation equipment may be classified under two general headings. The first of these is concerned with boiling at a submerged surface. A typical example of this is the horizontal tube evaporator considered in Section 5.7, where the basic heat transfer process is assumed to be nucleate boiling with convection induced predominantly by the growing and departing vapour bubbles. The second category includes two-phase forced-convection boiling processes occurring in closed conduits. In this case, convection is induced by the flow which results from natural or forced circulation effects.

As detailed in Volume 1, Chapter 9 and in Volume 6, the heat flux–temperature difference characteristic observed when heat is transferred from a surface to a liquid at its boiling point, is as shown in Fig. 5.2. In the range AB, although the liquid in the vicinity of the surface will be slightly superheated, there is no vapour formed and heat transfer is by natural convection with evaporation from the free surface. Boiling

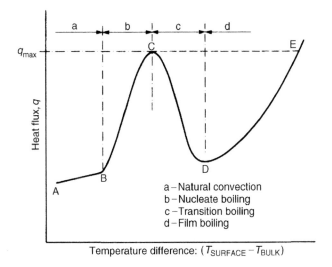

FIG. 5.2

Typical characteristic for boiling at a submerged surface.

commences at B with bubble columns initiated at preferred sites of nucleation centres on the surface. Over the nucleate boiling region, BC, the bubble sites become more numerous with increasing flux until, at C, the surface is completely covered. In the majority of commercial evaporation processes, the heating medium is a fluid, and therefore, the controlling parameter is the overall temperature difference. If an attempt is made to increase the heat flux beyond that at C, by increasing the temperature difference, the nucleate boiling mechanism will partially collapse and portions of the surface will be exposed to vapour blanketing. In the region of transition boiling CD, the average heat transfer coefficient, and frequently the heat flux, will decrease with increasing temperature difference, due to the increasing proportion of the surface exposed to vapour. This self-compensating behaviour is not exhibited if heat flux rather than temperature difference is the controlling parameter. In this case, an attempt to increase the heat flux beyond point C will cause the nucleate boiling regime to collapse completely, exposing the whole surface to a vapour film. The inferior heat transfer characteristics of the vapour mean that the surface temperature must rise to E in order to dissipate the heat. In many instances, this temperature exceeds the melting point of the surface and results can be disastrous. For obvious reasons the point C is generally known as *burnout*, although the terms *departure from nucleate boiling (DNB point)* and *maximum heat flux* are in common usage. In the design of evaporators, a method of predicting the heat transfer coefficient in nucleate boiling h_b, and the maximum heat flux which might be expected before h_b begins to decrease, is of extreme importance. The complexity of the nucleate boiling process has been the subject of many studies. In a review of the available correlations for nucleate boiling, Westwater [1] has presented some 14 equations. Palen and Taborek [2] reduced this list to seven and tested these against selected experimental data [3,4]. As a result of this study two equations, those due to McNelly [5] and Gilmour [6], were selected as the most accurate. Although the modified form of the Gilmour equation is somewhat more accurate, the relative simplicity of the McNelly equation is attractive and this equation is given in dimensionless form as:

$$\left[\frac{h_b d}{k}\right] = 0.225 \left[\frac{C_p \mu_L}{k}\right]^{0.69} \left[\frac{Pd}{\sigma}\right]^{0.31} \left[\frac{\rho_L}{\rho_v} - 1\right]^{0.31} \tag{5.2}$$

The inclusion of the characteristic dimension d is necessary dimensionally, though its value does not affect the result obtained for h_b.

This equation predicts the heat transfer coefficient for a single isolated tube and is not applicable to tube bundles, for which Palen and Taborek [2] showed that the use of this equation would have resulted in 50%–250% underdesign in a number of specific cases. The reason for this discrepancy may be explained as follows. In the case of a tube bundle, only the lowest tube in each vertical row is completely irrigated by the liquid with higher tubes being exposed to liquid–vapour mixtures. This partial vapour blanketing results in a lower average heat transfer coefficient for tube bundles than the value given by Eq. (5.2). In order to calculate these average values of h for a tube bundle, equations of the form $h = C_s h_b$ have been suggested [2] where the surface factor C_s is less than 1 and is, as might be expected, a function of the number of tubes in a vertical row, the pitch of the tubes, and the basic value of h_b. The factor C_s can only be determined by statistical analysis of experimental data and further work is necessary before it can be predicted from a physical model for the process.

The single tube values for h_b have been correlated by Eq. (5.2), which applies to the true nucleate boiling regime and takes no account of the factors which eventually lead to the maximum heat flux being approached. As discussed in Volume 1, Chapter 9, equations for *maximum flux*, often a limiting factor in evaporation processes, have been tested by Palen and Taborek [2], though the simplified equation of Zuber [7] is recommended. This takes the form:

$$ q_{max} = \frac{\pi}{24} \lambda \rho_v \left[\frac{\sigma g (\rho_L - \rho_v)}{\rho_v^2} \right]^{1/4} \left[\frac{\rho_L + \rho_v}{\rho_L} \right]^{1/2} \tag{5.3} $$

where q_{max} is the maximum heat flux, λ is the latent heat of vaporisation, ρ_L is the density of liquid, ρ_v is the density of vapour, σ is the interfacial tension, and g is the acceleration due to gravity.

5.2.3 **Forced convection boiling**

The performance of evaporators operating with forced convection depends very much on what happens when a liquid is vaporised during flow through a vertical tube. If the liquid enters the tube below its boiling point, then the first section operates as a normal heater and the heat transfer rates are determined by the well-established equations for single phase flow. When the liquid temperature reaches the boiling point corresponding to the local pressure, boiling commences. At this stage the vapour bubbles are dispersed in the continuous liquid phase although progressive vaporisation of the liquid gives rise to a number of characteristic flow patterns which are shown in Fig. 5.3. Over the initial boiling section convective heat transfer occurs with vapour bubbles dispersed in the liquid. Higher up, the tube bubbles become more numerous and elongated, and bubble coalescence occurs and eventually the bubbles form slugs which later collapse to give an annular flow regime in which vapour forms the central core with a thin film of liquid carried up the wall. In the final stage, dispersed flow with liquid entrainment in the vapour core occurs. In general, the conditions existing in the tube are those of annular flow. With further evaporation, the rising liquid film becomes progressively thinner and this thinning, together with the increasing vapour core velocity, eventually causes breakdown of the liquid film, leading to dry wall conditions.

For boiling in a tube, there is therefore a contribution from nucleate boiling arising from bubble formation, together with forced convection boiling due to the high velocity liquid–vapour mixture. Such a system is inherently complex since certain parameters influence these two basic processes in different ways.

Dengler and Addoms [8] measured heat transfer to water boiling in a 6 m tube and found that the heat flux increased steadily up the tube as the percentage of vapour increased, as shown in Fig. 5.4. Where convection was predominant, the data were correlated using the ratio of the observed two-phase heat transfer coefficient (h_{tp}) to

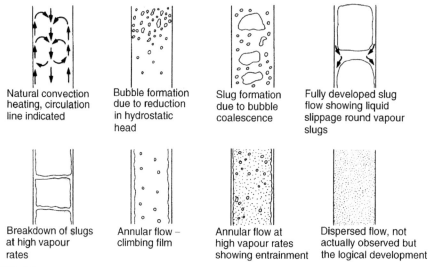

Natural convection heating, circulation line indicated

Bubble formation due to reduction in hydrostatic head

Slug formation due to bubble coalescence

Fully developed slug flow showing liquid slippage round vapour slugs

Breakdown of slugs at high vapour rates

Annular flow – climbing film

Annular flow at high vapour rates showing entrainment

Dispersed flow, not actually observed but the logical development

FIG. 5.3

The nature of two-phase flow in an evaporator tube.

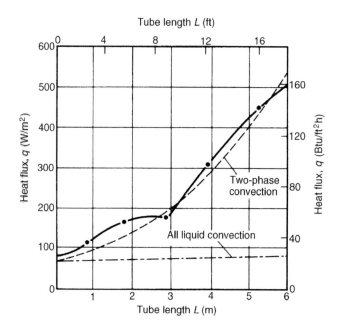

FIG. 5.4

Variation of the heat flux to water in an evaporator tube [8].

that which would be obtained had the same total mass flow been all liquid (h_L) as the ordinate. As discussed in Volume 6, Chapter 12, this ratio was plotted against the reciprocal of X_{tt}, the parameter for two-phase turbulent flow developed by Lockhart and Martinelli [9]. The liquid coefficient h_L is given by:

$$h_L = 0.023 \left[\frac{k}{d_t}\right] \left[\frac{4W}{\pi d_t \mu_L}\right]^{0.8} \left[\frac{C_p \mu_L}{k}\right]^{0.4} \tag{5.4}$$

where W is the total mass rate of flow. The parameter $1/X_{tt}$ is given by:

$$\frac{1}{X_{tt}} = \left[\frac{y}{1-y}\right]^{0.9} \left[\frac{\rho_L}{\rho_v}\right]^{0.5} \left[\frac{\mu_v}{\mu_L}\right]^{0.1} \tag{5.5}$$

$1/X_{tt}$ is strongly dependent on the mass fraction of vapour y. The density and viscosity terms give a quantitative correction for the effect of pressure in the absence of nucleate boiling.

Eighty-five per cent of the purely convective data for two-phase flow were correlated to within 20% by the expression:

$$\frac{h_{tp}}{h_L} = 3.5 \left[\frac{1}{X_{tt}}\right]^{0.5} \text{ where } 0.25 < \frac{1}{X_{tt}} < 70 \tag{5.6}$$

Similar results for a range of organic liquids are reported by Guerrieri and Talty [10], though, in this work, h_L is based on the point mass flow rate of the unvapourised part of the stream, that is, W is replaced by $W(1-y)$ in Eq. (5.4).

One unusual characteristic of Eq. (5.2) is the dependence of h_b on the heat flux q. The calculation of h_b presents no difficulty in situations where the controlling parameter is the heat flux, as is the case with electrical heating. If a value of q is selected, this together with a knowledge of operating conditions and the physical properties of the boiling liquid permits the direct calculation of h_b. The surface temperature of the heater may now be calculated from q and h_b and the process is described completely. Considering the evaluation of a process involving heat transfer from steam condensing at temperature T_c to a liquid boiling at temperature T_b, assuming that the condensing coefficient is constant and specified as h_c, and also that the thermal resistance of the intervening wall is negligible, an initial estimate of the wall temperature T_w may be made. The heat flux q for the condensing film may now be calculated since $q = h_c(T_c - T_w)$, and the value of h_b may then be determined from Eq. (5.2) using this value for the heat flux. A heat balance across the wall tests the accuracy of the estimated value of T_w since $h_c(T_c - T_w)$ must equal $h_b(T_w - T_b)$, assuming the intervening wall to be plane. If the error in this heat balance is unacceptable, further values of T_w must be assumed until the heat balance falls within specified limits of accuracy.

A more refined design procedure would include the estimation of the steam-side coefficient h_c by one of the methods discussed in Volume 1, Chapter 9. While such iterative procedures are laborious when carried out by hand, they are ideally handled by computers which enable a rapid evaluation to any degree of accuracy to be easily achieved.

Table 5.1 Advantages of vacuum operation.

	Atmospheric pressure (101.3 kN/m²)	Vacuum operation (13.5 kN/m²)
Boiling point	373 K	325 K
Temperature drop to liquor	7 K	55 K
Heat lost in condensate	419 kJ/kg	216 kJ/kg
Heat used	2266 kJ/kg	2469 kJ/kg

5.2.4 Vacuum operation

With a number of heat sensitive liquids it is necessary to work at low temperatures, and this is effected by boiling under a vacuum, as indeed is the case in the last unit of a multi-effect system. Operation under a vacuum increases the temperature difference between the steam and boiling liquid as shown in Table 5.1 and therefore tends to increase the heat flux. At the same time, the reduced boiling point usually results in a more viscous material and a lower film heat transfer coefficient.

For a standard evaporator using steam at 135 kN/m² and 380 K with a total heat content of 2685 kJ/kg, evaporating a liquor such as water, the capacity under vacuum is $(101.3/13.5) = 7.5$ times great than that at atmospheric pressure. The advantage in capacity for the same unit is therefore considerable, though there is no real change in the consumption of steam in the unit. In practice, the advantages are not as great as this since operation at a lower boiling point reduces the value of the heat transfer coefficient and additional energy is required to achieve and maintain the vacuum.

5.3 Single-effect evaporators

Single-effect evaporators are used when the throughput is low, when a cheap supply of steam is available, when expensive materials of construction must be used as is the case with corrosive feedstocks and when the vapour is so contaminated so that it cannot be reused. Single effect units may be operated in batch, semi-batch or continuous batch modes or continuously. In strict terms, batch units require that filling, evaporating and emptying are consecutive steps. Such a method of operation is rarely used since it requires that the vessel is large enough to hold the entire charge of feed and that the heating element is low enough to ensure that it is not uncovered when the volume is reduced to that of the product. Semi-batch is the more usual mode of operation in which feed is added continuously in order to maintain a constant level until the entire charge reaches the required product density. Batch-operated evaporators often have a continuous feed and, over at least part of the cycle, a continuous discharge. Often a feed drawn from a storage tank is returned until the entire contents of the tank reach the desired concentration. The final evaporation is then achieved by batch operation. In essence, continuous evaporators have a continuous feed and discharge and concentrations of both feed and discharge remain constant.

The heat requirements of single-effect continuous evaporators may be obtained from mass and energy balances. If enthalpy data or heat capacity and heat of solution data are not available, heat requirements may be taken as the sum of the heat needed to raise the feed from feed to product temperature and the heat required to evaporate the water. The latent heat of water is taken at the vapour head pressure instead of the product temperature in order to compensate, at least to some extent, for the heat of solution. If sufficient vapour pressure data are available for the liquor, methods are available for calculating the true latent heat from the slope of the Dühring line and detailed by Othmer [11]. The heat requirements in batch operation are generally similar to those in continuous evaporation. While the temperature and sometimes the pressure of the vapour will change during the course of the cycle which results in changes in enthalpy, since the enthalpy of water vapour changes only slightly with temperature, the differences between continuous and batch heat requirements are almost negligible for all practical purposes. The variation of the fluid properties, such as viscosity and boiling point rise, have a much greater effect on heat transfer, although these can only be estimated by a step-wise calculation. In estimating the boiling temperature, the effect of temperature on the heat transfer characteristics of the type of unit involved must be taken into account. At low temperatures some evaporator types show a marked drop in the heat transfer coefficient which is often more than enough to offset any gain in available temperature difference. The temperature and cost of the cooling water fed to the condenser are also of importance in this respect.

Example 5.1

A single-effect evaporator is used to concentrate 7 kg/s of a solution from 10% to 50% solids. Steam is available at 205 kN/m^2 and evaporation takes place at 13.5 kN/m^2. If the overall coefficient of heat transfer is 3 kW/m^2 K, estimate the heating surface required and the amount of steam used if the feed to the evaporator is at 294 K and the condensate leaves the heating space at 352.7 K. The specific heats of 10% and 50% solutions are 3.76 and 3.14 kJ/kg K respectively.

Solution

Assuming that the steam is dry and saturated at 205 kN/m^2, then from the Steam Tables in the Appendix, the steam temperature $=394$ K at which the total enthalpy $=2530$ kJ/kg.

At 13.5 kN/m^2, water boils at 325 K and, in the absence of data on the boiling point elevation, this will be taken as the temperature of evaporation, assuming an aqueous solution. The total enthalpy of steam at 325 K is 2594 kJ/kg.

Thus the feed, containing 10% solids, has to be heated from 294 to 325 K at which temperature the evaporation takes place.

In the feed, mass of dry solids $= (7 \times 10)/100 = 0.7$ kg/s and, for x kg/s of water in the product:

$$(0.7 \times 100)/(0.7 + x) = 50$$

from which:

$$x = 0.7 \text{ kg/s}$$

Thus:

$$\text{water to be evaporated} = (7.0 - 0.7) - 0.7 = 5.6 \text{ kg/s}$$

Summarising:

Stream	Solids (kg/s)	Liquid (kg/s)	Total (kg/s)
Feed	0.7	6.3	7.0
Product	0.7	0.7	1.4
Evaporation		5.6	5.6

Using a datum of 273 K:

$$\text{Heat entering with the feed} = (7.0 \times 3.76)(294 - 273) = 552.7 \text{ kW}$$
$$\text{Heat leaving with the product} = (1.4 \times 3.14)(325 - 273) = 228.6 \text{ kW}$$
$$\text{Heat leaving with the evaporated water} = (5.6 \times 2594) = 14,526 \text{ kW}$$

Thus:

$$\text{Heat transferred from the steam} = (14,526 + 228.6) - 552.7 = 14,202 \text{ kW}$$

The enthalpy of the condensed steam leaving at 352.7 K $=4.18(352.7-273)=333.2$ kJ/kg.

The heat transferred from 1 kg steam $=(2530-333.2)=2196.8$ kJ/kg and hence:

$$\text{Steam required} = (14,202/2196.8) = 6.47 \text{ kg/s}$$

As the preheating of the solution and the sub-cooling of the condensate represent but a small proportion of the heat load, the temperature driving force may be taken as the difference between the temperatures of the condensing steam and the evaporating water, or:

$$\Delta T = (394 - 325) = 69 \text{ K}$$

Thus:

$$\text{Heat transfer area}, A = Q/U\Delta T$$
$$= 14,202/(3 \times 69) = 68.6 \text{ m}^2. \tag{5.7}$$

5.4 **Multiple-effect evaporators**

The single effect evaporator uses rather more than 1 kg of steam to evaporate 1 kg of water. Three methods have been introduced which enable the performance to be improved, either by direct reduction in the steam consumption, or by improved energy efficiency of the whole unit. These are:

a) Multiple effect operation.
b) Recompression of the vapour rising from the evaporator.
c) Evaporation at low temperatures using a heat pump cycle.

The first of these is considered in this section and (b) and (c) are considered in Section 5.5.

5.4.1 **General principles**

If an evaporator, fed with steam at 399 K with a total heat of 2714 kJ/kg, is evaporating water at 373 K, then each kilogram of water vapour produced will have a total heat content of 2675 kJ. If this heat is allowed to go to waste, by condensing it in a tubular condenser or by direct contact in a jet condenser for example, such a system makes very poor use of steam. The vapour produced is, however, suitable for passing to the calandria of a similar unit, provided the boiling temperature in the second unit is reduced so that an adequate temperature difference is maintained. This, as discussed in Section 5.2.4, can be effected by applying a vacuum to the second effect in order to reduce the boiling point of the liquor. This is the principle reached in the multiple effect systems which were introduced by Rillieux in about 1830.

For three evaporators arranged as shown in Fig. 5.5, in which the temperatures and pressures are T_1, T_2, T_3, and P_1, P_2, P_3, respectively, in each unit, if the liquor has no boiling point rise, then the heat transmitted per unit time across each effect is:

$$\text{Effect 1: } Q_1 = U_1 A_1 \Delta T_1, \quad \text{where } \Delta T_1 = (T_0 - T_1),$$
$$\text{Effect 2: } Q_2 = U_2 A_2 \Delta T_2, \quad \text{where } \Delta T_2 = (T_1 - T_2),$$
$$\text{Effect 3: } Q_3 = U_3 A_3 \Delta T_3, \quad \text{where } \Delta T_3 = (T_2 - T_3).$$

Neglecting the heat required to heat the feed from T_f to T_1, the heat Q_1 transferred across where A_1 appears as latent heat in the vapour D_1 and is used as steam in the second effect, and:
So that:

$$Q_1 = Q_2 = Q_3$$
$$U_1 A_1 \Delta T_1 = U_2 A_2 \Delta T_2 = U_3 A_3 \Delta T_3$$

(5.8)

If, as is commonly the case, the individual effects are identical, $A_1 = A_2 = A_3$, and:

$$U_1 \Delta T_1 = U_2 \Delta T_2 = U_3 \Delta T_3$$

(5.9)

On this analysis, the difference in temperature across each effect is inversely proportional to the heat transfer coefficient. This represents a simplification, however, since:

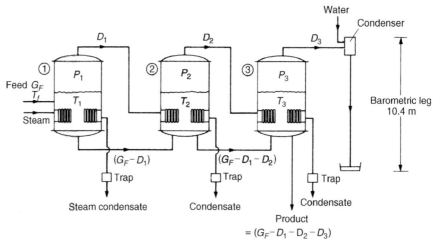

FIG. 5.5

Forward-feed arrangement for a triple-effect evaporator.

a) the heat required to heat the feed from T_f to T_1 has been neglected and
b) the liquor passing from stages ① to ② carries heat into the second effect, and this is responsible for some evaporation. This is also the case in the third effect.

The latent heat required to evaporate 1 kg of water in ①, is approximately equal to the heat obtained in condensing 1 kg of steam at T_0.

Thus 1 kg of steam fed to ① evaporates 1 kg of water in ①. Again the 1 kg of steam from ① evaporates about 1 kg of steam in ②. Thus, in a system of N effects, 1 kg of steam fed to the first effect will evaporate in all about N kg of liquid. This gives a simplified picture, as discussed later, although it does show that one of the great attractions of a multiple-effect system is that considerably more evaporation per kilogram of steam is obtained than in a single-effect unit. The economy of the system, measured by the kilograms of water vaporised per kilogram of steam condensed, increases with the number of effects.

The water evaporated in each effect is proportional to Q, since the latent heat is approximately constant. Thus the total capacity is:

$$Q = Q_1 + Q_2 + Q_3$$
$$= U_1 A_1 \Delta T_1 + U_1 A_2 \Delta T_2 + U_3 A_3 \Delta T_3 \qquad (5.10)$$

If an average value of the coefficients U_{av} is taken, then:

$$Q = U_{av}(\Delta T_1 + \Delta T_2 + \Delta T_3)A \qquad (5.11)$$

assuming the area of each effect is the same. A single-effect evaporator operating with a temperature difference $\sum \Delta T$, with this average coefficient U_{av}, would, however, have the same capacity $Q = U_{av}A \sum \Delta T$. Thus, it is seen that the capacity of a multiple-effect system is the same as that of a single effect, operating with the same total temperature difference and having an area A equal to that of one of the multiple-effect units. The value of the multiple-effect system is that better use is made of steam although, in order to achieve this, a much higher capital outlay is required for the increased number of units and accessories.

5.4.2 The calculation of multiple-effect systems

In the equations considered in Section 5.4.1, various simplifying assumptions have been made which are now considered further in the calculation of a multiple-effect system. In particular, the temperature distribution in such a system and the heat transfer area required in each effect are determined. The method illustrated in Example 5.2 is essentially based on that of Hausbrand [12].

Example 5.2A (Forward-feed)

4 kg/s (14.4 tonne/h) of a liquor containing 10% solids is fed at 294 K to the first effect of a triple-effect unit. Liquor with 50% solids is to be withdrawn from the third effect, which is at a pressure of 13 kN/m² (~0.13 bar). The liquor may be assumed to have a specific heat of 4.18 kJ/kg K and to have no boiling point rise. Saturated dry steam at 205 kN/m² is fed to the heating element of the first effect, and the condensate is removed at the steam temperature in each effect as shown in Fig. 5.5.

If the three units are to have equal areas, estimate the area, the temperature differences and the steam consumption. Heat transfer coefficients of 3.1, 2.0 and 1.1 kW/m² K for the first, second, and third effects, respectively, may be assumed.

Solution 1

A precise theoretical solution is neither necessary nor possible, since during the operation of the evaporator, variations of the liquor levels, for example, will alter the heat transfer coefficients and hence the temperature distribution. It is necessary to assume values of heat transfer coefficients, although, as noted previously, these will only be approximate and will be based on practical experience with similar liquors in similar types of evaporators.

Temperature of dry saturated steam at 205 kN/m² = 394 K.

At a pressure of 13 kN/m² (0.13 bar), the boiling point of water is 325 K, so that the total temperature difference $\Sigma\Delta T = (394-325) = 69$ K.

First approximation
Assuming that:

$$U_1\Delta T_1 = U_2\Delta T_2 = U_3\Delta T_3 \qquad (5.12)$$

then substituting the values of U_1, U_2 and U_3 and $\Sigma\Delta T = 69$ K gives:

$$\Delta T_1 = 13 \text{ K}, \quad \Delta T_2 = 20 \text{ K}, \quad \Delta T_3 = 36 \text{ K}$$

Since the feed is cold, it will be necessary to have a greater value of ΔT_1 than given by this analysis. It will be assumed that $\Delta T_1 = 18\,\mathrm{K}$, $\Delta T_2 = 17\,\mathrm{K}$, $\Delta T_3 = 34\,\mathrm{K}$.

If the latent heats are given by λ_0, λ_1, λ_2 and λ_3, then from the Steam Tables in the Appendix:

$$\text{For steam to 1:}\ T_0 = 394\ \mathrm{K} \quad \text{and} \quad \lambda_0 = 2200\ \mathrm{kJ/kg}$$
$$\text{For steam to 2:}\ T_1 = 376\ \mathrm{K} \quad \text{and} \quad \lambda_1 = 2249\ \mathrm{kJ/kg}$$
$$\text{For steam to 3:}\ T_2 = 359\ \mathrm{K} \quad \text{and} \quad \lambda_2 = 2293\ \mathrm{kJ/kg}$$
$$T_3 = 325\ \mathrm{K} \quad \text{and} \quad \lambda_3 = 2377\ \mathrm{kJ/kg}$$

Assuming that the condensate leaves at the steam temperature, then heat balances across each effect may be made as follows:

Effect 1:

$$D_0\lambda_0 = G_F C_p\left(T_1 - T_f\right) + D_1\lambda_1, \quad \text{or} \quad 2200D_0 = 4 \times 4.18(376 - 294) + 2249D_1$$

Effect 2:

$$D_1\lambda_1 + (G_F - D_1)C_p(T_1 - T_2) = D_2\lambda_2, \quad \text{or} \quad 2249D_1$$
$$+ (4 - D_1)4.18(376 - 359) + 2293D_2$$

Effect 3:

$$D_2\lambda_2 + (G_F - D_1 - D_2)C_p(T_2 - T_3) = D_3\lambda_3, \quad \text{or} \quad 2293D_2$$
$$+ (4 - D_1 - D_2)4.18(359 - 325)$$
$$= 2377D_3$$

where G_F is the mass flow rate of liquor fed to the system, and C_p is the specific heat capacity of the liquid, which is assumed to be constant.

A material balance over the evaporator is:

	Solids (kg/s)	Liquor (kg/s)	Total (kg/s)
Feed	0.4	3.6	4.0
Product	0.4	0.4	0.8
Evaporation		3.2	3.2

Making use of the previous equations and the fact that $(D_1 + D_2 + D_3) = 3.2\,\mathrm{kg/s}$, the evaporation in each unit is, $D_1 \approx 0.991$, $D_2 \approx 1.065$, $D_3 \approx 1.144$, and $D_0 \approx 1.635\,\mathrm{kg/s}$. The area of the surface of each calandria necessary to transmit the necessary heat under the given temperature difference may then be obtained as:

$$A_1 = \frac{D_0 \lambda_0}{U_1 \Delta T_1} = \frac{(1.635 \times 2200)}{(3.1 \times 18)} = 64.5 \text{ m}^2$$

$$A_2 = \frac{D_1 \lambda_1}{U_2 \Delta T_2} = \frac{(0.991 \times 2249)}{(2.0 \times 17)} = 65.6 \text{ m}^2$$

$$A_3 = \frac{D_2 \lambda_2}{U_3 \Delta T_3} = \frac{(1.085 \times 2293)}{(1.1 \times 34)} = 65.3 \text{ m}^2$$

These three calculated areas are approximately equal, so that the temperature differences assumed may be taken as nearly correct. In practice, ΔT_1 would have to be a little larger since A_1 is the smallest area. It may be noted that, on the basis of these calculations, the economy is given by $e = (3.2/1.635) = 2.0$. Thus, a triple effect unit working under these conditions gives a reduction in steam utilisation compared with a single effect, though not as large an economy as might be expected.

A simplified method of solving problems of multiple effect evaporation, suggested by Storrow [13], is particularly useful for systems with a large number of effects because it obviates the necessity for solving many simultaneous equations. Essentially the method depends on obtaining only an approximate value for those heat quantities which are a small proportion of the whole. Example 5.2A is now solved by this method.

Solution 2

From Fig. 5.5 it may be seen that for a feed G_F to the first effect, vapour D_1 and liquor $(G_F - D_1)$ are fed forward to the second effect. In the first effect, steam is condensed partly in order to raise the feed to its boiling point and partly to effect evaporation. In the second effect, further vapour is produced mainly as a result of condensation of the vapour from the first effect and to a smaller extent by flash vaporisation of the concentrated liquor which is fed forward. As the amount of vapour produced by the latter means is generally only comparatively small, this may be estimated only approximately. Similarly, the vapour produced by flash evaporation in the third effect will be a small proportion of the total and only an approximate evaluation is required.

Vapour production by flash vaporisation—Approximate evaluation. If the heat transferred in each effect is the same, then:

$$U_1 \Delta T_1 = U_2 \Delta T_2 = U_3 \Delta T_3 \tag{5.12}$$

or

$$3.1 \Delta T_1 = 2.0 \Delta T_2 = 1.1 \Delta T_3$$

Steam temperature $= 394$ K. Temperature in condenser $= 325$ K.
Thus:

$$\sum \Delta T = (394 - 325) = 69 \text{ K}$$

Solving:

$$\Delta T_1 = 13 \text{ K}, \quad \Delta T_2 = 20 \text{ K}, \quad \Delta T_3 = 36 \text{ K}$$

These values of ΔT will be valid provided the feed is approximately at its boiling point.

Weighting the temperature differences to allow for the fact that the feed enters at ambient temperature gives:

$$\Delta T_1 = 18 \text{ K}, \quad \Delta T_2 = 18 \text{ K}, \quad \Delta T_3 = 33 \text{ K}$$

and the temperatures in each effect are:

$$T_1 = 376 \text{ K}, \quad T_2 = 358 \text{ K}, \quad \text{and} \quad T_3 = 325 \text{ K}$$

The total evaporation $(D_1 + D_2 + D_3)$ is obtained from a material balance:

	Solids (kg/s)	Liquor (kg/s)	Total (kg/s)
Feed	0.4	3.6	4.0
Product	0.4	0.4	0.8
Evaporation		3.2	3.2

Assuming, as an approximation, equal evaporation in each effect, or $D_1 = D_2 = D_3 = 1.07 \text{ kg/s}$, then the latent heat of flash vaporisation in the second effect is given by:

$$4.18(4.0 - 1.07)(376 - 358) = 220.5 \text{ kW}$$

and latent heat of flash vaporisation in the third effect is:

$$4.18(4.0 - 2 \times 1.07)(358 - 325) = 256.6 \text{ kW}$$

Final calculation of temperature differences. Subsequent calculations are considerably simplified if it is assumed that the latent heat of vaporisation is the same at all temperatures in the multiple-effect system, since under these conditions the condensation of 1 kg of steam gives rise to the formation of 1 kg of vapour.

Thus:

$$\text{At 394 K, the latent heat} = 2200 \text{ kJ/kg}$$
$$\text{At 325 K, the latent heat} = 2377 \text{ kJ/kg}$$
$$\text{Mean value,} \lambda = 2289 \text{ kJ/kg}$$

The amounts of heat transferred in each effect (Q_1, Q_2, Q_3) and in the condenser (Q_c) are related by:

$$Q_1 - G_F C_p (T_1 - T_f) = Q_2 = (Q_3 - 220.5) = (Q_c - 220.5 - 256.6)$$

or

$$Q_1 - 4.0 \times 4.18(394 - \Delta T_1 - 294) = Q_2 = (Q_3 - 220.5) = (Q_c - 477.1) \text{ kW}$$

$$\text{Total evaporation} = (Q_2 + Q_3 + Q_c)/2289 = 3.2 \text{ kg/s}$$

Thus:

$$Q_2 + (Q_2 + 220.5) + Q_2 + (Q_2 + 477.1) = 7325 \text{ kW}$$

or

$$Q_2 = 2209 \text{ kW}$$

$$Q_2 = 2430 \text{ kW}$$

and

$$Q_1 = 2209 + 4.0 \times 4.18(394 - \Delta T_1 - 294)$$
$$= (3881 - 16.72\Delta T_1) \text{ kW}$$

Applying the heat transfer equations, then:

$$3881 - 16.72\Delta T_1 = 3.1A\Delta T_1, \quad \text{or} \quad A\Delta T_1 = (1252 - 5.4\Delta T_1) \text{ m}^2\text{K}$$
$$2209 = 2.0A\Delta T_2, \quad \text{or} \quad A\Delta T_2 = 1105 \text{ m}^2\text{K}$$
$$2430 = 1.1A\Delta T_3, \quad \text{or} \quad A\Delta T_3 = 2209 \text{ m}^2\text{K}$$

Further:

$$\Delta T_1 + \Delta T_2 + \Delta T_3 = 69 \text{ deg K}$$

Values of ΔT_1, ΔT_2, ΔT_3 are now chosen by trial and error to give equal values of A in each effect, as follows:

ΔT_1 (K)	A_1 (m^2)	ΔT_2 (K)	A_2 (m^2)	ΔT_3 (K)	A_3 (m^2)
18	64.2	18	61.4	33	66.9
19	60.5	17	65.0	33	66.9
18	64.2	17.5	63.1	33.5	65.9
18	64.2	17	65.0	34	64.9

The areas, as calculated in the last line, are approximately equal, so that the assumed temperature differences are acceptable and:

$$\text{Steam consumption} = (Q_1/2289) = (3580/2289) = 1.56 \text{ kg/s}$$
$$\text{Economy} = (3.2/1.56) \approx 2.0 \text{ kg/kg}$$

The calculation of areas in multiple-effect systems is relatively straightforward for one or two configurations, although it becomes tedious in the extreme where a wide range of operating conditions is to be investigated. Fortunately the calculations involved lend themselves admirably to processing by computer, and in this respect reference should be made to work such as that by Stewart and Beveridge [14].

5.4.3 Comparison of forward and backward feeds

In the unit considered in Example 5.2A, the weak liquor is fed to effect ① and flows on to ② and then to ③. The steam is also fed to ①, and the process is known as forward-feed since the feed is to the same unit as the steam and travels down the unit in the same direction as the steam or vapour. It is possible, however, to introduce the weak liquor to effect ③ and cause it to travel from ③ to ② to ①, while the steam and vapour still travel in the direction of ① to ② to ③. This system, shown in Fig. 5.6, is known as backward-feed. A further arrangement for the feed is known as parallel-feed, which is shown in Fig. 5.7. In this case, the liquor is fed to each of the three effects in parallel although the steam is fed only to the first effect. This arrangement is commonly used in the concentration of salt solutions, where the deposition of crystals makes it difficult to use the standard forward-feed arrangement. The effect of backward-feed on the temperature distribution, the areas of surface required, and the economy of the unit is of importance, and Example 5.2A is now considered for this flow arrangement.

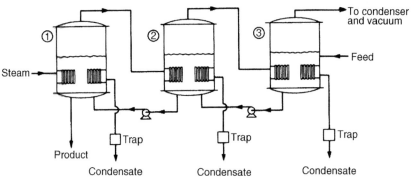

FIG. 5.6

Backward-feed arrangement for a triple-effect evaporator.

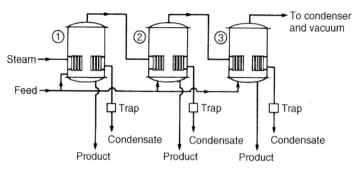

FIG. 5.7

Parallel-feed arrangement for a triple-effect evaporator.

Example 5.2B (Backward-feed).

Since the dilute liquor is now at the lowest temperature and the concentrated liquor at the highest, the heat transfer coefficients will not be the same as in the case of forward-feed. In effect ①, the liquor is now much more concentrated than in the former case, and hence U_1 will not be as large as before. Again, on the same argument, U_3 will be larger than before. Although it is unlikely to be exactly the same, U_2 will be taken as being unaltered by the arrangement. Taking values of $U_1 = 2.5$, $U_2 = 2.0$, and $U_3 = 1.6\,\text{kW/m}^2\,\text{K}$, the temperature distribution may be determined in the same manner as for forward feed, by taking heat balances across each unit.

Solution 1

In this case, it is more difficult to make a reasonable first estimate of the temperature differences because the liquid temperature is increasing as it passes from effect to effect $(3 \to 2 \to 1)$ and sensible heat must be added at each stage. It may therefore be necessary to make several trial and error solutions before achieving the conditions for equal areas. In addition, the values of U_1, U_2, and U_3 may be different from those in forward-feed, depending as they do on concentration as well as on temperature.

Taking:

$$\Delta T_1 = 20 \text{ K}, \quad \Delta T_2 = 24 \text{ K}, \quad \Delta T_3 = 25 \text{ K}$$

The temperatures in the effect and the corresponding latent heats are:

$$T_0 = 394 \text{ K} \quad \text{and} \quad \lambda_0 = 2200 \text{ kJ/kg}$$
$$T_1 = 374 \text{ K} \quad \text{and} \quad \lambda_1 = 2254 \text{ kJ/kg}$$
$$T_2 = 350 \text{ K} \quad \text{and} \quad \lambda_2 = 2314 \text{ kJ/kg}$$
$$T_3 = 325 \text{ K} \quad \text{and} \quad \lambda_3 = 2377 \text{ kJ/kg}$$

The heat balance equations are then:
Effect 3:

$$D_2\lambda_2 = G_F C_p(T_3 - T_f) + D_3\lambda_3, \quad \text{or} \quad 2314D_2 = 4 \times 4.18(325 - 294) + 2377D_3$$

Effect 2:

$$D_1\lambda_1 + (G_F - D_3)C_p(T_2 - T_3) = D_2\lambda_2, \quad \text{or} \quad 2254D_1$$
$$= (4 - D_3)4.18(350 - 325) + 2314D_2$$

Effect 1:

$$D_0\lambda_0 + (G_F - D_3 - D2)C_p(T_1 - T_2) = D_1\lambda_1, \quad \text{or} \quad 2200D_0$$
$$+ (4 - D_3 - D_2)4.18(374 - 350)$$
$$= 2254D_1$$

Again taking $(D_1 + D_2 + D_3) = 3.2\,\text{kg/s}$, these equations may be solved to give:

$$D_1 \approx 1.261, \quad D_2 \approx 1.086, \quad D_3 \approx 0.853, \quad D_0 \approx 1.387 \text{ kg/s}$$

The areas of transfer surface are then:

$$A_1 = \frac{D_0 \lambda_0}{U_1 \Delta T_1} = \frac{(1.387 \times 2200)}{(2.5 \times 20)} = 61.0 \text{ m}^2$$

$$A_2 = \frac{D_1 \lambda_1}{U_2 \Delta T_2} = \frac{(1.261 \times 2254)}{(2.00 \times 24)} = 59.2 \text{ m}^2$$

$$A_3 = \frac{D_2 \lambda_2}{U_3 \Delta T_3} = \frac{(1.086 \times 2314)}{(1.6 \times 25)} = 62.8 \text{ m}^2$$

These three areas are approximately equal, so that the temperature differences suggested are sufficiently acceptable for design purposes. The economy for this system is $(3.2/1.387) = 2.3$ kg/kg.

Solution 2

Using Storrow's method, as in Example 5.2A, the temperatures in the effects will be taken as:

$$T_1 = 374 \text{ K}, \quad T_2 = 350 \text{ K}, \quad T_3 = 325 \text{ K}$$

With backward-feed, as shown in Fig. 5.6, the liquid has to be raised to its boiling point as it enters each effect.

The heat required to raise the feed to the second effect to its boiling point is:

$$= 4.18(4.0 - 1.07)(350 - 325)$$
$$= 306.2 \text{ kW}$$

The heat required to raise the feed to the first effect to its boiling point is:

$$= 4.18(4.0 - 2 \times 1.07)(374 - 350)$$
$$= 186.6 \text{ kW}$$

Assuming a constant value of 2289 kJ/kg for the latent heat in all the stages, the relation between the heat transferred in each effect and in the condenser is:

$$Q_1 - 186.6 = Q_2 = (Q_3 + 306.2) = Q_c - 306.2 + 4 \times 4.18(325 - 294)$$
$$= Q_c + 824.5$$

$$\text{Total evaporation} = (Q_2 + Q_3 + Q_c)/2289 = 3.2 \text{ kg/s}$$

and

$$Q_2 + (Q_2 - 306.2) + (Q_2 - 824.5) = 7325 \text{ kW}$$

Thus:

$$Q_2 = 2819 = A \Delta T_2 \times 2.0 \text{ kW}$$

$$Q_3 = 2512 = A \Delta T_3 \times 1.6 \text{ kW}$$

sand

$$Q_1 = 3006 = A \Delta T_1 \times 2.5 \text{ kW}$$

or

$$A\Delta T_1 = 1202 \text{ m}^2 \text{ K}$$

$$A\Delta T_2 = 1410 \text{ m}^2 \text{ K}$$

$$A\Delta T_3 = 1570 \text{ m}^2 \text{ K}$$

and

$$\Delta T_1 + \Delta T_2 + \Delta T_3 = 69 \text{ K}$$

Thus:

ΔT_1 (K)	A_1 (m²)	ΔT_2 (K)	A_2 (m²)	ΔT_3 (K)	A_3 (m²)
20	60.1	24	58.9	25	62.8

The areas are approximately equal and the assumed values of ΔT are therefore acceptable.

$$\text{Economy} = \frac{3.2}{(3006/2289)} = 2.4 \text{ kg/kg}$$

On the basis of heat transfer area and thermal considerations, a comparison of the two methods of feed is:

	Forward	Backward
Total steam used D_0 (kg)	1.635	1.387
Economy (kg/kg)	2.0	2.3
Condenser load D_3 (kg)	1.44	0.853
Heat transfer surface per effect A (m²)	65.1	61.0

For the conditions of Example 5.2, the backward feed system shows a reduction in steam consumption, an improved economy, a reduction in condenser load, and a small reduction in heat transfer area.

Effect of feed system on economy

In the case of forward feed systems, all the liquor has to be heated from T_f to T_1 by steam although, in the case of backward feed, the heating of the feed in the last effect is done with steam that has already evaporated $(N-1)$ times its own mass of water, assuming ideal conditions. The feed temperature must therefore be regarded as a major feature in this class of problem. Webre [15] has examined the effect of feed

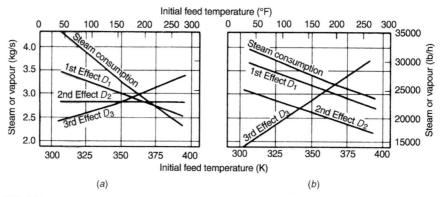

FIG. 5.8

Effect of feed temperature on the operation of a triple effect evaporator: (A) forward feed and (B) backward feed.

temperature on the economy and the evaporation in each effect, for the case of a liquor fed at the rate of 12.5 kg/s to a triple-effect evaporator in which a concentrated product was obtained at a flow rate of 8.75 kg/s. Neglecting boiling-point rise and working with a fixed vacuum on the third effect, the curves shown in Figs 5.8 and 5.9 for the three methods of forward, backward and parallel feed were prepared.

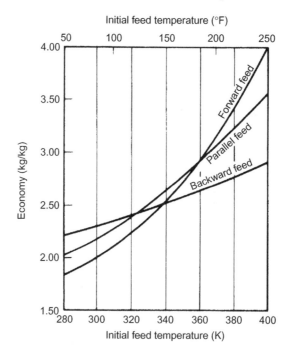

FIG. 5.9

Economy of triple-effect evaporators.

Fig. 5.8A illustrates the drop in steam consumption as the feed temperature is increased with forward feed. It may be seen that, for these conditions, D_1 falls, D_2 remains constant and D_3 rises with increase in the feed temperature T_f. With backward feed shown in Fig. 5.8B, the fall in steam consumption is not so marked and it may be seen that, whereas D_1 and D_2 fall, the load on the condenser D_3 increases. The results are conveniently interpreted in Fig. 5.9, which shows that the economy increases with T_f for a forward-feed system to a marked extent, while the corresponding increase with the backward-feed system is relatively small. At low values of T_f, the backward feed gives the higher economy. At some intermediate value, the two systems give the same value of economy, while for high values of T_f the forward-feed system is more economical in steam.

These results, while showing the influence of T_f on the economy, should not be interpreted too rigidly, since the values for the coefficients for the two systems and the influence of boiling-point rise may make a substantial difference to these curves. In general, however, it will be found that with cold feeds the backward-feed system is more economical. Despite this fact, the forward-feed system is the most common, largely because it is the simplest to operate, while backward feed requires the use of pumps between each effect.

The main criticism of the forward-feed system is that the most concentrated liquor is in the last effect, where the temperature is lowest. The viscosity is therefore high and low values of U are obtained. In order to compensate for this, a large temperature difference is required, and this limits the number of effects. It is sometimes found, as in the sugar industry, that it is preferable to run a multiple-effect system up to a certain concentration, and to run a separate effect for the final stage where the crystals are formed.

5.5 Improved efficiency in evaporation

5.5.1 Vapour compression evaporators

Considering an evaporator fed with saturated steam at 387 K, equivalent to $165 \, kN/m^2$, concentrating a liquor boiling at 373 K at atmospheric pressure, if the condensate leaves at 377 K, then:

1 kg of steam at 387 K has a total heat of 2698 kJ.

1 kg of condensate at 377 K has a total heat of 437 kJ and.

the heat given up is 2261 kJ/kg steam.

If this condensate is returned to the boiler, then at least 2261 kJ/kg must be added to yield 1 kg of steam to be fed back to the evaporator. In practice, of course, more heat per kilogram of condensate will be required. 2261 kJ will vaporise 1 kg of liquid at atmospheric pressure to give vapour with a total heat of 2675 kJ/kg. To regenerate 1 kg of steam in the original condition from this requires the addition of only 23 kJ. The idea of vapour compression is to make use of the vapour from the evaporator, and to upgrade it to the condition of the original steam. Such a system offers enormous advantages in thermal economy, though it is by no means easy to add the 23 kJ to each kilogram of vapour in an economical manner. The two methods available are:

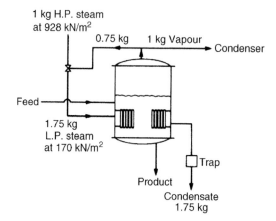

FIG. 5.10

Vapour compression evaporator with high-pressure steam-jet compression.

a) the use of steam-jet ejectors as shown in Fig. 5.10 and
b) the use of mechanical compressors as shown in Fig. 5.11.

In selecting a compressor for this type of operation, the main difficulty is the very large volume of vapour to be handled. Rotary compressors of the Rootes type, described in Volume 1, Chapter 8, are suitable for small and medium size units, though these have not often been applied to large installations. Mechanical compressors have been used extensively in evaporation systems for the purification of sea water.

The use of an ejector, fed with high-pressure steam, is illustrated in Fig. 5.10. High-pressure steam is injected through a nozzle and the low-pressure vapours are drawn in through a second inlet at right angles, the issuing jet of steam passing out to the calandria, as shown. These units are relatively simple in construction and can be made of corrosion-resistant material. They have no moving parts and for this reason will have a

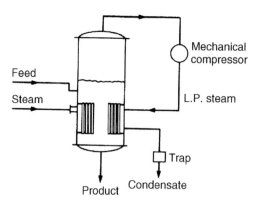

FIG. 5.11

Vapour compression evaporator with a mechanical compressor.

long life. They have the great advantage over mechanical compressors in that they can handle large volumes of vapour and can therefore be arranged to operate at very low pressures. The disadvantage of the steam-jet ejector is that it works at maximum efficiency at only one specific condition. Some indication of the performance of these units is shown in Fig. 5.12, where the pressure of the mixture, for different amounts of vapour compressed per kilogram of live steam, is shown for a series of different pressures. With an ejector of these characteristics using steam at $965\,kN/m^2$, 0.75 kg vapour/kg steam can be compressed to give 1.75 kg of vapour at $170\,kN/m^2$. An evaporator unit, as shown in Fig. 5.11, will therefore give 1.75 kg of vapour/kg high-pressure steam. Of the 1.75 kg of vapour, 0.75 kg is taken to the compressor and the remaining 1 kg to the condenser. Ideally, this single-effect unit gives an economy of 1.75, or approximately the economy of a double-effect unit.

Vapour compression may be applied to the vapour from the first effect of a multiple-effect system, thus giving increased utilisation of the steam. Such a device is not suitable for use with liquors with a high boiling-point rise, for in these cases the vapour, although initially superheated, has to be compressed to such a great degree, in order to give the desired temperature difference across the calandria, that the efficiency is reduced. The application of these compressors depends on the steam load of the plant. If there is plenty of low-pressure steam available, then the use of vapour compression can rarely be advocated. If, however, high-pressure steam is available, then it may be used to advantage in a vapour compression unit. It will, in fact, be far superior to the practice of passing high-pressure steam through a reducing valve to feed an evaporator.

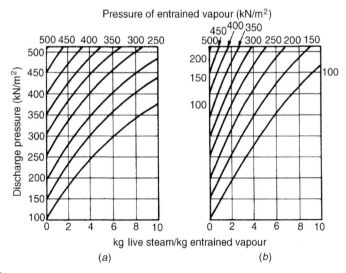

FIG. 5.12

Performance of a steam jet ejector: (A) $790\,kN/m^2$ operating pressure and (B) $1135\,kN/m^2$ operating pressure.

Example 5.3

Saturated steam leaving an evaporator at atmospheric pressure is compressed by means of saturated steam at $1135\,kN/m^2$ in a steam jet to a pressure of $135\,kN/m^2$. If 1 kg of the high-pressure steam compresses 1.6 kg of the vapour produced at atmospheric pressure, comment on the efficiency of the compressor.

Solution

The efficiency of an ejector η' is given by:

$$\eta' = (m_1 + m_2)(H_4 - H_3)/[m_1(H_1 - H_2)]$$

where m_1 is the mass of high-pressure steam (kg), m_2 is the mass of entrained steam (kg), H_1 is the enthalpy of high-pressure steam (kJ/kg), H_2 is the enthalpy of steam after isentropic expansion in the nozzle to the pressure of the entrained vapours (kJ/kg), H_3 is the enthalpy of the mixture at the start of compression in the diffuser section (kJ/kg), and H_4 is the enthalpy of the mixture after isentropic compression to the discharge pressure (kJ/kg).

The high-pressure steam is saturated at $1135\,kN/m^2$ at which $H_1 = 2780\,kJ/kg$. If this is allowed to expand isentropically to $101.3\,kN/m^2$, then from the entropy–enthalpy chart, given in the appendix, $H_2 = 2375\,kJ/kg$ and the dryness faction is 0.882.

Making an enthalpy balance across the system, then:

$$m_1 H_1 + m_2 H_e = (m_1 + m_2)H_4$$

where H_e is the enthalpy of entrained steam. Since this is saturated at $101.3\,kN/m^2$, then:

$$H_e = 2690\,kJ/kg \quad \text{and} \quad (1 \times 2780) + (1.6 \times 2690) = (1.0 + 1.6)H_4$$

from which:

$$H_4 = 2725\,kJ/kg$$

Again assuming isentropic compression from 101.3 to $135\,kN/m^2$, then:

$$H_3 = 2640\,kJ/kg\,(\text{from the chart})$$

and

$$\eta' = (1.0 + 1.6)(2725 - 2640)/[1.0(2780 - 2375)] = 0.55$$

This value is low, since in good design overall efficiencies approach 0.75–0.80. Obviously the higher the efficiency the greater the entrainment ratio or the higher the saving in live steam. The low efficiency is borne out by examination of Fig. 5.12B, which applies for an operating pressure of $1135\,kN/m^2$.

Since the pressure of entrained vapour $= 101.3\,kN/m^2$ and the discharge pressure $= 135\,kN/m^2$, the required flow of live steam $= 0.5\,kg/kg$ entrained vapour.

In this case the ratio is $(1.0/1.6) = \underline{0.63\,kg/kg}$.

Example 5.4

Distilled water is produced from sea water by evaporation in a single-effect evaporator working on the vapour compression system. The vapour produced is compressed by a mechanical compressor at 50% efficiency and then returned to the calandria of the evaporator. Additional steam, dry and saturated at $650\,kN/m^2$, is bled into the steam space through a throttling valve. The distilled water is withdrawn as condensate from the steam space. Fifty per cent of the sea water is evaporated in the plant. The energy supplied in addition to that necessary to compress the vapour may be assumed to appear as superheat in the vapour.

Using the following data, calculate the quantity of additional steam required in kg/s.

Production of distillate $=0.125\,kg/s$, pressure in vapour space $=101.3\,kN/m^2$, temperature difference from steam to liquor$=8\,K$, boiling point rise of sea water$=1.1\,K$, specific heat capacity of sea water$=4.18\,kJ/kg\,K$. The sea water enters the evaporator at 344 K from an external heater.

Solution

The pressure in the vapour space is $101.3\,kN/m^2$ at which pressure, water boils at 373 K. The sea water is therefore boiling at $(373+1.1)=374.1\,K$ and the temperature in the steam space is $(374.1+8)=382.1\,K$. At this temperature, steam is saturated at $120\,kN/m^2$ and has sensible and total enthalpies of 439 and 2683 kJ/kg, respectively.

Making a *mass balance*, there are two inlet streams—the additional steam, say G_x kg/s, and the sea water feed, say G_y kg/s. The two outlet streams are the distilled water product, 0.125 kg/s, and the concentrated sea water, $0.5\,G_y$ kg/s.

Thus:

$$(G_x + G_y) = (0.125 + 0.5G_y) \quad \text{or} \quad (G_x + 0.5G_y) = 0.125 \qquad (5.14)$$

Making an *energy balance*, energy is supplied by the compressor and in the steam and inlet sea water and is removed by the sea water and the product. At $650\,kN/m^2$, the total enthalpy of the steam $=2761\,kJ/kg$. Thus the energy in this stream $=2761\,G_x$ kW. The sea water enters at 344 K.

Thus:

$$\text{enthalpy of feed} = \left[G_y \times 4.18(344 - 273)\right] = 296.8\ G_y\ \text{kW}$$

The sea water leaves the plant at 374.1 K and hence:
the enthalpy of the concentrated sea water$=(0.5G_y \times 4.18)(374.1{-}273)=211.3\ G_y$ kW.

The product has an enthalpy of 439 kJ/kg or $(439 \times 0.125)=54.9\,kW$.

Making a balance:

and

$$(E + 2761G_x + 296.8G_y) = (211.3G_y + 54.9)$$
$$(E + 2761G_x + 58.5G_y) = 54.9 \qquad (5.14)$$

where E is the power supplied to the compressor.

Substituting from Eq. (5.14) into Eq. (5.15) gives:
and:

$$(E + 2761G_x) + 85.5(0.25 - 2G_x) = 54.9$$
$$(E + 2590G_x) = 33.5 \tag{5.14}$$

For a single-stage isentropic compression, the work done in compressing a volume V_1 of gas at pressure P_1 to a volume V_2 at pressure P_2 is given by eq. (8.32) in Volume 1 as:

$$[P_1V_1/(\gamma - 1)]\left[(P_2/P_1)^{\gamma-1/\gamma} - 1\right]$$

In the compressor, $0.5G_y$ kg/s vapour is compressed from $P_1 = 101.3 \text{ kN/m}^2$, the pressure in the vapour space, to $P_2 = 120 \text{ kN/m}^2$, the pressure in the calandria.

At 101.3 kN/m^2 and 374.1 K, the density of steam $= (18/22.4)(273/374.1) = 0.586 \text{ kg/m}^3$ and hence the volumetric flow rate at pressure P_1 is $(0.5G_y/0.586) = 0.853G_y \text{ m}^3/\text{s}$.

Taking $\gamma = 1.3$ for steam, then:

$$(E' \times 0.5G_y) = \left[(101.3 \times 0.853G_y)/(1.3 - 1)\right]\left[(120/101.3)^{0.3/1.3} - 1\right]$$
$$0.5E'G_y = 288.0G_y\left(1.185^{0.231} - 1\right) = 11.5G_y$$

and

$$E' = 23.0 \text{ kW/(kg/s)}$$

As the compressor is 50% efficient, then:

$$E = (E/0.5) = 46.0 \text{ kW/(kg/s)}$$
$$= (46.0 \times 0.5G_y) = 23.0G_y \text{ kW}$$

Substituting in Eq. (5.15) gives:

$$(E + 2761G_x) + 85.5(E/23.0) = 54.9$$

Thus:

$$2761G_x = (54.9 + 4.72E)$$

From Eq. (5.16):

$$E = (33.5 - 2590G_x)$$

and in Eq. (iv):

$$2761G_x = 54.9 + 4.72(33.5 - 2590G_x)$$

from which:

$$G_x = 0.014 \text{ kg/s}.$$

Example 5.5

An evaporator operating on the thermo-recompression principle employs a steam ejector to maintain atmospheric pressure over the boiling liquid. The ejector uses 0.14 kg/s of steam at 650 kN/m² and superheated by 100 degree and produces a pressure in the steam chest of 205 kN/m². A condenser removes surplus vapour from the atmospheric pressure line.

What is the capacity and economy of the system and how could the economy be improved?

Data

Properties of the ejector:

nozzle efficiency=0.95, efficiency of momentum transfer=0.80, efficiency of compression=0.90.

The feed enters the evaporator at 295 K and concentrated liquor is withdrawn at the rate of 0.025 kg/s. This concentrated liquor exhibits a boiling-point rise of 10 K. The plant is sufficiently well lagged so that heat losses to the surroundings are negligible.

Solution

It is assumed that P_1 is the pressure of live steam $=650$ kN/m² and P_2 is the pressure of entrained steam $=101.3$ kN/m².

The enthalpy of the live steam at 650 kN/m² and $(435+100)=535$ K, $H_1=2970$ kJ/kg.

Therefore, H_2, the enthalpy after isentropic expansion from 650 to 101.3 kN/m², using an enthalpy–entropy chart, is $H_2=2605$ kJ/kg and the dryness fraction, $x_2=0.97$. The enthalpy of the steam after actual expansion to 101.3 kN/m² is given by H_2', where:

$$(H - H_2') = 0.95(2970 - 2605) = 347 \ \text{kJ/kg}$$

and

$$H_2' = (2970 - 347) = 2623 \ \text{kJ/kg}$$

At $P_2=101.3$ kN/m², $\lambda=2258$ kJ/kg and the dryness after expansion but before entrainment x_2' is given by:

$$(x_2' - x_2)\lambda = (1 - e_1)(H_1 - H_2)$$

or

$$(x_2' - 0.97)2258 = (1 - 0.95)(2970 - 2605) \ \text{and} \ x_2' = 0.978.$$

If x_2'' is the dryness after expansion *and* entrainment, then:

$$(x_{22}'' - x_2')\lambda = (1 - e_3)(H_1 - H_2')$$

or

$$(x_{22}'' - 0.978)2258 = (1 - 0.80)(2970 - 2605) \ \text{and} \ x_{22}'' = 1.00$$

Assuming that the steam at the discharge pressure $P_3 = 205\,\text{kN/m}^2$ is also saturated, that is $x_3 = 1.00$, then from the steam chart in the Appendix, H_3 the enthalpy of the mixture at the start of compression in the diffuser section at $101.3\,\text{kN/m}^2$ is $H_3 = 2675\,\text{kJ/kg}$. Again assuming the entrained steam is also saturated, the enthalpy of the mixture after isentropic compression in the diffuser from 101.3 to $205\,\text{kN/m}^2$, $H_4 = 2810\,\text{kJ/kg}$.

The entrainment ratio is given by:

$$(m_2/m_1) = \{[(H_1 - H_2)/(H_4 - H_3)]\eta_1\eta_2\eta_3 - 1\}$$

where η_1, η_2 and η_3 are the efficiency of the nozzle, momentum transfer and compression, respectively.

Thus:

$$(m_2/m_1) = \{[(2970 - 2605)/(2810 - 2675)]0.95 \times 0.80 \times 0.90 - 1\}$$

It was assumed that $x_3 = 1.0$. This may be checked as follows:

$$x_3 = [x_2 + x_4(m_2/m_1)]/(1 + m_2/m_1)$$
$$= (1.0 + 1.0 \times 0.85)/(1 + 0.85) = 1.0$$

Thus with a flow of 0.14 kg/s live steam, the vapour entrained at $101.3\,\text{kN/m}^2$ is $(0.14 \times 0.85) = 0.12\,\text{kg/s}$, giving 0.26 kg/s steam saturated at $205\,\text{kN/m}^2$ to the calandria.

Allowing for a 10 K boiling-point rise, the temperature of boiling liquor in the unit is $T_1' = 383\,\text{K}$ and taking the specific heat capacity as 4.18 kJ/kg K, then:

$$D_0\lambda_0 = G_F C_p(T_1' - T_f) + D_1\lambda_1$$

or

$$0.26 \times 2200 = (G_F \times 4.18)(393 - 295) + (D_1 \times 2258)$$

$$572 = (368G_F + 2258D_1)$$

But:

$$(G_F - D_1) = 0.025 \text{ kg/s} \quad \text{and} \quad D_1 = 0.214 \text{ kg/s}$$

Thus:

$$\text{the economy of system} = (0.214/0.14) = 1.53$$

The capacity, in terms of the throughput of solution, is:

$$G_F = (0.214 + 0.025) = 0.239 \text{ kg/s}.$$

Apart from increasing the efficiency of the ejector, the economy of the system might be improved by operating with a higher live-steam pressure, increasing the pressure in the vapour space, and by using the vapour not returned to the ejector to preheat the feed solution.

5.5.2 **The heat pump cycle**

The evaporation of citrus juices at temperatures up to 328 K, or of pharmaceutical products at even lower temperatures, has led to the development of an evaporator incorporating a heat-pump cycle using a separate working fluid. The use of the heat pump cycle, with ammonia as the working fluid is shown in Fig. 5.13. In this arrangement, ammonia gas vaporises the feed liquor at 288–313 K. The ammonia is condensed and the liquid ammonia is then passed through an expansion valve, where it is cooled to a much lower temperature. The cooled liquid ammonia then enters the condenser where it condenses the vapour leaving the separator. The ammonia is vaporised and leaves as low-pressure gas, to be compressed in a mechanical compressor and then passed to the evaporator for a second cycle. The excess heat introduced by the compressor must be removed from the ammonia by means of a cooler.

The main advantage of this form of unit is the very great reduction in the volume of gas handled by the compressor. Thus, 1 kg of water vapour at, say, 311 K, with a volume of $22 \, \text{m}^3$ and latent heat about 2560 kJ/kg, passes this heat to ammonia at a

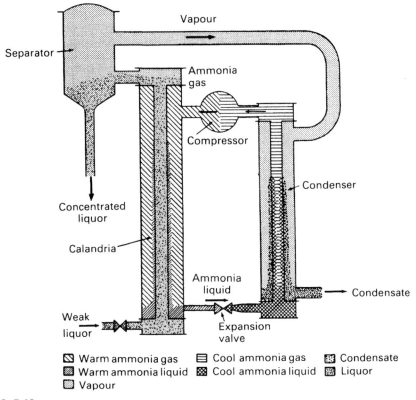

FIG. 5.13

Heat pump cycle using ammonia.

temperature of say 305 K. About 2.1 kg of ammonia will be vaporised to give a vapour with a volume of only about $0.22\,m^3$ at the high pressure used in the ammonia cycle.

Schwarz [16] gives a comparison of the various units used for low temperature evaporation. The three types in general use are the single-effect single-pass, the single-effect with recirculation, and the multiple-effect with recirculation. Each of these types may involve vapour compression or the addition of a second heat transfer medium. Schwarz suggests that multiple-effect units are the most economical, in terms of capital and operating costs. It is important to note that the single-effect, single-pass system offers the minimum holdup, and hence a very short transit time. With film-type units, there seems little to be gained by recirculation, since over 70% vaporisation can be achieved in one pass. The figures in Table 5.2 show the comparison between a double-effect unit with vapour compression on the first effect, and a unit with an ammonia refrigeration cycle, both units giving 1.25 kg/s (4.5 tonne/h) of evaporation.

The utilities required for the refrigeration system other than power are therefore very much less than for recompression with steam, although the capital cost and the cost of power will be much higher.

Reavell [17] has given a comparison of costs for the concentration of a feed of a heat-sensitive protein liquor at 1.70 kg/s from 10% to 50% solids, on the basis of a 288 ks (160 h) week. These data are shown in Table 5.3. It may be noted that, when

Table 5.2 Comparison of refrigeration and vapour compression systems.

System	Steam at 963 kN/m² (kg/s)	Water at 300 K (m³/s)	Power (kW)
Refrigeration cycle	0.062	0.019	320[a]
Vapour compression	0.95	0.076	20
Ratio of steam system to refrigeration	15.1	4	0.06

[a]Includes 300 kW compressor.

Table 5.3 Comparison of various systems for the concentration of a protein liquid.

Type	Approx. installed cost (£)	Cost of steam (£/year)	Net saving compared with single effect (£/year)
Single effect	50,000	403,000	–
Double effect	70,000	214,000	189,000
Double effect with vapour compression	90,000	137,000	266,000
Triple effect	100,000	143,000	260,000

using the double-effect evaporation with vapour compression, a lower temperature can be used in the first effect than when a triple-effect unit is used. In determining these figures, no account has been taken of depreciation, although if this is 15% of the capital costs it does not make a significant difference to the comparison.

The use of a heat pump cycle is the subject of Problem 5.22 at the end of this Volume, and a detailed discussion of the topic is given in the Solutions Manual.

Example 5.6

For the concentration of fruit juice by evaporation it is proposed to use a falling-film evaporator and to incorporate a heat pump cycle with ammonia as the medium. The ammonia in vapour form will enter the evaporator at 312K and the water will be evaporated from the juices at 287K. The ammonia in the vapour–liquid mixture will enter the condenser at 278K and the vapour will then pass to the compressor. It is estimated that the work for compressing the ammonia will be 150 kJ/kg of ammonia and that 2.28 kg of ammonia will be cycled/kg water evaporated. The following proposals are available for driving the compressor:
a) to use a diesel engine drive taking 0.4 kg of fuel/MJ; the calorific value being 42 MJ/kg and the cost £0.02/kg and
b) to pass steam, costing £0.01/10 kg through a turbine which operates at 70% isentropic efficiency, between 700 and 101.3 kN/m².

Explain by means of a diagram how this plant will work, and include all necessary major items of equipment required. Which method should be adopted for driving the compressor?

A simplified flow diagram of the plant is given in Fig. 5.14.

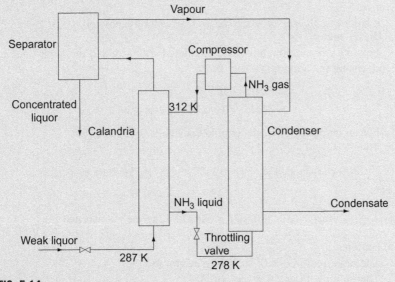

FIG. 5.14

Flow diagram for Example 5.6.

Solution

Considering the ammonia cycle

Ammonia gas will leave the condenser, probably saturated at low pressure, and enter the compressor which it leaves at high pressure and 312 K. In the calandria, heat will be transferred to the liquor, and the ammonia gas will be cooled to saturation, condense, and indeed may possibly leave the unit at 278 K as slightly sub-cooled liquid though still at high pressure. This liquid will then be allowed to expand adiabatically in the throttling valve to the lower pressure during which some vaporisation will occur and the vapour–liquid mixture will enter the condenser with a dryness fraction of, say, 0.1–0.2. In the condenser, heat will be transferred from the condensing vapours, and the liquid ammonia will leave the condenser, probably just saturated, though still at the low pressure. The cycle will then be repeated.

Considering the liquor stream

Weak liquor will enter the plant and pass to the calandria where it will be drawn up as a thin film by the partial vacuum caused by ultimate condensation of vapour in the condenser. Vaporisation will take place due to heat transfer from condensing ammonia in the calandria, and the vapour and concentrated liquor will then pass to a separator from which the concentrated liquor will be drawn off as product. The vapours will pass to the condenser where they will be condensed by heat transfer to the evaporating ammonia and leave the plant as condensate. A final point is that any excess heat introduced by the compressor must be removed from the ammonia by means of a cooler.

Fuller details of the cycle and salient features of operation are given in Section 5.5.2.

Choice of compressor drive (basis 1 kg water evaporated)

(a) Diesel engine

For 1 kg evaporation, ammonia circulated $= 2.28$ kg and the work done in compressing the ammonia

$$= (150 \times 2.28)$$
$$= 342 \text{ kJ or } 0.342 \text{ MJ/kg evaporation}$$

For an output of 1 MJ, the engine consumes 0.4 kg fuel.

Thus:

$$\text{fuel consumption} = (0.4 \times 0.342) = 0.137 \text{ kg/kg water evaporated}$$

and

$$\text{cost} = (0.02 \times 0.137) = 0.00274 \text{ £/kg water evaporated}$$

(b) Turbine

The work required is 0.342 MJ/kg evaporation.

Therefore, with an efficiency of 70%:

$$\text{energy required from steam} = (0.342 \times 100/70) = 0.489 \text{ MJ/kg}.$$

Enthalpy of steam saturated at $700MJ/kg^2 = 2764kJ/kg$.
Enthalpy of steam saturated at $101.3kN/m^2 = 2676kJ/kg$.

Thus

energy from steam $= (2764 - 2676) = 88$ kJ/kg or 0.088 MJ/kg

and

steam required $= (0.489/0.088) = 5.56$ kg/kg evaporation

at a cost of:

$(0.01 \times 5.56)/10 = 0.0056$ £/kg water evaporated

and hence:

the *Diesel engine would be used for driving the compressor*.

5.6 **Evaporator operation**

In evaporation, solids may come out of solution and form a deposit or scale on the heat transfer surfaces. This causes a gradual increase in the resistance to heat transfer and, if the same temperature difference is maintained, the rate of evaporation decreases with time and it is necessary to shut down the unit for cleaning at periodic intervals. The longer the boiling time, the lower is the number of shutdowns which are required in a given period although the rate of evaporation would fall to very low levels and the cost per unit mass of material handled would become very high. A far better approach is to make a balance which gives a minimum number of shutdowns while maintaining an acceptable throughput.

It has long been established [18] that, with scale formation, the overall coefficient of heat transfer may be expressed as a function of the boiling time by an equation of the form:

$$1/U^2 = at_b + b \tag{5.17}$$

where t_b is the boiling time. If Q_b is the total heat transferred in this time, then:

$$\frac{dQ_b}{dt_b} = UA\Delta T$$

and substituting for U from Eq. (5.17) gives:

$$\frac{dQ_b}{dt_b} = \frac{A\Delta T}{(at_b + b)^{0.5}} \tag{5.18}$$

Integrating between 0 and Q_b and 0 and t_b gives:

$$Q_b = (2A\Delta T/a) \leftarrow \left[(at_b + b)^{0.5} - b^{0.5}\right] \tag{5.19}$$

There are two conditions for which an optimum value of the boiling time may be sought—the time whereby the heat transferred and hence the solvent evaporated is a maximum and secondly, the time for which the cost per unit mass of solvent evaporated is a minimum. These are now considered in turn.

5.6.1 Maximum heat transfer

If the time taken to empty, clean and refill the unit is t_c, then the total time for one cycle is $t = (t_b + t_c)$ and the number of cycles in a period t_P is $t_P/(t_b + t_c)$. The total heat transferred during this period is the product of the heat transferred per cycle and the number of cycles in the period or:

$$Q_P = (2A\Delta T/a)\left[(at_b + b)^{0.5} - b^{0.5}\right][t_P/(t_b + t_c)] \tag{5.20}$$

The optimum value of the boiling time which gives the maximum heat transferred per cycle is obtained by differentiating Eq. (5.20) and equating to zero which gives:

$$t_{bopt} = t_c + (2/a)(abt_c)^{0.5} \tag{5.21}$$

5.6.2 Minimum cost

Taking C_c as the cost of a shutdown and the variable cost during operation including a labour component as C_b, then the total cost during period t_P is:

$$C_T = (C_c + t_bC_b)t_P/(t_b + t_c)$$

and substituting from Eq. (5.20):

$$C_T = [a \leftarrow Q_P(C_c + t_bC_b)]/2A\Delta T[at_b + b)^{0.5} - b^{0.5}] \tag{5.22}$$

The optimum value of the boiling time to give minimum cost is obtained by differentiating Eq. (5.22) and equating to zero to give:

$$t_{bopt} = (C_c/C_b) + 2(abC_cC_b)^{0.5}/(aC_b) \tag{5.23}$$

In using this equation, it must be ensured that the required evaporation is achieved. If this is greater than that given by Eq. (5.23), then it is not possible to work at minimum cost conditions. The use of these equations is illustrated in the following example which is based on the work of Harker [19].

Example 5.7

In an evaporator handling an aqueous salt solution, the overall coefficient U (kW/m^2 K) is given by a form of Eq. (5.20) as:

$$1/U^2 = 7 \times 10^{-5}t_b + 0.2,$$

the heat transfer area is $40\,\text{m}^2$, the temperature driving force is $40\,\text{K}$ and the latent heat of vaporisation of water is $2300\,\text{kJ/kg}$. If the down-time for cleaning is $15\,\text{ks}$ $(4.17\,\text{h})$, the cost of a shutdown is £600 and the operating cost during boiling is £18/ks $(£64.6/\text{h})$, estimate the optimum boiling times to give (a) maximum throughput and (b) minimum cost.

Solution
a) Maximum throughput

The boiling time to give maximum heat transfer and hence maximum throughput is given by Eq. (5.21):

$$t_{bopt} = \left(15 \times 10^3\right) + \left(2/\left(7 \times 10^{-5}\right)\right)\left(7 \times 10^{-5} \times 0.2 \times 15 \times 10^3\right)^{0.5}$$
$$= 2.81 \times 10^4 \text{ s or } 28.1 \text{ ks}(7.8 \text{ h})$$

The heat transferred during boiling is given by Eq. (5.19):

$$Q_b = (2 \times 40 \times 40)\left(7 \times 10^{-5}\right)\left[\left(\left(7 \times 10^{-5} \times 2.81 \times 10^4\right) + 0.2\right)^{0.5} - 0.2^{0.5}\right]$$
$$= 4.67 \times 10^7 \text{ kJ}$$

and the water vaporated $= (4.67 \times 10^7)/2300 = 2.03 \times 10^4\,\text{kg}$.

Rate of evaporation during boiling $= (2.03 \times 10^4)/(2.81 \times 10^4) = 0.723\,\text{kg/s}$.

Mean rate of evaporation during the cycle $= 2.03/[(2.8 \times 10^4) + (15 \times 10^3)] = 0.471\,\text{kg/s}$.

Cost of the operation $= ((2.81 \times 10^4 \times 18)/1000) + 600 = 1105.8$ £/cycle or

$$(1105.8/(2.03 \times 104)) = 0.055 \text{ £/kg}.$$

b) Minimum cost

The boiling time to give minimum cost is given by Eq. (5.23):

$$t_{bopt} = (600/0.018) + \left[2\left(7 \times 10^{-5} \times 0.2 \times 600 \times 0.018\right)^{0.5}\right]/\left(7 \times 10^{-5} \times 0.018\right)$$
$$= 5.28 \times 10^4 \text{ s or } 52.8 \text{ ks}(14.7 \text{ h})$$

The heat transferred during one boiling period is given by Eq. (5.19):

$$Q_b = \left[(2 \times 40 \times 40)/\left(7 \times 10^{-5}\right)\right]\left[\left(\left(7 \times 10^{-5} \times 5.28 \times 10^4 + 0.2\right)^{0.5} - 0.2^{0.5}\right]\right.$$
$$= 6.97 \times 10^7 \text{ kJ}$$

and the water evaporated $= (6.97 \times 10^7)/2300 = 3.03 \times 10^4\,\text{kg}$.

Rate of evaporation during boiling $= (3.03 \times 10^4)/(5.28 \times 10^4) = 0.574\,\text{kg/s}$.

Mean rate of evaporation during the cycle $= (3.03 \times 10^4)/[(5.28 \times 10^4) + (15 \times 10^3)] = 0.45\,\text{kg/s}$.

In this case, cost of one cycle $= (5.28 \times 10^4 \times 0.018) + 600 = £1550.4$ or

$$1550.4/(3.03 \times 104) = 0.0512 \text{ £/kg}.$$

Thus, the maximum throughput is 0.471 kg/s and the throughput to give minimum cost, 0.0512 £/kg, is 0.45 kg/s. If the desired throughput is between 0.45 and 0.471 kg/s, then this can be achieved although minimum cost operation is not possible. If a throughput of less than 0.45 kg/s is required, say 0.35 kg/s, then a total cycle time of $(3.03 \times 10^4)/0.35 = 8.65 \times 10^4$ s or 86.5 ks is required. This could be achieved by boiling at 0.423 kg/s for 71.5 ks followed by a shutdown of 15 ks, which gives a cost of 0.0624 £/kg. This is not the optimum boiling time for minimum cost and an alternative approach might be to boil for 52.8 ks at the optimum value, 0.45 kg/s, and, with a shutdown of 15 ks, a total cost of 0.0654 £/kg is estimated which is again higher than the minimum value. It would be, in fact, more cost effective to operate with the optimum boiling time of 52.8 ks and the down-time of 15 ks and to *close the plant down* for the remaining 18.7 ks of the 86.5 ks cycle. In this way, the minimum cost of 0.0512 £/kg would be achieved. In practice, the plant would probably not be closed down each cycle but rather for the equivalent period say once per month or indeed once a year. In all such considerations, it should be noted that, when a plant is shut down, there is no return on the capital costs and overheads which still have to be paid and this may affect the economics.

While calculated optimum cycle times may not exactly correspond to convenient operating schedules, this is not important as slight variations in the boiling times will not affect the economics greatly.

5.7 Equipment for evaporation

5.7.1 Evaporator selection

The rapid development of the process industries and of new products has provided many liquids with a wide range of physical and chemical properties all of which require concentration by evaporation. The type of equipment used depends largely on the method of applying heat to the liquor and the method of agitation. Heating may be either direct or indirect. Direct heating is represented by solar evaporation and by submerged combustion of a fuel. In indirect heating, the heat, generally provided by the condensation of steam, passes through the heating surface of the evaporator.

Some of the problems arising during evaporation include:

a) High product viscosity.
b) Heat sensitivity.
c) Scale formation and deposition.

Equipment has been developed in an attempt to overcome one or more of these problems. In view of the large number of types of evaporator which are available, the selection of equipment for a particular application can only be made after a detailed analysis of all relevant factors has been made. These will, of course, include the properties of the liquid to be evaporated, capital and running costs, capacity, holdup, and residence time characteristics. Evaporator selection considered in detail in Volume

6, has been discussed by Moore and Hesler [20] and Parker [21]. Parker has attempted to test the suitability of each basic design for dealing with the problems encountered in practice, and the basic information is presented in the form shown in Fig. 5.15. The factors considered include the ability to handle liquids in three viscosity ranges, to deal with foaming, scaling or fouling, crystal production, solids in suspension, and heat sensitive materials. A comparison of residence time and holding volume relative to the wiped film unit is also given. It is of interest to note that the agitated or wiped film evaporator is the only one which is shown to be applicable over the whole range of conditions covered.

5.7.2 Evaporators with direct heating

The use of solar heat for the production of Glauber's salt has been described by Holland [22,23]. Brine is pumped in hot weather to reservoirs of 100,000 m^2 in area to a depth of 3–5 m, and salt is deposited. Later in the year, the mother liquor is drained off and the salt is stacked mechanically, and conveyed to special evaporators in which hot gases enter at 1150–1250 K through a suitable refractory duct and leave at about 330 K. The salt crystals melt in their water of crystallisation and are then dried in the stream of hot gas. Bloch et al. [24], who examined the mechanism of evaporation of salt brines by direct solar energy, found that the rate of evaporation increased with the depth of brine. The addition of dyes, such as 2-naphthol green, enables the solar energy to be absorbed in a much shallower depth of brine, and this technique has been used to obtain a significant increase in the rate of production in the Dead Sea area.

The submerged combustion of a gas, such as natural gas, has been used for the concentration of very corrosive liquors, including spent pickle liquors, weak phosphoric and sulphuric acids. The depth of immersion of the burner is determined by the time of heat absorption and, for example, a 50 mm burner may be immersed by 250 mm and a 175 mm burner by about 450 mm. The efficiency of heat absorption is measured by the difference between the temperature of the liquid and that of the gases leaving the surface, values of 2–5 K being obtained in practice. The great attraction of this technique, apart from the ability to handle corrosive liquors, is the very great heat release obtained and the almost instantaneous transmission of the heat to the liquid, typically 70 MW/m^3.

5.7.3 Natural circulation evaporators

While each of the previous types of evaporator is of considerable importance in a given industry, it is the steam-heated evaporator that is the most widely used unit in the process industries and this is now considered in detail. In Chapter 9 of Volume 1, it is shown that the movement of the liquid over the heating surface has a marked influence on the rate of heat transfer, and it is thus convenient to classify evaporators according to the method of agitation or the nature of the circulation of the liquor over the heating surface. On this basis evaporators may be divided into three main types:

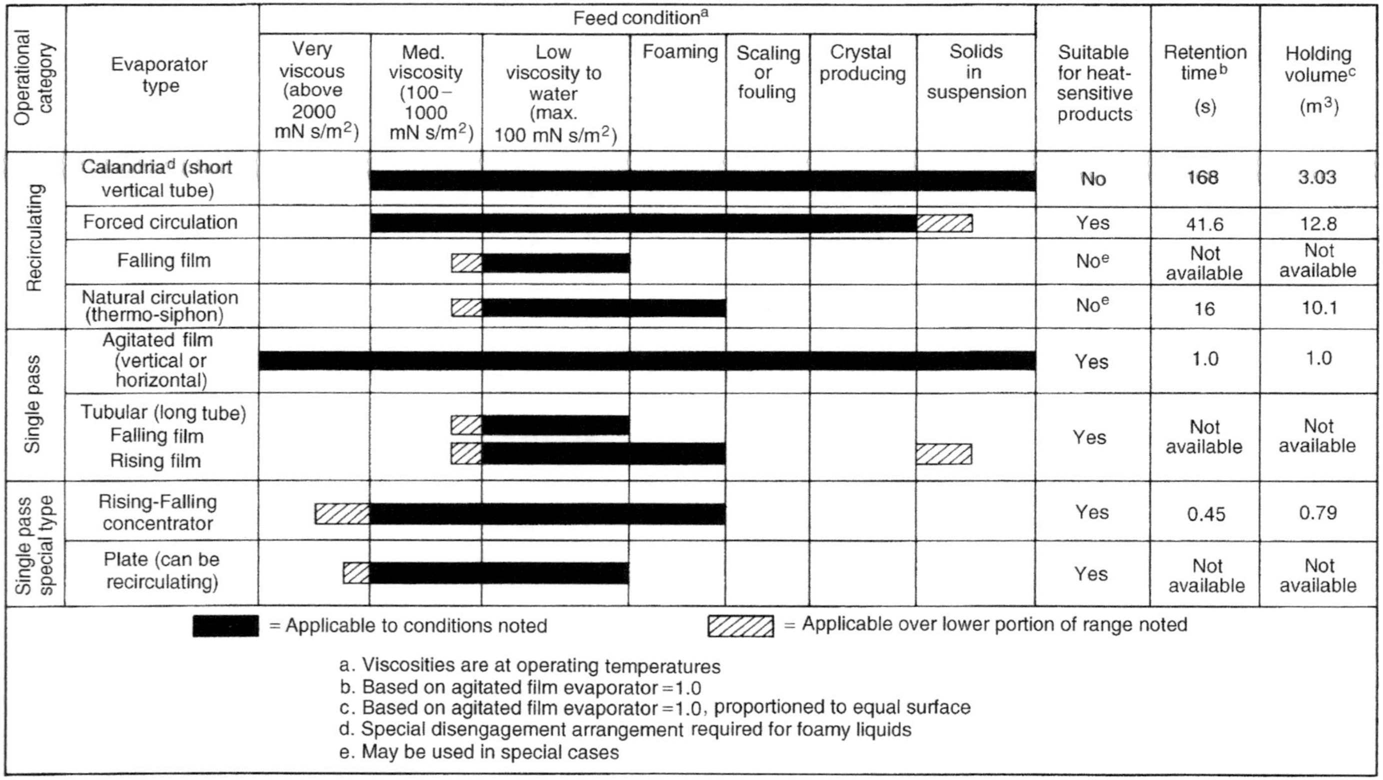

Operational category	Evaporator type	Feed condition[a]							Suitable for heat-sensitive products	Retention time[b] (s)	Holding volume[c] (m³)
		Very viscous (above 2000 mN s/m²)	Med. viscosity (100–1000 mN s/m²)	Low viscosity to water (max. 100 mN s/m²)	Foaming	Scaling or fouling	Crystal producing	Solids in suspension			
Recirculating	Calandria[d] (short vertical tube)								No	168	3.03
	Forced circulation								Yes	41.6	12.8
	Falling film								No[e]	Not available	Not available
	Natural circulation (thermo-siphon)								No[e]	16	10.1
Single pass	Agitated film (vertical or horizontal)								Yes	1.0	1.0
	Tubular (long tube) Falling film Rising film								Yes	Not available	Not available
Single pass special type	Rising-Falling concentrator								Yes	0.45	0.79
	Plate (can be recirculating)								Yes	Not available	Not available

■ = Applicable to conditions noted   ▨ = Applicable over lower portion of range noted

a. Viscosities are at operating temperatures
b. Based on agitated film evaporator = 1.0
c. Based on agitated film evaporator = 1.0, proportioned to equal surface
d. Special disengagement arrangement required for foamy liquids
e. May be used in special cases

FIG. 5.15

Evaporator selection.

After Parker, N.H., How to specify evaporators. Chem. Eng. 70 (1963) (15) 135–140.

a) Natural circulation units.
b) Forced circulation units.
c) Film-type units.

The developments that have taken place have, in the main, originated from the sugar and salt industries where the cost of evaporation represents a major factor in the process economics. In recent years, particular attention has been given to obtaining the most efficient use of the heating medium, and the main techniques that have been developed are the use of the multiple-effect unit, and of various forms of vapour compression units. With natural-circulation evaporators, circulation of the liquor is achieved by convection currents arising from the heating surface. This group of evaporators may be subdivided according to whether the tubes are horizontal with the steam inside, or vertical with the steam outside.

Rillieux is usually credited with first using *horizontal tubes*, and a unit of this type is shown in Fig. 5.16. The horizontal tubes extend between two tube plates to which they are fastened either by packing plates or, more usually, by expansion. Above the heating section is a cylindrical portion in which separation of the vapour from the liquid takes place. The vapour leaves through some form of de-entraining device to prevent the carry-over of liquid droplets with the vapour stream. The steam enters one steam chest, passes through the tubes and out into the opposite chest, and the condensate leaves through a steam trap. Horizontal evaporators are relatively cheap, require low head room, are easy to instal, and are suitable for handling liquors that do not crystallise. They can be used either as batch or as continuous units, and the shell is generally 1–3.5 m diameter and 2.5–4 m high. The liquor circulation is poor, and for this reason such units are unsuitable for viscous liquors.

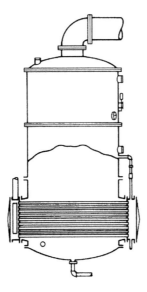

FIG. 5.16

Natural circulation evaporator with horizontal tubes.

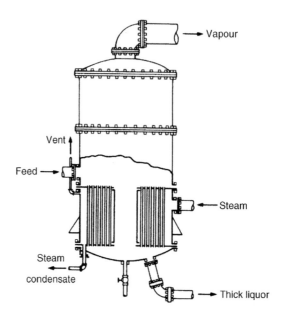

FIG. 5.17

Evaporator with vertical tubes and a large central downcomer.

The use of *vertical tubes* is associated with Robert, and this type is sometimes known as the Robert or Standard Evaporator. A typical form of vertical evaporator is illustrated in Fig. 5.17, in which a vertical cylindrical body is used, with the tubes held between two horizontal tube plates which extend right across the body. The lower portion of the evaporator is frequently spoken of as the calandria section shown in Fig. 5.18. Tubes are 1–2 m in length and 37–75 mm diameter, giving ratio of length to inside diameter of the tubes of 20–40. In the basket type shown in Fig. 5.19, vertical tubes are used with the steam outside, though the heating element is suspended in the body so as to give an annular downtake. The advantages claimed for this design are that the heating unit is easily removed for repairs, and that crystals formed in the downcomer do not break up. As the circulation of the liquor in the tubes is better, the vertical tube evaporator is used widely in the sugar and salt industries where through-puts are very large.

5.7.4 Forced circulation evaporators

Increasing the velocity of flow of the liquor through tubes results in a significant increase in the liquid-film transfer coefficient. This is achieved in the forced circulation units where a propeller or other impeller is mounted in the central downcomer, or a circulating pump is mounted outside the evaporator body. In the concentration of strong brines, for example, an internal impeller, often a turbine impeller, is fitted in the downtake, and this form of construction is particularly useful where crystallisation

FIG. 5.18

Calandria for an evaporator.

takes place. Forced circulation enables higher degrees of concentration to be achieved, since the heat transfer rate can be maintained in spite of the increased viscosity of the liquid. Because pumping costs increase roughly as the cube of the velocity, the added cost of operation of this type of unit may make it uneconomic, although many forced circulation evaporators are running with a liquor flow through the tubes of 2–5 m/s which is a marked increase on the value for natural circulation. Where stainless steel or expensive alloys such as Monel are to be used, forced circulation is favoured because the units can be made smaller and cheaper than those relying on natural circulation. In the type illustrated in Fig. 5.20, there is an external circulating pump, usually of the centrifugal type when crystals are present, though otherwise vane types may be used. The liquor is either introduced at the bottom and pumped straight through the calandria, or it is introduced in the separating section. In most units, boiling does not take place in the tubes, because the hydrostatic head of liquid raises the boiling point above that in the separating space. Thus the liquor enters the bottom of the tubes and is heated as it rises and at the same time the pressure falls. In the separator, the pressure is sufficiently low for boiling to occur. Forced circulation evaporators work well on

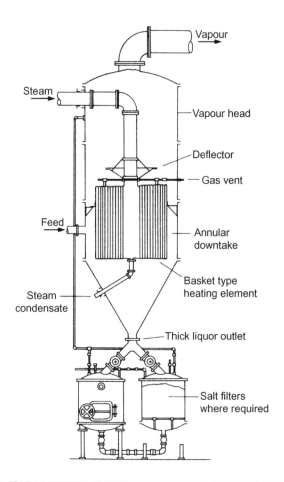

FIG. 5.19

Basket type of evaporator.

materials such as meat extracts, salt, caustic soda, alum, and other crystallising materials and also with glues, alcohols, and foam-forming materials.

For certain applications multi-pass arrangements are used. When a plate heat exchanger is used instead of the tubular unit, boiling on the heating surfaces is avoided by increasing the static head using a line restriction between the plate pack and the separator. Compared with tubular units, lower circulation rates and reduced liquid retention times are important advantages. Plate-type units are discussed further in Section 5.7.7.

For the handling of corrosive fluids, forced-circulation evaporators have been constructed in a variety of inert materials, and particularly in graphite where the unique combination of chemical inertness coupled with excellent thermal conductivity gives the material important advantages. Graphite differs from most constructional materials in its high anisotropy which results in directionally preferred

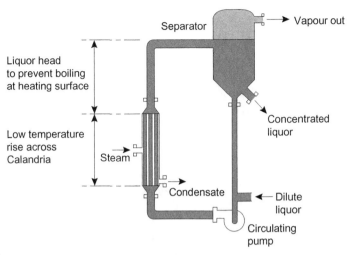

FIG. 5.20

Forced circulation evaporator with an external pump.

thermal conductivity, and in the difference between its relatively good compressive strength and its poor tensile or torsional strength. Although it is easily machinable, it is not ductile or malleable and cannot be cast or welded. The use of cements in assembly is undesirable because they are usually less chemically or thermally stable. There are also problems of differential expansion. In order to exploit the advantages of this material and to avoid the foregoing problems, special constructional techniques are necessary. The Polybloc system, which is described by Hilliard [25–27], is based on the use of robust blocks assembled exclusively under compression. Heat transfer occurs between fluids passing through holes drilled in the blocks and positioned so as to exploit preferred anisotropic crystal orientation for the highest thermal conductivity in the direction of heat flow. Inert gaskets eliminate the need for cements and enable units of varying size to be assembled simply by stacking the required number of blocks as shown in Fig. 5.21. A similar form of construction has been adopted by the Powell Duffryn Company. In commercial installations, high values of overall transfer coefficients have been achieved and, for example, a value of $1.1 \, kW/m^2K$ has been obtained for concentrating thick fruit juice containing syrup, and also for concentrating 40% sulphuric acid to 60%. A value of $0.8 \, kW/m^2K$ has been obtained for sulphuric acid concentration from 60% to 74% at a pressure of $1.5 \, kN/m^2$, and similar values have been obtained with spinning-bath liquors and some pharmaceuticals.

5.7.5 Film-type units

In all the units so far discussed, the liquor remains for some considerable time in the evaporator, up to several hours with batch operation, and this may be undesirable as many liquors decompose if kept at temperatures at or near their boiling points for any

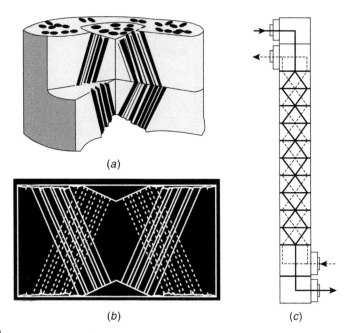

(a)

(b)   (c)

FIG. 5.21

The Polybloc system. (A) Cutaway section of *x*-flow block as used for two corrosive fluids, (B) section through *x*-flow block, and (C) stacked Polybloc exchanger.

length of time. The temperature can be reduced by operating under a vacuum, as discussed previously, though there are many liquors which are very heat-sensitive, such as orange juice, blood plasma, liver extracts and vitamins. If a unit is designed so that the residence time is only a few seconds, then these dangers are very much reduced. This is the principle of the Kestner long tube evaporator, introduced in 1909, which is fitted with tubes of 38 to 50mm diameter, mounted in a simple vertical steam chest. The liquor enters at the bottom, and a mixture of vapour and entrained liquor leaves at the top and enters a separator, usually of the tangential type. The vapour passes out from the top and the liquid from the bottom of the separator. In the early models, the thick liquid was recirculated through the unit, although the once-through system is now normally used.

An alternative name for the long-tube evaporator is the *climbing film evaporator*. *The* progressive evaporation of a liquid, while it passes through a tube, gives rise to a number of flow regimes discussed in Section 5.2.3. In the long-tube evaporator, the annular flow or climbing-film regime is utilised throughout almost all the tube length, the climbing film being maintained by drag induced by the vapour core which moves at a high velocity relative to the liquid film. With many viscous materials, however, heat transfer rates in this unit are low because there is little turbulence in the film, and the thickness of the film is too great to permit much evaporation from the film as a result of conduction through it. In evaporators of this type, it is essential that the feed should enter the tubes as near as possible to its boiling point. If the feed is subcooled, the initial sections will act merely as a feed heater thus reducing the overall performance of the unit. Pressure drop over the tube length will be attributable to the hydrostatic heads of the single-phase and two-phase regions, friction losses in these regions, and losses due to the acceleration of the vapour phase. The first published analysis of the operation of this type of unit was given by Badger and his associates [28–30] who fitted a small thermocouple inside the experimental tube, 32 mm outside diameter and 5.65 m long, so that the couple could be moved up and down the centre of the tube. In this way, it was found that the temperature rose slightly from the bottom of the tube to the point where boiling commenced, after which the change in temperature was relatively small. Applying this technique, it was possible to determine the heat transfer coefficients in the non-boiling and boiling sections of the tube.

A *falling-film evaporator* with the liquid film moving downwards, operates in a similar manner, as shown in Fig. 5.22. The falling-film evaporator is the simplest and most commonly used type of film-evaporator in which the liquid flows under gravitational force as a thin film on the inside of heated vertical tubes and the resulting vapour normally flows co-currently with the liquid in the centre of the tubes. A complete evaporator stage consists of the evaporator, a separator to separate the vapours from the residual liquid, and a condenser. Where high evaporation ratios are

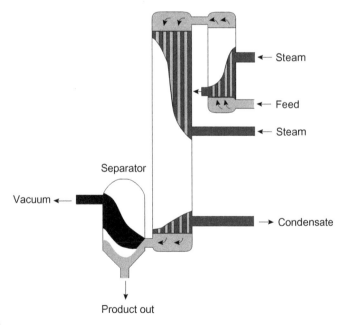

FIG. 5.22

Single-effect falling-film evaporator.

required, part of the concentrated liquid is recycled back to the evaporator inlet in order to ensure that the tubes are sufficiently wetted. An essential part of every falling-film evaporator is the liquid distribution system since the liquid feed must not only be evenly distributed to all the tubes, but also form a continuous film of the inner circumference of the tubes. Kuhni has developed a two-stage unit in which, after an initial pre-distribution, the liquid is directed tangentially onto the tubes through slits in specially designed tube inserts. The advantages of falling film evaporators include:

a) high heat transfer coefficients, 2000–5000 W/m^{2}K for water and 500–1000 W/m^{2}K for organics,
b) short residence times on the heated surface, 5–10 s without recirculation,
c) low pressure drops, 0.2–0.5 kN/m^2,
d) suitability for vacuum operation,
e) high evaporation ratios, c. 70% without and 95% with recirculation,
f) wide operating range, up to 400% of the minimum throughput,
g) low susceptibility to fouling,
h) minimum cost operation.

5.7.6 Thin-layer or wiped-film evaporators

This type of unit, known also as a thin-film evaporator is shown in Fig. 5.23. It consists of a vertical tube, the lower portion of which is surrounded by a jacket which contains the heating medium. The upper part of the tube is not jacketed and this acts as a separator. A rotor, driven by an external motor, has blades which extend nearly to the bottom of the tube, mounted so that there is a clearance of only about 1.3 mm between their tips and the inner surface of the tube. The liquor to be concentrated is picked up as it enters by the rotating blades and thrown against the tube wall. This action provides a thin film of liquid and sufficient agitation to give good heat transfer, even with very viscous liquids. The film flows down by gravity, becoming concentrated as it falls. The concentrated liquor is taken off at the bottom by a pump, and the vapour leaves the top of the unit where it is passed to a condenser. Development of this basic design has been devoted mainly to the modification of the blade system. An early alternative was the use of a hinged blade. In this type of unit, the blade is forced on to the wall under centrifugal action, the thickness of the film being governed by a balance between this force and the hydrodynamic forces produced in the liquid film on which the blade rides. The first experimental comparison of the fixed and hinged blade wiped-film evaporators was that of Bressler [31]. For each type of blade there appeared to be an optimum wiper speed beyond which an increase had no further effect on heat transfer. This optimum was reached at a lower speed with the hinged blade. Other agitator designs in which the blades, usually made from rubber, graphite or synthetic materials, actually scrape the wall have been studied. The use of nylon brushes as the active agitator elements has been investigated by McManus [32] using a small steam heated evaporator, 63 mm internal diameter and 762 mm long.

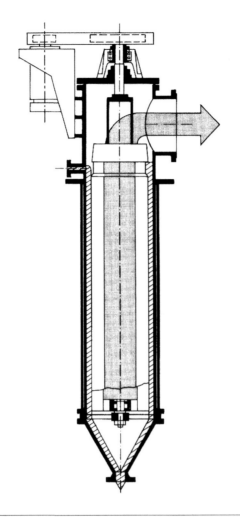

FIG. 5.23

Thin-film evaporator.

Water and various aqueous solutions of sucrose and glycerol were tested in the evaporator. A notable feature of the unit was the high heat fluxes obtained with the viscous solutions. Values as high as $70\,kW/m^2$ were obtained when concentrating a 60% sucrose feedstock to 73%, at a film temperature difference of 16.5 K with a wiper speed of 8.3 Hz. The fluxes obtained for the evaporation of water under similar conditions were nearly 4.5 times higher. A detailed analysis of the heat transfer mechanism, based on unsteady-state conduction to the rapidly renewed film, was presented. Similar analyses are to be found in the work of Harriott [33] and Kool [34]. Close agreement between the theory and experimental data confirmed the appropriateness of the model chosen to represent the heat transfer process. The theory has one main disadvantage, however, in that a satisfactory method for the estimation of liquid film thickness is not available. The most important factor influencing the evaporation coefficient is the thermal conductivity of the film material, and that the effects of viscosity and wiper speed which is inversely proportional to the heating time t, are of less significance.

A comprehensive discussion of the main aspects of the wiped-film evaporator technique covering thin-film technology in general, the equipment, and its economics and process applications is given by Mutzenburg [35], Parker [36], Fischer [37], and Ryley [38]. An additional advantage of wiped-film evaporators, especially those producing a scraped surface, is the reduction or complete suppression of scale formation though, in processes where the throughput is very high, this type of unit obviously becomes uneconomic and the traditional way of avoiding scale formation, by operating a flash evaporation process, is more suitable.

5.7.7 Plate-type units

A plate evaporator consists of a series of gasketted plates mounted within a support frame. Film-type plate evaporators can be climbing-film, falling-film or a combination of these. Fig. 5.24 shows the flow and plate arrangement of an APV falling-film plate evaporator. Each unit comprises a product plate and a steam plate, and this arrangement is repeated to provide the required heat transfer area. Product flow down each side of the plate may be in series where this is advantageous in terms of wetting rates.

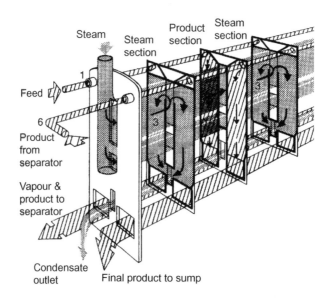

FIG. 5.24

Flow and plate arrangement for two-stage operation.

Both the vapour evaporated from the boiling film and the concentrated product are discharged from the evaporator to a vapour–liquid separator from which the product is pumped, the vapour passing to the next effect of the evaporator, or the condenser. Compared with tubular evaporators, plate evaporators can offer important advantages in terms of headroom, floorspace, accessibility and flexibility.

APV, whose 'Paraflow' plate heat exchanger is illustrated in Volume 1, Chapter 9, supply climbing and falling-film plate evaporators with evaporative capacities up to 10 kg/s. Such units offer the advantages of short contact and residence times and low liquor hold-up, and hence are widely used for the concentration of heat-sensitive materials.

For applications where viscosities or product concentrations are high, APV have developed the 'Paravap' evaporator in which corrugated-plate heat exchanger plates are used in a climbing-film arrangement, thereby increasing turbulence in the liquid film compared with standard plate evaporators. A typical arrangement of an APV 'Paravap' plant is shown in Fig. 5.25. Feed liquor from balance tank 1 is pumped 2 through the feed preheater 3 to the evaporator 4 where it boils. Concentrated product and evaporated vapour are discharged to the separator 5 from which product is pumped 6. Vapour passes to the condenser 7 from which condensate is pumped 8, vacuum being maintained by a liquid ring pump 9. Single-pass operation is used for low concentration ratios between feed and product, while higher ratios require

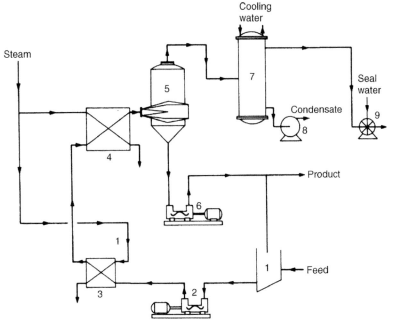

FIG. 5.25

Flowsheet for a typical 'Paravap' evaporator installation.

the recirculation of some of the product. This can be to the balance tank or feed pipework, although in some cases it is necessary to use in-line devices to achieve satisfactory mixing.

For some products it has been found advantageous to pressurise the plate heat exchanger, with an orifice or valve preventing boiling until the liquor enters the separator, in what is known as the APV 'Paraflash' system. This is a special case of the forced circulation evaporator described earlier.

APV 'Paravap' and 'Paraflash' evaporators are used for products with viscosities up to $5\,\mathrm{Ns/m^2}$ and concentrations in excess of 99% by mass. Evaporation rates are up to $4\,\mathrm{kg/s}$.

It may be noted that the gasket is a key component in plate heat exchangers, and this may limit the maximum temperature which can be used and may indeed prevent the use of this type of equipment with some corrosive fluids.

5.7.8 Flash evaporators

In the flash evaporator, boiling in the actual tubes is suppressed and the superheated liquor is flashed into a separator operating at reduced pressure. While the high heat transfer rates associated with boiling in tubes cannot be utilised, the thermodynamic and economical advantages of the system when operated in a multistage configuration outweigh this consideration. These advantages, stated independently by Frankel [39] and Silver [40], have been important in the past decade in the intensive effort to devise economic processes for the desalination of sea water. This topic is discussed further by Baker [41], who considers multistage flash evaporation with heat input supplied by a conventional steam boiler, by a gas-turbine cycle, or by vapour recompression. The combined power–water plant is also considered. Attempts to reduce scale formation in flash evaporators to even lower levels have resulted in a number of novel developments. In one unit described by Woodward [42], sea water is heated by a countercurrent spray of hot immiscible oil. In this respect, the process is similar to liquid–liquid extraction, the extracted quantity being heat in this case. The sea water is heated under pressure and subsequently flashed into a low-pressure chamber. A similar direct contact system is discussed by Wilke et al. [43]. Yet another arrangement which avoids the intervening metallic wall of the conventional heat exchanger is described by Othmer et al. [44]. In this process, direct mass transfer between brine and pure water is utilised in the desalination operation.

The formation of solids in evaporators is not always undesirable and, indeed, this is precisely what is required in the *evaporator-crystalliser* discussed in Chapter 15. The evaporator–crystalliser is a unit in which crystallisation takes place largely as a result of the removal of solvent by evaporation. Cooling of the liquor may, in some cases, produce further crystallisation thus establishing conditions similar to those in vacuum crystallisation. The true evaporator–crystalliser is distinguished, however, by its use of an external heat source. Crystallisation by evaporation is practised on salt solutions having a small change of solubility with temperature, such as sodium chloride and ammonium sulphate, which cannot be dealt with economically by other means, as well as those with inverted solubility curves. It is also widely used in the production of many other crystalline materials, as outlined by Bamforth [45]. The problem of design for crystallising equipment is extremely complicated and consequently design data are extremely meagre and unreliable. This topic is discussed further in Chapter 15.

The *development of unwanted foams* is a problem that evaporation has in common with a number of processes, and a considerable amount of effort has been devoted to the study of defoaming techniques using chemical, thermal, or mechanical methods. Chemical techniques involve the addition of substances, called antifoams, to foam-producing solutions to eliminate completely, or at least to reduce drastically, the resultant foam. Antifoams are, in general, slightly soluble in foaming solutions and can cause a decrease in surface tension. Their ability to produce an expanded surface film is, however, one explanation of their foam-inhibiting characteristic, as discussed by Beckerman [46]. Foams may be caused to collapse by raising or lowering the temperature. Many foams collapse at high temperature due to a decrease in surface tension, solvent evaporation, or chemical degradation of the foam-producing agents; at low temperatures freezing or a reduction in surface elasticity may be responsible. Other methods which are neither chemical nor thermal may be classified as mechanical. Tensile, shear, or compressive forces may be used to destroy foams, and such methods are discussed in some detail by Goldberg and Rubin [47]. The ultimate choice of defoaming procedure depends on the process under consideration and the convenience with which a technique may be applied.

5.7.9 Ancillary equipment

One important component of any evaporator installation is the *equipment for condensing the vapour* leaving the last effect of a multiple-effect unit, achieved either by direct contact with a jet of water, or in a normal tubular exchanger. If M is the mass of cooling water used per unit mass of vapour in a jet condenser, and H is the enthalpy per unit mass of vapour, then a heat balance gives:

$$\underset{\text{(Heat in)}}{H + M} \underset{\text{(Heat out)}}{\frac{C_p T_i}{}} = C_p T_e + M C_p T_e \qquad (5.24)$$

where T_i and T_e are the inlet and outlet temperatures of the water, above a standard datum temperature, and where the condensate is assumed to leave at the same temperature as the cooling water. From Eq. (5.24):

$$M = \frac{H - C_p T_e}{C_p (T_e - T_i)} \qquad (5.25)$$

If, for example, $T_e = 316\,\text{K}$, $T_i = 302\,\text{K}$, and the pressure $= 87.8\,\text{kN/m}^2$, then:

$$M = \frac{2596 - 4.18(316 - 273)}{4.18(316 - 302)} = 41.5 \ \text{kg/kg}$$

The water is then conveniently discharged at atmospheric pressure, without the aid of a pump, by allowing it to flow down a vertical pipe, known as a barometric leg, of sufficient length for the pressure at the bottom to be slightly in excess of atmospheric pressure. For a jet condenser with a barometric leg, a chart for determining the water requirement has been prepared by Arrowsmith [48].

Jet condensers may be either of the countercurrent or parallel flow type. In the countercurrent unit, the water leaves at the bottom through a barometric leg, and any entrained gases leave at the top. This provides what is known as a *dry vacuum system*, since the pump has to handle only the non-condensable gases. The cooling water will generally be heated to within 3–6 K of the vapour temperature. With the parallel flow system, the temperature difference will be rather greater and, therefore, more cooling water will be required. In this case, the water and gas will be withdrawn from the condenser and passed through a wet vacuum system. As there is no barometric leg, the unit can be mounted at floor level, although the pump displacement is about one and a half times that for the dry vacuum system.

Air is introduced into a jet condenser from the cooling water, as a result of the evolution of non-condensable gases in the evaporator, and as a result of leakages. The volume of air to be removed is frequently about 15% of that of the cooling water. The most convenient way of obtaining a vacuum is usually by means of a steam jet ejector. Part of the momentum of a high velocity steam jet is transferred to the gas entering the ejector, and the mixture is then compressed in the diverging portion of the ejector by conversion of kinetic energy into pressure energy. Good performance by the ejector is obtained largely by correct proportioning of the steam nozzle and diffuser, and poor ejectors will use much more high-pressure steam than a well-designed unit. The amount of steam required increases with the compression ratio. Thus, a single-stage ejector will remove air from a system at a pressure of $17 \, \text{kN/m}^2$ where a compression ratio of 6 is required. To remove air from a system at $3.4 \, \text{kN/m}^2$ would involve a compression ratio of 30, and a single-stage unit would be uneconomic in steam consumption. A two-stage ejector is shown in Fig. 5.26. The first stage withdraws air from the high vacuum vessel and compresses it to say $20 \, \text{kN/m}^2$, and the second stage compresses the discharge from the first ejector to atmospheric pressure. A further improvement is obtained if a condenser is inserted after the first stage, as this will reduce the amount of vapour to be handled in the final stage. The higher the steam pressure the smaller is the consumption, and pressures of $790–1135 \, \text{kN/m}^2$ are commonly used in multistage units.

In operating an evaporator, it is important to minimise entrainment of the liquid in the vapour passing over to the condenser. Entrainment is reduced by having a considerable headroom, of say 1.8 m, above the boiling liquid, though the addition of some form of de-entrainer is usually essential. Fig. 5.27 shows three *methods of reducing entrainment*. The simplest is to take the vapour from an upturned pipe as in *a*, and this has been found to give quite good results in small units. The deflector type *b* is a common form of de-entrainer and the tangential separator *c* is the type usually fitted to climbing-film units. This problem is particularly important in the concentration of radioactive waste liquors and has been discussed by McCullough [49], who cites the case of a batch evaporator of the forced-circulation type in which the vapours are passed to a 3.6 m diameter separator and then through four bubble cap trays to give complete elimination of entrained liquor. A good entrainment separator will reduce the amount of liquid carried over to $10–20 \, \text{kg}/10^6 \, \text{kg}$ of vapour.

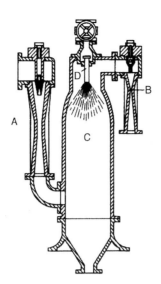

FIG. 5.26

Two-stage ejector with condenser: (A) First stage, (B) second stage, (C) condenser, and (D) water spray.

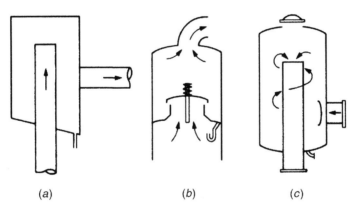

(a)	(b)	(c)

FIG. 5.27

Entrainment separators: (A) upturned pipe, (B) deflector type, and (C) tangential type.

References

[1] J.W. Westwater, Nucleate pool boiling, Petro/Chem. Eng. 33 (9) (1961) 186–189. and 33, No. 10 (Sept. 1961) 219–26.

[2] J.W. Palen, J.J. Taborek, Refinery kettle reboilers: proposed method for design and optimization, Chem. Eng. Prog. 58 (7) (1962) 37–46.

[3] D.S. Cryder, E.R. Gilliland, Heat transmission from metal surfaces to boiling liquids. I. Effect of physical properties of boiling liquid on liquid film coefficient, Ind. Eng. Chem. 24 (1932) 1382–1387.

[4] M.T. Cichelli, C.F. Bonilla, Heat transfer to liquids boiling under pressure, Trans. Am. Inst. Chem. Eng. 41 (1945) 755–787.

[5] M.J. Mcnelly, A correlation of the rates of heat transfer to nucleate boiling liquids, J. Imp. Coll. Chem. Eng. Soc. 7 (1953) 18.

[6] C.H. Gilmour, Performance of vaporizers: heat transfer analysis of plant data, Chem. Eng. Prog. 55 (1959) 67–78. Symp. Ser. No. 29.

[7] N. Zuber, On the stability of boiling heat transfer, Trans. Am. Soc. Mech. Eng. 80 (1958) 711–720.

[8] C.E. Dengler, J.N. Addoms, Heat transfer mechanism for vaporization of water in a vertical tube, Chem. Eng. Prog. 52 (1956) 95–103. Symp. Ser. No. 18.

[9] R.W. Lockhart, R.C. Martinelli, Proposed correlation of data for isothermal two-phase, two-component flow in pipes, Chem. Eng. Prog. 45 (1949) 39–48.

[10] S.A. Guerrieri, R.D. Talty, A study of heat transfer to organic liquids in single-tube, natural-circulation, vertical-tube boilers, in: Chem. Eng. Progress Symp. Ser., 52, 1956, pp. 69–77. No. 18.

[11] D.F. Othmer, Correlating vapor pressure and latent heat date. A new plot, Ind. Eng. Chem. 32 (1940) 841.

[12] E. Hausbrand, Evaporating, Condensing and Cooling Apparatus, Translated from the second revised German edition by A.C. Wright. Fifth English edition revised by B. Heastie, E. Benn, London, 1933.

[13] J.A. Storrow, Design calculations for multiple-effect evaporators—part 3, Ind. Chemist 27 (1951) 298.

[14] G. Stewart, G.S.G. Beveridge, Steady-state cascade simulation in multiple effect evaporation, Comput. Chem. Eng. 1 (1977) 3.

[15] A.L. Webre, Evaporation—a study of the various operating cycles in triple effect units, Chem. Met. Eng. 27 (1922) 1073.

[16] H.W. Schwarz, Comparison of low temperature (e.g. 15–24 citrus juice) evaporators, Food Technol. 5 (1951) 476.

[17] B.N. Reavell, Developments in evaporation with special reference to heat sensitive liquors, Ind. Chemist 29 (1953) 475.

[18] W.L. Mccabe, C.S. Robinson, Evaporator scale formation, Ind. Eng. Chem. 16 (1924) 478.

[19] J.H. Harker, Finding the economic balance in evaporator operation, Processing 12 (1978) 31–32.

[20] J.G. Moore, W.E. Hesler, Equipment for the food industry—2: evaporation of heat sensitive materials, Chem. Eng. Prog. 59 (2) (1963) 87–92.

[21] N.H. Parker, How to specify evaporators, Chem. Eng. 70 (15) (1963) 135–140.

[22] A.A. Holland, More Saskatchewan salt cake, Chem. Eng. 55 (xii) (1948) 121.

[23] A.A. Holland, New type evaporator, Chem. Eng. 58 (i) (1951) 106.

[24] M.R. Bloch, L. Farkas, K.S. Spiegler, Solar evaporation of salt brines, Ind. Eng. Chem. 43 (1951) 1544.

[25] A. Hilliard, Considerations on the design of graphite heat exchangers, Brit. Chem. Eng. 4 (1959) 138–143.

[26] A. Hilliard, The X-flow Polybloc system of construction for graphite, Br. Chem. Eng. 8 (1963) 234–237.

[27] A. Hilliard, Effect of anisotropy on design considerations for graphite, Ind. Chemist 39 (1963) 525–531.

[28] C.H. Brooks, W.L. Badger, Heat transfer coefficients in the boiling section of a long-tube, natural circulation evaporator, Trans. Am. Inst. Chem. Eng. 33 (1937) 392.

[29] G.W. Stroebe, E.M. Baker, W.L. Badger, Boiling film heat transfer coefficients in a long-tube vertical evaporator, Trans. Am. Inst. Chem. Eng. 35 (1939) 17.

[30] O.C. Cessna, J.R. Lientz, W.L. Badger, Heat transfer in a long-tube vertical evaporator, Trans. Am. Inst. Chem. Eng. 36 (1940) 759.

[31] R. Bressler, Versuche über die Verdampfung von dünnen Flüssigkeitsfilmen, Z. Ver. Deut. Ing. 100 (15) (1958) 630–638.

[32] T. McManus, The Influence of Agitation on the Boiling of Liquids in Tubes, University of Durham, 1963. PhD thesis.

[33] P. Harriott, Heat transfer in scraped surface exchangers, in: Chemical Engineering Progress Symposium Series, vol. 55, 1959, pp. 137–139. No. 29.

[34] J. Kool, Heat transfer in scraped vessels and pipes handling viscous materials, Trans. Inst. Chem. Eng. 36 (1958) 253–258.

[35] A.B. Mutzenburg, Agitated thin film evaporators. Part I. Thin film technology, Chem. Eng. 72 (19) (1965) 175–178.

[36] N. Parker, Agitated thin film evaporators. Part 2. Equipment and economics, Chem. Eng. 72 (19) (1965) 179–185.

[37] R. Fischer, Agitated thin film evaporators. Part 3. Process applications, Chem. Eng. 72 (19) (1965) 186–190.

[38] J.T. Ryley, Controlled film processing, Ind. Chemist 38 (1962) 311–319.

[39] A. Frankel, Flash evaporators for the distillation of sea-water, Proc. Inst. Mech. Eng. 174 (7) (1960) 312–324.

[40] R.S. Silver, Nominated lecture: fresh water from the sea, Proc. Inst. Mech. Eng. 179 (1964–5) 135–154. Pt. 1. No. 5.

[41] R.A. Baker, The flash evaporator, Chem. Eng. Prog. 59 (6) (1963) 80–83.

[42] T. Woodward, Heat transfer in a spray column, Chem. Eng. Prog. 57 (1) (1961) 52–57.

[43] C.R. Wilke, C.T. Cheng, V.L. Ledesma, J.W. Porter, Direct contact heat transfer for sea water evaporation, Chem. Eng. Prog. 59 (12) (1963) 69–75.

[44] D.F. Othmer, R.F. Benenati, G.C. Goulandris, Vapour reheat flash evaporation without metallic surfaces, Chem. Eng. Prog. 57 (1) (1961) 47–51.

[45] A.W. Bamforth, Industrial Crystallization, Leonard Hill, London, 1965.

[46] J.J. Beckerman, Foams, Theory and Industrial Applications, Reinhold, New York, 1953.

[47] M. Goldberg, E. Rubin, Mechanical foam breaking, Ind. Eng. Chem. Process. Des. Dev. 6 (1967) 195–200.

[48] G. Arrowsmith, Production of vacuum for industrial chemical processes, Trans. Inst. Chem. Eng. 27 (1949) 101.

[49] G.E. McCullough, Concentration of radioactive liquid waste by evaporation, Ind. Eng. Chem. 43 (1951) 1505.

Further reading

D.S. Azbel, Fundamentals of Heat Transfer in Process Engineering, Noyes, New York, 1984.

D.S. Azbel, Heat Transfer Applications in Process Engineering, Noyes, New York, 1984.

J.R. Backhurst, J.H. Harker, Process Plant Design, Heinemann, London, 1973.

W.L. Badger, Heat Transfer and Evaporation, Chemical Catalog Co, 1926.

R. Billet, Evaporation Technology: Principles, Applications, Economics, VCH, London, 1989.

E. Hausbrand, Evaporating, Condensing and Cooling Apparatus, Translated from the second revised German edition by A. C. Wright, Fifth English edition revised by B. Heastie, E. Benn, London, 1933.

F.A. Holland, R.M. Moores, F.A. Watson, J.K. Wilkinson, Heat Transfer, Heinemann, London, 1970.

D.Q. Kern, Process Heat Transfer, McGraw-Hill, New York, 1950.

C.J. King, Separation Processes, second ed., McGraw-Hill, New York, 1980.

F. Kreith, Principles of Heat Transfer, second ed., International Textbook Co., London, 1965.

W.H. Mcadams, Heat Transmission, third ed., McGraw-Hill, New York, 1954.

W.L. McCabe, J.C. Smith, P. Harriott, Unit Operations of Chemical Engineering, fourth ed., McGraw-Hill, New York, 1984.

B.M. Mckenna (Ed.), Engineering and Food, Vol. 2. Process Applications, Elsevier Applied Science, New York, 1984.

P.E. Minton, Handbook of Evaporation Technology, Noyes, New York, 1987.

R.H. Perry, D.W. Green (Eds.), Perry's Chemical Engineers' Handbook, sixth ed., McGraw-Hill Book Company, New York, 1984.

M.S. Peters, K.D. Timmerhaus, Plant Design and Economics for Chemical Engineers, third ed., McGraw-Hill, New York, 1984.

R.A. Smith, Vaporisors, Selection, Design and Operation, Longman, London, 1987.

M. Tyner, Process Engineering Calculations, Ronald Press, New York, 1960.

Crystallisation

Ajay Kumar Ray

Department of Chemical and Biochemical Engineering, University of Western Ontario, London, ON, Canada

Nomenclature

		Units in SI system	Dimensions in M, N, L, T, θ
A	crystal surface area	m^2	$\mathbf{L}^2$
A'	heat transfer area	m^2	$\mathbf{L}^2$
a_o	amount of component **A** in original solid	kg	$\mathbf{M}$
B	secondary nucleation rate	1/m^3 s	$\mathbf{L}^{-3}\mathbf{T}^{-1}$
b	kinetic order of nucleation or exponent in Eq. (6.11)	–	–
b_o	amount of component **B** in original solid	kg	$\mathbf{M}$
C	concentration of solute	kmol/m^3	$\mathbf{NL}^{-3}$
C^*	saturated concentration of solute	kmol/m^3	$\mathbf{NL}^{-3}$
C_P	specific heat capacity	J/kg K	$\mathbf{L}^2\mathbf{T}^{-2}\theta^{-1}$
c	solution concentration	kg/kg	–
c^*	equilibrium concentration	kg/kg	–
c_i	concentration at interface	kg/kg	–
c_{io}	initial concentration of impurity	kg/kg	–
c_{in}	impurity concentration after stage n	kg/kg	–
c_r	solubility of particles of radius r	kg/kg	–
c_s	slurry concentration or magma density	kg/m^3	$\mathbf{MT}^{-3}$
D	molecular diffusivity	m^2/s	$\mathbf{L}^2\mathbf{T}^{-1}$
d	characteristic size of crystal	m	$\mathbf{L}$

Continued

Coulson and Richardson's Chemical Engineering. https://doi.org/10.1016/B978-0-08-101097-6.00006-7

		Units in SI system	Dimensions in M, N, L, T, θ
d'	characteristic dimension of vaporisation chamber	m	**L**
d_D	dominant size of crystal size distribution	m	**L**
d_p	product crystal size	m	**L**
d_s	size of seed crystals	m	**L**
E	mass of solvent evaporated/mass of solvent in initial solution	kg/kg	–
e	evaporation coefficient (Eq. 6.51)	–	–
F	fraction of liquid removed after each decantation	–	–
F'	pre-exponential factor in Eq. (6.9)	$1/m^3$ s	$\mathbf{L^{-3}\,T^{-1}}$
ΔG	excess free energy	J	$\mathbf{ML^2T^{-2}}$
G'	flow rate of inert gas	kg/s	$\mathbf{MT^{-1}}$
G_d	overall growth rate	m/s	$\mathbf{LT^{-1}}$
ΔG_v	free energy change per unit volume	J/m^3	$\mathbf{ML^{-1}\,T^{-2}}$
H_f	heat of fusion	J/kg	$\mathbf{L^2T^{-2}}$
i	order of integration or relative kinetic order	–	–
J	rate of nucleation	$1/m^3$ s	$\mathbf{L^{-3}\,T^{-1}}$
j	exponent in Eq. (6.11)	–	–
K_b	birthrate constant	–	–
K_G	overall crystal growth coefficient	kg/s	$\mathbf{MT^{-1}}$
K_N	primary nucleation rate constant	$1/m^3$s	$\mathbf{L^{-3}\,T^{-1}}$
k	Boltzmann constant	1.38×10^{23} J/K	$\mathbf{ML^2T^{-2}\,\theta^{-1}}$
k_1	constant in Eq. (6.33)	$1/m^3$s	$\mathbf{L^{-3}\,T^{-1}}$
k_2	constant in Eq. (6.34)	m/s	$\mathbf{LT^{-1}}$
k_3	constant in Eq. (6.35)	$m^{-(i+3)}\,s^{(i-1)}$	$\mathbf{L^{-(i+3)}T^{(i-1)}}$
k_4	constant in Eq. (6.38)	$m^{-(i+3)}\,s^{(i-1)}$	$\mathbf{L^{-(i+3)}T^{(i-1)}}$
k_d	diffusion mass transfer coefficient	kg/m^2s	$\mathbf{ML^{-2}\,T^{-1}}$
k_r	integration or reaction mass transfer coefficient	kg/m^2s	$\mathbf{ML^{-2}\,T^{-1}}$
l	exponent in Eq. (6.11)	–	–
M	molecular weight of solute in solution	kg/kmol	$\mathbf{MN^{-1}}$
M'	molecular weight	kg/kmol	$\mathbf{MN^{-1}}$
M_g	molecular weight of inert gas	kg/kmol	$\mathbf{MN^{-1}}$
M_s	molecular weight of sublimed material	kg/kmol	$\mathbf{MN^{-1}}$
m	particle mass or mass deposited in time t	kg	**M**
m_s	mass of seeds	kg	**M**
N	rotational speed of impeller	Hz	$\mathbf{T^{-1}}$

Continued

		Units in SI system	Dimensions in M, N, L, T, θ
n	order of nucleation process or number of stages	–	–
n'	population density of crystals	m^{-4}	L^{-4}
n_i	moles of ions/mole of electrolyte	–	–
n^o	population density of nuclei	m^{-4}	L^{-4}
P	vapour pressure	N/m^2	$ML^{-1}T^{-2}$
P'	crystal production rate	kg/s	MT^{-1}
P_g	partial pressure of inert gas	N/m^2	$ML^{-1}T^{-2}$
P_s	vapour pressure at surface	N/m^2	$ML^{-1}T^{-2}$
P_s'	partial pressure of vaporised material	N/m^2	$ML^{-1}T^{-2}$
P_t	total pressure	N/m^2	$ML^{-1}T^{-2}$
Q	heat load	W	ML^2T^{-3}
q	heat of crystallisation	J/kg	L^2T^{-2}
R	universal gas constant	8314 J/kmol K	$MN^{-1}L^2T^{-2}\theta^{-1}$
R	molecular mass of hydrate/molecular mass of anhydrous salt	–	–
R_G	mass deposition rate	kg/m^2s	$ML^{-2}T^{-1}$
Re	Reynolds' number ($ud'\rho_v/\mu$)	–	–
r	radius of particle or equivalent sphere	m	L
r'	particle size in equilibrium with bulk solution	m	L
r_c	critical size of nucleus	m	L
S	supersaturation ratio	–	–
S'	mass sublimation rate	kg/s	MT^{-1}
Sc	Schmidt Number ($\mu/\rho_v D$)	–	–
s	order of overall crystal growth	–	–
T	temperature	K	θ
ΔT_m	logarithmic mean temperature difference	K	θ
T_M	melting point	K	θ
T_s	surface temperature	K	θ
t	time	s	T
t_b	batch time	s	T
t_i	induction period	s	T
t_r	residence time	s	T
U	overall coefficient of heat transfer	W/m^2K	$MT^{-3}\theta^{-1}$
u	gas velocity	m/s	LT^{-1}
u'	mean linear growth velocity	m/s	LT^{-1}
V'	volume of suspension of crystals	m^3	L^3
V_L	volume of liquor in vessel	m^3	L^3

Continued

		Units in SI system	Dimensions in M, N, L, T, θ
V_W	volume of wash water	m³	$\mathbf{L}^3$
v	molar volume	m³/kmol	$\mathbf{N}^{-1}\,\mathbf{L}^3$
v'	maximum theoretical vaporisation rate	kg/m²s	$\mathbf{ML}^{-2}\,\mathbf{T}^{-1}$
v_g	molar volume of gas	m³/kmol	$\mathbf{N}^{-1}\,\mathbf{L}^3$
v_1	molar volume of liquid	m³/kmol	$\mathbf{N}^{-1}\,\mathbf{L}^3$
V_s	molar volume of solid	m³/kmol	$\mathbf{N}^{-1}\,\mathbf{L}^3$
w_1	initial mass of solvent in liquor	kg	$\mathbf{M}$
w_2	final mass of solvent in liquor	kg	$\mathbf{M}$
x	solute concentration in terms of mole fraction	–	–
x'	effective film thickness	m	$\mathbf{L}$
x_c	amount of component **A** in crystallised solid	kg	$\mathbf{L}$
y	yield of crystals	kg	$\mathbf{M}$
y_c	amount of component **B** in crystallised solid	kg	$\mathbf{M}$
z	kmol of gas produced by 1 kmol of electrolyte	–	–
z_1, z_2	functions in Eq. (6.6)	1/K	θ^{-1}
α	shape factor in Eq. (6.16)	–	–
β	shape factor in Eq. (6.16)	–	–
ϕ	relative supersaturation	–	–
γ	ion activity coefficient	–	–
$\gamma \pm$	mean activity coefficient	–	–
φ	degree of supersaturation/ equilibrium saturation	–	–
λ	latent heat of vaporisation of solvent	J/kg	$\mathbf{L}^2\mathbf{T}^{-2}$
λ_f	latent heat of fusion	J/kg	$\mathbf{L}^2\mathbf{T}^{-2}$
λ_s	latent heat of sublimation	J/kg	$\mathbf{L}^2\mathbf{T}^{-2}$
λ_v	latent heat of vaporisation per unit mass	J/kg	$\mathbf{L}^2\mathbf{T}^{-2}$
λ_v'	latent heat of vaporisation per mole	J/kmol	$\mathbf{MN}^{-1}\mathbf{L}^2\mathbf{T}^{-2}$
μ	fluid viscosity	Ns/m²	$\mathbf{ML}^{-1}\mathbf{T}^{-1}$
ρ	density of crystal	kg/m³	$\mathbf{ML}^{-3}$
ρ_g	density of inert gas	kg/m³	$\mathbf{ML}^{-3}$
ρ_s	density of solid or sublimed material	kg/m³	$\mathbf{ML}^{-3}$
ρ_v	density of vapour	kg/m³	$\mathbf{ML}^{-3}$
σ	interfacial tension of crystallisation surface	J/m²	$\mathbf{MT}^{-2}$

6.1 **Introduction**

Crystallisation, one of the oldest of unit operations, is used to produce vast quantities of materials, including sodium chloride, sodium and aluminium sulphates, and sucrose which all have production rates in excess of 10^8 tonne/year on a world basis. Many organic liquids are purified by crystallisation rather than by distillation since, as shown by Mullin [1] in Table 6.1, enthalpies of crystallisation are generally much lower than enthalpies of vaporisation and crystallisation may be carried out closer to ambient temperature thereby reducing energy requirements. Against this, crystallisation is rarely the last stage in a process and solvent separation, washing and drying stages are usually required. Crystallisation is also a key operation in the freeze-concentration of fruit juices, the desalination of sea water, the recovery of valuable materials such as metal salts from electroplating processes, the production of materials for the electronic industries and in biotechnological operations such as the processing of proteins.

   Although crystals can be grown from the liquid phase—either a solution or a melt—and also from the vapour phase, a degree of supersaturation, which depends on the characteristics of the system, is essential in all cases for crystal formation or growth to take place. Some solutes are readily deposited from a cooled solution whereas others crystallise only after removal of solvent. The addition of a substance to a system in order to alter equilibrium conditions is often used in precipitation processes where supersaturation is sometimes achieved by chemical reaction between two or more substances and one of the reaction products is precipitated.

Table 6.1 Energy requirements for crystallisation and distillation [1].

Substance	Melting point (K)	Enthalpy of crystallisation (kJ/kg)	Boiling point (K)	Enthalpy of vaporisation (kJ/kg)
o-Cresol	304	115	464	410
m-Cresol	285	117	476	423
p-Cresol	306	110	475	435
o-Xylene	246	128	414	347
m-Xylene	225	109	412	343
p-Xylene	286	161	411	340
o-Nitrotoluene	268.9	120	495	344
m-Nitrotoluene	288.6	109	506	364
p-Nitrotoluene	325	113	511	366
Water	273	334	373	2260

6.2 Crystallisation fundamentals

In evaluating a crystallisation operation, data on phase equilibria are important as this indicates the composition of product which might be anticipated and the degree of supersaturation gives some idea of the driving force available. The rates of nuclei formation and crystal growth are equally important as these determine the residence time in, and the capacity of a crystalliser. These parameters also enable estimates to be made of crystal sizes, essential for the specification of liquor flows through beds of crystals and also the mode and degree of agitation required. It is these considerations that form the major part of this section.

6.2.1 Phase equilibria
One-component systems

Temperature and pressure are the two variables that affect phase equilibria in a one-component system. The phase diagram in Fig. 6.1 shows the equilibria between the solid, liquid, and vapour states of water where all three phases are in equilibrium at the *triple point*, $0.06 \, N/m^2$ and $273.3 \, K$. The *sublimation curve* indicates the vapour pressure of ice, the *vaporisation curve* the vapour pressure of liquid water, and the *fusion curve* the effect of pressure on the melting point of ice. The fusion curve for ice is unusual in that, in most one component systems, increased pressure increases the melting point, while the opposite occurs here.

A single substance may crystallise in more than one of seven crystal systems, all of which differ in their lattice arrangement, and exhibit not only different basic shapes but also different physical properties. A substance capable of forming more than one different crystal is said to exhibit *polymorphism*, and the different forms are called *polymorphs*. Calcium carbonate, for example, has three polymorphs—calcite

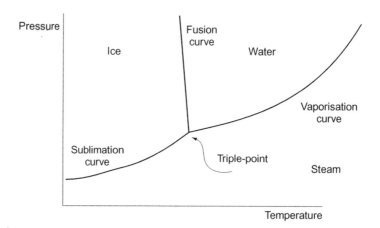

FIG. 6.1

Phase diagram for water.

(hexagonal), aragonite (tetragonal), and vaterite (trigonal). Although each poly-morph is composed of the same single substance, it constitutes a separate phase. Since only one polymorph is thermodynamically stable at a specified temperature and pressure, all the other polymorphs are potentially capable of being transformed into the stable polymorph. Some polymorphic transformations are rapid and revers-ible and polymorphs may be *enantiotropic* (interconvertible) or *monotropic* (incapa-ble of transformation). Graphite and carbon, for example, are monotropic at ambient temperature and pressure, whereas ammonium nitrate has five enantiotropic poly-morphs over the temperature range 255–398 K. Fig. 6.2A, taken from Mullin [2], shows the phase reactions exhibited by two enantiotropic forms, α and β, of the same

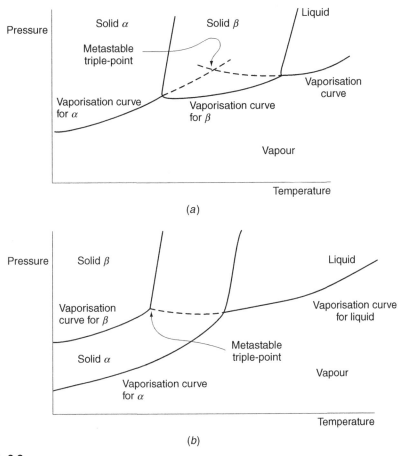

(a)

(b)

FIG. 6.2

Phase diagram for polymorphic substances [2]. Congruent melting point represents a definite temperature when the composition of the liquid that forms is the same as the composition of the solid.

substance. The point of intersection of the two vapour pressure curves is the transition point at which the two forms can coexist in equilibrium at the specified temperature and pressure. The triple point at which vapour, liquid and β solid can co-exist may be considered as the melting point of the β form. On slow heating, solid α changes into solid β and finally melts with the reverse process taking place on slow cooling. Rapid heating or cooling can, however, result in different behaviour where the vapour pressure of the α form follows a continuation of the vaporisation curve, and changes in the liquid are represented by the liquid vaporisation curve. The two curves intersect at a *metastable triple point* where the liquid, vapour, and a solid can coexist in metastable equilibrium. Fig. 6.2B shows the pressure–temperature curves for a monotropic substance for which the vapour pressure curves of the α and β forms do not intersect, and hence there is no transition point. In this case, solid β is the metastable form, and the metastable triple point is as shown.

Two-component systems

Temperature, pressure, and concentration can affect phase equilibria in a two-component or binary system, although the effect of pressure is usually negligible and data can be shown on a two-dimensional temperature–concentration plot. Three basic types of binary system – eutectics, solid solutions, and systems with compound formation – are considered and, although the terminology used is specific to melt systems, the types of behaviour described may also be exhibited by aqueous solutions of salts, since, as Mullin [3] points out, there is no fundamental difference in behaviour between a melt and a solution.

An example of a binary *eutectic system* **AB** is shown in Fig. 6.3A where the eutectic is the mixture of components that has the lowest crystallisation temperature in the system. When a melt at X is cooled along XZ, crystals, theoretically of pure **B**, will start to be deposited at point Y. On further cooling, more crystals of pure component **B** will be deposited until, at the eutectic point E, the system solidifies completely. At Z, the crystals C are of pure **B** and the liquid L is a mixture of **A** and **B** where the mass proportion of solid phase (crystal) to liquid phase (residual melt) is given by ratio of the lengths LZ to CZ; a relationship known as the *lever arm rule*. Mixtures represented by points above AE perform in a similar way, although here the crystals are of pure **A**. A liquid of the eutectic composition, cooled to the eutectic temperature, crystallises with unchanged composition and continues to deposit crystals until the whole system solidifies. While a eutectic has a fixed composition, it is not a chemical compound, but is simply a physical mixture of the individual components, as may often be visible under a low-power microscope.

The second common type of binary system is one composed of a continuous series of *solid solutions, where* the term *solid solution* or *mixed crystal* refers to an intimate mixture on the molecular scale of two or more components. The components of a solid-solution system cannot be separated as easily as those of a eutectic system. This is shown in Fig. 6.3B, where the liquidus represents the temperature at which mixtures of **A** and **B** begin to crystallise on cooling and the solidus represents temperatures at which mixtures begin to melt on heating. A melt at X begins to crystallise at Y and then at Z, the system consists of a mixture of crystals of a composition

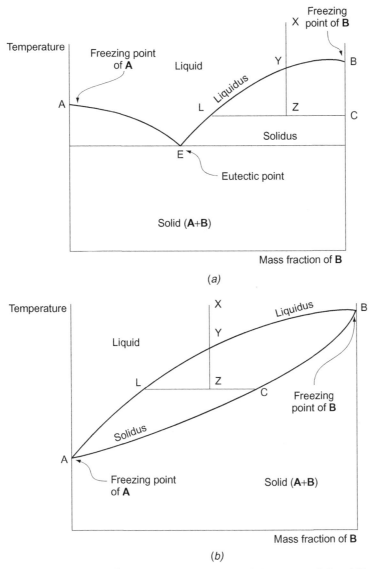

FIG. 6.3

Phase diagrams for binary systems: (A) eutectic and (B) solid solution.

represented by C and a liquid of a composition represented by L, where the ratio of crystals to liquid is again given by the lever arm rule. The crystals do not, however, consist of a single pure component as in a simple eutectic system but are an intimate mixture of components **A** and **B** which must be heated and re-crystallised, perhaps many times, in order to achieve further purification. In this way, a simple eutectic system may be purified in a single-stage crystallisation operation, whereas a solid-solution system always needs multistage operation.

The solute and solvent of a binary system can combine to form one or more different *compounds* such as, for example, hydrates in aqueous solutions. If the compound can co-exist in stable equilibrium with a liquid phase of the same composition, then it has a *congruent* melting point, that is where melting occurs without change in composition. If this is not the case, then the melting point is *incongruent*. In Fig. 6.4A, the heating–cooling cycle follows the vertical line through point D since melting and crystallisation occur without any change of composition. In

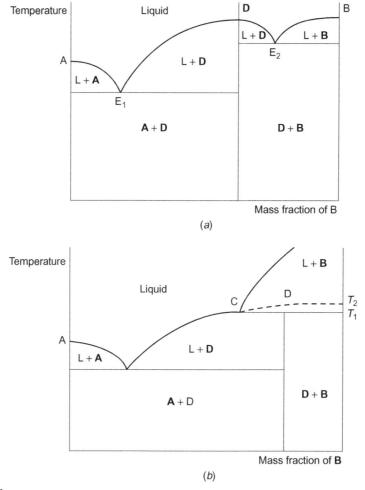

FIG. 6.4

Phase diagrams for binary systems (E—eutectic, L—liquid): (A) congruent melting point and (B) Incongruent melting point.

Fig. 6.4B, however, compound **D** decomposes at a temperature T_1 which is below its theoretical melting point T_2. Thus, if **D** is heated, melting begins at T_1, though is not complete. At T_1, a system of composition D contains crystals of pure **B** in a melt of composition C. If this mixture is cooled, then a solid mixture of **B** and that represented by point C is obtained and subsequent heating and cooling cycles result in further decomposition of the compound represented by D.

There is current interest in the use of inorganic-salt hydrates as heat-storage materials, particularly for storage of solar heat in domestic and industrial space heating, where, ideally, the hydrate should have a congruent melting point so that sequences of crystallisation–melting–crystallisation can be repeated indefinitely. Incongruently melting hydrate systems tend to stratify on repeated temperature cycling with a consequent loss of efficiency as melting gives a liquid phase that contains crystals of a lower hydrate or of the anhydrous salt, which settle to the bottom of the container and fail to re-dissolve on subsequent heating. Calcium chloride hexahydrate, while not a true congruently melting hydrate, appears to be one of the most promising materials [4,5] as are sodium sulphate deca-hydrate, sodium acetate tri-hydrate, and sodium thiosulphate penta-hydrate which all do have incongruent melting points [6,7].

Three-component systems

Phase equilibria in three-component systems are affected by temperature, pressure, and the concentrations of any two of the three components. Since the effect of pressure is usually negligible, phase equilibria may be plotted on an isothermal triangular diagram and, as an example, the temperature–concentration space model for o-, m-, and p-nitrophenol is shown in Fig. 6.5A [3]. The three components are **O**, **M**, and **P**, respectively and points O', M', and P' represent the melting points of the pure components o-(318 K), m- (370 K), and p-nitrophenol (387 K). The vertical faces of the prism represent temperature–concentration diagrams for the three binary eutectic systems **O-M**, **O-P**, and **M-P**, which are all similar to that shown in Fig. 6.4. The binary eutectics are represented by points A (304.7 K; 72.5% **O**, 27.5% **M**), B (306.7 K; 75.5% **O**, 24.5% **P**), and C (334.7 K; 54.8% **M**, 45.2% **P**) and AD within the prism represents the effect of adding **P** to the **O-M** binary eutectic at A. Similarly, curves BD and CD denote the lowering of freezing points of the binary eutectics represented by points B and C, respectively, upon adding the third component. Point D is a ternary eutectic point (294.7 K; 57.7% **O**, 23.2% **M**, 19.1% **P**) at which the liquid freezes to form a solid mixture of the three components. The section above the freezing point surfaces formed by the liquidus curves represents the homogeneous liquid phase, the section below these surfaces down to a temperature, D denotes solid and liquid phases in equilibrium and the section below this temperature represents a completely solidified system.

Fig. 6.5B is the projection of AD, BD, and CD in Fig. 6.5A on to the triangular base of the prism. Again **O**, **M**, and **P** are the pure components, points A, B, and C represent the three binary eutectic points and D is the ternary eutectic point. The diagram is divided by AD, BD, and CD into three regions which denote the three liquidus surfaces in the space model and the temperature falls from the apexes and

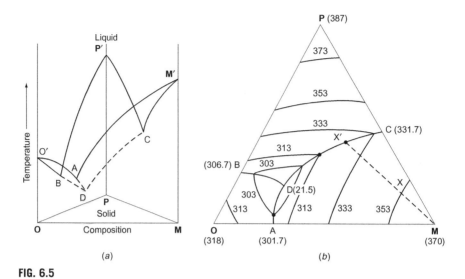

FIG. 6.5

Eutectic formation in the ternary system o-, m-, and p-nitrophenol [3].
(A) Temperature–concentration space model and (B) projection on a triangular diagram.
(Numerical values represent temperatures in K.)

sides of the triangle towards the eutectic point D. Several isotherms showing points on the liquidus surfaces are shown. When, for example, a molten mixture with a composition X is cooled, solidification starts when the temperature is reduced to 353 K and since X lies in the region ADCM, pure m-nitrophenol is deposited. The composition of the remaining melt changes along line MX' and at X', equivalent to 323 K, p-nitrophenol also starts to crystallise. On further cooling, both m and p-nitrophenol are deposited and the composition of the liquid phase changes in the direction $X'D$. When the melt composition and temperature reach point D, o-nitrophenol also crystallises out and the system solidifies without further change in composition.

Many different types of phase behaviour are encountered in ternary systems that consist of water and two solid solutes. For example, the system KNO_3–$NaNO_3$–H_2O which does not form hydrates or combine chemically at 323 K is shown in Fig. 6.6, which is taken from Mullin [3]. Point A represents the solubility of KNO_3 in water at 323 K (46.2 kg/100 kg solution), C the solubility of $NaNO_3$ (53.2 kg/100 kg solution), AB is the composition of saturated ternary solutions in equilibrium with solid KNO_3 and BC those in equilibrium with solid $NaNO_3$. The area above the line ABC is the region of unsaturated homogeneous solutions. At point B, the solution is saturated with both KNO_3 and $NaNO_3$. If, for example, water is evaporated isothermally from an unsaturated solution at X_1, the solution concentration increases along X_1X_2 and pure KNO_3 is deposited when the concentration reaches X_2. If more water is evaporated to give a system of composition X_3, the solution composition is represented by X'_3 on the saturation curve AB, and by point B when composition X_4 is reached.

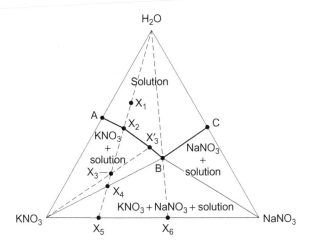

FIG. 6.6

Phase diagram for the ternary system KNO_3–$NaNO_3$–H_2O at 323 K [3].

Further removal of water causes deposition of $NaNO_3$. After this, all solutions in contact with solid have a constant composition B, which is referred to as the *eutonic point* or *drying-up point* of the system. After complete evaporation of water, the composition of the solid residue is indicated by X_5. Similarly, if an unsaturated solution, represented by a point to the right of B is evaporated isothermally, only $NaNO_3$ is deposited until the solution composition reaches B. KNO_3 is then also deposited and the solution composition remains constant until evaporation is complete. If water is removed isothermally from a solution of composition B, the composition of deposited solid is given by X_6 and it remains unchanged throughout the evaporation process.

Multi-component systems

The more components in a system, the more complex are the phase equilibria and it is more difficult to represent phases graphically. Descriptions of multi-component solid–liquid diagrams and their uses have been given by Mullin [3], Findlay and Campbell [8], Ricci [9], Null [10], and Nyvlt [11] and techniques for predicting multi-component solid–liquid phase equilibria have been presented by Hormeyer et al. [12], Kusik et al. [13], and Sander et al. [14].

Phase transformations

Metastable crystalline phases frequently crystallise to a more stable phase in accordance with Ostwald's rule of stages, and the more common types of phase transformation that occur in crystallising and precipitating systems include those between polymorphs and solvates. Transformations can occur in the solid state, particularly at temperatures near the melting point of the crystalline solid, and because of the intervention of a solvent. A stable phase has a lower solubility than a metastable

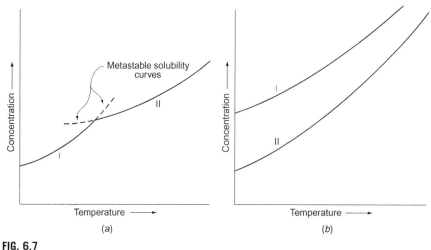

FIG. 6.7

Solubility curves for substances with two polymorphs I and II [2]: (A) enantiotropic and (B) monotropic.

phase, as indicated by the solubility curves in Fig. 6.7A and B for enantiotropic and monotropic systems respectively and, while transformation cannot occur between the metastable (I) and stable (II) phases in the monotropic system in the temperature range shown, it is possible above the transition temperature in an enantiotropic system. Polymorphic transformation adds complexity to a phase diagram, as illustrated by Nancollas et al. [15] and Nancollas and Reddy [16] who have studied dissolution–recrystallisation transformations in hydrate systems, and Cardew et al. [17] and Cardew and Davey [18] who have presented theoretical analyses of both solid state and solvent-mediated transformations in an attempt to predict their kinetics.

6.2.2 Solubility and saturation

Supersaturation

A solution that is in thermodynamic equilibrium with the solid phase of its solute at a given temperature is a saturated solution, and a solution containing more dissolved solute than that given by the equilibrium saturation value is said to be supersaturated. The degree of supersaturation may be expressed by:

$$\Delta c = c - c^* \tag{6.1}$$

where c and c^* are the solution concentration and the equilibrium saturation value, respectively. The supersaturation ratio, S, and the relative supersaturation, φ are then:

$$S = c/c^* \tag{6.2}$$

and

$$\varphi = \Delta c/c^* = S - 1 \qquad (6.3)$$

Solution concentrations may be expressed as mass of anhydrate/mass of solvent or as mass of hydrate/mass of free solvent, and the choice affects the values of S and φ as shown in the following example which is based on the data of Mullin [3].

Example 6.1

At 293 K, a supersaturated solution of sucrose contains 2.45 kg sucrose/kg water. If the equilibrium saturation value is 2.04 kg/kg water, what is the super-saturation ratio in terms of kg/kg water and kg/kg solution?

Solution

For concentrations in kg sucrose/kg water:

$$c = 2.45 \text{ kg/kg}, c^* = 2.04 \text{ kg/kg}$$

and

$$S = c/c^* = (2.45/2.04) = \underline{1.20.}$$

For concentrations in kg sucrose/kg solution:

$$c = 2.45/(2.45 + 1.0) = 0.710 \text{ kg/kg solution,}$$

$$c^* = 2.04/(2.04 + 1.0) = 0.671 \text{ kg/kg solution.}$$

and

$$S = (0.710/0.671) = \underline{1.06.}$$

While the fundamental driving force for crystallisation, the true thermodynamic supersaturation, is the difference in chemical potential, in practice supersaturation is generally expressed in terms of solution concentrations as given in Eqs. (6.1)–(6.3). Mullin and Söhnel [19] has presented a method of determining the relationship between concentration-based and activity-based supersaturation by using concentration-dependent activity-coefficients.

In considering the state of supersaturation, Ostwald [20] introduced the terms *labile* and *metastable* supersaturation to describe conditions under which spontaneous (primary) nucleation would or would not occur, and Miers and Isaac [21] have represented the metastable zone by means of a solubility–supersolubility diagram, as shown in Fig. 6.8.

While the (continuous) solubility curve can be determined accurately, the position of the (broken) supersolubility curve is less certain as it is influenced by factors such as the rate at which the supersaturation is generated, the degree of agitation and the presence of crystals or impurities. In the stable unsaturated zone, crystallisation is impossible. In the metastable supersaturated zone, spontaneous nucleation is improbable although a crystal would grow, and in

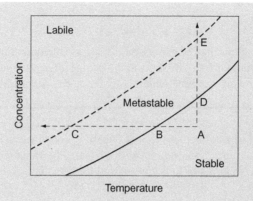

FIG. 6.8

Solubility supersolubility diagram.

the unstable or labile saturated zone, spontaneous nucleation is probable but not inevitable. If a solution at A is cooled without loss of solvent along ABC, spontaneous nucleation cannot occur until C is reached. Although the tendency to nucleate increases once the labile zone is reached, some solutions become too viscous to permit nucleation and set to a glass. Supersaturation can also be achieved by removing solvent and ADE represents such an operation carried out at constant temperature. Because the solution near the evaporating surface is more highly supersaturated than the bulk solution, penetration into the labile zone rarely occurs and crystals at the surface fall into the solution and induce nucleation, often before bulk conditions at E have been reached. Industrial crystallisers often combine cooling and evaporation. The width of the metastable zone is often expressed as a temperature difference, ΔT which is related to the corresponding concentration difference, Δc by the point slope of the solubility curve, dc^*/dT or:

$$\Delta c \simeq \frac{dc}{dT} \Delta T \tag{6.4}$$

The measurement of the width of the metastable zone is discussed in Section 6.2.4, and typical data are shown in Table 6.2. Provided the actual solution concentration and the corresponding equilibrium saturation concentration at a given temperature are known, the supersaturation may be calculated from Eqs. (6.1) to (6.3). Data on the solubility for two- and three-component systems have been presented by Seidell and Linke [22], Stephen et al. [23], and Broul et al. [24]. Supersaturation concentrations may be determined by measuring a concentration-dependent property of the system such as density or refractive index, preferably in situ on the plant. On industrial plant, both temperature

Table 6.2 Maximum allowable supercooling ΔT_{max} for aqueous salt solutions at 298 K [3].

Substance	K	Substance	K	Substance	K
NH_4Cl	0.7	$Na_2CO_3 \cdot 10H_2O$	0.6	$Na_2S_2O_3 \cdot 5H_2O$	1.0
NH_4NO_3	0.6	$Na_2CrO_4 \cdot 10H_2O$	1.6	K alum	4.0
$(NH_4)_2SO_4$	1.8	$NaCl$	4.0	KBr	1.1
$NH_4H_2PO_4$	2.5	$Na_2B_4O_7 \cdot 10H_2O$	4.0	KCl	1.1
$CuSO_4 \cdot 5H_2O$	1.4	NaI	1.0	KI	0.6
$FeSO_4 \cdot 7H_2O$	0.5	$NaHPO_4 \cdot 12H_2O$	0.4	KH_2PO_4	9.0
$MgSO_4 \cdot 7H_2O$	1.0	$NaNO_3$	0.9	KNO_3	0.4
$NiSO_4 \cdot 7H_2O$	4.0	$NaNO_2$	0.9	KNO_2	0.8
$NaBr \cdot 2H_2O$	0.9	$Na_2SO_4 \cdot 10H_2O$	0.3	K_2SO_4	6.0

Data measured in the presence of crystals with slow cooling and moderate agitation. The working value for a normal crystalliser may be 50% of these values or less.

and feedstock concentration can fluctuate, making the assessment of supersaturation difficult. Under these conditions, the use of a mass balance based on feedstock and exit-liquor concentrations and crystal production rates, averaged over a period of time, is usually an adequate approach.

Prediction of solubilities

Techniques are available for estimating binary and multi-component solubility behaviour. One example is the van't Hoff relationship which, as stated by Moyers and Rousseau [25], takes the following form for an ideal solution:

$$\ln \ x = \frac{H_f}{RT} \left(\frac{T}{T_M} - 1 \right) \tag{6.5}$$

where x is the mole fraction of solute in solution, H_f is the heat of fusion and T_M is the melting point of the pure component. One interesting consequence of this equation is that solubility depends only on the properties of the solute occurring in the equation. Another equation frequently used for ideal systems incorporates cryoscopic constants, values of which have been obtained empirically for a wide variety of materials in the course of the American Petroleum Research Project 44 [26]. This takes the form:

$$\ln (1/x) = z_1(T_M - T)[1 + z_2(T_M - T)...] \tag{6.6}$$

where

$$z_1 = \frac{H_f}{RT_M^2} \quad \text{and} \quad z_2 = \frac{1}{T_M} - \frac{C_p}{2H_f} / 2H_f.$$

Moyers and Rousseau [25] have used Eqs. (6.5) and (6.6) to calculate the freezing point data for *o*- and *p*-xylene shown in Table 6.3.

Table 6.3 Calculated freezing point curves for o- and p-xylene [25].

	Data	p-Xylene		o-Xylene
	T_M (K)	286.41		247.97
	ΔH_f (kJ/kmol)	17,120		13,605
	A (mole fraction/K)	0.02599		0.02659
	B (mole fraction/K)	0.0028		0.0030

	Mole fraction in solution			
	p-xylene		**o-xylene**	
Temperature (K)	**Eq. (6.5)**	**Eq. (6.6)**	**Eq. (6.5)**	**Eq. (6.6)**
286.41	1.00	1.00		
280	0.848	0.844		
270	0.646	0.640		
260	0.482	0.478		
249.97			1.00	1.00
240	0.249	0.256	0.803	0.805
235			0.695	0.699
230	0.172	0.183		

Crystal size and solubility

If *very small* solute particles are dispersed in a solution, the solute concentration may exceed the normal equilibrium saturation value. The relationship between particle size and solubility first applied to solid–liquid systems by Ostwald [20] may be expressed as:

$$\ln \frac{c_r}{c^*} = \frac{2M\sigma}{n_i RT\rho_s r} \tag{6.7}$$

where c_r is the solubility of particles of radius r, ρ_s the density of the solid, M the relative molecular mass of the solute in solution, σ the interfacial tension of the crystallisation surface in contact with its solution and n_i the moles of ions formed from one mole of electrolyte. For a non-electrolyte, $n_i = 1$ and for most inorganic salts in water, the solubility increase is really only significant for particles of less than 1 μm. The use of this equation is illustrated in the following example which is again based on data from Mullin [3].

Example 6.2

Compare the increase in solubility above the normal equilibrium values of 1, 0.1, and 0.01 μm particles of barium sulphate and sucrose at 298 K. The relevant properties of these materials are:

	Barium sulphate	Sucrose
Relative molecular mass (kg/kmol)	233	342
Number of ions (−)	2	1
Solid density (kg/m^3)	4500	1590
Interfacial tension (J/m^2)	0.13	0.01

Solution

Taking the gas constant, **R** as 8314 J/kmol K, then in Eq. (6.7):
For barium sulphate:

$$\ln(c_r/c^*) = (2 \times 233 \times 0.13)/(2 \times 8314 \times 298 \times 4500r) = 2.72 \times 10^{-9}/r.$$

For sucrose:

$$\ln(c_r/c^*) = (2 \times 342 \times 0.01)/(1 \times 8314 \times 298 \times 1590\,r) = 1.736 \times 10^{-9}/r.$$

Substituting 0.5×10^{-7}, 0.5×10^{-8} and 0.5×10^{-9} m for r gives the following data:

	Particle size d (μm)	r (μm)	c_r/c^*	Increase (per cent)
Barium sulphate	1	0.5	1.005	0.5
	0.1	0.05	1.06	6
	0.01	0.005	1.72	72
Sucrose	1	0.5	1.004	0.4
	0.1	0.05	1.035	3.5
	0.01	0.005	1.415	41.5

Effect of impurities

Industrial solutions invariably contain dissolved impurities that can increase or decrease the solubility of the prime solute considerably, and it is important that the solubility data used to design crystallisation processes relate to the actual system used. Impurities can also have profound effects on other characteristics, such as nucleation and growth.

6.2.3 Crystal nucleation

Nucleation, the creation of crystalline bodies within a supersaturated fluid, is a complex event, since nuclei may be generated by many different mechanisms. Most nucleation classification schemes distinguish between *primary nucleation*—in the absence of crystals and *secondary nucleation*—in the presence of crystals. Strickland-Constable [27] and Kashchiev [28] have reviewed nucleation, and Garside and Davey [29] have considered secondary nucleation in particular.

Primary nucleation

Classical theories of primary nucleation are based on sequences of bimolecular collisions and interactions in a supersaturated fluid that result in the build-up of lattice-structured bodies which may or may not achieve thermodynamic stability. Such primary nucleation is known as *homogeneous*, although the terms *spontaneous* and *classical* have also been used. As discussed by Ubbelhode [30] and Garten and Head [31], ordered solute-clustering can occur in supersaturated solutions prior to the onset of homogeneous nucleation, and Berglund et al. [32] have detected the presence of quasi-solid-phase species even in unsaturated solutions. Mullin and Leci [33] discussed the development of concentration gradients in supersaturated solutions of citric acid under the influence of gravity, and Larson and Garside [34] estimated the size of the clusters at 4–10 nm. Primary nucleation may also be initiated by suspended particles of foreign substances, and this mechanism is generally referred to as *heterogeneous* nucleation. In industrial crystallisation, most primary nucleation is almost certainly heterogeneous, rather than homogeneous, in that it is induced by foreign solid particles invariably present in working solutions. Although the mechanism of heterogeneous nucleation is not fully understood, it probably begins with adsorption of the crystallising species on the surface of solid particles, thus creating apparently crystalline bodies, larger than the critical nucleus size, which then grow into macro-crystals.

Homogeneous nucleation. A consideration of the energy involved in solid-phase formation and in creation of the surface of an arbitrary spherical crystal of radius r in a supersaturated fluid gives:

$$\Delta G = 4\pi r^2 \sigma + (4\pi/3)r^3 \Delta G_v \qquad (6.8)$$

where ΔG is the overall excess free energy associated with the formation of the crystalline body, σ is the interfacial tension between the crystal and its surrounding supersaturated fluid, and ΔG_v is the free energy change per unit volume associated with the phase change. The term $4\pi r^2 \sigma$, which represents the surface contribution, is positive and is proportional to r^2 and the term $(4\pi/3)r^3 \Delta G_v$ which represents the volume contribution, is negative and is proportional to r^3. Any crystal smaller than the critical nucleus size r_c is unstable and tends to dissolve while any crystal larger than r_c is stable and tends to grow. Combining Eqs. (6.7) and (6.8), and expressing the rate of nucleation J in the form of an Arrhenius reaction rate equation, gives the nucleation rate as:

$$J = F \ \exp \left[-\frac{16\pi \sigma^3 v^2}{3 \ \mathbf{k}^3 T^3 (\ln \ S)^2} \right] \qquad (6.9)$$

where F is a pre-exponential factor, v is molar volume, $\mathbf{k}$ is the Boltzmann constant and S is the supersaturation ratio. Since Eq. (6.9) predicts an explosive increase in the nucleation rate beyond some so-called critical value of S, it not only demonstrates the powerful effect of supersaturation on homogeneous nucleation, but also indicates the possibility of nucleation at any level of supersaturation.

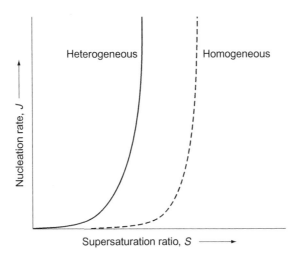

FIG. 6.9

Effect of supersaturation on the rates of homogeneous and heterogeneous nucleation.

Heterogeneous nucleation. The presence of foreign particles or heteronuclei enhances the nucleation rate of a given solution, and equations similar to those for homogeneous nucleation have been proposed to express this enhancement. The result is simply a displacement of the nucleation rate against supersaturation curve, as shown in Fig. 6.9, indicating that nucleation occurs more readily at a lower degree of supersaturation. For primary nucleation in industrial crystallisation, classical relationships similar to those based on Eq. (6.9) have little use, and all that can be justified is a simple empirical relationship such as:

$$J = K_N (\Delta c)^n \tag{6.10}$$

which relates the primary nucleation rate J to the supersaturation Δc from Eq. (6.1). The primary nucleation rate constant K_N, and the order of the nucleation process n, which is usually greater than 2, depend on the physical properties and hydrodynamics of the system.

Secondary nucleation

Secondary nucleation can, by definition, take place only if crystals of the species under consideration are already present. Since this is usually the case in industrial crystallisers, secondary nucleation has a profound influence on virtually all industrial crystallisation processes.

Apart from deliberate or accidental introduction of tiny seed crystals to the system, and productive interactions between existing crystals and quasi-crystalline embryos or clusters in solution, the most influential mode of new crystal generation in an industrial crystalliser is contact secondary nucleation between the existing crystals themselves, between crystals and the walls or other internal parts of the crystalliser, or between crystals and the mechanical agitator. Secondary nucleation rates (in m^{-3} s^{-1}) are most commonly correlated by empirical relationships such as:

$$B = K_b \rho_m^j N^l \Delta c^b \tag{6.11}$$

where B is the rate of secondary nucleation or birthrate, K_b is the birthrate constant, ρ_m is the slurry concentration or magma density and N is a term that gives some measure of the intensity of agitation in the system such as the rotational speed of an impeller. The exponents j, l, and b vary according to the operating conditions.

Nucleation measurements

One of the earliest attempts to derive nucleation kinetics for solution crystallisation was proposed by Nyvlt [35] and Nyvlt et al. [36] whose method is based on the measurement of metastable zone widths shown in Fig. 6.8, using a simple apparatus, shown in Fig. 6.10, consisting of a 50 mL flask fitted with a thermometer and a magnetic stirrer, located in an external cooling bath. Nucleation is detected visually and both primary and secondary nucleation can be studied in this way. Typical results [3] shown in Fig. 6.11 demonstrate that seeding has a considerable influence on the nucleation process, and the difference between the slopes of the two lines indicates that primary and secondary nucleation occur by different mechanisms. Solution turbulence also affects nucleation and, in general, agitation reduces the metastable zone width. For example, the metastable zone width for gently agitated potassium sulphate solutions is about 12 K while vigorous agitation reduces this to around 8 K. The

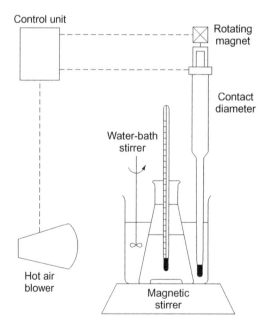

FIG. 6.10

Simple apparatus for measuring metastable zone widths [36].

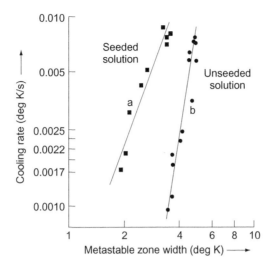

FIG. 6.11

Metastable zone width of aqueous ammonium [3].

presence of crystals also induces secondary nucleation at a supercooling of around 4 K. The relation between supercooling ΔT and supersaturation Δc is given by Eq. (6.4). Useful information on secondary nucleation kinetics for crystalliser operation and design can be determined only from model experiments that employ techniques such as those developed for MSMPR (mixed-suspension mixed-product removal) crystallisers. As discussed by Nyvlt et al. [36] and Randolph and Larson [37], in a real crystalliser, both nucleation and growth proceed together and interact with other system parameters in a complex manner.

Induction periods

A delay occurs between attainment of supersaturation and detection of the first newly created crystals in a solution, and this so-called *induction period*, t_i is a complex quantity that involves both nucleation and growth components. If it is assumed that t_i is essentially concerned with nucleation, that is $t_i \propto 1/J$, then Mullin [3] has shown, from Eq. (6.9), that:

$$\frac{1}{t_i} \propto \exp \frac{\sigma^3}{T^3(\log S)^2} \tag{6.12}$$

Thus, for a given temperature, a logarithmic plot of t_i against $(\log S)^{-2}$ should yield a straight line which, if the data truly represent homogeneous nucleation, will allow the calculation of the interfacial tension σ and the evaluation of the effect of temperature on σ. Nielsen and Söhnel [38] has attempted to derive a general correlation between interfacial tension and the solubility of inorganic salts as shown in Fig. 6.12A, although the success of this method depends on precise measurement of the induction period t_i, which presents problems if t_i is less than a few seconds.

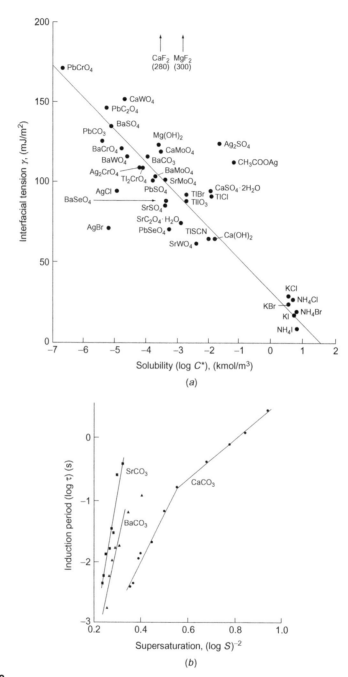

FIG. 6.12

(A) Interfacial tension as a function of solubility [38] and (B) induction period as a function of initial supersaturation [39].

Söhnel and Mullin [39] have shown that short induction periods can be determined by a technique that detects rapid changes in the conductivity of a supersaturated solution. Typical results for $CaCO_3$, $SrCO_3$, and $BaCO_3$, produced by mixing an aqueous solution of Na_2CO_3 with a solution of the appropriate chloride, are shown in Fig. 6.12B. The slopes of the linear, high-supersaturation regions are used to calculate the interfacial tensions $(0.08–0.12\,J/m^2)$, which compare reasonably well with the values predicted from the interfacial tension–solubility relationship in Fig. 6.12A. Although interfacial tensions evaluated from experimentally measured induction periods are somewhat unreliable, measurements of the induction period can provide useful information on other crystallisation phenomena, particularly the effect of impurities.

6.2.4 **Crystal growth**

Fundamentals

As with nucleation, classical theories of crystal growth [3,20,21,35,40–42] have not led to working relationships, and rates of crystallisation are usually expressed in terms of the supersaturation by empirical relationships. In essence, overall mass deposition rates, which can be measured in laboratory fluidised beds or agitated vessels, are needed for crystalliser design, and growth rates of individual crystal faces under different conditions are required for the specification of operating conditions.

In simple terms, the crystallisation process may be considered to take place in two stages—a diffusional step in which solute is transported from the bulk fluid through the solution boundary layer adjacent to the crystal surface, and a deposition step in which adsorbed solute ions or molecules at the crystal surface are deposited and integrated into the crystal lattice. These two stages which are shown in Fig. 6.13, may be described by:

$$dm/dt = k_d A(c - c_i) = k_r A(c_i - c^*)^i \tag{6.13}$$

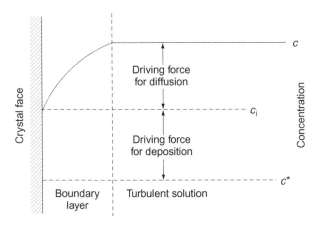

FIG. 6.13

Concentration driving forces for crystal growth from solution.

where m is the mass deposited in time t, A is the crystal surface area, c, c_i and c^* are the solute concentrations in the bulk solution, at the interface and at equilibrium saturation and k_d and k_r are the diffusion and deposition or reaction mass transfer coefficients. Because it is not possible to determine the interfacial concentration, this is eliminated by using the overall concentration driving force $\Delta c = (c - c^*)$, where:

$$(c - c^*) = \left(\frac{1}{k_d}\right) \frac{dm}{dt} + \left(\frac{1}{k_r}\right) \left(\frac{dm}{dt}\right)^{\frac{1}{i}}$$

Eliminating c and introducing an overall crystal growth coefficient, K_G gives the approximate relation:

$$dm/dt = K_G(\Delta c)^s \tag{6.14}$$

The exponents i and s in Eqs. (6.13) and (6.14), referred to as the order of integration and overall crystal growth process, should not be confused with their more conventional use in chemical kinetics where they always refer to the power to which a concentration should be raised to give a factor proportional to the rate of an elementary reaction. As Mullin [3] points out, in crystallisation work, the exponent has no fundamental significance and cannot give any indication of the elemental species involved in the growth process. If $i = 1$ and $s = 1$, c_i may be eliminated from Eq. (6.13) to give:

$$\frac{1}{K_G} = \frac{1}{K_D} + \frac{1}{k_r} \tag{6.15}$$

where the rate of integration is very high, K_G is approximately equal to k_d and the crystallisation is diffusion controlled. When the diffusional resistance is low, K_G is approximately equal to k_r and the process is controlled by the deposition step. While the diffusional step is generally proportional to the concentration driving force, the *integration* process is rarely first-order and for many inorganic salts crystallising from aqueous solution, s lies in the range 1–2.

Comprehensive reviews of theories of crystal growth have been presented by Garside [43], Nielsen [44], Pamplin [45], and Kaldis and Scheel [46].

Measurement of growth rate

Methods used for the measurement of crystal growth rates are either (a) direct measurement of the linear growth rate of a chosen crystal face or (b) indirect estimation of an overall linear growth rate from mass deposition rates measured on individual crystals or on groups of freely suspended crystals [3,35,41,47,48].

Face growth rates. Different crystal faces grow at different rates and faces with a high value of s grow faster than faces with low values. Changes in growth environment such as temperature, supersaturation pH, and impurities can have a profound effect on growth, and differences in individual face growth rates give rise to habit changes in crystals. For the measurement of individual crystal-face growth-rates, a fixed crystal in a glass cell is observed with a travelling microscope under precisely controlled conditions of solution temperature, supersaturation and liquid velocity [3]. The solution velocity past the fixed crystal is often an important growth-determining parameter, sometimes responsible for the so-called size-dependent growth effect often observed in agitated and fluidised-bed crystallisers. Large crystals have higher settling velocities than small crystals and, if their growth is diffusion-controlled, they tend to grow faster. Salts that exhibit solution velocity dependent growth rates include the alums, nickel ammonium sulphate, and potassium sulphate, although salts such as ammonium sulphate and ammonium or potassium dihydrogen phosphate are not affected by solution velocity.

Overall growth rates. In the laboratory, growth rate data for crystalliser design can be measured in fluidised beds or in agitated vessels, and crystal growth rates measured by growing large numbers of carefully sized seeds in fluidised suspension under strictly controlled conditions. A warm undersaturated solution of known concentration is circulated in the crystalliser and then supersaturated by cooling to the working temperature. About 5 g of closely sized seed crystals with a narrow size distribution and a mean size of around 500 μm is introduced into the crystalliser, and the upward solution velocity is adjusted so that the crystals are maintained in a reasonably uniform fluidised state in the growth zone. The crystals are allowed to grow at a constant temperature until their total mass is some 10 g, when they are removed, washed, dried, and weighed. The final solution concentration is measured, and the mean of the initial and final supersaturations is taken as the average for the run, an assumption which does not involve any significant error because the solution concentration is usually not allowed to change by more than about 1% during a run. The overall crystal growth rate is then calculated in terms of mass deposited per unit area per unit time at a specified supersaturation.

Expression of growth rate

Because the rate of growth depends, in a complex way, on temperature, supersaturation, size, habit, system turbulence, and so on, there is no simple way of expressing the rate of crystal growth, although, under carefully defined conditions, growth may be expressed as an overall mass deposition rate, R_G (kg/m^2 s), an overall linear growth rate, G_d ($=dd/dt$) (m/s) or as a mean linear velocity, $u' = dr/dt$ (m/s). Here d is some characteristic size of the crystal such as the equivalent aperture size, and r is the radius corresponding to the equivalent sphere where $r=0.5d$. The relationships between these quantities are:

$$R_G = K_G \Delta c^s = \left(\frac{1}{A}\right)\frac{dm}{dt} = \frac{3\alpha\rho\frac{dd}{dt}}{\beta} = \left(\frac{6\alpha\rho}{\beta}\right)\frac{dr}{dt} = \frac{6\alpha\rho u'}{\beta} \qquad (6.16)$$

where ρ is the density of the crystal and the volume and surface shape factors, α and β, are related to particle mass m and surface area A, respectively, by:

$$m = \alpha\rho d^3 \qquad (6.17)$$

and

$$A = \beta d^2 \qquad (6.18)$$

Table 6.4 Mean over-all crystal growth rates expressed as a linear velocity [3].

Substance	Supersaturation ratio		u' (m/s)
	K	S	
NH_4NO_3	313	1.05	8.5×10^{-7}
$(NH_4)_2SO_4$	303	1.05	2.5×10^{-7}
	333	1.05	4.0×10^{-7}
$MgSO_4 \cdot 7H_2O$	293	1.02	4.5×10^{-8}
	303	1.01	8.0×10^{-8}
	303	1.02	1.5×10^{-7}
KCl	313	1.01	6.0×10^{-7}
KNO_3	293	1.05	4.5×10^{-8}
	313	1.05	1.5×10^{-7}
K_2SO_4	293	1.09	2.8×10^{-8}
	293	1.18	1.7×10^{-7}
	303	1.07	4.2×10^{-8}
	323	1.06	7.0×10^{-8}
	323	1.12	3.2×10^{-7}
NaCl	323	1.002	2.5×10^{-8}
	323	1.003	6.5×10^{-8}
	343	1.002	9.0×10^{-8}
	343	1.003	1.5×10^{-7}

Values of $6\alpha/\beta$ are 1 for spheres and cubes and 0.816 for octahedra and typical values of the mean linear growth velocity, $u'(=0.5\,G_d)$ for crystals 0.5–1 mm growing in the presence of other crystals are given in Table 6.4 which is taken from Mullin [3].

Dependence of growth rate on crystal size
Experimental evidence indicates that crystal growth kinetics often depend on crystal size, possibly because the size depends on the surface deposition kinetics and different crystals of the same size can also have different growth rates because of differences in surface structure or perfection. In addition, as discussed by White et al. [49], Jones and Mullin [50], Janse and de Jong [51], and Garside and Jančič [52], small crystals of many substances grow much more slowly than larger crystals, and some do not grow at all. The behaviour of very small crystals has considerable influence on the performance of continuously operated crystallisers because new crystals with a size of 1–10 μm are constantly generated by secondary nucleation and these then grow to populate the full crystal size distribution.

Growth–nucleation interactions

Crystal nucleation and growth in a crystalliser cannot be considered in isolation because they interact with one another and with other system parameters in a complex manner. For a complete description of the crystal size distribution of the product in a continuously operated crystalliser, both the nucleation and the growth processes must be quantified, and the laws of conservation of mass, energy, and crystal population must be applied. The importance of population balance, in which all particles are accounted for, was first stressed in the pioneering work of Randolph and Larson [37].

Crystal habit modification

Differences in the face growth-rates of crystals give rise to changes in their habit or shape. Although the growth kinetics of individual crystal faces usually depend to various extents on supersaturation so that crystal habit can sometimes be controlled by adjusting operating conditions, the most common cause of habit modification is the presence of impurities. Although a soluble impurity will often remain in the liquid phase so that pure crystals are formed, in many cases, both the rate of nucleation and the crystal growth rate are affected. More usually, the effect is one of retardation, thought to be due to the adsorption of the impurity on the surface of the nucleus or crystal. Materials with large molecules such as tannin, dextrin or sodium hexametaphosphate, added in small quantities to boiler feed water, prevent the nucleation and growth of calcium carbonate crystals and hence reduce scaling. In a similar way, the addition of 0.1% of HCl and 0.1% $PbCl_2$ prevent the growth of sodium chloride crystals. In some cases the adsorption occurs preferentially on one particular face of the crystal, thus modifying the crystal shape. One example is that sodium chloride crystallised from solutions containing traces of urea forms octahedral instead of the usual cubic crystals. In a similar way, dyes are preferentially adsorbed on inorganic crystals [53], although the effect is not always predictable. Garrett [54] has described a number of uses of additives as habit modifiers, and industrial applications of habit modification are reported in several reviews [3,55,56] in which the factors that must be considered in selecting a suitable habit modifier are discussed. In the main, solid impurities act as condensation nuclei and cause dislocations in the crystal structure.

Inclusions in crystals

Inclusions are small pockets of solid, liquid, or gaseous impurities trapped in crystals that usually occur randomly although a regular pattern may be sometimes observed. As described by Mullin [3], a simple technique for observing inclusions is to immerse the crystal in an inert liquid of similar refractive index or, alternatively, in its own saturated solution when, if the inclusion is a liquid, concentration streamlines will be seen as the two fluids meet and, if it is a vapour, a bubble will be released. Industrial crystals may contain significant amounts of included mother liquor that can significantly affect product purity and stored crystals may cake because of liquid seepage from inclusions in broken crystals. In order to minimise inclusions, the crystallising system should be free of dirt and other solid debris, vigorous agitation or boiling should be avoided, and ultrasonic irradiation may be used to suppress adherence of bubbles to a growing crystal face. As fast crystal growth is probably the most common cause of inclusion formation, high supersaturation levels should be avoided. Deicha [57], Powers [58], Wilcox and Kuo [59], and Saska and Myerson [60] have published detailed accounts of crystal inclusion.

6.2.5 Crystal yield

The yield of crystals produced by a given degree of cooling may be estimated from the concentration of the initial solution and the solubility at the final temperature, allowing for any evaporation, by making solvent and solute balances. For the solvent, usually water, the initial solvent present is equal to the sum of the final solvent in the mother liquor, the water of crystallisation within the crystals and any water evaporated, or:

$$w_1 = w_2 + y\frac{R-1}{R} + w_1 E \tag{6.19}$$

where w_1 and w_2 are the initial and final masses of solvent in the liquor, y is the yield of crystals, R is the ratio (molecular mass of hydrate/molecular mass of anhydrous salt) and E is the ratio (mass of solvent evaporated/mass of solvent in the initial solution). For the solute:

$$w_1 c_1 = w_2 c_2 + y/R \tag{6.20}$$

where c_1 and c_2 are the initial and final concentrations of the solution expressed as (mass of anhydrous salt/mass of solvent). Substituting for w_2 from Eq. (6.19):

$$w_1 c_1 = c_2\left[w_1(1-E) - y\frac{R-1}{R}\right] + \frac{y}{R} \tag{6.21}$$

from which the yield for aqueous solutions is given by:

$$y = Rw_1\frac{c_1 - c_2(1-E)}{1 - c_2(R-1)} \tag{6.22}$$

The actual yield may differ slightly from that given by this equation since, for example, when crystals are washed with fresh solvent on the filter, losses may occur through dissolution. On the other hand, if mother liquor is retained by the crystals, an extra quantity of crystalline material will be deposited on drying. Since published solubility data usually refer to pure solvents and solutes that are rarely encountered industrially, solubilities should always be checked against the actual working liquors.

Before Eq. (6.22) can be applied to vacuum or adiabatic cooling crystallisation, the quantity E must be estimated, where, from a heat balance:

$$E = \frac{qR(c_1 - c_2) + C_p(T_1 - T_2)(1 + c_1)[1 - c_2(R-1)]}{\lambda[1 - c_2(R-1)] - qRc_2} \tag{6.23}$$

In this equation, λ is the latent heat of evaporation of the solvent (J/kg), q is the heat of crystallisation of the product (J/kg), T_1 is the initial temperature of the solution (K), T_2 is the final temperature of the solution (K) and C_p is the specific heat capacity of the solution (J/kg K).

Example 6.3

What is the theoretical yield of crystals which may be obtained by cooling a solution containing 1000 kg of sodium sulphate (molecular mass = 142 kg/kmol) in 5000 kg water to 283 K? The solubility of sodium sulphate at 283 K is 9 kg anhydrous salt/100 kg water and the deposited crystals will consist of the deca-hydrate (molecular mass = 322 kg/kmol). It may be assumed that 2% of the water will be lost by evaporation during cooling.

Solution

The ratio, $R = (322/142) = 2.27$.

The initial concentration, $c_1 = (1000/5000) = 0.2$ kg Na_2SO_4/kg water.

The solubility, $c_2 = (9/100) = 0.09$ kg Na_2SO_4/kg water.

The initial mass of water, $w_1 = 5000$ kg and the water lost by evaporation, $E = (2/100) = 0.02$ kg/kg.

Thus, in Eq. (6.22):

$$\text{yield,} \quad y = (5000 \times 2.27)[0.2 - 0.09(1 - 0.02)]/[1 - 0.09(2.27 - 1)]$$
$$= 1432 \text{ kg } \underline{Na_2SO_4.10H_2O}$$

Example 6.4

What is the yield of sodium acetate crystals ($CH_3COONa \cdot 3H_2O$) obtainable from a vacuum crystalliser operating at 1.33 kN/m² when it is supplied with 0.56 kg/s of a 40% aqueous solution of the salt at 353 K? The boiling point elevation of the solution is 11.5 K.

Data

Heat of crystallisation, $q = 144$ kJ/kg trihydrate.

Heat capacity of the solution, $C_p = 3.5$ kJ/kg K.

Latent heat of water at 1.33 kN/m², $\lambda = 2.46$ MJ/kg.

Boiling point of water at 1.33 kN/m² = 290.7 K.

Solubility of sodium acetate at 290.7 K, $c_2 = 0.539$ kg/kg water.

Solution

Equilibrium liquor temperature $= (290.7 + 11.5) = 302.2$ K.

Initial concentration, $c_1 = 40/(100 - 40) = 0.667$ kg/kg water.

Final concentration, $c_2 = 0.539$ kg/kg water.

Ratio of molecular masses, $R = (136/82) = 1.66$.

Thus, in Eq. (6.23):

$$E = \{144 \times 1.660.667 - 0.539 + 3.5353 - 302.21 + 0.667[1 - 0.5391.66 - 1]\}/$$
$$\{2460[1 - 0.5391.66 - 1] - 144 \times 1.66 \times 0.539\}.$$
$$= 0.153 \text{ kg/kg water originally present.}$$

The yield is then given by Eq. (6.22) as:

$$y = (0.56(100 - 40)/100)1.66[0.667 - 0.539(1 - 0.153)]/[1 - 0.539(1.66 - 1)].$$
$$= 0.183 \, \text{kg/s}.$$

6.2.6 Caking of crystals

Crystalline materials frequently cake or cement together on storage and crystal size, shape, moisture content, and storage conditions can all contribute to the caking tendency. In general, caking is caused by a dampening of the crystal surfaces in storage because of inefficient drying or an increase in atmospheric humidity above some critical value that depends on both substance and temperature. The presence of a hygroscopic trace impurity in the crystals, can also greatly influence their tendency to absorb atmospheric moisture. Moisture may also be released from inclusions if crystals fracture under storage conditions and, if crystal surface moisture later evaporates, adjacent crystals become firmly joined together with a cement of recrystallised solute. Caking may be minimised by efficient drying, packaging in airtight containers, and avoiding compaction on storage. In addition, crystals may be coated with an inert dust that acts as a moisture barrier. Small crystals are more prone to cake than large crystals because of the greater number of contact points per unit mass, although actual size is less important than size distribution and shape and the narrower the size distribution and the more granular the shape, the lower is the tendency of crystals to cake. Crystal size distribution can be controlled by adjusting operating conditions in a crystalliser and crystal shape may be influenced by the use of habit modifiers. A comprehensive account of the inhibition of caking by trace additives has been given by Phoenix [61].

6.2.7 Washing of crystals

The product from a crystalliser must be subjected to efficient solid–liquid separation in order to remove mother liquor and, while centrifugal filtration can reduce the liquor content of granular crystals to 5%–10%, small irregular crystals may retain more than 50%. After filtration, the product is usually washed to reduce the amount of liquor retained still further and, where the crystals are very soluble in the liquor, another liquid in which the substance is relatively insoluble is used for the washing, although this two-solvent method means that a solvent recovery unit is required. When simple washing is inadequate, two stages may be required for the removal of mother liquor with the crystals removed from the filter, re-dispersed in wash liquor and filtered again. This may cause a loss of yield although this is much less than the loss after a complete re-crystallisation.

If, for simplicity, it is assumed that the soluble impurity is in solution and that solution concentrations are constant throughout the dispersion vessel, then wash liquor requirements for decantation washing may be estimated as follows.

If, in *batch operation*, c_{io} and c_{in} denote the impurity concentrations in the crystalline material (kg impurity/kg product) initially and after washing stage n respectively, and F is the fraction of liquid removed at each decantation, then a mass balance gives:

$$c_{in} = c_{io}(1 - F)^n \qquad (6.24)$$

or

$$\ln(c_{in}/c_{io}) = n \ \ln(1 - F) \qquad (6.25)$$

For *continuous operation*, where fresh wash-liquid enters the vessel continuously and liquor is withdrawn through a filter screen, then a mass balance gives:

$$V_L \ dc = -c_i \ d \ V_W \qquad (6.26)$$

or

$$\ln(c_{in}/c_{io}) = -V_W/V_L \qquad (6.27)$$

where c_{io} and c_{in} are the initial and final concentrations and V_L and V_W are the volumes of liquor in the vessel and of the wash-water respectively. Combining Eqs. (6.25) and (6.27):

$$n \ \ln(1 - F) = -V_W/V_L \qquad (6.28)$$

or

$$\frac{1}{nF}\frac{V_w}{V_L} = -\ln\frac{1-F}{F} \qquad (6.29)$$

As Mullin [3] points out, this equation can be used for comparing batch and continuous processing since V_W and nFV_L represent the wash liquor requirements for both cases.

6.3 Crystallisation from solutions

Solution crystallisers are usually classified according to the method by which supersaturation is achieved, that is by cooling, evaporation, vacuum, reaction and salting out. The term *controlled* denotes supersaturation control while *classifying* refers to classification of product size.

6.3.1 Cooling crystallisers

Non-agitated vessels

The simplest type of cooling crystalliser is an unstirred tank in which a hot feedstock solution is charged to an open vessel and allowed to cool, often for several days, by natural convection. Metallic rods may be suspended in the solution so that large crystals can grow on them thereby reducing the amount of product that sinks to the bottom of the unit. The product is usually removed manually. Because cooling is slow, large interlocked crystals are usually produced. These retain mother liquor and thus the dried crystals are generally impure. Because of the uncontrolled nature of the process, product crystals range from a fine dust to large agglomerates. Labour costs are high, but the method is economical for small batches since capital, operating, and maintenance costs are low, although productivity is low and space requirements are high.

Agitated vessels

Installation of an agitator in an open-tank crystalliser gives smaller and more uniform crystals and reduces batch times. Because less liquor is retained by the crystals after filtration and more efficient washing is possible, the final product has a higher purity. Water jackets are usually preferred to coils for cooling because the latter often become encrusted with crystals and the inner surfaces of the crystalliser should be smooth and flat to minimise encrustation. Operating costs of agitated coolers are higher than for simple tanks and, although the productivity is higher, product handling costs are still high. Tank crystallisers vary from shallow pans to large cylindrical tanks.

The typical agitated cooling crystalliser, shown in Fig. 6.14A, has an upper conical section which reduces the upward velocity of liquor and prevents the crystalline product from being swept out with the spent liquor. An agitator, located in the lower region of a draught tube circulates the crystal slurry through the growth zone of the crystalliser; cooling surfaces may be provided if required. External circulation, as shown in Fig. 6.14B, allows good mixing inside the unit and promotes high rates of heat transfer between liquor and coolant, and an internal agitator may be installed in the crystallisation tank if required. Because the liquor velocity in the tubes is high, low temperature differences are usually adequate, and encrustation on heat transfer surfaces is reduced considerably. Batch or continuous operation may be employed.

Scraped-surface crystallisers

The Swenson-Walker scraped-surface unit, which is used for processing inorganic salts that have a high temperature solubility coefficient with water, is a shallow semi-cylindrical trough, about 600 mm wide and 3–12 m long, fitted with a water-cooled jacket. A helical scraper rotating at 0.8–1.6 Hz, keeps the cooling surfaces clean and enhances growth of crystals by moving them through the solution which flows down the sloping trough. Several units may be connected in series and the capacity is determined by the heat transfer rate which should exceed 60 kW [1] for economic operation, with heat transfer coefficients in the range 50–150 W/m^2 K. High coefficients and hence high production rates are obtained with double-pipe, scraped-surface units such as Votator and Armstrong crystallisers in which spring-loaded internal agitators scrape the heat transfer surfaces. With turbulent flow in the tube, coefficients of 50–700 W/m^2 K are achieved. Such units range from 75 to 600 mm in diameter and 0.3–3 m long. They are used mainly for processing fats, waxes and other organic melts, as outlined in Section 6.4, although the

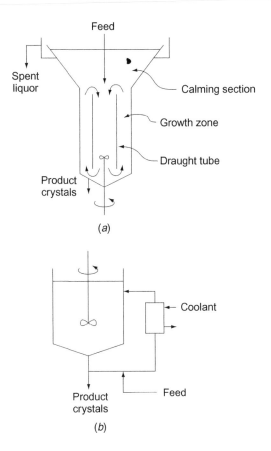

FIG. 6.14

Cooling crystallisers: (A) internal circulation through a draught tube and (B) external circulation through a heat exchanger.

processing of inorganic solutions such as sodium sulphate from viscose spin-bath liquors, has been reported by Armstrong [62].

Example 6.5

A solution containing 23% by mass of sodium phosphate is cooled from 313 to 298 K in a Swenson-Walker crystalliser to form crystals of $Na_3PO_4 \cdot 12H_2O$. The solubility of Na_3PO_4 at 298 K is 15.5 kg/100 kg water, and the required product rate of crystals is 0.063 kg/s. The mean heat capacity of the solution is 3.2 kJ/kg K and the heat of crystallisation is 146.5 kJ/kg. If cooling water enters and leaves at 288 and 293 K, respectively, and the overall coefficient of heat transfer is 140 W/m^2 K, what length of crystalliser is required?

Solution

The molecular mass of hydrate/molecular mass of anhydrate, $R = (380/164) = 2.32$.

It will be assumed that the evaporation is negligible and that $E = 0$.

The initial concentration, $c_1 = 0.23$ kg/kg solution or $0.23/(1 - 0.23) = 0.30$ kg/kg water.

The final concentration, $c_2 = 15.5$ kg/kg water or 0.155 kg/kg water.

In 1 kg of the initial feed solution, there is 0.23 kg salt and 0.77 kg water and hence $w_1 = 0.77$ kg The yield is given by Eq. (6.22):

$$y = 2.32 \times 0.77[0.30 - 0.155(1 - 0)]/[1 - 0.155(2.32 - 1)]. = 0.33 \, \text{kg}.$$

In order to produce 0.063 kg/s of crystals, the required feed is:

$$= (1 \times 0.063/0.33) = 0.193 \, \text{kg/s}.$$

The heat required to cool the solution $= 0.193 \times 3.2(313 - 298) = 9.3$ kW.

Heat of crystallisation $= (0.063 \times 146.5) = 9.2$ kW; a total of $(9.3 + 9.2) = 18.5$ kW.

Assuming countercurrent flow, $\Delta T_1 = (313 - 293) = 20$ K

$$\Delta T_2 = (298 - 288) = 10 \, \text{K}.$$

and the logarithmic mean, $\Delta T_m = (20 - 10)/\ln(20/10) = 14.4$ K.

The heat transfer area required, $A' = Q/U\Delta T_m = 18.5/(0.14 \times 14.4) = 9.2 \, \text{m}^2$.

Assuming that the area available is, typically, $1 \, \text{m}^2$/m length, the length of exchanger required $= 9.2$ m. In practice 3 lengths, each of 3 *m* length would be specified.

Direct-contact cooling

The occurrence of crystal encrustation in conventional heat exchangers can be avoided by using direct-contact cooling (DCC) in which supersaturation is achieved by allowing the process liquor to come into contact with a cold heat-transfer medium. Other potential advantages of DCC include better heat transfer and lower cooling loads, although disadvantages include product contamination from the coolant and the cost of extra processing required to recover the coolant for further use. Since a solid, liquid, or gaseous coolant can be used with transfer of sensible or latent heat, the coolant may or may not boil during the operation, and it can be either miscible or immiscible with the process liquor, several types of DCC crystallisation are possible:

(a) immiscible, boiling, solid or liquid coolant where heat is removed mainly by transfer of latent heat of sublimation or vaporisation;

(b) immiscible, non-boiling, solid, liquid, or gaseous coolant with mainly sensible heat transfer;

(c) miscible, boiling, liquid coolant with mainly latent heat transfer; and

(d) miscible, non-boiling, liquid coolant with mainly sensible heat transfer.

Crystallisation processes employing DCC have been used successfully in the de-waxing of lubricating oils [63], the desalination of water [64], and the production of inorganic salts from aqueous solution [65].

6.3.2 Evaporating crystallisers

If the solubility of a solute in a solvent is not appreciably decreased by lowering the temperature, the appropriate degree of solution supersaturation can be achieved by evaporating some of the solvent and the oldest and simplest technique, the use of solar energy, is still employed commercially throughout the world [66]. Common salt is produced widely from brine in steam-heated evaporators, multiple-effect evaporator-crystallisers are used in sugar refining and many types of forced-circulation evaporating crystallisers are in large-scale use [3,40,67]. Evaporating crystallisers are usually operated under reduced pressure to aid solvent removal, minimise heat consumption, or decrease the operating temperature of the solution, and these are described as *reduced-pressure evaporating crystallisers.*

6.3.3 Vacuum (adiabatic cooling) crystallisers

A vacuum crystalliser operates on a slightly different principle from the reduced-pressure unit since supersaturation is achieved by simultaneous evaporation and adiabatic cooling of the feedstock. A hot, saturated solution is fed into an insulated vessel maintained under reduced pressure. If the feed liquor temperature is higher than the boiling point of the solution under the low pressure existing in the vessel, the liquor cools adiabatically to this temperature and the sensible heat and any heat of crystallisation liberated by the solution evaporate solvent and concentrate the solution.

6.3.4 Continuous crystallisers

The majority of continuously operated crystallisers are of three basic types: forced-circulation, fluidised-bed and draught-tube agitated units.

Forced-circulation crystallisers

A *Swenson forced-circulation crystalliser* operating at reduced pressure is shown in Fig. 6.15. A high recirculation rate through the external heat exchanger is used to provide good heat transfer with minimal encrustation. The crystal magma is circulated from the lower conical section of the evaporator body, through the vertical tubular heat exchanger, and reintroduced tangentially into the evaporator below the liquor level to create a swirling action and prevent flashing. Feed-stock enters on the pump inlet side of the circulation system and product crystal magma is removed below the conical section.

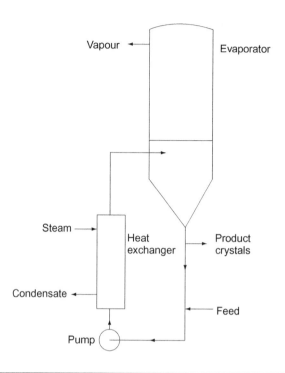

FIG. 6.15

Forced-circulation Swenson crystalliser.

Fluidised-bed crystallisers

In an *Oslo fluidised-bed crystalliser*, a bed of crystals is suspended in the vessel by the upward flow of supersaturated liquor in the annular region surrounding a central downcomer, as shown in Fig. 6.16. Although originally designed as classifying crystallisers, fluidised-bed Oslo units are frequently operated in a mixed-suspension mode to improve productivity, although this reduces product crystal size [68]. With the classifying mode of operation, hot, concentrated feed solution is fed into the vessel at a point directly above the inlet to the circulation pipe. Saturated solution from the upper regions of the crystalliser, together with the small amount of feedstock, is circulated through the tubes of the heat exchanger and cooled by forced circulation of water or brine. In this way, the solution becomes supersaturated although care must be taken to avoid spontaneous nucleation. Product crystal magma is removed from the lower regions of the vessel.

Draught-tube agitated vacuum crystallisers

A *Swenson draught-tube-baffled (DTB) vacuum unit* is shown in Fig. 6.17. A relatively slow-speed propellor agitator is located in a draught tube that extends to a small distance below the liquor level. Hot, concentrated feed-stock, enters at the base of the draught tube, and the steady movement of magma and feed-stock to the surface

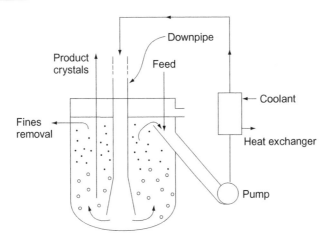

FIG. 6.16

Oslo cooling crystalliser.

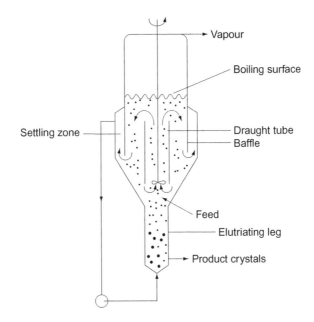

FIG. 6.17

Swenson draught-tube-baffled (DTB) crystalliser.

of the liquor produces a gentle, uniform boiling action over the whole cross-sectional area of the crystalliser. The degree of supercooling thus produced is less than 1 K and, in the absence of violent vapour flashing, both excessive nucleation and salt build-up on the inner walls are minimised. The internal baffle forms an annular space free of agitation and provides a settling zone for regulating the

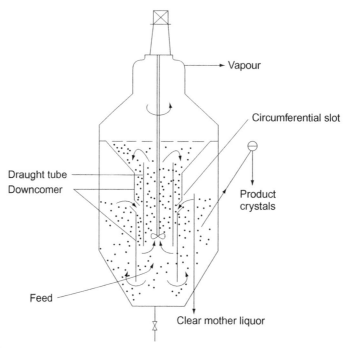

Vapour

Circumferential slot

Draught tube

Downcomer

Product crystals

Feed

Clear mother liquor

FIG. 6.18

Standard-Messo turbulence crystalliser.

magma density and controlling the removal of excess nuclei. An integral elutri-ating leg may be installed below the crystallisation zone to effect some degree of product classification.

The *Standard-Messo turbulence crystalliser*, Fig. 6.18, is a draught-tube vac-uum unit in which two liquor flow circuits are created by concentric pipes: an outer ejector tube with a circumferential slot, and an inner guide tube in which circulation is effected by a variable-speed agitator. The principle of the Oslo crys-talliser is utilised in the growth zone, partial classification occurs in the lower regions, and fine crystals segregate in the upper regions. The primary circuit is created by a fast upward flow of liquor in the guide tube and a downward flow in the annulus. In this way, liquor is drawn through the slot between the ejector tube and the baffle, and a secondary flow circuit is formed in the lower region of the vessel. Feedstock is introduced into the draught tube and passes into the vaporiser section where flash evaporation takes place. In this way, nucleation occurs in this region, and the nuclei are swept into the primary circuit. Mother liquor may be drawn off by way of a control valve that provides a means of con-trolling crystal slurry density.

The *Escher–Wyss Tsukishima double-propeller* (DP) *crystalliser*, shown in Fig. 6.19, is essentially a draught-tube agitated crystalliser. The DP unit contains

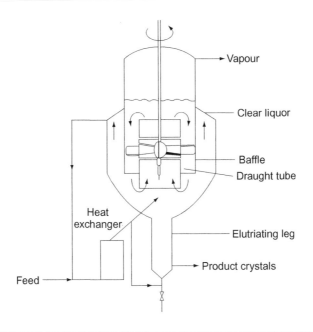

FIG. 6.19

Escher–Wyss Tsukishima double-propeller (DP) crystalliser.

an annular baffled zone and a double-propellor agitator which maintains a steady upward flow inside the draught tube and a downward flow in the annular region, thus giving very stable suspension characteristics.

6.3.5 Controlled crystallisation

Carefully selected seed crystals are sometimes added to a crystalliser to control the final product crystal size. The rapid cooling of an unseeded solution is shown in Fig. 6.20A in which the solution cools at constant concentration until the limit of the metastable zone is reached, where nucleation occurs. The temperature increases slightly due to the release of latent heat of crystallisation, but on cooling more nucleation occurs. The temperature and concentration subsequently fall and, in such a process, nucleation and growth cannot be controlled. The slow cooling of a seeded solution, in which temperature and solution composition are controlled within the metastable zone throughout the cooling cycle, is shown in Fig. 6.20B. Crystal growth occurs at a controlled rate depositing only on the added seeds and spontaneous nucleation is avoided because the system is never allowed to become labile. Many large-scale crystallisers are operated on this batch operating method that is known as *controlled crystallisation*.

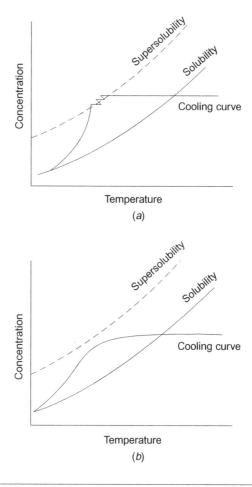

FIG. 6.20

Effect of seeding on cooling crystallisation: (A) rapid cooling-unseeded and (B) slow cooling-seeded.

If crystallisation occurs only on the added seeds, the mass m, of seeds of size d_s that can be added to a crystalliser depends on the required crystal yield y and the product crystal size d_p, as follows:

$$m_s = y \left(\frac{d_s^3}{d_p^3 - d_s^3} \right) \tag{6.30}$$

The product crystal size from a batch crystalliser can also be controlled by adjusting the rates of cooling or evaporation. Natural cooling, for example, produces a super-saturation peak in the early stages of the process when rapid, uncontrolled heavy nucleation inevitably occurs, although nucleation can be controlled within acceptable limits by following a cooling path that maintains a constant low level of supersaturation. As Mullin and Nyvlt [69] has pointed out, the calculation of optimum cooling curves for different operating conditions is complex, although the following simplified relationship is usually adequate for general application:

$$T_t = T_0 - (T_0 - T_f)(t/t_b)^3 \qquad (6.31)$$

where T_0, T_f, and T_t are the temperatures at the beginning, end and any time t during the process, respectively, and t_b is the overall batch time.

6.3.6 Batch and continuous crystallisation

Continuous, steady-state operation is not always the ideal mode for the operation of crystallisation processes, and batch operation often offers considerable advantages such as simplicity of equipment and reduced encrustation on heat-exchanger surfaces. While only a batch crystalliser can, in certain cases, produce the required crystal form, size distribution, or purity, the operating costs can be significantly higher than those of a comparable continuous unit, and problems of product variation from batch to batch may be encountered. The particular attraction of a continuous crystalliser is its built-in flexibility for control of temperature, supersaturation nucleation, crystal growth, and other parameters that influence the size distribution of the crystals. The product slurry may have to be passed to a holding tank, however, to allow equilibrium between the crystals and the mother liquor to be reached if unwanted deposition in the following pipelines and effluent tanks is to be avoided. One important advantage of batch operation, especially in the pharmaceutical industry, is that the crystalliser can be cleaned thoroughly at the end of each batch, thus preventing contamination of the next charge with any undesirable products that might have been formed as a result of transformations, rehydration, dehydration, air oxidation and so on during the batch cycle. In continuous crystallisation systems, undesired self-seeding may occur after a certain operating time, necessitating frequent shutdowns and washouts.

Semi-continuous crystallisation processes which often combine the best features of both batch and continuous operation are described by Nyvlt [35], Randolph [37], Robinson and Roberts [70], and Abbeg and Balakrishnam [71]. It may be possible to use a series of tanks which can then be operated as individual units or in cascade. Mullin [3] suggests that for production rates in excess of 0.02 kg/s (70 kg/h) or liquor feeds in excess of 0.005 m³/s, continuous operation is preferable although sugar may be produced batch-wise at around 0.25 kg/s (900 kg/h) per crystalliser.

6.3.7 Crystalliser selection

The temperature–solubility relationship for solute and solvent is of prime importance in the selection of a crystalliser and, for solutions that yield appreciable amounts of crystals on cooling, either a simple cooling or a vacuum cooling unit is appropriate. An evaporating crystalliser would be used for solutions that change little in composition on cooling and salting-out would be used in certain cases. The shape, size and size distribution of the product is also an important factor and for large uniform crystals, a controlled suspension unit fitted with suitable traps for fines, permitting the discharge of a partially classified product, would be suitable. This simplifies washing and drying operations and screening of the final product may not be necessary. Simple cooling-crystallisers are relatively inexpensive, though the initial cost of a mechanical unit is fairly high although no costly vacuum or condensing equipment is required. Heavy crystal slurries can be handled in cooling units without liquor circulation, though cooling surfaces can become coated with crystals thus reducing the heat transfer efficiency. Vacuum crystallisers with no cooling surfaces do not have this disadvantage but they cannot be used when the liquor has a high boiling point elevation. In terms of space, both vacuum and evaporating units usually require a considerable height.

Once a particular class of unit has been decided upon, the choice of a specific unit depends on initial and operating costs, the space available, the type and size of the product, the characteristics of the feed liquor, the need for corrosion resistance and so on. Particular attention must be paid to liquor mixing zones since the circulation loop includes many regions where flow streams of different temperature and composition mix. These are all points at which temporary high supersaturations may occur causing heavy nucleation and hence encrustation, poor performance and operating instabilities. As Toussaint and Donders [72] stresses, it is essential that the compositions and enthalpies of mixer streams are always such that, at equilibrium, only one phase exists under the local conditions of temperature and pressure.

6.3.8 Crystalliser modelling and design
Population balance
Growth and nucleation interact in a crystalliser in which both contribute to the final crystal size distribution (CSD) of the product. The importance of the population balance [37] is widely acknowledged. This is most easily appreciated by reference to the simple, idealised case of a mixed-suspension, mixed-product removal (MSMPR) crystalliser operated continuously in the steady state, where no crystals are present in the feed stream, all crystals are of the same shape, no crystals break down by attrition, and crystal growth rate is independent of crystal size. The crystal size distribution for steady state operation in terms of crystal size d and population density n' (number of crystals per unit size per unit volume of the system), derived directly from the population balance over the system [37] is:

$$n' = n° \exp\left(-d/G_d t_r\right) \tag{6.32}$$

where $n°$ is the population density of nuclei and t_r is the residence time. Rates of nucleation B and growth $G_d(=dd/dt)$ are conventionally written in terms of supersaturation as:

$$B = k_1 \Delta c^b \tag{6.33}$$

and

$$G_d = k_2 \Delta c^s \tag{6.34}$$

These empirical expressions may be combined to give:

$$B = k_3 G^i \tag{6.35}$$

where

$$i = b/s \quad \text{and} \quad k_3 = k_1/k_2^i \tag{6.36}$$

where b and s are the kinetic orders of nucleation and growth, respectively, and i is the relative kinetic order. The relationship between nucleation and growth may be expressed as:

$$B = n^\circ = G_d \tag{6.37}$$

or

$$n^\circ = k_4 G_d^{i-1} \tag{6.38}$$

In this way, experimental measurement of crystal size distribution, recorded on a number basis, in a steady-state MSMPR crystalliser can be used to quantify nucleation and growth rates. A plot of log n against d should give a straight line of slope $-(G_d t_r)^{-1}$ with an intercept at $d = 0$ equal to n° and, if the residence time t_r is known, the crystal growth rate G_d can be calculated. Similarly, a plot of log n° against log G_d should give a straight line of slope $(i - 1)$ and, if the order of the growth process s is known, the order of nucleation b may be calculated. Such plots are shown in Fig. 6.21.

The mass of crystals per unit volume of the system, the so-called magma density, ρ_m is given by:

$$\rho_m = 6\alpha\rho n^\circ \, (G_d t_r)^4 \tag{6.39}$$

where α is the volume shape factor defined by $\alpha = \text{volume}/d^3$ and ρ is the crystal density.

The peak of the mass distribution, the dominant size d_D of the CSD, is given by Mullin [1] as:

$$d_D = 3 \, G_d t_r \tag{6.40}$$

and this can be related to the crystallisation kinetics by:

$$d_D \propto t_r^{(i-1)/(i+3)} \tag{6.41}$$

This interesting relationship [37] enables the effect of changes in residence time to be evaluated. For example, if $i = 3$, a typical value for many inorganic salt systems, a doubling of the residence time would increase the dominant product crystal size by only 26%. This could be achieved either by doubling the volume of the crystalliser or by halving the volumetric feed rate, and hence the production rate. Thus, residence time adjustment is usually not a very effective means of controlling product crystal size.

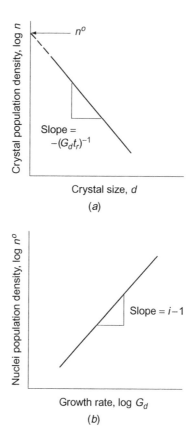

Crystal population density, log n

n^o

Slope $= -(G_d t_r)^{-1}$

Crystal size, d

(a)

Nuclei population density, log n^o

Slope $= i-1$

Growth rate, log G_d

(b)

FIG. 6.21

Population plots for a continuous mixed-suspension mixed-product removal (MSMPR) crystalliser: (A) crystal size distribution and (B) nucleation and growth kinetics.

CSD modelling based on population balance considerations may be applied to crystalliser configurations other than MSMPR [37] and this has become a distinct, self-contained branch of reaction engineering [56,59,60,73].

Example 6.6

An MSMPR crystalliser operates with a steady nucleation rate of $n^o = 10^{13}/m^4$, a growth rate, $G_d = 10^{-8}$ m/s and a mixed-product removal rate, based on clear liquor of 0.00017 m^3/s. The volume of the vessel, again based on clear liquor, is 4 m^3, the crystal density is 2660 kg/m^3 and the volumetric shape factor is 0.7. Determine:
(a) the solids content in the crystalliser,
(b) the crystal production rate,

(c) the per cent of nuclei removed in the discharge by the time they have grown to 100 μm, and

(d) the liquor flow rate which passes through a trap which removes 90% of the original nuclei by the time they have grown to 100 μm.

Solution

Draw-down time $=(4/0.00017)=23,530$ s.

(a) From a mass balance, the total mass of solids is:

$$c_s = 6a\rho n^\circ (G_d t_r)^4$$
$$= \left(6 \times 0.6 \times 2660 \times 10^{13}\right)\left(10^{-8} \times 23,530\right)^4 \qquad (6.42)$$
$$= \underline{343 \ \text{kg/m}^3}$$

(b) The production rate $= (343 \times 0.00017) = \underline{0.058\text{kg/s} \ (200 \ \text{kg/h})}$

(c) The crystal population decreases exponentially with size or:

$$n/n^\circ = \exp(-L/G_d t_r)$$
$$= \exp\left[(-100 \times 10^{-6})/(10^{-8} \times 23,\ 530)\right] \qquad (6.43)$$
$$= 0.66 \ \text{or 66 per cent}$$

Thus : $(100 - 66) = \underline{34\% \text{ have been discharged by the time they reach } 100\mu m}$.

(d) If $(100 - 90) = 10\%$ of the nuclei remain and grow to $>100\,\mu$m, then in Eq. (6.32):

$$(1/0.10) = \exp.\left[(-100 \times 10^{-6})/(10^{-8} \ t_r)\right]$$

and $t_r = 4343$ s

Thus:

$$4343 = 4/(0.00017 + Q_F)$$

and $Q_F = 0.00075 \ \text{m}^3/\text{s} \ (2.68 \ \text{m}^3/\text{h})$.

Design procedures

Mullin [3] has given details of a procedure for the design of classifying crystallisers in which the calculation steps are as follows:

(a) The maximum allowable supersaturation is obtained and hence the working saturation, noting that this is usually about 30% of the maximum.

(b) The solution circulation rate is obtained from a materials balance.

(c) The maximum linear growth-rate is obtained based on the supersaturation in the lowest layer which contains the product crystals and assuming that $(\beta/\alpha) = 6$.

(d) The crystal growth time is calculated from the growth rate for different relative desupersaturations (100% desupersaturation corresponding to the reduction of the degree of supersaturation to zero).

(e) The mass of crystals in suspension and the suspension volume are calculated assuming a value for the voidage which is often about 0.85.

(f) The solution up-flow velocity is calculated for very small crystals ($<0.1\,$mm) using Stokes' Law although strictly this procedure should not be used for particles other than spheres or for $Re>0.3$. In a real situation, laboratory measurements of the velocity are usually required.

(g) The crystalliser area and diameter are first calculated and then the height which is taken as (volume of suspension/cross-sectional area).

(h) A separation intensity (SI), defined by Griffiths [74] as the mass of equivalent 1 mm crystals produced in $1\,m^3$ of crystalliser volume in 1 s, is calculated. Typical values are $0.015\,kg/m^3$ s at 300 K and up to 0.05 at higher temperatures and, for crystals $>1\,$mm, the intensity is given by:

$$SI = d_p P'/V \qquad (6.44)$$

where d_p is the product crystal size, P' (kg/s) is the crystal production rate and $V\,(m^3)$ is the suspension volume.

Mullin [3] has used this procedure for the design of a unit for the crystallisation of potassium sulphate at 293 K. The data are given in Table 6.5 from which it will be

Table 6.5 Design of a continuous classifying crystalliser [3].

Basic data				

Substance: potassium sulphate at 293 K
Product: 0.278 kg/s of 1 mm crystals
Growth constant: $k_d = 0.75\Delta c^{-2}$ kg/m²s
Nucleation constant: $k_n = 2\times10^8\,\Delta c^{-7.3}$ kg/s
Crystal size: smallest in fluidised bed = 0.3 mm, (free settling velocity = 40 mm/s)
Smallest in system = 0.1 mm
Crystal density = 2660 kg/m³, solution density = 1082 kg/m³
Solution viscosity = 0.0012 Ns/m², solubility, $c^* = 0.1117$ kg/kg water

Desupersaturation	1.0	0.9	0.5	0.1
Maximum growth rate (µm/s)	5.6	5.6	5.6	5.6
Up-flow velocity (m/s)	0.04	0.04	0.04	0.04
Circulation rate (m³/s)	0.029	0.032	0.058	0.286
Crystal residence time (ks)	1469	907	51.8	12.6
Mass of crystals (Mg)	145	90	5.1	1.25
Volume of crystal suspension (m³)	364	225	12.8	3.15
Cross-sectional area of crystalliser (m²)	0.72	0.80	1.45	7.2
Crystalliser diameter (m)	0.96	1.01	1.36	3.02
Crystalliser height (m)	505	281	8.8	0.44
Height/diameter	525	280	6.5	0.15
Separation intensity	3.0	4.5	78	320
Economically possible	No	No	Yes	No

noted that the cross-sectional area depends linearly on the relative degree of de-supersaturation and the production rate depends linearly on the area but is independent of the height. If the production rate is fixed, then the crystalliser height may be adjusted by altering the sizes of the seed or product crystals. Mullin and Nyvlt [75] have proposed a similar procedure for mixed particle-size in a crystalliser fitted with a classifier at the product outlet which controls the minimum size of product crystals.

Scale-up problems

Crystalliser design is usually based on data measured on laboratory or pilot-scale units or, in difficult cases, both. One of the main problems in scaling up is characterisation of the particle–fluid hydrodynamics and the assessment of its effects on the kinetics of nucleation and crystal growth. In fluidised-bed crystallisers, for example, the crystal suspension velocity must be evaluated—a parameter which is related to crystal size, size distribution, and shape, as well as bed voidage and other system properties—such as density differences between particles and liquid and viscosity of the solution. Possible ways of estimating suspension velocity are discussed in the literature [3,41,43,76,77], although, as Mullin [3] points out, determination of suspension velocities on actual crystal samples is often advisable. In agitated vessels, the 'just-suspended' agitator speed N_{JS}, that is the minimum rotational speed necessary to keep all crystals in suspension, must be determined since, not only do all the crystals have to be kept in suspension, but the development of 'dead spaces' in the vessel must also be avoided as these are unproductive zones and regions of high supersaturation in which vessel surfaces can become encrusted. Fluid and crystal properties, together with vessel and agitator geometries, are important in establishing N_{JS} values [3,43]. As discussed in Volume 1, Section 7.3, agitated vessel crystallisers are often scaled up successfully on the crude basis of either *constant power input per unit volume* or *constant agitator tip speed*, although Bennett et al. [78] have suggested that, in draught-tube agitated vessels, the quantity (tip speed)^2/(vessel volume/volumetric circulation rate) should be kept constant.

6.4 **Crystallisation from melts**
6.4.1 **Basic techniques**

A *melt* is a liquid or a liquid mixture at a temperature near its freezing point and *melt crystallisation* is the process of separating the components of a liquid mixture by cooling until crystallised solid is deposited from the liquid phase. Where the crystallisation process is used to separate, or partially separate, the components, the composition of the crystallised solid will differ from that of the liquid mixture from which it is deposited. The ease or difficulty of separating one component from a multicomponent mixture by crystallisation may be represented by a phase diagram as shown in Figs 6.4 and 6.5, both of which depict binary systems – the former depicts a eutectic, and the latter a continuous series of solid solutions. These two systems behave quite differently on freezing since a eutectic system can deposit a pure component, whereas a solid solution can only deposit a mixture of components.

Two basic techniques of melt crystallisation are:

(a) gradual deposition of a crystalline layer on a chilled surface in a static or laminar-flowing melt and

(b) fast generation of discrete crystals in the body of an agitated vessel.

Gradual deposition (a) occurs in the Proabd refiner [79] which essentially utilises a batch cooling process in which a static liquid feedstock is progressively crystallised on to extensive cooling surfaces, such as fin-tube heat-exchangers, supplied with a cold heat-transfer fluid located inside a crystallisation tank. As crystallisation proceeds, the liquid becomes increasingly impure and crystallisation may be continued until virtually the entire charge has solidified. When the crystallised mass is then slowly melted by circulating a hot fluid through the heat exchanger, the impure fraction melts first and drains out of the tank. As melting proceeds, the melt run-off becomes progressively richer in the desired component, and fractions may be taken off during the melting stage. A typical flow diagram, based on a scheme for the purification of naphthalene, is shown in Fig. 6.22 where the circulating fluid is usually cold water that is heated during the melting stage by steam injection. Another example of gradual deposition occurs in the rotary drum crystalliser which consists of a horizontal cylinder, partially immersed in the melt, or otherwise supplied with feedstock. The coolant enters and leaves the inside of the hollow drum through trunnions and, as the drum rotates, a crystalline layer forms on the cold surface and is removed with a scraper knife. Two feed and discharge arrangements are shown in Fig. 6.23. Rotary drum behaviour and design have been discussed by Gelperin and Nosov [80] and Ponomarenko et al. [81]. Wintermantel [82] has shown that the structure and impurity levels of growing crystal layers are determined primarily by mass-transfer effects at the layer front.

Fast-melt crystallisation (b) takes place in the scraped-surface heat exchanger, which consists of a cylindrical tube surrounded by a heat-exchange jacket. The tube

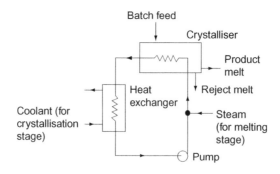

FIG. 6.22

Batch cooling crystallisation of melts: flow diagram for the Proabd refiner.

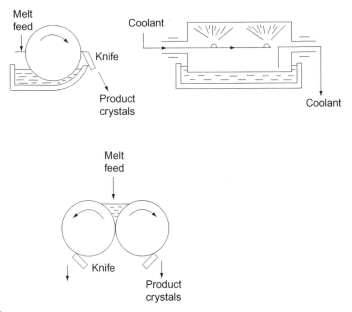

FIG. 6.23

Feed and discharge arrangements for drum crystallisers.

is surrounded by close-clearance scraper blades and rotates at relatively low speed. Two basic types are available: the large (>200 mm in diameter, >3 m long) slow-speed (<0.15 Hz) unit, and the small (<150 mm in diameter, <1.5 m long) high-speed (>8 Hz) machine. Both types can handle viscous magmas, operating at temperatures as low as 190 K, and are widely used in the manufacture of margarine (crystallisation of triglycerides), de-waxing of lubricating oils (crystallisation of higher n-alkanes), and large-scale processing of many organic substances, such as naphthalene, p-xylene, chlorobenzenes, and so on. The magma emerging from a scraped-surface crystalliser generally contains very small crystals, often less than 10 µm which are difficult to separate and can subsequently cause reprocessing problems unless the crystals are first grown to a larger size in a separate holdup tank.

6.4.2 **Multistage-processes**

A single-stage crystallisation process may not always achieve the required product purity and further separation, melting, washing, or refining may be required. Two approaches are used:

a) a repeating sequence of crystallisation, melting, and re-crystallisation and
b) a single crystallisation step followed by countercurrent contacting of the crystals with a relatively pure liquid stream.

The first approach is preferred if the concentration of impurities in the feedstock is high, and is essential if the system forms a continuous series of solid solutions. The second approach is used where the concentration of impurities is low, although some industrial operations require a combination of both systems. Atwood [83] has offered an analysis of different types of multistage crystallisation schemes.

As described by Mullin [3], Bennett et al. [78], and Rittner and Steiner [84], many industrial melt-crystallisation processes have been developed, and further interest is being stimulated by the energy-saving potential in large scale processing, as compared with distillation. One example is the *Newton–Chambers process*, described by Molinari et al. [79], in which benzene is produced from a coal-tar benzole fraction by contacting the impure feedstock with brine. The slurry is centrifuged, yielding benzene crystals (freezing point 278.6 K) and a mixture of brine and mother liquor which is then allowed to settle. The brine is returned for refrigeration and the mother liquor is reprocessed to yield motor fuel. The process efficiency depends to a large extent on the efficient removal of impure mother liquor that adheres to the benzene crystals, and several modes of operation are possible. In the thaw–melt method shown in Fig. 6.24, benzene crystals are washed in the centrifuge with brine at a temperature above 279 K. Some of the benzene crystals partially melt, which helps to wash away the adhering mother liquor. The thawed liquor may then be recycled. Multistage operation can be employed, in which the first crop of crystals is removed as product and the second, from the liquor, is melted for recycle. The *Sulzer MWB*

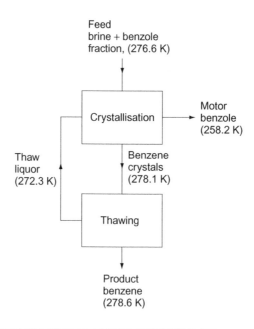

FIG. 6.24

Newton–Chambers process for the purification of benzene.

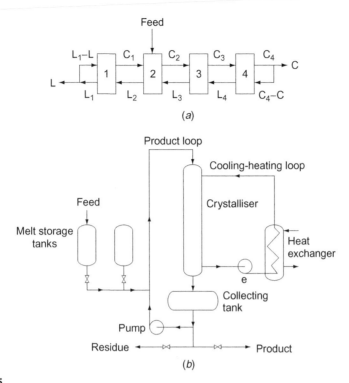

FIG. 6.25

Sulzer MWB process: (A) flow diagram (C—crystals, L—liquid) and (B) layout.

process, described by Fischer et al. [85], involves crystallisation on a cold surface and, since it may be operated effectively as a multistage separation device, it can be used to purify solid solutions. In an effective multistage countercurrent scheme, only one crystalliser, a vertical multi-tube heat exchanger, is required and the crystals do not have to be transported since they remain deposited on the internal heat-exchange surfaces in the vessel until they are melted for further processing. The intermediate storage tanks and crystalliser are linked by a control system consisting of a programme timer, actuating valves, pumps, and cooling loop, as shown in Fig. 6.25. This process has been used on a large scale for the purification of organics, such as chloronitrobenzenes, nitrotoluenes, cresols, and xylenols, and in the separation of fatty acids.

6.4.3 Column crystallisers

Because the components of the melt feedstock components can form both eutectic and solid-solution systems with one another, sequences of washing, partial or complete melting, and re-crystallisation are often necessary to produce one of the

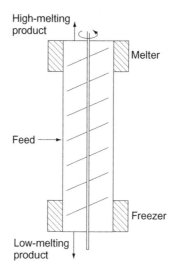

High-melting product

Melter

Feed

Freezer

Low-melting product

FIG. 6.26

Schildknecht column.

components in near-pure form. Because the operation of a sequence of melt crystallisation steps can be time-consuming and costly, many attempts have been made to carry out some of these operations in a single unit, such as a column crystalliser. One example is the *Schildknecht column* developed in the 1950s which is shown in Fig. 6.26. Liquid feedstock enters the column continuously at an intermediate point and freezing at the bottom of the column and melting at the top are achieved using, respectively, a suitable refrigerant and a hot fluid or an electrical heating element. Crystals and liquid pass through the column countercurrently, and the solid phase is transported downward by a helical conveyor fixed on a central shaft. The purification zone is usually operated at a virtually constant temperature, intermediate between the temperatures of the freezing and melting sections. Crystals are formed mainly in the freezing section, although some may also be deposited on the inner surface of the column and removed by the helical conveyor. During this operation, crystals make contact with the counter-flowing liquid melt and are thereby surface-washed. A system in which an upward flowing liquid is in contact with crystals being conveyed downward has also been used and, in this case, the locations of the freezer and melter are the reverse of those shown in Fig. 6.26. Gates and Powers [86] and Henry and Powers [87] have discussed the modelling of column crystallisers.

While the Schildknecht column is essentially a laboratory-scale unit, a melt-crystalliser of the wash-column type was developed by Phillips Petroleum Company in the 1960s for large-scale production of *p*-xylene. The key features of this *Phillips pulsed-column crystalliser*, as described by McKay et al. [88], are shown in Fig. 6.27. A cold slurry feed, produced in a scraped-surface chiller, enters at the top of the column and crystals are pulsed downward in the vertical bed by a piston,

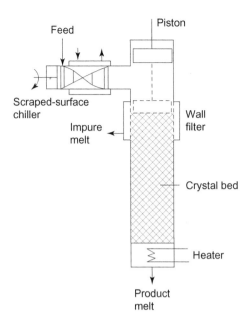

FIG. 6.27

Phillips pulsed-column crystalliser [88].

while impure mother liquor leaves through a filter. The upward-flowing wash liquor is generated at the bottom by a heater that melts pure crystals before they are removed from the column.

While the *Brodie purifier* [89] developed in the late 1960s incorporates several features of the column crystallisers described previously, it also has the potential to deal effectively with solid-solution systems. As shown in Fig. 6.28, it is essentially a centre-fed column that can convey crystals from one end to the other. As the crystals move through the unit, their temperature is gradually increased along the flow path and thus they are subjected to partial melting which encourages the release of low-melting impurities. The interconnected scraped-surface heat exchangers are of progressively smaller diameter so as to maintain reasonably constant axial flow velocities and to prevent back-mixing. The vertical purifying column acts as a countercurrent washer in which descending, nearly pure, crystals meet an upward stream of pure melt. The Brodie purifier has been used in the large-scale production of high-purity 1,4-dichlorobenzene and naphthalene.

The *Tsukishima Kikai (TSK) countercurrent cooling crystallisation process* described by Takegami [90] is, in effect, a development of Brodie technology. A typical system, consisting of three conventional cooling crystallisers connected in series is shown in Fig. 6.29. Feed enters the first-stage vessel and partially crystallises, and the slurry is then concentrated in a hydrocyclone before passing into a Brodie purifying column. After passage through a settling zone in the crystalliser, clear liquid

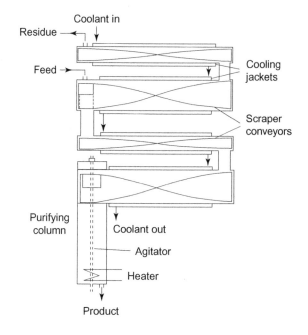

FIG. 6.28

Brodie purifier [89].

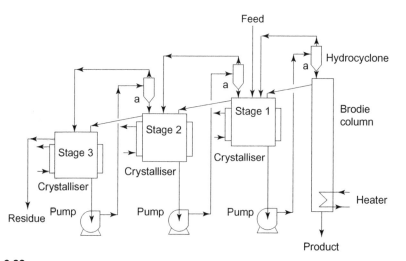

FIG. 6.29

Tsukishima Kikai (TSK) countercurrent cooling crystallisation process [90].

overflows to the next stage. Slurry pumping and overflow of clear liquid in each stage result in countercurrent flow of liquid and solid. This process has been applied in the large-scale production of *p*-xylene.

Units have been developed by Sulzer Chemtech which consist of vertical tubes where the product flows as a film down the inside surface of the tubes, and the liquid used for cooling and heating is distributed so as to wet the external surface of the tubes. During the initial freezing stage, the heat transfer medium chills the tubes, partial melting is then induced by raising the temperature of the heat transfer medium and higher temperatures are then applied for the final melting stage. A distribution system equalises the flow through the tubes and optimum performance is achieved by accurate control of the heating and cooling profiles. Sulzer also produce a unit for static melt crystallisation which employs cooled plates immersed in a stagnant melt. After a crystal layer has formed, sweating is induced, as with the falling-film device, and the sweated fraction is removed. The remaining crystal layer is then melted and passed to storage. A higher degree of purity may be obtained by using the intermediate product as feedstock and repeating the procedure and, in a similar way, the residue drained from the first phase of operation may be further depleted by additional melt-freezing processes to give an enhanced yield. A relatively new development is the use of a heat pump in which one crystalliser operates in the crystallisation mode as an evaporator, and a further identical unit operates in the sweating or melting mode as a condenser. In this way, energy costs are reduced due to the use of the enthalpy of condensation for crystal melting. Auxiliary exchangers are required only for the elimination of excess energy and for the start-up operation.

A further development, discussed by Moritoki [91], is *high-pressure crystallisation*, which is considered here.

6.4.4 Prilling and granulation

Prilling, a melt-spray crystallisation process in which solid spherical granules are formed, is used particularly in the manufacture of fertilisers such as ammonium nitrate and urea. Shearon and Dunwoody [92] describe the prilling of ammonium nitrate, in which a very concentrated solution containing about 5% of water is sprayed at 415 K into the top of a 30 m high, 6 m diameter tower, and the droplets fall countercurrently through an upwardly flowing air stream that enters the base of the tower at 293 K. The solidified droplets, which leave the tower at 353 K and contain about 4% water, must be dried to an acceptable moisture content at a temperature below 353 K in order to prevent any polymorphic transitions. Nunnelly and Cartney [93] describe a melt granulation technique for urea in which molten urea is sprayed at 420 K on to cascading granules in a rotary drum. Seed granules of less than 0.5 mm diameter can be built up to the product size of 2–3 mm. Heat released by the solidifying melt is removed by the evaporation of a fine mist of water sprayed into the air as it passes through the granulation drum.

An important application of granulation is in improving the 'flowability' of very fine (submicron) particles which stick together because of the large surface forces acting in materials with very high surface/volume ratios.

6.5 Crystallisation from vapours
6.5.1 Introduction

The term *sublimation* strictly refers to the phase change: solid → vapour, with no intervention of a liquid phase. In industrial applications, however, the term usually includes the reverse process of condensation or *desublimation*: solid → vapour → solid. In practice, it is sometimes desirable to vaporise a substance from the liquid state and hence the complete series of phase changes is then: solid → liquid → vapour → solid, and, on the condensation side of the process, with the supersaturated vapour condensing directly to the crystalline solid state without the creation of a liquid phase.

Common organic compounds that can be purified led by sublimation include [94]:

> *2-aminophenol, anthracene, anthranilic acid, anthraquinone, benzanthrone, benzoic acid, 1,4-benzoquinone, camphor, cyanuric chloride, isophthalic acid, naphthalene, 2-napththol, phthalic anhydride, phthalimide, pyrogallol, salicylic acid, terephthalic acid and thymol.*

and the following elemental and inorganic substances for which the process is suitable include:

> *aluminium chloride, arsenic, arsenic(III) oxide, calcium, chromium(III) chloride, hafnium tetrachloride, iodine, iron(III) chloride, magnesium, molybdenum trioxide, sulphur, titanium tetrachloride, uranium hexafluoride and zirconium tetrachloride.*

In addition, the sublimation of ice in freeze-drying, discussed in Chapter 7, has become an important operation particularly in the biological and food industries. The various industrial applications of sublimation techniques are discussed by several authors [3,40,95–97], and the principles underlying vaporisation and condensation [98] and the techniques for growing crystals from the vapour phase [99] are also presented in the literature.

6.5.2 Fundamentals
Phase equilibria

A sublimation process is controlled primarily by the conditions under which phase equilibria occur in a single-component system, and the phase diagram of a simple one-component system is shown in Fig. 6.30 where the *sublimation curve* is dependent on the vapour pressure of the solid, the *vaporisation curve* on the vapour

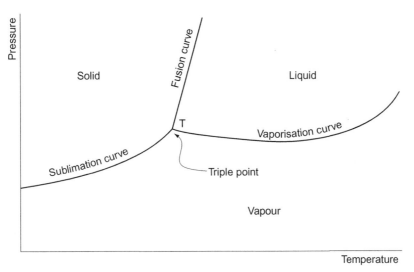

FIG. 6.30

Phase diagram for a single-component system.

pressure of the liquid, and the *fusion curve* on the effect of pressure on the melting point. The slopes of these three curves can be expressed quantitatively by the Clapeyron equation:

$$(dP/dT)_{sup} = \lambda_s/T(v_g - v_s) \qquad (6.45)$$

$$(dP/dT)_{vap} = \lambda_v/T(v_g - v_l) \qquad (6.46)$$

$$(dP/dT)_{fus} = \lambda_f/T(v_l - v_s) \qquad (6.47)$$

where P is the vapour pressure, and v_s, v_l, and v_g are the molar volumes of the solid, liquid, and gas phases, respectively. The molar latent heats (enthalpies) of sublimation, vaporisation, and fusion (λ_s, λ_v, and λ_f, respectively) are related at a given temperature by:

$$\lambda_s = \lambda_v + \lambda_f \qquad (6.48)$$

Although there are few data available on sublimation-desublimation, a considerable amount of information can be calculated using the Clausius–Clapeyron equation provided that information on vapour pressure is available at two or more temperatures. In this way:

$$\ln \frac{P_1}{P_2} = \frac{\lambda'_v}{R} \left(\frac{1}{T_2} - \frac{1}{T_1} \right) \qquad (6.49)$$

where λ'_v is latent heat of vaporisation per mole.

Example 6.7

The vapour pressures of naphthalene at 463 and 433 K are 0.780 and 0.220 kN/m^2 respectively. If λ'_v does not vary greatly over the temperature range considered, what is the vapour pressure at 393 K?

Solution

In Eq. (6.49):

$$\ln(780/220) = [\lambda'_v(463 - 433)]/[8314 \times 463 \times 433]$$

and

$$\lambda'_v = 70,340 \, \text{kJ/kmol}.$$

Thus:

$$\ln(220/P) = [70,340(433 - 393)]/[8.314 \times 433 \times 393]$$

and

$$P = \underline{\underline{30 \, \text{kN/m}^2}}.$$

The position of the *triple point T*, which represents the temperature and pressure at which the solid, liquid, and gas phases co-exist in equilibrium, is of the utmost importance in sublimation-desublimation processes. If it occurs at a pressure above atmospheric, the solid cannot melt under normal atmospheric conditions, and true sublimation (solid → vapour) is easily achieved. For example, since the triple point for carbon dioxide is at 216 K and 500 kN/m^2, liquid CO_2 is not formed when solid CO_2 is heated at atmospheric pressure and the solid simply vaporises. If the triple point occurs at a pressure less than atmospheric, certain precautions are necessary if the phase changes solid → vapour and vapour → solid are to be controlled. For example, since the triple point for water is 273.21 K and 0.6 kN/m^2, ice melts when it is heated above 273.2 K at atmospheric pressure. For ice to sublime, both the temperature and the pressure must be kept below the triple-point values. If the solid → liquid stage takes place before vaporisation, the operation is often called *pseudo-sublimation*. Both true sublimation and pseudo-sublimation cycles are depicted in Fig. 6.31. For a substance with a triple point at a pressure greater than atmospheric, true sublimation occurs. The solid at A is heated to a temperature B and the increase in vapour pressure is given by AB. The condensation is given by $BCDE$. Since the vapour passing to the condenser may cool slightly and be diluted with an inert gas such as air, C can be taken as the condition at the condenser inlet. After entering the condenser, the vapour is mixed with more inert gas, and the partial pressure and temperature drop to D. The vapour then cools at essentially constant pressure to E which is the condenser temperature. When the triple point of the substance occurs at a pressure lower than atmospheric, heating may result in the temperature and vapour pressure of the solid exceeding the values at the triple-point, and the solid will then melt in

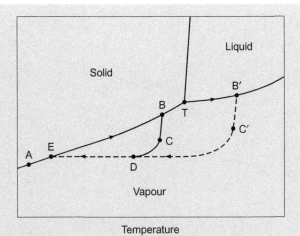

Solid

Liquid

B′

B

T

C′

A

E

C

D

Vapour

Temperature

FIG. 6.31

Phase diagram showing true sublimation (*ABCDE*) and pseudosublimation (*AB′C′DE*) cycles [94].

the vaporiser along *AB′*. In the condensation stage, the partial pressure in the vapour stream entering the condenser must be reduced below the pressure at the triple-point to prevent initial condensation to a liquid by diluting the vapour with an inert gas, although the frictional pressure drop in the vapour line is often sufficient to effect the required drop in partial pressure. *C′* then represents the conditions at the entry into the condenser and the condensation path is *C′DE*.

Fractional sublimation. If two or more sublimable substances form true solid solutions, their separation by fractional sublimation is theoretically possible. The phase diagram for a binary solid-solution system at a pressure below the triple-point pressures of the two components is shown in Fig. 6.32, where points *A* and *B* represent the equilibrium sublimation temperatures of pure components **A** and **B**, respectively, at a given pressure. The lower curve represents the sublimation temperatures of mixtures of **A** and **B**, while the upper curve represents the solid-phase condensation temperatures, generally called *snow points*. Fig. 6.33 shows that if a solid solution at *S* is heated to some temperature *X*, the resulting vapour phase at *Y* and residual solid solution at *Z* contain different proportions of the original components, quantified by the *lever arm rule*. The sublimate and the residual solid may then each be subjected again to this procedure and, therefore, the possibility of fractionation exists, although the practical difficulties may be considerable. Gillot and Goldberger [100], Vitovec et al. [101], and Eggers et al. [102] have described experimental studies of fractional sublimation, and nucleation and growth rates, of organic condensates. Matsouka et al. [103] have also applied the procedure to the fractionation of mixed vapours.

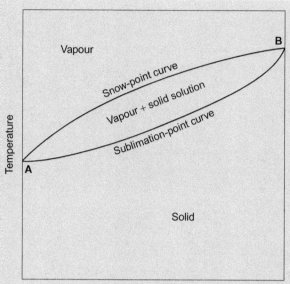

FIG. 6.32

Phase diagram for a two-component solid-solution system at a pressure below the triple points of the two components **A** and **B**.

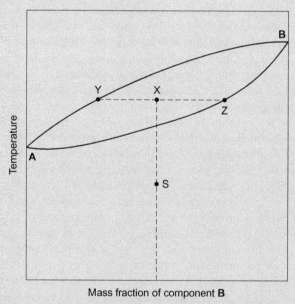

FIG. 6.33

Fractional sublimation of two components **A** and **B** demonstrated on a solid-solution phase diagram.

Vaporisation and condensation

Vaporisation. The maximum theoretical vaporisation rate v' (kg/m^2 s) from the surface of a pure liquid or solid is limited by its vapour pressure and is given by the Hertz–Knudsen equation [104], which can be derived from the kinetic theory of gases:

$$v' = P_s(M'/2\pi \mathbf{R} T_s)^{0.5} \tag{6.50}$$

where P_s is the vapour pressure at the surface temperature T_s, M' is the molecular weight and $\mathbf{R}$ is the gas constant. In practice, the actual vaporisation rate may be lower than predicted by Eq. (6.50), and a correction factor e, generally referred to as an evaporation coefficient, is included to give:

$$v' = e P_s (M'/2\pi \mathbf{R} T_s)^{0.5} \tag{6.51}$$

A laboratory technique used to measure values of e for sublimable solid materials is described by Plewes and Klassen [105].

Sublimation rates of pure solids into turbulent air streams have been successfully correlated by the Gilliland–Sherwood equation [102]:

$$d'/x' = 0.023 \, Re^{0.38} Sc^{0.44} \tag{6.52}$$

where d' *is* a characteristic dimension of the vaporisation chamber, x' is the effective film thickness for mass transfer at the vapour–solid interface, and Re and Sc are the dimensionless Reynolds and Schmidt numbers, respectively.

Condensation is generally a transient operation in which, as discussed by Ueda and Takashima [106], simultaneous heat and mass transfer are further complicated by the effects of spontaneous condensation in the bulk gaseous phase. After the creation of supersaturation in the vapour phase, nucleation normally occurs which may be homogeneous in special circumstances, but more usually heterogeneous. This process is followed by both crystal growth and agglomeration which lead to the formation of the final crystal product. As a rate process, the condensation of solids from vapours is less well understood than vaporisation [98]. Strickland-Constable [107] has described a simple laboratory technique for measuring kinetics in subliming systems which has been used to compare the rates of solid evaporation and crystal growth of benzophenone under comparable conditions. The two most common ways of creating the supersaturation necessary for crystal nucleation and subsequent growth are: (a) cooling by a metal surface to give either a glassy or multi-crystalline deposit that requires mechanical removal and (b) dilution with an inert gas to produce a loose crystalline mass that is easy to handle. Bilik and Krupiczka [108] have described the measurement and correlation of heat transfer rates during condensation of phthalic anhydride in a pilot plant connected to an industrial desublimation unit, Ciborowski and Wrenski [109] have reported on the condensation of several sublimable materials in a fluidised bed, and Knuth and Weinspach [110] have summarised an extensive study on heat- and mass-transfer processes in a fluidised-bed desublimation unit.

6.5.3 Sublimation processes

Simple and vacuum sublimation

Simple sublimation is a batch-wise process in which the solid material is vaporised and then diffuses towards a condenser under the action of a driving force attributable to difference in partial pressures at the vaporising and condensing surfaces. The vapour path between the vaporiser and the condenser should be as short as possible in order to reduce mass-transfer resistance. Simple sublimation has been used for centuries, often in very crude equipment, for the commercial production of ammonium chloride, iodine, and flowers of sulphur.

Vacuum sublimation is a development of simple sublimation, which is particularly useful if the pressure at the triple-point is lower than atmospheric, where the transfer of vapour from the vaporiser to the condenser is enhanced by the increased driving force attributable to the lower pressure in the condenser. Iodine, pyrogallol, and many metals have been purified by vacuum sublimation processes in which the exit gases from the condenser are usually passed through a cyclone or scrubber to protect the vacuum equipment and to minimise product loss.

Entrainer sublimation

In entrainer sublimation, an entrainer gas is blown into the vaporisation chamber of a sublimer in order to increase the vapour flow rate to the condensing equipment, thereby increasing the yield. Air is the most commonly used entrainer, though superheated steam can be employed for substances such as anthracene that are relatively insoluble in water. If steam is used, the vapour may be cooled and condensed by direct contact with a spray of cold water. Although the recovery of the sublimate is efficient, the product is wet. The use of an entrainer gas in a sublimation process also provides the heat needed for sublimation and an efficient means of temperature control. If necessary, it may also provide dilution for the fractional condensation at the desublimation stage. Entrainer sublimation, whether by gas flow over a static bed of solid particles or through a fluidised bed, is ideally suited to continuous operation.

A general-purpose, continuous entrainer–sublimation plant is shown in Fig. 6.34. The impure feedstock is pulverised in a mill and, if necessary, a suitable entrainer

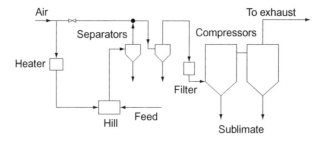

FIG. 6.34

General-purpose continuous sublimation unit [3].

gas, such as hot air, is used to blow the fine particles, which volatilise readily, into a series of separators, such as cyclones, where nonvolatile solid impurities are removed. A filter may also be located in the vapour line to remove final traces of inert, solid impurities. The vapour then passes to a series of condensers from which the sublimate is subsequently discharged. The exhaust gases may be recycled, or passed to the atmosphere through a cyclone or wet scrubber to recover any entrained product.

Although the application of fluidisation techniques to sublimation–desublimation processes was first proposed by Matz [111], the technique has not yet been widely adopted for large-scale commercial use, despite its obvious advantage of improving both heat and mass transfer rates. Cedro [112] has, however, reported on a fluidised-bed de-sublimation unit operating in the United States for the production of aluminium chloride at the rate of 3 kg/s (11 t/h).

The *product yield* from an entrainer–sublimation process may be estimated as follows. The mass flow rate G' of the inert gas and the mass sublimation rate S' are related by:

$$\frac{G'}{S'} = \frac{\rho_g P_g}{\rho_g P'_s} \tag{6.53}$$

where P_g and P'_s are the partial pressures of the inert gas and vaporised material, respectively, in the vapour stream, and ρ_g and ρ_s are their respective vapour densities. The total pressure P_t of the system is the sum of the partial pressures of the components or:

From Eq. (6.53):

$$P_t = P_g + P'_s \tag{6.54}$$

$$S' = G' \left(\frac{\rho_s}{\rho_g}\right) \left(\frac{P'_s}{P_t - P'_s}\right) \tag{6.55}$$

or, in terms of the molecular weights of the inert gas, M_g and of the material being sublimed, M_s:

$$S' = G' \left(\frac{M_s}{M_g}\right) \left(\frac{P'_s}{P_t - P'_s}\right) \tag{6.56}$$

The theoretical maximum yield from an entrainer sublimation process is the difference between the calculated sublimation rates corresponding to the conditions in the vaporisation and condensation stages.

Example 6.8

Salicylic acid ($M_s = 138$ kg/kmol) is to be purified by entrainer sublimation with air ($M_g = 29$ kg/kmol) at 423 K. The vapour is fed at 101.5 kN/m² to a series of condensers, the internal temperature and pressure of the last condenser being 313 K and 100 kN/m², respectively. The air flow rate is

0.56 kg/s and the pressure drop between the vaporiser and the last condenser is 1.5 kN/m^2. The vapour pressures of salicylic acid at 423 and 313 K are 1.44 and 0.0023 kN/m^2, respectively. What are the mass sublimation rates in the vaporiser and condenser?

Solution

Under saturated conditions:

Vaporisation stage: $P_t = 101.5$ kN/m^2, $P_s' = 1.44$ kN/m^2.
Thus, in Eq. (6.56):

$$S_v' = 0.56(138/29)(1.44/(101.5 - 1.144)) = 0.038 \text{ kg/s } (38 \text{ g/s})$$

Condensation stage: $P_t = 100$ kN/m^2, $P_s' = 0.0023$ kN/m^2.
Thus, in Eq. (6.56):

$$S' = 0.56(138/29)(00023/(100 - 0.0023)) = \underline{0.000061 \text{ kg/s } (0.061 \text{ g/s})}.$$

In this example, the loss from the condenser exit gases is only 0.061 g/s while the theoretical maximum yield is 38 g/s. This maximum yield is obtained, however, only if the air is saturated with salicylic acid vapour at 423 K, and saturation is approached only if the air and salicylic acid are in contact for a sufficiently long period of time at the required temperature. A fluidised-bed vaporiser may achieve these optimum conditions though, if air is simply blown over bins or trays containing the solid, saturation will not be achieved and the actual rate of sublimation will be lower than that calculated. In some cases, the degree of saturation achieved may be as low as 10% of the calculated value. The actual loss of product in the exit gases from the condenser is then greater than the calculated value. There are other losses which can be minimised by using an efficient scrubber.

Comparison of entrainer-sublimation and crystallisation

An analysis by Kudela and Sampson [97] of the processes for commercial purification of naphthalene suggests that sublimation is potentially more economical than conventional melt crystallisation. In the sublimation method, the feedstock is completely vaporised in a nitrogen stream and then partially condensed. Heat is removed by vaporising water at the top of the condenser and, in order to prevent deposition of sublimate on the vessel walls, the inner wall of the condenser is sufficiently permeable to allow it to pass some of the entrainer gas. Impurities remain in the vapour stream and are subsequently condensed in a cooler located after the compressor used to circulate the entrainer. The stream carrying impurities and wastewater from the separator is washed with benzene in an extractor. Sublimation gives a higher yield of naphthalene at a lower cost, and with a smaller space requirement, than crystallisation [94]. Although steam and electricity consumption is higher for sublimation, this is offset by a much lower cooling-water requirement.

Fractional sublimation

As discussed in Section 6.5.2, the separation of two or more sublimable substances by fractional sublimation is theoretically possible if the substances form true solid solutions. Gillot and Goldberger [100] have reported the development of a laboratory-scale process known as thin-film fractional sublimation which has been applied successfully to the separation of volatile solid mixtures such as hafnium and zirconium tetrachlorides, 1,4-dibromobenzene and 1-bromo-4-chlorobenzene, and anthracene and carbazole. A stream of inert, non-volatile solids fed to the top of a vertical fractionation column falls countercurrently to the rising supersaturated vapour which is mixed with an entrainer gas. The temperature of the incoming solids is maintained well below the snow-point temperature of the vapour, and thus the solids become coated with a thin film (10 μm) of sublimate which acts as a reflux for the enriching section of the column above the feed entry point.

6.5.4 **Sublimation equipment**

Very few standard forms of sublimation or de-sublimation equipment are in common use and most industrial units, particularly on the condensation side of the process, have been developed on an ad hoc basis for a specific substance and purpose. The most useful source of information on sublimation equipment is the patent literature, although as Holden and Bryant [95] and Kemp et al. [96] point out, it is not clear whether a process has been, or even can be, put into practice.

Vaporisers

A variety of types of vaporisation units has been used or proposed for large-scale operation [95], the design depending on the manner in which the solid feedstock is to be vaporised. These include:

(a) a bed of dry solids without entrainer gas,
(b) dry solids suspended in a dense non-volatile, liquid;
(c) solids suspended in a boiling (entrainer) liquid where the entrainer vapour is formed in situ;
(d) entrainer gas flowing through a fixed bed of solid particles;
(e) entrainer gas bubbling through molten feedstock such that vaporisation takes place above the triple-point pressure;
(f) entrainer gas flowing through a dense phase of solid particles in a fluidised bed;
(g) entrainer gas flowing through a dilute phase of solid particles, such as in a transfer-line vaporiser where the solid and gas phases are in co-current flow, or in a raining solids unit where the solids and entrainer may be in countercurrent flow.

Condensers

Sublimate condensers are usually large, air-cooled chambers which tend to have very low heat-transfer coefficients (5–10 W/m² K) because sublimate deposits on the condenser walls act as an insulator, and vapour velocities in the chambers are generally very low. Quenching the vapour with cold air in the chamber may increase the rate of heat removal although excessive nucleation is likely and the product crystals will be very small. Condenser walls may be kept free of solid by using internal scrapers, brushes, and other devices, and all vapour lines in sublimation units should be of large diameter, be adequately insulated, and if necessary, be provided with supplementary heating to minimise blockage due to the buildup of sublimate. One of the main hazards of air-entrainment sublimation is the risk of explosion since many solids that are considered safe in their normal state can form explosive mixtures with air. All electrical equipment should therefore be flame-proof, and all parts of the plant should be efficiently earthed to avoid build-up of static electricity.

The method of calculating the density of deposited layers of sublimate and of other variables and the optimisation of sublimate condenser design, has been discussed by Wintermantel et al. [113] It is generally assumed that the growth rate of sublimate layers is governed mainly by heat and mass transfer. The model which is based on conditions in the diffusion boundary layer takes account of factors such as growth rate, mass transfer, and concentrations in the gas. The model shows a reasonably good fit to experimental data.

In a variant of the large-chamber de-sublimation condenser, the crystallisation chamber may be fitted with gas-permeable walls as described by Vitovec et al. [101]. The vapour and the entrainer gas are cooled by evaporation of water dispersed in the pores of the walls, and an inert gas passes through the porous walls into the cooling space and protects the internal walls from solid deposits. Crystallisation takes place in the bulk vapour–gas mixture as a result of direct contact with the dispersed water. This arrangement has been used, for example, for the partial separation of a mixture of phthalic anhydride and naphthalene by using nitrogen as the entrainer. Although fluidised-bed condensers have been considered for large-scale application, most of the published reports are concerned with laboratory-scale investigations [110].

6.6 Fractional crystallisation

A single crystallisation operation performed on a solution or a melt may fail to produce a pure crystalline product for a variety of reasons including:

(a) the impurity may have solubility characteristics similar to those of the desired pure component, and both substances consequently co-crystallise,
(b) the impurity may be present in such large amounts that the crystals inevitably become contaminated, and
(c) a pure substance cannot be produced in a single crystallisation stage if the impurity and the required substance form a solid solution.

Re-crystallisation from a solution or a melt is, therefore, widely employed to increase crystal purity.

Example 6.9

Explain how fractional crystallisation may be applied to a mixture of sodium chloride and sodium nitrate given the following data. At 293 K, the solubility of sodium chloride is 36 kg/100 kg water and of sodium nitrate 88 kg/100 kg water. While at this temperature, a saturated solution comprising both salts will contain 25 kg sodium chloride and 59 kg of sodium nitrate per 100 kg of water. At 373 K, these values, again per 100 kg of water, are 40 and 176 and 17 and 160, respectively.

Solution

The data enable a plot of kg NaCl/100 kg of water to be drawn against kg $NaNO_3$/100 kg of water as shown in Fig. 6.35. On the diagram, points C and E represent solutions saturated with respect to both NaCl and $NaNO_3$ at 293 and 373 K, respectively. Fractional crystallisation may then be applied to this system as follows:

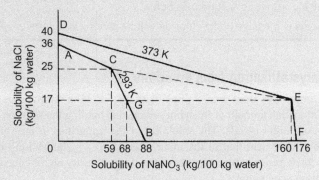

FIG. 6.35

Effect of sodium chloride on the solubility of sodium nitrate.

(a) A solution saturated with both NaCl and $NaNO_3$ is made up at 373 K. This is represented by point E, and, on the basis of 100 kg water, this contains 17 kg NaCl and 160 kg $NaNO_3$.

(b) The solution is separated from any residual sold and then cooled to 293 K. In so doing, the composition of the solution moves along EG.

(c) Point G lies on CB which represents solutions saturated with $NaNO_3$ but not with NaCl. Thus the solution still contains 17 kg NaCl and in addition is saturated with 68 kg $NaNO_3$. That is $(168 - 68) = 92$ kg of pure $NaNO_3$ crystals have come out of solution and this may be drained and washed.

In this way, relatively pure $NaNO_3$, depending on the choice of conditions and particle size, has been separated from a mixture of $NaNO_3$ and NaCl.

The amount of $NaNO_3$ recovered from the saturated solution at 373 K is:

$$(92 \times 100)/160 = \underline{57.5\%.}$$

An alternative approach is to note that points C and B represent 59 and 88 kg $NaNO_3$/100 kg water and assuming CB to be a straight line, then by similar triangles:

concentration of $NaNO_3 = 59 + [(88 - 59)\,(25 - 17)]/24$

$$= 68.3\,kg/100\,kg\ water$$

and

yield of $NaNO_3 = (160 - 68.3) = \underline{91.7\,kg/100\,kg\ water.}$

while all the sodium chloride remains in solution.

If the cycle is then repeated, during the evaporation stage the sodium chloride is precipitated (and removed!) while the concentration of the nitrate re-attains 160 kg/100 kg water. On cooling again, the amount of sodium nitrate which crystallises out is 91.7 kg/100 kg water, or: $(91.7 \times 100)/160 = \underline{57.3\%\ of\ the\ nitrate\ in\ solution}$, as before.

The same percentage of the chloride will be precipitated on re-evaporation.

6.6.1 Recrystallisation from solutions

Most of the impurities from a crystalline mass can often be removed by dissolving the crystals in a small amount of fresh hot solvent and cooling the solution to produce a fresh crop of purer crystals. The solubility of the impurities in the solvent must, however, be greater than that of the main product. *Re*-crystallisation may have to be repeated many times before crystals of the desired purity are obtained. A simple recrystallisation scheme is:

$$
\begin{array}{ccccccc}
S & & S & & & & \\
\downarrow & & \downarrow & & & & \\
\mathbf{AB} & \rightarrow & X_1 & \rightarrow & X_2 & \rightarrow & X_3 \\
& & \downarrow & & \downarrow & & \downarrow \\
& & L_1 & & L_2 & & L_3
\end{array}
$$

An impure crystalline mass **AB**, where **A** is the less soluble, desired component, is dissolved in the minimum amount of hot solvent **S** and then cooled. The first crop of crystals X_1 will contain less impurity **B** than the original mixture, and **B** is concentrated in the liquor L_1. To achieve a higher degree of crystal purity, the procedure can be repeated. In each stage of such a sequence, losses of the desired component **A** can be considerable, and the final amount of 'pure' crystals may easily be a small fraction of the starting mixture **AB**. Many schemes have been designed to increase both the yield and the separation efficiency of fractional re-crystallisation. The choice of solvent depends on the characteristics of the required substance **A** and the impurity **B**. Ideally, **B** should be very soluble in the solvent at the lowest temperature employed and A should have a high temperature coefficient of solubility, so that high yields of **A** can be obtained from operation within a small temperature range.

6.6.2 Recrystallisation from melts

Schemes for recrystallisation from melts are similar to those for solutions, although a solvent is not normally added. Usually, simple sequences of heating (melting) and cooling (partial crystallisation) are followed by separation of the purified crystals from the residual melt. Selected melt fractions may be mixed at intervals according to the type of scheme employed, and fresh feed-stock may be added at different stages if necessary. As Bailey [114] reports, several such schemes have been proposed for purification of fats and waxes.

As described in Section 6.2.1, eutectic systems can be purified in theory by single-stage crystallisation, whereas solid solutions always require multistage operations. Countercurrent fractional crystallisation processes in column crystallisers are described in Section 6.4.3.

6.6.3 Recrystallisation schemes

A number of fractional crystallisation schemes have been devised by Mullin [3] and Gordon et al. [115], and the use of such schemes has been discussed by Joy and Payne [116] and Salutsky and Sites [117].

6.7 Freeze crystallisation

Crystallisation by freezing, or *freeze crystallisation*, is a process in which heat is removed from a solution to form crystals of the solvent rather than of the solute. This is followed by separation of crystals from the concentrated solution, washing the crystals with near-pure solvent, and finally melting the crystals to produce virtually pure solvent. The product of freeze crystallisation can be either the melted crystals, as in water desalination, or the concentrated solution, as in the concentration of fruit juice or coffee extracts. Freeze crystallisation is applicable in principle to a variety of solvents and solutions although, because it is most commonly applied to aqueous systems, the following comments refer exclusively to the freezing of water.

One of the more obvious advantages of freezing over evaporation for removal of water from solutions is the potential for saving heat energy resulting from the fact that the enthalpy of crystallisation of ice, 334 kJ/kg, is only one-seventh of the enthalpy of vaporisation of water, 2260 kJ/kg, although it has to be acknowledged that the cost of producing 'cold' is many times more than the cost of producing 'heat'. Process energy consumption may be reduced below that predicted by the phase-change enthalpy, however, by utilising energy recycle methods, such as multiple-effect or vapour compression, as commonly employed in evaporation as discussed in Chapter 5. In freeze-crystallisation plants operating by direct heat exchange, vapour compression has been used to recover refrigeration energy by using the crystals to condense the refrigerant evaporated in the crystalliser. Another advantage of freeze crystallisation, important in many food applications, is that the volatile flavour components normally lost during conventional evaporation can be retained in the freeze-concentrated product. Despite earlier enthusiasm, large-scale applications in desalination, effluent treatment, dilute liquor concentration and solvent recovery and so on have not been developed as yet.

All freeze separation processes depend on the formation of pure solvent crystals from solution, as described for eutectic systems in Section 6.2.1. which allows single-stage operation. Solid-solution systems, requiring multistage-operation, are not usually economic. Several types of freeze crystallisation processes may be designated according to the kind of refrigeration system used as follows:.

(a) In *indirect-contact freezing*, the liquid feedstock is crystallised in a scraped-surface heat exchanger as described in Section 6.4, fitted with internal rotating scraper blades and an external heat-transfer jacket through which a liquid refrigerant is passed. The resulting ice-brine slurry passes to a wash column where the ice crystals are separated and washed before melting. van Pelt and van Nistelrooy [118] have described one of the commercial systems which are based on this type of freezing process.

(b) *Direct-contact freezing* processes utilise inert, immiscible refrigerants and are suitable for desalination. A typical scheme taken from Barduhn [64] is shown in Fig. 6.36. Sea-water, at a temperature close to its freezing point, is fed continuously into the crystallisation vessel where it comes in direct contact with

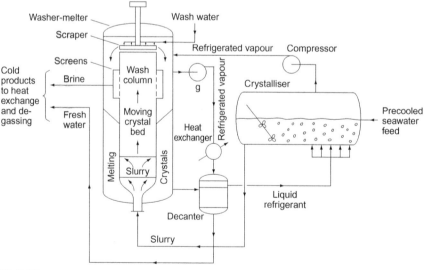

FIG. 6.36

Desalination of seawater by freezing [64].

a liquid refrigerant such as n-butane which vaporises and causes ice crystals to form due to the exchange of latent heat. The ice-brine slurry is fed to a wash column where it is washed countercurrently with fresh water. The emerging brine-free ice is melted by the enthalpy of the condensation of the vapour released from the compressed refrigerant. A major part of the energy input is that required for the compressors.

(c) *Vacuum freezing processes* do not require a conventional heat exchanger, and the problems of scale formation on heat-transfer surfaces are avoided. Cooling is effected by flash evaporating some of the solvent as the liquid feedstock enters a crystallisation vessel maintained at reduced pressure. Although vacuum freezing is potentially attractive for aqueous systems it has not, as yet, achieved widespread commercial success.

Thijssen and Spicer [119] has given a general review of freeze concentration as an industrial separation process and Bushnell and Eagen [63] have discussed the status of freeze desalination. The potential of freeze crystallisation in the recycling and re-use of wastewater has been reviewed by Heist [120], and the kinetics of ice crystallisation in aqueous sugar solutions and fruit juice are considered by Omran and King [121].

6.8 **High-pressure crystallisation**

As noted previously, high-pressure crystallisation in which an impure liquid feedstock is subjected to pressures of up to 300 MN/m^2 in a relatively small chamber, 0.001 m^3 in volume, under adiabatic conditions is a relatively recent development. As the pressure and temperature of the charge increase, fractional crystallisation takes place and the impurities are concentrated in the liquid phase which is then discharged from the pressure chamber. At the end of the cycle, further purification is possible since residual impurities in the compressed crystalline plug may then be 'sweated out' when the pressure is released. Moritoki [91] has claimed that a single-cycle operation lasting less than 300 s is capable of substantially purifying a wide range of organic binary melt systems.

The principle of operation is illustrated in Fig. 6.37 which shows the pressure-volume relationship. Curve a shows the phase change of a pure liquid as it is pressurised isothermally. Crystallisation begins at point A_1 and proceeds by compression without any pressure change until it is complete at point A_2. Beyond this point, the solid phase is compressed resulting in a very sharp rise in pressure. If the liquid contains impurities, these nucleate at point B_1. As the crystallisation of the pure substance progresses, the impurities are concentrated in the liquid phase and a higher pressure is required to continue the crystallisation process. As a result, the equilibrium pressure of the liquid–solid system rises exponentially with increase of the solid fraction, as shown by curve b which finally approaches the solidus curve. A liquid–solid equilibrium line in terms of pressure and temperature is shown in

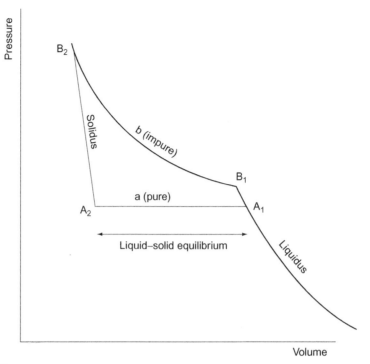

FIG. 6.37

Relationship between pressure and volume for isothermal conditions.

Fig. 6.38A. The liquid–solid equilibrium line moves from line a to line b with increase in impurities and line c represents the liquid–solid equilibrium for eutectic composition. On the industrial scale, a liquid is adiabatically compressed first from point A to point B, accompanied by heat generation and then to point C at which nucleation occurs accompanied by a temperature rise due to the release of the latent heat. Again, it is during this step that the impurities are concentrated into the mother liquor. At this stage, the liquid is separated from the solid phase and removed from the vessel at point C which is at a slightly lower pressure than the eutectic line. When the greater part of the liquid has been removed, its pressure decreases at first gradually and then rapidly to atmospheric pressure while the crystals are maintained at the initial separation pressure. In this way, the crystals are compacted and their surfaces are purified by slight melting, or by the so-called 'sweating' phenomenon. After the separation at point D, the crystals are highly purified and the line representing the equilibrium state gradually approaches line a. The basic pattern of operation as a function of time is shown in Fig. 6.38B.

Pilot scale investigations by Kobe Steel have shown, for example, that the impurity level in a feed of mesitylene is reduced from 0.52% to 0.002% in a single operating cycle at $15 \, MN/m^2$ and a concentration of greater than 99% p-xylene is

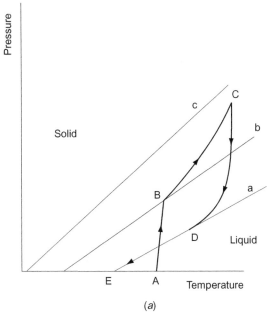

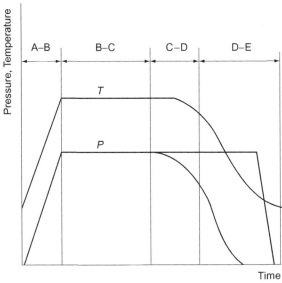

A–B–C: Pressurising step
C: Liquid-phase discharge
C–D–E: Solid phase compaction/sweating
         step

(*b*)

FIG. 6.38

High-pressure crystallisation: (A) adiabatic application of pressure and (B) pressure and temperature variation during a cycle.

obtained from a mixture of p-xylene and mesitylene containing 80% p-xylene. It has also been shown that while crystals of cumenealdehyde are very difficult to obtain by cooling, nucleation and crystal growth occur at pressures of $50–70\,MN/m^2$ and, where the crystals obtained are then used as seed material, cumenealdehyde is then easily crystallised and purified at pressures below $20\,MN/m^2$. In this work, even though the capacity of the pilot unit was only $0.0015\,m^3$, some 360 tonne/year of raw material could be processed in 120 s cycles over a period of 8000 h.

Kobe Steel claim that, in terms of running costs, not only is the energy consumption low, being 10%–50% compared with conventional processes, but high-pressure operation is ideally suited to the separation of isomers which are difficult to purify by other processes. These substances may have close boiling points or may be easily decomposed by temperature elevation. In this respect, recent work on supercritical fluids, as described by Poloakoff et al. [122], is of great importance.

As discussed in Chapter 14, supercritical fluids are gases that are compressed until their densities are close to those of liquids. They are extremely non-ideal gases in which interactions between molecules of a supercritical fluid and a potential solute can provide a 'solvation energy' for many solids to dissolve. The higher the pressure and hence the density of the supercritical fluid, the greater are the solvent–solute interactions and hence the higher the solubility of the solid. In other words the solvent power of a supercritical fluid is 'tunable' and this is the key factor in the use of supercritical fluids for a wide range of processes. As discussed in Chapter 14, supercritical fluids in common use include ethene, ethane and propane together with supercritical carbon dioxide and these have all been widely discussed in the literature [123–127]. Although supercritical fluids have considerable advantages in the field of process intensification and can also replace environmentally undesirable solvents and indeed most organics, their most important property as far as crystallisation is concerned is that they can be tuned to dissolve the desired product, or indeed any impurities, which are then separated by crystallisation at high pressure. It is in this way that crystallisation is moving from the simple production of hydrates from salt solutions towards a fully fledged separation technique which has and will have many advantages in comparison with more traditional operations in the years to come.

References

[1] J.W. Mullin, Crystallization, in: Kirk-Othmer: Encyclopedia of Chemical Technology, third ed., vol. 7, John Wiley & Sons, New York, 1979.

[2] J.W. Mullin, Crystallization and precipitation, in: Ullmann's Encyclopedia of Industrial Chemistry, vol. B2, VCH Verlagsgesellschaft mbH, Weinheim, 1988.

[3] J.W. Mullin, Crystallization, fourth ed., Butterworth-Heinemann, Oxford, 2001.

[4] H. Feilchenfeld, S. Sarig, Calcium chloride hexahydrate: a phase-changing material for energy storage, Ind. Eng. Chem. Process. Prod. Res. Dev. 24 (1985) 130–133.

[5] H. Kimurah, Impurity effect on growth rates of $CaCl2.6H2O$ crystals, J. Cryst. Growth 73 (1985) 53–62.

[6] F. Gronvold, K.K. Meisingset, Thermodynamic properties and phase transitions of salt hydrates between 270 and 400K: NH4Al(SO4)2.12H2O, KAl(SO4)2.12H2O, Al2(SO4)3.17H2O, ZnSO4.7H2O, NaSO4.10H2O and Na2S2O3.5H2O, J. Chem. Thermodyn. 14 (1982) 1083–1098.

[7] H. Kimura, J. Kai, Phase change stability of sodium acetate trihydrate and its mixtures, Sol. Energy 35 (1985) 527–534.

[8] A. Findlay, A.N. Campbell, The Phase Rule and Its Applications, ninth ed., Longman, London, 1951.

[9] J.E. Ricci, The Phase Rule and Heterogeneous Equilibrium, van Nostrand, New York, 1951.

[10] H.R. Null, Phase Equilibrium in Process Design, Wiley-Interscience, New York, 1970.

[11] J. Nyvlt, Solid-Liquid Phase Equilibrium, Academia, Prague, Elsevier, Amsterdam, 1977.

[12] H. Hörmeyer, Calculation of crystallisation equilibria, Germ. Chem. Eng. 6 (1983) 277–281.

[13] C.L. Kusik, H.P. Meissner, E.L. Field, Estimation of phase diagrams and solubilities for aqueous multi-ion systems, AIChE J. 25 (1979) 759–762.

[14] B. Sander, P. Rasmussen, A. Fredenslund, Calculation of solid-liquid equilibria in aqueous solutions of nitrate salts using an extended uniquac equation, Chem. Eng. Sci. 41 (1986) 1197–1202.

[15] G.H. Nancollas, M.M. Reddy, F. Tsai, Crystal growth kinetics of minerals encountered in water treatment processes, J. Cryst. Growth 20 (1973) 125–134.

[16] G.H. Nancollas, M.M. Reddy, The kinetics of crystallisation of scale-forming materials, Soc. Pet. Eng. J. (1974) 117–123.

[17] P.T. Cardew, R.J. Davey, A.J. Ruddick, Kinetics of polymorphic solid-state transformations, J. Chem. Soc., Faraday Trans. 2 80 (1984) 659–668.

[18] P.T. Cardew, R.J. Davey, The kinetics of solvent-mediated phase transformations, Proc. R. Soc. London Ser. A 398 (1985) 415–428.

[19] J.W. Mullin, O. Söhnel, Expressions of supersaturation in crystallisation studies, Chem. Eng. Sci. 32 (1977) 683–686.

[20] W. Ostwald, Über die vermeintliche Isomeric des roten und gelben Quecksilberoxyds und die Oberflaschen spannung fester Körper, Z. Phys. Chem. (Leipzig) 34 (1900) 493–503.

[21] H.A. Miers, F. Isaac, Refractive indices of crystallizing solutions, J. Chem. Soc. 89 (1906) 413–454.

[22] A. Seidell, in: W.F. Linke (Ed.), Solubilities of Inorganic and Metal Organic Compounds, fourth ed., vol. 1, van Nostrand, New York, 1958. vol. 2. Am. Chem. Soc., Washington, DC, 1965.

[23] H. Stephen, T. Stephen, Solubilities of Inorganic and Organic Compounds, Pergamon Press, London, 1963.

[24] M. Broul, J. Nyvlt, O. Söhnel, Solubilities in Binary Aqueous Solutions, Academia, Prague, 1981.

[25] C.G. Moyers, R.W. Rousseau, in: R.W. Rousseau (Ed.), Handbook of Separation Process Technology, John Wiley, New York, 1987.

[26] American Petroleum Institute, Selected Values of Properties of Hydrocarbons Research Project No. 44, American Petroleum Institute, 1976.

[27] R.F. Strickland-Constable, Kinetics and Mechanism of Crystallization, Academic Press, London, 1968.

[28] D. Kashchiev, Nucleation—Basic Theory With Applications, Butterworth-Heinemann, Oxford, 2000.

[29] J. Garside, R.J. Davey, Secondary contact nucleation: kinetics, growth and scale-up, Chem. Eng. Commun. 4 (1980) 393–424.

[30] A.R. Ubbleohde, Melting and Crystal Structure, Oxford University Press, Oxford, 1965.

[31] V.A. Garten, R.B. Head, Philos. Mag. 8 (1963) 1793–1803. V.A. Garten, R.B. Head, Homogeneous nucleation and the phenomenon of crystalloluminescence, Philos. Mag. 14 (1966) 1243–1253.

[32] K.A. Berglund, in: S.J. Jančič, E.J. de Jong (Eds.), Industrial Crystallisation, 84, Elsevier, Amsterdam, 1984.

[33] J.W. Mullin, C.L. Lecl, Evidence of molecular cluster formation in supersaturated solutions of citric acid, Philos. Mag. 19 (1969) 1075–1077.

[34] M.A. Larson, J. Garside, Solute clustering in supersaturated solutions, Chem. Eng. Sci. 41 (1986) 1285–1289.

[35] J. Nyvlt, Industrial Crystallization From Solutions, Butterworth, London, 1971.

[36] J. Nyvlt, O. Söhnel, M. Matuchova, M. Broul, The Kinetics of Industrial Crystallization, Academia, Prague, 1985.

[37] A.D. Randolph, M.A. Larson, Theory of Particulate Processes, Academic Press, New York, 1971.

[38] A.E. Nielsen, O. Söhnel, Interfacial tensions, electrolyte crystal-aqueous solution, from nucleation data, J. Cryst. Growth 11 (1971) 233–242.

[39] O. Söhnel, J.W. Mullin, A method for the determination of precipitation induction periods, J. Cryst. Growth 44 (1978) 377–382.

[40] G. Matz, Kristallisation, second ed., Springer Verlag, Berlin, 1969.

[41] S.J. Jančič, P.A.M. Grootscholten, Industrial Crystallization, Reidel, Dordecht, 1984.

[42] R.A. Laudice, The Growth of Single Crystals, Prentice-Hall, Englewood Cliffs, NJ, 1970.

[43] J. Garside, Industrial crystallisation from solution, Chem. Eng. Sci. 40 (1985) 3–26.

[44] A.E. Nielsen, Electrolyte crystal growth mechanisms, J. Cryst. Growth 67 (1984) 289–310.

[45] B.R. Pamplln (Ed.), Crystal Growth, second ed., Pergamon Press, Oxford, 1980.

[46] E. Kaldis, H.J. Scheel, Crystal Growth and Materials, North-Holland, Amsterdam, 1977.

[47] T. Tengler, A. Mersmann, Influence of temperature, saturation and flow velocity on crystal growth from solutions, Germ. Chem. Eng. 7 (1984) 248–259.

[48] W. Wöhlk, Meszanordnungen zur Bestimmung von Kristallwachstumsgeschwindigkeiten, Fortschr. Ber. VDI Z., Reihe, 3, 1982. 71.

[49] E.T. White, L.L. Bendig, M.A. Larson, Analysis and design of crystallization systems, AIChE Symp. Ser. 72 (153) (1976) 41–47.

[50] A.G. Jones, J.W. Mullin, Programmed cooling crystallisation of potassium sulphate solutions, Chem. Eng. Sci. 29 (1974) 105–118.

[51] A.H. Janse, E.J. de Jong, in: J.W. Mullin (Ed.), Industrial Crystallization, 75, Plenum Publishing, London, 1976.

[52] J. Garside, S.J. Jančič, Measurement and scale-up of secondary nucleation kinetics for the potash alum-water system, AIChE J. 25 (1979) 948–958.

[53] J. Whetstone, The crystal habit modification of inorganic salts with dyes (in two parts), Trans. Faraday Soc. 51 (1955) 973–980. 1142–53.

[54] D.E. Garrett, Industrial crystallisation: influence of chemical environment, Br. Chem. Eng. 4 (1959) 673–677.

[55] G.D. Botsaris, in: S.J. Jančič, E.J. de Jong (Eds.), Industrial Crystallization, 81, North-Holland, Amsterdam, 1982.

[56] R.J. Davey, in: S.J. Jančič, E.J. de Jong (Eds.), Industrial Crystallization, 78, North-Holland, Amsterdam, 1979.

[57] G. Deicha, Lacunes Des Cristeaux et Leurs Inclusions Fluides, Masson, Paris, 1955.

[58] H.E.C. Powers, Sucrose crystals: inclusions and structure, Sugar Technol. Rev. 1 (1970) 85–190.

[59] W.R. Wilcox, V.H.S. Kuo, Nucleation of monosodium urate crystals, J. Cryst. Growth 19 (1973) 221–229.

[60] M. Saska, A.S. Myerson, A crystal-growth model with concentration dependent diffusion, J. Cryst. Growth 67 (1984) 380–382.

[61] L. Phoenix, How trace additives inhibit the caking of inorganic salts, Br. Chem. Eng. 11 (1966) 34–38.

[62] A.J. Armstrong, Scraped surface crystallisers, Chem. Process. Eng. 51 (11) (1970) 59.

[63] J.D. Bushnell, J.F. Eagen, Dewax process produces benefits, Oil Gas J. 73 (42) (1975) 80–84.

[64] A.J. Barduhn, The status of freeze-desalination, Chem. Eng. Prog. 71 (11) (1975) 80–87.

[65] J.W. Mullin, J.R. Williams, A comparison between indirect contact cooling methods for the crystallisation of potassium sulphate, Chem. Eng. Res. Des. 62 (1984) 296–302.

[66] E. Finkelstein, On solar ponds: critique, physical fundamentals and engineering aspects, J. Heat Recovery Syst. 3 (1983) 431–437.

[67] A.W. Bamforth, Industrial Crystallisation, Leonard Hill, London, 1965.

[68] W.C. Saeman, Crystal size distribution in mixed systems, AICHE J. 2 (1956) 107–112.

[69] J.W. Mullin, J. Nyvlt, Programmed cooling of batch crystallizers, Chem. Eng. Sci. 26 (1971) 369–377.

[70] J. Robinson, J.E. Roberts, A mathematical study of crystal growth in a cascade of agitators, Can. J. Chem. Eng. 35 (1957) 105–112.

[71] C.F. Abegg, N.S. Balakrishnam, The tanks-in-series concept as a model for imperfectly mixed crystallizers, AIChE. Symp. Ser. 72 (153) (1976) 88–94.

[72] A.G. Toussaint, A.J.M. Donders, The mixing criterion in crystallisation by cooling, Chem. Eng. Sci. 29 (1974) 237–245.

[73] E.T. White, A.D. Randolph, Graphical solution of the material balance constraint for MSMPR crystallizers, AIChE J. 33 (1984) 686–689.

[74] H. Griffiths, Crystallisation, Trans. Inst. Chem. Eng. 25 (1947) 14–18.

[75] J.W. Mullin, J. Nyvlt, Design of classifying crystalisers, Trans. Inst. Chem. Eng. 48 (1970) 7–14.

[76] A. Mersmann, M. Kind, Modeling of chemical process equipment: the design of crystallizers, Int. Chem. Eng. 29 (1989) 616–626.

[77] E.J. de Jong, Development of crystallizers, Int. Chem. Eng. 24 (1984) 419–431.

[78] R.C. Bennett, H. Fiedelman, A.D. Randolph, Crystallizer influenced nucleation, Chem. Eng. Prog. 69 (1973) 86–93.

[79] J.G.D. Molinari, in: M. Zeif, W.R. Wilcox (Eds.), Fractional Solidification, Dekker, New York, 1967, pp. 393–400.

[80] N.I. Gel'perin, G.A. Nosov, Fractional crystallisation of melts in drum crystallizers, Sov. Chem. Ind. (Eng. Transl.) 9 (1977) 713–717.

[81] K. Ponomarenko, G.F. Potebyna, V.I. Bei, Crystallization of melts on a thin moving wall, Theor. Found. Chem. Eng. 13 (1980) 724–729.

[82] K. Wintermantel, Effective separation power in freezing out layers of crystals from melts and solutions—a unified presentation, Chem. Ing. Tech. 58 (6) (1986) 498–499.

[83] G.R. Atwood, The progressive mode of multistage crystallisation, in: Crystallization From Solutions and Melts, Springer, 1969, pp. 112–121 (in Ref. 42a).

[84] S. Rittner, R. Steiner, Melt crystallisation of organic substances and its large scale application, Chem. Ing. Tech. 57 (2) (1985) 91–102.

[85] O. Fischer, S.J. Jančič, K. Saxer, in: S.J. Jančič, E.J. De Jong (Eds.), Industrial Crystallization, 84, Elsevier, Amsterdam, 1984.

[86] W.C. Gates, J.E. Powers, Determination of the mechanics causing and limiting separation by column crystallization, AIChE J. 16 (1970) 648–657.

[87] J.D. Henry, J.E. Powers, Experimental and theoretical investigation of continuous flow column crystallisation, AIChE J. 16 (1970) 1055–1063.

[88] D.L. McKay, in: F. Zeit, W.R. Wilcox (Eds.), Fractional Solidification, Marcel Dekker, New York, 1967, pp. 427–439.

[89] J.A. Brodie, A continuous multistage melt purification process, Mech. Chem. Trans. Inst. Eng. Aust. 7 (1) (1971) 37–44.

[90] K. Takegami, N. Nakamaru, M. Morita, in: S.J. Jančič, E.J. de Jong (Eds.), Industrial Crystallization, 84, Elsevier, Amsterdam, 1984, pp. 143–146.

[91] M. Moritoki, Crystallization and sweating of p-cresol by application of high pressure, Int. Chem. Eng. 20 (1980) 394–401.

[92] W.H. Shearon, W.B. Dunwoody, Ammonium nitrate, Ind. Eng. Chem. 45 (1953) 496–504.

[93] L.M. Nunelly, F.T. Cartney, Granulation of urea by the falling-curtain process, Ind. Eng. Chem. Prod. Res. Dev. 21 (1982) 617–620.

[94] J.W. Mullin, Sublimation in Ullmann's Encyclopedia of Industrial Chemistry, VCH Verlagsgesellschaft mbH, Weinheim, 1988.

[95] C.A. Holden, H.S. Bryant, Purification by sublimation, Sep. Sci. 4 (1) (1969) 1–13.

[96] S.D. Kemp, Sublimation and vacuum freeze drying, in: H.W. Cremer, T. Davies (Eds.), Chemical Engineering Practice, 6, Butterworth, London, 1958, pp. 567–600.

[97] L. Kudela, M.J. Sampson, Understanding sublimation technology, Chem. Eng. 93 (12) (1986) 93–98.

[98] E. Rutner, P. Goldfinger, J.P. Hirth (Eds.), Condensation and Evaporation of Solids, Gordon and Breach, New York, 1964.

[99] M.M. Faktor, I. Garrett, Growth of Crystals From the Vapour, Chapman and Hall, London, 1974.

[100] J. Gillot, W.M. Goldberger, Separation by thin-film fractional sublimation, Chem. Eng. Prog. Symp. Ser. 65 (91) (1969) 36–42.

[101] J. Vitovec, J. Smolik, K. Kugler, Separation of sublimable compounds by partial crystallisation from their vapour mixture, Collect. Czechoslov. Chem. Commun. 43 (1978) 396–400.

[102] H.H. Eggers, D. Ollmann, D. Heinz, D.W. Drobot, A.W. Nikolajew, Sublimation und desublimation im system AlCl3–FeCl3, Z. Phys. Chem. (Leipzig) 267 (1986) 353–363.

[103] M. Matsouka, Rates of nucleation and growth of organic condensates of mixed vapours, in: S.J. Jančič, E.J. de Jong (Eds.), Industrial Crystallization, 84, Elsevier, Amsterdam, 1984, pp. 357–360.

[104] T.K. Sherwood, C. Johannes, The maximum rate of sublimation of solids, AIChE J. 8 (1962) 590–593.

[105] A.C. Plewes, J. Klassen, Mass transfer rates from the solid to the gas phase, Can. J. Technol. 29 (1951) 322.

[106] H. Ueda, Y. Takashima, Desublimation of two-component systems, J. Chem. Eng. Jpn. 10 (1977) 6–12.

[107] R.F. Strickland-Constable, Kinetics and Mechanism of Crystallization, Academic Press, London, 1968.

[108] R.J. Bilik, R. Krupiczka, Heat transfer in the desublimation of phthalic anhydride, Chem. Eng. J. 26 (1983) 169–180.

[109] J. Ciborowski, S. Wrenski, Condensation of sublimable materials in a fluidized bed, Chem. Eng. Sci. 17 (1962) 481–489.

[110] M. Knuth, P.M. Weinspach, Experimentelle Untersuchung des Wärme und Stoffübergangs an die Partikeln einer Wirbelschicht bei der Desublimation, Chem. Ing. Tech. 48 (1976) 893.

[111] G. Matz, Fleitzbett-sublimation, Chem. Ing. Tech. 30 (1958) 319–329.

[112] V. Cedro, Fluid beds and sublimation, Chem. Eng. 93 (21) (1986) 5.

[113] K. Wintermantel, H. Holzknech, P. Thoma, Density of sublimed layers, Chem. Eng. Technol. 10 (1987) 205–210.

[114] A.E. Bailey, Solidification of Fats and Waxes, Interscience, New York, 1950.

[115] L. Gordon, M.L. Salutsky, H.H. Willard, Precipitation From Homogeneous Solution, Wiley-Interscience, New York, 1959.

[116] E.F. Joy, J.H. Payne, Fractional precipitation or crystallisation systems, Ind. Eng. Chem. 47 (1955) 2157–2161.

[117] M.L. Salutsky, J.G. Sites, Ra-Ba separation process, Ind. Eng. Chem. 47 (1955) 2162–2166.

[118] W.H.S.M. van Pelt, M.G.J. van Nistelrooy, Procedure for the concentration of beer, Food Eng. (1975) 77–79.

[119] H.A.C. Thijssen, in: A. Spicer (Ed.), Advances in Preconcentration and Dehydration of Foods, Wiley-Interscience, New York, 1974.

[120] J.A. Heist, Freeze crystallisation: waste water recycling and reuse, AIChE. Symp. Ser. 77 (209) (1984) 259–272.

[121] A.M. Omran, C.J. King, Kinetics of ice crystallisation in sugar solutions and fruit juices, AIChE J. 20 (1974) 799–801.

[122] M. Poloakoff, N.J. Meehan, S.K. Ross, A supercritical success story, Chem. Ind. 10 (1999) 750–752.

[123] P.G. Jessop, W. Leitner (Eds.), Chemical Synthesis Using Supercritical Fluids, Wiley-VCH, Weinheim, 1999.

[124] M.A. McHugh, V.J. Krukonis, Supercritical Fluid Extraction: Principles and Practice, second ed., Butterworth-Heinemann, Boston, 1994.

[125] J.A. Darr, M. Poliakoff, New directions in inorganic and metal-organic coordination chemistry in supercritical fluids, Chem. Rev. 99 (1999) 495.

[126] P.G. Jessop, T. Ikariya, R. Noyori, Homogeneous catalysis in supercritical fluids, Chem. Rev. 99 (1999) 475.

[127] A. Baiker, Supercritical fluids in heterogeneous catalysis, Chem. Rev. 99 (2) (1999) 453–473.

Further reading

A.E. Bailey, Solidification of Fats and Waxes, Interscience, New York, 1950.

A.W. Bamforth, Industrial Crystallisation, Leonard Hill, London, 1965.

A. Bartolomai (Ed.), Food Factories: Processes, Equipment and Costs, VCH, New York, 1987.

H.E. Buckley, Crystal Growth, Wiley, New York, 1951.

S. Bruln (Ed.), Preconcentration and Drying of Food Materials, Elsevier, New York, 1988.

M.M. Faktor, D.E. Garrett, Growth of Crystals From the Vapour, Chapman and Hall, London, 1974.

A. Findlay, A.N. Campbell, The Phase Rule and Its Applications, ninth ed., Longman, London, 1951.

S.J. Jancic, A.M. Grootscholten, Industrial Crystallisation, Reidel, Dordrecht, 1984.

E. Kaldis, H.J. Scheel, Crystal Growth and Materials, North-Holland, Amsterdam, 1977.

R. Larson, Constitutive Equations for Polymer Melts and Solutions, Butterworth, London, 1988.

W.D. Lawson, S. Nielsen, Preparation of Single Crystals, Butterworths, London, 1958.

Y.A. Lui, H.A. Mcgee, W.R. Epperly (Eds.), Recent Developments in Chemical Process and Plant Design, Wiley, New York, 1987.

G. Matz, Kristallisation, second ed., Springer Verlag, Berlin, 1969.

A. Merstmann (Ed.), Crystallization Technology Handbook, Marcel Dekker, New York, 1995.

C.G. Moyers, R.W. Rousseau, in: R.W. Rousseau (Ed.), Handbook of Separation Process Technology, John Wiley, New York, 1987.

J.W. Mullin, Crystallization, third ed., Butterworth-Heinemann, Oxford, 1993. 1997.

J.W. Mullin, Crystallization, fourth ed., Butterworth-Heinemann, Oxford, 2001.

J.W. Mullin, Crystallisation, in: Kirk-Othmer: Encyclopedia of Chemical Technology, third ed., vol. 7, John Wiley & Sons, New York, 1979.

J.W. Mullin, Crystallisation and precipitation, in: Ullmann's Encyclopedia of Industrial Chemistry, vol. B2, VCH Verlagsgesellschaft mbH, Weinheim, 1988.

A.S. Myerson, Handbook of Industrial Crystallization, second ed., Butterworth-Heinemann, Oxford, 2000.

H.R. Null, Phase Equilibrium in Process Design, Wiley-Interscience, New York, 1970.

J. Nyvlt, Industrial Crystallisation From Solutions, Butterworths, London, 1971.

J. Nyvlt, Solid–Liquid Phase Equilibria, Elsevier-North Holland, New York, 1977.

J. Nyvlt, Industrial Crystallisation, Verlag Chemie, Weinheim, NY, 1978. B.R. Pamplln (Ed.), Crystal Growth, second ed., Pergamon Press, Oxford, 1980.

D. Randolph, M.A. Larson, Theory of Particulate Processes, Academic Press, New York, 1971.

E. Rutner, P. Goldfinger, J.P. Hirth (Eds.), Condensation and Evaporation of Solids, Gordon and Breach, New York, 1964.

P.A. Schweitzer (Ed.), Handbook of Separation Techniques for Chemical Engineers, second ed., McGraw-Hill, New York, 1988.

R.F. Strickland-Constable, Kinetics and Mechanism of Crystallisation, Academic Press, London, 1968.

H.A. Thijssen, in: A. Spicer (Ed.), Advances in Preconcentration and Dehydration of Foods, Wiley-Interscience, New York, 1974.

A.R. Ubbelohde, Melting and Crystal Structure, OUP, Oxford, 1965.

A.G. Walton, The Formation and Properties of Precipitates, Interscience, New York, 1967.

S.M. Walas, Chemical Process Equipment: Selection and Design, Butterworth, London, 1989.

J. Wisniak, Phase Diagrams. Physical Sciences Data, 10, Elsevier-North Holland, New York, 1981.

A.C. Zettlemoyer (Ed.), Nucleation, Marcel Dekker, New York, 1969.

(A) Crystallization from solutions and melts. AIChE Symp. Ser. 65 (1969) (95); (B) Factors affecting size distribution, 67 (1971) (110); (C) Crystallization from solutions: Nucleation phenomena in growing crystal systems, 68 (1972) (121); (D) Analysis and design of crystallisation processes, 72 (1976) (153); (E) Design, control and analysis of crystallisation processes, 76 (1980) (193); (F) Nucleation, growth and impurity effects in crystallisation process engineering, 78 (1982) (215); (G) Advances in crystallisation from solutions, 80 (1984) (240).

Drying

7

Ajay Kumar Ray

*Department of Chemical and Biochemical Engineering, University of Western Ontario,
London, ON, Canada*

Nomenclature

		Units in SI system	Dimensions in M, N, L, T, θ
A	area for heat transfer or evaporation	m^2	L^2
A_N	area of nozzle or jet normal to direction of flow	m^2	L^2
a	surface area per unit volume	m^2/m^3	L^{-1}
B	width of surface	m	L
C	coefficient	$kg^{0.2}/m^{0.4}s^{0.2}$	$M^{0.2}\,L^{-0.4}\,T^{-0.2}$
C_D	coefficient of discharge	$-$	$-$
C_w	water concentration of a point in solid	kg/m^3	ML^{-3}
C_f	heat capacity of wet material	$J/kg\,K$	$L^2\,T^{-2}\,\theta^{-1}$
C_m	mean heat capacity of gas	$J/kg\,K$	$L^2\,T^{-2}\,\theta^{-1}$
C_s	heat capacity of dry solid	$J/kg\,K$	$L^2\,T^{-2}\,\theta^{-1}$
C_x	heat capacity of liquid evaporated	$J/kg\,K$	$L^2\,T^{-2}\,\theta^{-1}$
D	diameter of drum or disc	m	L
D_L	diffusion coefficient (liquid phase)	m^2/s	$L^2\,T^{-1}$
d	particle diameter	m	L
d_j	nozzle or jet diameter	m	L
d_m	main drop diameter	m	L
d_s	surface-mean diameter of drop	m	L

Continued

Coulson and Richardson's Chemical Engineering. https://doi.org/10.1016/B978-0-08-101097-6.00007-9

		Units in SI system	Dimensions in M, N, L, T, θ
d_t	diameter of evaporating drop at time t	m	**L**
d_o	initial diameter of drop	m	**L**
F	wet solid feed rate	kg/s	**MT^{-1}**
F'	volumetric rate of feed per unit cross-section	m^3/sm^2	**LT^{-1}**
f	free moisture content	kg	**M**
f_c	free moisture content at critical condition	kg	**M**
f_1	initial free moisture content	kg	**M**
FN	flow number of nozzle (Eq. 7.34)	–	–
G	mass flow in jet or nozzle *or* of gas	kg/s	**MT^{-1}**
G_v	mass rate of evaporation	kg/s	**MT^{-1}**
G'	mass rate of flow of air per unit cross-section	kg/m^2 s	**ML^{-2} T^{-1}**
g	acceleration due to gravity	m/s^2	**LT^{-2}**
H	humidity	kg/kg	–
H_s	humidity of saturated air at surface temperature	kg/kg	–
H_w	humidity of saturated air at temperature θ_w	kg/kg	–
H_0	humidity of saturated air	kg/kg	–
h	heat transfer coefficient	W/m^{2}K	**MT$^{-3}\theta^{-1}$**
h_D	mass transfer coefficient	m/s	**LT^{-1}**
h_f	friction head over a distance z_1 from surface	m	**L**
h_s	suction potential immediately below meniscus	m	**L**
h_t	theoretical suction potential of pore or waist	m	**L**
h_1	suction potential at distance z_1 below meniscus	m	**L**
j	volumetric gas/liquid ratio	–	–
K	coefficient	m^{2}s/kg	**M^{-1} L^2 T**
K'	coefficient	s$^{1.8}$/m$^{1.5}$	**L$^{-1.8}$ T$^{1.8}$**
K''	transfer coefficient ($h_{D\rho A}$)	kg/m^2 s	**ML^{-2} T^{-1}**
k_G	mass transfer coefficient	s/m	**L^{-1} T**
k_f	thermal conductivity of gas film at interface	W/mK	**MLT$^{-3}\theta^{-1}$**
L	length of surface	m	**L**

Continued

		Units in SI system	Dimensions in M, N, L, T, θ
l	half thickness of slab	m	**L**
M_A	molecular weight of air	kg/kmol	**MN**$^{-1}$
M_w	molecular weight of water	kg/kmol	**MN**$^{-1}$
m	ratio of rate of drying per unit area to moisture content	m^{-2}s^{-1}	**L**$^{-2}$ **T**$^{-1}$
N	revolutions per unit time	Hz	T^{-1}
n	index	–	–
n_f	number of flights	–	–
P	total pressure	N/m^2	**ML**$^{-1}$ **T**$^{-2}$
P_s	vapour pressure of water at surface of material	N/m^2	**ML**$^{-1}$ **T**$^{-2}$
P_w	partial pressure of water vapour	N/m^2	**ML**$^{-1}$ **T**$^{-2}$
P_{wo}	partial pressure at surface of material at wet bulb temperature	N/m^2	**ML**$^{-1}$ **T**$^{-2}$
$-\Delta P$	pressure drop across nozzle	N/m^2	**ML**$^{-1}$ **T**$^{-2}$
Q	rate of heat transfer	W	**ML**^{2} **T**$^{-3}$
R_c	rate of drying per unit area for constant rate period	kg/m^2 s	**ML**$^{-2}$ **T**$^{-1}$
R	universal gas constant	8314 J/kmol K	**MN**$^{-1}$ **L**^{2} **T**$^{-2}\theta^{-1}$
R	exit gas relative humidity (per cent)	–	–
r	radius of sphere	m	**L**
r'	radius of capillary	m	**L**
S	slope of drum	–	–
T	absolute temperature	K	θ
T_b	bed temperature	K	θ
T_f	temperature of wet solids feed	K	θ
T_m	mean temperature of inlet gas	K	θ
ΔT	temperature difference	K	θ
t	time	s	**T**
t_c	time of constant rate period of drying	s	**T**
t_f	time of drying in falling rate period	s	**T**
U	overall heat transfer coefficient	W/m^2 K	**MT**$^{-3}\theta^{-1}$
u	gas velocity	m/s	**LT**$^{-1}$
u_f	fluidising velocity	m/s	**LT**$^{-1}$
u_l	liquid velocity in jet or spray	m/s	**LT**$^{-1}$
u_r	velocity of gas relative to liquid	m/s	**LT**$^{-1}$
V	volume	m^3	**L**^{3}
W	mass rate of evaporation	kg/s	**MT**$^{-1}$
w	total moisture	kg	**M**

Continued

		Units in SI system	Dimensions in M, N, L, T, θ
w_c	critical moisture content	kg	**M**
w_e	equilibrium moisture content	kg	**M**
w_1	initial moisture content	kg	**M**
X	hold-up of drum	–	–
X_a	hold-up of drum with air flow	–	–
X_f	moisture content of wet feed	kg/kg	–
x	factor depending on type of packing	–	–
y	distance in direction of diffusion	m	**L**
z_1	distance below meniscus	m	**L**
α	angle of contact	–	–
$\bar{\kappa}$	coefficient	$kg^{1-n}\ m^{2n}\ s^{n-3}$/K	$\mathbf{M}^{1-n}\ \mathbf{L}^{2n}\ \mathbf{T}^{n-3\theta-1}$
θ	gas temperature	K	θ
θ_s	surface temperature	K	θ
θ_w	wet bulb temperature	K	θ
λ	latent heat vaporisation per unit mass	J/kg	$\mathbf{L}^2\ \mathbf{T}^{-2}$
λ_b	mean latent heat vaporisation per unit mass of T_b	kJ/kg	$\mathbf{L}^2\ \mathbf{T}^{-2}$
λ_{opt}	optimum wavelength for jet disruption	m	**L**
μ	viscosity	Ns/m^2	$\mathbf{M}\mathbf{L}^{-1}\ \mathbf{T}^{-1}$
ρ	density	kg/m^3	$\mathbf{M}\mathbf{L}^{-3}$
ρ_A	density of air at its mean partial pressure	kg/m^3	$\mathbf{M}\mathbf{L}^{-3}$
σ	surface tension	J/m^2	$\mathbf{M}\mathbf{T}^{-2}$
ϕ	half-angle of spray cone or sheet	–	–

7.1 Introduction

The drying of materials is often the final operation in a manufacturing process, carried out immediately prior to packaging or dispatch. Drying refers to the final removal of water, or another solute, and the operation often follows evaporation, filtration, or crystallisation. In some cases, drying is an essential part of the manufacturing process, as for instance in paper making or in the seasoning of timber, although, in the majority of processing industries, drying is carried out for one or more of the following reasons:

(a) To reduce the cost of transport.
(b) To make a material more suitable for handling as, for example, with soap powders, dyestuffs, and fertilisers.

(c) To provide definite properties, such as, for example, maintaining the free-flowing nature of salt.

(d) To remove moisture which may otherwise lead to corrosion. One example is the drying of gaseous fuels or benzene prior to chlorination.

With a crystalline product, it is essential that the crystals are not damaged during the drying process, and, in the case of pharmaceutical products, care must be taken to avoid contamination. Shrinkage, as with paper, cracking, as with wood, or loss of flavour, as with fruit, must also be prevented. With the exception of the partial drying of a material by squeezing in a press or the removal of water by adsorption, almost all drying processes involve the removal of water by vaporisation, which requires the addition of heat. In assessing the efficiency of a drying process, the effective utilisation of the heat supplied is the major consideration.

7.2 **General principles**

The moisture content of a material is usually expressed in terms of its water content as a percentage of the mass of the dry material, though moisture content is sometimes expressed on a wet basis, as in Example 7.3. If a material is exposed to air at a given temperature and humidity, the material will either lose or gain water until an equilibrium condition is established. This equilibrium moisture content varies widely with the moisture content and the temperature of the air, as shown in Fig. 7.1. A non-porous insoluble solid, such as sand or China clay, has an equilibrium

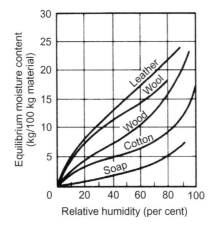

FIG. 7.1

Equilibrium moisture content of a solid as a function of relative humidity at 293 K.

moisture content approaching zero for all humidities and temperatures, although many organic materials, such as wood, textiles, and leather, show wide variations of equilibrium moisture content. Moisture may be present in two forms:

Bound moisture. This is water retained so that it exerts a vapour pressure less than that of free water at the same temperature. Such water may be retained in small capillaries, adsorbed on surfaces, or as a solution in cell walls.

Free moisture. This is water which is in excess of the equilibrium moisture content.

The water removed by vaporisation is generally carried away by air or hot gases, and the ability of these gases to pick up the water is determined by their temperature and humidity. In designing dryers using air, the properties of the air–water system are essential, and these are detailed in Volume 1, Chapter 13, where the development of the humidity chart is described. For the *air–water system*, the following definitions are of importance:

Humidity H, mass of water per unit mass of dry air.
Since:

$$\frac{\text{moles of water vapour}}{\text{moles of dry air}} = \frac{P_w}{(P - P_w)}$$

then:

$$H = \frac{18 P_w}{29(P - P_w)}$$

where P_w is the partial pressure of water vapour and P is the total pressure.

Humidity of saturated air H_0. This is the humidity of air when it is saturated with water vapour. The air then is in equilibrium with water at the given temperature and pressure.

Percentage humidity

$$= \frac{\text{Humidity of air}}{\text{Humidity of saturated air}} \times 100 = \frac{H}{H_0} \times 100$$

Percentage relative humidity, R

$$= \frac{\text{Partial pressure of water vapour in air}}{\text{Vapour pressure of water at the same temperature}} \times 100$$

The distinction between *percentage humidity* and *percentage relative humidity* is of significance though, the difference in the values of the two quantities does not usually exceed 7% to 8%. Reference may be made here to Volume 1, Section 13.2.1.

Humid volume is the volume of unit mass of dry air and its associated vapour. Then, under ideal conditions, at atmospheric pressure:

$$\text{humid volume} = \frac{22.4}{29}\left(\frac{T}{273}\right) + \frac{22.4.H}{18}\left(\frac{T}{273}\right) \text{ m}^3/\text{kg}$$

where T is in degrees K,

or:

$$\frac{359}{29}\left(\frac{T}{492}\right) + \frac{359.H}{18}\left(\frac{T}{492}\right) \text{ ft}^3/\text{lb}$$

where T is in degrees Rankine.

Saturated volume is the volume of unit mass of dry air, together with the water vapour required to saturate it.

Humid heat is the heat required to raise unit mass of dry air and associated vapour through 1 degree K at constant pressure or $1.00 + 1.88H$ kJ/kg K.

Dew point is the temperature at which condensation will first occur when air is cooled.

Wet bulb temperature. If a stream of air is passed rapidly over a water surface, vaporisation occurs, provided the temperature of the water is above the dew point of the air. The temperature of the water falls and heat flows from the air to the water. If the surface is sufficiently small for the condition of the air to change inappreciably and if the velocity is in excess of about 5 m/s, the water reaches the wet bulb temperature θ_w at equilibrium.

The rate of heat transfer from gas to liquid is given by:

$$Q = hA(\theta - \theta_w) \tag{7.1}$$

The mass rate of vaporisation is given by:

$$\begin{aligned}G_v &= \frac{h_D A M_w}{RT}(P_{w0} - P_w) \\ &= \frac{h_D A M_A}{RT}\left[(P - P_w)_{\text{mean}}(H_w - H)\right] \\ &= h_D A_{\rho A}(H_w - H)\end{aligned} \tag{7.2}$$

The rate of heat transfer required to effect vaporisation at this rate is given by:

$$G_v = h_D A_{\rho A}(H_w - H)\lambda \tag{7.3}$$

At equilibrium, the rates of heat transfer given by Eqs (7.1), (7.3) must be equal, and hence:

$$H - H_w = -\frac{h}{h_D \rho_A \lambda}(\theta - \theta_w) \tag{7.4}$$

In this way, it is seen that the wet bulb temperature θ_w depends only on the temperature and humidity of the drying air.

In these equations:

h is the heat transfer coefficient,
h_D is the mass transfer coefficient,
A is the surface area,
θ is the temperature of the air stream,
θ_w is the wet bulb temperature,
P_{w0} is the vapour pressure of water at temperature θ_w,

M_A is the molecular weight of air,
M_w is the molecular weight of water,
R is the universal gas constant,
T is the absolute temperature,
H is the humidity of the gas stream,
H is the humidity of saturated air at temperature θ_w,
ρ_A is the density of air at its mean partial pressure, and
λ is the latent heat of vaporisation of unit mass of water.

Eq. (7.4) is identical with Eq. (13.8) in Volume 1, and reference may be made to that chapter for a more detailed discussion.

7.3 Rate of drying
7.3.1 Drying periods

In drying, it is necessary to remove free moisture from the surface and also moisture from the interior of the material. If the change in moisture content for a material is determined as a function of time, a smooth curve is obtained from which the rate of drying at any given moisture content may be evaluated. The form of the drying rate curve varies with the structure and type of material, and two typical curves are shown in Fig. 7.2. In curve 1, there are two well-defined zones: AB, where the rate of drying is constant and BC, where there is a steady fall in the rate of drying as the moisture content is reduced. The moisture content at the end of the constant rate period is represented by point B, and this is known as the *critical moisture content*. Curve 2 shows three stages, DE, EF, and FC. The stage DE represents a constant rate period, and EF and FC are falling rate periods. In this case, the Section EF is a straight line, however, and only the portion FC is curved. Section EF is known as the first falling rate period and the final stage, shown as FC, as the second falling rate period. The drying of soap gives rise to a curve of type 1, and sand to a curve of type 2. A number

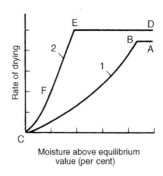

FIG. 7.2

Rate of drying of a granular material.

Table 7.1 Evaporation rates for various materials under constant conditions [8].

Material	Rate of evaporation	
	(kg/m² h)	(kg/m² s)
Water	2.7	0.00075
Whiting pigment	2.1	0.00058
Brass filings	2.4	0.00067
Brass turnings	2.4	0.00067
Sand (fine)	2.0–2.4	0.00055–0.00067
Clays	2.3–2.7	0.00064–0.00075

of workers, including Sherwood [1] and Newitt and co-workers [2–7], have contributed various theories on the rate of drying at these various stages.

Constant rate period

During the constant rate period, it is assumed that drying takes place from a saturated surface of the material by diffusion of the water vapour through a stationary air film into the air stream. Gilliland [8] has shown that the rates of drying of a variety of materials in this stage are substantially the same as shown in Table 7.1.

In order to calculate the rate of drying under these conditions, the relationships obtained in Volume 1 for diffusion of a vapour from a liquid surface into a gas may be used. The simplest equation of this type is:

$$W = k_G A (P_s - P_w) \tag{7.5}$$

where k_G is the mass transfer coefficient.

Since the rate of transfer depends on the velocity u of the air stream, raised to a power of about 0.8, then the mass rate of evaporation is:

$$W = k_G A (P_s - P_w) u^{0.8} \tag{7.6}$$

where: A is the surface area,

P_s is the vapour pressure of the water, and
P_w is the partial pressure of water vapour in the air stream.

This type of equation, used in Volume 1 for the rate of vaporisation into an air stream, simply states that the rate of transfer is equal to the transfer coefficient multiplied by the driving force. It may be noted, however, that $(P_s - P_w)$ is not only a driving force, but it is also related to the capacity of the air stream to absorb moisture.

These equations suggest that the rate of drying is independent of the geometrical shape of the surface. Work by Powell and Griffiths [9] has shown, however, that the ratio of the length to the width of the surface is of some importance, and that the evaporation rate is given more accurately as:

(a) For values of $u = 1-3$ m/s:

$$W = 5.53 \times 10^{-9} L^{0.77} B(P_s - P_w)(1 + 61u^{0.85}) \text{ kg/s} \tag{7.7}$$

(b) For values of $u < 1$ m/s:

$$W = 3.72 \times 10^{-9} L^{0.73} B^{0.8}(P_s - P_w)(1 + 61u^{0.85}) \text{ kg/s} \tag{7.8}$$

where:

P_s, the saturation pressure at the temperature of the surface (N/m^2),
P_w, the vapour pressure in the air stream (N/m^2), and
L and B are the length and width of the surface, respectively (m).

For most design purposes, it may be assumed that the rate of drying is proportional to the transfer coefficient multiplied by $(P_s - P_w)$. Chakravorty [10] has shown that, if the temperature of the surface is greater than that of the air stream, then P_w may easily reach a value corresponding to saturation of the air. Under these conditions, the capacity of the air to take up moisture is zero, whilst the force causing evaporation is $(P_s - P_w)$. As a result, a mist will form and water may be redeposited on the surface. In all drying equipment, care must therefore be taken to ensure that the air or gas used does not become saturated with moisture at any stage.

The rate of drying in the constant rate period is given by:

$$W = \frac{dw}{dt} = \frac{hA\Delta T}{\lambda} = k_G A(P_s - P_w) \tag{7.9}$$

where:

W is the rate of loss of water,
h is the heat transfer coefficient from air to the wet surface,
ΔT is the temperature difference between the air and the surface,
λ is the latent heat of vaporisation per unit mass,
k_G is the mass transfer coefficient for diffusion from the wet surface through the gas film,
A is the area of interface for heat and mass transfer, and
$(P_s - P_w)$ is the difference between the vapour pressure of water at the surface and the partial pressure in the air.

It is more convenient to express the mass transfer coefficient in terms of a humidity difference, so that $k_G A(P_s - P_w) \simeq kA(H_s - H)$. The rate of drying is thus determined by the values of h, ΔT and A, and is not influenced by the conditions inside the solid. h depends on the air velocity and the direction of flow of the air, and it has been found that $h = CG'^{0.8}$ where G' is the mass rate of flow of air in kg/s m^2. For air flowing parallel to plane surfaces, Shepherd et al. [11] have given the value of C as 14.5 where the heat transfer coefficient is expressed in W/m^2 K.

If the gas temperature is high, then a considerable proportion of the heat will pass to the solid by radiation, and the heat transfer coefficient will increase. This may result in the temperature of the solid rising above the wet bulb temperature.

First falling-rate period

The points B and E in Fig. 7.2 represent conditions where the surface is no longer capable of supplying sufficient free moisture to saturate the air in contact with it. Under these conditions, the rate of drying depends very much on the mechanism by which the moisture from inside the material is transferred to the surface. In general, the curves in Fig. 7.2 will apply, although for a type 1 solid, a simplified expression for the rate of drying in this period may be obtained.

Second falling-rate period

At the conclusion of the first falling rate period, it may be assumed that the surface is dry and that the plane of separation has moved into the solid. In this case, evaporation takes place from within the solid and the vapour reaches the surface by molecular diffusion through the material. The forces controlling the vapour diffusion determine the final rate of drying, and these are largely independent of the conditions outside the material.

7.3.2 Time for drying

If a material is dried by passing hot air over a surface which is initially wet, the rate of drying curve in its simplest form is represented by BCE, shown in Fig. 7.3
    where:

w is the total moisture,
w_e is the equilibrium moisture content (point E),
$w - w_e$ is the free moisture content, and
w_c is the critical moisture content (point C).

Constant-rate period

During the period of drying from the initial moisture content w_1 to the critical moisture content w_c, the rate of drying is constant, and the time of drying t_c is given by:

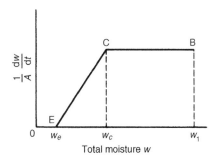

FIG. 7.3

The use of a rate of drying curve in estimating the time for drying.

$$t_c = \frac{w_1 - w_c}{R_c A}$$

(7.10)

where:

R_c is the rate of drying per unit area in the constant rate period, and
A is the area of exposed surface.

Falling-rate period

During this period, the rate of drying is, approximately, directly proportional to the free moisture content $(w - w_e)$, or:

$$-\left(\frac{1}{A}\right)\frac{dw}{dt} = m(w - w_e) = mf \quad (\text{say})$$

(7.11)

Thus:

$$-\frac{1}{mA}\int_{w_e}^{w}\frac{dw}{w - w_e} = \int_0^{t_f} dt$$

or:

$$\frac{1}{mA}\ln\left[\frac{w_c - w_e}{w - w_e}\right] = t_f$$

and:

$$t_f = \frac{1}{mA}\ln\left(\frac{f_c}{f}\right)$$

(7.12)

Total time of drying

The total time t of drying from w_1 to w is given by $t = (t_c + t_f)$.

The rate of drying R_c over the constant rate period is equal to the initial rate of drying in the falling rate period, so that $R_c = mf_c$.

Thus:

$$t_c = \frac{w_1 - w_c}{mAf_c}$$

(7.13)

and the total drying time,

$$
\begin{aligned}
t &= \frac{(w_1 - w_c)}{mAf_c} + \frac{1}{mA}\ln\left(\frac{f_c}{f}\right) \\
&= \frac{1}{mA}\left[\frac{(f_1 - f_c)}{f_c} + \ln\left(\frac{f_c}{f}\right)\right]
\end{aligned}
$$

(7.14)

Example 7.1

A wet solid is dried from 25% to 10% moisture under constant drying conditions in 15 ks (4.17 h). If the critical and the equilibrium moisture contents are 15% and 5%, respectively, how long will it take to dry the solid from 30 to 8% moisture under the same conditions?

Solution

> *For the first drying operation*:
> Thus:

$$w_1 = 0.25\,\text{kg/kg}, w = 0.10\,\text{kg/kg}, w_c = 0.15\,\text{kg/kg and } w_e = 0.05\,\text{kg/kg}$$
$$f_1 = (w_1 - w_e) = (0.25 - 0.05) = 0.20\,\text{kg/kg}$$
$$f_c = (w_c - w_e) = (0.15 - 0.05) = 0.10\,\text{kg/kg}$$
$$f = (w - w_e) = (0.10 - 0.05) = 0.05\,\text{kg/kg}$$

From Eq. (7.14), the total drying time is:

$$t = (1/mA)[(f_1 - f_c)/f_c + \ln(f_c/f)]$$

or:

$$15 = (1/mA)[(0.20 - 0.10)/0.10 + \ln(0.10/0.05)]$$

and:

$$mA = 0.0667(1.0 + 0.693) = 0.113\,\text{kg/s}$$

> *For the second drying operation*:
> Thus:

$$w_1 = 0.30\,\text{kg/kg}, w = 0.08\,\text{kg/kg}, w_c = 0.15\,\text{kg/kg and } w_e = 0.05\,\text{kg/kg}$$
$$f_1 = (w_1 - w_e) = (0.30 - 0.05) = 0.25\,\text{kg/kg}$$
$$f_c = (w_c - w_e) = (0.15 - 0.05) = 0.10\,\text{kg/kg}$$
$$f = (w - w_e) = (0.08 - 0.05) = 0.03\,\text{kg/kg}$$

> *The total drying time is then*:

$$t = (1/0.113)[(0.25 - 0.10)/0.10 + \ln(0.10/0.03)]$$
$$= 8.856(1.5 + 1.204)$$
$$= 23.9\,\text{ks}\,(6.65\,\text{h})$$

Example 7.2

Strips of material 10 mm thick are dried under constant drying conditions from 28% to 13% moisture in 25 ks (7 h). If the equilibrium moisture content is 7%, what is the time taken to dry 60 mm planks from 22% to 10% moisture under the same conditions assuming no loss from the edges? All moistures are given on a wet basis.

The relation between E, the ratio of the average free moisture content at time t to the initial free moisture content, and the parameter J is given by:

E	1	0.64	0.49	0.38	0.295	0.22	0.14
J	0	0.1	0.2	0.3	0.5	0.6	0.7

It may be noted that $J = kt/l^2$, where k is a constant, t the time in ks and $2L$ the thickness of the sheet of material in millimetres.

Solution

For the 10 mm strips
Initial free moisture content $=(0.28-0.07)=0.21$ kg/kg.
Final free moisture content $=(0.13-0.07)=0.06$ kg/kg.
Thus: when $t = 25$ ks, $E = (0.06/0.21) = 0.286$
and from Fig. 7.4, a plot of the given data,
Thus:

$$J = 0.52$$

$$0.52 = (k \times 25)/(10/2)^2$$

and:

$$k = 0.52$$

For the 60 mm planks
Initial free moisture content $=(0.22-0.07)=0.15$ kg/kg.
Final free moisture content $=(0.10-0.07)=0.03$ kg/kg.

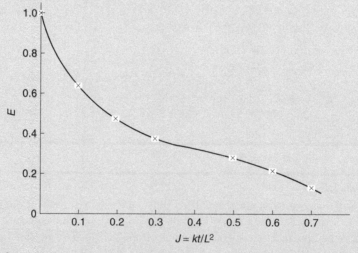

FIG. 7.4

Drying data for Example 6.2.

$$E = (0.03/0.15) = 0.20$$

From Fig. 7.4:

$$J = 0.63$$

and hence:

$$t = Jl^2/k$$

$$= 0.63 \, (60/2)^2/0.52 = \underline{1090 \, ks} \, (12.6 \, \text{days})$$

Example 7.3

A granular material containing 40% moisture is fed to a countercurrent rotary dryer at a temperature of 295 K and is withdrawn at 305 K, containing 5% moisture. The air supplied, which contains 0.006 kg water vapour/kg dry air, enters at 385 K and leaves at 310 K. The dryer handles 0.125 kg/s wet stock.

Assuming that radiation losses amount to 20 kJ/kg dry air used, determine the mass flow rate of dry air supplied to the dryer and the humidity of the exit air.

The latent heat of water vapour at 295 K = 2449 kJ/kg, specific heat capacity of dried material = 0.88 kJ/kg K, the specific heat capacity of dry air = 1.00 kJ/kg K, and the specific heat capacity of water vapour = 2.01 kJ/kg K.

Solution

This example involves a heat balance over the system. 273 K will be chosen as the datum temperature, and it will be assumed that the flow rate of dry air = G kg/s.

Heat in:

(a) *Air*

G kg/s dry air enter with 0.006G kg/s water vapour and hence the heat content of this stream

$$= [(1.00G) + (0.006G \times 2.01)](385 - 273) = 113.35G \, kW$$

(b) *Wet solid*

0.125 kg/s enter containing 0.40 kg water/kg wet solid, assuming the moisture is expressed on a wet basis.

Thus:

$$\text{mass flowrate of water} = (0.125 \times 0.40) = 0.050 \, \text{kg/s}$$

and:

$$\text{mass flowrate of dry solid} = (0.125 - 0.050) = 0.075 \, \text{kg/s}$$

Hence:

the heat content of this stream $= [(0.050 \times 4.18) + (0.075 \times 0.88)]$ $(295 - 273) = 6.05$ kW

Heat out:

(a) *Air*

Heat in exit air $= [(1.00\,G) + (0.006\,G \times 2.01)](310 - 273) = 37.45G\,\text{kW}.$

Mass flow rate of dry solids $= 0.075$ kg/s containing 0.05 kg water/kg wet solids.

Hence:

water in the dried solids leaving $= (0.05 \times 0.075)/(1 + 0.05) = 0.0036\,\text{kg/s}$

and

the water evaporated into gas steam $= (0.050 - 0.0036) = 0.0464\,\text{kg/s}.$

Assuming evaporation takes place at 295 K, then:

heat in the water vapour $= 0.0464[2.01(310 - 295) + 2449 + 4.18(295 - 273)]$
$= 119.3\,\text{kW}$

and:

the total heat in this stream $= (119.30 + 37.45G)\,\text{kW}.$

(b) *Dried solids*

The dried solids contain 0.0036 kg/s water and hence heat content of this stream is:

$= [(0.075 \times 0.88) + (0.0036 \times 4.18)/(305 - 273)] = 2.59\,\text{kW}$

(c) *Losses*

These amount to 20 kJ/kg dry air or 20 *m* kW.

Heat balance

$$(113.35G + 6.05) = (119.30 + 37.45G + 2.59 + 20G)$$

and:

$$G = 2.07\ \text{kg/s}$$

Water in the outlet air stream $= (0.006 \times 2.07) + 0.0464 = 0.0588\,\text{kg/s}$

and:

the humidity $H = (0.0588/2.07) = 0.0284\ \text{kg/kg dry air}$

7.4 **The mechanism of moisture movement during drying**

7.4.1 **Diffusion theory of drying**

In the general form of the curve for the rate of drying of a solid shown in Fig. 7.2, there are two and sometimes three distinct sections. During the constant-rate period, moisture vaporises into the air stream and the controlling factor is the transfer coefficient for diffusion across the gas film. It is important to understand how the moisture moves to the drying surface during the falling-rate period, and two models have been used to describe the physical nature of this process, the diffusion theory and the capillary theory. In the diffusion theory, the rate of movement of water to the air interface is governed by rate equations similar to those for heat transfer, whilst in the capillary theory the forces controlling the movement of water are capillary in origin, arising from the minute pore spaces between the individual particles.

Falling rate period, diffusion control

In the falling-rate period, the surface is no longer completely wetted and the rate of drying steadily falls. In the previous analysis, it has been assumed that the rate of drying per unit effective wetted area is a linear function of the water content, so that the rate of drying is given by:

$$\left(\frac{1}{A}\right)\frac{dw}{dt} = -m(w - w_e) \tag{7.15}$$

In many cases, however, the rate of drying is governed by the rate of internal movement of the moisture to the surface. It was initially assumed that this movement was a process of diffusion and would follow the same laws as heat transfer. This approach has been examined by a number of workers, and in particular by Sherwood [12] and Newman [13].

Considering a slab with the edges coated to prevent evaporation, which is dried by evaporation from two opposite faces, the Y-direction being taken perpendicular to the drying face, the central plane being taken as $y = 0$, and the slab thickness $2L$, then on drying, the moisture movement by diffusion will be in the Y-direction, and hence from Volume 1, Eq. (10.66):

$$\frac{\partial C_w}{\partial t} = D_L \frac{\partial^2 C_w}{\partial y^2}$$

where C_w is the concentration of water at any point and any time in the solid, and D_L is the coefficient of diffusion for the liquid.

If w is the liquid content of the solid, integrated over the whole depth, w_1 the initial content, and w_e the equilibrium content, then:

$$\frac{(w - w_e)}{(w_1 - w_e)} = \frac{\text{Free liquid at any time}}{\text{Initial free liquid content}}$$

Sherwood [12] and Newman [13] have presented the following solution assuming an initially uniform water distribution and zero water-concentration at the surface once drying has started:

$$\frac{(w - w_e)}{w_1 - w_e} = \frac{8}{\pi^2}\left\{e^{-D_L t(\pi/2l)^2} + \frac{1}{9}e^{-9D_L t(\pi/2l)^2} + \frac{1}{25}e^{-25D_L t(\pi/2l)^2} + \cdots\right\} \tag{7.16}$$

This equation assumes an initially uniform distribution of moisture, and that the drying is from both surfaces. When drying is from one surface only, then l is the total thickness. If the time of drying is long, then only the first term of the equation need be used and thus, differentiating Eq. (7.16) gives:

$$\frac{dw}{dt} = -\frac{2D_L}{l^2}e^{-\frac{D_L t\pi^2}{4l^2}}(w_1 - w_e) \tag{7.17}$$

In the drying of materials such as wood or clay, the moisture concentration at the end of the constant rate period is not uniform, and is more nearly parabolic. Sherwood has presented an analysis for this case, and has given experimental values for the drying of brick clay.

In this case, it is assumed that the rate of movement of water is proportional to a concentration gradient, and capillary and gravitational forces are neglected. Water may, however, flow from regions of low concentration to those of high concentration if the pore sizes are suitable, and for this and other reasons, Ceaglske and Hougen [14,15] have proposed a capillary theory which is now considered.

7.4.2 Capillary theory of drying
Principles of the theory
The capillary theory of drying has been proposed in order to explain the movement of moisture in the bed during surface drying. The basic importance of the pore space between granular particles was first pointed out by Slichter [16] in connection with the movement of moisture in soils, and this work has been modified and considerably expanded by Haines [17]. The principles are now outlined and applied to the problem of drying. Considering a systematic packing of uniform spherical particles, these may be arranged in six different regular ways, ranging from the most open to the closest packing. In the former, the spheres are arranged as if at the corners of a cube with each sphere touching six others. In the latter arrangement, each sphere rests in the hollow of three spheres in adjacent layers, and touches twelve other spheres. These configurations are shown in Fig. 7.5. The densities of packing of the other four arrangements will lie between those illustrated.

In each case, a regular group of spheres surrounds a space which is called a pore, and the bed is made up of a series of these elemental groupings. The pores are connected together by passages of various sizes, the smallest portions of which are known as *waists*. The size of a pore is defined as the diameter of the largest sphere which can be fitted into it, and the size of a waist as the diameter of the inscribed

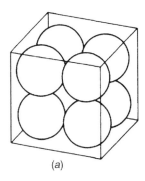 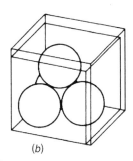

(a)   (b)

FIG. 7.5

Packing of spherical particles. (A) Cubic arrangement, one sphere touching six others. (B) Rhombohedral packing, one sphere touching twelve others, with layers arranged in rhombic formation.

circle. The sizes of the pores and waists will differ for each form of packing, as shown in Table 7.2.

The continuous variation in the diameter of each passage is the essential difference between a granular packing and a series of capillary tubes. If a clean capillary of uniform diameter $2r'$ is placed in a liquid, the level will rise in the capillary to a height h_s given by:

$$h_s = \left(\frac{2\sigma}{r'\rho g}\right) \cos \alpha \qquad (7.18)$$

where:

ρ is the density of the liquid,
σ is the surface tension, and
α is the angle of contact.

Table 7.2 Properties of packing of spheres of radius r.

Packing arrangement	Pore space (per cent total volume)	Radius of pore	Radius of waist	Value of x in Eq. (7.21) for:	
				limiting suction potential of pores	entry suction potential of waists
Cubical	47.64	0.700r	0.414r	2.86	4.82
Rhombohedral	25.95	0.288r	0.155r	6.90	12.90

A negative pressure, known as a *suction potential*, will exist in the liquid in the capillary. Immediately below the meniscus, the suction potential will be equivalent to the height of the liquid column h_s and, if water is used, this will have the value:

$$h_s = \frac{2\sigma}{r'\rho g} \tag{7.19}$$

If equilibrium conditions exist, the suction potential h_1 at any other level in the liquid, a distance z_1 below the meniscus, will be given by:

$$h_s = h_1 + z_1 \tag{7.20}$$

Similarly, if a uniform capillary is filled to a height greater than h_s, as given by Eq. (7.18), and its lower end is immersed, the liquid column will recede to this height.

The non-uniform passages in a porous material will also display the same characteristics as a uniform capillary, with the important difference that the rise of water in the passages will be limited by the pore size, whilst the depletion of saturated passages will be controlled by the size of the waists. The height of rise is controlled by the pore size, since this approximates to the largest section of a varying capillary, whilst the depletion of water is controlled by the narrow waists which are capable of a higher suction potential than the pores.

The theoretical suction potential of a pore or waist containing water is given by:

$$h_t = \frac{x\sigma}{r\rho g} \tag{7.21}$$

where:

x is a factor depending on the type of packing, shown in Table 7.2, and
r is the radius of the spheres.

For an idealised bed of uniform rhombohedrally packed spheres of radius r, for example, the waists are of radius $0.155r$, from Table 7.2, and the maximum theoretical suction potential of which such a waist is capable is:

$$\frac{2\sigma}{0.155r\rho g} = \frac{12.9\sigma}{r\rho g}$$

from which $x = 12.9$.

The maximum suction potential that can be developed by a waist is known as the *entry suction potential*. This is the controlling potential required to open a saturated pore protected by a meniscus in an adjoining waist and some values for x are given in Table 7.2.

When a bed is composed of granular material with particles of mixed sizes, the suction potential cannot be calculated and it must be measured by methods such as those given by Haines [17] and Oliver and Newitt [3].

Drying of a granular material according to the capillary theory

If a bed of uniform spheres, initially saturated, is to be surface dried in a current of air of constant temperature, velocity and humidity, then the rate of drying is given by:

$$\frac{\mathrm{d}w}{\mathrm{d}t} = k_G A (P_{w0} - P_w) \tag{7.22}$$

where P_{w0} is the saturation partial pressure of water vapour at the wet bulb temperature of the air, and P_w is the partial pressure of the water vapour in the air stream. This rate of drying will remain constant so long as the inner surface of the 'stationary' air film remains saturated.

As evaporation proceeds, the water surface recedes into the waists between the top layer of particles, and an increasing suction potential is developed in the liquid. When the menisci on the cubical waists, that is the largest, have receded to the narrowest section, the suction potential h_s at the surface is equal to $4.82\sigma/r\rho g$, from Table 7.2. Further evaporation will result in h_s increasing so that the menisci on the surface cubical waists will collapse, and the larger pores below will open. As h_s steadily increases, the entry suction of progressively finer surface waists is reached, so that the menisci collapse into the adjacent pores which are thereby opened.

In considering the conditions below the surface, the suction potential h_1 a distance z_1 from the surface is given by:

$$h_s = h_1 + z_1 \tag{7.23}$$

The flow of water through waists surrounding an open pore is governed by the size of the waist as follows:

(a) If the size of the waist is such that its entry suction potential exceeds the suction potential at that level within the bed, it will remain full by the establishment of a meniscus therein, in equilibrium with the effective suction potential to which it is subjected. This waist will then protect adjoining full pores which cannot be opened until one of the waists to which it is connected collapses.

(b) If the size of the waist is such that its entry suction potential is less than the suction potential existing at that level, it will in turn collapse and open the adjoining pore. In addition, this successive collapse of pores and waists will progressively continue so long as the pores so opened expose waists having entry suction potentials of less than the suction potentials existing at that depth.

As drying proceeds, two processes take place simultaneously:

(a) The collapse of progressively finer surface waists, and the resulting opening of pores and waists connected to them, which they previously protected, and

(b) The collapse of further full waists within the bed adjoining opened pores, and the consequent opening of adjacent pores.

Even though the effective suction potential at a waist or pore within the bed may be in excess of its entry or limiting suction potential, this will not necessarily collapse or open. Such a waist can only collapse if it adjoins an opened pore, and the pore in question can only open upon the collapse of an adjoining waist.

Effect of particle size. Reducing the particle size in the bed will reduce the size of the pores and the waists, and will increase the entry suction potential of the waists.

This increase means that the percentage variation in suction potentials with depth is reduced, and the moisture distribution is more uniform with small particles.

As the pore sizes are reduced, the frictional forces opposing the movement of water through these pores and waists may become significant, so that Eq. (7.23) is more accurately represented by:

$$h_s = h_1 + z_1 + h_f \tag{7.24}$$

where h_f, the frictional head opposing the flow over a depth z_1 from the surface, will depend on the particle size. It has been found [2] that, with coarse particles when only low suction potentials are found, the gravity effect is important though h_f is small, whilst with fine particles h_f becomes large.

(a) For particles of 0.1 to 1 mm radius, the values of h_1 are independent of the rate of drying, and vary appreciably with depth. Frictional forces are, therefore, negligible whilst capillary and gravitational forces are in equilibrium throughout the bed and are the controlling forces. Under such circumstances the percentage moisture loss at the critical point at which the constant rate period ends is independent of the drying rate, and varies with the depth of bed.

(b) For particles of 0.001 to 0.01 mm radius, the values of h_1 vary only slightly with rate of drying and depth, indicating that both gravitational and frictional forces are negligible whilst capillary forces are controlling. The critical point here will be independent of drying rate and depth of bed.

(c) For particles of less than 0.001 mm (1 μm) radius, gravitational forces are negligible, whilst frictional forces are of increasing importance and capillary and frictional forces may then be controlling. In such circumstances, the percentage moisture loss at the critical point diminishes with increased rate of drying and depth of bed. With beds of very fine particles an additional factor comes into play. The very high suction potentials which are developed cause a sufficient reduction of the pressure for vaporisation of water to take place inside the bed. This internal vaporisation results in a breaking up of the continuous liquid phase and a consequent interruption in the free flow of liquid by capillary action. Hence, the rate of drying is still further reduced.

Some of the experimental data of Newitt et al. [2] are illustrated in Fig. 7.6.

Freeze drying

Special considerations apply to the movement of moisture in freeze drying. Since the water is frozen, liquid flow under capillary action is impossible, and movement must be by vapour diffusion, analogous to the 'second falling rate period' of the normal case. In addition, at very low pressures the mean free path of the water molecules may be comparable with the pore size of the material. In these circumstances the flow is said to be of the 'Knudsen' type, referred to in Volume 1, Section 10.1.

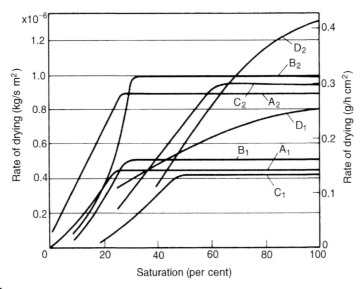

FIG. 7.6

Rates of drying of various materials as a function of percentage saturation. A – 60 μm glass spheres, bed 51 mm deep. B – 23.5 μm silica flour, bed 51 mm deep. C – 7.5 μm silica flour, bed 51 mm deep. D – 2.5 μm silica flour, bed 65 mm deep. Subscripts: 1. Low drying rate 2. High drying rate.

7.5 Drying equipment

7.5.1 Classification and selection of dryers

Because of the very wide range of dryer designs available, classification is a virtually impossible task. Parker [18] takes into account, however, the means by which material is transferred through the dryer as a basis of his classification, with a view to presenting a guide to the selection of dryers. Probably the most thorough classification of dryer types has been made by Kröll [19] who has presented a decimalised system based on the following factors:

(a) Temperature and pressure in the dryer,
(b) The method of heating,
(c) The means by which moist material is transported through the dryer,
(d) Any mechanical aids aimed at improving drying,
(e) The method by which the air is circulated,
(f) The way in which the moist material is supported,
(g) The heating medium, and
(h) The nature of the wet feed and the way it is introduced into the dryer.

In selecting a dryer for a particular application, as Sloan [20] has pointed out, two steps are of primary importance:

(a) A listing of the dryers which are capable of handling the material to be dried,
(b) Eliminating the more costly alternatives on the basis of annual costs, capital charges + operating costs. A summary of dryer types, together with cost data, has been presented by Backhurst and Harker [21] and the whole question of dryer selection is discussed further in Volume 6.

Once a group of possible dryers has been selected, the choice may be narrowed by deciding whether batch or continuous operation is to be employed and, in addition to restraints imposed by the nature of the material, whether heating by contact with a solid surface or directly by convection and radiation is preferred.

In general, continuous operation has the important advantage of ease of integration into the rest of the process coupled with a lower unit cost of drying. As the rate of throughput of material becomes smaller, however, the capital cost becomes the major component in the total running costs and the relative cheapness of batch plant becomes more attractive. This is illustrated in Fig. 7.7 which is taken from the work of Keey [22]. In general, throughputs of 5000 kg/day (0.06 kg/s) and over are best handled in batches, whilst throughputs of 50,000 kg/day (0.06 kg/s) and over are

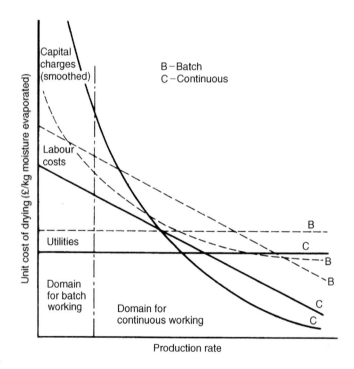

FIG. 7.7

Variation of unit costs of drying with production rate [22].

better handled continuously. The ease of construction of a small batch dryer compared with the sophistication of the continuous dryer should also be taken into account. In addition, a batch dryer is much more versatile and it can often be used for different materials. The humidity may be controlled during the drying operation, and this is especially important in cases where the humidity has to be maintained at different levels for varying periods of time.

Direct heating, in which the material is heated primarily by convection from hot gases has several advantages. Firstly, directly heated dryers are, in general, less costly, mainly because of the absence of tubes or jackets within which the heating medium must be contained. Secondly, it is possible to control the temperature of the gas within very fine limits, and indeed it is relatively simple to ensure that the material is not heated beyond a specified temperature. This is especially important with heat-sensitive materials. Against this, the overall thermal efficiency of directly heated dryers is generally low due to the loss of energy in the exhaust gas and, where an expensive solvent is evaporated from the solid, the operation is often difficult and costly. Losses also occur in the case of fluffy and powdery materials, and further problems are encountered where either the product or the solvent reacts with oxygen in the air.

A major cost in the operation of a dryer is in heating the air or gas. Frequently, the hot gases are produced by combustion of a fuel gas or atomised liquid, and considerable economy may be effected by using a combined heat and power system in which the hot gases are first passed through a turbine connected to an electrical generator.

Many of these disadvantages may be overcome by modifications to the design, although these increase the cost, and often an indirectly heated dryer may prove to be more economical. This is especially the case when thermal efficiency, solvent recovery or maximum cleanliness is of paramount importance and, with indirectly heated dryers, there is always the danger of overheating the product, since the heat is transferred through the material itself.

The maximum temperature at which the drying material may be held is controlled by the thermal sensitivity of the product and this varies inversely with the time of retention. Where lengthy drying times are employed, as for example in a batch shelf dryer, it is necessary to *operate under vacuum* in order to maintain evaporative temperatures at acceptable levels. In most continuous dryers, the retention time is very low, however, and operation at atmospheric pressure is usually satisfactory. As noted previously, dryer selection is considered in some detail in Volume 6.

7.5.2 Tray or shelf dryers

Tray or shelf dryers are commonly used for granular materials and for individual articles. The material is placed on a series of trays which may be heated from below by steam coils and drying is carried out by the circulation of air over the material. In some cases, the air is heated and then passed once through the oven, although, in the majority of dryers, some recirculation of air takes place, and the air is reheated before it is passed over each shelf. As air is passed over the wet material, both its temperature and its humidity change. This process of air humidification is discussed in Volume 1, Chapter 13.

If air of humidity H_1 is passed over heating coils so that its temperature rises to θ_1, this operation may be represented by the line AB on the humidity chart shown in Fig. 7.8. This air then passes over the wet material and leaves at, say 90% relative humidity, with its temperature falling to some value θ_2. This change in the condition of the air is shown by the line BC, and the humidity has risen to H_2. The wet bulb temperature of the air will not change appreciably and therefore BC will coincide with an adiabatic cooling line. Each kg of air removes $(H_2 - H_1)$ kg of water, and the air required to remove a given amount of water from the material may easily be found. If the air at θ_2 is now passed over a second series of heating coils and is heated to the original temperature θ_1, the heating operation is shown by the line CD. This reheated air can then be passed over wet material on a second tray in the dryer, and pick up moisture until its relative humidity rises again to 90% at a temperature θ_3. This is at point E. In this way, each kilogram of air has picked up water amounting to $(H_3 - H_1)$ kg of water. Reheating in this way may be effected a number of times, as shown in Fig. 7.8, so that the moisture removed per kilogram of air can be considerably increased compared with that for a single pass. Thus, for three passes of air over the material, the total moisture removed is $(H_4 - H_1)$ kg/kg air.

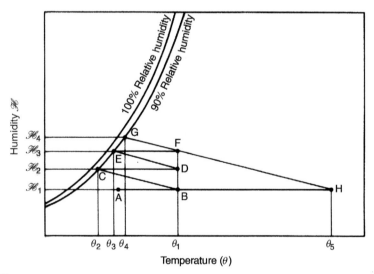

FIG. 7.8

Drying with reheating of air.

If the air of humidity H_1 had been heated initially to a temperature θ_5, the same amount of moisture would have been removed by a single passage over the material, assuming that the air again leaves at a relative humidity of 90%.

This reheating technique has two main advantages. Firstly, very much less air is required, because each kilogram of air picks up far more water than in a single stage system, and secondly, in order to pick up as much water in a single stage, it would be necessary to heat the air to a very much higher temperature. This reduction in the amount of air, needed simplifies the heating system, and reduces the tendency of the air to carry away any small particles.

A modern tray dryer consists of a well-insulated cabinet with integral fans and trays which are stacked on racks, or loaded on to trucks which are pushed into the dryer. Tray areas are 0.3 to 1 m^2 with a depth of material of 10 to 100 mm, depending on the particle size of the product. Air velocities of 1 to 10 m/s are used and, in order to conserve heat, 85% to 95% of the air is recirculated. Even at these high values, the steam consumption may be 2.5 to 3.0 kg/kg moisture removed. The capacity of tray dryers depends on many factors including the nature of the material, the loading and external conditions, although for dyestuffs an evaporative capacity of 0.03 to 0.3 kg/m^2 ks (0.1 to 1 kg/m^2 h) has been quoted with air at 300 to 360 K [22].

Example 7.4

A 100 kg batch of granular solids containing 30% moisture is to be dried in a tray drier to 15.5% of moisture by passing a current of air at 350 K tangentially across its surface at a velocity of 1.8 m/s. If the constant rate of drying under these conditions is 0.0007 kg/sm^2 and the critical moisture content is 15%, calculate the approximate drying time. Assume the drying surface to be 0.03 m^2/kg dry mass.

Solution

$$\text{In 100 kg feed, mass of water} = (100 \times 30/100) = 30 \text{ kg}$$

and:

$$\text{mass of dry solids} = (100 - 30) = 70 \text{ kg}$$

For b kg water in the dried solids: $100b/(b+70)=15.5$.
and the water in the product, $b=12.8$ kg
Thus: initial moisture content, $w_1 = (30/70)=0.429$ kg/kg dry solids
final moisture content, $w_2 = (12.8/70)=0.183$ kg/kg dry solids.
and water to be removed$=(30-12.8)=17.2$ kg.
The surface area available for drying$=(0.03 \times 70)=2.1\,m^2$ and hence the rate of drying during the constant period$=(0.0007 \times 2.1)=0.00147$ kg/s.

As the final moisture content is above the critical value, all the drying is at this constant rate and the time of drying is:

$$t = (17.2/0.00147) = 11,700 \text{ s or } \underline{11.7 \text{ ks}} \text{ (3.25 h)}$$

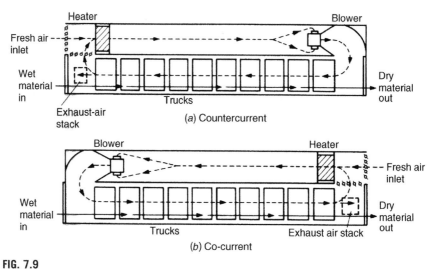

FIG. 7.9

Arrangements for tunnel dryers.

7.5.3 Tunnel dryers

In tunnel dryers, a series of trays or trolleys is moved slowly through a long tunnel, which may or may not be heated, and drying takes place in a current of warm air. Tunnel dryers are used for drying paraffin wax, gelatine, soap, pottery ware, and wherever the throughput is so large that individual cabinet dryers would involve too much handling. Alternatively, material is placed on a belt conveyor passing through the tunnel, an arrangement which is well suited to vacuum operation. In typical tunnel arrangements, shown in Fig. 7.9, the construction is of block or sheet metal and the size varies over a wide range, with lengths sometimes exceeding 30 m.

7.5.4 Rotary dryers

For the continuous drying of materials on a large scale, 0.3 kg/s (1 t/h) or greater, a rotary dryer, which consists of a relatively long cylindrical shell mounted on rollers and driven at a low speed, up to 0.4 Hz is suitable. The shell is supported at a small angle to the horizontal so that material fed in at the higher end will travel through the dryer under gravity, and hot gases or air used as the drying medium are fed in either at the upper end of the dryer to give co-current flow or at the discharge end of the machine to give countercurrent flow. One of two methods of heating is used:

(a) Direct heating, where the hot gases or air pass through the material in the dryer.
(b) Indirect heating, where the material is in an inner shell, heated externally by hot gases. Alternatively, steam may be fed to a series of tubes inside the shell of the dryer.

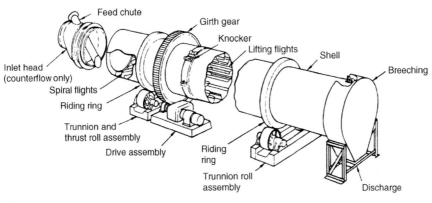

FIG. 7.10

Rotary dryer, 0.75 m diameter × 4.5 m long for drying desiccated coconut [21].

The shell of a rotary dryer is usually constructed by welding rolled plate, thick enough for the transmission of the torque required to cause rotation, and to support its own weight and the weight of material in the dryer. The shell is usually supported on large tyres which run on wide rollers, as shown in Fig. 7.10, and although mild steel is the usual material of construction, alloy steels are used, and if necessary the shell may be coated with a plastics material to avoid contamination of the product.

With countercurrent operation, since the gases are often exhausted by a fan, there is a slight vacuum in the dryer, and dust-laden gases are in this way prevented from escaping. This arrangement is suitable for sand, salt, ammonium nitrate, and other inorganic salts, and is particularly convenient when the product is discharged at a high temperature. In this case, gas or oil firing is used and, where air is used as a drying medium, this may be filtered before heating, in order to minimise contamination of the product. As the gases leaving the dryer generally carry away very fine material, some form of cyclone or scrubber is usually fitted. Since the hot gases come into immediate contact with the dried material, the moisture content may be reduced to a minimum, though the charge may become excessively heated. Further, since the rate of heat transfer is a minimum at the feed end, a great deal of space is taken up with heating the material.

With co-current flow, the rate of passage of the material through the dryer tends to be greater since the gas is travelling in the same direction. Contact between the wet material and the inlet gases gives rise to rapid surface drying, and this is an advantage if the material tends to stick to the walls. This rapid surface drying is also helpful with materials containing water of crystallisation. The dried product leaves at a lower temperature than with countercurrent systems, and this may also be an advantage. The rapid lowering of the gas temperature as a result of immediate contact with the wet material also enables heat sensitive materials to be handled rather more satisfactorily.

Since the drying action arises mainly from direct contact with hot gases, some form of lifter is essential to distribute the material in the gas stream. This may take the form of flights, as shown in Fig. 7.11, or of louvres. In the former case, the flights lift the

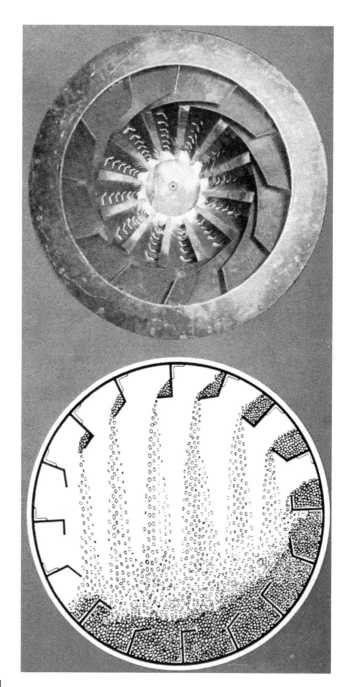

FIG. 7.11

Helical and angled lifting flights in a rotary dryer.

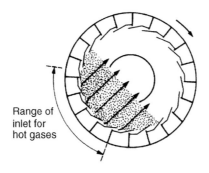

FIG. 7.12

Air flow through a rotary louvre dryer.

material and then shower it across the gas stream, whilst in the latter, the gas stream enters the shell along the louvres. In Fig. 7.12, it may be seen that, in the rotary louvre dryer, the hot air enters through the louvres, and carries away the moisture at the end of the dryer. This is not strictly a co-current flow unit, but rather a through circulation unit, since the material continually meets fresh streams of air. The rotation of the shell, at about 0.05 Hz (3 rpm), maintains the material in agitation and conveys it through the dryer. Rotary dryers are 0.75 to 3.5 m in diameter and up to 9 m in length [23].

The thermal efficiency of rotary dryers is a function of the temperature levels, and ranges from 30% in the handling of crystalline foodstuffs to 60% to 80% in the case of inert materials. Evaporative capacities of 0.0015 to 0.0080 kg/m^3 s may be achieved and these are increased by up to 50% in louvre dryers.

In one form of *indirectly heated dryer*, shown in Fig. 7.13, hot gases pass through the innermost cylinder, and then return through the annular space between the outer cylinders. This form of dryer can be arranged to give direct contact with the wet

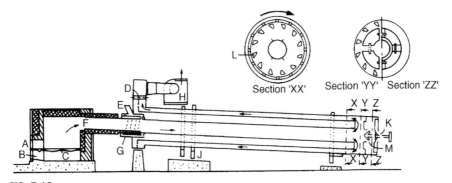

FIG. 7.13

Indirectly heated rotary dryer. A – Firing door. B – Air regulator. C – Furnace. D – Control valves. E – Feed chute. F – Furnace flue. G – Feed screw. H – Fan. J – Driving gear. K – Discharge bowl. L – Duct lifters.ss M – By-pass valve.

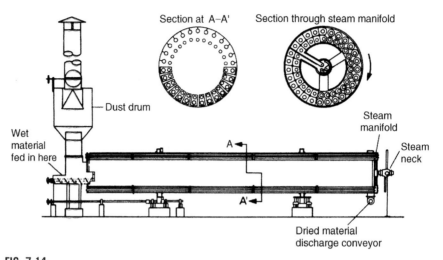

FIG. 7.14

Steam-tube rotary dryer.

material during the return passage of the gases. Flights on the outer surface of the inner cylinder, and the inner surface of the outer cylinder, assist in moving the material along the dryer. This form of unit gives a better heat recovery than the single flow direct dryer, though it is more expensive. In a simpler arrangement, a single shell is mounted inside a brickwork chamber, through which the hot gases are introduced.

The steam-tube dryer, shown in Fig. 7.14, incorporates a series of steam tubes, fitted along the shell in concentric circles and rotating with the shell. These tubes may be fitted with fins to increase the heat transfer surface although material may then stick to the tubes. The solids pass along the inclined shell, and leave through suitable ports at the other end. A small current of air is passed through the dryer to carry away the moisture, and the air leaves almost saturated. In this arrangement, the wet material comes in contact with very humid air, and surface drying is therefore minimised. This type of unit has a high thermal efficiency, and can be made from corrosion resisting materials without difficulty.

Design considerations

Many of the design problems associated with rotary dryers have been discussed by Friedman and Marshall [24], by Prutton et al. [25] and by Miller et al. [26]

Heat from the air stream passes to the solid material during its fall through the air stream, and also from the hot walls of the shell, although the first mechanism is much the more important. The heat transfer equation may be written as:

$$Q = UaV\Delta T \tag{7.25}$$

where:

Q is the rate of heat transfer,
U is the overall heat transfer coefficient,

V is the volume of the dryer,

a is the area of contact between the particles and the gas per unit volume of dryer, and

ΔT is the mean temperature difference between the gas and material.

The combined group Ua has been shown [25,26] to be influenced by the feed rate of solids, the air rate and the properties of the material, and a useful approximation is given by:

$$Ua = \bar{\kappa}\, G'^n / D \qquad (7.26)$$

where $\bar{\kappa}$ is a dimensional coefficient. Typical values for a 300 mm diameter dryer revolving at 0.08 to 0.58 Hz (5–35 rpm) show that $n = 0.67$ for specific gas rates in the range 0.37 to 1.86 kg/m^2, as given by Saeman [27]. The coefficient $\bar{\kappa}$ is a function of the number of flights and, using SI units, this is given approximately by:

$$\bar{\kappa} = 20\big(n_f - 1\big) \qquad (7.27)$$

Eq. (7.27) was derived for a 200 mm diameter dryer with between 6 and 16 flights [28]. Combining Eqs (7.26), (7.27) gives:

$$Ua = 20\big(n_f - 1\big) G'^{0.67} / D \qquad (7.28)$$

and hence for a 1 m diameter dryer with 8 flights, Ua would be about 140 W/m^3 K for a gas rate of 1 kg/m^2 s.

Saeman [27] has investigated the countercurrent drying of sand in a dryer of 0.3 m diameter and 2.0 m long with 8 flights rotating at 0.17 Hz (10 rpm) and has found that the volumetric heat transfer coefficient may be correlated in terms of the hold-up of solids, as shown in Fig. 7.15, and is independent of the gas rate in the range 0.25 to 20 kg/m^2 s.

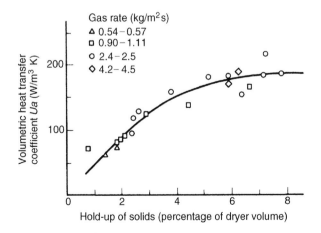

FIG. 7.15

Correlation of volumetric heat-transfer coefficients with hold-up [27].

Other typical values of heat transfer coefficients achieved in drying operations are:

Application	Parameter		Value	Reference
Heated surface (fairly dry solid)	U	5–12	W/m²K	28
Unagitated dryers	U	5–29	W/m²K	29
Moderate agitation	U	29–85	W/m²K	29
High agitation	U	85–140	W/m²K	29
Cake-encrusted heating surface	U	5–29	W/m²K	29
Light powdery materials	$U\Delta T$	950	W/m²	24
Coarse granular materials	$U\Delta T$	6300	W/m²	24

The hold-up in a rotary dryer varies with the feed rate, the number of flights, the shell diameter and the air rate. For zero air flow, Friedman and Marshall [24] give the hold-up as:

$$X = \frac{25.7F'}{SN^{0.9}D} \text{ percent of dryer volume} \qquad (7.29)$$

where:

D is the diameter of the drum (m),
F' is the feed rate (m³/s m²),
S is the slope of the dryer (m/m length),
N is the rate of rotation (Hz), and
X is the holdup, expressed as a percentage of the drum volume.

As the air flow rate is increased, X changes and an empirical relation for the hold-up X_a with air flow is:

$$X_a = X \pm KG' \qquad (7.30)$$

although the values of K are poorly defined. X_a usually has a value of about 3%, when working with a slope of about 0.1. In Eq. (7.30), the positive sign refers to counter-current flow and the negative sign to co-current flow.

A more general approach has been made by Sharples et al. [29] who have solved differential moisture and heat-balance equations coupled with expressions for the forward transport of solids, allowing for solids being cascaded out of lifting baffles. Typical results were obtained using a 2.6 m diameter dryer, 16 m long with a 1° slope (60 mm/m). Commercial equipment is available with diameters up to 3 m and lengths up to 30 m, and hence the correlations outlined in this section must be used with caution beyond the range used in the experimental investigations.

In general, the allowable mass velocity of the gas in a direct-contact rotary dryer depends on the dust content of the solids and is 0.55 to 7.0 kg/m² s with coarse particles. Inlet gas temperatures are typically 390 to 450 K for steam-heated air and 800 to 1100 K for flue gas, and the peripheral speed of the shell is 0.3 to 0.4 m/s.

Example 7.5

A flow of 0.35 kg/s of a solid is to be dried from 15% to 0.5% moisture on a dry basis. The mean heat capacity of the solids is 2.2 kJ/kg deg. K and it is proposed that a co-current adiabatic dryer should be used with the solids entering at 300 K and, because of the heat sensitive nature of the solids, leaving at 325 K. Hot air is available at 400 K with a humidity of 0.01 kg/kg dry air and the maximum allowable mass velocity of the air is 0.95 kg/m²s. What diameter and length should be specified for the proposed dryer?

Solution

With an inlet air temperature and humidity of 400 K and 0.01 kg/kg dry air respectively from Fig. 13.4 in Volume 1, the inlet wet bulb temperature $=312$ K. If, not unreasonably, it is assumed that the number of transfer units is 1.5, then for adiabatic drying the outlet air temperature, T_0 is given by:

$$1.5 = \ln\left((400 - 312)/(T_0 - 312)\right) \text{ and } T_0 = 331.5 \text{ K.}$$

The solids outlet temperature will be taken as the maximum allowable, 325 K.

From the steam tables in the Appendix, the latent heat of vaporisation of water at 312 K is 2410 kJ/kg. Again from steam tables, the specific heat capacity of water vapour $=1.88$ kJ/kg K and that of the solids will be taken as 2.18 kJ/kg K.

For a mass flow of solids of 0.35 kg/s and inlet and outlet moisture contents of 0.15 and 0.005 kg/kg dry solids respectively, the mass of water evaporated $=0.35(0.15{-}0.005)=0.0508$ kg/s.

For unit mass of solids, the heat duty includes:

heat to the solids $=2.18(325{-}300)=54.5$ kJ/kg.

heat to raise the moisture to the dew point $=(0.15 \times 4.187(312-300))=$ 7.5 kJ/kg.

heat of vaporisation $=2410(0.15{-}0.005)=349.5$ kJ/kg.

heat to raise remaining moisture to the solids outlet temperature

$$= (0.005 \times 4.187)(325 - 312) = 0.3 \text{ kJ/kg}$$

and heat to raise evaporated moisture to the air outlet temperature

$$= (0.15 - 0.005)1.88(331.5 - 312) = 5.3 \text{ kJ/kg}$$

a total of $(54.5{+}7.5{+}349.5{+}0.3{+}5.3)=417.1$ kJ/kg.

or:

$(417.1 \times 0.35) = 146 \, \text{kW}$

From Fig. 13.4 in Volume 1, the humid heat of the entering air is 1.03 kJ/kg K and making a heat balance:

$$G_1(1+H_1) = Q/C_{p1}(T_1 - T_2)$$

where:

G_1 (kg/s) is the mass flow rate of inlet air,

H_1 (kg/kg) is the humidity of inlet air,

Q (kW) is the heat duty,

C_{p1} (kJ/kg K) is the humid heat of inlet air

and: T_1 and T_2 (K) are the inlet and outlet air temperatures respectively.

In this case:

$$G_1(1+0.01) = 146/(1.03(400 - 331.5)) = 2.07 \, \text{kg/s}$$

and:

$$\text{mass flowrate of dry air}, G_a = (2.07/1.01) = 2.05 \, \text{kg/s}$$

The humidity of the outlet air is then:

$$H_2 = 0.01 + (0.0508/2.05) = 0.0347 \, \text{kg/kg}$$

At a dry bulb temperature of 331.5 K, with a humidity of 0.0347 kg/kg, the wet bulb temperature of the outlet air, from Fig. 13.4 in Volume 1, is 312 K, the same as the inlet, which is the case for adiabatic drying.

The dryer diameter is then found from the allowable mass velocity of the air and the entering air flow and for a mass velocity of 0.95 kg/m²s, the cross sectional area of the dryer is:

$$(2.07/0.95) = 2.18 \, \text{m}^2$$

equivalent to a diameter of $[(4 \times 2.18)/\pi]^{0.5} = \underline{1.67 \, \text{m}}$

With a constant drying temperature of 312 K:

at the inlet:

$$\Delta T_1 = (400 - 312) = 88 \, \text{deg K}$$

and at the outlet:

$$\Delta T_2 = (331.5 - 312) = 19.5 \, \text{deg K}$$

and the logarithmic mean, $\Delta T_m = (88 - 19.5)/\ln(88/19.5) = 45.5 \, \text{deg K}$.

The length of the dryer, L is then:

$$L = Q/\left(0.0625\pi DG'^{0.67}\Delta T_m\right)$$

where D (m) is the diameter and G' (kg/m^{2}s) is the air mass velocity.
In this case:

$$L = 146/[0.0625\pi \times 1.67(0.95)^{0.67} \times 45.5)] = \underline{10.1\ \text{m}}$$

This gives a length/diameter ratio of $(10.1/1.67)=6$, which is a reasonable value for rotary dryers.

7.5.5 Drum dryers

If a solution or slurry is run on to a slowly rotating steam-heated drum, evaporation takes place and solids may be obtained in a dry form. This is the basic principle used in all drum dryers, some forms of which are illustrated in Fig. 7.16. The feed to a single drum dryer may be of the dip, pan, or splash type. The dip-feed system, the earliest design, is still used where liquid can be picked up from a shallow pan. The agitator prevents settling of particles, and the spreader is sometimes used to produce a uniform coating on the drum. The knife, which is employed for removing the dried material, functions in a similar manner to the doctor blade on a rotary filter. If the material is dried to give a free-flowing powder, this comes away from the drum quite easily. The splash-feed type is used for materials, such as calcium arsenate, lead arsenate, and iron oxide, where a light fluffy product is desired. The revolving cylinder throws the liquor against the drum, and a uniform coating is formed with materials which do not stick to the hot surface of the drum.

Double drum dryers may be used in much the same way, and Fig. 7.16 shows dip-feed and top-feed designs. Top-feed gives a larger capacity as a thicker coating is obtained. It is important to arrange for a uniform feed to a top-feed machine, and this may be effected by using a perforated pipe for solutions and a travelling trough for suspensions.

Drums are usually made from cast iron, although chromium-plated steel or alloy steel is often used where contamination of the product must be avoided, such as with pharmaceuticals or food products. Arrangements must be made for accurate adjustment of the separation of the drums, and the driving gears should be totally enclosed. A range of speeds is usually obtained by selecting the gears, rather than by fitting a variable speed drive. Removal of the steam condensate is important, and an internal syphon is often fitted to keep the drum free of condensate. In some cases, it is better for the drums to be rotated upwards at the point of closest proximity, and the knives are then fitted at the bottom. By this means, the dry material is kept

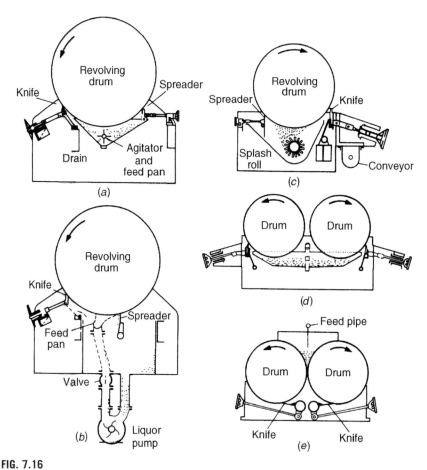

FIG. 7.16

Methods of feeding drum dryers. (A) Single drum, dip-feed. (B) Single drum, pan-feed. (C) Single drum, splash-feed. (D) Double drum, dip-feed. (E) Double drum, top-feed.

away from the vapour evolved. Some indication of the sizes of this type of dryer is given in Table 7.3, and it may be noted that the surface of each drum is limited to about 35 m^2. This, coupled with developments in the design of spray dryers, renders the latter more economically attractive, especially where large throughputs are to be handled. For steam-heated drum dryers, normal evaporative capacities are proportional to the active drum area and are 0.003 to 0.02 kg/m^2 s, although higher rates are claimed for grooved drum dryers.

The solid is usually in contact with the hot metal for 6 to 15 s, short enough to prevent significant decomposition of heat sensitive materials, and heat transfer coefficients are 1 to 2 kW/m^2 K.

Table 7.3 Sizes of double drum dryers.

Drum dimensions Diameter (m) × Length (m)	Length (m)	Width (m)	Height (m)	Mass (kg)
0.61 × 0.61	3.5	2.05	2.3	3850
0.61 × 0.91	3.8	2.05	2.4	4170
0.81 × 1.32	5.0	2.5	2.8	7620
0.81 × 1.83	5.5	2.5	2.8	8350
0.81 × 2.28	5.9	2.5	2.8	8890
0.81 v 2.54	6.25	2.5	2.8	9300
0.81 × 3.05	6.7	2.5	2.8	10,120
1.07 × 2.28	6.4	3.0	3.0	14,740
1.07 × 2.54	6.7	3.0	3.0	15,420
1.07 × 3.05	7.2	3.0	3.0	16,780
1.52 × 3.66	7.9	4.1	4.25	27,220

When the temperature of the drying material must be kept as low as possible, vacuum drying is used, and one form of vacuum dryer is shown in Fig. 7.17. The dried material is collected in two screw conveyors and carried usually to two receivers so that one can be filled whilst the other is emptied.

7.5.6 Spray dryers

Water may be evaporated from a solution or a suspension of solid particles by spraying the mixture into a vessel through which a current of hot gases is passed. In this way, a large interfacial area is produced and consequently a high rate of evaporation is obtained. Drop temperatures remain below the wet bulb temperature of the drying gas until drying is almost complete, and the process thus affords a convenient means of drying substances which may deteriorate if their temperatures rise too high, such as milk, coffee, and plasma. Furthermore, because of the fine state of subdivision of the liquid, the dried material is obtained in a finely divided state.

In spray drying, it is necessary to atomise and distribute, under controlled conditions, a wide variety of liquids, the properties of which range from those of solutions, emulsions, and dispersions, to slurries and even gels. Most of the atomisers commonly employed are designed for simple liquids, that is mobile Newtonian liquids. When atomisers are employed for slurries, pastes, and liquids having anomalous properties, there is a great deterioration in performance and, in many cases, atomisers may be rapidly eroded and damaged so as to become useless. There is therefore much to be gained by considering various types and designs of atomiser so that a suitable selection can be made for the given duty.

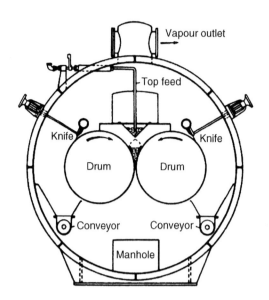

FIG. 7.17

Vacuum drum dryer.

The performance of a spray dryer or reaction system is critically dependent on the drop size produced by the atomiser and the manner in which the gaseous medium mixes with the drops. In this context an atomiser is defined as a device which causes liquid to be disintegrated into drops lying within a specified size range, and which controls their spatial distribution.

Atomisers

Atomisers are classified in Table 7.4 [30] according to the three basic forms of energy commonly employed – pressure energy, centrifugal kinetic energy, and gaseous energy. Where greater control is required over disintegration or spatial dispersion, combinations of atomiser types may be employed, and, for example, swirl-spray nozzles or spinning discs may be incorporated in a blast atomiser, their primary functions being to produce thin liquid sheets which are then eventually atomised by low-, medium-, or high-velocity gas streams.

The fundamental principle of disintegrating a liquid consists in increasing its surface area until it becomes unstable and disintegrates. The theoretical energy requirement is the increase in surface energy plus the energy required to overcome viscous forces, although in practice this is only a small fraction of the energy required. The process by which drops are produced from a liquid stream depends upon the nature of the flow in the atomiser, that is whether it is laminar or turbulent, the way in which energy is imparted to the liquid, the physical properties of the liquid, and the

Table 7.4 Classification of atomisers [30].

Pressure used in pressure atomizers	Centrifugal energy used in rotary atomizers	Gaseous energy used in twin-fluid or blast atomizers
Fan-spray nozzles 0.25–1.0 MN/m²	Spinning cups 6–30 m/s	External mixing (fluid pressures independent)
Impact nozzles Impinging jet nozzles 0.25–1.0 MN/m²	Spinning discs Flat discs Saucer-shaped discs	Internal mixing (fluid pressures interdependent)
Impact plate nozzles Up to 3.0 MN/m²	Radial-vaned discs Multiple discs	Low velocity
Deflector nozzles 7.0 MN/m²	30–180 m/s	Gas velocity 30–120 m/s
		Gas–liquid ratio
Swirl-spray nozzles Hollow cone, full cone 0.4–70 MN/m²		2–25 kg/kg
		Medium velocity Gas velocity 120–300 m/s
Divergent pintle nozzles 0.25–7.0 MN/m²		Gas–liquid ratio
Fixed or vibrating pintle		0.2–1 kg/kg
		High velocity Gas velocity sonic or above Gas–liquid ratio 0.2–1 kg/kg

properties of the ambient atmosphere. The basic mechanism is, however, unaffected by these variables and consists essentially of the breaking down of unstable threads of liquid into rows of drops and conforms to the classical mechanism postulated by Lord Rayleigh [31]. This theory states that a free column of liquid is unstable if its length is greater than its circumference, and that, for a non-viscous liquid, the wavelength of that disturbance which will grow most rapidly in amplitude is 4.5 times the diameter. This corresponds to the formation of droplets of diameter approximately 1.89 times that of the jet d_j. Weber [32] has shown that the optimum wavelength for the disruption of jets of viscous liquid is:

$$\lambda_{opt} = \sqrt{(2\pi d_j)} \left[1 + \frac{3\mu}{\sqrt{(\rho\sigma\, d_j)}} \right]^{0.5} \tag{7.31}$$

Rim   Perforated sheet   Wavy sheet

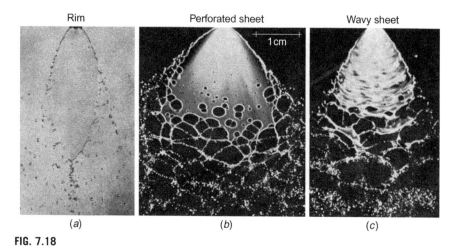

(a)   (b)   (c)

FIG. 7.18

Modes of disintegration of liquid films.

A uniform thread will break down into a series of drops of uniform diameter, each separated by one or more satellite drops. Because of the heterogeneous character of the atomisation process, however, non-uniform threads are produced and this results in a wide range of drop sizes. An example of a part of a laminar sheet collapsing into a network of threads and drops is shown in Fig. 7.18. Only when the formation and disintegration of threads are controlled can a homogeneous spray cloud be produced. One method by which this can be achieved is by using a rotary cup atomiser operating within a critical range of liquid flow rates and rotor speeds, although, as shown later, this range falls outside that normally employed in practice.

Although certain features are unique to particular atomiser types, many of the detailed mechanisms of disintegration are common to most forms of atomiser [33]. The most effective way of utilising energy imparted to a liquid is to arrange that the liquid mass has a large specific surface before it commences to break down into drops. Thus, the primary function of an atomiser is to transpose bulk liquid into thin liquid sheets. Three modes of disintegration of such spray sheets have been established [34], namely *rim*, *perforated sheet*, and *wave*. Because of surface tension, the free edge of any sheet contracts into a thick rim, and *rim disintegration* occurs as it breaks up by instabilities analogous to those of free jets. Fig. 7.18A illustrates a fan spray sheet and shows that, as the liquid in each edge moves along the curved boundary, the latter becomes disturbed and disintegrates. When this occurs, the resulting drops sustain the direction of flow of the edge at the point at which the drops are formed, and remain attached to the receding surface by thin threads which rapidly disintegrate into streams of drops.

In *perforated-sheet disintegration* shown in Fig. 7.18B, small holes suddenly appear in the sheet as it advances into the atmosphere. They rapidly grow in size [35] until the thickening rims of adjacent holes coalesce to form threads of varying diameter. The threads finally break down into drops.

Disintegration can also occur through the superimposition of a wave motion on the sheet, as shown in Fig. 7.18C. Sheets of liquid corresponding to half or full wavelengths of liquid are torn off and tend to draw up under the action of surface tension, although these may suffer disintegration by air action or liquid turbulence before a regular network of threads can be formed.

In a *pressure atomiser*, liquid is forced under pressure through an orifice, and the form of the resulting liquid sheet can be controlled by varying the direction of flow towards the orifice. By this method, flat and conical spray sheets can be produced. From the Bernoulli equation, given in Volume 1, Chapter 6, the mass rate of flow through a nozzle may be derived as:

$$G = C_{D\rho} A_N \sqrt{[2(-\Delta P)]\rho} \tag{7.32}$$

For a given nozzle and fluid, and an approximately constant coefficient of discharge, C_D, then:

$$G = \text{constant}\sqrt{(-\Delta P)} \tag{7.33}$$

The capacity of a nozzle is conveniently described by the *flow number*, **FN** a dimensional constant defined by:

$$\mathbf{FN} \equiv \frac{\text{Volumetric flow (gal/h)}}{\sqrt{\text{Pressure (lb/in}^2)}} = 2.08 \times 10^6 \frac{\text{Volume flow (m}^3/\text{s)}}{\sqrt{\text{Pressure (kN/m}^2)}} \tag{7.34}$$

In the fan-spray drop nozzle shown in Fig. 7.19A, two streams of liquid are made to impinge behind an orifice by specially designed approach passages, and a sheet is formed in a plane perpendicular to the plane of the streams. The orifice runs full and, since the functional portion is sharp-edged, high discharge coefficients are obtained which are substantially constant over wide ranges of Reynolds number.

The influence of conditions on the droplet size where the spray sheet disintegrates through aerodynamic wave motion may be represented by the following expression proposed by Dombrowski and Munday [33] for ambient densities around normal atmospheric conditions:

$$d_m = \left(\frac{0.000156}{C_D}\right) \left[\frac{\mathbf{FN}\sigma\rho}{\sin\phi(-\Delta P)}\right]^{1/3} \rho_A^{-1/6} \tag{7.35}$$

where **FN** is given in Eq. (7.34), ϕ is the half-angle of the sheet, $-\Delta P$ is in kN/m^2 and all other quantities are expressed in SI units.

The principle of operation of the impinging jet nozzle resembles that of the fan spray nozzle with the exception that two or more independent jets are caused to impinge in the atmosphere. In impact atomisers, one jet is caused to strike against a solid surface, and for two jets impinging at 180° [33], using SI units:

$$d_m = 1.73 \left(\frac{d_j^{0.75}}{u_l^{0.5}}\right) \left[\frac{\sigma}{\rho}\right]^{0.25} \tag{7.36}$$

With this atomiser, the drop size is effectively independent of viscosity, and the size spectrum is narrower than with other types of pressure nozzle.

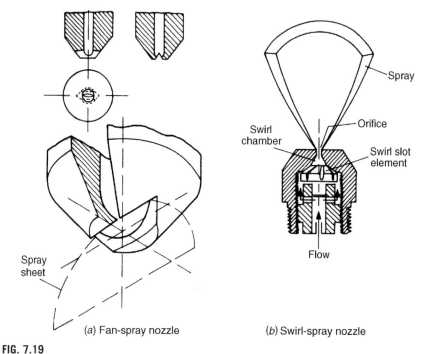

FIG. 7.19

Pressure atomisers.

When liquid is caused to flow through a narrow divergent annular orifice or around a pintle against a divergent surface on the end of the pintle, a conical sheet of liquid is produced where the liquid is flowing in radial lines. Such a sheet generally disintegrates by an aerodynamic wave motion. The angle of the cone and the root thickness of the sheet may be controlled by the divergence of the spreading surface and the width of the annulus. For small outputs, this method is not so favourable because of difficulties in making an accurate annulus. A conical sheet is also produced when the liquid is caused to emerge from an orifice with a tangential or swirling velocity resulting from its path through one or more tangential or helical passages before the orifice. Fig. 7.19B shows a typical nozzle used for a spray dryer. In such swirl-spray nozzles the rotational velocity is sufficiently high to cause the formation of an air core throughout the nozzle, resulting in low discharge coefficients for this type of atomiser.

Several empirical relations have been proposed to express drop size in terms of the operating variables. One suitable for small atomisers with 85° spray cone angles, at atmospheric pressure is [33]:

$$d_m = 0.0134 \left(\frac{FN^{0.209}(\mu/\rho)^{0.215}}{(-\Delta P)^{0.348}} \right) \tag{7.37}$$

where $-\Delta P$ is in kN/m^2, **FN** is given in Eq. (7.34), and all other quantities are in SI units.

Pressure nozzles are somewhat inflexible since large ranges of flow rate require excessive variations in differential pressure. For example, for an atomiser operating satisfactorily at 275 kN/m^2, a pressure differential of 17.25 MN/m^2 is required to increase the flow rate to 10 times its initial value. These limitations, inherent in all pressure-type nozzles, have been overcome in swirl-spray nozzles by the development of spill, duplex, multi-orifice, and variable port atomisers, in which ratios of maximum to minimum outputs in excess of 50 can be easily achieved [33].

In a *rotary atomiser*, liquid is fed on to a rotating surface and spread out by centrifugal force. Under normal operating conditions the liquid extends from the periphery in the form of a thin sheet which breaks down some distance away from the periphery, either freely by aerodynamic action or by the action of an additional gas blast. Since the accelerating force can be controlled independently, this type of atomiser is extremely versatile and it can handle successfully a wide range of feed rates with liquids having a wide range of properties. The rotating component may be a simple flat disc though slippage may then occur, and consequently it is more usual to use bowls, vaned discs, and slotted wheels, as shown in Fig. 7.20. Diameters are 25 to 460 mm and small discs rotate at up to 1000 Hz whilst the larger discs rotate at up to 200 Hz with capacities of 1.5 kg/s. Where a coaxial gas blast is used to effect atomisation, lower speeds of the order of 50 Hz may be used. At very low flow rates, such as 30 mg/s, the liquid spreads out towards the cup lip where it forms a ring. As liquid continues to flow into the ring, its inertia increases, overcomes the restraining surface tension and viscous forces and is centrifuged off as discrete drops of

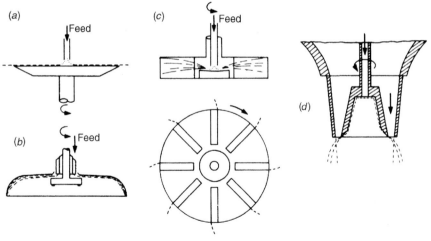

FIG. 7.20

Characteristic rotary atomisers: (A) Sharp-edge flat disc; (B) Bowl; (C) Vaned disc; (D) Air-blast bowl atomiser.

uniform size, which initially remain attached to the rim by a fine attenuating thread. When the drop is finally detached, the thread breaks down into a chain of small satellite drops. Since the satellite drops constitute only a small proportion of the total liquid flow rate, a cup operating under these conditions effectively produces a monodisperse spray. Under these conditions, the drop size from sharp-edged discs has been given by Walton and Prewett [36] as:

$$d_m = \left(\frac{0.52}{N}\right)\sqrt{\left(\frac{\sigma}{D\rho}\right)} \qquad (7.38)$$

When the liquid flow rate is increased, the retaining threads grow in thickness and form long jets. As they extend into the atmosphere, these jets are stretched and finally break down into strings of drops. Under more practical ranges of flow rate, the jets are unable to remove all the liquid, the ring is forced away and a thin sheet extends around the cup lip and eventually breaks up into a polydisperse spray, as shown in Fig. 7.21.

A far greater supply of energy for disintegration of the liquid jet can be provided by using a high-speed gas stream which impinges on a liquid jet or film. By this means a greater surface area is formed and drops of average size less than $20\,\mu m$ can be produced. This is appreciably smaller than is possible by the methods previously described, although the energy requirements are much greater. The range of flow rates which can be used is wide, because the supply of liquid and the energy for atomisation can be controlled independently. Gas velocities ranging from $50\,m/s$ to sonic velocity are common, and sometimes the gas is given a vortex motion.

Break-up of the jet occurs as follows. Ligaments of liquid are torn off, which collapse to form drops. These may be subsequently blown out into films, which in turn further collapse to give a fine spray. Generally, this spray has a small cone angle and is capable of penetrating far greater distances than the pressure nozzle. Small atomisers of this type have been used in spray-drying units of low capacities.

Where the gas is impacted on to a liquid jet, the mean droplet size is given approximately by [33]:

$$d_s = \frac{0.585}{u_r}\sqrt{\left(\frac{\sigma}{\rho}\right) + 0.0017\left(\frac{\mu}{\sqrt{(\sigma\rho)}}\right)^{0.45}\left(\frac{1000}{j}\right)^{1.5}} \qquad (7.39)$$

where:

d_s is the surface mean diameter (m),
j is the volumetric ratio of gas to liquid rates at the pressure of the surrounding atmosphere,
u_r is the velocity of the gas relative to the liquid (m/s),
μ is the liquid viscosity (Ns/m^2),
ρ is the liquid density (kg/m^3), and
σ is the surface tension (N/m).

As Fraser et al. point out, more efficient atomisation is achieved if the liquid is spread out into a film before impact [35].

(a) low speed

(b) high speed

FIG. 7.21

Spray sheet from a rotating cup.

Drying of drops

The amount of drying that a drop undergoes depends upon the rate of evaporation and the contact time, the latter depending upon the velocity of fall and the length of path through the dryer. The terminal velocity and the transfer rate depend upon the flow conditions around the drop. Because of the nature of the flow pattern, the latter also varies with angular position around the drop, although no practical design method has incorporated such detail and the drop is always treated as if it evaporates uniformly from all its surface.

There are two main periods of evaporation. When a drop is ejected from an atomiser its initial velocity relative to the surrounding gas is generally high and very high rates of transfer are achieved. The drop is rapidly decelerated to its terminal velocity, however, and the larger proportion of mass transfer takes place during the free-fall period. Little error is therefore incurred in basing the total evaporation time on this period.

An expression for the evaporation time for a pure liquid drop falling freely in air has been presented by Marshall [37,38]. For drop diameters less than $100\,\mu m$, this may be simplified to give:

$$t = \frac{\rho\lambda\left(d_0^2 - d_t^2\right)}{8k_f\Delta T} \tag{7.40}$$

Sprays generally contain a wide range of drop sizes, and a stepwise procedure can be used [38] to determine the size spectrum as evaporation proceeds.

When single drops containing solids in suspension or solutions are suspended in hot gas streams, it is found that evaporation initially proceeds in accordance with Eq. (7.40) for pure liquid, although when solids deposition commences, a crust or solid film is rapidly formed which increases the resistance to transfer. Although this suggests the existence of a falling rate period similar to that found in tray-drying, the available published data indicate that it has little effect on the total drying time. As a result of crust formation, the dried particles may be in the form of hollow spheres.

Industrial spray dryers

Spray dryers are used in a variety of applications where a fairly high-grade product is to be made in granular form. In the drying chamber, the gas and liquid streams are brought into contact, and the efficiency of mixing depends upon the flow patterns induced in the chamber. Rotating disc atomisers are most commonly used. Counter-current dryers give the highest thermal efficiencies although product temperatures are higher in these units. This limits their use to materials which are not affected by overheating. Co-current dryers suffer from relatively low efficiencies, although they have the advantage of low product temperatures unless back-mixing occurs. In the case of materials which are extremely sensitive to heat, great care has to be taken in the design of the chamber to avoid overheating. Combustion gases are frequently used directly although, in some cases, such as the preparation of food products, indirectly heated air is used. Maximum temperatures are then normally limited to lower values than those with direct heating. Typical flow arrangements in spray drying are shown in Fig. 7.22.

The drying time and size of the particles are directly related to the droplet size, and therefore the initial formation of the spray is of great importance. The factors which govern the choice of atomisers for any specific drying application are principally dependent upon the characteristics of the liquid feed and upon the required drying characteristics of the drying chamber. A general guide is given in Table 7.5.

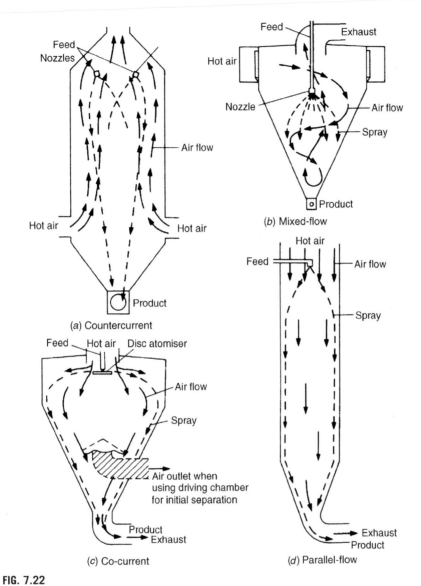

FIG. 7.22

Flow arrangements in spray-drying [21].

Pressure nozzles are most suited to low viscosity liquids and, where possible, viscous liquids should be preheated to ensure the minimum viscosity at the nozzle. Because of their simplicity, pressure nozzles are also employed to atomise viscous liquids with a kinematic viscosity up to $0.01\,\text{m}^2/\text{s}$, depending upon the nozzle capacity. Under these conditions, injection pressures of up to $50\,\text{MN/m}^2$ may be required to

Table 7.5 Choice of atomizer.

	Atomiser			
	Pressure	**Rotary**	**Twin-fluid**	**Spinning-cup plus gas blast**
Drying chamber				
Co-current	*	*	*	*
Countercurrent	*		*	
Feed				
Low viscosity solutions	*		*	
High viscosity solutions			*	*
Slurries		*	*	*
Pastes		*	*	*
Flexibility	Flow rate $\alpha v(-\Delta P)$	Flow rate independent of cup speed	Liquid flow independent of gas energy	Liquid flow independent of cup speed and gas energy

produce the required particle size. With slurries, the resulting high liquid velocities may cause severe erosion of the orifice and thus necessitate frequent replacement.

Spinning discs are very suitable for slurries and pastes, whilst high-viscosity liquids tend to produce a stringy product. Care must also be taken in design to minimise incrustation around the lip and subsequent out-of-balance as drying takes place.

The simple gas atomiser is inherently fairly flexible although it has not yet found widespread application. This is a result of its tendency to produce a dusty product containing a large proportion of very small particles.

Often, little difficulty is experienced in removing the majority of the dried product, though in most cases the smaller particles that may be carried over in the exit gases must be reclaimed. Cyclones are the simplest form of separator though bag filters or even electrostatic precipitators may be required. With heat-sensitive materials, and in cases where sterility is of prime importance, more elaborate methods are required. For example, cooling streams of air may be used to aid the extraction of product whilst maintaining the required low temperature. Mechanical aids are often incorporated to prevent particles adhering to the chamber walls, and, in one design, the cooling air also operates a revolving device which sweeps the walls.

In some cases all the product is conveyed from the dryer by the exhaust gases and collected outside the drying chamber. This method is liable to cause breakage of the particles though it is particularly suited for heat-sensitive materials which may deteriorate if left in contact with hot surfaces inside the dryer.

Spray drying has generally been regarded as a relatively expensive process, especially when indirect heating is used. The data given in Table 7.6 taken from Grose

Table 7.6 Operating costs for a spray dryer evaporating 0.28 kg/s (1 t/h) of water [39].

Basis: Air outlet temperature, 353 K

Cost of steam, £10/Mg

Cost of fuel oil, £150/Mg

Cost of power, £0.054/kWh (£15.0×10^{-6}/kJ)

	Air inlet temperature (K)	Steam flow (kg/s)	Oil flow (kg/s)	Power consumption (kW)	Cost (£/100 h)			Cost (£/Mg)	
					Fuel	Power	Total	Total	
Air heated:									
Indirectly	453	0.708	–	71	2550	384	2934	8150	
Direct combustion	523	–	0.033	55	1800	296	2096	5820	
Direct combustion	603	–	0.031	47	1650	254	1904	5300	

and Duffield [39] using 1990 costs illustrate the cost penalties associated with indirect heating or with low inlet temperatures in direct heating.

Quinn [40] has drawn attention to the advantages with larger modern units using higher air inlet temperatures, 675 K for organic products and 925 K for inorganic products.

In spray dryers, using either a nozzle or rotating disc as the atomiser [40], volumetric evaporative capacities are 0.0003 to 0.0014 kg/m^3 s for cross-and co-current flows, with drying temperatures of 420 to 470 K. For handling large volumes of solutions, spray dryers are unsurpassed, and it is only at feed rates below 0.1 kg/s, that a drum dryer becomes more economic. Indeed the economy of spray drying improves with capacity until, at evaporative capacities of greater than 0.6 kg/s, the unit running cost is largely independent of scale.

In the jet spray dryer, cold feed is introduced [41] into preheated primary air which is blown through a nozzle at velocities up to 400 m/s. Very fine droplets are obtained with residence times of around 0.01 s, and an air temperature of 620 K. This equipment has been used for evaporating milk without adverse effect on flavour and, although operating costs are likely to be high, the system is well suited to the handling of heat-sensitive materials.

7.5.7 Pneumatic dryers

In pneumatic dryers, the material to be dried is kept in a state of fine division, so that the surface per unit volume is high and high rates of heat transfer are obtained. Solids are introduced into the dryer by some form of mechanical feeder, such as a rotating star wheel, or by an extrusion machine arranged with a high-speed guillotine to give short lengths of material, such as 5 to 10 mm. Hot gases from a furnace, or more frequently from an oil burner, are passed into the bottom of the dryer, and these pick up the particles and carry them up the column. The stream of particles leaves the dryer through a cyclone separator and the hot gases pass out of the system. In some instances, final collection of the fine particles is by way of a series of bag filters. The time of contact of particles with the gases is small, typically 5 s, even with a lengthy duct – and the particle temperature does not approach the temperature of the hot gas stream. In some cases, the material is recycled, especially where bound moisture is involved. Evaporative capacities, which are high, are greatly affected by the solids–air ratio. Typical thermal and power requirements are given by Quinn [42] as 4.5 MJ/kg moisture evaporated and 0.2 MJ/kg, respectively.

A typical installation with the associated equipment is shown in Fig. 7.23. Wet feed is delivered on to a bed of previously dried material in a double-paddle mixer to produce a friable mixture. This then passes to a cage mill where it comes into contact with hot gases from the furnace. Surface and some inherent moisture is immediately flash evaporated. The cage mill breaks up any agglomerates to ensure uniform drying of each individual particle. Gases and product are drawn up the drying column, where inherent moisture continues to be evaporated, before passing into the cyclone collector. Separated solids discharge through a rotary air lock into a dry divider which is set

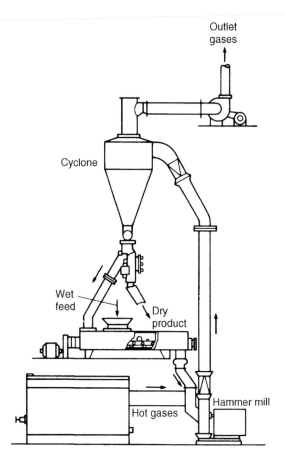

FIG. 7.23

Air-lift dryer with an integral mill.

to recycle an adequate percentage of solids for conditioning the new wet feed in the double paddle mixer. Gases from the cyclone are vented to atmosphere through a suitable dust collector or wet scrubber. The system operates under suction and dust is therefore reduced to a minimum. A direct oil or gas fired furnace is generally employed and the heat input is controlled according to the vent stack gas temperature. Indirect heating may be used where contamination of the product is undesirable.

Materials handled include food products, chalk, coal, organic chemicals, clays, spent coffee grounds, sewage sludge, and chicken manure. Where exhaust gases have unpleasant odours, after-burners can be supplied to raise the temperature and burn off the organic and particulate content causing the problem.

Convex dryers are continuously operating pneumatic dryers with an inherent classifying action in the drying chamber which gives residence times for the individual particles differing according to particle size and moisture content. Such units offer the processing advantages of short-time, co-current dryers and are used primarily for drying reasonably to highly free-flowing moist products that can be conveyed pneumatically and do not tend to stick together. By virtue of the pronounced classifying action, such dryers are also well suited to the drying of thermally sensitive moist materials with widely differing particle sizes where the large particles have to be completely dried without any overheating of the small ones. Basically, this form of pneumatic dryer consists of a truncated cyclone with a bottom outlet that acts as a combined classifier and dryer.

7.5.8 Fluidised bed dryers

The principles of fluidisation, discussed in Chapter 6, are applied in this type of dryer, shown typically in Fig. 7.24. Heated air, or hot gas from a burner, is passed by way of a plenum chamber and a diffuser plate, fitted with suitable nozzles to prevent any back-flow of solids, into the fluidised bed of material, from which it passes to a dust separator. Wet material is fed continuously into the bed through a rotary valve, and this mixes immediately with the dry charge. Dry material overflows through a downcomer to an integral after-cooler. An alternative design of this type of dryer is one in which a thin bed is used.

Quinn [42] points out that, whilst it would seem impossible to obtain very low product moisture levels when the incoming feed is very wet and at the same time ensure that the feed point is well away from the discharge point, this is not borne

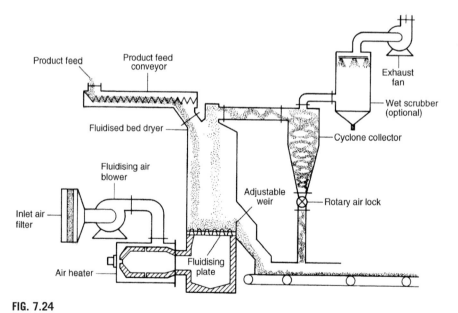

FIG. 7.24

Flow diagram for a typical continuous fluidised-bed dryer.

out by operating experience. Mixing in the bed is so rapid that it may be regarded as homogeneous, and baffles or physical separation between feed and discharge points are largely ineffective. The very high mass-transfer rates achieved make it possible to maintain the whole bed in a dry condition. Some rectangular fluidised-bed dryers have separately fluidised compartments through which the solids move in sequence. The residence time is very similar for all particles and the units, known as *plug-flow* dryers, often employ cold air to effect product cooling in the last stage.

Many large-scale uses include the drying of fertilisers, plastics materials, foundry sand, and inorganic salts, and Agarwal et al. [43] describe a plant consisting of two units, each drying 10.5 kg/s of fine coal. Small fluidised-bed dryers also find use in, for example, the drying of tablet granulations in the pharmaceutical industry [44].

Dryers with grid areas of up to 14 m^2 have been built [45] and evaporative capacities vary from 0.02 kg/s m^2 grid area for the low-temperature drying of food grains to 0.3 kg/s m^2 for the drying of pulverised coal by direct contact with flue gases [43]. Specific air rates are usually 0.5 to 2.0 kg/s m^2 grid area and the total energy demand is 2.5 to 7.5 MJ/kg moisture evaporated. The exit gas is nearly always saturated with vapour for all allowable fluidisation velocities.

The choice between spray, pneumatic and fluidised dryers depends very much on the properties of the particles and some guidance in this respect is given in Table 7.7.

Design considerations

A simple, concise method for the preliminary sizing of a fluidised bed dryer has been proposed by Clark [46] and this is now considered.

The minimum bed diameter is a function of the operating velocity, the particle characteristics and the humidity of the drying gas. The hot gas at the inlet rapidly

Table 7.7 Selection of spray, pneumatic and fluidised-bed dryers.

Particle property	Spray dryer	Pneumatic dryer	Fluidised-bed dryer
Initial moisture greater than 80%	Yes	No	No
Too dry to pump	No	Yes	Yes
Wet enough to pump but moisture less than 80%	Yes	Yes	No
Solids in dissolved state	Yes	No	No
Partially dry but sticky particles	Yes	No	No
Fragile particles	Yes	No	Possible
Very small particles	Yes	Yes	No
Residence time (s)	3–10	1–10, often less than 1	Widely variable, greater than 10
Heat sensitive material	Yes	Yes	No
Relative drying speed	Third	First	Second

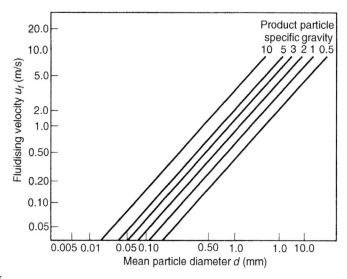

FIG. 7.25

Superficial operating velocity in fluidised bed dryers [46].

loses heat and gains moisture as it passes through the bed which it eventually leaves at the bed temperature T_b and with a relative humidity R, which is approximately equal to the relative humidity which would be in equilibrium with the dried product at the bed temperature. The operating velocity may be taken as twice the minimum fluidising velocity, obtained from the equations in Section 6.1.3, by laboratory tests, or more conveniently from Fig. 7.25.

For drying media other than air at approximately atmospheric pressure, the velocity obtained from Fig. 7.25 should be multiplied by $0.00975\,\rho^{-0.29}\mu^{-0.43}$ where ρ and μ are the density (kg/m^3) and the viscosity (Ns/m^2) of the fluidising gas.

From a mass balance across the bed:

$$(R100)(P_w/P) = [W + (G/(1 + H)H]/[W + (G/(1 + H)(0.625 + H)] \qquad (7.41)$$

where:

 $R =$ exit gas relative humidity (per cent),
 $P_w =$ vapour pressure of water at the exit gas temperature (N/m^2),
 $P =$ total static pressure of gases leaving the bed (N/m^2),
 $W =$ evaporative capacity (kg/s),
 $G =$ inlet flow of air (kg/s), and
 $H =$ humidity of inlet air (kg/kg dry air).

The term $G/(1 + H)$ is in effect the flow of dry air, and 0.625 is approximately the ratio of molecular masses in the case of water and air.

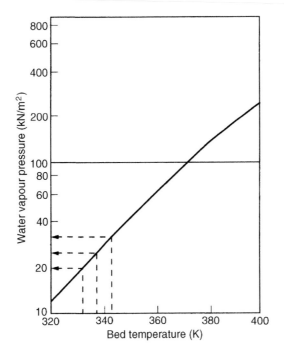

FIG. 7.26

Water vapour pressure at the bed exit [46].

Values of P_w may be obtained from Fig. 7.26 and, for indirect heating, H is the humidity of the ambient air. For direct heating, H may be obtained from Fig. 7.27 which assumes that the air at 293 K, with a humidity of 0.01 kg/kg, is heated by the combustion of methane.

A heat balance (in W) across the bed gives:

$$C_m G(T_m - T_b) = \lambda_b W + C_f F(T_b - T_f) \tag{7.42}$$

where:

C_m = mean thermal capacity of the gas between T_m and T_b (J/kg deg. K) (1000 J/kg deg. K for air),

$C_f = [(X_f c_x + c_s)/(X_f + 1)]$ average thermal capacity of the wet solid between T_f and T_b (J/kg deg. K),

X_f = moisture content of wet feed (kg/kg dry solid),

C_x = heat capacity of liquid being evaporated (J/kg deg. K),

C_s = heat capacity of dry solid (J/kg deg. K),

λ_b = mean latent heat of liquid at T_b (J/kg),

T_m, T_b, T_f = temperatures of inlet gas, bed and wet feed respectively (K), and

F = rate of wet solid feed (kg/s).

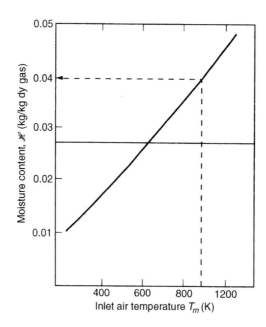

FIG. 7.27

Moisture content of the inlet air [46].

Values of G and T_b are obtained from Eqs (7.41), (7.42) and the bed diameter is then obtained from:

$$D^2 = \frac{(G + 1.58 \ W)T_b}{278 u_f} \ \text{m}^2 \qquad (7.43)$$

noting that: $\frac{\text{Mean molecular weight of inlet air}}{\text{molecular weight of water}} = \frac{29}{18} \approx 1.58$

and:

$$\frac{\pi}{4} \left\{ \frac{(\text{molecular weight or air}) \times (\text{datum temperature})}{\text{kilomolecular volume}} \right\} = \frac{\pi}{4} \left(\frac{29 \times 273}{22.4} \right)$$

$$= 278 \ \text{kg deg K} / \text{m}^3$$

Example 7.6

A granular material of density $5000 \ \text{kg/m}^3$ is to be dried in a fluidised bed dryer using directly heated air at $811 \ \text{K}$. The particle size is $0.5 \ \text{mm}$ and $1.26 \ \text{kg/s}$ of water is to be removed from $12.6 \ \text{kg/s}$ of solid feed at $293 \ \text{K}$. What diameter of bed should be specified?

$\lambda_b = 2325 \ \text{kJ/kg}$, ambient air is at $293 \ \text{K}$, $H = 0.01 \ \text{kg/kg}$ and $C_f = 1.67 \ \text{kJ/kg deg. K}$.

Solution

From Fig. 7.27: $H = 0.036 \ \text{kg/kg}$ at $811 \ \text{K}$.

and the right-hand side of Eq. (7.41) is:

$$= (1.26 + 0.0347G)/(1.26 + 0.638\,G)$$

Taking R as 90% and P as 101.3 kN/m^2, then, for assumed values of T_b of 321, 333, and 344 K:

$P_w = 13$, 20, and 32 kN/m^2, respectively.

and: $G = 27.8$, 12.9, and 6.02 kg/s, respectively.

Using Eq. (7.42), for $T_b = 321$, 333, and 344 K,

$G = 7.16$, 7.8, and 7.54 kg/s, respectively.

Plotting G against T_b for each equation on the same axis, then:

$G = 8.3$ kg/s when $T_b = 340$ K.

From Fig. 7.26, $u_f = 0.61$ m/s. Hence, in Eq. (7.43):

$$D_2 = 340(8.3 + (1.58 \times 1.26))/(278 \times 0.61)$$

and:

$$\underline{D = 4.60\ \text{m}}$$

This is a very large diameter bed and it may be worthwhile increasing the fluidising velocity provided any increased elutriation is acceptable. If u_f is increased to 1.52 m, the diameter then becomes 2.88 m with a subsequent reduction in capital costs.

One development in the field of fluidised bed drying is what is known as *sub-fluidised conditioning*. A fluidised-bed dryer normally works with a maximum residence time of some 1200 s. If this is increased, then the spread of product residence times increases excessively due to the fact that axial and longitudinal mixing occur in the bed during the fluidisation process and, the more vigorous the fluidisation, the greater the spread of residence times. If fluidisation continues for too long, breakage and other product damage is likely to occur. A solution to this problem, developed by Ventilex, is sub-fluidised conditioning where forward movement of the product is created by a shaking mechanism in which the product is kept at the threshold of fluidisation. As a result, there is little or no axial or longitudinal mixing and the spread of residence time is kept to a minimum. In essence, this process provides a situation which approaches that of ideal plug flow, and many products can be conditioned by this combination of fluidised and sub-fluidised techniques. In addition, even less air is required for drying and thus small auxiliary equipment is required and energy requirements are less. Nominal bed thicknesses are 150 to 250 mm and long residence times, of up to 7 ks, are possible with a minimal time spread. In the drying of drugs, nuts, meats and rice, no degradation takes place and, in this respect, the process is comparable with traditional installations such as conveyor dryers, with the added benefits of fluidisation. The system is also applicable to the drying of minerals such as China clay and sand, and also to animal wastes and sludges. A typical installation together with a flow diagram is shown in Fig. 7.28.

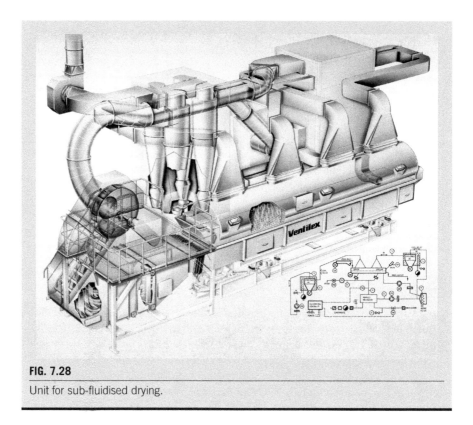

FIG. 7.28

Unit for sub-fluidised drying.

7.5.9 Turbo-shelf dryers

The handling of sticky materials can present difficulties, and one type of dryer which is useful for this type of material is the turbo-dryer. As shown in Fig. 7.29, wet solid is fed in a thin layer to the top member of a series of annular shelves each made of a number of segmental plates with slots between them. These shelves rotate and, by means of suitably placed arms, the material is pushed through a slot on to a shelf below. After repeated movements, the solid leaves at the bottom of the dryer. The shelves are heated by a row of steam pipes, and in the centre there are three or more fans which suck the hot air over the material and remove it at the top.

The accelerated drying induced by the raking of the material gives evaporative capacities of 0.0002 to 0.0014 kg/s m^2 shelf area which are comparable with those obtained by through circulation on perforated belts. Shelf areas are 0.7 to 200 m^2 in a single unit and the dryer may easily be converted to closed-circuit operation, either to prevent emission of fumes or in order to recover valuable solvents. Typical air

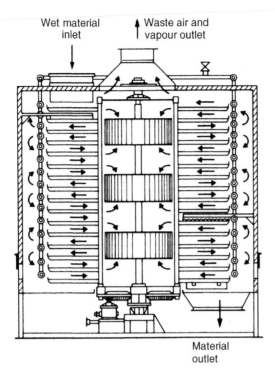

Wet material inlet

Waste air and vapour outlet

Material outlet

FIG. 7.29

Turbo-shelf dryer.

velocities are 0.6 to 2.5 m/s, and the lower trays are often used to cool the dry solids. A turbo-dryer combines cross-circulation drying, as in a tray dryer, with drying by showering the particles through the hot air as they tumble from one tray to another.

7.5.10 Disc dryers

A disc dryer provides a further way in which pasty and sticky materials may be handled, and it can also cope satisfactorily with materials which tend to form a hard crust or pass through a rheologically difficult phase during the drying operation.

As shown in Fig. 7.30, a single-agitator disc contact dryer, consists of a heated cylindrical housing (1) assembled from unit sections, and a heated hollow agitating rotor (2) which has a simultaneous rotating and oscillating movement produced by means of a rotating drive (3) and a reciprocating drive (4). The drive (3 and 4) and stuffing box (8) are located at the dry product end, and the stuffing box is protected by a reverse acting flight (12). The wet product is introduced, drawn in by the screw flight (13) and continuously conveyed through the dryer. The vapour passes through the vapour filter (15) to the condenser. This vapour filter system has a

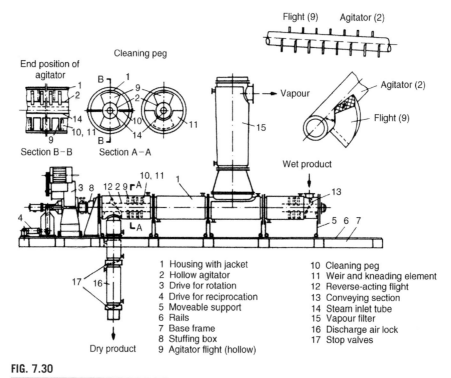

FIG. 7.30

Arrangement of a disc dryer.

back-scavenging action, and is specially designed for removing dust from vapour in vacuum drying plants. The hollow agitator (2) is fitted, over its whole length, with heated flights, which are arranged equidistantly in pairs. Between every two axially neighbouring agitator flights, there are, projecting inwards from the housing fixed wiping pegs (10) or annular weir/kneading elements (11) which extend inward to the agitator core.

The self-cleaning of the heating surfaces is achieved by the combined rotating-reciprocating movement of the agitator. The stationary elements (10 and 11) clean the faces of the agitator flights, at the end of each oscillating movement, during one rotation of the agitator. By the forward and backward movement, the edges of the agitator flights (9) clean the inner surface of the housing and the fixed elements projecting inwards to the core of the agitator clean the agitator (2). In all, about 95% of the heating surface is cleaned.

The rotating and reciprocating motions need not be synchronised, because the individual agitator flights (9) oscillate only between two adjacent fixed elements (10 and 11). This is so that the speed of rotation, the frequency of reciprocation and the forward and backward speed of reciprocation can be adjusted independently of one another over a wide range of settings. The housing sections (1) which are supported by frames (5) on rails, can be drawn forward when cleaning is necessary.

The transport of pasty products through the dryer may be achieved by the differential forward and backward oscillatory motion, combined with the action of the bevelled edges of the agitator flights.

Contivac dryers, for example, have heating surfaces of 4 to 60 m^2 and volumes of 0.1 to 3 m^3. They may be operated under vacuum or up to 400 kN/m^2 with heating fluids at 330 K to 670 K. The evaporative capacity is 0.03 to 0.55 kg/s and the agitator speed ranges from 0.1 Hz (6 rpm) for rheologically difficult materials, to 1 Hz (60 rpm) for easier applications.

Operating on a similar principle is the *Buss paddle* dryer which effects batch drying of liquids, pasty and sandy materials. The paddle dryer, which is the main element of the drying unit. It consists of a horizontal, cylindrical housing which contains a paddle agitator in the form of a hollow shaft carrying agitator arms. The jacket, the hollow shaft and the agitator arms are steam-heated. The paddle agitator is driven by an electric motor and gear unit. A built-on torque bracket with a microswitch protects the paddle agitator against overloading. The product to be dried is passed through the charging nozzle into the paddle dryer, where it is distributed uniformly by the rotating paddle agitator. The drying proceeds under vacuum whilst intensive intermixing by the agitator causes continual renewal of the product particles in contact with the heated surfaces. This guarantees efficient heat transfer and uniform product quality. Vapours are purged of dust in passing through a vapour filter and are then fed to a condenser. Non-condensable gases are drawn off by the vacuum system. The vapour filter is equipped with a removable filter insert and, to prevent excessive pressure drop across the fabric from thick, and possibly moist dust build-up on the filter sacks, these are provided with a reverse jet arrangement, that is the individual filter sacks are cleaned in turn automatically during operation by short though powerful countercurrent blasts of steam blown through the filter sack. This serves to blow and shake off the dust layer and keep the filter sack dry. After the drying process is complete, the dried material may be cooled by applying cooling water to the dryer jacket and paddle agitator. The dryer is emptied by the arms of the paddle agitator, which are designed to move the material in the vessel towards the outlet when rotated backwards. The discharge outlet is specially constructed to prevent the formation of a plug of material. At 373 K, an evaporation rate of over 4 g/s m^2 can be achieved for a steam consumption of 1.5 kg/kg of evaporated water.

7.5.11 Centrifuge dryers

Where a product of a high quality and purity is required, this can, in many cases, be achieved only by washing the solids during a centrifugation process in order to flush out the suspension mother liquor or to dissolve salt crystals. The washed crystals are then usually dried in a separate physical unit operation involving different equipment and problems can be encountered due to the exposure to the atmosphere of the wet cake as it is transferred between the centrifuge and the dryer. These problems may be largely eliminated by using a combined centrifuge–dryer which it is claimed, has the following advantages:

(a) It is hermetically sealed from the environment and is therefore easy to inert.
(b) There is no human contact with the product at any stage of the operation and no possibility of foreign materials being introduced.
(c) The handling of the product is simple and gentle.
(d) Drying times can be reduced by using a jet-pulsed bed drying technique.
(e) The unit can be emptied completely with no residual cake in the basket or on the wall of the drum.

All these advantages are inherent in the FIMA, TZT centrifuge–dryer which is operated as follows.

The feed suspension is introduced into the centrifuge through a hollow drive shaft, and the liquid from the suspension is separated from the crystals and discharged from a sealed filter basket by passing it through a metal filter which retains the solids inside a sealed chamber. Any contaminants are then washed away from the filling pipe, the centrifuge basket and the separated solids by introducing wash liquid through the hollow drive shaft. This operation may be repeated as required. One advantage of the TZT system is that, with no internal pipes or structures that might accumulate unwashed solids, all the separated solids make contact with the wash water. The ring of separated solids is then removed from the wall of the centrifuge by introducing powerful gas blasts beneath the filter medium and the wet en-masse solids accumulate at the bottom of the filter drum. The drying process is achieved by rotating the drum at slow speeds and injecting heated drying gas as process into the closed centrifuge chamber. The moisture-laden gas leaves passing through the filter medium so that the solids are retained within the basket. Such a process is extremely gentle to the solids and dries at low product temperatures even when using higher gas temperatures as a result of the cooling effect of the surface evaporation of the moisture. Dry powder is discharged from the unit by opening the sealing centrifuge housing and fluidising the solids which then enter an integrated powder conveying system. In this system, the gas used for drying may be recycled in a closed gas loop, or it may be discharged to atmosphere in an open system.

Units of this type have a capacity of 20 to 400 kg in terms of filling mass, filter areas of 0.37 to 2.4 m^2 and drum diameters of 400 to 1300 mm and lengths of 300 to 600 mm.

Although this is, in essence a batch operation, close reproducibility between batches overcomes many of the problems associated with batch identification which can be a problem with more conventional drying equipment.

7.6 Specialised drying methods

7.6.1 Solvent drying

Two processes are used here:

Superheated-solvent drying in which a material containing non-aqueous moisture is dried by contact with superheated vapours of its own associated liquid, and,

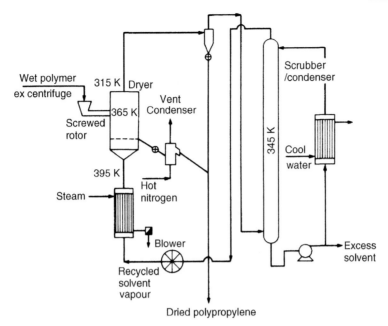

FIG. 7.31

Flowsheet for drying polypropylene by fluidising with a hot stream of solvent vapour [48].

Solvent dehydration in which water-wet substances are exposed to an atmosphere of a saturated organic solvent vapour.

The first of these has advantages where a material containing a flammable liquid such as butanol is involved. Drying is effected with a gas with an air–moisture ratio of 90 kg/kg moisture in order to ensure that the composition is well below the lower explosive limit. The heat requirement is as great as those for superheated solvent drying at the same gas outlet temperature of 400 K [47]. Superheated solvent drying in a fluidised bed has been used for the drying of polypropylene pellets to eliminate the need for water-washing and for fractionating. A flowsheet of the installation is shown [48] in Fig. 7.31.

The most important applications of solvent dehydration lie within the field of kiln drying for seasoning timber where substantial reductions in drying times have been achieved [49].

7.6.2 Superheated steam drying

The replacement of air by superheated steam to take up evaporating moisture is attractive in that it provides a high temperature heat source which also gives rise to a much higher driving force for mass transfer since it does not become saturated at relatively low moisture contents as is the case with air. In the drying of foodstuffs, a further advantage is the fact that the steam is completely clean and there is much

Steam

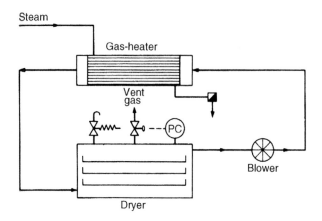

FIG. 7.32

Superheated-steam dryer.

less oxidation damage. In the seasoning of timber, for example, drying times can be reduced quite significantly. Although the principles involved have been understood for some considerable time, as Basel and Grey [48] points out, applications have been limited to due to corrosion problems and the lack of suitable equipment. A flowsheet of a batch dryer using superheated steam is shown in Fig. 7.32. The dryer is initially filled with air circulated by a blower, together with evaporated moisture. Any excess moisture is vented to atmosphere so that the air is gradually replaced by steam. For an evaporation rate of $10 \, kg/m^3$ volume of the dryer chamber, the composition of the gas phase would reach about 90% steam in about 600 s.

As Luikov [50] reports, superheated steam drying may also be used to dry wet material by heating it in a sealed autoclave, and periodically releasing the steam which is generated. This pressure release causes flash evaporation of moisture throughout the whole extent of the material, thus avoiding drying stresses and severe moisture gradients. Such an operation has been applied to the drying of thermal insu-lating materials by Bond et al. [51] who have investigated the drying of paper. In this work, impinging jets of superheated steam were used at 293 to 740 K during the con-stant drying period, with jet Reynolds numbers, 100 to 12,000. Above 450 K, steam drying was found to be much faster than air drying for the same mass velocity of gas. The specific blower power was found to be much lower than for air drying at tem-peratures of industrial importance. It was concluded that steam-impingement drying can lead to significant reductions in both capital investment and energy costs.

In tests on the drying of sand, Wenzel and White [52] found that the use of steam rather than air did not alter the general characteristics of the drying process, and that the drying rate during the constant rate period was determined by the heat transfer rate. In these tests, the heat transferred by radiation from the steam and surrounding surfaces was 7.5 to 31% of the total heat transferred, and coefficients of convective heat transfer were 13 to $100 \, W/m^2 K$. It was concluded that higher drying rates and greater thermal efficiencies are possible when drying with superheated steam as

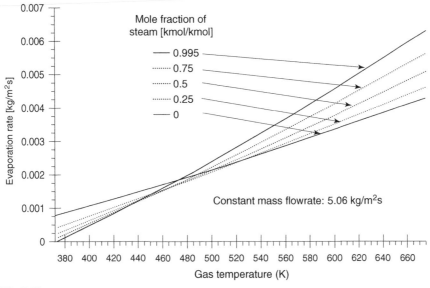

FIG. 7.33

Inversion temperature with superheated steam drying [53].

opposed to air, and that the choice of steam drying must depend on a balance between the savings in operating costs and the higher capital investment attributable to the higher temperatures and pressures.

Schwartze and Bröcker [53], who have made a theoretical study of the evaporation of water into mixtures of superheated steam and air, have calculated the inversion temperature above which the evaporation rate into pure superheated steam is higher than that into dry air under otherwise similar conditions. The data obtained are given in Fig. 7.33 which shows quite clearly the enhanced evaporation rates at gas temperatures above about 475 K. This inversion temperature is given by the point of intersection of the curves for evaporation rate with dry air and superheated steam. The Nusselt and Sherwood equations for heat and mass transfer coefficients for the relevant geometrical configuration, given in Volume 1, were used to calculate the evaporation rates, and these were found to be in excellent agreement with experimental data in the literature. Taylor and Krishna [54] point out that this approach may be used for a wide range of applications, and Vidaurre and Martinez [55] show that the model may be extended to include specialised applications, such as the evaporation of multicomponent liquids.

7.6.3 **Freeze drying**

In this process, the material is first frozen and then dried by sublimation in a very high vacuum, 10 to 40 N/m^2, at a temperature of 240 to 260 K. During the sublimation of the ice, a dry surface layer is left, though this is not free to move because it has been frozen, and a honeycomb structure is formed. With normal vacuum drying of biological solutions containing dissolved salts, a high local salt concentration is formed at the skin, although with freeze drying this does not occur because of the freezing of the solid. Thus, freeze drying is useful not only as a method of working at low temperatures, but also as a method of avoiding surface hardening.

As described by Chambers [56], heat has to be supplied to the material to provide the heat of sublimation. The rate of supply of heat should be such that the highest possible water vapour pressure is obtained at the ice surface without danger of melting the material. During this stage, well over 95% of the water present should be removed, and in order to complete the drying, the material should be allowed to rise in temperature, to say ambient temperature. The great attraction of this technique is that it does not harm heat-sensitive materials, and it is suitable for the drying of penicillin and other biological materials. Initially, high costs restricted the application of the process though economic advances have reduced these considerably and foodstuffs are now freeze-dried on a large scale. Maguire [57] has estimated a total cost of £0.06/kg of water evaporated, in a plant handling 0.5 to 0.75 kg/s (2–3 t/h) of meat.

A typical layout of a freeze-drying installation is shown in Fig. 7.34. Heat is supplied either by conduction, or by radiation from hot platens which interleave with trays containing the product, and the sublimed moisture condenses on to a refrigerated coil at the far end of the drying chamber. The use of dielectric-heating has been investigated [58], though uneven loading of the trays can lead to scorching, and ionisation of the residual gases in the dryer results in browning of the food.

Continuous freeze-drying equipment has been developed [59], and chopped meat and vegetables may be dried in a rotating steam-jacketed tube enclosed in the vacuum chamber. A model of the freeze-drying process has been presented by Dyer and Sunderland [60], and further details are given in Section 15.8 on freeze crystallisation.

7.6.4 Flash drying

Conventional drying consists of mixing and heating the solids to achieve even drying, and simultaneously transporting the vapours away from the surface of the solids so as to maintain a high rate of mass transfer. Although these two stages can be

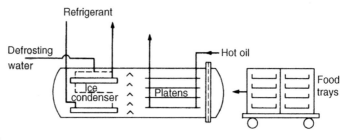

FIG. 7.34

Flowsheet of freeze-drying plant.

achieved simultaneously by contacting the solids with a hot gas stream, heat-sensitive materials, such as foods and pharmaceuticals, may suffer thermal degradation, and other solids may lose some of their water of hydration when subjected to high temperatures. This problem may often be overcome by using a flash drying technique to ensure that the solids are in contact with hot gas in a highly turbulent environment for only a very short time, perhaps only for a few seconds.

A typical flash-drying process consists of a modified pneumatic conveyor in which the wet solids are introduced into a pipe through which they are transported in a high-velocity hot gas stream. In such a process, sticky sludges must first be mixed with dry solids so that the resultant mixture breaks up into smaller particles in the gas stream. Such a process can be used for a wide range of applications [61]. In the drying of the gypsum by-product of flue gas desulphurisation, for example, operating at a minimum solids temperature allows the gypsum to retain its water of hydration, and in the conversion of sewerage sludge into dry fertiliser flash drying prevents the fertiliser from oxidising. Because waste flue gases can often be used as the drying medium, little or no external heat energy is required and, when the dried sludge is burned as a fuel, the heat generated provides much of the energy required for drying.

Even when back-mixed with dry solids, some materials are too sticky or paste-like to feed to a flash dryer, though these can often be handled in a spin dryer consisting of a high shear agitator inside a specially designed drying chamber. In this way, the wet paste may be subjected simultaneously to high shear to break up the solid mass, and to gas at a high velocity to dry the smaller particles. As the particles become small enough and dry enough, they are carried out of the chamber with the spent air and, as with conventional flash drying, the contact time is but a few seconds.

Neither flash nor spin-drying produces highly uniform particles and, even though the bulk of the particles remain at or near the wet bulb temperature, small particles are often over-dried and overheated. Particles of uniform size can be formed by spray drying, however, although it is then necessary that the feed should be in the form of a pumpable liquid which can be atomised so that the particles are evenly exposed to the hot air.

7.6.5 Partial-recycle dryers

The majority of drying operations depend on direct heating using a high flow rate of hot air and/or combustion products which is passed once through or over the wet material and then exhausted to the atmosphere. A variation of this arrangement is the closed-loop system where the entire air or gas stream is confined and recycled after condensing out the vapour. Such a system is justified only where the vaporised liquid or gas has to be recovered for economic or environmental reasons. In addition to the once-through and the closed-loop systems, there is a third class of dryers which incorporates the partial-recycle mode of operation. In this system, a substantial proportion, typically 40% to 60%, of the outlet air is returned to the dryer in order to minimise the heating requirements and the amount of effluent treatment required. Conveyor, flash-pneumatic conveying-, fluidised-bed, rotary, spray and tray types of direct dryers can all be designed for closed-loop or partial-recycle operation and Cook [62] has claimed that recycling part of the exhaust stream has the following advantages:

(a) smaller collectors are needed for removing dried product from the exhaust gases,
(b) less heat is lost in the exhausts which leave, typically, at 340 to 420 K,
(c) since the total volume of exhaust gases is reduced, the cost of preventing undesirable gases or particles from entering the atmosphere is reduced,
(d) when using direct burners to heat the air, the level of oxygen content is minimised, thus enabling sensitive products to be handled and reducing potential fire and/or explosion hazards.

In conveyor and tray types of dryer, air is often recirculated inside the drying vessel in an attempt to save energy or to maintain a relatively high moisture content in the drying air. In other direct dryers such as flash, fluidised-bed, rotary and spray units, any recycle of exhaust air must be returned to the dryer using external ducting, the cost of which is offset by the net savings from the lower volumes of exhaust streams which have to be handled.

As shown in Fig. 7.35, in a partial-recycle drying process, the total airflow leaving the dryer is first passed to a particulate or dust collector in order to remove entrained product or fines, and is then split into two streams – a bleed stream which is vented to atmosphere and a recycle stream which is returned to the system. The bleed stream is large enough to carry all of the moisture that has entered the system, and the recycle portion is returned to the inlet of either the heater or the drying vessel. In the condensing operation, shown in Fig. 7.35, the exhaust stream from the dust

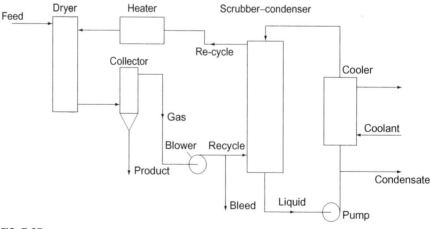

FIG. 7.35

Partial-recycle drying systems.

collector is passed through a condenser, usually a wet scrubber, to remove most of the entrained moisture before it is returned to the inlet of the heater or the dryer. The bleed stream may also be passed through the condenser for gas-cleaning prior to discharge to atmosphere, thus removing moisture and thereby reducing the load on the condenser. Although condensers minimise the amount of bleed to the atmosphere and the moisture content of the drying air, they also cool the recycle stream and this increases costs due to the need to reheat the stream. The total volume of the bleed must be balanced by the heater output and any leakage into the system, noting that moisture can enter the dryer from three sources – the feed, combustion air and the supply air – and can leave in the bleed in the product and as condensate, if any. It may be noted that, in Fig. 7.35, the scrubber–condenser also serves as a collector for particulates. If an oxygen content of less than 21% by volume or a temperature above 270 to 640 K is required, then a direct-fired heater may be used. If non-aqueous solvents are to be evaporated in the dryer, the effluent may require incineration or other treatment prior to discharge to the atmosphere, and dual blowers are often needed to maintain the required pressures throughout the system.

Cook [62] has presented data on the performance of a partial-recycle units based on a direct-fired burner using natural gas. These are compared with the once-through operation in Table 7.8.

In these systems, the total collection efficiencies of the dry product are 85% for the drying vessel, 90% for the cyclone collector and 98% for the scrubber-condenser. The net efficiency of the system may be as high as 99.97% if the scrubber effluent is considered as product. All the runs are based on 1.25 kg/s product and 0.75 kg/s evaporation at an elevation of 300 m above sea level. The total air flow is measured at the outlet before the stream is split into the recycle and bleed portions and, for such flows, the design of suitable fans is outlined by Jorgensen [63]. The calculations outlined here may be confirmed by the use of psychometric charts, and this procedure has been considered in some detail by Cook and Dumount [64].

In recent years, air-recycling in drying systems has become more widely adopted as a means of reducing total dryer discharges to the atmosphere. It would seem that this trend is likely to continue because of environmental concerns and the ever-increasing costs of treating emissions.

7.7 The drying of gases

The drying of gases is carried out on a very large scale. The most important applications of the process are:

(a) in order to reduce the tendency of the gas to cause corrosion,
(b) preparation of dry gas for use in a chemical reaction, and
(c) reduction of the humidity of air in air-conditioning plants.

Table 7.8 Comparison of once-through with partial-recycle operation.

Example	Recycle mode	Burner outlet temperature (K)	Recycle (per cent)	Bleed (per cent)	Moisture (kg/ kg air)	Heat load (MW)	Air flow (m³/s)
1	condensing	1090	46.0	54.0	0.206	2.74	8.14
2	condensing	810	19.7	80.3	0.173	2.83	8.22
3	condensing	1370	58.9	41.1	0.222	2.68	8.11
4	bleed only	1160	46.0	54.0	0.325	2.52	7.92
5	bleed only	858	25.4	74.6	0.206	2.68	8.14
6	once-through	700	0.0	100.0	0.148	2.90	8.28

Dryer inlet = 700K, dryer outlet = 365K.

The problems involved in the drying or dehumidification of gases are referred to in Volume 1, Chapter 13, and the most important methods available are now summarised.

Cooling. A gas stream may be dehumidified by bringing it into contact with a cold liquid or a cold solid surface. If the temperature of the surface is lower than the dew point of the gas, condensation will take place, and the temperature of the surface will tend to rise by virtue of the liberation of latent heat: It is therefore necessary to remove heat constantly from the surface. Because a far larger interfacial surface can be produced with a liquid, it is usual to spray a liquid into the gas and then to cool it again before it is recycled. In many cases, countercurrent flow of the gas and liquid is obtained by introducing the liquid at the top of a column and allowing the gas to pass upwards.

Compression. The humidity of a gas may be reduced by compressing it, cooling it down to near its original temperature, and then draining off the water which has condensed. During compression, the partial pressure of the water vapour increases and condensation occurs as soon as the saturation value is exceeded.

Liquid absorbents. If the partial pressure of the water in the gas is greater than the equilibrium partial pressure at the surface of a liquid, condensation will take place as a result of contact between the gas and liquid. Thus, water vapour is frequently removed from a gas by bringing it into contact with concentrated sulphuric acid, phosphoric acid, or glycerol. Concentrated solutions of salts, such as calcium chloride, are also effective. The process may be carried out either in a packed column or in a spray chamber. Regeneration of the liquid is an essential part of the process, and this is usually effected by evaporation.

Solid adsorbents and absorbents. The use of silica gel or solid calcium chloride to remove water vapour from gases is a common operation in the laboratory. Moderately large units can be made, although the volume of packed space required is generally large because of the comparatively small transfer surface per unit volume. If the particle size is too small, the pressure drop through the material becomes excessive. The solid desiccants are regenerated by heating.

Gas is frequently dried by using a calcium chloride liquor containing about 0.56 kg calcium chloride/kg solution. The extent of recirculation of the liquor from the base of the tower is governed by heating effects, since the condensation of the water vapour gives rise to considerable heating. It is necessary to instal heat exchangers to cool the liquor leaving the base of the tower.

In the contact plant for the manufacture of sulphuric acid, sulphuric acid is itself used for drying the air for the oxidation of the sulphur. When drying hydrocarbons such as benzene, it is sometimes convenient to pass the material through a bed of solid caustic soda, although, if the quantity is appreciable, this method is expensive.

The great advantage of materials such as silica gel and activated alumina is that they enable the gas to be almost completely dried. Thus, with silica gel, air may be dried down to a dew point of 203 K. Small silica gel containers are frequently used to prevent moisture condensation in the low pressure lines of pneumatic control installations.

References

[1] T.K. Sherwood, The air drying of solids, Trans. Am. Inst. Chem. Eng. 32 (1936) 150.

[2] J.F. Pearse, T.R. Oliver, D.M. Newitt, The mechanism of the drying of solids. Part I. The forces giving rise to movement of water in granular beds, during drying, Trans. Inst. Chem. Eng. 27 (1949) 1.

[3] T.R. Oliver, D.M. Newitt, The mechanism of drying of solids. Part II. The measurement of suction potentials and moisture distribution in drying granular solids, Trans. Inst. Chem. Eng. 27 (1949) 9.

[4] D.M. Newitt, M. Coleman, The mechanism of drying of solids. Part III. The drying characteristics of China clay, Trans. Inst. Chem. Eng. 30 (1952) 28.

[5] R.W. Corben, D.M. Newitt, The mechanism of drying of solids. Part VI. The drying characteristics of porous granular material, Trans. Inst. Chem. Eng. 33 (1955) 52.

[6] A.R. King, D.M. Newitt, The mechanism of drying of solids. Part VII. Drying with heat transfer by conduction, Trans. Inst. Chem. Eng. 33 (1955) 64.

[7] R.W. Corben, The Mechanism of Drying of Solids. Part IV. A Study of the Effect of Granulation on the Rate of Drying of a Porous Solid (Ph.D. thesis), University of London, 1955.

[8] E.R. Gilliland, Fundamentals of drying and air conditioning, Ind. Eng. Chem. 30 (1938) 506.

[9] R.W. Powell, E. Griffiths, The evaporation of water from plane and cylindrical surfaces, Trans. Inst. Chem. Eng. 13 (1935) 175.

[10] K.R. Chakravorty, Evaporation from free surfaces, J. Imp. Coll. Chem. Eng. Soc. 3 (1947) 46.

[11] C.B. Shepherd, C. Hadlock, R.C. Brewer, Drying materials in trays. Evaporation of surface moisture, Ind. Eng. Chem. 30 (1938) 388.

[12] T.K. Sherwood, The drying of solids I, II, Ind. Eng. Chem. 21 (1929) 12. 976.

[13] A.B. Newman, The drying of porous solids: diffusion and surface emission equations, Trans. Am. Inst. Chem. Eng. 27 (1931) 203.

[14] N.H. Ceaglske, O.A. Hougen, The drying of granular solids, Trans. Am. Inst. Chem. Eng. 33 (1937) 283.

[15] N.H. Ceaglske, O.A. Hougen, Drying granular solids, Ind. Eng. Chem. 29 (1937) 805.

[16] C.S. Slichter, Theoretical investigation of the motion of ground waters, in: U.S. Geol. Survey, 19th Annual Report, 1897–1898. Part 2, 301.

[17] W.B. Haines, Studies in the physical properties of soils. IV. A further contribution to the theory of capillary phenomena in soil, J. Agric. Sci. 17 (1927) 264.

[18] N.H. Parker, Aids to dryer selection, Chem. Eng. Albany 70 (13) (1963) 115.

[19] K. Kröll, Trockner, Einteilen, ordnen, benennen, benummern, Schilde Schriftenreihe 6, Schilde Bad-Hersfeld, 1965.

[20] C.E. Sloan, Drying systems and equipment, Chem Eng. Albany 74 (13) (1967) 169.

[21] J.R. Backhurst, J.H. Harker, Process Plant Design, Heinemann, London, 1973.

[22] R.B. Keey, Drying Principles and Practice, Pergamon Press, Oxford, 1972.

[23] J.L. Erisman, Roto-louvre dryer, Ind. Eng. Chem. 30 (1938) 996.

[24] S.J. Friedman, W.R. Marshall, Studies in rotary drying, Chem. Eng. Prog. 45 (1949) 482. 573.

[25] C.F. Prutton, C.O. Miller, W.H. Schuette, Factors influencing the performance of rotary dryers, Trans. Am. Inst. Chem. Eng. 38 (1942) 123. 251.

[26] C.O. Miller, B.A. Smith, W.H. Schuette, Factors influencing the performance of rotary dryers, Trans. Am. Inst. Chem. Eng. 38 (1942) 841.

[27] W.C. Saeman, Air-solids interaction in rotary dryers and coolers, Chem. Eng. Prog. 58 (6) (1962) 49–56.

[28] J.J. Fisher, Low temperature drying in vacuum tumblers, Ind. Eng. Chem. 55 (2) (1963) 18.

[29] K. Sharples, P.G. Gliken, R. Warne, Complete simulation of rotary dryers, Trans. Inst. Chem. Eng. 42 (1964) T275.

[30] R.P. Fraser, P. Eisenklam, N. Dombrowski, Liquid atomisation in chemical engineering, Br. Chem. Eng. 2 (1957) 414–417. 496–501, 536–43, and 610–13.

[31] L. Rayleigh, On the instability of jets, Proc. Lond. Math. Soc. 10 (1878–1879) 4–13.

[32] C. Weber, Zum Zerfall eines Flüssigkeitsstrahles, Z. Angew. Math. Mech. 11 (1931) 136–154.

[33] N. Dombrowski, G. Munday, Spray drying, in: Biochemical and Biological Engineering Science, vol. 2, Academic Press, 1967 (Chapter 16).

[34] N. Dombrowski, R.P. Fraser, A photographic investigation into the disintegration of liquid sheets, Philos. Trans. 247A (1954) 101.

[35] R.P. Fraser, P. Eisenklam, N. Dombrowski, D. Hasson, Drop formation from rapidly moving liquid sheets, AIChE J 8 (1962) 672–680.

[36] W.H. Walton, W.C. Prewett, The production of sprays and mists of uniform drop size by means of spinning disc type sprayers, Proc. Phys. Soc. 628 (1949) 341.

[37] W.R. Marshall, Atomization and spray drying, Chem. Eng. Prog. Monogr. Ser. (2) (1954) 50.

[38] W.R. Marshall, Heat and mass transfer in spray drying, Trans. Am. Soc. Mech. Eng. 77 (1955) 1377.

[39] J.W. Grose, G.H. Duffield, Chemical engineering methods in the food industry, Chem. Ind. (1954) 1464.

[40] J.J. Quinn, The economics of spray drying, Ind. Eng. Chem. 57 (1) (1965) 35–37.

[41] P. Bradford, S.W. Briggs, Equipment for the food industry — 3. Jet spray drying, Chem. Eng. Prog. 59 (3) (1963) 76.

[42] M.F. Quinn, Fluidized bed dryers, Ind. Eng. Chem. 55 (7) (1963) 18–24.

[43] J.C. Agarwal, W.L. Davis, D.T. King, Fluidized-bed coal dryer, Chem. Eng. Prog. 58 (11) (1962) 85–90.

[44] M.W. Scott, H.A. Lieberman, A.S. Rankell, F.S. Chow, G.W. Johnston, Drying as a unit operation in the pharmaceutical industry. I. Drying of tablet granulations in fluidized beds, J. Pharm. Sci. 52 (3) (1963) 284–291.

[45] V. Vaněček, M. Markvart, R. Drbohlav, Fluid Bed Drying, Leonard Hill, 1966.

[46] W.E. Clark, Fluid-bed drying, Chem. Eng. Albany 74 (6) (1967) 177.

[47] J.C. Chu, A.M. Lane, D. Conker, Evaporation of liquids into their superheated vapours, Ind. Eng. Chem. 45 (1953) 1586.

[48] L. Basel, E. Gray, Superheated solvent drying in a fluidized bed, Chem. Eng. Prog. 58 (1962) 67.

[49] E.L. Ellwood, J.W. Gottstein, W.G. Kauman, A Laboratory Study of the Vapour Drying Process. CSIRO Forest Prod. Div., Paper 14, 1961, p. 111.

[50] A.V. Luikov, Heat and Mass Transfer in Capillary-Porous Bodies, Pergamon Press, Oxford, 1966.

[51] J.F. Bond, A.S. Mujumdar, A.R.P. van Heiningen, W.J.M. Douglas, Drying paper by impinging jets of superheated steam, Can. J. Chem. Eng. 72 (1994) 452–456.

[52] L. Wenzel, R.R. White, Drying granular solids by superheated steam, Ind. Eng. Chem. Process Des. Dev. 9 (1970) 207–214.

[53] J.P. Schwartze, S. Bröcker, The evaporation of water into air of different humidities and the inversion temperature phenomenon, Int. J. Heat Mass Transf. 43 (2000) 1791–1800.

[54] R. Taylor, R. Krishna, Multicomponent Mass Transfer, Wiley, New York, 1993.

[55] M. Vidaurre, J. Martinez, Continuous drying of a solid wetted with ternary mixtures, AIChE J 43 (3) (1997) 681–692.

[56] H.H. Chambers, Vacuum freeze drying, Trans. Inst. Chem. Eng. 27 (1949) 19.

[57] J.F. Maguire, Freeze drying moves ahead in U.S, Food Eng. 34 (8) (1962) 54–55. 34, No. 9 (Sept. 1962) 48–52.

[58] J.C. Harper, C.O. Chichester, T.E. Roberts, Freeze-drying of foods, Agric. Eng. 43 (1962) 78, 90.

[59] H.G. Maister, E.N. Heger, W.M. Bogard, Continuous freeze-drying of *Serratia marcescens*, Ind. Eng. Chem. 50 (1958) 623.

[60] D.F. Dyer, J.E. Sunderland, The role of convection in drying, Chem. Eng. Sci. 23 (1968) 965.

[61] S. France, Delicately drying solids - in a flash, Chem. Eng. Albany (1996) 83–84.

[62] E.M. Cook, Process calculations for partial-recycle dryers, Chem. Eng. Albany (1996) 82–89.

[63] R. Jorgensen (Ed.), Fan Engineering, eighth ed., Buffalo Forge Co., Buffalo, New York, 1983.

[64] E.M. Cook, H.D. Dumount, Process Drying Practice, McGraw-Hill, New York, 1991.

Further reading

J.R. Backhurst, J.H. Harker, Process Plant Design, Heinmann, London, 1973.

J.R. Backhurst, J.H. Harker, J.E. Porter, Problems in Heat and Mass Transfer, Edward Arnold, London, 1974.

A.G. Bailey, Electrostatic Spraying of Liquids, John Wiley, New York, 1988.

S.F. Barclay, A Study of Drying, Inst. Energy, London, 1957.

M.J. Bridgman, Drying—Principles and Practice in the Process Industries, Caxton, Christchurch, 1966.

W.H. Brown, An Introduction to the Seasoning of Timber, Pergamon Press, Oxford, 1965.

S. Bruin (Ed.), Preconcentration and Drying of Food Materials, Elsevier, London, 1988.

C. Butcher, Fluid bed dryers, Chem. Eng. Lond. (450) (1988) 16.

R.J. Clarke, R. Macrae (Eds.), Coffee, Vol. 2: Technology, Elsevier, London, 1987.

R.B. Keey, Drying Principles and Practice, Pergamon Press, Oxford, 1972.

R.B. Keey, Introduction to Industrial Drying Operations, Pergamon Press, Oxford, 1978.

R.B. Keey, Drying of Loose and Particulate Materials, Hemisphere Publishing Corporation, London and Washington DC, 1991.

A.V. Luikov, Heat and Mass Transfer in Capillary-porous Bodies, Pergamon Press, Oxford, 1966.

K. Masters, Spray Drying Handbook, George Godwin, London, 1985.

W.L. McCabe, J.C. Smith, P. Harriott, Unit Operations of Chemical Engineering, fourth ed., McGraw-Hill, New York, 1984.

A.S. Mujumder (Ed.), Handbook of Industrial Drying, Marcel Dekker, London, 1987.

G. Nonhebel, Gas Purification Processes, Newnes, London, 1964.

G. Nonhebel, A.A.H. Msoss, Drying of Solids in the Chemical Industry, Butterworth, London, 1971.

H.A.C. Thijssen, W.H. Rulkens, Recent Developments in Freeze Drying, International Institute of Refrigeration, Lausanne, 1969.

J. Vanécék, M. Markvart, R. Drbohlav, Fluidized Bed Drying, Leonard Hill, London, 1966.

S.M. Walas, Chemical Process Equipment: Selection and Design, Butterworth, London, 1989.

A. Williams-Gardner, Industrial Drying, George Godwin, London, 1971.

Adsorption

Ajay Kumar Ray

Department of Chemical and Biochemical Engineering, University of Western Ontario,
London, ON, Canada

Nomenclature

		Units in SI system	Dimensions in M, N, L, T, θ
A	superficial cross-sectional area of bed	m^2	L^2
A_p	interfacial area for condensate in a pore	m^2	L^2
A_s	area of an adsorbed film	m^2	L^2
$A_1{}', A_2{}'$	Arrhenius frequency factors for desorption	$kmol/m^2 s$	$NL^{-2} T^{-1}$
a_m	area occupied by one molecule in an adsorbed film	m^2	L^2
a_p	external area of adsorbent per unit volume of adsorbent	m^{-1}	L^{-1}
$a_p{}'$	external surface of a pellet	m^2	L^2
a_z	external area of adsorbent per unit volume of bed	m^{-1}	L^{-1}
a_0, a_1, a_2	fraction of the adsorbent surface covered by no adsorbate, one, two layers	–	–
B_0, B_1, B_i, B_j	constants in Langmuir-type equations	various	
B_2	α_0/β in Eq. (8.9)	–	–

Continued

Coulson and Richardson's Chemical Engineering. https://doi.org/10.1016/B978-0-08-101097-6.00008-0

		Units in SI system	Dimensions in M, N, L, T, θ
C	concentration of adsorbate in the fluid	kmol/m^3	$\mathbf{NL}^{-3}$
C_B	concentration of adsorbate **B** in the fluid	kmol/m^3	$\mathbf{NL}^{-3}$
C_c	concentration of adsorbate after 1 cold stage	kmol/m^3	$\mathbf{NL}^{-3}$
C_c'	concentration of adsorbate within a pore after 1 cold stage	kmol/m^3	$\mathbf{NL}^{-3}$
C_{cc}	concentration of adsorbate after two cold stages	kmol/m^3	$\mathbf{NL}^{-3}$
C_E	concentration of adsorbate at the emergence of the adsorption zone	kmol/m^3	$\mathbf{NL}^{-3}$
C_F	concentration of adsorbate in feed conditions	kmol/m^3	$\mathbf{NL}^{-3}$
C_i	concentration of adsorbate at the exterior surface of a pellet	kmol/m^3	$\mathbf{NL}^{-3}$
C_H, C_{HH}	concentration of adsorbate after one, two hot stages	kmol/m^3	$\mathbf{NL}^{-3}$
C_0	concentration of adsorbate initially or at inlet	kmol/m^3	$\mathbf{NL}^{-3}$
C'_p	concentration of adsorbate in feed conditions within a pore	kmol/m^3	$\mathbf{NL}^{-3}$
C_r, C_{sr}	concentration of adsorbate at radius r in the fluid, or in adsorbed phases	kmol/m^3	$\mathbf{NL}^{-3}$
C_s	concentration of adsorbed phase	kmol/m^3	$\mathbf{NL}^{-3}$
C'_s	concentration of adsorbed phase within a pore	kmol/m^3	$\mathbf{NL}^{-3}$
C''_s	mass concentration of adsorbed phase referred to adsorbate free adsorbent	kg/m^3	$\mathbf{ML}^{-3}$

Continued

		Units in SI system	Dimensions in M, N, L, T, θ
$\bar{C}_s$	mean concentration of adsorbed phase	kmol/m^3	$\mathbf{NL^{-3}}$
C'_{sc}	concentration of adsorbed phase in a pore after one cold stage	kmol/m^3	$\mathbf{NL^{-3}}$
C'_{SF}	concentration of adsorbed phase in a pore in feed conditions	kmol/m^3	$\mathbf{NL^{-3}}$
C_{sm}	concentration of adsorbed phase in a monolayer	kmol/m^3	$\mathbf{NL^{-3}}$
$C_{s\infty}$	ultimate or maximum concentration of adsorbed phase	kmol/m^3	$\mathbf{NL^{-3}}$
C^*	concentration of adsorbate in equilibrium with $\bar{C}_s$	kmol/m^3	$\mathbf{NL^{-3}}$
c_{pg}	specific heat capacity of the gas phase	J/kg K	$\mathbf{L^2T^{-2}\theta^{-1}}$
c_{ps}	specific heat capacity of the adsorbent with adsorbate	J/kg K	$\mathbf{L^2T^{-2}\theta^{-1}}$
c_{pw}	specific heat capacity of the wall	J/kg K	$\mathbf{L^2T^{-2}\theta^{-1}}$
D	diffusivity	m^2/s	$\mathbf{L^2T^{-1}}$
D_{av}	average diffusivity	m^2/s	$\mathbf{L^2T^{-1}}$
D_e	effective diffusivity	m^2/s	$\mathbf{L^2T^{-1}}$
D_k	Knudsen diffusivity	m^2/s	$\mathbf{L^2T^{-1}}$
D_L	longitudinal diffusivity	m^2/s	$\mathbf{L^2T^{-1}}$
D_M	molecular diffusivity	m^2/s	$\mathbf{L^2T^{-1}}$
D_s	surface diffusivity	m^2/s	$\mathbf{L^2T^{-1}}$
D_{s0}	surface diffusivity in standard conditions	m^2/s	$\mathbf{L^2T^{-1}}$
D_T	total diffusivity	m^2/s	$\mathbf{L^2T^{-1}}$
d_b	bed diameter	m	$\mathbf{L}$
d_p	pellet diameter	m	$\mathbf{L}$
E	argument of an error function	–	–
E_s	energy of activation in surface diffusion	J/kmol	$\mathbf{MN^{-1}L^2T^{-2}}$
$E_0, E_1 \dots E_n$	energy of activation of desorption from empty surface, monolayer etc.	J/kmol	$\mathbf{MN^{-1}L^2T^{-2}}$
F	a function in Eq. (8.92)	–	–

Continued

		Units in SI system	Dimensions in M, N, L, T, θ
$f'(C)$	slope of an adsorption isotherm	–	–
G	function in Eq. (8.82)	–	–
G	Gibbs free energy of an adsorbed film *or*	J	ML^2T^{-2}
	mass flow rate of fluid	kg/s	MT^{-1}
G', G_s'	mass flow rate per unit cross sectional area of fluid, solid	kg/m^{2}s	$ML^{-2}T^{-1}$
G_1, G_2	functions in Eq. (8.83)	–	–
H	enthalpy per kmol	J/kmol	$MN^{-1}L^2T^{-2}$
h	film heat transfer coefficient	W/m^2 K	$MT^{-3}\theta^{-1}$
i	unsaturated fraction of an adsorption zone	–	–
J	a factor in Eq. (8.58)	–	–
K	an equilibrium constant based on activities	various	
K_a, K_c, K_H	adsorption equilibrium constants	–	–
K_0	adsorption equilibrium constant at a standard condition	–	–
k	thermal conductivity of fluid	W/mK	$MLT^{-3}\theta^{-1}$
k	reaction rate constant (1st order)	s^{-1}	T^{-1}
k_B	Boltzmann constant	1.3805×10^{-23} J/K	$ML^2T^{-2}\theta^{-1}$
k_e	effective thermal conductivity of solid	W/mK	$MLT^{-3}\theta^{-1}$
k_g	film mass transfer coefficient	m/s	LT^{-1}
k_g^o	overall mass transfer coefficient	m/s	LT^{-1}
k_p	mass transfer coefficient based on 'solid film'	m/s	LT^{-1}
k_0, k_1	adsorption velocity constants for empty surface, monolayer	kmol/Ns	$M^{-1}NL^{-1}T$
k_1', k_2'	desorption velocity constants for monolayer, bilayer	kmol/m^{2}s	$NL^{-2}T^{-1}$
L	length of porous medium	m	L
L'	constant in Eq. (8.26)	–	–

Continued

		Units in SI system	Dimensions in M, N, L, T, θ
L_e	length of porous path	m	**L**
L_p	mean length of a pore	m	**L**
l	distance into a pore	m	**L**
M	molecular weight	kg/kmol	**MN**$^{-1}$
M'	constant in Eq. (7.26)	m^{-6}	**L**$^{-6}$
M_A, M_B	molecular weights of **A, B**	kg/kmol	**MN**$^{-1}$
m	inter-pellet void ratio $\varepsilon/(1-\varepsilon)$	–	–
m'	slope of an operating line	–	–
m_M	mass of one molecule	kg	**M**
N	Avogadro number (6.023×10^{26} molecules per kmol *or* 6.023×10^{23} molecules per mol)	kmol^{-1}	**N**$^{-1}$
N_A, N_B	flux of molecular species **A, B**	kmol/m^{2}s	**NL**$^{-2}$ **T**$^{-1}$
N_{AC}	interstitial flux of **A**	kmol/m^{2}s	**NL**$^{-2}$ **T**$^{-1}$
N_{Ai}	flux along a tortuous pore	kmol/m^{2}s	**NL**$^{-2}$ **T**$^{-1}$
N_u	Nusselt number (hd_p/k)	–	–
n	number of adsorbed layers	–	–
n	index in the Freundlich Eq. (8.35)	–	–
n'	number of molecules per unit volume of gas	m^{-3}	**L**$^{-3}$
n''	number of components in Eq. (8.33)	–	–
n_0	initial number of mols of adsorbate	kmol	**N**
n_p	number of pores	–	–
n_s	number of mols of adsorption in a film or capillary	kmol	**N**
n_{s1}	number of mols of adsorbate at pressure P_1	kmol	**N**
P	total pressure	N/m^2	**ML**$^{-1}$ **T**$^{-2}$
$\bar{P}$	mean pressure	N/m^2	**ML**$^{-1}$ **T**$^{-2}$
P^0	vapour pressure of the adsorbed phase	N/m^2	**ML**$^{-1}$ **T**$^{-2}$
P_A	partial pressure of **A**	N/m^2	**ML**$^{-1}$ **T**$^{-2}$
P_e	Peclet number (ud_p/D_L)	–	–
P_H/P_L	high/low pressure ratio	–	–

Continued

		Units in SI system	Dimensions in M, N, L, T, θ
P_m	vapour pressure over a meniscus	N/m^2	$\mathbf{ML^{-1}T^{-2}}$
P_r	Prandtl number ($c_p\mu/k$)	–	–
P_1	pressure at which the adsorbed film becomes mobile	N/m^2	$\mathbf{ML^{-1}T^{-2}}$
$\mathbf{R}$	gas constant	8314 J/kmol K	$\mathbf{MN^{-1}L^2T^{-2}\theta^{-1}}$
r	radius within a spherical pellet or pore	m	$\mathbf{L}$
r_i	outer radius of a spherical pellet	m	$\mathbf{L}$
r_p	radius of a pore	m	$\mathbf{L}$
r_p'	net radius of a pore allowing for adsorbed layers	m	$\mathbf{L}$
Sc	Schmidt number ($\mu/\rho D$)	–	–
Sh	Sherwood number ($k_g d_p/D$)	–	–
T	absolute temperature	K	$\boldsymbol{\theta}$
T_a	ambient temperature	K	$\boldsymbol{\theta}$
T_0	initial or reference temperature	K	$\boldsymbol{\theta}$
t	time	s	$\mathbf{T}$
t_a	time for an adsorption zone to move its own length	s	$\mathbf{T}$
t_b	time to breakpoint	s	$\mathbf{T}$
t_1	time parameter in Eq. (8.71)	s	$\mathbf{T}$
t_{min}	minimum time to saturate unit cross section of a bed of length, z	s	$\mathbf{T}$
U_0	overall heat transfer coefficient	W/m^2 K	$\mathbf{MT^{-3}\theta^{-1}}$
u	interpellet velocity of fluid	m/s	$\mathbf{LT^{-1}}$
u_c, u_H	velocity of a point on an adsorption wave at lower/ higher temperatures	m/s	$\mathbf{LT^{-1}}$
u_T	velocity of a point on a thermal wave	m/s	$\mathbf{LT^{-1}}$
V	volume of fluid *or* volume per unit mass of adsorbent	m^3 (m^3/kg)	$\mathbf{L^3\ (M^{-1}L^3)}$

Continued

		Units in SI system	Dimensions in M, N, L, T, θ
V^1	volume in the gas phase equivalent to an adsorbed monolayer	m^3	L^3
V_i	volume of ith component	m^3	L^3
V_0	initial volume of solution	m^3	L^3
V_p	volume of a pore	m^3	L^3
V_s	volume of the adsorbed phase	m^3	L^3
V_M	molar volume of adsorbate	$m^3/kmol$	$N^{-1}L^3$
V'_m	molar volume of liquid adsorbate	$m^3/kmol$	$N^{-1}L^3$
V_s^1	volume of the adsorbed phase in a monolayer	m^3	L^3
v_s	volume of adsorbate on unit area of surface	m^3/m^2	L
v_s^1	volume of adsorbate on unit area of monolayer	m^3/m^2	L
W	mass per unit length of containing vessel	kg/m	ML^{-1}
x_f	thickness of boundary film	m	L
x_r	mass ratio of adsorbed phase to adsorbent	–	–
y	mole fraction of adsorbate in the fluid phase	–	–
y_H, y_L	mole fractions in the effluent during high, low pressure stages of PSA	–	–
y_r	mass ratio of adsorbate to carrier fluid	–	–
y_r^B, y_r^E	y_r at breakpoint, after emergence of adsorption zone	–	–
y_r^*	y_r in equilibrium with mean concentration of adsorbed phase	–	–
z	distance along a bed	m	L
z'	position of concentration C	m	L
z_a	length of an adsorption zone	m	L
z_e	length of a fixed bed	m	L

Continued

		Units in SI system	Dimensions in M, N, L, T, θ
z_H	distance moved by an adsorption zone at high pressure	m	**L**
z_L, z_L'	distances moved by an adsorption zone in successive low pressure stages	m	**L**
z_0	initial position of a point on the adsorption wave	m	**L**
α	intra-pellet void fraction	–	–
$\alpha_0, \alpha_1, \alpha_2$	constants in Eqs (8.7), (8.24), (8.35)	various	
β, β_1, β_2	constants in Eqs (8.8), (8.24), (8.30)	various	**MT^{-2}**
Γ	two dimensional spreading pressure	N/m	
Δ	change in property	–	–
ε	interpellet void fraction	–	–
ε_p	adsorption potential	J/kmol	**MN^{-1} L^{2}T^{-2}**
θ_1	constant in Eq. (8.36)	m^3	**L^3**
Ω	collision integral	–	–
λ	mean free path of a gas molecule	m	**L**
λ_A	average distance between adsorption sites	m	**L**
λ_M	molar latent heat	J/kmol	**MN^{-1} L^{2}T^{-2}**
λ'	length parameter in Eq. (8.93)	–	–
μ	viscosity	Ns/m^2	**ML^{-1} T^{-1}**
μ_g	chemical potential of a gas	J/kmol	**MN^{-1} L^{2}T^{-2}**
$\mu\rho_s$	chemical potential of an adsorbed film	J/kmol	**MN^{-1} L^{2}T^{-2}**
ρ_B	mass or molar density of a packed bed	kg/m^3 (kmol/m^3)	**ML^{-3} (NL^{-3})**
ρ_g	mass (or molar) density of a gas	kg/m^3 (kmol/m^3)	**ML^{-3} (NL^{-3})**
ρ_p	mass (or molar density) of an adsorbent pellet	kg/m^3 (kmol/m^3)	**ML^{-3} (NL^{-3})**
ρ_s	mass (or molar) density of a solid	kg/m^3 (kmol/m^3)	**ML^{-3} (NL^{-3})**
σ_{AB}	collision diameter between **A** and **B** molecules	m	**L**

Continued

		Units in SI system	Dimensions in M, N, L, T, θ
$\sigma_{sl}, \sigma_{sv}, \sigma_{lv}$	surface tensions at three interfaces in a solid–liquid–gas system	N/m	MT^{-2}
ϕ	interfacial angle between liquid and solid	–	–
ϕ_1	constant in Eq. (8.36)	m^3	L^3
τ	tortuosity, L/L_e	–	–
τ'	time parameter in Eq. (8.94)	–	–
φ	a resistance parameter in Eq. (8.95)	–	–
χ	time parameter in Eqs (8.71) and (8.83)	s	T

8.1 Introduction

Although adsorption has been used as a physical–chemical process for many years, it is only over the last four decades that the process has developed to a stage where it is now a major industrial separation technique. In adsorption, molecules distribute themselves between two phases, one of which is a solid whilst the other may be a liquid or a gas. The only exception is in adsorption on to foams, a topic which is not considered in this chapter.

Unlike *absorption*, in which solute molecules diffuse from the bulk of a gas phase to the bulk of a liquid phase, in *adsorption*, molecules diffuse from the bulk of the fluid to the surface of the solid adsorbent forming a distinct adsorbed phase.

Typically, gas adsorbers are used for removing trace components from gas mixtures. The commonest example is the drying of gases in order to prevent corrosion, condensation, or an unwanted side reaction. For items as diverse as electronic instruments and biscuits, sachets of adsorbent may be included in the packaging in order to keep the relative humidity low. In processes using volatile solvents, it is necessary to guard against the incidental loss of solvent carried away with the ventilating air, and recovery may be affected by passing the air through a packed bed of adsorbent.

Adsorption may be equally effective in removing trace components from a liquid phase and may be used either to recover the component or simply to remove a noxious substance from an industrial effluent.

Any potential application of adsorption has to be considered along with alternatives such as distillation, absorption, and liquid extraction. Each separation process exploits a difference between a property of the components to be separated. In distillation, it is volatility. In absorption, it is solubility. In extraction, it is a distribution coefficient. Separation by adsorption depends on one component being more readily adsorbed than another. The selection of a suitable process may also depend on the ease with which the separated components can be recovered. Separating *n*- and iso-paraffins by distillation requires a large number of stages because of the low relative volatility of the components. It may be economic, however, to use a selective adsorbent which separates on the basis of slight differences in mean molecular diameters, where for example, *n*- and isopentane have diameters of 0.489 and 0.558 nm, respectively. When an adsorbent with pore size of 0.5 nm is exposed to a mixture of the gases, the smaller molecules diffuse to the adsorbent surface and are retained whilst the larger molecules are excluded. In another stage of the process, the retained molecules are desorbed by reducing the total pressure or increasing the temperature.

Most commercial processes for producing nitrogen and oxygen from air use the cryogenic distillation of liquefied air. There is also interest in adsorptive methods, particularly for moderate production rates. Some adsorbents take up nitrogen preferentially and can be used to generate an oxygen-rich gas containing 95 mol% of oxygen. Regeneration of the adsorbent yields a nitrogen-rich gas. It is also possible to separate the gases using an adsorbent with 0.3 nm pores. These allow oxygen molecules of 0.28 nm in size to diffuse rapidly on to the adsorption surface, whereas nitrogen molecules of 0.30 nm will diffuse more slowly. The resulting stream is a nitrogen-rich gas of up to 99% purity. An oxygen stream of somewhat lower purity is obtained from the desorption stage.

Other applications of commercial adsorbents are given in Table 8.1, taken from the work of Crittenden [1]. Some typical solvents which are readily recovered by adsorptive techniques are listed in Table 8.2, taken from information supplied by manufacturers.

All such processes suffer one disadvantage in that the capacity of the adsorbent for the adsorbate in question is limited. The adsorbent has to be removed at intervals from the process and regenerated, that is, restored to its original condition. For this reason, the adsorption unit was considered in early industrial applications to be more difficult to integrate with a continuous process than, say, a distillation column. Furthermore, it was difficult to manufacture adsorbents which had identical adsorptive properties from batch to batch. The design of a commercial adsorber and its operation had to be sufficiently flexible to cope with such variations.

These factors, together with the rather slow thermal regeneration that was common in early applications, resulted in the adsorber being an unpopular option with plant designers. Since a greater variety of adsorbents has become available, each tailor-made for a specific application, the situation has changed, particularly as faster alternatives to thermal regeneration are often possible.

Adsorption occurs when molecules diffusing in the fluid phase are held for a period of time by forces emanating from an adjacent surface. The surface represents a gross discontinuity in the structure of the solid, and atoms at the surface have a residue of molecular forces which are not satisfied by surrounding atoms such as those in the body of the structure. These residual or van der Waals forces are common to all surfaces and the only reason why certain solids are designated 'adsorbents' is

Table 8.1 Typical applications of commercial adsorbents [1].

Type	Typical applications
Silica gel	Drying of gases, refrigerants, organic solvents, transformer oils; desiccant in packings and double glazing; dew point control of natural gas.
Activated alumina	Drying of gases, organic solvents, transformer oils; removal of HCl from hydrogen; removal of fluorine and boron-fluorine compounds in alkylation processes.
Carbons	Nitrogen from air; hydrogen from syn-gas and hydrogenation processes; ethene from methane and hydrogen; vinyl chloride monomer (VCM) from air; removal of odours from gases; recovery of solvent vapours; removal of SO_x and NO_x; purification of helium; clean-up of nuclear off-gases; decolourising of syrups, sugars and molasses; water purification, including removal of phenol, halogenated compounds, pesticides, caprolactam, chlorine.
Zeolites	Oxygen from air; drying of gases; removing water from azeotropes; sweetening sour gases and liquids; purification of hydrogen; separation of ammonia and hydrogen; recovery of carbon dioxide; separation of oxygen and argon; removal of acetylene, propane and butane from air; separation of xylenes and ethyl benzene; separation of normal from branched paraffins; separation of olefins and aromatics from paraffins; recovery of carbon monoxide from methane and hydrogen; purification of nuclear off-gases; separation of cresols; drying of refrigerants and organic liquids; separation of solvent systems; purification of silanes; pollution control, including removal of Hg, NO_x and SO_x from gases; recovery of fructose from corn syrup.

Continued

Table 8.1 Typical applications of commercial adsorbents [1]—*cont'd*

Type	Typical applications
Polymers and resins	Water purification, including removal of phenol, chlorophenols, ketones, alcohols, aromatics, aniline, indene, polynuclear aromatics, nitro- and chlor-aromatics,
	PCB, pesticides, antibiotics, detergents, emulsifiers, wetting agents, kraftmill effluents, dyestuffs;
	recovery and purification of steroids, amino acids and polypeptides;
	separation of fatty acids from water and toluene;
	separation of aromatics from aliphatics;
	separation of hydroquinone from monomers;
	recovery of proteins and enzymes;
	removal of colours from syrups;
	removal of organics from hydrogen peroxide.
Clays (acid-treated and pillared)	Treatment of edible oils;
	removal of organic pigments;
	refining of mineral oils;
	removal of polychlorobiphenyl (PCB).

that they can be manufactured in a highly porous form, giving rise to a large internal surface. In comparison, the external surface makes only a modest contribution to the total, even when the solid is finely divided. Iron oxide particles with a radius of 5 μm and a density of 5000 kg/m^3 have an external surface of 12,000 m^2/kg. A typical value for the total surface of commercial adsorbents is 400,000 m^2/kg.

The adsorption which results from the influence of van der Waals forces is essentially physical in nature. Because the forces are not strong, the adsorption may be easily reversed. In some systems, additional forces bind absorbed molecules to the solid surface. These are chemical in nature involving the exchange or sharing of electrons, or possibly molecules forming atoms or radicals. In such cases, the term *chemisorption* is used to describe the phenomenon. This is less easily reversed than physical adsorption, and regeneration may be a problem. Chemisorption is restricted to just one layer of molecules on the surface, although it may be followed by additional layers of physically adsorbed molecules.

When molecules move from a bulk fluid to an adsorbed phase, they lose degrees of freedom and the free energy is reduced. Adsorption is always accompanied by the liberation of heat. For physical adsorption, the amount of heat is similar in magnitude to the heat of condensation. For chemisorption it is greater and of an order of magnitude normally associated with a chemical reaction. If the heat of adsorption cannot be dispersed by cooling, the capacity of the adsorbent will be reduced as its temperature increases.

Table 8.2 Properties of some solvents recoverable by adsorptive techniques.

	Molecular weight (kg/kmol)	Specific heat capacity at 293K (kJ/kg K)	Density at 293K (kg/m³)	Latent heat evaporation (kJ/kg)	Boiling point (K)	Explosive limits in air at 293K (per cent by volume) low	high	Solubility in water at 293K (kg/m³)
Acetone	58.08	2.211	791.1	524.6	329.2	2.15	13.0	∞
Allyl alcohol	58.08	2.784	853.5	687	369.9	2.5	18	∞
n-Amyl acetate	130.18	1.926	876	293	421.0	3.6	16.7	1.8
Isoamyl acetate	130.18	1.9209	876	289	415.1	3.6	–	2.5
n-Amyl alcohol	88.15	2.981	817	504.9	410.9	1.2	–	68
Isoamyl alcohol	88.15	2.872	812	441.3	404.3	1.2	–	32
Amyl chloride	106.6	–	883	–	381.3	–	–	Insol.
Amylene	70.13	1.181	656	314.05	309.39	1.6	–	Insol.
Benzene	78.11	1.720	880.9	394.8	353.1	1.4	4.7	0.8
n-Butyl acetate	116.16	1.922	884	309.4	399.5	1.7	15	10
Isobutyl acetate	116.16	1.921	872	308.82	390.2	2.4	10.5	6.7
n-Butyl alcohol	74.12	2.885	809.7	600.0	390.75	1.45	11.25	78
Isobutyl alcohol	74.12	2.784	805.7	578.6	381.8	1.68	–	85
Carbon disulphide	76.13	1.005	1267	351.7	319.25	1.0	50	2
Carbon tetrachloride	153.84	0.846	1580	194.7	349.75	Non-flammable		0.084
Cellosolve	90.12	2.324	931.1	–	408.1	–	–	∞
Cellosolve methyl	76.09	2.236	966.3	565	397.5	–	–	∞
Cellosolve acetate	132.09	–	974.8	–	426.0	–	–	230
Chloroform	119.39	0.942	1480	247	334.26	Non-flammable		8
Cyclohexane	84.16	2.081	778.4	360	353.75	1.35	8.35	Insol.
Cyclohexanol	100.16	1.746	960	452	433.65	–	–	60

Continued

Table 8.2 Properties of some solvents recoverable by adsorptive techniques—*cont'd*

	Molecular weight (kg/kmol)	Specific heat capacity at 293 K (kJ/kg K)	Density at 293 K (kg/m³)	Latent heat evaporation (kJ/kg)	Boiling point (K)	Explosive limits in air at 293 K (per cent by volume)		Solubility in water at 293 K (kg/m³)
						low	high	
Cyclohexanone	98.14	1.813	947.8	–	429.7	3.2	9.0	50
Cymene	134.21	1.666	861.2	283.07	449.7	–	–	Insol.
n-Decane	142.28	2.177	730.1	252.0	446.3	0.7	–	Insol.
Dichloroethylene	96.95	1.235	1291	305.68	333.0	9.7	12.8	Insol.
Ether (diethyl)	74.12	2.252	713.5	360.4	307.6	1.85	36.5	69
Ether (di-n-butyl)	130.22	–	769.4	288.1	415.4	–	–	3
Ethyl acetate	88.10	2.001	902.0	366.89	350.15	2.25	11.0	79.4
Ethyl alcohol	56.07	2.462	789.3	855.4	351.32	3.3	19.0	∞
Ethyl bromide	108.98	0.812	1450	250.87	311.4	6.7	11.2	9.1
Ethyl carbonate	118.13	1.926	975.2	306	399.0	–	–	V.sl.sol.
Ethyl formate	74.08	2.135	923.6	406	327.3	2.7	16.5	100
Ethyl nitrite	75.07	–	900	–	290.0	3.0	–	Insol.
Ethylene Dichloride	96.97	1.298	1255.0	323.6	356.7	6.2	15.6	8.7
Ethylene oxide	44.05	–	882	580.12	283.5	3.0	80	∞
Furfural	96.08	1.537	1161	450.12	434.8	2.1	–	83
n-Heptane	100.2	2.123	683.8	318	371.4	1.0	6.0	0.052
n-Hexane	86.17	2.223	659.4	343	341.7	1.25	6.9	0.14
Methyl acetate	74.08	2.093	927.2	437.1	330.8	4.1	13.9	240
Methyl alcohol	32.04	2.500	792	1100.3	337.7	6.72	36.5	67.2
Methyl cyclohexanol	114.18	–	925	–	438.0	–	–	11

Methyl cyclohexanone	112.2	1.842	924	—	438.0	—	—	30
Methyl cyclohexane	98.18	1.855	768	323	373.9	1.2	—	Insol.
Methyl ethyl ketone	72.10	2.085	805.1	444	352.57	1.81	11.5	265
Methylene chloride	84.93	1.089	1336	329.67	313.7	Non-flammable		20
Monochlorobenzene	112.56	1.256	1107.4	324.9	404.8	—	—	0.49
Naphthalene	128.16	1.683	1152	316.1	491.0	0.8	—	0.04
Nonane	128.25	2.106	718	274.2	422.5	0.74	2.9	Insol.
Octane	114.23	2.114	702	296.8	398.6	0.95	3.2	0.015
Paraldehyde	132.16	1.825	904	104.75	397.0	1.3	—	120
n-Pentane	72.15	2.261	626	352	309.15	1.3	8.0	0.36
Isopentane	72.09	2.144	619	371.0	301.0	1.3	7.5	Insol.
Pentachloroethane	202.33	0.900	1678	182.5	434.95	Non-flammable		0.5
Perchloroethylene	165.85	0.879	1662.6	209.8	393.8	Non-flammable		0.4
n-Propyl acetate	102.3	1.968	888.4	336.07	374.6	2.0	8.0	18.9
Isopropyl acetate	102.3	2.181	880	332.4	361.8	2.0	8.0	29
n-Propyl alcohol	60.09	2.453	803.6	682	370.19	2.15	13.5	∞
Isopropyl alcohol	60.09	2.357	786.3	667.4	355.4	2.02	—	∞
Propylene dichloride	112.99	1.398	1159.3	302.3	369.8	3.4	14.5	2.7
Pyridine	79.10	1.637	978	449.59	388.3	1.8	12.4	∞
Tetrachloroethane	167.86	1.130	1593	231.5	419.3	Non-flammable		3.2
Tetrahydrofuran	72.10	1.964	888	410.7	339.0	1.84	11.8	∞
Toluene	92.13	1.641	871	360	383.0	1.3	7.0	0.47
Trichloroethylene	131.4	0.934	1465.5	239.9	359.7	Non-flammable		1.0
Xylene	106.16	1.721	897	347.1	415.7	1.0	6.0	Insol.
Water	18	4.183	998	2260.9	373.0	Non-flammable		∞

It is often convenient to think of adsorption as occurring in three stages as the adsorbate concentration increases. Firstly, a single layer of molecules builds up over the surface of the solid. This monolayer may be chemisorbed and associated with a change in free energy which is characteristic of the forces which hold it. As the fluid concentration is further increased, layers form by physical adsorption and the number of layers which form may be limited by the size of the pores. Finally, for adsorption from the gas phase, capillary condensation may occur in which capillaries become filled with condensed adsorbate, and its partial pressure reaches a critical value relative to the size of the pore.

Although the three stages are described as taking place in sequence, in practice, all three may be occurring simultaneously in different parts of the adsorbent since conditions are not uniform throughout. Generally, concentrations are higher at the outer surface of an adsorbent pellet than in the centre, at least until equilibrium conditions have been established. In addition, the pore structure will consist of a distribution of pore sizes and the spread of the distribution depends on the origin of the adsorbent and its conditions of manufacture.

8.2 The nature of adsorbents

Adsorbents are available as irregular granules, extruded pellets and formed spheres. The size reflects the need to pack as much surface area as possible into a given volume of bed and at the same time minimise pressure drop for flow through the bed. Sizes of up to about 6 mm are common.

To be attractive commercially, an adsorbent should embody a number of features:

(a) it should have a large internal surface area.
(b) the area should be accessible through pores big enough to admit the molecules to be adsorbed. It is a bonus if the pores are also small enough to exclude molecules which it is desired not to adsorb.
(c) the adsorbent should be capable of being easily regenerated.
(d) the adsorbent should not age rapidly, that is lose its adsorptive capacity through continual recycling.
(e) the adsorbent should be mechanically strong enough to withstand the bulk handling and vibration that are a feature of any industrial unit.

8.2.1 Molecular sieves

An increase in the use of adsorbers as a means of separation on a large scale is the result of the manufacturers' skill at developing and producing adsorbents which are tailored for specific tasks. First, by using naturally occurring zeolites and, later, synthesised members of that family of minerals, it has been possible to manufacture a range of adsorbents known collectively as *molecular sieves*. These have lattice structures composed of tetrahedra of silica and alumina arranged in various ways. The net effect is the formation of a cage-like structure with windows which admit only molecules less than a certain size as shown in Fig. 8.1. By using different source

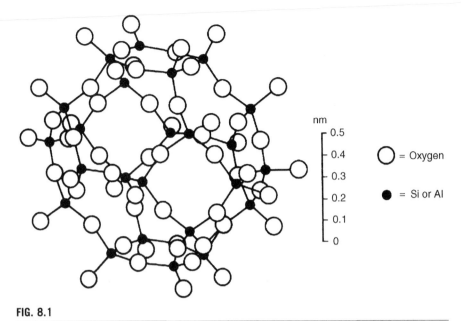

FIG. 8.1

A cubo-octahedral unit composed of SiO_4 and AlO_4 tetrahedra [2].

materials and different conditions of manufacture, it is possible to produce a range of molecular sieves with access dimensions of 0.3–1 nm. The dimensions are precise for a particular sieve because they derive from the crystal structure of that sieve. Some of the molecules admitted by different molecular sieves are given in Table 8.3 which is taken from the work of Barrer [2]. The crystallites of a sieve are about 10 µm in size and are aggregated for commercial use by mixing with a clay binder and extruding as pellets or rolling into spheres. The pelletising creates two other sets of pores, between crystallites and between pellets. Neither may add significantly to the adsorptive surface though each will influence rates of diffusion and pressure drop. It has been estimated by Yang [3] that there are about forty naturally occurring zeolites and that some 150 have been synthesised.

The manufacture of molecular sieves has been reviewed in the literature, and particularly by Breck [4], Barrer [5], and Roberts [6].

8.2.2 Activated carbon

In some of the earliest recorded examples of adsorption, activated carbon was used as the adsorbent. Naturally occurring carbonaceous materials such as coal, wood, coconut shells or bones are decomposed in an inert atmosphere at a temperature of about 800 K. Because the product will not be porous, it needs additional treatment or *activation* to generate a system of fine pores. The carbon may be produced in the activated state by treating the raw material with chemicals, such as zinc chloride or phosphoric acid before carbonising. Alternatively, the carbon from the carbonising

Table 8.3 Classification of some molecular sieves [2].

Molecular size increasing →

(1)	(2)	(3)	(4)	(5)	(6)	(7)	(8)	(9)	(10)
He, Ne, A, CO H_2, O_2, N_2,	Kr, Xe CH_4	C_3H_8 $n\text{-}C_4H_{10}$	CF_4 C_2F_6	SF_6 $isoC_4H_{10}$	$(CH_3)_3N$ $(C_2H_5)_3N$	C_6H_6 $C_6H_5CH_3$	Naphthalene quinoline,	1, 3, 5 Triethyl benzene	$(n\text{-}C_4F_9)_3N$
NH_3, H_2O	C_2H_6	$n\text{-}C_7H_{16}$	CF_2Cl_2	$isoC_5H_{12}$	$C(CH_3)_4$	$C_6H_4(CH_3)_2$	6-decyl-	1, 2, 3, 4, 5, 6,	
	CH_3OH	$n\text{-}C_{14}H_{30}$	CF_3Cl	$isoC_8H_{18}$	$C(CH_3)_3Cl$	Cyclohexane	1, 2, 3, 4-	7, 8, 13, 14, 15,	
	CH_3CN CH_3NH_2 CH_3Cl CH_3Br CO_2 C_2H_2	etc. C_2H_5Cl C_2H_5Br C_2H_5OH $C_2H_5NH_2$ CH_2Cl_2	$CHFCl_2$	etc. $CHCl_3$ $CHBr_3$ CHI_3 $(CH_3)_2CHOH$ $(CH_3)_2CHCl$	C $(CH_3)_3Br$ C $(CH_3)_3OH$ CCl_4 CBr_4 $C_2F_2Cl_4$	Thiophen Furan Pyridine Dioxane $B_{10}H_{14}$	tetrahydro-naphthalene, 2-butyl-1-hexyl indan $C_6F_{11}CF_3$	16- Decahydro-chrysene	
	CS_2	CH_2Br_2 CHF_2Cl CHF_3		$n\text{-}C_3F_8$ $n\text{-}C_4F_{10}$ $n\text{-}C_7F_{16}$ B_5H_9					
		$(CH_3)_2NH$ CH_3I							
		B_2H_6							

Size limit for Ca- and Ba-mordenites and levynite about here (~0.38 nm)

Size limit for Na-mordenite and Linde sieve 4 A about here (~0.4 nm)

Size limit for Ca-rich chabazite, Linde sieve 5 A, Ba-zeolite and gmelinite about here (~0.49 nm)

Size limit for Linde sieve 10 × about here (~0.8 nm)

Size limit for Linde sieve 13 × about here (~1.0 nm)

Type 5

Type 4

Type 3

Type 2

Type 1

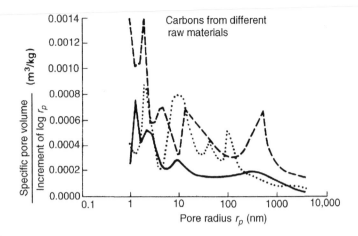

FIG. 8.2

Typical pore volume distributions for three activated carbons used for liquid-phase processes [7].

stage may be selectively oxidised at temperatures in excess of 1000 K in atmospheres of materials such as steam or carbon dioxide.

Activated carbon has a typical surface area of 10^6 m^2/kg, mostly associated with a set of pores of about 2 nm in diameter. There is likely to be another set of pores of about 1000 nm in diameter, which do not contribute to the surface area. There may even be a third, intermediate set of pores which is particularly developed in carbons intended for use with liquids, as shown in Fig. 8.2 which is taken from the work of Schweitzer [7].

Activated carbon may be used as a powder, in which form it is mixed in with the liquid to be treated, and then removed by filtration. It may also be used in granular form. When the use of carbon is low, it is normally economic to regenerate it, and this is usually the case with powder. Granular carbon is normally regenerated after use. Because it has a low affinity for water, activated carbon may preferentially adsorb components from aqueous solution or from moist gases.

By carefully choosing the starting material and the activating process, it has been possible in recent years to generate in carbon a pore system with a narrow span of pore sizes. With a mean pore diameter of perhaps 0.6 nm, such products are known as carbon molecular sieves.

8.2.3 Silica gel

When a silicate solution such as sodium silicate is acidified, a gel of polymeric colloidal salicylic acid is formed as an agglomerate of micro-particles. When the gel is heated, water is expelled leaving a hard, glassy structure with voids between the micro-particles equivalent to a mean pore diameter of about 3 nm and an internal surface of about $500,000\,m^2/kg$. As discussed by Everett and Stone [8] these properties may be varied by controlling the pH of the solution from which the gel is precipitated.

Silica gel is probably the adsorbent which is best known. Small sachets of gel are often included in packages of material that might deteriorate in a damp atmosphere. Sometimes a dye is added in the manufacturing process so that the gel changes colour as it becomes saturated.

Unlike the activated carbons, the surface of silica gel is hydrophilic and it is commonly used for drying gases and also finds applications where there is a requirement to remove unsaturated hydrocarbons. Silica gels are brittle and may be shattered by the rapid release of the heat of adsorption that accompanies contact with liquid water. For such applications, a tougher variety is available with a slightly lower surface area.

8.2.4 Activated alumina

When an adsorbent is required which is resistant to attrition and which retains more of its adsorptive capacity at elevated temperatures than does silica gel, activated alumina may be used. This is made by the controlled heating of hydrated alumina. Water molecules are expelled and the crystal lattice ruptures along planes of structural weakness. A well-defined pore structure results, with a mean pore diameter of about $4\,nm$ and a surface area of some $350,000\,m^2/kg$. The micrograph shown in Fig. 8.3 taken from the work of Bowen et al. [9] shows, at the higher magnification, the regular hexagonal disposition of the pores at the thin edges of particles of alumina.

Activated alumina has a high affinity for water in particular, and for hydroxyl groups in general. It cannot compete in terms of capacity or selectivity with molecular sieves although its superior mechanical strength is important in moving-bed applications. As a powder, activated alumina may be used as a packing for chromatographic columns, as described in Chapter 19.

8.3 Adsorption equilibria

Much of the early work on the nature of adsorbents sought to explain the equilibrium capacity and the molecular forces involved. Adsorption equilibrium is a dynamic concept achieved when the rate at which molecules adsorb on to a surface is equal to the rate at which they desorb. The physical chemistry involved may be complex and no single theory of adsorption has been put forward which satisfactorily explains all systems. Fortunately for the engineer, what is needed is an accurate representation of equilibrium, the theoretical minutiae of which is not of concern. For this reason, some of the earliest theories of adsorption are still the most useful, even though the assumptions on which they were based were seen in later years to be not entirely valid. Most theories have been developed for gas–solid systems because the gaseous state is better understood than the liquid. Statistical theories are being developed which should apply equally well to gas–solid and liquid–solid equilibria, though

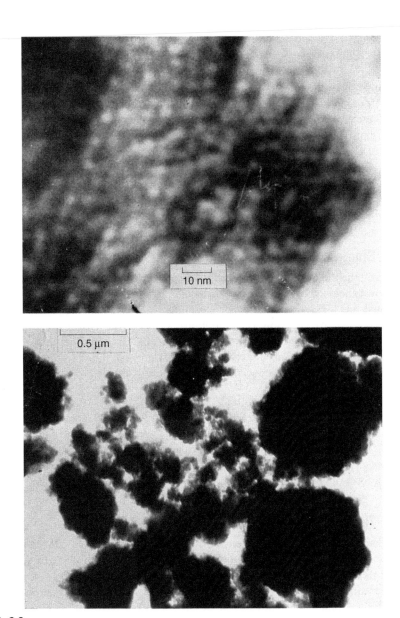

FIG. 8.3

Electron micrographs of a commercial activated alumina at two magnifications [9].

these are not yet at a stage when they can be applied easily and confidently to the design of equipment.

The capacity of an adsorbent for a particular adsorbate involves the interaction of three properties—the concentration C of the adsorbate in the fluid phase, the concentration C_s of the adsorbate in the solid phase and the temperature T of the system. If one of these properties is kept constant, the other two may be graphed to represent the equilibrium. The commonest practice is to keep the temperature constant and to plot C against C_s to give an adsorption isotherm. When C_s is kept constant, the plot of C against T is known as an adsorption isostere. In gas–solid systems, it is often convenient to express C as a pressure of adsorbate. Keeping the pressure constant and plotting C_s against T gives adsorption isobars. The three plots are shown for the ammonia-charcoal system in Fig. 8.4 which is taken from the work of Brunauer [10].

8.3.1 Single component adsorption

Most early theories were concerned with adsorption from the gas phase. Sufficient was known about the behaviour of ideal gases for relatively simple mechanisms to be postulated, and for equations relating concentrations in gaseous and adsorbed phases to be proposed. At very low concentrations the molecules adsorbed are widely spaced over the adsorbent surface so that one molecule has no influence on another. For these limiting conditions it is reasonable to assume that the concentration in one phase is proportional to the concentration in the other, that is:

$$C_s = K_a C \tag{8.1}$$

This expression is analogous to Henry's Law for gas–liquid systems even to the extent that the proportionality constant obeys the van't Hoff equation and $K_a = K_0 e^{-\Delta H/RT}$, where ΔH is the enthalpy change per mole of adsorbate as it transfers from gaseous to adsorbed phase. At constant temperature, Eq. (8.1) becomes the simplest form of adsorption isotherm. Unfortunately, few systems are so simple.

8.3.2 The Langmuir isotherm

At higher gas-phase concentrations, the number of molecules absorbed soon increases to the point at which further adsorption is hindered by lack of space on the adsorbent surface. The rate of adsorption then becomes proportional to the empty surface available, as well as to the fluid concentration. At the same time as molecules are adsorbing, other molecules will be desorbing if they have sufficient activation energy. At a fixed temperature, the rate of desorption will be proportional to the surface area occupied by adsorbate. When the rates of adsorption and desorption are equal, a dynamic equilibrium exists. For adsorption which is confined to a monomolecular layer, the equilibrium may be written as:

$$k_0 a_0 C = k_0 (1 - a_1) C = k_1' a_1$$

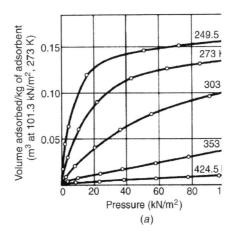

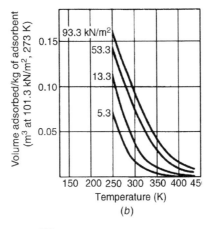

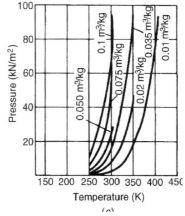

FIG. 8.4

Equilibrium data for the adsorption of ammonia on charcoal [10] (A) Adsorption isotherm
(B) Adsorption isobar (C) Adsorption isostere.

or:

$$a_1 = \frac{B_0 C}{1 + B_0 C} \tag{8.2}$$

where:

a_0 is the fraction of empty surface,

a_1 is the fraction of surface occupied by a monolayer of adsorbed molecules,

$B_0\, C_s = K_a C$.

k_0 is the velocity constant for adsorption on to empty surface, and.

$k_1{}'$ is the velocity constant for desorption from a monolayer.

Equation (8.2) has been developed for adsorption from the gas phase. It is convenient to also express it in terms of partial pressures, which gives:

$$\frac{C_s}{C_{sm}} = \frac{B_1 P}{1 + B_1 P} \tag{8.3}$$

where:

C_s is the concentration of the adsorbed phase,

C_{sm} is the concentration of the adsorbed phase when the monolayer is complete,

$B_1 = B_0/\mathbf{R}T$, and.

P is the partial pressure of adsorbate in the gas phase.

Equations (8.2) and (8.3) have the form of the Langmuir [11] equation, developed in 1916, which describes the adsorption of gases on to plane surfaces of glass, mica and platinum. A number of assumptions is implicit in this development. As well as being limited to monolayer adsorption, the Langmuir equation assumes that:

(a) these are no interactions between adjacent molecules on the surface.

(b) the energy of adsorption is the same all over the surface.

(c) molecules adsorb at fixed sites and do not migrate over the surface.

When $B_1 P << 1$, Eq. (8.3) reverts to the form of Henry's Law, as given in Eq. (8.1). Equation (8.3) can be rewritten in linear form to give:

$$\frac{P}{C_s} = \frac{P}{C_{sm}} + \frac{1}{B_1 C_{sm}} \tag{8.4}$$

so that a plot of P/C_s against P will be a straight line when applied to a system that behaves in accordance with the Langmuir isotherm.

It has been shown experimentally, however, that many systems do not follow this isotherm and efforts continue to find an improved equation.

8.3.3 The BET isotherm

In 1938, Brunauer, Emmett and Teller [12] and Emmett and de Witt [13] developed what is now known as the BET theory. As in the case in Langmuir's isotherm, the theory is based on the concept of an adsorbed molecule which is not free to move over the surface, and which exerts no lateral forces on adjacent molecules of adsorbate. The BET theory does, however, allow different numbers of adsorbed layers to build up on different parts of the surface, although it assumes that the net amount of surface which is empty or which is associated with a monolayer, bilayer and so on is constant for any particular equilibrium condition. Monolayers are created by adsorption on to empty surface and by desorption from bilayers. Monolayers are lost both through desorption and through the adsorption of additional layers. The rate of adsorption is proportional to the frequency with which molecules strike the surface and the area of that surface. From the kinetic theory of gases, the frequency is proportional to the pressure of the molecules and hence:

The rate of adsorption on to empty surface $= k_0 a_0 P$, and the rate of desorption from a monolayer $= k'_1 a_1$.

Desorption is an activated process. If E_1 is the excess energy required for 1 mol in the monolayer to overcome the surface forces, the proportion of molecules possessing such energy is $e^{-E_1/RT}$. Hence the rate of desorption from a monolayer may be written as:

$$A'_1 e^{-E_1/RT} a_1$$

where A'_1 is the frequency factor for monolayer desorption.

The dynamic equilibrium of the monolayer is given by:

$$k_0 a_0 P + A'_2 e^{-E_2/RT} a_2 = k_1 a_1 P + A'_1 e^{-E_1/RT} a_1 \tag{8.5}$$

where A_2' is the frequency factor for description from a bilayer, thus creating a monolayer.

Applying similar arguments to the empty surface, then:

$$k_0 a_0 P = A'_1 e^{-E_1/RT} a_1 \tag{8.6}$$

From Eqs. (8.5) and (8.6):

$$k_1 a_1 P = A'_2 e^{-E_2/RT} a_2$$

$$a_1 = \frac{k_0}{A'_1} e^{E_1/RT} a_0 P = a_0 a_{01} \tag{8.7}$$

and:

$$a_2 = \frac{k_1}{A'_2} e^{E_2/RT} a_1 P = \beta a_1 \tag{8.8}$$

The BET theory assumes that the reasoning used for one or two layers of molecules may be extended to n layers. It argues that energies of activation after the first layer are all equal to the latent heat of condensation, so that:

$$E_2 = E_3 = E_4 = \cdots = E_n = \lambda_M.$$

Hence it may be assumed that β is constant for layers after the first and:

$$a_i = \beta^{i-1} a_1 = B_2 \beta^i a_0$$

where $B_2 = \alpha_0/\beta$, and a_i is the fraction of the surface area containing i layers of adsorbate.

Since a_0, a_1, ... are fractional areas, their summation over n layers will be unity and:

$$1 = a_0 + \sum_{i=1}^{n} a_i$$

$$= a_0 + \sum_{i=1}^{n} B_2 \beta^i a_0 \qquad (8.9)$$

The total volume of adsorbate associated with unit area of surface is given by:

$$v_s = v_s^1 \sum_{i=1}^{n} i a_i = v_s^1 \sum_{i=1}^{n} i B_2 \beta^i a_0 \qquad (8.10)$$

where v_s^1 is the volume of adsorbate in a unit area of each layer.

Since v_s^1 does not change with n, a geometrically plane surface is implied. Strictly, Eq. (8.10) is not applicable to highly convex or concave surfaces. Equations (8.9) and (8.10) may be combined to give:

$$\frac{v_s}{v_s^1} = \frac{\sum_{i=1}^{n} i B_2 \beta^i a_0}{a_0 + \sum_{i=1}^{n} B_2 \beta^i a_0} \qquad (8.11)$$

The numerator of Eq. (8.11) may be written as:

$$B_2 a_0 \beta \frac{d}{d\beta} \left(\sum_{i=1}^{n} \beta^i \right) = B_2 a_0 \beta \frac{d}{d\beta} \left\{ \left(\frac{1 - \beta^n}{1 - \beta} \right) \beta \right\}$$

and the denominator as:

$$a_0 \left[1 + B_2 \beta \left(\frac{1 - \beta^n}{1 - \beta} \right) \right]$$

Substituting these values into Eq. (8.11) and rearranging gives:

$$\frac{v_s}{v_s^1} = \frac{B_2 \beta}{1 - \beta} \frac{\left[1 - (n + 1)\beta^n + n\beta^{n+1} \right]}{\left[1 + (B_2 - 1)\beta - B_2 \beta^{n+1} \right]} \qquad (8.12)$$

On a flat unrestricted surface, there is no theoretical limit to the number of layers that can build up. When $n = \infty$, Eq. 8.12 becomes:

$$\frac{v_s}{v_s^1} = \frac{B_2 \beta}{(1 - \beta)(1 - \beta + B_2 \beta)} \qquad (8.13)$$

When the pressure of the adsorbate in the gas phase is increased to the saturated vapour pressure, condensation occurs on the solid surface and v_s/v_s^1 approaches infinity. In Eq. (8.13), this condition corresponds to putting $\beta = 1$. It may be noted that putting $\beta = 1/(1 - B_2)$ is not helpful.

Hence, from Eq. (8.8):

$$1 = \frac{k_1}{A_2'} e^{\lambda M / RT} \cdot P^0$$

where: λ_M is the molar latent heat and.

P^0 is the saturated vapour pressure.

Hence, from Eq. (8.8), $\beta = P/P^0$.

Equation (8.12) may be rewritten for unit mass of adsorbent instead of unit surface. This is known as the limited form of the BET equation which is:

$$\frac{V_s}{V_s^1} = B_2 \frac{P}{P^0} \frac{\left[1 - (n+1)(P/P^0)^n + n(P/P^0)^{n+1}\right]}{(1 - P/P^0)\left[1 + (B_2 - 1)(P/P^0) - B_2(P/P^0)^{n+1}\right]} \tag{8.14}$$

where V_s^1 is the volume of adsorbate contained in a monolayer spread over the surface area present in unit mass of adsorbent.

When $n = 1$, adsorption is confined to a monolayer and Eq. (8.14) reduces to the Langmuir equation.

When $n = \infty$, $(P/P^0)^n$ approaches zero and Eq. (8.13) may be rearranged in a convenient linear form to give:

$$\frac{P/P_0}{V(1 - P/P_0)} = \frac{1}{V^1 B_2} + \frac{B_2 - 1}{V^1 B_2}\left(\frac{P}{P^0}\right) \tag{8.15}$$

where V and V^1 are the equivalent gas-phase volumes of V_s and V_s^1.

If a plot of the left-hand term against P/P^0 is linear, the experimental data may be said to fit the infinite form of the BET equation. From the slope and the intercept, V^1 and B_2 may be calculated.

The advantage of Eq. (8.14) is that it may be fitted to all known shapes of adsorption isotherm. In 1938, a classification of isotherms was proposed which consisted of the five shapes shown in Fig. 8.5 which is taken from the work of Brunauer et al. [14]

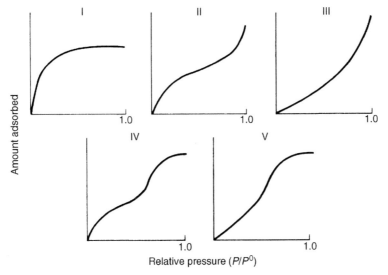

FIG. 8.5

Classification of isotherms into five types of Brunauer, Deming, Deming, and Teller [14].

Only gas–solid systems provide examples of all the shapes, and not all occur frequently. It is not possible to predict the shape of an isotherm for a given system, although it has been observed that some shapes are often associated with a particular adsorbent or adsorbate properties. Charcoal, with pores just a few molecules in diameter, almost always gives a Type I isotherm. A non-porous solid is likely to give a Type II isotherm. If the cohesive forces between adsorbate molecules are greater than the adhesive forces between adsorbate and adsorbent, a Type V isotherm is likely to be obtained for a porous adsorbent and a Type III isotherm for a non-porous one.

In some systems, three stages of adsorption may be discerned. In the activated alumina–air–water vapour system at normal temperature, the isotherm is found to be of Type IV. This consists of two regions which are concave to the gas concentration axis separated by a region which is convex. The concave region that occurs at low gas concentrations is usually associated with the formation of a single layer of adsorbate molecules over the surface. The convex portion corresponds to the build-up of additional layers, whilst the other concave region is the result of condensation of adsorbate in the pores – so-called *capillary condensation* as discussed earlier in this section.

At low gas concentrations, whilst the monolayer is still incomplete, the absorbed molecules are relatively immobile. In the multilayer region, the adsorbed molecules behave more like a liquid film. The amount of capillary condensation that occurs depends on the pore sizes and their distribution, as well as on the concentration in the gas phase.

When $n = 1$, Eq. (8.14) represents a Type I isotherm.

When $n = \infty$, Eq. (8.14) represents a Type II, and the rarer Type III isotherm by choosing a suitable value for B_2. As B_2 is increased, the point of inflexion or 'knee' of Type II becomes more prominent. This corresponds to an increasing tendency for the monolayer to become complete before a second layer starts. In the extreme case of an adsorbent whose surface is very uniform from an energy point of view, the adsorbate builds up in well-defined layers. This gives rise to a stepped isotherm, in which each step corresponds to another layer. When B_2 is less than 2, there is no point of inflexion and Type III isotherms are obtained. The condition $1 > B_2 > 0$ often corresponds to a tendency for molecules to adsorb in clusters rather than in complete layers.

The success of the BET equation in representing experimental data should not be regarded as a measure of the accuracy of the model on which it is based. Its capability of modelling the mobile multilayers of a Type IV isotherm is entirely fortuitous because, in the derivation of the equation, it is assumed that adsorbed molecules are immobile.

Example 8.1

Spherical particles of 15 nm diameter and density 2290 kg/m³ are pressed together to form a pellet. The following equilibrium data were obtained for the sorption of nitrogen at 77 K. Obtain estimates of the surface area of the pellet from the adsorption isotherm and compare the estimates with the geometric surface. The density of liquid nitrogen at 77 K is 808 kg/m³.

P/P^0	0.1	0.2	0.3	0.4	0.5	0.6	0.7	0.8	0.9
m^3 liq $N_2 \times 10^6/$ kg solid	66.7	75.2	83.9	93.4	108.4	130.0	150.2	202.0	348.0

where P is the pressure of the sorbate and P^0 is its vapour pressure at 77 K. Use the following data:

$$\text{density of liquid nitrogen} = 808\,\text{kg/m}^3$$

$$\text{area occupied by one adsorbed molecule of nitrogen} = 0.162\,\text{nm}^2$$

$$\text{Avogadro Number} = 6.02 \times 10^{26}\,\text{molecules/kmol}$$

Solution

For $1\,\text{m}^3$ of pellet with a voidage ε, then:

$$\text{Number of particles} = (1 - \varepsilon)/(\pi/6)\left(15 \times 10^{-9}\right)^3$$

$$\text{Surface area per unit volume} = (1 - \varepsilon)\pi\left(15 \times 10^{-9}\right)^2/(\pi/6)\left(15 \times 10^{-9}\right)^3$$

$$= 6(1 - \varepsilon)/\left(15 \times 10^{-9}\right)\,\text{m}^2/\text{m}^3$$

$1\,\text{m}^3$ of pellet contains 2290 $(1 - \varepsilon)$ kg solid and hence:

$$\text{specific surface} = 6(1 - \varepsilon)/[(15 \times 10^{-9}(1 - \varepsilon)2290]$$

$$= 1.747 \times 10^5\,\text{m}^2/\text{kg}$$

(a) Using the BET isotherm

$$(P/P^0)/[V(1 - P/P^0)] = 1/V'B + (B - 1)(P/P^0)/V'B \quad (8.15)$$

where V and V' are the liquid volumes of adsorbed nitrogen.
From the adsorption data given:

(P/P^0)	0.1	0.2	0.3	0.4	0.5	0.6
V (m^3 liquid N_2/kg solid $\times 10^6$)	66.7	75.2	83.9	93.4	108.4	130.0
$((P/P^0)/V) \times 10^{-6}$	1500	2660	3576	4283	4613	4615
$(P/P^0)/[V(1 - P/P^0)]$	1666	3333	5109	7138	9226	11,538

Plotting $(P/P^0)/[V(1 - P/P^0)]$ against (P/P^0), as shown in Fig. 8.6, then: intercept, $1/V'B = 300$, and slope, $(B - 1)/V'B = 13,902$.
from which:
$B = (13,902/300) + 1 = 47.34$.

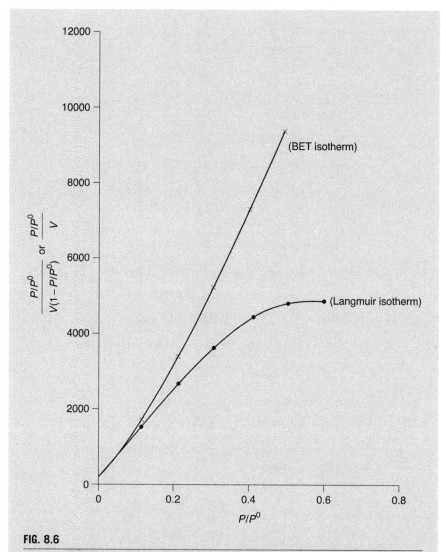

FIG. 8.6

Adsorption isotherms for Example 8.1.

and:

$$V' = 1/(300 \times 47.34) = 70.4 \times 10^{-6}\,\text{m}^3/\text{kg}.$$

The total surface area $= \left[(70.4 \times 10^{-6} \times 808 \times 6.2 \times 10^{26} \times 0.162 \times 10^{-18})\right]/28$

$$= \underline{\underline{2.040 \times 10^5\,\text{m}^2/\text{kg}.}}$$

(b) Using the Langmuir form of the isotherm:

Assuming this applies at low concentrations then, expressing pressure as the ratio P/P^0, and the amount adsorbed as a volume of liquid adsorbate, Eq. (8.4) becomes:

$$(P/P^0)/V = (P/P^0)/V' + 1/(B_2V')$$

Thus, a plot of $(P/P^0)/V$ against (P/P^0) will have a slope of $(1/V')$.
Thus, from Fig. 8.6:
and:

$$1/V' = 13,902$$
$$V' = 71.9 \times 10^{-6} m^3/kg$$

which agrees with the value from the BET isotherm.

It may be noted that areas calculated from the isotherm are some 20% greater than the geometric surface, probably due to the existence of some internal surface within the particles.

8.3.4 The Gibbs isotherm

An entirely different approach to equilibrium adsorption is to assume that adsorbed layers behave like liquid films, and that the adsorbed molecules are free to move over the surface. It is then possible to apply the equations of classical thermodynamics. The properties which determine the free energy of the film are pressure and temperature, the number of molecules contained and the area available to the film. The Gibbs free energy G may be written as:

$$G = F(P, T, n_s, A_s) \tag{8.16}$$

Hence:

$$dG = \left(\frac{\partial G}{\partial P}\right)_{T,ns,A_s} ; dP + \left(\frac{\partial G}{\partial T}\right)_{P,n_s,A_s} ; dT + \left(\frac{\partial G}{\partial n_s}\right)_{T,P,A_s} .; dn_s + \left(\frac{\partial G}{\partial A_s}\right)_{T,P,n_s} dA_s \tag{8.17}$$

At constant temperature and pressure this becomes:

$$dG = \left(\frac{\partial G}{\partial n_s}\right) dn_s + \left(\frac{\partial G}{\partial A_s}\right) dA_s \tag{8.18}$$

$$= \mu_s dn_s - \Gamma dA_s \tag{8.19}$$

where: μ_s is the free energy per mole or chemical potential of the film, and.

Γ is defined as a two-dimensional or spreading pressure.

The total Gibbs free energy may be written as:

$$G = \mu_s n_s - \Gamma A_s \tag{8.20}$$

so that:

$$dG = \mu_s dn_s + n_s d\mu_s - \Gamma dA_s - A_s d\Gamma \tag{8.21}$$

A comparison of Eqs (8.19) and (8.21) shows that:

$$d\Gamma = \frac{n_s}{A_s} d\mu_s$$

If the gas phase is ideal and equilibrium exists between it and the sorbed phase then, by definition:

$$d\mu_s = d\mu_g = \mathbf{R}T d(\ln P) \tag{8.22}$$

where μ_g is the chemical potential of the gas.

Substituting for $d\mu_s$ gives:

$$d\Gamma = \frac{n_s}{A_s} \mathbf{R}T d(\ln P) \tag{8.23}$$

Equation (8.23) has the form of an adsorption isotherm since it relates the amount adsorbed to the corresponding pressure. This is known as the *Gibbs Adsorption Isotherm*. For it to be useful, an expression is required for Γ. Assuming an analogy between adsorbed and liquid films, Harkins and Jura [15] have proposed that:

$$\Gamma = \alpha_1 - \beta_1 a_m \tag{8.24}$$

where α_1, β_1, are constants, $a_m = A_s/\mathbf{N}n_s$, the area per molecule of adsorbate and $\mathbf{N}$ is the Avogadro number. Substituting for $d\Gamma$ in Eq. (8.23) and integrating, at constant A_s, from some condition P_1, n_{s1} at which the adsorbed film becomes mobile, to an arbitrary coverage n_s at pressure P, gives:

$$\ln \frac{P}{P_1} = \frac{1}{\mathbf{R}T} \frac{\beta_1 A_s^2}{2\mathbf{N}} \left(\frac{1}{n_{s_1}^2} - \frac{1}{n_s^2} \right) \tag{8.25}$$

Equation (8.25) may be rewritten as:

$$\ln \frac{P}{P^0} = L' - \frac{M'}{V^2} \tag{8.26}$$

where V is the volume occupied in the gas phase by n_s moles of sorbate at a temperature T and pressure P.

Thus:

$$L' = \ln \frac{P_1}{P^0} + \frac{1}{\mathbf{R}T} \frac{\beta_1 A_s^2}{2\mathbf{N}} \frac{1}{n_{s_1}^2} \tag{8.27a}$$

and:

$$M' = \frac{1}{\mathbf{R}T} \frac{\beta_1 A_s^2}{2\mathbf{N}} \frac{1}{\rho_g^2} \tag{8.27b}$$

where ρ_g is the molar density of the gas phase. Equation (8.26) is the *Harkins-Jura (H–J) equation*, which may be used to correlate adsorption data and to obtain an estimate of the surface area of an adsorbent.

The H–J model may be criticised because the comparison between an adsorbed film on a solid surface and a liquid film on a liquid surface does not stand detailed examination. Other equations of state have been used instead of Eq. (8.24). These are discussed in more detail by Ruthven [16].

Although making fundamentally different assumptions about the static or mobile nature of an adsorbed layer, the BET and H–J equations may often represent a set of experimental data equally well. As pointed out by Garg and Ruthven [17], their respective assumptions lead one to expect that the BET equation may be applied to low surface coverage when it is thought that adsorbed molecules do not move. The H–J equation may be applied to the middle range of relative pressures (P/P^0) corresponding to the completion of a monolayer and the formation of additional mobile layers. Neither theory, it seems, is equipped to deal with the filling of micropores that occurs with the onset of capillary condensation at higher relative pressures. For this condition, a third theory of adsorption has been shown to be useful.

8.3.5 The potential theory

In this theory the adsorbed layers are considered to be contained in an *adsorption space* above the adsorbent surface. The space is composed of equipotential contours, the separation of the contours corresponding to a certain adsorbed volume, as shown in Fig. 8.7. The theory was postulated in 1914 by Polanyi [18], who regarded the potential of a point in adsorption space as a measure of the work carried out by surface forces in bringing one mole of adsorbate to that point from infinity, or a point at such a distance

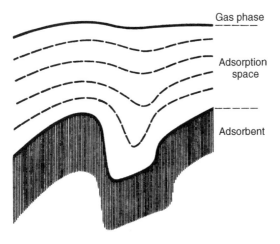

FIG. 8.7

The concept of adsorption space used by Polanyi [18].

from the surface that those forces exert no attraction. The work carried out depends on the phases involved. Polanyi considered three possibilities: (a) that the temperature of the system was well below the critical temperature of the adsorbate and the adsorbed phase could be regarded as liquid, (b) that the temperature was just below the critical temperature and the adsorbed phase was a mixture of vapour and liquid, (c) that the temperature was above the critical temperature and the adsorbed phase was a gas. Only the first possibility, the simplest and most common, is considered here.

At temperatures below the critical point, the pressure at the adsorbent surface is the saturated vapour pressure of the liquid adsorbate, P^0. At the limit of adsorption space, the pressure exerted by the adsorbate molecules in the gas phase is P. For an ideal gas, the work done bringing a mole to the adsorbent surface is given by:

$$\varepsilon_p = \mathbf{R}T \ln P^0/P \tag{8.28}$$

The potential theory postulates a unique relationship between the adsorption potential ε_p and the volume of adsorbed phase contained between that equipotential surface and the solid. It is convenient to express the adsorbed volume as the corresponding volume in the gas phase.

Hence:

$$\varepsilon_p = f(V) \tag{8.29}$$

where the function is assumed to be independent of temperature. Equations (8.28) and (8.29) may be combined to give an adsorption isotherm equation, although this is not explicit until the form of f(V) has been specified. The form may be expressed graphically by plotting $\mathbf{R}T \ln P^0/P$ against V, giving what is called a characteristic curve as shown in Fig. 8.8. The curve is valid for a particular adsorbent and adsorbate at any temperature and its validity may often be extended to other adsorbates on the same adsorbent. The adsorption potential must then be rewritten as:

$$\varepsilon_p = \beta_2 f_1(V) \tag{8.30}$$

where β_2 is a coefficient of affinity. For many adsorbates, β_2 is proportional to the molar volume V'_M of the liquid adsorbate at a temperature T. Values of β_2 for some hydrocarbons, relative to benzene, are given in Table 8.4 which is taken from the work of Young and Crowell [19], with relative molar volumes listed for comparison. For the same volume V of adsorbates 1 and 2:

$$\frac{\varepsilon_{p1}}{\varepsilon_{p2}} = \frac{\beta_{21}}{\beta_{22}} = \frac{V'_{M1}}{V'_{M2}} \tag{8.31}$$

so that the pressures exerted by a particular volume of adsorbate on the same adsorbent are related by:

$$\left(\frac{T}{V'_M} \ln \frac{P^0}{P}\right)_1 = \left(\frac{T}{V'_M} \ln \frac{P^0}{P}\right)_2 \tag{8.32}$$

Thus, from adsorption data for one gas, data for other gases on the same adsorbent may be found. Several other methods of plotting the characteristic curve have been proposed [3].

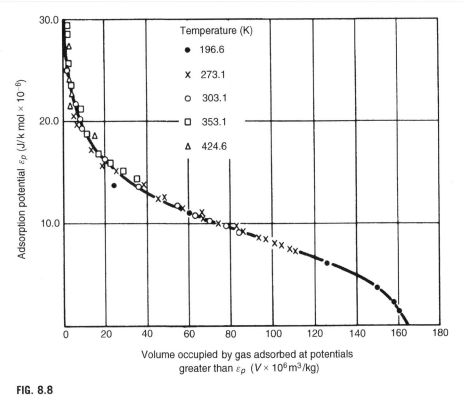

FIG. 8.8

Characteristic curve for the adsorption of carbon dioxide on to charcoal [10].

8.4 Multicomponent adsorption

The three isotherms discussed, BET, (H–J based on Gibbs equation) and Polanyi's potential theory involve fundamentally different approaches to the problem. All have been developed for gas–solid systems and none is satisfactory in all cases. Many workers have attempted to improve these and have succeeded for particular systems. Adsorption from gas mixtures may often be represented by a modified form of the single adsorbate equation. The Langmuir equation, for example, has been applied to a mixture of n'' components [11].

The volume of the ith component V_i which is adsorbed at a partial pressure P_i is given by:

$$\frac{V_i}{V_i^1} = \frac{B_i P_i}{1 + \sum_{j=1}^{n''} B_j P_j} \tag{8.33}$$

where V_i^1 is the volume of i contained in a monolayer.

Table 8.4 Values of coefficients of affinity relative to $\beta_2 = 1$ for benzene, and of $V'_M/V_{m(benzene)}'$.

Vapour	β_2 expt	$V'_M/V_{m(benzene)}'$
C_6H_6	1	1
C_5H_{12}	1.12	1.28
C_6H_{12}	1.04	1.21
C_7H_{16}	1.50	1.65
$C_6H_5CH_3$	1.28	1.19
CH_3Cl	0.56	0.59
CH_2Cl_2	0.66	0.71
$CHCl_3$	0.88	0.90
CCl_4	1.07	1.09
C_2H_5Cl	0.78	0.80
CH_3OH	0.40	0.46
C_2H_5OH	0.61	0.65
$HCOOH$	0.60	0.63
CH_3COOH	0.97	0.96
$(C_2H_5)_2O$	1.09	1.17
CH_3COCH_3	0.88	0.82
CS_2	0.70	0.68
CCl_3NO	1.28	1.12
NH_3	0.28	0.30

Thermodynamic arguments indicate that V^1 should be the same for all species. A mean value V^1 is used, which for a binary mixture is given by:

$$\frac{1}{V^1} = \frac{y_1}{V_1^1} + \frac{y_2}{V_2^1} \tag{8.34}$$

where y_1, y_2 are mole fractions.

Equation (8.33) has a limited application. A fuller discussion of adsorption from gas mixtures can be found in the literature [3,16,19,20].

8.5 Adsorption from liquids

Adsorption from liquids is less well understood than adsorption from gases. In principle, the equations derived for gases ought to be applicable to liquid systems, except when capillary condensation is occurring. In practice, some offer an empirical fit of the equilibrium data. One of the most popular adsorption isotherm equations used for liquids was proposed by Freundlich [21] in 1926. Arising from a study of the adsorption of organic compounds from aqueous solutions on to charcoal, it was shown that the data could be correlated by an equation of the form:

$$C_s' = \alpha_2 (C^*)^{1/n} \tag{8.35}$$

where: C_s' is mass of adsorbed solute per unit mass of charcoal, and.

C^* is the concentration of solute in solution in equilibrium with that on the solid.

α_2 and n are constants, the latter normally being greater than unity.

Although this was proposed originally as an empirical equation, the Freundlich isotherm was shown later to have some thermodynamic justification by Glueckauf and Coates [22].It has also been modified for binary mixtures. In gas–solid systems, it is convenient to measure the amount adsorbed by noting the changes in pressures and volume of the gas phase, or the change in the mass of the solid. Neither approach is practicable in liquid–solid systems because the volume changes in the liquid phase are small, and because of the difficulty of distinguishing between the adsorbed phase and liquid clinging to the solid that is removed from solution. Instead, the amount adsorbed is calculated from the changes in concentration of a known volume of liquid after it has attained equilibrium with a known amount of adsorbent. Adsorption from solutions of non-electrolytes has been discussed by Kipling [23].

8.6 Structure of adsorbents

The equilibrium capacity of an adsorbent for different molecules is one factor affecting its selectivity. Another is the structure of the system of pores which permeates the adsorbent. This determines the size of molecules that can be admitted and the rate at which different molecules diffuse towards the surface. Molecular sieves, with their precise pore sizes, are uniquely capable of separating on the basis of molecular size. In addition, it is sometimes possible to exploit the different rates of diffusion of molecules to bring about their separation. A particularly important example referred to earlier, concerns the production of oxygen and nitrogen from air.

Although it is sometimes possible to view pores with an electron microscope and to obtain a measure of their diameter, it is difficult by this means to measure the distribution of sizes and impossible to measure the associated surface area. Adsorptive methods are used instead, employing some of the theories of adsorption explained previously.

8.6.1 Surface area

The simplest measure of surface area is that obtained by the so-called *point B method*. Many isotherms of Type II or IV show a straight section at intermediate relative pressures. The more pronounced the section, the more complete is the adsorbed monolayer before multilayer adsorption begins. As may be seen from Fig. 8.9, the lower limit of the straight section is the point B which was identified by Emmett

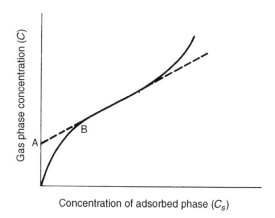

FIG. 8.9

The 'point B' method of estimating surface area.

[24] and Yang [3] as corresponding to a complete monolayer. This interpretation is accepted as a convenient empiricism.

The BET isotherm, represented by Eq. (8.14), may be rearranged as: where:

$$\frac{\varphi_1}{V_s} = \frac{1}{V_s^1 B_2} + \frac{\theta_1}{V_s^1}$$

$$\varphi_1 = \frac{P/P^0 \left[1 - \left(P/P^0\right)^n - n\left(P/P^0\right)^n\left(1 - P/P^0\right)\right]}{\left(1 - P/P^0\right)^2} \tag{8.36}$$

$$\theta_1 = \frac{P}{P^0}\frac{\left[1 - \left(P/P^0\right)^n\right]}{\left(1 - P/P^0\right)^n}$$

For a given value of n, a plot of ϕ_1/V_s against θ_1 should give a straight line from the slope of which V_s^1 may be calculated. The simpler, infinite, form of the BET isotherm is given by Eq. (8.15). The appropriate plot gives a straight line, from the slope and intercept of which V_1 may be calculated. Equation (8.15) is most likely to apply at low relative pressures.

Equation (8.26), derived by Harkins and Jura [15], may be plotted as ln (P/P_0) against $1/V^2$ to give a straight line. The slope is proportional to The constant of proportionality may be found by using the same adsorbate on a solid of known surface area. Since the equation was derived for mobile layers and makes no provision for capillary condensation, it is most likely to fit data in the intermediate range of relative pressures.

8.6.2 Pore sizes

Having obtained a measure of surface area, a mean pore size may be calculated by simplifying the pore system into n_p cylindrical pores per unit mass of adsorbent, of mean length L_p and mean pore radius r_p.

Hence:

$$\text{Pore volume } V_p = \pi r_p^2 L_p$$
$$\text{Surface area } A_p = 2n_p \pi r_p L_p$$

and:

$$r_p = \frac{2V_p}{A_p} \tag{8.37}$$

This expression has been generalised by Everett and Stone [8] for any shape of capillary by including a shape factor γ which takes a value that depends on the geometry of the capillary. This is unity for parallel sided fissures as well as for cylindrical pores.

Where there is a wide distribution of pore sizes and, possibly, quite separately developed pore systems, a mean size is not a sufficient measure. There are two methods of finding such distributions. In one a porosimeter is used, and in the other the hysteresis branch of an adsorption isotherm is utilised. Both require an understanding of the mechanism of capillary condensation.

The regions of Type IV and Type V isotherms which are concave to the gas-concentration axis at high concentrations correspond to the bulk condensation of adsorbate in the pores of the adsorbent. An equation relating volume condensed to partial pressure of adsorbate and pore size, may be found by assuming transfer of adsorbate to occur in three stages:

(i) transfer from the gas to a point above the meniscus in the capillary.
(ii) condensation on the meniscus.
(iii) transfer from a plane surface source of liquid adsorbate to the gas phase in order to maintain the first stage.

At equilibrium the free energy changes associated with the second and third stages are zero. For the first stage at, constant temperature T:

$$\Delta G = \int_{P^0}^{P_m} V \, dP = n_s \mathbf{R} T \ln \frac{P_m}{P^0} \tag{8.38}$$

where: n_s is the number of moles condensed,

P_m is the vapour pressure over the meniscus, and.

P^0 is the vapour pressure over the plane surface.

Another expression for ΔG may be obtained from consideration of the interfacial changes that occur as the capillaries fill. This takes the form:

$$\Delta G = \Delta A_p (\sigma_{sl} - \sigma_{sv}) = -\Delta A_p \sigma_{lv} \cos \varphi \tag{8.39}$$

where:

σ represents surface tensions at the three interfaces of the solid (s), liquid (l) and vapour (v),

ϕ is the interfacial angle between liquid and solid, and.

ΔA_p is the change in interfacial area.

Hence:

$$n_s RT \ln \frac{P_m}{P^0} = -\Delta A_p \sigma_{lv} \cos \varphi \qquad (8.40)$$

If the transfer of dn_s moles results in a change in interfacial area dA_p, then:

$$\frac{dn_s}{dA_p} = -\frac{\sigma_{lv} \cos \varphi}{RT \ln (P_m/P^0)} = \frac{1}{V_M} \frac{dV}{dA_p} \qquad (8.41)$$

where: dV is the volume occupied by dn_s moles in the vapour phase, and.
V_M is the molar volume.
Thus:

$$\frac{dV}{dA_p} = \frac{V_M \sigma_{lv} \cos \varphi}{RT \ln (P^0/P_m)} \qquad (8.42)$$

The left-hand side of Eq. (8.42) is a characteristic dimension of the pores in which condensation or evaporation occurs at an adsorbate pressure P_m. This depends on the geometry of the pores, as well as whether adsorption or desorption is occurring. Fig. 8.10 illustrates the conditions in a cylindrical pore. For desorption occurring from the free surface of condensate:

$$dV = d\left(\pi r_p^2 L_p\right) = \pi r_p^2 dL_p$$
$$dA_p = \left(2\pi r_p L_p\right) = 2\pi r_p dL_p$$

Hence:

$$\frac{dV}{dA_p} = \frac{r_p}{2} \qquad (8.43)$$

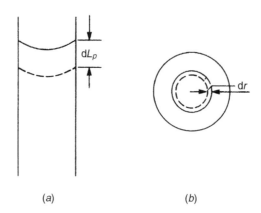

(a)   (b)

FIG. 8.10

The capillary condensation equation applied to a cylindrical pore (A) for desorption (B) for adsorption.

and:

$$r_p = \frac{2V_M}{RT}\frac{\sigma_{lv}\cos\varphi}{\ln P^0/P} \tag{8.44}$$

Equation (8.44) is known as the *Kelvin equation*, and is considered to be valid for desorption.

For adsorption on to existing layers of adsorbate on the cylindrical surface of the pores:

$$dV = d\left[\pi\left(r_p^2 - r\right)L_p\right] = -2\pi r L_p dr$$
$$dA_p = d\left[2\pi r L_p\right] = 2\pi L_p dr$$

Therefore,

$$r'_p = \frac{V_M}{RT}\frac{\sigma_{lv}\cos\varphi}{\ln P^0/P} \tag{8.45}$$

where r'_p is the net radius of the pore.

Equation (8.45) is known as the *Cohan equation* [25]. Together with Eq. (8.44), this offers an explanation of the hysteresis effect which many isotherms exhibit at high relative concentrations of 0.4 to 0.95. Fig. 8.11 shows an isotherm for nitrogen on activated-alumina. A difference between adsorption and desorption conditions

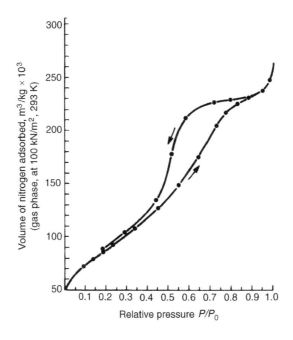

FIG. 8.11

Nitrogen isotherm for activated alumina determined at the boiling point of liquid nitrogen.

over the middle-to-high range of relative pressures may be noted. When applied to this range, Eq. (8.44) or Eq. (8.45) may be used to estimate the distribution of pore sizes contained within the adsorbent. A method, such as that of Cranston and Inkley [26] leads to the distribution shown in Fig. 8.12.

Interpretation of the pore radii calculated in this way is complicated by the fact that adsorption is occurring on the surface of pores which are, as yet, too big for condensation to occur. Radii given by Eqs (8.44) or (8.45) are net values, and an allowance has to be made for any adsorbed layers present when condensation occurs.

The actual structure of an adsorbent may be very different from a model assuming independent cylindrical pores. It may be possible to allow for the difference by including a geometric factor in the simple equations, or it may be necessary to 'view' the surface using, for example, a scanning electron microscope. Sometimes the results are a pleasing confirmation of the simple assumptions. Fig. 8.3 shows a transmission electron micrograph of an activated alumina similar to that which gave rise to Figs. 8.11 and 8.12. The thin edges of particles are sufficiently transparent to the electron beam to show up parallel sets of pores on a hexagonal array, and about 4 nm in diameter, very close to the peak in Fig. 8.12. The physical structure of an adsorbent

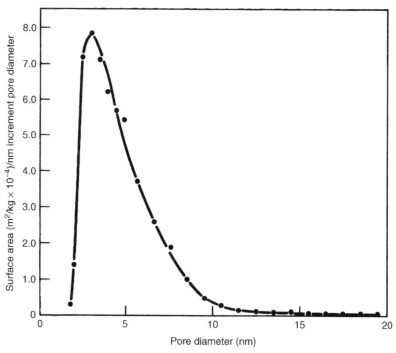

FIG. 8.12

Pore size distribution of an activated alumina calculated from the isotherm by the method of Cranston and Inkley [26].

is an important factor in determining the rate at which adsorbate molecules diffuse to the adsorbent surface.

Example 8.2

Data taken from the adsorption leg of the isotherm of Fig. 8.11 are listed in the first two columns of the following table. Test the applicability of the following equilibrium theories: (a) Langmuir (b) infinite BET and (c) Harkins and Jura. From (a) and (b) obtain estimates of the surface area of the adsorbent and compare the values with that obtained by the 'point B' method. One molecule of nitrogen adsorbed on alumina occupies $0.162\,nm^2$.

The Avogadro Number is 6.02×10^{26} molecules/kmol.

P/P^0 (—)	$V \times 10^2$ (m^3N_2/kg alumina at 293 K and 10^5 N/m^2)	$\dfrac{P}{P^0}\dfrac{1}{V}$ (kg/m^3)	$\dfrac{P}{P^0}\dfrac{1}{V}\dfrac{1}{(1-P/P^0)}$ (kg/m^3)	$\dfrac{1}{V^2}$ (kg^2/m^6)
0.05	6.6	0.76	0.80	230
0.10	7.4	1.35	1.50	183
0.15	8.1	1.85	2.18	152
0.20	8.8	2.27	2.88	129
0.25	9.4	2.66	3.55	113
0.30	10.2	2.94	4.20	96
0.35	10.9	3.21	4.94	84
0.40	11.7	3.42	5.73	73
0.50	13.8			53
0.60	16.5			37
0.87	19.6			26
0.80	22.1			20

Solution

(a) For $n = 1$, Eq. (8.14) may be written in a Langmuir form to give:

$$\frac{P/P^0}{V} = \frac{P/P^0}{V^1} + \frac{1}{B_2 V^1} \tag{8.15}$$

where V is the gas-phase volume equivalent to the amount adsorbed.

The data, which are plotted as $(P/P^0)/V$ against P/P^0 in Fig. 8.13, may be seen to conform to a straight line only at low values of P/P^0, suggesting that more than one layer of molecules is adsorbed.

From the slope of the line, $1/V^1 = 12.56$, hence $V^1 = 7.96 \times 10^{-2}$ m^3/kg.

The surface area occupied by this adsorbed volume $=$

$$\frac{7.96 \times 10^{-2}}{24} \times 6.02 \times 10^{26} \times 0.162 \times 10^{-18} = 323{,}000\,m^2/kg$$

$$\frac{P/P^0}{V(1-P/P^0)} = \frac{1}{V^1 B_2} + \frac{B_2 - 1}{V^1 B_2}(P/P^0) \tag{8.15}$$

The left-hand side of Eq. (8.15) is plotted against P/P^0 in Fig. 8.13. The data conform to a straight line over a wider range of relative pressure than was the case in (a),

$$\text{Slope,} \left(\frac{B_2 - 1}{V^1 B_2}\right) = 14.05$$

(8.15)

$$\text{Intercept,} \left(\frac{1}{V^1 B_2}\right) = 0.2$$

Hence: $B_2 = 71.2$ and $V^1 = 7.02 \times 10^{-2}$ m^3/kg.
Surface area $= 285{,}000$ m^2/kg.
From Fig. 8.11, point B corresponds to about 7.5×10^{-2} m^3/kg or $305{,}000$ m^2/kg.

(c) According to Eq. (8.26), a plot of $\ln(P/P^0)$ against $1/V^2$ should be linear. Fig. 8.14 shows that agreement in the middle range of relative pressure is good.

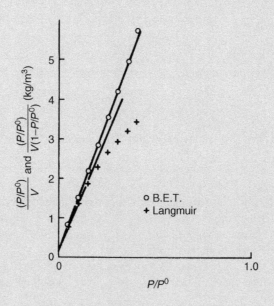

FIG. 8.13

Langmuir and BET plots for Example 8.2.

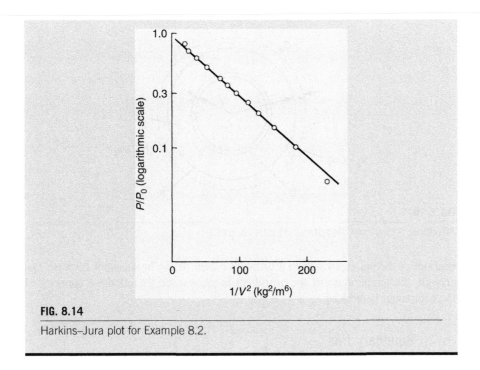

FIG. 8.14

Harkins–Jura plot for Example 8.2.

8.7 **Kinetic effects**

In much of the early theory used to describe adsorption in different kinds of equipment, it was assumed that equilibrium was achieved instantly between the concentrations of adsorbate in the fluid and in the adsorbed phases. Whilst it may be useful to make this assumption because it leads to relatively straightforward solutions and shows the interrelationship between system parameters, it is seldom true in practice. In large-scale plant particularly, performance may fall well short of that predicted by the equilibrium theory.

There are several resistances which may hinder the movement of a molecule of adsorbate from the bulk fluid outside a pellet to an adsorption site on its internal surface, as shown in Fig. 8.15. Some of these are sequential and have to be traversed in series, whilst others derive from possible parallel paths. In broad terms, a molecule, under the influence of concentration gradients, diffuses from the turbulent bulk fluid through a laminar boundary layer around a solid pellet to its external surface. It then diffuses, by various possible mechanisms, through the pores or the lattice vacancies in the pellet until it is held by an adsorption site. During desorption the process is reversed.

In the process of being adsorbed, molecules lose degrees of freedom and a heat of adsorption is released. All the resistances to mass transfer also affect the transfer of

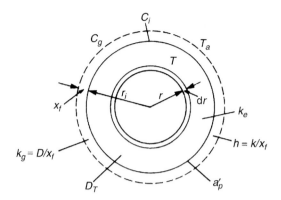

FIG. 8.15

Adsorbate and energy transport in a spherical pellet.

heat out of the pellet, though to a different extent. If the heat cannot disperse fast enough, the temperature of the adsorbent increases and the effective capacity of the equipment is reduced.

8.7.1 Boundary film

There have been many studies of the effect of boundary films on mass and heat transfer to single pellets and in packed beds, including the work of Ranz and Marshall [27] and Dwivedi and Upadhey [28]. Other theories of mass and heat transfer are discussed in Volume 1, Chapter 10, although only the steady-state film-theory is considered here. It is assumed that the difference in concentration and temperature between the bulk fluid and the external surface of a pellet is confined to a narrow laminar boundary-layer in which the possibility of accumulation of adsorbate or of heat is neglected.

Adsorbate is transferred to an adsorbent at a rate which depends upon a_p' the external surface area of the solid and upon ΔC, the concentration difference across the boundary film.

Thus:

$$\text{Rate of mass transfer} = k_g a_p' \Delta C$$

where k_g is a mass transfer coefficient.

The rate at which heat is generated by mass transfer into the pellet is given approximately by:

$$\text{Rate of liberation of heat} = k_g a_p' \Delta C \Delta H$$

where ΔH is the heat of adsorption. When the temperature of the pellet has increased to the extent that heat is lost as fast as it is generated, then:

$$k_g a_p' \Delta C \Delta H = h a_p' \Delta T \tag{8.46}$$

where h is a heat transfer coefficient.

Since $k_g = D/x_f$ and $h = k/x_f$ then:

$$\Delta T = \frac{D}{k} \Delta H \Delta C \qquad (8.47)$$

which may be written as:

$$\Delta T = \frac{1}{\rho c_{pg}} \frac{Pr}{Sc} \Delta H \Delta C \qquad (8.48)$$

where:

D = fluid diffusivity,
k = fluid thermal conductivity,
x_f = thickness of boundary film,
Pr = Prandtl Number $c_{pg}\mu/k$, and.
Sc = Schmidt Number $\mu/\rho D$.

For gas mixtures, the ratio Pr/Sc is about 0.2 to 5. $\Delta H/\rho c_{pg}$ is normally large. It follows that a small value of ΔC corresponds to a large value of ΔT. Generally, the boundary film does not offer much resistance to mass transfer, except in relative terms during the unsteady state conditions that exist when a pellet is first exposed to a fluid. The boundary film may, however, offer a significant resistance to heat transfer.

A re-evaluation of published data for mass transfer carried out by Wakao and Funazkri [29] indicates that, for the range $3 < Re < 10^4$:

$$Sh = 2.0 + 1.1 Sc^{1/3} Re^{0.6} \qquad (8.49)$$

where Sh is the Sherwood number ($k_g d_p/D$).

Re is the Reynolds number based on the superficial velocity through the bed, and pellet diameter. Equation (8.49) assumes that axial diffusion in the bed has been allowed for, which is not the case in most of the earlier work.

The corresponding equation for heat transfer, neglecting radiation, is:

$$Nu = 2.0 + 1.1 Pr^{1/3} Re^{0.6} \qquad (8.50)$$

where Nu is the Nusselt number $h d_p/k$.

8.7.2 Intra-pellet effects

After passing through the boundary layer, the molecules of adsorbate diffuse into the complex structure of the adsorbent pellet, which is composed of an intricate network of fine capillaries or interstitial vacancies in a solid lattice. The problem of diffusion through a porous solid has attracted a great deal of interest over the years and there is a fairly good understanding of the mechanisms involved, at least for gas-phase diffusion. Here, diffusion within a single cylindrical pore is considered and, then, the pore is related to the pellet as a whole.

Molecules entering a pore move randomly between the pore-wall and the pore-space. Sometimes molecules are more closely associated with the wall and may be thought of as diffusing over the surface, subject to a surface concentration-difference and a surface diffusion-coefficient. At other times, molecules move in pore-space, where their behaviour depends on the mean free path of the gas and on the pore diameter. The mean free path λ, which is the average distance between collisions of gas molecules, may be predicted using the kinetic theory of gases, and given by Glasstone and Lewis [30] as:

$$\lambda = \frac{1}{\sqrt{2}n'\pi\sigma^2} \qquad (8.51)$$

where: n' = number of molecules in unit volume of gas phase,
σ = collision diameter.
At atmospheric pressure and 293 K:

$$n' = \frac{6 \times 10^{26}}{24.0} = 2.5 \times 10^{25}\,\text{molecules/m}^3$$

The collision diameter of most molecules is about 0.2 nm.
Hence: $\lambda = 225$ nm.
In pores that are appreciably smaller than the mean free path, molecules tend to collide with the pore walls rather than with other molecules. Having collided with the wall, the molecules are momentarily retained and then released in a random direction. The coefficient, D_k, which controls this *Knudsen diffusion*, considered by Satterfield [31], and in Volume 1, Chapter 3, may be derived from the kinetic theory to give:

$$D_k = \frac{2}{3}\left(\frac{8'_B T}{\pi m_M}\right) r_p \qquad (8.52)$$

or:

$$D_k = 1.0638\left(\frac{\mathbf{R}T}{M}\right) r_p \qquad (8.53)$$

where: r_p is the pore radius.
m_M is the mass of a molecule, and.
M is the molecular weight.
The flux due to Knudsen diffusion is given by:

$$N_{Ak} = -D_k \frac{\partial C_A}{\partial l} \qquad (8.54)$$

where l is the distance into the pore.
When the mean free path is small compared with pore diameter, the dominating experience of molecules is that of collision with other molecules in the gas phase. In that respect, the situation is much the same as that which exists in the bulk gas. The appropriate diffusion coefficient D_m may be obtained from published experimental values, or calculated from a theoretical expression. For molecular diffusion in a binary gas, the Chapman and Enskog equation may be used, as discussed by Bird, Stewart and Lightfoot [32]. This takes the form:

$$D_m = \frac{1.88 \times 10^{-4}T^{3/2}}{P\sigma_{AB}^2\Omega_{AB}}\left(\frac{1}{M_A} + \frac{1}{M_B}\right)^{1/2} \qquad (8.55)$$

where: D_m is the diffusion coefficient (m^2/s),

M_A, M_B are molecular weights of **A** and **B**,
P is the pressure (N/m^2),
T is the temperature (K),
σ_{AB} is a collision diameter (nm), and.
Ω_{AB} is a dimensionless collision integral.
Bird Stewart and Lightfoot have published tables which give values of Ω and σ.
The contribution to the total flux that comes from bulk diffusion is given by:

$$N_{Am} = -D_m \frac{\partial C_A}{\partial l} \tag{8.56}$$

The total diffusion in any one adsorbent is complicated by the fact that there is no sudden transition from Knudsen to molecular diffusion at a particular pore size, and most adsorbents contain a range of pore sizes. Methods of obtaining a value for an average coefficient have been suggested. For example, Evans et al. [33] have suggested that, in a binary system:

$$\frac{1}{D_{av}} = \frac{1}{D_k} + \frac{1}{D_m}\left[1 - \left(1 + \frac{N_B}{N_A}\right)y_A\right] \tag{8.57}$$

where y_A is the mole fraction of one component. The final bracket allows for deviations from equimolecular counterdiffusion. Whilst both Knudsen and molecular diffusion may be regarded as alternative mechanisms in a particular pore, a third mechanism, namely surface diffusion, is a possible additional route by which molecules may enter a pellet. Surface diffusion is sometimes likened to movement in a liquid film, the surface diffusion coefficient D_s increasing with surface coverage until the monolayer is complete and then remaining constant. Since it exists in parallel with other forms of diffusion, surface diffusion is additive in its effect and:

$$\alpha D_T = \alpha D_{av} + (1 - \alpha)J D_s \tag{8.58}$$

where: α is pore volume per unit volume of pellet [34], and.
J is a factor which has been shown to decrease with temperature.
Surface diffusion is only significant when the pores are small and $D_{av} = D_k$.
There is an energy of activation E_s for surface diffusion which leads to a temperature dependence of an Arrhenius kind and:

$$D_s = D_{s_0} e^{-E_s/RT} \tag{8.59}$$

Equation (8.59) has been confirmed experimentally, suggesting that molecules move over a surface by 'hopping' to adjacent adsorption sites. It may be assumed that this process involves a lower energy of activation than that required for complete desorption. The molecule will continue to 'hop' until it finds a vacant adsorption site, thus explaining the increase of surface diffusion coefficient with coverage.

The Einstein equation for surface diffusion gives $D_s = \lambda_A^2/2\delta t$, where λ_A is the average distance between vacant adsorption sites and δt is the residence time on a site.

Experimental measurements of surface diffusion are usually calculated by subtracting from the measured total diffusion that predicted theoretically for Knudsen and molecular diffusion.

Using the total diffusion coefficient defined in Eq. (8.58), the flux into a pore may be written as:

$$N_A = -D_T \frac{\partial C_A}{\partial l} \tag{8.60}$$

The total diffusional flow into a pellet must allow for the fraction of pellet which is pore, and the tortuous path through the pore system.

In terms of an effective diffusivity D_e and a mean concentration gradient across a porous medium of thickness L, the flux through the medium may be written as:

$$N_A = D_e(-\Delta C_A)/L \tag{8.61}$$

N_A, which refers to unit superficial area, may be expressed in terms of an interstitial flow N_{Ai} and voidage ε as:

$$N_A = \varepsilon N_{Ai} \tag{8.62}$$

A third flux N_{Ac} may be defined which refers to the actual flow of molecules along the tortuous path of a pore of length L_e.

Thus:

$$N_{Ac} = D_T(-\Delta C_A/L_e) \tag{8.63}$$

The velocity of **A** along the direct axial path $= N_{Ai}/\overline{C}_A$.

The velocity of **A** through a tortuous pore $= N_{Ac}/\overline{C}_A$.

where $\overline{C}_A$ is the mean molar concentration of **A**.

The tortuous and direct paths are equivalent if the time taken for **A** to traverse each is the same.

Thus:

$$\frac{L}{N_{Ai}/\overline{C}_A} = \frac{L_e}{N_{Ac}/\overline{C}_A} \tag{8.64}$$

Hence:

$$N_{Ac} = N_{Ai} \frac{L_e}{L} = \frac{N_A}{\varepsilon} \frac{L_e}{L}$$

Substituting in Eq. (8.63):

$$N_A = D_T \varepsilon \frac{(-\Delta C_A)}{L} \left(\frac{L}{L_e}\right)^2 \tag{8.65}$$

From Eqs (8.61) and (8.65):

$$D_e = \frac{D_T \varepsilon}{(L_e/L)^2} = \frac{D_T \varepsilon}{\tau^2} \tag{8.66}$$

where τ is the tortuosity (actual pore length/superficial diffusion path) and τ^2 is a tortuosity factor.

Values of τ^2 of 2–6 are common. Occasionally, much higher values are quoted which are more likely to indicate shortcomings in the theory rather than highly tortuous paths. There is some confusion in the literature between τ^2 and τ, as discussed by Epstein [35].

8.7.3 Adsorption

The final stage in getting a molecule from the bulk phase outside a pellet on to the interior surface is the adsorption step itself. Where adsorption is physical in nature, this step is unlikely to affect the overall rate. An equilibrium state is likely to exist between adsorbate molecules immediately above a surface and those on it. An adsorption isotherm such as Eq. (8.4) or (8.35) may be applied.

When chemisorption is involved, or when some additional surface chemical reaction occurs, the process is more complicated. The most common combinations of surface mechanisms have been expressed in the *Langmuir–Hinshelwood* relationships [36]. Since the adsorption process results in the net transfer of molecules from the gas to the adsorbed phase, it is accompanied by a bulk flow of fluid which keeps the total pressure constant. The effect is small and usually neglected. As adsorption proceeds, diffusing molecules may be denied access to parts of the internal surface because the pore system becomes blocked at critical points with condensate. Complex as the situation may be in theory, it may often be simplified by selecting what is termed the *controlling resistance*. This relates to the step whose resistance is overwhelmingly large in comparison with that for the other layers. The whole of the difference in adsorbate concentration, from its value in the bulk gas to its value at the adsorbent surface, occurs across the controlling resistance. It may be, however, that rate control is mixed, involving two or more resistances.

The relative importance of the boundary-film and intra-pellet diffusion in mass transfer is measured by the *Biot number*. On the assumption that there is no accumulation of adsorbate at the external surface of a pellet, then:

$$k_g (C_g - C_i) = D_e \frac{\partial C}{\partial r}\bigg|_{r=r_i} \qquad \frac{k_g r_i}{D_e} = \frac{1}{(1 - C_i/C_g)} \frac{\partial (C/C_g)}{\partial (r/r_i)}\bigg|_{r=r_i} \qquad (8.67)$$

where the left-hand dimensionless group Bi is the *Biot number*. A high value of Bi indicates that intra-pellet diffusion is controlling the rate of transport.

The discussion so far has concentrated on mass transfer. The transfer of the heat liberated on adsorption or consumed on desorption may also limit the rate process or the adsorbent capacity. Again the possible effects of the boundary-film and the intra-pellet thermal properties have to be considered. A Biot number for heat transfer is hr_i/k_e. In general, this is less than that for mass transfer because the boundary layer offers a greater resistance to heat transfer than it does to mass transfer, whilst the converse is true in the interior of the pellet.

8.8 Adsorption equipment

The scale and complexity of an adsorption unit varies from a laboratory chromatographic column a few millimetres in diameter, as used for analysis, to a fluidised bed several metres in diameter, used for the recovery of solvent vapours, from a simple container in which an adsorbent and a liquid to be clarified are mixed, to a highly-automated moving-bed of solids in plug-flow.

All such units have one feature in common in that in all cases the adsorbent becomes saturated as the operation proceeds. For continuous operation, the spent adsorbent must be removed and replaced periodically and, since it is usually an expensive commodity, it must be regenerated, and restored as far as possible to its original condition.

In most systems, regeneration is carried out by heating the spent adsorbent in a suitable atmosphere. For some applications, regeneration at a reduced pressure without increasing the temperature is becoming increasingly common. The precise way in which adsorption and regeneration are achieved depends on the phases involved and the type of fluid–solid contacting employed. It is convenient to distinguish three types of contacting:

(a) Those in which the adsorbent and containing vessel are fixed whilst the inlet and outlet positions for process and regenerating streams are moved when the adsorbent becomes saturated. The fixed bed adsorber is an example of this arrangement. If continuous operation is required, the unit must consist of at least two beds, one of which is on-line whilst the other is being regenerated.
(b) Those in which the containing vessel is fixed, though the adsorbent moves with respect to it. Fresh adsorbent is fed in and spent adsorbent removed for regeneration at such a rate as to confine the adsorption within the vessel. This type of arrangement includes fluidised beds and moving beds with solids in plug flow.
(c) Those in which the adsorbent is fixed relative to the containing vessel which moves relative to fixed inlet and outlet positions for process and regenerating fluids. The rotary-bed adsorber is an example of such a unit.

8.8.1 Fixed or packed beds

When used as part of a commercial operation with gas or liquid mixtures, the single pellets discussed in the context of *rate processes* are consolidated in the form of packed beds. Usually the beds are stationary and the feed is switched to a second bed when the first becomes saturated. Whilst there are applications for moving-beds, as discussed later, only fixed-bed equipment will be considered, here as this is the most widely used type.

Fig. 8.16A depicts the way in which adsorbate is distributed along a bed, during an adsorption cycle. At the inlet end of the bed, the adsorbent has become saturated and is in equilibrium with the adsorbate in the inlet fluid. At the exit end, the adsorbate content of the adsorbent is still at its initial value. In between, there is a reasonably well-defined mass-transfer zone in which the adsorbate concentration drops from the inlet to the exit value. This zone progresses through the bed as the run proceeds. At t_1 the zone is fully formed, t_2 is some intermediate time, and t_3 is the breakpoint time t_b at which the zone begins to leave the column. For efficient operation the run must be stopped just before the breakpoint. If the run extends for too long, the

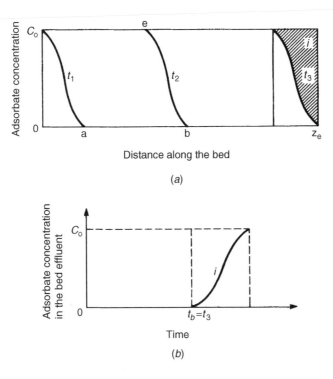

FIG. 8.16

The distribution of adsorbate concentration in the fluid phase through a bed (A) Development and progression of an adsorption wave along a bed. (B) Breakthrough curve.

breakpoint is exceeded and the effluent concentration rises sharply, as is shown in the breakthrough curve of Fig. 8.16B.

A mass balance of adsorbate in the fluid flowing through an increment dz of bed as shown in Fig. 8.17 gives:

$$uA\varepsilon C - \left[uA\varepsilon C + \frac{\partial(uA\varepsilon C)}{\partial z}\,\mathrm{d}z\right] = \frac{\partial(\varepsilon AC\mathrm{d}z)}{\partial t} + \text{Loss} \tag{8.68}$$

INPUT − OUTPUT = ACCUMULATION + LOSSBYADSORPTION

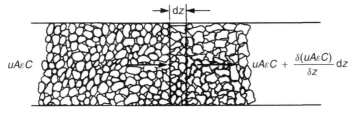

FIG. 8.17

Conservation of adsorbate across an increment of a packed bed.

The rate of loss by adsorption from the fluid phase equals the rate of gain in the adsorbed phase and:

$$\text{Rate of adsorption} = \frac{\partial[(1 - \varepsilon)AC_s dz]}{\partial t}$$

The equations may be rearranged to give:

$$\left(\frac{\partial(uC)}{\partial z}\right)_t + \left(\frac{\partial C}{\partial t}\right)_z = -\frac{1}{m}\left(\frac{\partial C_s}{\partial t}\right)_z \tag{8.69}$$

where:

$$m = \frac{\varepsilon}{1 - \varepsilon}$$

When the adsorbate content of the inlet stream is small, the fluid velocity is virtually constant along the bed.

Therefore,

$$u\left(\frac{\partial C}{\partial z}\right)_t + \left(\frac{\partial C}{\partial t}\right)_z = -\frac{1}{m}\left(\frac{\partial C_s}{\partial t}\right)_z \tag{8.70}$$

which may be simplified further by substituting to give:

$$\chi = \frac{z}{mu} \quad \text{and} \quad t_1 = \left(t - \frac{z}{u}\right)$$

so that:

$$\frac{\partial C}{\partial \chi} = -\frac{\partial C_s}{\partial t_1} \tag{8.71}$$

Equation (8.70) includes explicitly the inter-pellet voidage. The intra-pellet voidage α contributed by the pores is subsumed in the term C_s, which is the mean adsorbate content on the pellet. C_s varies along the bed although it is assumed to be constant at any radius at a particular distance from the inlet.

If α is included in the conservation equation, then this becomes:

$$u\frac{\partial C}{\partial z} + \frac{\partial C}{\partial t} = -\frac{1}{m}\frac{\partial}{\partial t}\left[(1 - \alpha)C_s' + \alpha C'\right] \tag{8.72}$$

where: C' is the mean adsorbate concentration in the fluid phase which is present in the pore volume of the pellet, and.

C_s' is the mean adsorbate concentration in the adsorbed phase in a pellet,

C' is not equal to C, except in equilibrium operation, although it is likely to be in equilibrium with C_s' through an adsorption isotherm $C' = f(C_s')$. C is normally expressed in moles per unit volume of fluid. In Eq. (8.70), consistent units for C_s are moles per unit volume of pellet. In practice, C_s is often quoted as mass of adsorbate per unit mass of adsorbate-free adsorbent (C_s'').

It then follows that:

$$C_s'' = \frac{M}{\rho_p}C_s$$

where: M is the molecular weight of adsorbate, and.

ρ_p is the pellet density.

The latter may be related to true solid density ρ_s by:

$$\rho_p = (1 - \alpha)\rho_s$$

and to the bed density, ρ_B, by:

$$\rho_B = (1 - \varepsilon)\rho_p$$

Whilst it is convenient to refer to pellet mean concentrations when analysing the performance of agglomerates of pellets such as is found in a packed bed, in reality, both C_s and C decrease from the outside of the pellet to the centre, as shown in Fig. 8.18.

Another important approximation implicit in Eq. (8.70) is that radial and longitudinal dispersions may be neglected. Radial concentration gradients are likely to be small. It has been shown that because of the greater bed voidage at the wall, and within about three pellet diameters of it, a peak in longitudinal velocity occurs near the wall and the breakthrough for wall-flow is earlier. For a bed/pellet diameter ratio of greater than 20, the effect is small. At low Reynolds numbers longitudinal

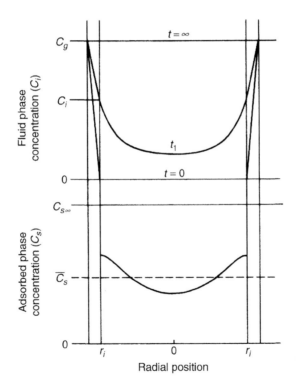

FIG. 8.18

Distribution of adsorbate through a spherical pellet.

dispersion may become important. This gives rise to axial mixing, which elongates the mass transfer zone and reduces separation efficiency. When it is necessary to take longitudinal dispersion into account, a form of Fick's Law is assumed to apply and the term $D_L(\partial^2 C/\partial z^2)$ is added to the left-hand side of Eq. (8.70). Values of D_L may be calculated from published correlations of Peclet Numbers. For gases, Edwards and Richardson [37] have shown that:

$$\frac{1}{Pe} = \frac{0.73\varepsilon}{Re\ Sc} + \frac{1}{2\left(1 + \frac{9.5\varepsilon}{Re\ Sc}\right)} \tag{8.73}$$

where:

$$Pe = \frac{2ur_i}{D_L}$$

In liquids the effects of longitudinal dispersion are small, even at low Reynolds Numbers.

Many solutions are available for Eq. (8.70) and its refinements. Three cases are considered to illustrate the range of solutions. Firstly, it is assumed that the bed operates isothermally and that equilibrium is maintained between adsorbate concentrations in the fluid and on the solid. Secondly, the non-equilibrium isothermal case is considered and, finally, the non-equilibrium non-isothermal case.

8.8.2 Equilibrium, isothermal adsorption in a fixed bed, single adsorbate

At all positions in the bed, concentrations in the fluid and adsorbed phases are related by the adsorption isotherm. This implies that there is no resistance to the transfer of molecules of adsorbate from bulk fluid to adsorption site.

If the adsorption isotherm is written as $C_s = f(C)$, Eq. (8.69) may be rewritten as:
Also:

$$u\frac{\partial C}{\partial z} + \frac{\partial C}{\partial t} = -\frac{1}{m}\frac{\partial C_s}{\partial C}\frac{\partial C}{\partial t}$$

$$= -\frac{1}{m}f'(C)\frac{\partial C}{\partial t} \tag{8.74}$$

$$\left(\frac{\partial C}{\partial t}\right)_z = -\left(\frac{\partial C_s}{\partial z}\right)_t \cdot \left(\frac{\partial z}{\partial t}\right)_c$$

and:

$$\left\{u - \left[1 + \frac{1}{m}f'(C)\right]\frac{\partial z}{\partial t}\right\}\frac{\partial C}{\partial t} = 0$$

Hence:

$$\left(\frac{\partial z}{\partial t}\right)_c = \frac{u}{1 + \frac{1}{m}f'(C)} \tag{8.75}$$

Equation (8.75) is important as it illustrates, for the equilibrium case, a principle that applies also to the non-equilibrium cases more commonly encountered. The principle concerns the way in which the shape of the adsorption wave changes as it moves along the bed. If an isotherm is concave to the fluid concentration axis it is termed *favourable*, and points of high concentration in the adsorption wave move more rapidly than points of low concentration. Since it is physically impossible for points of high concentration to overtake points of low concentration, the effect is for the adsorption zone to become narrower as it moves along the bed. It is, therefore, termed *self-sharpening*.

An isotherm which is convex to the fluid concentration axis is termed *unfavourable*. This leads to an adsorption zone which gradually increases in length as it moves through the bed. For the case of a linear isotherm, the zone goes through the bed unchanged. Fig. 8.19 illustrates the development of the zone for these three conditions.

Whilst the simple theory predicts that the adsorption zone associated with a favourable isotherm reduces to a step change in concentration, in practice finite resistance to mass transfer and the effect of longitudinal diffusion will result in a zone of finite and constant width being propagated. The property is important because it leads to simplified methods of sizing fixed beds. Fig. 8.20 taken from the work of

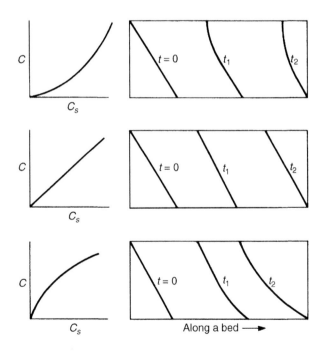

FIG. 8.19

Effect of the shape of the isotherm on the development of an adsorption wave through a bed with the initial distribution of adsorbate shown at $t = 0$.

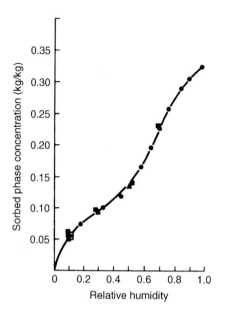

FIG. 8.20

Adsorption isotherm for water vapour on activated alumina, plotted as a function of the relative humidity. Temperatures: ● 303 K, ■ 308 K, ▲ 315 K, □ 325 K, ○ 335 K.

Bowen and Rimmer [38] shows a typical isotherm for activated alumina and water. This has sections concave to the gas concentration axis at high and low concentrations and the middle section is convex. From the predictions of the equilibrium solutions, it might be expected that the portions of the adsorption wave corresponding to the extremes of the concentration range would sharpen as they moved through the column. The middle of the range should become longer. These were obtained from a narrow, jacketed, laboratory column operating essentially isothermally, though not under equilibrium conditions. The figure does show the tendency for high and low concentrations to develop a constant pattern and for the middle range to spread as the wave progresses. In order to predict the breakpoint, only the leading zone has to be considered. Equation (8.75) may be integrated at constant C to give:

$$\frac{ut}{1 + \frac{1}{m} f'(C)} z - z_0 \tag{8.76}$$

where z_0 is the position of C initially. For a bed initially free of adsorbate, z_0 is zero for all values of C.

The condition for which the bed is likely to operate at near equilibrium is when the feed rate is low. This is also the condition when longitudinal dispersion may be significant. Equilibrium solutions have been found by Lapidus and Amundson [39] and by Levenspiel and Bischoff [40] for this case. These take the form:

$$\frac{C}{C_0} = \frac{1}{2} \cdot \left\{ 1 + \mathrm{erf} \left[\left(\frac{uz}{4D_L} \right)^{1/2} \frac{(t - t_{\min})}{(tt_{\min})^{1/2}} \right] \right\} \tag{8.77}$$

where $t_{\min}$ is the minimum time, under given flow conditions, to saturate a bed of unit cross-section and length z. C_0 is the constant concentration of adsorbate in the fluid entering the bed. The minimum time is given by:

$$t_{\min} = \left(1 + \frac{1}{m} \frac{C_{s\infty}}{C_0} \right) \frac{z}{u} \tag{8.78}$$

where $C_{s\infty}$ is the concentration of adsorbed phase in equilibrium with C_0.

Example 8.3

A solvent contaminated with $0.03 \, \mathrm{kmol/m^3}$ of a fatty acid is to be purified by passing it through a fixed bed of activated carbon which will adsorb the acid but not the solvent. If the operation is essentially isothermal and equilibrium is maintained between the liquid and the solid, calculate the length of a bed of $0.15 \, \mathrm{m}$ diameter to give $3600 \, \mathrm{s}$ ($1 \, \mathrm{h}$) of operation when the fluid is fed at $1 \times 10^{-4} \, \mathrm{m^3/s}$. The bed is initially free of adsorbate, and the inter-granular voidage is 0.4. Use an equilibrium, fixed-bed theory to obtain an answer for three types of isotherm:

(a) $C_s = 10C$
(b) $C_s = 3.0 \, C^{0.3}$ – use the mean slope
(c) $C_s = 10^4 \, C^2$ – take the breakthrough concentration as $0.003 \, \mathrm{kmol/m^3}$.
    C and C_s refer to concentrations ($\mathrm{kmol/m^3}$) in the gas phase and the adsorbent, respectively.

Solution
From Eq. (8.76):

$$ut / \left(1 + \frac{1}{m} f'(C) \right) = (z - z_0)$$

For case (a):
$C_s = 10C$ which represents a linear isotherm.
    All concentrations move at the same velocity. If $z_0 = 0$ at $t = 0$ for all concentrations, the adsorption wave propagates as a step change from the inlet to the outlet concentration,

$$f'(C_s) = 10$$
$$u = 1 \times 10^{-4} / \left[(\pi/4)(0.15^2 \varepsilon) \right] \mathrm{m/s}$$

where ε is the inter-granular voidage $= 0.4$.

Thus:

$$m = \varepsilon/(1-\varepsilon) = (0.4/0.6)$$

$$t = 3600\,s$$

$$z = \left(\frac{4 \times 10^{-4}}{\pi \times 0.15^2 \times 0.4}\right)\left(\frac{3600}{1 + 10(0.6/0.4)}\right)$$

$$= \underline{3.18\,m}$$

It may be noted that, when the adsorption wave begins to emerge from the bed, the bed is saturated in equilibrium with the inlet concentration.
Hence:

$$uA\varepsilon tC_0 = zA\left[\varepsilon C_0 + 1(1-\varepsilon)C_s^*\right]$$

which is the same as that obtained by applying Eq. (8.76) to a linear isotherm.
For case (b):
$C_s = 3C^{0.3}$ which represents a favourable isotherm.

As C increases, f(C) decreases and points of higher concentrations are predicted to move a greater distance in a given time than lower concentrations. It is not possible for points of higher concentrations to overtake lower concentrations, and if $z_0 = 0$ for all concentrations, the adsorption wave will propagate as a step change similar to case a.

Hence:

$$z = ut\Big/\left[1 + \frac{1}{m}\frac{C_s^*}{C_0}\right]$$

$$= \left(\frac{4 \times 10^{-4}}{\pi \times 0.15^2 \times 0.4}\right)\left(\frac{3600}{1 + (0.6/0.4)(3/C_0^{0.7})}\right)$$

$$= \underline{0.95\,m}$$

For case (c):

$$Cs = 10^4 C^2 \text{ which represents an unfavourable isotherm.}$$
$$f'(C) = 2 \times 10^4\, C.$$

As C increases, f'(C) increases such that, in a given time, z for lower concentrations is greater than for higher concentrations. Following the progress of the breakpoint concentration, $C = 0.003\,kmol/m^3$, then:
Hence:

$$f'(0.003) = 60$$

$$z = \left(\frac{4 \times 10^{-4}}{\pi \times 0.15^2 \times 0.4}\right)\left(\frac{3600}{1 + (0.60/0.40)60}\right)$$

$$= \underline{0.55\,m}$$

At breakpoint, the bed is far from saturated and:

$$\text{saturation} = \frac{100uA\varepsilon t C_0}{zAC_0\varepsilon\left[1 + \dfrac{1}{m}\cdot\dfrac{C_s^*}{C_0}\right]} = \frac{100ut}{z\left[1 + \dfrac{1}{m}\cdot\dfrac{C_s^*}{C_0}\right]}$$

$$= 100\left(\frac{4\times10^{-4}\times3600}{\pi\times0.15^2\times0.4}\right)\Big/[0.55(1+(0.6/0.4)(9/0.03))]$$

$$= \underline{\underline{20.5\,\text{percent}}}$$

8.8.3 Non-equilibrium adsorption—Isothermal operation

Isothermal operation in a fixed bed may be achieved in a well-cooled laboratory column and also in large-scale equipment if the concentration of adsorbate is low and the release of heat of adsorption is not great. A third and rather specialised situation in which isothermal conditions may exist is that in which a component is adsorbed on to a surface already covered with a second component. If this second component is displaced by the first, its heat of desorption will 'consume' the heat released when the adsorption of the first component occurs.

Constant patterns analysis

A constant wave-pattern develops when adsorption is governed by a favourable isotherm. In Fig. 8.16A, a typical wave is assumed to move a distance dz in a time dt. If the wave is already fully developed, it will retain its shape. A mass balance gives:

$$\varepsilon uAC_0 dt = A[(1-\varepsilon)C_{s\infty} + \varepsilon C_0]dz$$

$$\frac{dz}{dt} = \frac{u}{1 + \dfrac{1}{m}\dfrac{C_{s\infty}}{C_0}} \tag{8.79}$$

where Eq. (8.79) is similar in form to the equilibrium Eq. (8.75), and identical to it if the isotherm is linear.

The mass balance may be carried out at any level of concentration within the zone to give:

$$\frac{dz}{dt} = \frac{u}{1 + \dfrac{1}{m}\dfrac{C_s}{C}} \tag{8.80}$$

For a constant pattern wave, all concentrations within the wave have the same velocity.

Therefore,

$$\frac{C_s}{C}\frac{C_{s\infty}}{C_0} \tag{8.81}$$

This is the *constant-pattern simplification* that enables many solutions to be obtained from what might otherwise be complex rate equations. It represents a condition that is approached as the wave becomes fully developed and leads to what are termed *asymptotic solutions*.

Representing the mass balance in a fixed bed by:

$$\left(\frac{\partial C_s}{\partial \chi}\right)_{t1} = -\left(\frac{\partial C_s}{\partial t_1}\right)_\chi \tag{8.71}$$

and assuming a general rate expression:

$$\frac{\partial C_s}{\partial t_1} = G(C, C_s) \tag{8.82}$$

where G denotes a function.

The constant pattern assumption gives:

$$\frac{\partial C_s}{\partial t_1} = G_1(C_s) = G_2(C) = -\frac{\partial C}{\partial \chi} \tag{8.83}$$

Equation (8.83) may be integrated to give, at constant χ to give:

or, at constant t_1 to give :

$$\int \frac{dC_s}{G_1(C_s)} = \left[\int \frac{dC_s}{G_1(C_s)}\right]_{t_1=0} + t_1$$

$$\int \frac{dC_s}{G_2(C)} = \left[\int \frac{dC_s}{G_2(C)}\right]_{x=0} - \chi \tag{8.84}$$

If, for example, a rate expression may be written as:

$$\frac{\partial C_s}{\partial t_1} = k_g^0(C - C^*) \tag{8.85}$$

where C^* is the fluid concentration in equilibrium with the mean solid concentration C_s, then, assuming a Langmuir equilibrium relationship similar to Eq. (8.3):

$$C_s = \frac{C_{sm}B_3 C^*}{1 + B_3 C^*} \tag{8.86}$$

Substituting in Eq. (8.85) for C and C^*, from Eqs (8.81) and (8.86), respectively, gives the expression for $G_1(C_s)$ for the particular case.

Rosen's solutions

Rate equations, such as Eq. 8.85, make no attempt to distinguish mechanisms of transfer within a pellet. All such mechanisms are taken into account within the rate constant k. A more fundamental approach is to select the important factors and combine them to form a rate equation, with no regard to the mathematical complexity of the equation. In most cases, this approach will lead to the necessity for numerical solutions although for some limiting conditions, useful analytical solutions are possible, particularly that presented by Rosen [41].

A mass balance for diffusion of adsorbate into a spherical pellet may be written as:

$$\underbrace{\left[-4\pi r^2 D_e \frac{\partial C_r}{\partial r}\right]}_{\text{IN}} - \underbrace{\left[-4\pi r^2 D_e \frac{\partial C_r}{\partial r} + \frac{\partial\left(-4\pi r^2 D_e \frac{\partial C_r}{\partial r}\right)dr}{\partial r}\right]}_{\text{OUT}}$$

$$= \underbrace{\left[4\pi r^2 dr \frac{\partial C_r}{\partial t} + 4\pi r^2 dr \frac{\partial C_{sr}}{\partial t}\right]}_{\text{ACCUMULATION}}$$

(8.87)

where C_r and C_{sr} are concentrations of adsorbate at a radius r.

If there is equilibrium between fluid in the pores and the adjacent surface, and if the equilibrium is linear then:

$$C_{sr} = K_a C_r \tag{8.88}$$

Hence the mass balance may be written as:

$$\frac{\partial C_r}{\partial t} = \frac{D_e}{1 + K_a}\left(\frac{\partial^2 C_r}{\partial r^2} + \frac{2}{r}\frac{\partial C_r}{\partial r}\right) \tag{8.89}$$

The total adsorbate in the pellet at any particular time may be written as:

$$4\pi \int \left(r^2 C_r + r^2 K_a C_r\right) dr$$

so that the mean concentration C_s is given by:

$$C_s = \frac{3}{r_i^3}(1 + K_a)\int r^2 C_r dr \tag{8.90}$$

The rate at which adsorbate enters a pellet may be expressed in terms of the concentration driving force across the boundary film to give:

$$\frac{4}{3}\pi r_i^3 \frac{\partial C_s}{\partial t} = 4\pi r_i^2 k_g (C - C)$$

Thus:

$$\frac{\partial C_s}{\partial t} = \frac{3k_g}{r_i}(C - C_i) \tag{8.91}$$

where C_i is in equilibrium with the solid concentration at $r = r_i$.

Solutions have been found by Rosen, using the fixed-bed Eq. (8.70) together with Eqs (8.90) and (8.91). The general solution takes the form:

$$\frac{C}{C_0} = \frac{1}{2} + F(\lambda', \tau', \psi) \tag{8.92}$$

where the length parameter is:

$$\lambda' = \frac{3D_e K_a z}{mur_i^2}$$ (8.93)

the time parameter is:

$$\tau' = \frac{2D_e}{r_i^2}\left(t - \frac{z}{u}\right)$$ (8.94)

and the resistance parameter is:

$$\psi = \frac{D_e K_a}{r_i k_g}$$ (8.95)

Solutions are available in tabulated and graphical form. Except for small values of λ', the solution has the following convenient form:

$$\frac{C}{C_0} = \frac{1}{2}\left\{1 + \mathrm{erf}\left[\frac{(3\frac{\tau'}{2\lambda'} - 1)}{2\sqrt{[(1 + 5\psi)/5\lambda']}}\right]\right\}$$ (8.96)

Ruthven [16] gives a useful summary of other solutions.

Example 8.4

A bed is packed with dry silica gel beads of mean diameter 1.72 mm to a density of 671 kg of gel/m³ of bed. The density of a bead is 1266 kg/m³ and the depth of packing is 0.305 m. Humid air containing 0.00267 kg of water/kg of dry air enters the bed at the rate of 0.129 kg of dry air/m² s. The temperature of the air is 300 K and the pressure is 1.024×10^5 N/m². The bed is assumed to operate isothermally. Use the method of Rosen to find the effluent concentration as a function of time. Equilibrium data for the silica gel are given by the curve in Fig. 8.21. An appropriate value of k_g, the film mass transfer coefficient, is 0.0833 m/s.

Solution

The length parameter λ' is calculated from Eq. (8.83):

$$\lambda' = \frac{3D_e K_a z}{mur_i^2}$$

$$K_a = \left(\frac{0.084}{0.00267}\right) \times \left(\frac{1266}{1.186}\right) = 3.36 \times 10^4$$

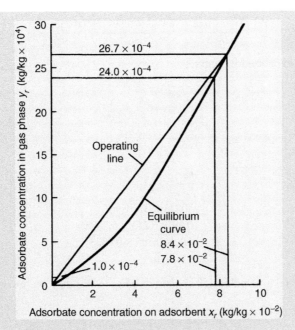

FIG. 8.21

Adsorption isotherm for water vapour in air on silica gel.

where K_a is obtained from the mean slope of the isotherm between its origin and the point corresponding to the inlet concentration. This slope has been multiplied by the ratio of bead to gas densities to give K_a in the same units as Eq. (8.88).

$$m = \frac{\varepsilon}{1 - \varepsilon} = \frac{1266 - 671}{671} = 0.89$$

where ε is the volume (and area) voidage between beads.

$$\varepsilon = \frac{1266 - 671}{1266} = 0.47$$

$$u = \frac{0.129}{0.47 \times 1.186} = 0.233 \text{ m/s}$$

where u is the interstitial velocity of the air.

$$r_i = 0.86 \text{ mm}$$
$$z = 0.305 \text{ m}$$

D_e is the diffusivity of sorbate referred to the sorbed phase (and this has a value of $10^{-10}-10^{-11}$ m^2/s). Substituting values gives $\lambda' = 18{,}500-1850$. The use of Eq. (8.96) is valid for large values of λ'. From Eqs (8.93) and (8.94):

$$\frac{\tau'}{\lambda'} = \frac{2\left(t - \frac{z}{u}\right)mu}{3K_a z} = \left(\frac{2 \times 0.89}{3 \times 3.36 \times 10^4}\right)\left(\frac{0.233t}{0.305} - t\right)$$

where t is expressed in hours.

Thus:

$$t = 20.55\frac{\tau'}{\lambda'} + \frac{1}{2750}$$

From Eqs (8.93) and (8.95):

Thus:

$$\frac{\psi}{\lambda'} = \frac{mur_i}{3zk_g}$$

$$k_g = 0.0833 \, \text{m/s (given)}$$

$$\frac{\psi}{\lambda'} = \frac{(0.89 \times 0.233 \times 0.00086)}{(3 \times 0.305 \times 0.0833)}$$

$$= 2.36 \times 10^{-3}$$

For the relative values of λ' and ψ/λ', Eq. (8.96) may be rewritten as:

where:

$$\frac{C}{C_0} = \frac{1}{2}\left[1 + \text{erf}\left(\frac{(3\tau'/2\lambda' - 1)}{2\sqrt{(\psi/\lambda)}}\right)\right]$$

$$= \frac{1}{2}[1 + \text{erf}E]$$

$$E = \frac{(3\tau'/2\lambda' - 1)}{2\sqrt{(\psi/\lambda')}}$$

For selected values of C/C_0, the value of E may be found from the tables of error functions given in the Appendix in Volume 1. From E a ratio τ/λ' may be calculated and hence a corresponding time. These calculations are summarised as follows:

$\frac{C}{C_0}$	erf E	E	$\frac{\tau'}{\lambda'}$	t(h)
0.024	−0.952	−1.40	0.576	11.8
0.045	−0.910	−1.20	0.589	12.1
0.079	−0.842	−1.00	0.602	12.4
0.24	−0.520	−0.50	0.635	13.0
0.50	0	0	0.667	13.7
0.715	0.430	0.40	0.693	14.2
0.92	0.840	1.00	0.732	15.0

8.8.4 **Non-equilibrium adsorption—Non-isothermal operation**

When the effects of heats of adsorption cannot be ignored – the situation in most industrial adsorbers – equations representing heat transfer have to be solved simultaneously with those for mass transfer. All the resistances to mass transfer will also affect heat transfer although their relative importance will be different. Normally, the greatest resistance to mass transfer is found within the pellet and the smallest in the external boundary film. For heat transfer, the thermal conductivity of the pellet is normally greater than that of the boundary film so that temperatures through a pellet are fairly uniform. The temperature difference between bulk conditions outside a pellet and conditions within it occurs almost wholly across the boundary film.

The effects of non-isothermal adsorption in fixed beds have been examined by Leavitt [42]. Non-isothermal adsorption of a single adsorbate from a carrier fluid leads to the complex wave front shown in Fig. 8.22. As adsorption proceeds, the

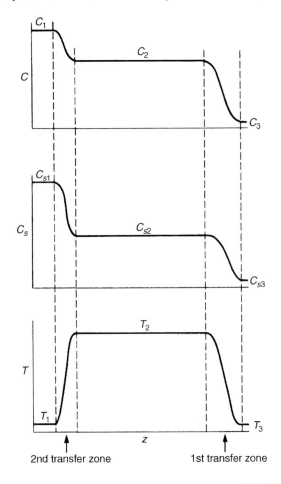

FIG. 8.22

Idealised distributions of temperature and concentrations during adiabatic adsorption.

leading edge of the adsorption wave meets adsorbent that is essentially free of adsorbate. Adsorption occurs and the temperature of the adsorbent increases until a dynamic equilibrium is established between fluid and adsorbed phase at the prevailing temperature. These conditions persist through a first plateau zone until the rate of adsorption falls to a point where the temperature of the plateau zone cannot be sustained. As the temperature falls, more adsorption occurs until the second plateau zone is formed, in equilibrium with the incoming stream. The net result is that adiabatic, or near adiabatic, conditions can lead to the formation of two transfer zones separated by a zone in which conditions remain constant.

The adiabatic profile may be further complicated by the shape of the isotherm. Under isothermal conditions, a favourable isotherm produces a single transfer zone, although an isotherm with favourable and unfavourable sections may generate a more complex profile.

A heat balance over an increment of bed dz may be written as:

$$(\text{IN} - \text{OUT}) = -\frac{\partial}{\partial z}\left(u\varepsilon A\rho_g c_{pg}(T - T_0)\right)dz - U_0\pi d_b dz(T - T_a) \qquad (8.97)$$

where: d_b is the diameter of the containing vessel,

T_a is the temperature of the surroundings,

T_0 is the reference temperature, and.

U_0 is the overall coefficient for heat transfer between the containing vessel and the surroundings.

Accumulation:

$$= \frac{\partial}{\partial t}\left(\varepsilon A\,dz\rho_g c_{pg}(T - T_0) + (1 - \varepsilon)A\,dz\rho_p c_{ps}(T - T_0)\right) + \frac{\partial}{\partial t}\left(W\,dzc_{pw}(T - T_0)\right) \qquad (8.98)$$

where W is the mass per unit length of containing vessel.

Loss by adsorption:

$$= (1 - \varepsilon)A\,dz\frac{\partial(C_s\Delta H)}{\partial t} = (1 - \varepsilon)A\,dz\frac{\partial(C_s\Delta H)}{\partial T}\frac{\partial T}{\partial t} \qquad (8.99)$$

where ΔH is the heat of adsorption, a negative quantity.

Hence:

$$\frac{\partial\left(u\varepsilon A\rho_g c_{pg}(T - T_0)\right)}{\partial z}dz + U_0\pi d_b dz(T - T_a) + \frac{\partial}{\partial t}\left(\varepsilon A\,dz\rho_g C_{pg}(T - T_0)\right)$$

$$+ (1 - \varepsilon)A dz\rho_p c_{ps}(T - T_0) + W\,dzc_{pw}(T - T_0)) + (1 - \varepsilon)A\,dz\frac{\partial(C_s\Delta H)}{\partial T}\frac{\partial T}{\partial t} = 0$$

For a cylindrical bed:

$$u\frac{\partial T}{\partial z} + \frac{4U_0}{\varepsilon d_b}\frac{(T - T_a)}{\rho_g c_{pg}} + \left(1 + \frac{1 - \varepsilon}{\varepsilon}\frac{\rho_p c_{ps}}{\rho_g c_{pg}}\right.$$

$$\left. + \frac{4Wc_{pw}}{\varepsilon\pi d_b^2\rho_g c_{pg}} + \frac{1 - \varepsilon}{\varepsilon}\frac{1}{c_{pg}}\frac{\partial(C_s\Delta H)}{\partial T}\right)\frac{\partial T}{\partial t} = 0 \qquad (8.100)$$

For the particular case of adiabatic operation with no sinks for heat in the wall, $U_0 = 0$ and $W = 0$.

Since:

$$\frac{\partial T}{\partial z} = -\left(\frac{\partial T}{\partial t}\right)\left(\frac{\partial t}{\partial z}\right)$$

Then:

$$\left.\frac{\partial z}{\partial t}\right|_T = u_T = \frac{u}{1 + \frac{1}{m} \cdot \left(\frac{\rho_p c_{ps}}{\rho_g c_{pg}} + \frac{1}{\rho_g c_{pg}} + \frac{\partial(C_s \Delta H)}{\partial T}\right)} \tag{8.101}$$

where u_T is the velocity of a point of constant temperature. If the thermal wave is coherent, that is all points travel at the same velocity, then u_T is the thermal wave velocity. This may be compared with the concentration wave velocity u_c, where:

$$u_c = \frac{u}{1 + \frac{1}{m}\frac{\partial C_s}{\partial C}} \tag{8.75}$$

If the velocities of the thermal and concentration waves are equal, then from Eqs (8.75) and (8.101):

$$\frac{\partial C_s}{\partial C} = \frac{\rho_p c_{ps}}{\rho_g c_{pg}} + \frac{1}{\rho_g c_{pg}} + \frac{\partial(C_s \Delta H)}{\partial T} \tag{8.102}$$

When Eq. (8.102) is applied to the finite difference between inlet concentration and plateau, and between plateau and exit concentration, as shown in Fig. 8.22, it becomes:

$$\frac{C_{s1} - C_{s2}}{C_1 - C_2} = \frac{\rho_p c_{ps}}{\rho_g c_{pg}} + \frac{1}{\rho_g c_{pg}}\left(\frac{(C_s \Delta H)_1 - (C_s \Delta H)_2}{T_1 - T_2}\right) \tag{8.103}$$

$$\frac{C_{s2} - C_{s3}}{C_2 - C_3} = \frac{\rho_p c_{ps}}{\rho_g c_{pg}} + \frac{1}{\rho_g c_{pg}}\left(\frac{(C_s \Delta H)_2 - (C_s \Delta H)_3}{T_2 - T_3}\right) \tag{8.104}$$

In Eqs (8.103) and (8.104), c_{ps} and c_{pg} are the mean specific heats over the ranges of temperature and concentrations encountered. Properties with subscripts 1 and 3 are known from inlet and exit conditions, respectively. If the plateau values represented by subscript 2 are in equilibrium, then the values C_2, C_{s2}, and T_2 may be found from the equations for any known form of the adsorption isotherm $C_{s2} = f(C_2)$.

Equation (8.102) was derived on the assumption that concentration and thermal waves propagated at the same velocity. Amundson et al. [43] showed that it was possible for the temperatures generated in the bed to propagate as a pure thermal wave leading the concentration wave. A simplified criterion for this to occur can be obtained from Eqs (8.75) and (8.101). Since there is no adsorption term associated with a pure thermal wave, and if changes within the bed voids are small, then:

$$u_T = \frac{m\rho_g c_{pg}}{\rho_p c_{ps}} u \tag{8.105}$$

Also:

$$u_c = m\frac{\Delta C}{\Delta C_s}u \tag{8.106}$$

For a bed initially free of adsorbate, the thermal wave propagates more quickly than the concentration wave if:

$$\frac{\rho_g}{\rho_p}\frac{c_{pg}}{c_{ps}} > \frac{C_1}{C_{s1}} \tag{8.107}$$

Since C_1/C_{s1} increases with temperature, u_c (non-isothermal) $> u_c$ (isothermal). It has been estimated that the rear wave moves at about two-thirds of the velocity of the front wave, so a more cautious criterion than Eq. (8.107) is given by:

$$\frac{\rho_g c_{pg}}{\rho_p c_{ps}} > \frac{3}{2}\frac{C_1}{C_{s1}} \tag{8.108}$$

where C_{s1} is a function of C_1 and the maximum temperature $T_{2\text{max}}$ of the plateau zone.

Equation (8.101) may be rearranged to give:

$$T_2 = T_1 + \frac{[(\Delta H)C_s]_2 - [(\Delta H)C_s]_1}{\rho_g c_{pg}\frac{\Delta C_s}{\Delta C} - \rho_p c_{ps}} \tag{8.109}$$

T_2 is a maximum when C_{s2} is zero.
Hence:

$$T_{2\text{max}} = T_1 + \frac{[(-\Delta H)C_s]_1}{\rho_g c_{pg}(C_s/C)_1 - \rho_p c_{ps}} \tag{8.110}$$

When equilibrium between the fluid and the solid cannot be assumed, it may still be possible to obtain analytical solutions for beds operating non-isothermally. In general, however, it will be necessary to look for numerical solutions. This problem has been summarised by Ruthven [16].

8.9 Regeneration of spent adsorbent

Most theory concerning the dynamics of an industrial adsorption unit is directed at sizing a bed for a given adsorption duty. In the design of a complete adsorption unit, however, it is most important to ensure that the spent adsorbent can be regenerated in a given time and that the total inventory of adsorbent is kept to a minimum. If spent adsorbent could be regenerated instantly, all units would consist of a single cylindrical-bed packed with a depth of pellets just sufficient to accommodate the adsorption zone. In practice, units vary in size and configuration because regeneration takes a finite time and what is an optimum arrangement for one application is not necessarily the optimum case for another.

In an ideal fixed-bed adsorber, the adsorption stage continues until the adsorbate wave is about to emerge from the bed and the effluent concentration begins to rise, as shown in Fig. 8.16B. Conditions behind the adsorption zone are such that the adsorbent is more or less in equilibrium with the feed. The equilibrium condition has then to be changed for regeneration to occur. This is usually brought about by changing the temperature or the pressure so that the driving force, which had previously resulted in the movement of adsorbate from fluid to solid, is now reversed.

8.9.1 Thermal swing

The simplest and the most common way of regenerating an adsorbent in industrial applications is by heating. The vapour pressure exerted by the adsorbed phase increases with temperature, so that molecules desorb until a new equilibrium with the fluid phase is established. Fig. 8.23 depicts adsorption isotherms for a lower temperature T_1 and a higher temperature T_2. For a fixed concentration C in the fluid phase, the adsorbate concentration falls from C_{s1} to C_{s2} when the temperature is increased.

An adsorption unit using thermal swing regeneration usually consists of two packed beds, one on-line and one regenerating, as shown in Fig. 8.24. Regenerating consists of heating, and purging to remove adsorbate. The arrangement is flexible and robust. The desorption temperature depends on the properties of the adsorbent and the adsorbates. Manufacturers normally recommend the best regenerating temperature for their particular adsorbent. Exceeding this temperature may accelerate the ageing processes which cause pores to coalesce and capacity to be reduced. Too low a temperature may result in incomplete regeneration so that the effluent concentration in subsequent adsorption stages will be higher than its design value. Hot spots may develop in the operation of a fixed bed, so particular care has to be taken to control temperature when handling flammable materials.

The relatively poor conductivity of a packed bed makes it difficult to get the heat of regeneration into the bed, either from a jacket or from coils embedded in the

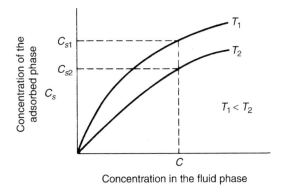

FIG. 8.23

Thermal regeneration utilises the change in concentration that follows from a change in temperature.

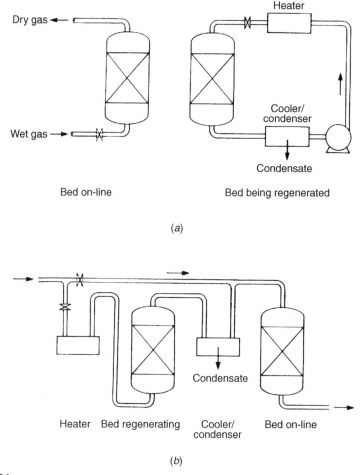

FIG. 8.24

Typical arrangement of a two bed dryer. (A) Separate regeneration (B) Integrated regeneration.

packing. This is more easily achieved by preheating the purge stream. Even in the best conditions, it takes time for the temperature of the bed to rise to the required level. Thermal regeneration is normally associated with long cycle times, measured in hours. Such cycles require large beds and, since the adsorption wave occupies only a small part of the bed on-line, the utilisation of the total adsorbent in the unit is low.

It is good operating practice to regenerate a bed in the reverse direction to that followed during adsorption. This ensures that the adsorbent at the end of the bed, which controls the quality of the treated stream, is that which is most thoroughly regenerated. Carter [44] has quantified the effect and showed that regeneration is achieved in a shorter time in this way.

8.9.2 **Plug-flow of solids**

Better utilisation of adsorbent would be achieved if a unit could be designed in which adsorbent were removed for regeneration as soon as it became saturated, and even better if the advancing adsorption wave were presented with only sufficient fresh adsorbent to contain the wave. Both characteristics are possible if the adsorbent moves countercurrent to the fluid at such a rate that the adsorption zone is stationary.

The earliest application of a moving bed in which solids moved with respect to the containing vessel was reported in the late 1940s. A typical application was the recovery of ethylene from gas composed mainly of hydrogen and methane, and with some propane and butane. The unit shown diagrammatically in Fig. 8.25, taken from the work of Berg [45], is known as the *hypersorber*.

The mixture to be separated is fed to the centre of the column down which activated carbon moves slowly. Immediately above the feed, the rising gas is stripped of ethylene and heavier components leaving hydrogen, methane and any non-adsorbed gases to be discharged as a top-product. The adsorbent with its adsorbate continues down the column into an enriching section where it meets an upwards stream of recycled top-product. The least-strongly adsorbed ethylene is desorbed and recovered in a side stream. The heavy components continue downwards on the carbon until these are also

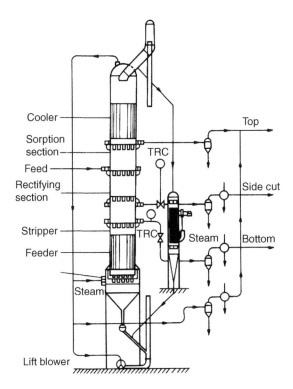

FIG. 8.25

The hypersorber [45].

desorbed by steam, to be recovered as a bottom product. The carbon, now stripped of all the adsorbates, is lifted to the top of the column where it is cooled before the cycle starts again.

The rate at which the carbon is circulated may be controlled at entry to the lift-pipe. In normal operation, regenerating conditions do not maintain the carbon in its initial highly active state. Consequently, a small proportion of the regenerated carbon is steam-treated in a small column at a temperature of about 870 K. Some large hypersorbers, about 25 m high and 1.4 m in diameter, have been built for commercial operation. It seems that the units were beset from the beginning with problems of solids-handling. There was difficulty in maintaining an even flow of adsorbent and the problem of solids attrition and their subsequent loss as fines. Recently developed adsorbents, which are more selective and therefore more attractive as separating agents are, if anything, less resistant to attrition, and are unsuitable for moving-solid applications as a result.

In the 1960s, there were attempts to use a moving bed of carbon to remove sulphur dioxide from flue gas on a pilot scale. As described by Katell [46] and Cartelyou [47], this *Reinluft* process was abandoned because of the problems caused by the carbon igniting in the presence of oxygen.

In order to design moving-bed equipment, the velocity of the adsorption zone relative to the solid has to be calculated. This gives the velocity at which the solids must move in plugflow in order that the zone remains within the equipment. The depth of packing, z_a, should be sufficient to contain the zone. A mass balance across an increment dz gives:

$$u \varepsilon A \, dC = k_g A (1 - \varepsilon) a_p (C - C_i) dz \tag{8.111}$$

C_i is an interfacial concentration which, in general, will not be known. It is often possible to express the rate of transfer of adsorbate in terms of an overall driving force $(C - C^*)$. In this case, Eq. (8.111) may be rearranged and integrated to give:

$$z_a = \frac{mu}{k_g^0 a_p} \int_{C_B}^{C_E} \frac{dC}{C - C^*} \tag{8.112}$$

where C^* is the fluid concentration in equilibrium with the mean concentration of the adsorbed phase, at any instant, and k_g^0 is an overall mass transfer coefficient.

The integral in Eq. (8.112) may be evaluated numerically or graphically as shown in Fig. 8.26. This is the *number of transfer units*, and the group outside the integral is the *height of a transfer unit*. The integral in Eq. (8.112) covers the whole concentration span of the adsorption process. If, instead, the limits are taken as from C_B to an arbitrary concentration C, then the length z' corresponding to C is given by:

$$z' = \frac{mu}{k_g^0 a_p} \int_{C_B}^{C} \frac{dC}{C - C^*} \tag{8.113}$$

and:

$$\frac{z'}{z_a} = \frac{\int_{C_B}^{C} \frac{dC}{C - C^*}}{\int_{C_B}^{C_E} \frac{dC}{C - C^*}} \tag{8.114}$$

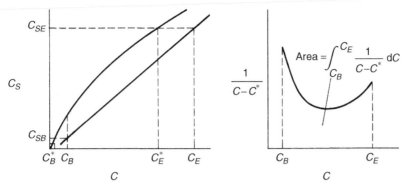

FIG. 8.26

Graphical calculation of the number of transfer units.

In Fig. 8.27, C/C_E is plotted as a function of z'/z_a, and the unsaturated fraction i of the adsorption zone may then be found. Since the zone in a moving bed is essentially the same as that which develops in a fixed bed, the time to break-point for the latter case, t_b, may be found from:

$$uA\varepsilon C_0 t_b = (z' - iz_a)C_{s\infty}A(1 - \varepsilon) \qquad (8.115)$$

The shape of the breakthrough wave, subsequent to the breakpoint, may then be determined from:

$$z' = \frac{(t - t_b)}{t_a}z_a \qquad (8.116)$$

where t_a is the time for the adsorption zone to move its own length, and $z!$ is measured from the inlet of the adsorption zone.

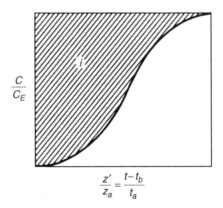

FIG. 8.27

Dimensionless breakthrough curve showing fractional unsaturation of the adsorption zone.

The method, which is illustrated in Example 8.5, applies only to isothermal beds in which the zone becomes fully-developed quite soon after the flow begins.

Example 8.5

An adsorption unit is to be designed to dry air using silica gel. A moving-bed design is considered in which silica gel moves down a cylindrical column in plug flow whilst air flows up the column. Air enters the unit at the rate of $0.129\,kg$ of dry air/m²s and with a humidity of $0.00267\,kg$ water/kg dry air. It leaves essentially bone dry. There is equilibrium between air and gel at the entrance to and the exit from the adsorption zone. Experiments were carried out to find the relative resistances of the external gas film and pellet diffusion. Referred to a driving force expressed as mass ratios then:

(a) for the gas film, the coefficient $k_g a_z = 31.48 G'^{0.55}$ kg/m³s, where G' is the mass flow rate of dry air per unit cross-section of bed.
(b) for pellet diffusion, solid-film coefficient $k_p a_z = 0.964\,kg/m^3 s$

where a_z is the external area of adsorbent per unit volume of bed

(a) Using the transfer-unit concept, calculate the minimum length of packing which will reduce the moisture content of the air to $0.0001\,kg$ water/kg dry air. At what rate should the gel travel through the bed?
The properties of the gel and the condition of the air are as given in Example 8.4.
(b) After operating for some time, the gel jams and the unit continues operating as a fixed bed. How long after jamming will it be before the moisture content of the effluent rises to half the inlet value?

Solution

(a) The bed must be long enough to contain the adsorption zone. From Eq. (8.112), the number of transfer units may be written as: $\int dy_r/(y_r - y_r^*)$ and the height of a transfer unit in appropriate units is $G'/k_g^0 a_z$, where k_g^0 is the overall mass transfer coefficient.

The integral may be evaluated graphically from a plot of $1/(y_r - y_r^*)$ against y_r over the concentration range of the adsorption zone.
A mass balance over a part of the bed gives an operating line:

$$y_r = \frac{G'_s}{G'} x_r + \left(y_{rin} - \frac{G'_s}{G'} x_{rout} \right)$$

This line, together with the equilibrium line, is similar to that shown in Fig. 8.26. Corresponding values of y and y^* may be measured and the integral evaluated. The pinches between the operating and equilibrium lines which occur at each end of the zone prevent the end concentration from being used as the limits of the integration. If the lower limit of $y_r = 0.0001$ and the upper limit 0.0024, then from a graphical construction:

$$\int_{0.0001}^{0.0024} \frac{dy_r}{y_r - y_r^*} = 10.95\,\text{transfer units}$$

The height of a transfer unit is: $\frac{G'}{k_g^0 a_z}$ (8.112)

k_g^0 may be evaluated from the film coefficients, as discussed in Chapter 12.

When the zone is fully developed, each part will move at the same constant velocity. If f'(C) is the mean slope of the isotherm over the range of concentrations of interest, then, in appropriate units:

$$[f'(C)]_{mean} = \frac{(8.4 \times 10^{-2} \times 1266)}{(2.67 \times 10^{-3} \times 1.186)}$$

$$m = \frac{\varepsilon}{1-\varepsilon} = \left(\frac{0.47}{0.53}\right) = 0.89$$

The inter-pellet air velocity $=0.233$ m/s. The velocity u_c with which the adsorption wave moves through the column may be obtained from Eq. (8.79).

Hence:

$u_c = 6.2 \times 10^{-6}$ m/s.

$G_s' = u_c(1-\varepsilon)\rho_p$

and:

$= 6.2 \times 10{-6}(0.53)1266$

$= \underline{4.16 \times 10{-3} \text{ kg/m2s}}$

The rate at which clean adsorbent must be added and spent adsorbent removed in order to maintain steady state may also be found from an overall balance:

$$G_s'(0.084 - 0) = 0.129(0.00267)$$
$$G_s' = 4.11 \times 10^{-3} \text{ kg/m}^2\text{s}$$

(b) When the gel stops moving, the bed behaves as a fixed-bed already at its breakpoint. The concentration of water in the effluent begins to rise.

The time t_a for the adsorption zone to move its own length z_a is given by:

$$t_a = \frac{z_a}{u_c} = 8.3\text{h}$$

The time taken for a point at a distance z' into the zone to emerge is given by:

$$t = \frac{z'}{z_a}t_a$$

where:

$$\frac{z'}{z_a} = \int_{y_r^B}^{y_r} \frac{dy_r}{y_r - y_r^*} \bigg/ \int_{y_r^B}^{y_r^E} \frac{dy_r}{y_r - y_r^*}$$

The results for graphical integration are tabulated below.

$$\frac{1}{k_g^0 a_z} = \frac{1}{k_g a_z} + \frac{m'}{k_p a_z}$$

where m' is the slope of the operating line $= \left(\frac{0.00267}{0.084}\right) = 0.0318$, as shown in Fig. 8.21.

$$k_g a_z = 31.48 G'^{0.55} = 10.21 \, \text{kg/m}^3 \, \text{s}$$

Hence:

$$k_g^0 a_z = 7.64 \, \text{kg/m}^3 \, \text{s}$$

and:

$$\frac{G'}{k_g^0 a_z} = \left(\frac{0.129}{7.64}\right) = 0.0169 \, \text{m}$$

The length of the adsorption zone $= (10.95 \times 0.0169) = 0.185 \, \text{m}$.

Hence the minimum length of bed to contain the adsorption zone is 0.185 m. In practice, a somewhat greater length would be used to allow for variations in the length of the zone that might result from fluctuations in operating conditions. The data are summarised as follows:

y_r	y_r^*	$\frac{1}{y_r - y_r^*}$	$\int_{y_r^B}^{y_r} \frac{dy_r}{y_r - y_r^*}$	$\frac{z'}{z_a}$	$\frac{y_r}{y_{r0}}$	t (h)
0.0001	0.00005	20,000	0	0	0.038	0
0.0002	0.00010	10,000	1.50	0.137	0.075	0.8
0.0006	0.00032	3570	4.00	0.362	0.225	3.1
0.0010	0.00062	2630	5.18	0.473	0.374	4.1
0.0014	0.00100	2500	6.13	0.560	0.525	4.5
0.0018	0.00133	3700	7.38	0.674	0.674	5.8
0.0022	0.00204	6250	9.33	0.852	0.825	7.1
0.0024	0.00230	10,000	10.95	1.000	0.899	7.7

By interpolation, $y_r/y_r 0 = 0.5$ when $t \approx 4.4 \, \text{h}$.

8.9.3 Rotary bed

Because of the difficulty of ensuring that the solid moves steadily and at a controlled rate with respect to the containing vessel, other equipment has been developed in which solid and vessel move together, relative to a fixed inlet for the feed and a fixed outlet for the product. Fig. 8.28 shows the principle of operation of a rotary-bed adsorber used, for example, for solvent recovery from air on to activated-carbon. The activated-carbon is contained in a thick annular layer, divided into cells by radial partitions. Air can enter through most of the drum circumference and passes through the carbon layer to emerge free of solvent. The clean air leaves the equipment through a duct connected along the axis of rotation. As the drum rotates, the carbon

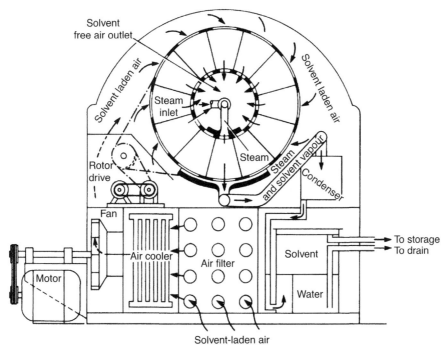

FIG. 8.28

Rotary-bed adsorber.

enters a section in which it is exposed to steam. Steam flows from the inside to the outside of the annulus so that the inner layer of carbon, which determines the solvent content of effluent air, is regenerated as thoroughly as possible. Steam and solvent pass to condensers and the solvent recovered, either by decanting or by a process such as distillation. In the particular equipment shown, there is no separate provision for cooling the regenerated adsorbent; instead it is allowed to cool in contact with vapour-laden air and the adsorptive capacity may be lower as a result [16].

8.9.4 Moving access

In an interesting alternative to a moving-bed or a moving-container adsorber, a multi-way valve effectively changes the position of the inlet and outlet valves relative to a fixed bed. Such a system is shown diagrammatically in Fig. 8.29 which shows a unit consisting of 12 small beds housed in one column. It may be assumed that it is fed with a mixture containing components **A** and **B,** the former being the more strongly absorbed of the two, with the desorption carried out using a third component **D,** the most strongly adsorbed component of all, and therefore capable of displacing **B** and

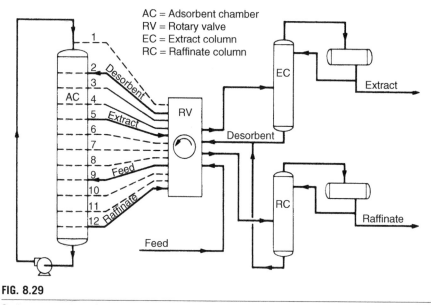

AC = Adsorbent chamber
RV = Rotary valve
EC = Extract column
RC = Raffinate column

FIG. 8.29

Sorbex: a number of small beds used with a rotary valve to simulate a moving bed.

A in that order. The valve rotates in a stepwise fashion, at regular intervals, with the positions of the inlets and the outlets for the process and regenerating streams moving to each numbered position in turn, thus simulating the behaviour of a moving bed as shown in Fig. 8.30. As described by Broughton [48] and Johnson and Kabza [49], the arrangement was developed by Universal Oil Products under the general name

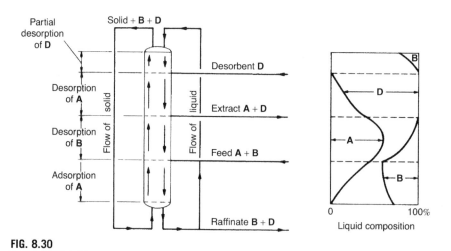

FIG. 8.30

The moving-bed equivalent of the Sorbex process.

'Sorbex'. The unit currently operates in the liquid phase, chiefly for separating *p*-xylene from C_8 aromatics, normal from branched and cycloparaffins, or olefins from mixture with paraffins. In principle, the unit may be used for gas-phase separations, although, in either phase, success depends crucially on the proper working of the rotary valve.

8.9.5 Fluidised beds

Although the moving packed-bed has not yet achieved commercial success, another arrangement in which the solid moves with respect to the containing vessel, the fluidised bed, has fared better. Behaving essentially as a stirred tank reactor, the fluidised bed does not gives the ideal configuration for use as an adsorber. The solid is thoroughly mixed, so that the condition of the effluent is controlled by the mean adsorbate concentration on the solid, rather than by the initial concentration as in the case of a fixed bed. Nevertheless, there are other considerations which outweigh this disadvantage. Solid is easily added to and removed from fluidised bed. The pressure drop through a fluidised bed is effectively constant over a wide range of fluid flow rates, and this makes it possible to treat materials at high flow rates in relatively compact equipment.

If the mean residence time in the fluidised bed is sufficiently long, it may be regarded as a single stage, from which streams of fluid and solid leave in equilibrium.

8.9.6 Compound beds

There is sometimes an advantage in using two kinds of adsorbent in an adsorption bed. Near the inlet would be an adsorbent with a high capacity at high concentrations, although it may have an unfavourable isotherm so that, on its own, the adsorption zone would then be unduly long, particularly if large pellets were used to minimise the pressure drop. If it is followed by a second bed of adsorbent with a highly favourable isotherm and a low mass transfer resistance, a short mass transfer zone will be sufficient to effect the required separation.

8.9.7 Pressure-swing regeneration

In thermal-swing regeneration, the bed may need a substantial time to reach the regeneration temperature. The high temperatures may also affect the product and accelerate the ageing processes in the adsorbent.

An alternative is to use pressure rather than temperature as the thermodynamic variable to be changed with adsorption taking place at high pressure and desorption at low pressure—hence the description *pressure–swing adsorption*. An arrangement utilising this principle was proposed by Skarstrom [50,51]. Fig. 8.31 shows a typical arrangement of a unit consisting of two fixed beds, one adsorbing and one regenerating. These functions are later reversed. A simple cycle consists of four steps. In step 1, high-pressure feed flows through bed A. Part of its effluent is expanded to the

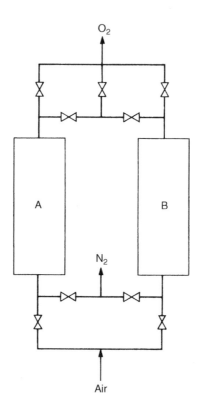

FIG. 8.31

A two-bed unit using pressure swing regeneration for separating oxygen and nitrogen on a small scale.

lower pressure, and then passed through bed B which it regenerates. In step 2, B is repressurised to the feed pressure using feed gas, whilst A is blown down to the purge pressure. Steps 3 and 4 follow the sequence of 1 and 2 except that the functions of beds A and B are reversed.

In large-scale equipment, more than two beds may be used so that pressure energy is better utilised. The regenerating effect of the purge stream depends on its volume rather than on its mass, so only a fraction of the high pressure effluent, say 20%, is needed to achieve effective regeneration. Because changes in pressure can be brought about more rapidly than changes in temperature, pressure-swing regeneration can be used with shorter cycle-times than was possible with thermal-swing. This, in turn, allows smaller beds to be used and consequently a smaller inventory of adsorbent is needed in the system.

Pressure-swing regeneration is useful when the stream to be treated is needed at pressures above atmospheric, as for example in the case of instrument air. Pressure-swing units are compact and can readily be made portable. When the process stream is at atmospheric pressure or below, it may be possible to regenerate using a partial vacuum. When Skarstrom was patenting his 'heatless' adsorber, Guerin de Montgareuil, and Domine [52] were patenting a system using vacuum regeneration. The original aim of both patents was to separate oxygen and nitrogen from air. With the range of adsorbents then available, neither process was particularly successful for that particular application. Skarstrom's equipment was, however, found to be suitable for the drying of gases, and, after some modifications, the Guerin-Domine process was applied successfully to the separation of air on a large scale.

A mathematical description of a pressure swing system has been presented by Shendalman and Mitchell [53] who assumed that isothermal equilibrium adsorption takes place and that the isotherm is linear, with the feed consisting of a single adsorbate at low concentration in a non-adsorbed carrier gas.

The mass conservation equation for the adsorber, over an increment dz of bed may be written as:

$$\frac{\partial(uC)}{\partial z} + \frac{\partial C}{\partial t} + \frac{1}{m}\frac{\partial C_s}{\partial t} = 0 \tag{8.69}$$

where m is the inter-pellet void ratio,

For an ideal gas:

$$C = \frac{yP}{RT}$$

where y is the mole fraction of the adsorbate and P is the total pressure.

If the adsorbed phase is in linear equilibrium with the gas, then:

$$C_s = \frac{K_a yP}{RT}$$

where K_a is the equilibrium constant.

Substituting in Eq. (8.69):

$$\frac{\partial(uyP)}{\partial z} + \frac{\partial P}{\partial t} + \frac{K_a}{m}\frac{\partial(uP)}{\partial t} = 0 \tag{8.118}$$

Neglecting the pressure gradient $\partial P/\partial z$:

$$\frac{\partial u}{\partial z} + u\frac{\partial \ln y}{\partial z} + \left(1 + \frac{K_a}{m}\right)\frac{\partial \ln P}{\partial t} + \left(1 + \frac{K_a}{m}\right)\frac{\partial \ln y}{\partial t} = 0 \tag{8.119}$$

For a pure carrier gas, $y = 1$ and $K_a = 0$, and therefore,

$$\frac{\partial u}{\partial z} + \frac{\partial \ln P}{\partial t} = 0 \tag{8.120}$$

Substituting in Eq. (8.119):

$$u\frac{\partial \ln y}{\partial z} + \frac{K_a}{m}\frac{\partial \ln P}{\partial t} + \left(1 + \frac{K_a}{m}\right)\frac{\partial \ln y}{\partial t} = 0 \tag{8.121}$$

By the method of characteristics [54], Eq. (8.121) may be reduced to an ordinary differential equation to give:

$$\frac{dz}{u} = \frac{dt}{\left(1 + \frac{K_a}{m}\right)} = -\frac{d\ln y}{\left(\frac{K_a}{m}\right)\left(\frac{d\ln P}{dt}\right)} \tag{8.122}$$

from which:

$$\frac{dz}{dt} = \frac{u}{1 + \frac{1}{m}K_a} \tag{8.123}$$

Equation (8.123) is a particular case of Eq. (8.75).

The left-hand side of Eq. (8.123) is the velocity of a point of fixed concentration on the adsorption wave. For a linear isotherm and if longitudinal diffusion is neglected, all points of concentration will move at the same velocity. Changing the pressure will affect u and, to a lesser extent, K_a.

Pressure-swing regeneration is achieved by using a part of the high-pressure adsorber effluent for purging. The volume of purge must be such that the distance the adsorption wave moves at high pressure is completely reversed in the same time at low pressure. The requirement is normally satisfied by using a fraction of the high pressure effluent which is equal to the ratio of the low pressure to the high pressure.

The second characteristic that follows from Eq. (8.122) is:

$$\frac{d\ln y}{d\ln P} = \frac{-\frac{1}{m}K_a}{1 + \frac{K_a}{m}} \tag{8.124}$$

where m is the inter-pellet void ratio $\varepsilon/(1 - \varepsilon)$.

Integrating between the high and low pressure gives:

$$\frac{y_H}{y_L} = \left(\frac{P_L}{P_H}\right)^{(K_a/m)/[1+(K_a/m)]} \tag{8.125}$$

This relates the change in effluent concentration to the pressure ratio.

The change in the position of the characteristic that results from a pressure-swing is given by integrating Eq. (8.120) from the closed end of the bed where $u = 0$, $z = 0$. Thus:

$$u = -\left(\frac{\partial \ln P}{\partial t}\right)z \tag{8.126}$$

From Eq. (8.123):

$$u = \left(1 + \frac{K_a}{m}\right)\frac{dz}{dt} = -\left(\frac{d\ln P}{dt}\right)z$$

which, on integration gives:

$$\frac{z_H}{z_L} = \left(\frac{P_L}{P_H}\right)^{1/(1+K_a/m)} \tag{8.127}$$

where z_H represents the distance moved by the adsorption front during the high pressure stage.

The net distance moved by the front, from the beginning of one position of low pressure to the next is given by:

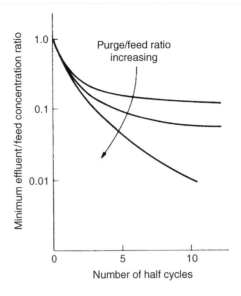

FIG. 8.32

Effect of cycling and purge/feed volumetric ratio on minimum effluent/feed concentration ratio, using pressure-swing regeneration.

$$\Delta z = z_L - z'_L = z_L - \left(\frac{P_H}{P_L}\right)^{1/(1+K_a/m)} \tag{8.128}$$

If Δz is negative, insufficient regeneration is occurring to sustain a condition of cyclic steady-state.

Fig. 8.32 shows how the purge–feed volumetric ratio and cycling affect effluent concentration. Ratios of 1.1 to 1.5 are normal.

8.9.8 Parametric pumping

When operated in a conventional mode, a fixed bed is fed with the stream to be processed until the breakpoint is reached. Thus, maximum use is made of the adsorptive capacity of the bed, without exceeding it. Regeneration is accomplished by changing a variable, such as temperature, pressure or concentration, and purging the bed in a countercurrent manner.

As described by Wilhelm et al. [55], an alternative operating procedure has been developed in order to improve the separation obtained, where separation is defined as the ratio of concentrations in the upper and lower reservoirs, or in a reservoir and the feed. The technique has become known as *parametric pumping* because changing an operating parameter, such as temperature, may be considered as *pumping* the adsorbate into a reservoir at one end of a bed and, by difference, depleting the adsorbate in a reservoir at the other end.

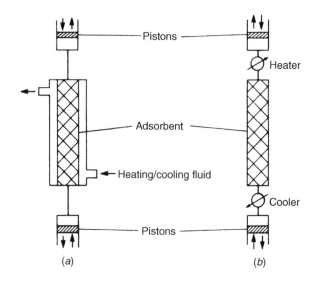

FIG. 8.33

Parametric pumping, batch operation (A) Direct thermal mode (B) Recuperative thermal mode.

Direct mode of operation

Fig. 8.33A shows a simple one-bed unit, operating in batch mode, and heated and cooled through a jacket. The arrangement is known as the *direct thermal mode* because heat is supplied to the whole length of bed at the same instant. Finite resistances to heat transfer will mean, however, that, in practice, the bed takes a finite time to reach the required temperature.

To illustrate the principle, a number of assumptions will be made which are not realised in practice, although they enable the source of the separation to be identified. It is assumed that there is equilibrium at all times between fluid and the solid with which it is in contact. It is assumed that changes in temperature can be achieved instantaneously. It is further assumed that the adsorption front travels at such a velocity at the higher temperature that it traverses the length of the bed in the time allowed for upwards flow of the fluid. At the lower temperature, it is assumed that the adsorption wave travels half the distance in the same time with flow downwards. Each change may be regarded as taking place in two steps—first the temperature is changed and the new equilibrium established and then the pistons move to reverse the direction of flow. This simplified process is illustrated in Fig. 8.34.

When the temperature of the bed is changed, the mass of adsorbate in an increment of bed dz is conserved. Hence, for the first change from hot to cold:

$$A\,dz[\varepsilon C_F + (1 - \varepsilon)C_{sF}] = A\,dz[\varepsilon C_c + (1 - \varepsilon)C_{sc}] \qquad (8.129)$$

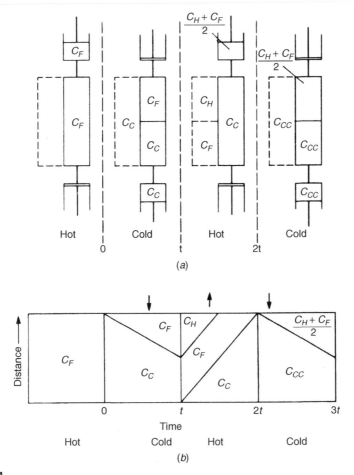

FIG. 8.34

Ideal parapump, direct heating (A) Dotted section represents the condition immediately after a temperature change but before the pistons move (B) Position of the wave-front against time.

For equilibrium operation and a linear isotherm, $C_{sc}=K_cC_c$ and $C_{sF}=K_HC_F$, and hence:

$$\frac{C_F}{C_c}=\frac{1+\frac{1}{m}K_c}{1+\frac{1}{m}K_H} \tag{8.130}$$

It may be seen from Eq. (8.123) that the right-hand side of Eq. (8.130) is the ratio of zone velocities, and:

$$\frac{C_F}{C_c}=\frac{u_H}{u_c} \tag{8.131}$$

It was assumed that $u_H/u_c=2$, and hence $C_F=2C_c$, $C_H=2C_F$, $C_c=2C_{cc}$, and so on.

In Eq. (8.129), C_s is the mean adsorbate concentration over a pellet of adsorbent. It may be desirable to include the intra-pellet voidage α and intra-pellet concentrations C'_F, C'_{sF}, and so on to give:

$$\varepsilon C_F + (1 - \varepsilon)\alpha C'_F + (1 - \varepsilon)(1 - \alpha)C'_{sF} = \varepsilon C_c + (1 - \varepsilon)\alpha C'_c + (1 - \varepsilon)(1 - \alpha)C'_{sc}$$

At equilibrium: $C_F = C'_F$, $C_c = C'_c$,

$$C'_{sF} = K_H C_F, \quad C'_{sc} = K_c C_c$$

Hence, Eq. (8.130) becomes:

$$\frac{C_F}{C_c} = \frac{1 + \frac{1}{m}[a + (1 - \alpha)K_c]}{1 + \frac{1}{m}[a + (1 - \alpha)K_H]} \tag{8.132}$$

It may be seen that even with only the two cycles shown, a significant difference has been achieved between the concentrations of the material in the two reservoirs $[0.5(C_H + C_F)/C_{cc} = 6]$. Separation is sometimes defined as the ratio of the upper reservoir concentration to that of the feed. In this case, a value of 3/2 is obtained. Continuing the cycling process will increase the degree of separation without limit in this ideal case. In practice, thermal lags and diffusional processes make it impossible to sustain sharp differences of concentration, though separations giving a 100,000-fold change in concentrations have been achieved.

The process described so far has made maximum use of the adsorptive capacity of the bed in upwards flow. If the flow were continued beyond that point, the contents of the lower reservoir would pass unchanged into the top reservoir and the separation would be reduced. It may be more convenient, however, to use a shorter time for upwards flow. The effect of allowing the process to continue only until the concentration wave has reached two-thirds of the way along the bed is shown in Fig. 8.35. As in Fig. 8.34, $u_H = 2u_c$. After the second up-flow stage, the average concentration in the material that has left the top of the bed is given by:

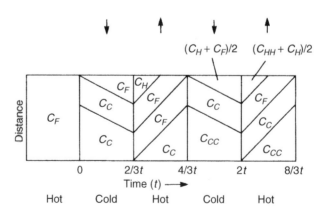

FIG. 8.35

Effect of cycle-time on separation.

$$\frac{1}{2}[(C_{HH} + C_H)/2 + C_F] = 2C_F \tag{8.133}$$

The separation between upper and lower reservoirs in this case is 8. This compares with a value of 7 shown in Fig. 8.34B after the same number of flow-reversals. The degree of separation depends on frequency of cycling as well as the total number of cycles.

Recuperative mode

Fig. 8.33B shows an alternative method of supplying heat in thermal parametric-pumping. Heat is supplied in upwards flow by passing feed from the lower reservoir through a heat exchanger. Cold operation in downward flow is achieved by cooling the feed from the top reservoir. Clearly, even when this method is idealised, the thermal wave takes a finite time to travel the length of the bed. The method is known as the *indirect* or *recuperative mode*, and is shown in Fig. 8.33B as applied to a batch process.

The velocity with which a pure thermal wave travels through an insulated packed bed may be obtained from Eq. (8.100) by putting $U_0 = 0$ and $(\Delta/\partial T)(C_s \Delta H) = 0$ to give:

$$\frac{u}{u_T} = \left[1 + \left(\frac{1}{m}\frac{\rho_p c_{ps}}{\rho_g c_{pg}}\right) + \left(\frac{4W c_{pw}}{\varepsilon \pi d_p^2 \rho_g c_{pg}}\right)\right] \tag{8.134}$$

It has been assumed that the gas and solid have the same temperature at any point, and that the fluid concentration is constant throughout a pellet at a value equal to that immediately outside the pellet. Within the limits of these assumptions, the thermal wave velocity u_T is independent of temperature. As discussed in Section 8.8.4, the velocity of the thermal wave relative to that of the concentration wave can be positive, as it normally is in liquids, negative or zero.

Fig. 8.36 shows a thermal wave plotted as a dotted line of distance against time. The velocity u_c of the concentration wave depends on where it is in relation to the thermal wave, as can be seen by comparison with the full line in the Fig. 8.36.

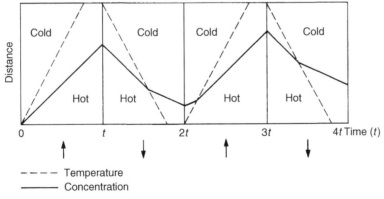

--- - Temperature
——— Concentration

FIG. 8.36

Wave propagation in recuperative mode.

It may be shown that the ratio of the concentration in a hot zone to that in a cold zone for recuperative parametric pumping is given by:

$$\frac{C_H}{C_c} = \frac{1/u_c - 1/u_T}{1/u_H - 1/u_T} \qquad (8.135)$$

For 'instant' heating and cooling, u_T equals infinity and Eq. (8.135) becomes equivalent to Eq. (8.131) for the direct-heating mode.

The net movement upwards of a concentration wave is greater in the direct mode. Fewer cycles are needed to achieve a given separation. Nevertheless, the recuperative mode is probably the more convenient method to use on a commercial scale. Indeed, its equivalent is the only mode that can be used when other parameters, such as pH or pressure, are changed instead of temperature.

Many workers have demonstrated the effectiveness of parametric pumping in order to achieve separations in laboratory-scale equipment. It is mainly liquid systems that have been studied, using either temperature or pH as the control variable. Pressure parametric-pumping is described in a US patent and is discussed by Yang [3].

The principles of separation have been discussed using equilibrium theory. Finite resistances to heat and mass transfer will reduce the separation achieved.

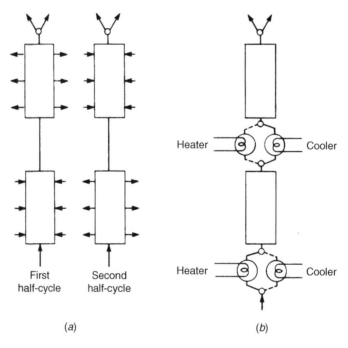

First half-cycle   Second half-cycle

Heater   Cooler

Heater   Cooler

(a)   (b)

FIG. 8.37

A two-bed cycling zone adsorption unit (A) Direct heating mode (B) Recuperative heating mode.

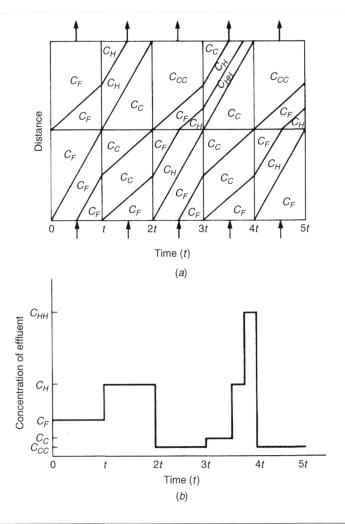

FIG. 8.38

Direct mode cycling zone adsorption (A) Progression of concentration bands through a two-bed unit (B) Effluent concentration.

8.9.9 Cycling-zone adsorption (CZA)

When parametric pumping was being developed, an alternative parameter-swinging technique was proposed by Pigford et al. [56] which is called *cycling-zone adsorption*. Instead of reversing the flow through a single bed as temperature is changed, a number of beds are used, connected in series, alternatively hot and cold. As with parametric pumping, heat may be supplied in the feed stream to each bed or through a jacket. The reversals of temperature remain, though each reversal of direction of the parametric pump corresponds to an additional stage of CZA. Fig. 8.37 illustrates a two-bed unit. Given the same assumptions of ideality and, in particular, equilibrium and instant temperature changes, the sequence of events giving rise to concentration peaks and troughs in the effluent is shown in Fig. 8.38. The effluent is switched between high and low concentration reservoirs.

In the example considered, the separation between highest and lowest effluent concentrations, after four temperature reversals, is $C_{HH}/C_{cc} = 16$, on the basis of the earlier assumptions. A single bed operating in a similar way would produce a separation of C_H/C_c equal to only 4. There is no theoretical limit to the separation that may be achieved by adding further stages. Clearly, there are practical considerations which will limit the number, such as pressure drop and total capital cost.

The other factor affecting separation will be the frequency with which the temperatures are changed. The maximum time for one stage will be the time taken for the feed to break through a hot bed. The minimum time will be determined by the fact that, if there are too many temperature changes, the concentration bands will pass through unchanged.

The principle temperature-cycling of separation has been described. Pressure-cycling has been described by Platt and Lavie [57].

Further discussions of pressure swing adsorption, parametric pumping and cycling-zone adsorption have been presented by Yang [3], Schweitzer [7] and Wankat [58].

References

[1] B. Crittenden, Selective adsorption, Chem. Eng. 452 (1988) 21.
[2] R.M. Barrer, New selective sorbents: porous crystals as molecular filters, Br. Chem. Eng. 4 (1959) 267.
[3] R.T. Yang, Gas Separation by Adsorption Processes, Butterworth, London, 1987.
[4] D.W. Breck, Zeolite Molecular Sieves, Wiley, New York, 1974.
[5] R.M. Barrer, Zeolites and Clay Minerals, Academic Press, London, 1978.
[6] C.W. Roberts, Properties and Applications of Zeolites, Chemical Society, London, 1979.
[7] P.A. Schweitzer (Ed.), Handbook of Separation Techniques for Chemical Engineers, second ed., McGraw-Hill, New York, 1988.
[8] D.H. Everett, F.S. Stone (Eds.), The Structure and Properties of Porous Materials, Butterworth, Oxford, 1958.
[9] J.H. Bowen, R. Bowrey, A.S. Malin, A study of the surface area and structure of activated alumina by direct observation, J. Catal. 7 (1967) 457.
[10] S. Brunauer, The Adsorption of Gases and Vapours, Oxford U.P, Oxford, 1945.
[11] I. Langmuir, The adsorption of gases on plane surfaces of glass, mica and platinum, J. Am. Chem. Soc. 40 (1918) 1361.
[12] S. Brunauer, P.H. Emmett, E. Teller, Adsorption of gases in multimolecular layers (for errata see reference 13), J. Am. Chem. Soc. 60 (1938) 309.
[13] P.H. Emmett, T. de Witt, Determination of surface areas, Ind. Eng. Chem. Anal. Ed. 13 (1941) 28.

[14] S. Brunauer, L.S. Deming, W.E. Deming, E. Teller, On a theory of the van der Waals adsorption of gases, J. Am. Chem. Soc. 62 (1940) 1723.

[15] W.D. Harkins, G. Jura, J. Chem. Phys. 11 (1943) 431. W.D. Harkins, G. Jura, Surface of solids. Part XIII. An adsorption method for the determination of the area of a solid without the assumption of a molecular area and the area occupied by nitrogen molecules on the surface of solids, J. Am. Chem. Soc. 66 (1944) 1366.

[16] D.M. Ruthven, Principles of Adsorption and Adsorption Processes, Wiley, New York, 1984.

[17] D.R. Garg, D.M. Ruthven, Linear driving force approximation for diffusion controlled adsorption in molecular sieve columns, AICHE J. 21 (1975) 200.

[18] M. Polanyi, Theory of adsorption of gases — introductory paper, Trans. Faraday Soc. 28 (1932) 316.

[19] D.M. Young, A.D. Crowell, Physical Adsorption of Gases, Butterworth, 1962.

[20] A. Mersmann, U. Munstermann, J. Schadl, Separation of gas mixtures by adsorption, Germ. Chem. Eng. 7 (1984) 137.

[21] H. Freundlich, Colloid and Capillary Chemistry, Methuen, London, 1926.

[22] E. Glueckauf, J.I. Coates, Theory of chromatography, J. Chem. Soc. 241 (1947) 1315.

[23] J.J. Kipling, Adsorption from Solutions of Non-Electrolytes, Academic Press, 1965.

[24] P.H. Emmett, The measurement of the surface areas of finely divided or porous solids by low temperature adsorption isotherms, in: Advances in Colloid Science, 1, Interscience Publishers, 1942, p. 1.

[25] L.H. Cohan, Sorption hysteresis and vapour pressure of concave surfaces, J. Am. Chem. Soc. 60 (1938) 433.

[26] R.W. Cranston, F.A. Inkley, The determination of pore structures from nitrogen adsorption isotherms, Adv. Catal. 9 (1957) 143.

[27] W.E. Ranz, W.R. Marshall, Evaporation from drops. Part II, Chem. Eng. Prog. 48 (4) (1952) 173.

[28] P.N. Dwivedi, S.N. Upadhey, Particle–fluid mass transfer in fixed and fluidized beds, Ind. Eng. Chem. Process. Des. Dev. 16 (1977) 157.

[29] N. Wakao, T. Funazkri, Effect of fluid dispersion coefficients on particle–fluid mass transfer coefficients in packed beds, Chem. Eng. Sci. 33 (1978) 1375.

[30] S. Glasstone, D. Lewis, Elements of Physical Chemistry, Macmillan, 1970.

[31] C.N. Satterfield, Mass Transfer in Heterogeneous Catalysis, MIT Press, Boston, 1970.

[32] R.B. Bird, W.E. Stewart, E.N. Lightfoot, Transport Phenomena, Wiley, New York, 1960.

[33] R.B. Evans, G.M. Watson, E.A. Mason, Gaseous diffusion in porous media at uniform pressure, J. Chem. Phys. 33 (1961) 2076.

[34] P. Schneider, J.M. Smith, Adsorption rate constants from chromatography, AICHE J. 14 (1968) 762.

[35] N. Epstein, On tortuosity and the tortuosity factor in flow and diffusion through porous media, Chem. Eng. Sci. 44 (1989) 777.

[36] C.N. Hinshelwood, The Kinetics of Chemical Change, Clarendon Press, Oxford, 1940.

[37] M.F. Edwards, J.F. Richardson, Gas dispersion in packed beds, Chem. Eng. Sci. 23 (1968) 109.

[38] J.H. Bowen, P.G. Rimmer, Design of fixed bed sorbers using a quadratic driving force equation, Trans. Inst. Chem. Eng. 50 (1972) 168.

[39] L. Lapidus, N.R. Amundson, Mathematics of adsorption in fixed beds – the rate determining steps in radial adsorption analysis, J. Phys. Chem. 56 (1952) 373. L. Lapidus, N. R. Amundson, The effect of longitudinal diffusion in ion exchange and chromatographic columns, J. Phys. Chem. 56 (1952) 984.

[40] O. Levenspiel, K.B. Bischoff, Patterns of flow in chemical process vessels, in: T.B. Drew, J.W. Hoopes, T. Vermeulen (Eds.), Advances in Chemical Engineering, vol. 4, Academic Press, 1963 (Chapter 95).

[41] J.B. Rosen, Kinetics of fixed bed systems for solid diffusion into spherical particles, J. Chem. Phys. 20 (1965) 387.

[42] F.W. Leavitt, Nonisothermal adsorption in large fixed beds, Chem. Eng. Prog. 58 (8) (1962) 54.

[43] N.R. Amundson, R. Aris, R. Swanson, On simple exchange waves in fixed beds, Proc. R. Soc. Lond. A 286 (1965) 129.

[44] J.W. Carter, On the regeneration of fixed adsorber beds, AICHE J. 21 (1975) 380.

[45] C. Berg, Hypersorption design, Chem. Eng. Prog. 47 (11) (1951) 585.

[46] S. Katell, Removing Sulphur dioxide from flue gases, Chem. Eng. Prog. 62 (1966) 67.

[47] C.G. Cartelyou, Commercial processes for SO2 removal, Chem. Eng. Prog. 65 (9) (1969) 69.

[48] D.B. Broughton, Molex, history of a process, Chem. Eng. Prog. 64 (8) (1968) 60.

[49] Johnson, J.A. and Kabza, R.G.: I. Chem. E. Annual Research Meeting, Swansea (1990) Sorbex: Industrial Scale Adsorptive Separation.

[50] C.W. Skarstrom, Use of adsorption phenomena in automatic plant-type gas analysis, Ann. N. Y. Acad. Sci. 72 (1959) 751.

[51] C.W. Skarstrom, Method and Apparatus for Fractionating a Gaseous Mixture by Adsorption. US Patent 2944627, 1960.

[52] P. Guerin de Montgareuil, D. Domine, Process for Separating a Binary Gaseous Mixture by Adsorption. US Patent 3155468, 1964.

[53] L.H. Shendalman, J.E. Mitchell, A study of heatless adsorption in the model system CO2 in He, Chem. Eng. Sci. 27 (1972) 1449.

[54] A. Acrivos, Method of characteristics technique. Application to heat and mass transfer problems, Ind. Eng. Chem. 45 (1956) 703.

[55] R.H. Wilhelm, A.W. Rice, R.W. Rolke, N.H. Sweed, Parametric pumping, Ind. Eng. Chem. Fundam. 7 (1968) 337.

[56] R.L. Pigford, B. Baker, D. Blum, An equilibrium theory of the parametric pump, Ind. Eng. Chem. Fundam. 8 (1969) 144.

[57] D. Platt, R. Lavie, Pressure cycle zone adsorption, Chem. Eng. Sci. 40 (1985) 733.

[58] P.C. Wankat, Cyclic separations—parametric pumping, pressure swing adsorption and cycling zone adsorption, in: A.I.Ch.E. Modular Instruction Series, Module B6.11, 1986.

Further reading

Adsorption and Ion Exchange: A.I.Ch.E. Symposium Series No 259, 83, 1987 (see also numbers 14, 24, 69, 74, 80, 96, 117, 120, 134, 152, 165, 219, 233, 242).

Adsorption and its Applications in Industry and Environmental Protection Vol 1 Applications in Industry Vol II Applications in Environmental Protection, Elsevier, Amsterdam, 1999.

D.O. Cooney, Adsorption Design for Wastewater Treatment, CRC Press, Lewis Publishers, Boca Raton, 1998.

N.P. Cheremisinoff, P.N. Cheremisinoff, Carbon Adsorption for Pollution Control, Prentice Hall, Englewood Cliff, New Jersey, 1993.

J. Karger, D.M. Ruthven, Diffusion in Zeolites and Other Microporous Solids, John Wiley & Sons, New York, 1992.

M.D. le Van (Ed.), Fundamentals of Adsorption V, Kluwer Academic Publishers, Norwell, 1996.

R.H. Perry, D.W. Green, J.O. Maloney (Eds.), Perry's Chemical Engineers' Handbook, seventh ed., McGraw-Hill Book Company, New York, 1997.

D.M. Ruthven, Principles of Adsorption and Adsorption Processes, Wiley, 1984.

D.M. Ruthven, S. Farooq, K.S. Knaebel, Pressure Swing Adsorption, VCH Publishers, New York, 1994.

F.L. Slejko, Adsorption Technology, Marcel Dekker, New York, 1985.

M. Suzuki (Ed.), Fundamentals of Adsorption IV, Kodansha, Tokyo, 1993.

M. Suzuki, Adsorption Engineering, Elsevier, Amsterdam, 1990.

W.J. Thomas, B.D. Crittenden, Adsorption, Technology and Design, Butterworth-Heinemann, Oxford, 1998.

C. Tien, Adsorption Calculations and Modeling, Butterworth, Boston, 1994.

D.Y. Valenzuela, A.I. Myers, Adsorption Equilibrium Data Handbook, Prentice Hall, Englewood Cliffs, 1989.

P.C. Wankat, Large Scale Adsorption and Chromatography (2 Vols), CRC Press, Boca Raton, 1986.

R.T. Yang, Gas Separation by Adsorption Processes, Butterworth, London, 1987.

R.T. Yang, Gas Separation by Adsorption Processes. Series on Chemical Engineering, vol. 1, World Scientific Publishing Co, 1997.

Ion exchange

Ajay Kumar Ray

Department of Chemical and Biochemical Engineering, University of Western Ontario,
London, ON, Canada

Nomenclature

		Units in SI system	Dimensions in M, N, L, T, θ, A
a_z	external surface area of resin per unit volume of bed	m^{-1}	$\mathbf{L}^{-1}$
b_A	slope of a linear sorption isotherm in Eq. (9.18)	–	–
C	concentration of counter-ions in the liquid phase	$kmol/m^3$	$\mathbf{NL}^{-3}$
$\overline{C}$	mean concentration of counter-ions in the liquid phase	$kmol/m^3$	$\mathbf{NL}^{-3}$
C_{Ab}	concentration of counter-ions **A** in the bulk liquid	$kmol/m^3$	$\mathbf{NL}^{-3}$
C_A^*	concentration of counter-ions **A** in equilibrium with the mean concentration of counter-ions in the resin	$kmol/m^3$	$\mathbf{NL}^{-3}$
C_i^*	concentration of counter-ions in the liquid phase, in equilibrium at the external surface of the resin	$kmol/m^3$	$\mathbf{NL}^{-3}$
C_n	concentration of counter-ions in the liquid leaving the nth stage	$kmol/m^3$	$\mathbf{NL}^{-3}$
C_0	concentration of counter-ions in the liquid initially, or in the feed	$kmol/m^3$	$\mathbf{NL}^{-3}$
C_s	concentration of counter-ions in the resin phase	$kmol/m^3$	$\mathbf{NL}^{-3}$

Continued

Coulson and Richardson's Chemical Engineering. https://doi.org/10.1016/B978-0-08-101097-6.00009-2

		Units in SI system	Dimensions in M, N, L, T, θ, A
C_s^*	concentration of counter-ions in the resin phase in equilibrium with the bulk liquid	kmol/m^3	$\mathbf{NL}^{-3}$
$\overline{C}_s$	mean concentration of counter-ions in the resin	kmol/m^3	$\mathbf{NL}^{-3}$
C_{s1}, $C_{s(n+1)}$	concentrations of counter-ions in resin streams leaving 1st, (n+1)th stage	kmol/m^3	$\mathbf{NL}^{-3}$
C_{s0}	concentration of counter-ions in the resin initially	kmol/m^3	$\mathbf{NL}^{-3}$
$C_{s\infty}$	ultimate concentration of counter-ions in the resin	kmol/m^3	$\mathbf{NL}^{-3}$
D_A	diffusivity of species $\mathbf{A}$	m^2/s	$\mathbf{L}^2\,\mathbf{T}^{-1}$
D_R	diffusivity in the resin	m^2/s	$\mathbf{L}^2\,\mathbf{T}^{-1}$
D^*	distribution coefficient $VC_0/R_vC_{s\infty}$	–	–
D_1^*	distribution coefficient $\dot{V}C_0/\dot{R}_vC_{s1}$	–	–
$\mathbf{F}$	faraday constant	9.6487×10^7 C/kmol	$\mathbf{N}^{-1}\,\mathbf{TA}$
$f()$	various functions	–	–
K	ion exchange equilibrium constant	–	–
K_c	selectivity coefficient	–	–
K_i	equilibrium constant in Eq. (9.36)	–	–
KCa^+Na^+	selectivity between calcium ions and sodium ions in a cationic resin	–	–
k	velocity constant in Eq. (9.21)	m^3/kmol s	$\mathbf{N}^{-1}\,\mathbf{L}^3\,\mathbf{T}^{-1}$
k_l	liquid-film mass transfer coefficient	m/s	$\mathbf{LT}^{-1}$
k_p	hypothetical solid 'film' mass transfer coefficient	m/s	$\mathbf{LT}^{-1}$
m	$\varepsilon/(1-\varepsilon)$	–	–
N_A, N_B	molar fluxes of $\mathbf{A}$, $\mathbf{B}$	kmol/m^{2}s	$\mathbf{NL}^{-2}\,\mathbf{T}^{-1}$
n	an index (Eq. 9.11) *or* number of stages (Fig. 9.5)	–	–
$\mathbf{R}$	gas constant	8314 J/kmol K	$\mathbf{MN}^{-1}\,\mathbf{L}^2\,\mathbf{T}^{-2}\dot{\mathbf{e}}^{-1}$
R_c^-	cationic resin	–	–
R_v	volume of resin	m^3	$\mathbf{L}^3$
$\dot{R}_v$	volumetric flow rate of resin	m^3/s	$\mathbf{L}^3\,\mathbf{T}^{-1}$
r	radius within a spherical pellet	m	$\mathbf{L}$

Continued

		Units in SI system	Dimensions in M, N, L, T, θ, A
r_i	outside radius of a spherical pellet	m	**L**
r^*	equilibrium parameter	–	–
T	temperature	K	θ
t	time	s	**T**
$t_{f1/2}$	time for half saturation assuming film diffusion control	s	**T**
$t_{p1/2}$	time for half saturation assuming pellet diffusion control	s	**T**
V	volume of liquid	m^3	L^3
$\dot{V}$	volume flow rate of liquid	m^3/s	**L**^{3}**T**$^{-1}$
x	ionic fraction in the liquid, C/C_0	–	–
y	ionic fraction in the resin, $C_S/C_{s\infty}$	–	–
y'	ionic ratio in the resin, C_S/C_{s1}	–	–
Z	distance along a fixed bed	m	**L**
a_B^A	separation factor of **A** relative to **B**	–	–
γ, γ_s	activity coefficient for liquid, resin	–	–
ε	inter-pellet voidage	–	–
v	valence	–	–
ϕ	electric potential	V	**ML**^{2}**T**$^{-3}$**A**$^{-1}$
κ	constant in Eq. (9.14)	–	–
χ	distance parameter in Eq. (9.34)	–	–
τ	time parameter in Eq. (9.34)	–	–

9.1 **Introduction**

Ion exchange is a unit operation in its own right, often sharing theory with adsorption or chromatography, although it has its own special areas of application. The oldest and most enduring application of ion exchange is in water treatment, to soften or demineralise water before industrial use, to recover components from an aqueous effluent before it is discharged or recirculated, and this is discussed by Arden [1]. Ion exchange may also be used to separate ionic species in various liquids as discussed by Helffrich [2] and Schweitzer [3]. Ion exchangers can catalyse specific reactions or be suitable to use for chromatographic separations, although these last two applications are not discussed in this chapter. Applications of ion exchange membranes are considered in Chapter 8.

The modern history of ion exchange began in about 1850 when two English chemists, Thompson [4] and Way [5], studied the exchange between ammonium ions in fertilisers and calcium ions in soil. The materials responsible for the exchange were shown later to be naturally occurring alumino-silicates [6]. History records very much earlier observations of the phenomenon and, for example, Aristotle [7], in 330 BC, noted that sea-water loses some of its salt when allowed to percolate through some sands. Those who claim priority for Moses [8] should note however that the process described may have been adsorption!

In the present context, the *exchange* is that of equivalent numbers of similarly charged ions, between an immobile phase, which may be a crystal lattice or a gel, and a liquid surrounding the immobile phase. If the exchanging ions are positively charged, the ion exchanger is termed *cationic*, and *anionic* if they are negatively charged. The rate at which ions diffuse between an exchanger and the liquid is determined, not only by the concentration differences in the two phases, but also by the necessity to maintain electroneutrality in both phases.

As well as occurring naturally, alumino-silicates are manufactured. Their structure is that of a framework of silicon, aluminium and oxygen atoms. If the framework contains water, then this may be driven off by heating, leaving a porous structure, access to which is controlled by 'windows' of precise molecular dimensions. Larger molecules are excluded, hence the description 'molecular sieve' as discussed in Chapter 17.

If the framework had originally contained not only water but also a salt solution, the drying process would leave positive and negative ions in the pores created by the loss of water. When immersed in a polar liquid, one or both ions may be free to move. It is often found that only one polarity of ion moves freely, the other being held firmly to the framework. An exchange of ions is then possible between the mobile ions in the exchanger and ions with like-charge in the surrounding liquid, as long as those ions are not too large and electro-neutrality is maintained.

9.2 Ion exchange resins

A serious obstacle to using alumino-silicates as ion exchangers, is that they become unstable in the presence of mineral acids. It was not possible, therefore, to bring about exchanges involving hydrogen ions until acid-resisting exchangers had been developed. First, sulphonated coal and, later, synthetic phenol formaldehyde were shown to be capable of cation exchange. Nowadays, cross-linked polymers, known as resins, are used as the basic framework for most ion exchange processes, both cationic and anionic.

The base resin contains a styrene–divinylbenzene polymer, DVB. If styrene alone were used, the long chains it formed would disperse in organic solvents. The divinylbenzene provides cross-linking between the chains. When the cross-linked structure is immersed in an organic solvent, dispersion takes place only to the point at which the osmotic force of solvation is balanced by the restraining force of the stretched polymer structure.

When the styrene–DVB polymer is sulphonated, it becomes the cation exchanger which is polystyrene sulphonic acid, with exchangeable hydrogen ions. The framework of the resin has a fixed negative charge, so that no exchange can occur with the mobile negative ions outside the resin. Ions of the same polarity as the framework are termed *co-ions*. Those of opposite polarity have the potential to exchange and are called *counter-ions*. Resins for anion exchange may also be manufactured from polystyrene as a starting material, by treating with monochloroacetone and trimethylamine, for example. The structures of these particular resins are shown in Fig. 9.1.

Both resins can be described as strongly ionic. Each is fully ionised so that all the counter-ions within the gel may be exchanged with similarly charged ions outside the gel, whatever the concentration of the latter.

9.3 Resin capacity

Various measures of the capacity of a resin for ion exchange are in common use. The *maximum capacity* measures the total number of exchangeable ions per unit mass of resin, commonly expressed in milliequivalents per gram (meq/g). The base unit of a polystyrenesulphonic-acid polymer, as shown in Fig. 9.1, has a molecular weight of approximately 184 kg/kmol. Each unit has one exchangeable hydrogen ion, so its maximum capacity is (1000/184) or 5.43 milliequivalents per gram.

The capacities of styrene-based anion exchangers are not so easily calculated because there may not be an anionic group on every benzene ring. Values of 2.5 to 4.0 meq/g are typical for strong anion resins.

It is the number of fixed ionic groups which determines the maximum exchange capacity of a resin although the extent to which that capacity may be exploited depends also on the chemical nature of those groups. Weak acid groups such as the carboxyl ion, COO–, ionise only at high pH. At low pH, that is a high concentration of hydrogen ions, they form undissociated COOH. Weak base groups such as NH_3^+ lose a proton when the pH is high, forming uncharged NH_2 ions. Consequently, for resins which are weakly ionic, the exploitable capacity depends on the pH of the liquid being treated. Fig. 9.2 illustrates the expected dependence.

When the resin is incompletely ionised, its *effective capacity* will be less than the maximum. If equilibrium between resin and liquid is not achieved, a *dynamic capacity* may be quoted which will depend on the contact time. When equipment is designed to contain the resin, it is convenient to use unit volume of water-swollen resin as the basis for expressing the capacity. For fixed-bed equipment, the *capacity at breakpoint* is sometimes quoted. This is the capacity per unit mass of bed, averaged over the whole bed, including the ion exchange zone, when the breakpoint is reached.

9.4 Equilibrium

The equilibrium distribution of ions between resin and liquid phases depends on many factors. As well as temperature, the degree of ionisation of solvent, solute and resin may be important although, to simplify the discussion, it is assumed that the resin is fully ionised (strong acid or strong base), that the solvent is not ionised

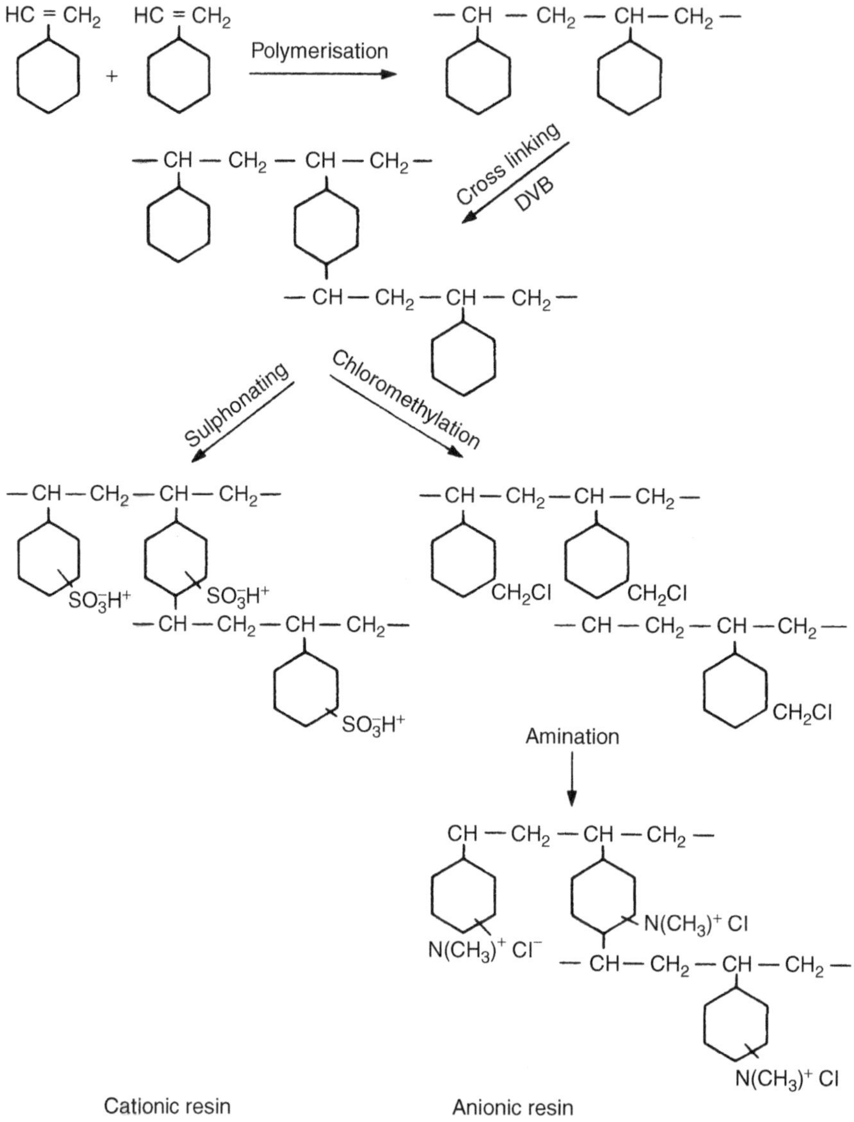

Cationic resin   Anionic resin

FIG. 9.1

Formation of styrene-based cationic and anionic resins.

and that the solute is completely ionised. Only ion exchange itself is considered, although adsorption on to the resin surface is possible as well as diffusion into the resin of neutral groups of ions and of uncharged molecules.

Freed of other restrictions, a mobile ion may be expected to diffuse down any concentration gradient that exists between porous solid and liquid. In the particular

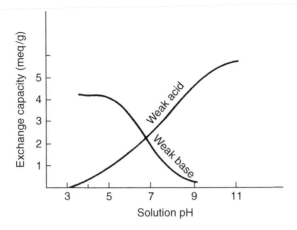

FIG. 9.2

Exchange capacity of weak resins.

case of ion exchange, there is an additional requirement that the resin and liquid phases should remain electrically neutral. Any tendency for molecules to move in such a way as to disturb this neutrality will generate a large electrostatic potential opposing further movement, known as the *Donnan potential*.

Mobile co-ions are confined almost entirely to the liquid phase. A few, however, may have diffused into the resin accompanied by neutralising counter-ions. The net effect is an increase in the number of ions in the resin, causing it to swell and increasing its exchange capacity slightly above that which arises from the fixed ionic groups alone. Swelling is a reproducible equilibrium characteristic of a resin, depending on its degree of cross-linking and its ion exchange capacity, as well as temperature and the solution composition. Polyvalent exchanging ions create more cross-linking within the resin and, therefore, produce less swelling than monovalent ions. $Al^{3+} < Ca^{2+} < Na^+$. Water swelling is generally the result of the hydration of ionic groups.

In considering ionic equilibria, it is convenient to write the exchange process in the form of a chemical equation. For example, a water-softening process designed to remove calcium ions from solution may be written as:

$$Ca^{++} + 2Cl^- + 2Na^+R_c^- = 2Na^+ + 2Cl^- + Ca^{++}(R_c^-)_2$$

where R_c is a cationic exchange resin. When its capacity is depleted, the resin may be regenerated by immersing it in sodium chloride solution so that the reverse reaction takes place.

In general, the exchange of an ion **A** of valency v_A in solution, for an ion **B** of valency v_B on the cationic resin may be written as:

$$v_B C_A + v_A C_{SB}(R_c)_{v_B} = v_A C_B + v_B C_{SA}(R_c)_{v_A} \qquad (9.1)$$

Including activity coefficients γ, the thermodynamic equilibrium constant K becomes:

$$K = \frac{(\gamma_B C_B)^{\upsilon_A}(\gamma_{SA} C_{SA})^{\upsilon_B}}{(\gamma_A C_A)^{\upsilon_B}(\gamma_{SB} C_{SB})^{\upsilon_A}}$$
$$= \frac{(\gamma_B)^{\upsilon_A}(\gamma_{SA})^{\upsilon_B}}{(\gamma_A)^{\upsilon_B}(\gamma_{SB})^{\upsilon_A}} K_c \qquad (9.2)$$

where K_c is the selectivity coefficient, a measure of preference for one ionic species. Defining ionic fractions as:

then:

$$x_A = C_A/C_0 \text{ and } y_A = C_{SA}/C_{S\infty}$$
$$K_c = \frac{C_B^{\upsilon_A} C_{SA}^{\upsilon_B}}{C_A^{\upsilon_B} C_{SB}^{\upsilon_A}}$$
$$= \frac{(y_A/x_A)^{\upsilon_B}}{(y_B/x_B)^{\upsilon_A}} \left(\frac{C_0}{C_{S\infty}}\right)^{\upsilon_A - \upsilon_B} \qquad (9.3)$$

where: C_0 is the total ionic strength of the solution, and:

$C_{S\infty}$ is the exchangeable capacity of the resin.

In a dilute solution, the activity coefficients approach unity and K_c approaches K. For the water softening, represented by the equation given previously, $\upsilon_A = 2$ and $\upsilon_B = 1$.

Hence:

$$K_c = \frac{y_A/x_A}{(y_B/x_B)^2} \frac{C_0}{C_{S\infty}}$$
$$= \frac{y_A/x_A}{[(1-y_A)/(1-x_A)]^2} \frac{C_0}{C_{S\infty}} \qquad (9.4)$$

Except when $\upsilon_A = \upsilon_B$, the selectivity coefficient depends on the total ionic concentrations of the resin and the liquid phases.

Another measure of the preference of an ion exchanger for one other ionic species is the *separation factor* α. This is defined in a similar way to relative volatility in vapour–liquid binary systems, and is independent of the valencies of the ions.

Thus:

$$\alpha_B^A = \frac{y_A/x_A}{y_B/x_B} \qquad (9.5)$$

When K_c is greater than unity, the exchanger takes up ion **A** in preference to ion **B**. In general, the value of K_c depends on the units chosen for the concentrations although it is more likely than α to remain constant when experimental conditions change. When **A** and **B** are monovalent ions, $K_c = \alpha$.

Equilibrium relationships may be plotted as y against x diagrams using Eq. (9.4). Fig. 9.3 shows such a plot for $\upsilon_A = 2$ and $\upsilon_B = 1$. The group $K_c(C_{S\infty}/C_0)^{\upsilon_A - \upsilon_B}$ has values of 0.01 to 100.

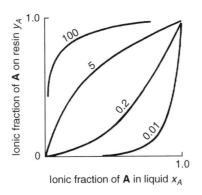

FIG. 9.3

Equilibrium isotherm for the ion exchange $A + 2B(S) = A(S) + 2B$. The parameters are values of the group $K_c C_{s8}/C_1$.

As is the case with adsorption isotherms, those curves in Fig. 9.3 which are concave to the concentration axis for the mobile phase are termed *favourable* and lead to self-sharpening ion exchange waves.

Deciding which of several counter-ions will be preferably exchanged may be difficult without experimental work, although some general guidance may be given. The Donnan potential results in counter-ions with a high valency being exchanged preferentially. If there is a specific interaction between a counter-ion and a fixed ionic group, that ion will be preferred. Ions may be preferred because of their small size or shape.

Approximate selectivity coefficients for the exchange of various cations for lithium ions on a sulphonated polystyrene, a typically strong acid resin, are given in Table 9.1. The values are relative to $Li = 1.0$. The selectivity coefficient between two ions is the ratio of their selectivities relative to lithium. Hence, for a sodium–hydrogen exchange:

$$K_{Na^+ H^+} = \left(\frac{2}{1.3}\right) \approx 1.5$$

Because these ions have the same valency, the selectivity coefficient is equal to the separation factor and a value greater than unity indicates that Na^+ adheres to the resin in preference to H^+. The same procedure may be used for exchange between di- and mono-valent ions although its validity is more questionable.

Thus:

$$K_{Ca^{++} Na^+} = \left(\frac{5.2}{2.0}\right) = 2.6$$

A similar table, shown in Table 9.2, is available for anion exchange although this is based on fewer data [3]. The problem is complicated by the fact that there are two

Table 9.1 Selectivities on 8% cross-linked strong acid resin for cations. Values are relative to lithium [3].

Li^+	1.0	Zn^{2+}	3.5
H^+	1.3	Co^{2+}	3.7
Na^+	2.0	Cu^{2+}	3.8
NH_4^+	2.6	Cd^{2+}	3.9
K^+	2.9	Ba^{2+}	4.0
Rb^+	3.2	Mn^{2+}	4.1
Cs^+	3.3	Ni^{2+}	3.9
Ag^+	8.5	Ca^{2+}	5.2
UO_2^{2+}	2.5	Sr^{2+}	6.5
Mg^{2+}	3.3	Pb^{2+}	9.9
		Ba^{2+}	11.5

Table 9.2 Selectivities on strong base resin [3].

I^-	8	HCO_3^-	0.4
NO_3^-	4	CH_3COO^-	0.2
Br^-	3	F^-	0.1
HSO_4^-	1.6	OH^- (II)	0.06
NO_2^-	1.3	SO_4^{2-}	0.15
CN^-	1.3	CO_3^{2-}	0.03
Cl^-	1.0	HPO_4^{2-}	0.01
BrO_3^-	1.0		
OH^- (I)	0.65		

types of functional structure used in strong anion exchange resins and that anions in solution may exist in complex form. Nevertheless, the table provides some guidance when several systems are being compared.

9.4.1 Ion exclusion and retardation

As well as being used for ion exchange, resins may be used to separate ionic and non-ionic solutes in aqueous solution. A packed bed of resin is then filled with water and a sample of solution added. If water is then drained from the bed as more water is added to the top, the sample is eluted through the column. If the Donnan potential prevents the ionic components from entering the resin, there will be no effect on the non-ionic species. When a solution of HCl and CH_3COOH is eluted through a bed of hydrogen and chloride resin, the HCl appears first in the effluent, followed by the CH_3COOH.

The process described is referred to as *ion-exclusion* as discussed by Asher and Simpson [9]. The resins used are normal and the non-ionic molecules are assumed to be small enough to enter the pores. When large non-ionic molecules are involved, an alternative process called *ion-retardation* may be used, as discussed by Hatch et al. [10]. This requires a special resin of an amphoteric type known as a *snake cage polyelectrolyte. The* polyelectrolyte consists of a cross-linked polymer physically entrapping a tangle of linear polymers. For example, an anion exchange resin which is soaked in acrylic acid becomes entrapped when the acrylic acid is polymerised. The intricacy of the interweaving is such that counter-ions cannot be easily displaced by other counter-ions. On the other hand, ionic mobility within the resin maintains the electro-neutrality. The ionic molecule as a whole is absorbed by the resin in preference to the non-ionic molecule. When a solution of NaCl and sucrose is treated by the method of ion-retardation, the sucrose appears first in the effluent.

9.5 **Exchange kinetics**

It is insufficient to have data on the extent of the ion exchange at equilibrium only. The design of most equipment requires data on the amount of exchange between resin and liquid that will have occurred in a given contact time. The resistances to transfer commonly found in such a system are discussed in Chapter 17. It is necessary to consider the counter-diffusion of ions through a boundary film outside the resin and through the pores of the resin. The ion exchange process on the internal surface does not normally constitute a significant resistance.

It is theoretically possible that equilibrium between liquid and resin will be maintained at all points of contact. Liquid and solid concentrations are then related by the sorption isotherm. It is usual, however, that pellet or film diffusion will dominate or 'control' the rate of exchange. It is also possible that control will be mixed, or will change as the ion exchange proceeds. In the latter case, the initial film-diffusion control will give way to pellet-diffusion control at a later stage.

9.5.1 **Pellet diffusion**

Ions moving through the body of an exchanger are subject to more constraints than are molecules moving through an uncharged porous solid. If the exchanging ions are equivalent and are of equal mobility, the complications are relatively minor and are associated with the tendency of the exchanger to swell and of some neutral groups of ions to diffuse. In the general case of ions with different valencies and mobilities, however, allowances have to be made for a diffusion potential arising from electrostatic differences, as well as the usual driving force due to concentration differences.

Exchange between counter-ions **A** in beds of resin and counter-ions **B** in a well-stirred solution may be represented by the Nernst–Planck equation as:

$$N_A = (N_A)_{\text{diff}} + (N_A)_{\text{elec}} = -D_A \left(\text{grad } C_A + \frac{v_A C_A \mathbf{F}}{RT} \text{grad } \varphi \right) \quad (9.6)$$

and similarly for **B**, where ϕ is the electrical potential and **F** is the Faraday constant. The requirements of maintaining electroneutrality and no net electric current may be expressed as:
Thus:

$$
\left.\begin{array}{l}
v_A N_A + v_B N_B = 0 \\[4pt]
v_A C_A + v_B C_B = \text{constant} \\[4pt]
v_A \,\mathrm{grad}\, C_A + v_B \,\mathrm{grad}\, C_B = 0
\end{array}\right\}
\tag{9.7}
$$

and Eq. (9.6) may be written as:

$$
N_A = -\left[\frac{D_A D_B \left(v_A^2 C_A + v_B^2 C_B\right)}{v_A^2 C_A D_A + v_B^2 C_B D_B}\right] \mathrm{grad}\, C_A
\tag{9.8}
$$

The term in the square bracket is an effective diffusion coefficient D_{AB}. In principle, this may be used together with a material balance to predict changes in concentration within a pellet. Algebraic solutions are more easily obtained when the effective diffusivity is constant. The conservation of counter-ions diffusing into a sphere may be expressed in terms of resin-phase concentration C_{Sr}, which is a function of radius and time.
Thus:

$$
\frac{\partial C_{Sr}}{\partial t} = \frac{1}{r^2} \frac{\partial}{\partial r}\left(r^2 D_R \frac{\partial C_{Sr}}{\partial r}\right)
\tag{9.9}
$$

where D_R is the diffusivity referred to concentrations in the resin phase. If this is constant, Eq. (9.9) can be rewritten as:

$$
\frac{\partial C_{Sr}}{\partial t} = D_R\left(\frac{\partial^2 C_{Sr}}{\partial r^2} + \frac{2}{r}\frac{\partial C_S}{\partial r}\right)
\tag{9.10}
$$

This equation has been solved by Eagle and Scott [11] for conditions of constant concentration outside the sphere and negligible resistance to mass transfer in the boundary film. The solution may be written in terms of a mean concentration through a sphere C_s, which is a function of time only, to give:

$$
\frac{C_S - C_{S0}}{C_S^* - C_{S0}} = 1 - \frac{6}{\pi^2}\sum_{n=1}^{\infty}\frac{1}{n^2}\exp\left[-\left(D_R \pi^2 t\right)/n^2 r_i^2\right]
\tag{9.11}
$$

where C_{S0} is the initial concentration on the resin and C_S^* is the concentration on the resin in equilibrium with C_0, the constant concentration in the solution.

When t is large, the summation may be restricted to one term and the equation becomes:

$$
\frac{C_S - C_{S0}}{C_S^* - C_{S0}} = 1 - \frac{6}{\pi^2}\exp\left[-\left(D_R \pi^2 t\right)/r_i^2\right]
\tag{9.12}
$$

The corresponding rate equation may be found by taking the derivative with respect to time and rearranging to give:

$$\frac{dC_S}{dt} = \frac{\pi^2 D_R}{r_i^2}\left(C_S^* - C_S\right) \tag{9.13}$$

Equation (9.13) is a *linear driving force* equation. Vermeulen [12] has suggested the following form that more accurately represents experimental data:

$$\frac{dC_S}{dt} = \frac{\kappa D_R}{r_i^2}\frac{\left(C_S^{*2} - C_S^2\right)}{2(C_S - C_{S0})} \tag{9.14}$$

This is known as the *quadratic driving force* equation. A plot of $\ln[1 - (C_S/C_S^*)^2]$ against t gives a straight line and the diffusion factor $\kappa D_S/r_i^2$ may be obtained from the slope.

When ion exchange involves ions of different mobilities, the rate depends also on the relative positions of the ions. If the more mobile ion is diffusing out of the resin, the rate will be greater than if it is diffusing into the resin, when pellet-diffusion controls.

Example 9.1

A single pellet of alumina is exposed to a flow of humid air at a constant temperature. The increase in mass of the pellet is followed automatically, yielding the following results:

t (min)	2	4	10	20	40	60	120
x_r (kg/kg)	0.091	0.097	0.105	0.113	0.125	0.128	0.132

Assuming the effect of the external film is negligible, predict time t against x_r values for a pellet of twice the radius, where x_r is the mass of adsorbed phase per unit mass of adsorbent.

Solution

When pellet diffusion is dominant, assuming $C_{s0}=0$ and the fluid concentration outside the pellet is constant, then:

$$\frac{C_S}{C_S^*} = 1 - \frac{6}{\pi^2}\exp\left[-\left(D_R\pi^2 t/r_i^2\right)\right] \tag{9.12}$$

Hence a plot of $\ln\left[1 - \frac{C_s}{C_s^*}\right]$ against t should be linear.

For $r = r_i$, with $C_s^* = 0.132\,\text{kg/kg}$ then:

t (min)	C_s (kg/kg)	(C_s/C_s^*)	$1-(C_s/C_s^*)$
2	0.091	0.69	0.31
4	0.097	0.73	0.27
10	0.105	0.80	0.20
20	0.113	0.86	0.14
40	0.125	0.95	0.05
60	0.128	0.97	0.03
120	0.132	1.0	0

These data are plotted in Fig. 9.4, which confirms the linearity and from which:

$$\pi^2 D_R / r_i^2 = 0.043$$

For a pellet of twice the radius, that is $r = 2r_i$.
and the slope $= (-0.043/4) = -0.011$.
Thus, when the radius $= 2r_i$:

$$C_s/C_s^* = 1 - \left(6/\pi^2\right) \exp\left(0.011t\right) \tag{i}$$

Alternatively, use may be made of the quadratic driving force equation:

$$\frac{dC_s}{dt} = \frac{\kappa D_R}{r_i^2} \frac{\left(C_S^{*2} - C_S^2\right)}{2(C_S - C_{S0})} \tag{9.14}$$

Integrating from the initial condition, $t = 0$ and $C_s = 0$, then:

$$C_S/C_S^* = \left[1 - \exp\left(-\frac{\kappa D_R t}{r_i^2}\right)\right]^{0.5}$$

indicating that a plot of $\ln[1 - (C_s/C_s^*)^2]$ against t should also be linear. Thus:

t(min)	C_s (kg/kg)	(C_s/C_s^*)	$1 - (C_s/C_s^*)^2$
2	0.091	0.69	0.52
4	0.097	0.73	0.47
10	0.105	0.80	0.36
20	0.113	0.86	0.26
40	0.125	0.95	0.10
60	0.128	0.97	0.06
120	0.132	1.0	0

This is shown in Fig. 9.4 from which:

$$\kappa D_R / r_i^2 = 0.04$$

For a pellet twice the size:

$$C_s/C_S^* = [1 - \exp(-0.01t)]^{0.5} \tag{ii}$$

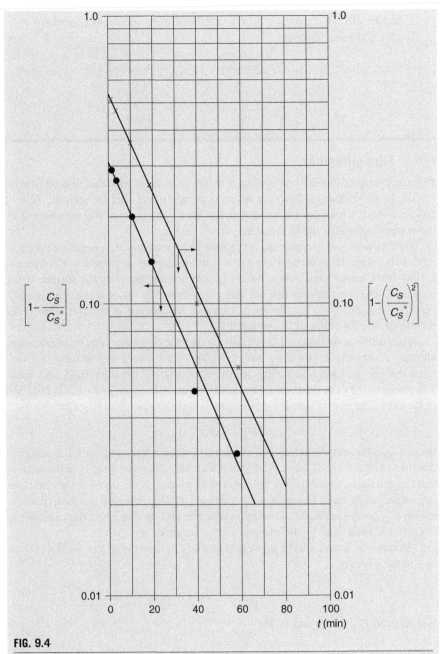

FIG. 9.4

Data for Example 9.1.

Values of C_s for radius $= 2r_i$ are calculated from equations (i) and (ii) to give the following results:

t(min)	C_S (kg/kg) Equation (i)	Equation (ii)
4	0.055	0.026
20	0.068	0.056
60	0.091	0.088

9.5.2 Film diffusion

Diffusion through liquid films is usually better understood than that through porous bodies. In ion exchange, however, there is an additional flux through the film of mobile co-ions which are not present in the resin. The co-ions will be affected by the relative mobilities of the counter-ions.

If a cationic resin contains the more mobile counter-ion **A,** a negative potential tends to build up at the outer surface of the resin and co-ions are repelled. Conversely, a slow resin counter-ion will result in co-ion concentration at the surface being increased. The net effect is that the rate of exchange is faster if the more mobile counter-ion is diffusing into the resin, that is if film-diffusion controls. This is the reverse of that for pellet-diffusion control.

When diffusion is assumed to be controlled by the boundary film, by implication, all other resistances to diffusion are negligible. Therefore, concentrations are uniform through the solid and local equilibrium existing between fluid and solid. The whole of the concentration difference between bulk liquid and solid is confined to the film. The rate of transfer into a spherical pellet may then be expressed as:

$$4\pi r_i^2 k_l \left(C_{Ab} - C_A^*\right) \tag{9.15}$$

where C_{Ab} is the concentration of the molecular species **A** in the liquid. It is assumed that the volume of liquid is large compared with the exchange capacity of the resin so that C_{Ab} remains constant. C_A^* is the concentration of **A** in the liquid at the outer surface of the pellet and it is assumed that this is in equilibrium with the mean concentration C_{SA} on the pellet, an assumption which is strictly true only when transfer to the pellet is controlled by the external film-resistance.

The rate which may also be expressed as a rate of increase of that molecular species in the pellet is:

$$\frac{\mathrm{d}}{\mathrm{d}t} \left(\frac{4}{3}\pi r_i^3 C_{SA}\right) \tag{9.16}$$

Hence, from Eqs (9.15) and (9.16):

$$\frac{\mathrm{d}C_{SA}}{\mathrm{d}t} = \frac{3k_l}{r_i}\left(C_{Ab} - C_A^*\right). \tag{9.17}$$

and:

$$C_{SA} = \mathrm{f}(C_A), \text{the sorption isotherm.}$$

If the equilibrium relationship between C_A and C_{SA} is linear so that $C_A^* = (1/b_A)C_{SA}$ and $C_{Ab} = (1/b_A)C_{S\infty}$, then Eq. (9.17) becomes:

$$\frac{dC_{SA}}{dt} = \frac{3k_l}{r_i b_A}(C_{s\infty} - C_{SA})$$
(9.18)

For a solid initially free of **A,** Eq. 9.18 may be integrated to give:

$$\ln\left[\frac{C_{S\infty} - C_{SA}}{C_{S\infty}}\right] = \frac{-3k_l}{r_i b_A}t$$
(9.19)

The situation is more complicated when charged ions rather than uncharged molecules are transferring. In this case, a Nernst–Planck equation which includes terms for both counter-ions and mobile co-ions must be applied. The problem may be simplified by assuming that the counter-ions have equal mobility, when the relationship is:

$$\ln\left[\frac{C_{S\infty} - C_{SA}}{C_{S\infty}}\right] + \left(1 - \frac{1}{\alpha_B^A}\right)\left(\frac{C_{SA}}{C_{S\infty}}\right) = \frac{-3k_l}{r_i b_A}t$$
(9.20)

where α_B^A is the separation factor, which equals b_A/b_B. Equations 9.19 and 9.20 are identical when $\alpha_B^A = 1$. More complex systems are discussed by Crank [13].

9.5.3 Ion exchange kinetics

In a theory of fixed bed performance for application to ion exchange columns, Thomas [14] assumed that the rate was controlled by the ion-exchange step itself. A rate equation may be written as:

$$\frac{dC_S}{dt} = k\left[C(C_{S\infty} - C_S) - \frac{1}{K_i}C_S(C_0 - C)\right]$$
(9.21)

where: k is the forward velocity constant of the exchange,
C_{S8} is the total concentration of exchangeable ion in the resin,
C, C_S are fluid and resin concentrations of counter-ion,
C_0 is the initial concentration in the fluid, and.
K_i is an equilibrium constant.

Although, in practice, ion exchange kinetics are unlikely to limit the rate, the solutions proposed by Thomas may be adapted to represent other controlling mechanisms, as discussed later.

9.5.4 Controlling diffusion

Whether film or pellet diffusion is rate-determining may be found experimentally. A pellet is immersed in an ionic solution and the change in concentration of the solution is measured with time. Before the exchange is complete, the pellet is taken out of the solution and held in air for a short period. If, after returning the pellet to the solution, the change in concentration continues smoothly from where it had stopped, then the rate of ion exchange is controlled by film-diffusion. If the resumed rate is higher than when the pellet was removed, the process is pellet-diffusion controlled. The buildup of concentration at the outer edges of the pellet which occurs when diffusion through the pellet is difficult, has been given time to disperse.

Various criteria have been developed to indicate whether film- or pellet-diffusion will be controlling. In one, proposed by Helffrich and Plesset [15], the times are compared for a pellet to become half-saturated under the hypothetical conditions of either film-diffusion control, $t_{f(1/2)}$, or pellet-diffusion control, $t_{p(1/2)}$.

From Eq. (9.20):

$$t_{f(1/2)} = (0.167 + 0.064\,\alpha_B^A) \frac{r_i b_A}{k_l \alpha_B^A} \tag{9.22}$$

From Eq. (9.11) and limiting the summation to the first order term in t, then:

$$t_{p(1/2)} = 0.03 \frac{r_i^2}{D_R} \tag{9.23}$$

If $t_{f(1/2)} > t_{p(1/2)}$, then film-diffusion controls, and conversely. In an alternative approach proposed by Rimmer and Bowen [16], it was recommended that film-diffusion should be assumed to control until the rate predicted by Eq. (9.14) is less than that predicted by Eq. (9.17) when $C_A^* = 0$.

9.6 Ion exchange equipment

Equipment for ion exchange is selected on the basis of the method to be used for regenerating the spent resin. Regeneration has to be carried out with the minimum disruption of the process and at a minimum cost. At its simplest, equipment may consist of a vessel containing the liquid to be treated, possibly fitted with stirrer to ensure good mixing. Ion exchange beads, a few millimetres in diameter, are added. Counter-ions diffuse from the liquid to the resin against a counterflow of ions diffusing from resin to liquid. Rates are such as to keep both resin and solution electrically neutral.

9.6.1 Staged operations

In the simple batch process, conservation of counter-ions leaving the liquid may be written as:

$$V(C_0 - C) = R_v(C_S - C_{S0}) \tag{9.24}$$

where V and R_v refer to initial volumes of liquid and resin.

Hence:

$$C_S = \frac{-V}{R_v} C + \left(\frac{V}{R_v} C_0 + C_{S0} \right) \tag{9.25}$$

If the batch process behaves as an equilibrium stage, the phases in contact will achieve equilibrium.

If the equilibrium relationship is known, then:

$$C_S^* = f(C) \tag{9.26}$$

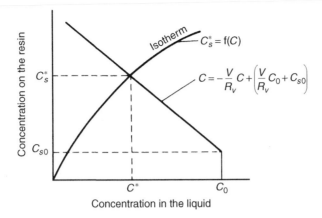

FIG. 9.5

Graphical solution for a single batch stage.

and Eqs (9.25) and (9.26) may be solved. In Fig. 9.5, it is assumed that V and R_v remain constant. It is sometimes convenient to use the fractional concentrations:

$$y = \frac{C_S}{C_{S\infty}} \quad \text{and} \quad x = \frac{C}{C_0} \tag{9.27}$$

where $C_{S\infty}$ is the maximum concentration of counter-ions on the resin.

Thus:

$$y = -D^* x + D^* + y_0 \tag{9.28}$$

where a distribution coefficient D^* is defined as:

$$D^* = VC_0/R_v C_{S\infty} \tag{9.29}$$

A plot of y against x gives a straight line of slope $-D^*$, passing through the point $(1, y_0)$. The intercept of the line with the equilibrium curve $y^* = f(x)$ gives the equilibrium condition that will be achieved in a single stage of mixing, starting from concentrations $(1, y_0)$, using volumes V and R_v of liquid and resin respectively.

If liquid and resin flow through a series of equilibrium stages at constant rates $\dot{V}$ and $\dot{R}_v$ respectively, as shown in Fig. 9.6, a mass balance over the first n stages gives:

$$\dot{V}C_0 + \dot{R}_v C_{S(n+1)} = \dot{R}_v C_{S1} + \dot{V}C_n \tag{9.30}$$

Equation (9.30) may be written as:

where:

$$y'_{n+1} = D_1^* x_n + \left(1 - D_1^*\right)$$

$$y' = \frac{C_S}{C_{S1}} \tag{9.31}$$

$$D_1^* = \dot{V}C_0/\dot{R}_v C_{S1}$$

If the ion exchange equilibrium isotherm is written in the form:

$$y'^* = f(x) \tag{9.32}$$

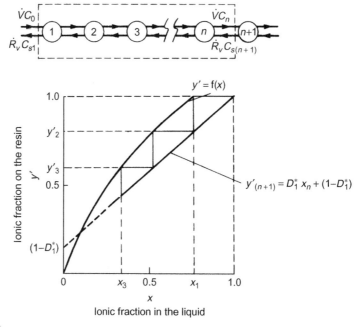

FIG. 9.6

Multistage countercurrent ion-exchange.

the total exchange obtained from a series of countercurrent equilibrium stages may be found by stepping off between the operating line, Eq. (9.31), and the equilibrium line, Eq. (9.32), as shown in Fig. 9.6.

9.6.2 Fixed beds

Most ion exchange operations are carried out in fixed beds of resin contained in vertical cylindrical columns. The resin is supported on a grid fine enough to retain the pellets of resin, but sufficiently open so as not to hinder liquid flow. Sizes range from laboratory scale to industrial units 1–3 m in diameter and height. Liquid is fed to the top of the column through a distributor carefully designed and fitted to ensure an even flow over the whole cross-section of the bed. A mass transfer zone develops at the inlet to the bed in which the ion exchange takes place. As more feed is added, the zone travels through the bed and operation continues until unconverted material is about to emerge with the effluent. The bed has now reached the limit of its working capacity, its breakpoint, so the run is stopped and the resin regenerated.

Regenerating liquid is normally arranged to flow countercurrently to the feed direction thus ensuring that the end of the bed, which controls the effluent condition, is the most thoroughly regenerated. Fine particles from impurities entering with the feed and from attrition of resin pellets may accumulate in the bed. Such fines have to be removed from time to time by backwashing, so that the ion-exchange capacity is not reduced.

Backwashing may also be used for re-arranging the components of a mixed-resin bed. To demineralise water, for example, it is convenient to use a bed containing a random mixture of cationic and anionic resins. When either becomes exhausted, the bed is taken off-line and a back flow of untreated water is used to separate the resins into two layers according to their densities. Fig. 9.7 shows such a process in which an anion layer, of density $1100\,kg/m^3$, rests on the cation layer, of density $1400\,kg/m^3$. The former is regenerated using a 5% solution of caustic soda. After rinsing to remove residual caustic soda, 5% hydrochloric acid is introduced above the cation layer for its regeneration. Finally, the resins are remixed using a flow of air and the bed is ready to be used again [1].

The design of fixed-bed ion exchangers shares a common theory with fixed-bed adsorbers, which are discussed in Chapter 17. In addition, Thomas [14] has developed a theory of fixed-bed ion exchange based on Eq. (9.21). It assumed that diffusional resistances are negligible. Though this is now known to be unlikely, the general form of the solutions proposed by Thomas may be used for film- and pellet-diffusion control.

A material balance of counter-ions across an increment of bed may be written as:

$$u\frac{\partial C}{\partial z} + \frac{\partial C}{\partial t} = -\frac{1}{m}\frac{\partial C_S}{\partial t} \tag{9.33}$$

Defining distance and time variables, then:

$$\chi = \frac{kC_{S\infty}z}{mu}; \tau = kC_0\left(t - \frac{z}{u}\right) \tag{9.34}$$

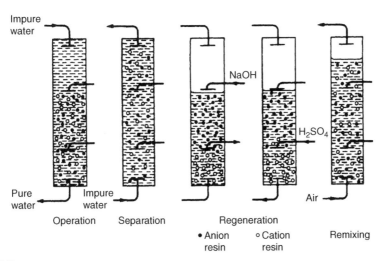

Impure water

NaOH

H$_2$SO$_4$

Pure water   Impure water   Air

Operation   Separation   Regeneration   Remixing

• Anion resin   ○ Cation resin

FIG. 9.7

A mixed demineralising bed [1].

Equations (9.33) and (9.21) may be written as:

$$\frac{\partial(C/C_0)}{\partial\chi} = -\frac{\partial(C_S/C_{S\infty})}{\partial\tau}$$ (9.35)

$$\frac{\partial(C_S/C_{S\infty})}{\partial\tau} = \frac{C}{C_0}\left(1 - \frac{C_S}{C_{S\infty}}\right) - \frac{1}{K_i}\frac{C_S}{C_{S\infty}}\left(1 - \frac{C}{C_0}\right)$$ (9.36)

The solutions given by Thomas are expressed as complex functions of χ, τ, and K_i. For design purposes, these are more conveniently presented in graphical form. Vermeulen [12] and Hester and Vermeulen [17] have extended their use to include diffusion control.

When film-diffusion controls, the kinetics step, given by Eq. (9.36), is essentially at equilibrium. Hence:

$$\frac{C_i^*}{C_0}\left(1 - \frac{C_S}{C_{S\infty}}\right) = \frac{1}{K_i}\frac{C_S}{C_{S\infty}}\left(1 - \frac{C_i^*}{C_0}\right)$$ (9.37)

and:

$$C_i^* = \frac{C_S}{C_{S\infty}}\frac{r^*C_0}{[1 + (r^* - 1)C_S/C_{S\infty}]}$$ (9.38)

where $r^* = 1/K_i$, is an equilibrium parameter.

The rate-controlling step may be written as:

$$\frac{\partial C_S}{\partial t} = \frac{k_l a_z}{1 - \varepsilon}(C - C_i^*)$$ (9.39)

where C_i^* is the fluid concentration at the exterior surface of the resin. This is assumed to be in equilibrium with the resin and uniform throughout the pellet.

Substituting in Eq. (9.37), then:

$$\frac{\partial(C_S/C_{S\infty})}{\partial\tau} = \frac{C}{C_S}\left(1 - \frac{C_S}{C_{S\infty}}\right) - r^*\left(1 - \frac{C}{C_S}\right)\frac{C_S}{C_{S\infty}}$$ (9.40)

where:

$$\tau = \frac{k_l a_z C_0}{(1 - \varepsilon)[1 + (r^* - 1)\overline{C}_S/C_{S\infty}]C_{S\infty}}\left(t - \frac{z}{u}\right)$$ (9.41)

If a mean value $\overline{C}_s/C_{s\infty}$ is taken for $C_S/C_{S\infty}$ in the definition of τ, Eq. (9.41) is a time parameter similar in form to Eq. (9.34).

Therefore, the solutions found for the kinetics-controlling-condition may be used with the new time parameter for the case of film-diffusion control.

When *pellet-diffusion* controls the exchange rate, the rate is often expressed in terms of a hypothetical solid-film coefficient k_p and a contrived driving force $(C_S^* - C_S)$ where C_S^* is the concentration of the resin phase in equilibrium with C, and C_S is the concentration of resin phase, averaged over the pellet.

Thus:

$$\frac{\partial C_S}{\partial t} = \frac{k_p a_z}{(1-\varepsilon)}\left(C_S^* - C_S\right) \tag{9.42}$$

The ion exchange step is at equilibrium, and hence:

$$C_S^* = \frac{C_{S\infty}}{(1-r^*) + r^*(C_0/C)} \tag{9.43}$$

Substituting for C_S^* in Eq. (9.42) gives:

$$\frac{\partial(C_S/C_{S\infty})}{\partial t} = \frac{k_p a_z}{\left[\overline{C}/C_0(1-r^*) + r^*\right](1-\varepsilon)}\left[\frac{C}{C_0}\left(1 - \frac{C_S}{C_{S\infty}}\right) - r^*\frac{C_S}{C_{S\infty}}\left(1 - \frac{C}{C_0}\right)\right] \tag{9.44}$$

For a mean value of $\overline{C}/C_0$ outside the bracket, or several mean values for different concentration ranges, Eq. (9.44) has the same form as Eq. (9.36) with a new time parameter given by:

$$\tau = \frac{k_p a_z}{\left[(\overline{C}/C_0)(1-r^*) + r^*\right](1-\varepsilon)}\left(t - \frac{Z}{u}\right) \tag{9.45}$$

The application to the design of fixed beds using graphed versions of the solution given by Thomas is discussed by Heister and Vermeulin [17]. Other solutions for fixed beds, including those that apply to equilibrium operation, are discussed in Chapter 17.

9.6.3 Moving beds

In principle, all the moving bed devices discussed in Chapter 17 may be used for ion exchange. Ion exchange is largely a liquid-phase phenomenon in which the solids may be made to flow relatively easily when immersed in liquid. A method of moving resin discontinuously has been developed by Higgins, as shown in Fig. 9.8, and described in Kirk-Othmer [18] and by Setter et al. [19] For a short period in a cycle time, ranging from a few minutes to several hours, the resin is moved by pulses of liquid generated by a double-acting piston which simultaneously sucks liquid from the top of the column and delivers it to the bottom. The operational problems encountered with this equipment are mechanical, resulting from wear and tear on the valves and attrition of the resin. Replacement of the resin can be as high as 30% each year when the equipment is used for water-softening. Other commercial applications include recovery of phosphoric acid from pickle liquor and the recovery of ammonium nitrate.

Fluidised beds are used commercially for ion exchange. These generally consist of a compartmented column with fluidisation in each compartment. The solid is moved periodically downwards from stage to stage and leaves at the bottom from which it passes to a separate column for regeneration. An arrangement of the Himsley-type is shown in Fig. 9.9 [18].

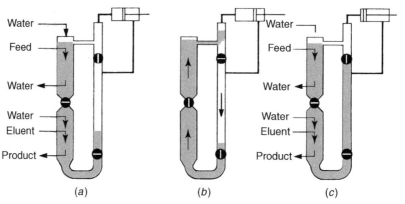

FIG. 9.8

Principles of the Higgins contactor (A) Solution pumping (several minutes) (B) Resin movement (3–5 s) (C) Solution pumping (several minutes).

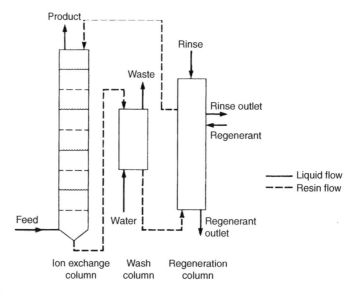

FIG. 9.9

A staged fluidised-bed ion-exchange column of the Himsley type [18].

Example 9.2

An acid solution containing 2% by mass of $NaNO_3$ and an unknown concentration of HNO_3 is used to regenerate a strong acid resin. After sufficient acid had been passed over the resin for equilibrium to be attained, analysis showed that 10% of resin sites were occupied by sodium ions. What was the concentration of HNO_3 in the solution, if its density were $1030 \, kg/m^3$.

Solution

In the solution:

$$_{NaNO} \text{ (Molecular weight} = 85 \text{ kg/kmol)}$$
$$\text{Concentration} = 2 \text{ per cent by mass}$$
$$= (20/85)(103/1000) = 0.242 \text{ kg/m}^3$$
$$_{HNO} \text{ (Molecular weight} = 63 \text{ kg/kmol)}$$
$$\text{Concentration} = p \text{ per cent}$$
$$\text{Concentration} = (10p/63)(1030/1000)$$
$$= 0.163p \text{ kg/m}^3$$
$$x_{Na^+} = 0.242/(0.242 + 0.163p)$$

For univalent ion exchange, Eq. (9.3) becomes:
But:

$$y_{Na^+}/(1 - y_{Na^+}) = K_{Na^+H^+}[x_{Na^+}/(1 - x_{Na^+})]$$
$$y_{Na^+} = 0.1$$

and from Table 9.1:

$$K_{Na^+H^+} = \left(\frac{2.0}{1.3}\right) = 1.5$$

Thus:

$$(0.1/0.9) = 1.5[0.242/(0.242 + 0.163p)]/[0.163p/(0.242 + 0.163p)]$$
$$p = \underline{20\, per cent}$$

References

[1] T.V. Arden, Water Purification by Ion Exchange, Butterworth, London, 1968.

[2] F. Helffrich, Ion Exchange, McGraw-Hill, New York, 1962.

[3] P.A. Schweitzer (Ed.), Handbook of Separation Techniques for Chemical Engineers, second ed., McGraw-Hill, New York, 1988.

[4] H.S. Thompson, On the absorbent power of soils, J. R. Agric. Soc. Eng. 11 (1850) 68.

[5] J.T. Way, On the power of soils to absorb manure, J. R. Agric. Soc. Eng. 11 (1850) 313 (13 (1852) 123).

[6] J. Lemberg, Z. Deut. Geol. Ges. 22 (1870) 355. Ueber einige Umwandlungen Finländischer Feldspat, 28, Ueber Siliciumumwandlungen, 1876, p. 519.

[7] Aristotle, Works, vol. 7, Clarendon Press, London, 1977, p. 933b (About 330 BC).

[8] Moses: Exodus 15 vv 23–25.

[9] D.R. Asher, D.W. Simpson, Glycerol purification by ion exclusion, J. Phys. Chem. 60 (1956) 518.

[10] M.J. Hatch, J.A. Dillon, H.B. Smith, The preparation and use of snake cage polyelectro-lytes, Ind. Chem. Eng. 49 (1957) 1812.

[11] S. Eagle, J.W. Scott, Liquid phase adsorption equilibrium and kinetics, Ind. Eng. Chem. 42 (1950) 1287.

[12] T. Vermeulen, Separation by adsorption methods, in: T.B. Drew, J.W. Hoopes (Eds.), Advances in Chemical Engineering, vol. 2, Academic Press, 1958, p. 148.

[13] J. Crank, Diffusion coefficients in solids, their measurement and significance, Discuss. Faraday Soc. 23 (1957) 99.

[14] H.C. Thomas, Heterogeneous ion exchange in a flowing system, J. Am. Chem. Soc. 66 (1944) 1664.

[15] F. Helffrich, M.S. Plesset, Ion exchange kinetics. A non-linear diffusion problem, J. Chem. Phys. 28 (1958) 418.

[16] P.G. Rimmer, J.H. Bowen, The design of fixed bed adsorbers using the quadratic driving force equation, Trans. Inst. Chem. Eng. 50 (1972) 168.

[17] N.K. Heister, T. Vermeulen, Saturated performance of ion exchange and adsorption columns, Chem. Eng. Prog. 48 (1952) 505.

[18] Kirk-Othmer, Encyclopedia of Chemical Technology, Ion Exchange, vol. 13, Wiley, New York, 1981.

[19] N.J. Setter, J.M. Googin, G.B. Marrow, The Recovery of Uranium From Reduction Residues by Semi-Continuous Ion Exchange USAEC Report Y-1257 (9th July 1959), 1959.

Further reading

K. Dorfner, Ion Exchangers, Walter De Gruyter, 1991.

Kirk-Othmer, Encyclopedia of Chemical Technology, Ion Exchange, vol. 13, Wiley, New York, 1981.

L. Libirti, J.R. Millar (Eds.), Fundamentals and applications of Ion Exchange (NATO Asi Series' Series E Applied Sciences No. 98), 1985.

J.A. Morinski, V. Marcus (Eds.), Ion Exchange and Solvent Extraction (A Series of Advances), Marcel Dekker, New York, 2001.

P.A. Schweitzer (Ed.), Handbook of Separation Techniques for Chemical Engineers, second ed., McGraw-Hill, New York, 1988.

A.K. Sengupta (Ed.), Ion Exchange Technology: Advances in Environmental Pollution Control, Technomic Publishing Co, Lancaster, Pennsylvania, 1995.

M.J. Slater (Ed.), Ion Exchange Advances SCI Conference IEX 92-Ion, 1995.

Chromatographic separations

10

Arvind Rajendran

Department of Chemical and Materials Engineering, University of Alberta, Edmonton, AB, Canada

Nomenclature

Symbol	Quantity	Units
Roman letters		
c	fluid-phase concentration	NL^{-3}
D	desorbent flow rate	L^3T^{-1}
D_L	axial dispersion	MT^{-2}
D_m	molecular diffusion	MT^{-2}
d_p	particle diameter	L
E	extract flow rate	L^3T^{-1}
F	feed flow rate	L^3T^{-1}
H	Henry constant	–
$HETP$	height equivalent to a theoretical plate	L
K	equilibrium constant	L^3N^{-1}
k	mass transfer coefficient	T^{-1}
m	dimensionless flow-rate ratio	–
N	number of theoretical plates	–
Q	volumetric flow-rate	L^3T^{-1}
q^*	equilibrium solid-phase concentration	NL^{-3}
q^*	solid-phase concentration	NL^{-3}
R	raffinate flow rate	L^3T^{-1}
t	time	T
t^*	switch time	T
t_R	retention time	T
v	interstitial velocity	LT^{-1}
V_{inj}	injection volume	L^3

Continued

Coulson and Richardson's Chemical Engineering. https://doi.org/10.1016/B978-0-08-101097-6.00010-9

Symbol	Quantity	Units
w	peak-width at half peak-height	T
z	axial coordinate	L
Greek letters		
γ_1, γ_2	constants in Eq. (10.14)	–
ε	bed voidage	–
μ	first moment of a pulse	T
σ^2	second-moment of the pulse	T^2
ω	roots of Eq. (10.26)	–
ψ	wave velocity	LT^{-1}

10.1 Introduction

Chromatography is an advanced separation processes that is capable of producing very high purity products and finds extensive application in the food, pharmaceutical, and petrochemical industries [1,2]. The basis of a chromatographic separation lies in the ability of certain solids to retain one or more components preferentially over others [2–4]. In most practical applications, the solid, which is referred either as an adsorbent or as a stationary phase, is packed in a column, and the mixture to be separated is dissolved in a suitable solvent and introduced as the column inlet. Since various components in the mixture show different affinities to the stationary phase, they travel at distinct velocities. This provides the ability to separate a mixture.

Chromatography is perhaps the most commonly used analytical technique. The goal of analytical chromatography, a schematic of which is shown in Fig. 10.1A, is to separate the contents of a mixture and quantify their presence. Analytical chromatography consists of an injection system where a small pulse of the mixture to be separated in injected into a solvent that flows through the column. A suitable detector is positioned at the column outlet which provides the trace of the 'eluted' solutes. This trace (concentration or signal vs time) is referred to as a chromatogram. The peaks in the chromatogram are usually related to a specific species. Analytical chromatography can be operated using a solvent/carrier that can be any one of gas, liquid, and supercritical fluid. In gas chromatography (GC), a gas-phase carrier is used, and the injected material can either be in a gas phase or in a liquid phase that is vapourised prior to injection. In GC, the column can either be a long capillary with the adsorbent coated on the inner surface or with a solid packing. Liquid chromatography (LC) is an operating mode where the solvent is a liquid and the injected solute is either a liquid or a solid that is dissolved in a solvent. LC is typically operated using a packed column containing very small particles. The reduction of particle size results in higher resolution between the peaks and this technique is also referred to as high-performance liquid chromatography (HPLC). The recent years have also seen the blossoming of supercritical fluid chromatography (SFC) in which the solvent is a

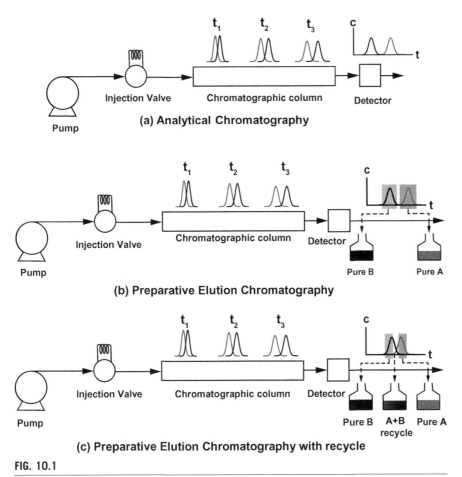

FIG. 10.1

Various modes of operation of single-column chromatographic processes.

supercritical fluid, i.e., a dense fluid. SFC traditionally employed capillary columns and has progressed to the use of packed columns [5].

Chromatography used for the industrial separation of mixtures is termed as preparative chromatography [6,7]. For large-scale separations, LC and SFC are more commonly used compared to GC. A schematic of preparative chromatographic systems is shown in Fig. 10.1B. Preparative separations differ from analytical separations in many ways. First, from an instrumentation perspective, preparative systems are scaled-up versions of the analytical systems with the addition of fraction collection units downstream of the detector. Second, in analytical systems, dilute samples are used in order to improve the resolution. In preparative separations, injection concentrations are fairly high, often close to solubility limits. In chromatography parlance, this is referred to an overloaded condition. Third, while the analytical separations are almost always performed under an elution mode; i.e., a pulse of the mixture to be analysed is injected and 'eluted' using the solvent, preparative separations can be operated in more complex configurations as will be discussed in this chapter. One of the key highlights of chromatographic separations, compared to other separation processes lies in the simplicity of scale-up procedures. Once a suitable stationary phase and solvent combination can be established in an analytical system, the rules for scale-up are quite straightforward. This chapter will focus on the fundamentals required for the engineering design of preparative separations.

10.2 Adsorption equilibria

Chromatography, as mentioned earlier, relies on the ability of certain solids (natural or synthetic) to selectively retain one species over another. The retention mechanism in most practical separations tends to rely on the differences in equilibrium between two species. Hence describing the thermodynamic equilibrium is critical. The most simplest form of representation is the linear isotherm in which the fluid-phase concentration of a species i, c_i, is related to the equilibrium concentration on the solid phase q_i^* as:

$$q_i^* = H_i c_i \qquad (10.1)$$

where H_i is the Henry constant [1,3]. The Henry constant for a particular solute and an adsorbent is a function of the solvent (composition) and the temperature. Most systems tend to the linear isotherm at low concentrations. At higher fluid-phase concentrations, since the adsorption sites on the solid surface are finite, the adsorption isotherm deviates from the linear form. Adsorption isotherms at high concentrations tend to be non-linear and their specific form depends on the specific system studied. One of the simplest and elegant descriptions of what is called as Type-I systems, is the Langmuir isotherm which is written as

$$q_i^* = \frac{H_i c_i}{1 + K_i c_i}; \; H_i = q_{s,i} K_i \qquad (10.2)$$

where $q_{s,i}$ is the saturation loading and K_i is the equilibrium constant. As can be seen at low concentrations, the Langmuir isotherm tends to the linear form and at very high concentrations reaches the saturation limit $q_{s,i}$. In order for the Langmuir adsorption to be thermodynamically consistent, all solutes should have identical saturation capacities [3]. However, in many cases, either because of the nature of the solid phase (e.g. stationary phases that contain functionalised entities on solid supports, or when pore sizes could potentially exclude some solutes), this required cannot be satisfied. In such cases, the Langmuir isotherm is often fitted to experimental data and is used for the purposes of preliminary design. While this approach could be reasonable, caution should be exercised when using such formulations beyond regions that have been confirmed by experiments. When more than one solute species is present in high concentrations, they compete for adsorption sites and can be conveniently represented by the extended Langmuir isotherm or the competitive-Langmuir isotherm for a binary system of solutes A and B as:

$$q_i^* = \frac{H_i c_i}{1 + K_A c_A + K_B c_B}; \; H_i = q_{s,i} K_i \qquad (10.3)$$

In some stationary phases, e.g., chiral stationary phases, the solute molecules adsorb both on the support and the functionalised chiral selector, the bi-Langmuir isotherm is commonly used [1].

$$q_i^* = \frac{H_{1,i} c_i}{1 + K_{1,i} c_i} + \frac{H_{2,i} c_i}{1 + K_{2,i} c_i}; \; H_{1,i} = q_{s1,i} K_{1,i} \text{ and } H_{2,i} = q_{s2,i} K_{2,i} \qquad (10.4)$$

As it can be seen, the bi-Langmuir isotherm is a simple extension of the Langmuir isotherm by assuming two distinct sites '1' and '2' exist on the solid.

The determination of adsorption isotherms is typically the first step in developing a separation. Although computational modelling is emerging as a reliable tool, in a practical situation, it is straightforward to characterise the isotherms thorough suitable experiments. Both batch and dynamic methods are available [8]. It is worth noting that adsorption isotherms should be measured over the entire concentration range that is anticipated in the processes. Further, suitable multi-component experiments need to be performed in order to validate the isotherms that will be used for process design.

10.3 Solute retention theory

Chromatographic separations exploit the potential of certain solids to selectively retain one of the species in a mixture over the other(s). Hence, it is critical to understand how the affinity of a component to the stationary phase affects its transport across a chromatographic column. In order to understand this, consider a column of length L, diameter d, and void fraction, ϵ. Note that the void fraction in this case is defined as the ration of the void volume to the empty column volume. Also consider a non-adsorbing solvent that is flowing through the column with an interstitial velocity v. The flow of the solvent is assumed to be plug-flow, i.e., there are no radial velocity gradients. Under these conditions, the mass balance of the solute is given by

$$\frac{\partial c_i}{\partial t} + v \frac{\partial c_i}{\partial z} + \frac{1 - \epsilon}{\epsilon} \frac{\partial q_i}{\partial t} = 0 \qquad (10.5)$$

q_i in the above equation denotes the solid-phase concentration of the solute i [9,10]. If it can be assumed that at every point in the column the fluid-phase and solid-phase concentrations are in equilibrium, i.e., mass transfer is fast enough, q_i in the above equation can be replaced by the solid-phase concentration that in equilibrium with the fluid-phase concentration, i.e., q_i^*. Then, Eq. (10.5) now becomes:

$$\frac{\partial c_i}{\partial t} + v \frac{\partial c_i}{\partial z} + \frac{1 - \epsilon}{\epsilon} \frac{\partial q_i^*}{\partial t} = 0 \qquad (10.6)$$

which can further be written as

$$\frac{\partial c_i}{\partial t} + v\frac{\partial c_i}{\partial z} + \frac{1-\epsilon}{\epsilon}\frac{dq_i^*}{dc_i}\frac{\partial c_i}{\partial t} = 0 \tag{10.7}$$

or

$$\left[1 + \frac{1-\epsilon}{\epsilon}\frac{dq_i^*}{dc_i}\right]\frac{\partial c_i}{\partial t} + v\frac{\partial c_i}{\partial z} = 0 \tag{10.8}$$

Solving the above mass balance equation with suitable initial and boundary conditions allows us to track how a concentration of the solute introduced at the column inlet propagates through the column. The wave velocity of the solute, ψ, can now be written as

$$\frac{dz}{dt}\bigg|_{c_i} = \psi_i = \frac{v}{\left[1 + \frac{1-\epsilon}{\epsilon}\frac{dq_i^*}{dc_i}\right]} \tag{10.9}$$

Integrating the above equation, for a specific initial condition, gives a line in the physical plane called as a characteristic.

10.3.1 Solute retention for linear isotherms

For the simplest case of equilibrium relationship, i.e., the linear isotherm, as given by Eq. (10.1), the wave velocity is given by

$$\psi_i = \frac{v}{\left[1 + \frac{1-\epsilon}{\epsilon}H_i\right]} \tag{10.10}$$

The above expression indicates that the wave velocity for a system that follows the linear isotherm is independent of the concentration. It also indicates that a component with a stronger affinity, i.e., with a larger value of H_i will travel slower than the one with a weaker affinity. This provides the fundamental principle upon which chromatographic separation processes are built upon.

The solution of Eq. (10.9) can be conveniently obtained using the method of characteristics, specifically for a class of problems called as Riemann problems (constant initial value problems). While the formal derivations and the mathematics are beyond the scope of this chapter, the solution itself serves the goal of understanding solute propagation. The interested reader is referred to more detailed treatments that can be found elsewhere [9,11–13]. The solution of a simple case where the column is initially filled with the solvent and a step change in the concentration of the solute is introduced at the inlet is shown in Fig. 10.2A. The surface spanned by the axial length on the abscissa and the time on the ordinate is referred to as the physical plane. Using Eq. (10.10) the retention time of the solute $t_{R, i}$, can hence be calculated as

$$t_{R,i} = \frac{L}{\psi_i} = \frac{L}{v}\left[1 + \frac{1-\epsilon}{\epsilon}H_i\right] \tag{10.11}$$

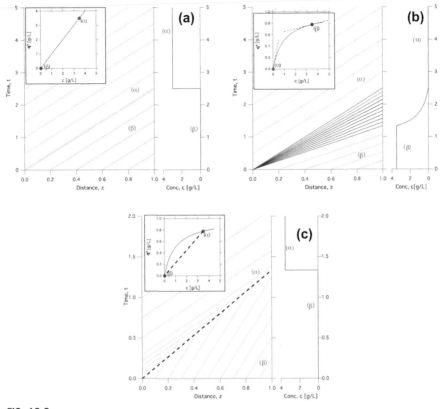

FIG. 10.2

Propagation of step inputs for various isotherm types. Characteristics plotted in the physical plane along with the corresponding elution curve at the column outlet. (A) Adsorption step for a linear isotherm; (B) adsorption step for a Langmuir isotherm; and (C) desorption step for a Langmuir isotherm. In all of these plots, states α and β represent the inlet and initial states, respectively.

Reproduced from Mazzotti and Rajendran, Annual Reviews.

10.3.2 Solute retention for non-linear isotherms

In preparative separations, in order to increase the throughput, the injected concentrations are rather high, and under these conditions, the adsorption isotherm takes a non-linear form. A common, but effective relationship between the fluid-phase and the solid-phase concentrations is given by the Langmuir isotherm, i.e., Eq. (10.2). For the case of the Langmuir isotherm, the wave velocity is

$$\psi_i = \frac{v}{\left[1 + \frac{1-\epsilon}{\epsilon}\frac{H_i}{(1 + K_i c_i)^2}\right]} \qquad (10.12)$$

Note that, unlike the case of linear isotherms, the wave velocity is no longer independent of the concentration. This is a key result that illustrates the principle that a higher concentration propagates faster than a lower concentration.

In order to illustrate this, consider a column that is originally equilibrated (initial state) with a certain concentration of the solute. At time $t=0$, pure solvent is introduced at the inlet, in order to regenerate (or desorb) the column. The solution of this problem, that can be obtained using the retention theory described above, is illustrated in Fig. 10.2B. As it can be seen, owing to the variation in propagation velocities, a step input at the column inlet spreads as it travels through the column. It is important to emphasise that this spread is caused because of the nature of the adsorption isotherm, and not by either axial dispersion or mass transfer resistances that will be discussed later. This type of a transition is called as a simple wave [12]. The elution curve at the column outlet, also shown in Fig. 10.2B, clearly illustrates this phenomenon. Another aspect that is evident from Eq. (10.12) is that time required to completely regenerate the column can be obtained by substituting $c_i=0$ which then reduces to the well know expression Eq. (10.11). This is an important result that shows that the time required to completely regenerate the column, for a system that follows the Langmuir isotherm, is governed by the Henry constant [14].

Now let us consider the opposite case, i.e., when the column is initially filled with the solvent and at time $t=0$, a step change in the concentration is introduced at the column inlet. Applying the solute retention theory directly creates a unique situation in which the characteristics corresponding to the initial state, feed state and the final state of the column will interfere with each other leading to an unrealistic solution. This very interesting situation, observed very commonly in practice, has been solved by mathematicians by introducing a discontinuity in the solution [12,13,15]. This leads to formation of a 'shock' that propagates with a constant velocity which is now given by

$$\psi_i = \frac{v}{\left[1 + \frac{1-\epsilon}{\epsilon}\frac{\Delta q_i^*}{\Delta c_i}\right]} \tag{10.13}$$

where the term $\frac{\Delta q_i^*}{\Delta c_i}$ represents the slope of the chord on the isotherm that connects the initial concentration in the column to the final. The solution of the problem is illustrated in Fig. 10.2C. This is again a key result that illustrates that for a Langmuir-type isotherm 'adsorption' steps lead to a shock transition, while 'desorption' steps lead to a simple wave transition.

It is worth noting that a pulse input can be modelled as a combination of an adsorption step followed by a desorption step. While in the case of linear isotherms, all transitions propagate at the same velocity, the shape of the pulse is preserved as it moves through the column. However, for the case of high concentration pulse injections, the shock wave fronts can intersect within the column and this leads to pulse elution curves that resemble a (degenerated form) right triangle, i.e., a sharp front followed by a spread-out tail [1,10].

10.4 Mass transfer and band broadening in chromatographic systems

In the solute retention theory mentioned above, band broadening was shown to be caused owing to the nature of the adsorption isotherm. However, in real systems band broadening is accentuated by dispersion and mass transfer effects and recognising their importance is critical to the understanding of chromatography. In this section, we will focus on linear systems, for which band-broadening is absent within the framework of local-equilibrium and no axial-dispersion [1].

When a dilute solute travels through the column, the causes for band-broadening can be classified those that are caused by phenomena that take place in the fluid phase and those that take place within the stationary phase. The band-broadening phenomena that occurs in the fluid phase is collectively charac-terised as axial dispersion [1,3]. Three mechanisms contribute to axial dispersion: (1) Non-uniform flow in the fluid phase that can be caused specifically when the diameter of the particle is not significantly smaller than the column diameter; (2) molecular diffusion caused by the concentration gradients that exist when a pulse travels through the column; and (3) mixing that takes place in the inter-particle voids. The simplest characterisation of the axial dispersion coefficient D_L is given by

$$D_L = \gamma_1 D_m + \gamma_2 d_p v \qquad (10.14)$$

where D_m is the molecular diffusion of the solute in the solvent, d_p is the particle diameter. The two constants in the above equation are usually in the range of $\gamma_1 = 0.7$ and $\gamma_2 = 0.5$. More rigorous correlations are available in the literature [1].

The second key contribution comes from the mass transfer resistances that reside within the particle. They resistances that a solute molecule could face as it diffuses from the fluid phase into the solid phase can be classified as (1) *Film resistance*: the resistance that is presents at the surface of the particle owing to a thin film of the solvent that surrounds the particle; (2) *macropore diffusion*: the resistance that lies in the macropores; (3) *micropore resistance*: the resistance that lies in the small pores that contain the bulk of the adsorption surface area; (4) the kinetics of adsorption/desorption. The specific location where the controlling resistance lies is often deter-mined experimentally. Very often, the transfer from the fluid to the solid phase is represented by simple linear driving force (LDF) expression where the mass transfer rate is given by

$$\frac{dq_i}{dt} = k_i \left(q_i^* - q_i \right) \qquad (10.15)$$

where k_i is called as mass transfer or the LDF coefficient [16].

Martin and Synge presented what is called as a the plate theory of chromatogra-phy [17]. The origins of this theory lie in the concept of number of theoretical plates that is used in classical separation processes such as distillation and absorption. This theory can be understood by considering the injection of an infinitesimally small

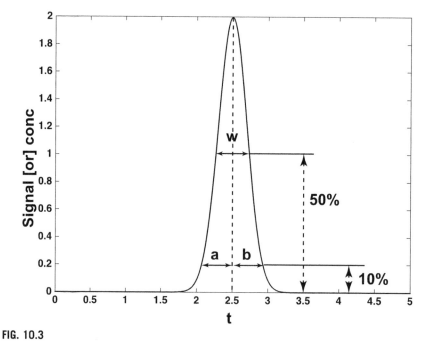

FIG. 10.3

Example of a response of a chromatographic column for a pulse injection.

pulse. The elution curve of such a pulse injection is shown in Fig. 10.3. Under these conditions, the first (μ), and the second (σ^2) moments of the elution curve can be written as:

$$\mu = \frac{\int_0^\infty ct\,dt}{\int_0^\infty c\,dt} \tag{10.16}$$

$$\sigma^2 = \frac{\int_0^\infty c(t-\mu)^2\,dt}{\int_0^\infty c\,dt} \tag{10.17}$$

Based on the two moments measured, the number of theoretical plates, N, was defined as

$$N = \frac{L}{\text{HETP}} = \frac{\mu^2}{\sigma^2} \tag{10.18}$$

where HETP refers to the height equivalent to a theoretical plate. For a Gaussian response, N can be calculated using the expression:

$$N = 5.54 \left(\frac{\mu}{w} \right)^2 \tag{10.19}$$

where μ represents the elution time of the peak while w represents the width of the peak at half-peak height. From the definition, it is clear that N is a measure of the 'spread' of the pulse. A larger value of N represents minimal band-broadening. By performing a moment analysis on a transport model that accounts for axial dispersion and the mass transfer into the particle, the HETP can be described by:

$$\text{HETP} = \frac{\sigma^2}{\mu^2} L = 2\frac{D_L}{v} + 2v \left(\frac{\epsilon}{1-\epsilon} \right) \frac{1}{kH} \tag{10.20}$$

By substituting Eq. (10.14) into Eq. (10.20), we obtain a very simple form for calculating the HETP:

$$\text{HETP} = \frac{A}{v} + B + Cv \tag{10.21}$$

In the above equation, 'A' represents the contribution from molecular diffusion; 'B' to the contribution from the inter-particle mixing; and 'C' to the contribution of inter-particle mass transfer. The plot of HETP as a function of velocity, shown in Fig. 10.4, is commonly referred to as a van-Deemter plot [18,19]. As it can be seen from Fig. 10.4, with increasing velocity, the HETP drops, reaches a minima and increases with increasing velocity. This plot allows for the comparison of various stationary phases and for the identification of the operating velocity (corresponding to minimum HETP) that will result in minimal band-broadening.

From the van-Deemter equation it can be deduced that HETP can be reduced by lowering particle size, however at the cost of increased pressure drops. There has been a continuing effort to reduce particle sizes in both analytical and preparative chromatography. Developments in particle technology and instrumentation have

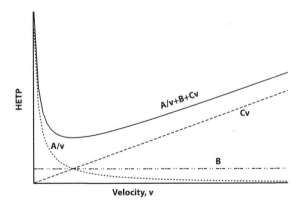

FIG. 10.4

A qualitative plot of a van-Deemter curve showing the variation of the height equivalent to a theoretical plate (HETP) as a function of fluid velocity.

now made it possible to use sub-micron scale particles in analytical columns. Core-shell particles have also contributed to the reduction of the inter-particle diffusional paths [20]. All these developments have together resulted in minimising band broadening, increase separation efficiency and leading to greener chromtography [21].

10.5 Elution chromatography

Elution chromatography is not only the simplest of chromatographic separations, but also a very powerful one [1,2]. The principle of the separation is rather straightforward. In this method, the mixture to be separated is injected into a column in the form of a pulse. If the column has suitable resolution for the components of the mixture, it is possible to separate the components as they elute from the column. Based on the times at which they elute, product fractions can be collected. This technique has two main advantages. The first one is the simple mode of operation which allows for a quick estimate of the separation potential. In practical situations, it is not uncommon to screen a host of stationary-phase solvent combination for each mixture and identify the best combination. The second and the most important advantage of elution chromatography is its ability to perform multi-component separations. Provided a suitable column can be found, the number of pure fractions that can be collected is theoretically infinite. This distinguishes elution chromatography from most other techniques which tend to be binary separators, e.g., distillation.

A variant of standard elution chromatography is to inject a highly concentrated pulse, collect the pure fractions and recycle the mixed fraction to the feed. This type of operation is referred to as steady-state recycle-chromatography (SSR), a technique that is used in small to medium scale separations [22–24]. The key advantage of this operation is to increase the performance of the process without increasing the complexity of instrumentation.

10.5.1 Design for linear isotherms

In order to illustrate the key principles of the design of elution chromatography, consider a binary mixture, with both solutes being describable by the linear isotherm, being separated in an ideal column. There are two key design variables, namely the injection time, t_{inj}, and the wait-time between two injections, t_{wait}, that will provide base-line separation, i.e., the components elute from the column without overlapping with each other and there will be no gap in between the elution of subsequent injections. Let us consider the elution characteristics of a pulse that is injected over a duration of t_{inj} units of time. Note that $t_{inj} = \frac{V_{inj}}{Q}$, where V_{inj} is the injection volume and Q is the volumetric flow rate of the solvent. If A and B represent the more and less retained components, the characteristic retention times of the leading and the lagging edges of the elution curves for the two solutes can be computed.

$$t_{R,i,\text{lead}} = \frac{L}{\psi_i}; \quad t_{R,i,\text{lag}} = \frac{L}{\psi_i} + t_{\text{inj}} \qquad (10.22)$$

From the above expressions, the relationship to calculate t_{inj} can be calculated by imposing the condition of 'touching-band-separation' that requires

$$t_{R,B,\text{lag}} = t_{R,A,\text{lead}} \qquad (10.23)$$

By substituting the appropriate expressions, it is quite straightforward to show that

$$t_{\text{inj}} = \frac{L}{v} \frac{1-\epsilon}{\epsilon} [H_A - H_B] \qquad (10.24)$$

Further, the time at which the next injection can be performed, t_{wait} can also be shown to be equal to t_{inj}. This type of injection strategy is often referred to a stacked injections and is shown in Fig. 10.5.

Naturally, it is worth pointing out that the calculations shown here should be used with caution, as band broadening is bound to happen in real systems which will prevent the collection of pure products with complete recovery based on the injection times calculated here. However, the injection time calculated by the ideal model provides an estimate of the largest injection that can provide 100% pure products.

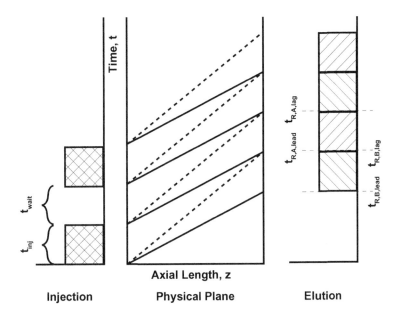

Injection   Physical Plane   Elution

FIG. 10.5

Illustration of touching band separation for a pulse injection of a mixture following linear isotherms. The solid and dotted lines in the physical plane refer to the weakly adsorbed (B) and strongly adsorbed (A) components. The illustration on the left indicates the injections, while those on the right illustrate the elution curves.

10.5.2 Design for Langmuir isotherms

The mathematics describing the propagation of pulse injections of concentrated systems whose equilibrium is described by non-linear isotherms is very complex. Nevertheless, this problem at least for a set of well-described Langmuir and Langmuir-type isotherms has been solved analytically [12,25]. While the mathematics of such a system is quite complex, it is illustrative to consider the progression of the two components. Fig. 10.6 shows the propagation of a pulse injection of a binary mixture that obey the Langmuir adsorption isotherm into a column that contains just the solvent (state O) [25]. Note that the injection concentrations of the weakly retained and strongly retained components are 3.5 and 2.5 g/L (state F), respectively. Due to the nature of the non-linear equilibria, two new states, i.e., plateaus, namely

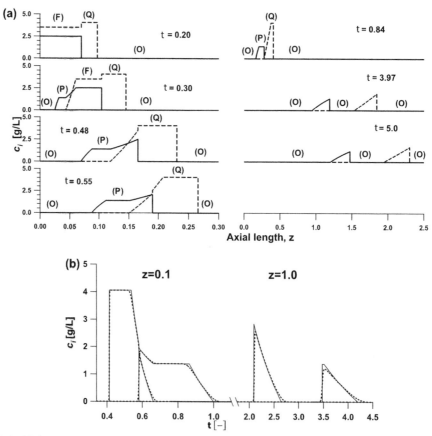

FIG. 10.6

Elution of a pulse injection for a system following a competitive Langmuir isotherm. (A) Shows the axial profiles at various instances of time and (B) elution curve at two positions in the column. Note the formation of two intermediate states Q and P.

Reproduced with permission from Industrial and Engineering Chemistry, ACS Publications.

P and Q are formed. Note that Q represents a concentration of B that is larger than the injected state. This is traditionally referred to as a 'roll-up', i.e., an enrichment of the weaker component. The state P, refers to a lower concentration of A compared to its feed value. As the pulse propagates through the column, all of these states are eroded, and if the column is long enough, the two components are separated. The final elution curve, in most cases, appear like right triangles, i.e., with a very sharp front, followed by a spread-out rear (see Fig. 10.6B). These are signatures of Langmuir adsorption isotherms. Under these conditions the injection time that will provide baseline separation is given by

$$t_{inj} = \frac{L}{v} \frac{1-\epsilon}{\epsilon} \left[\frac{\omega_1^F \omega_2^F (H_A - H_B)^2}{H_A H_B (\omega_2^F - H_B)} \right] \tag{10.25}$$

where ω_1^F and ω_2^F are the roots of the equation:

$$\left(1 + K_A c_A^F + K_B c_B^F\right)\omega^2 - \left[H_B\left(1 + K_A c_A^F\right) + H_A\left(1 + K_B c_B^F\right)\right]\omega + H_A H_B = 0 \tag{10.26}$$

and obeying the following condition:

$$0 < \omega_1 \leq H_1 \leq \omega_2 \leq H_2 \tag{10.27}$$

10.6 Simulated countercurrent processes

In the previous section, we discussed the separation using elution chromatography. While it is a powerful separation technique, it has a few shortcomings. First, it becomes difficult to collect pure products when the column efficiency is low. This means that the unit cannot achieve high productivity. Second, in order to achieve high purity and recovery, the injection size should be small enough to enable baseline separation. Third, the solvent consumption, which accounts for bulk of the operating cost of the process is high. Most of these limitations arise from the fixed-bed nature of the process along with the intermitted feed injection. Traditional separation processes such as distillation and absorption, overcome a few of these limitations by contacting the two phases (liquid and gas in the case of the two processes) in a continuous countercurrent manner. Although, there were a few attempts to achieve true countercurrent contact between the solid and fluid phases for performing chromatographic separations, it is quite challenging to avoid dispersion because of the movement of the solid. This aspect prevents the separation of systems with low selectivity.

Broughton and Gerhold overcame the limitations of a true moving bed by developing a rather creative way of achieving a simulated counter-current contact [26]. This process is called simulated moving bed chromatography (SMB) [2,27,28]. The simplest manifestation of this process is shown in Fig. 10.7. The SMB consists of four sections with the possibility of multiple columns per section. Pure solvent (desorbent) is introduced into Section 10.1 and the feed containing the mixture dissolved in a solvent is introduced between Sections 10.2 and 10.3. The extract containing a product enriched in the strongly adsorbing component is collected between

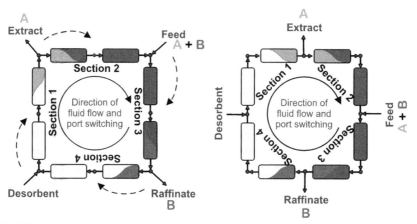

FIG. 10.7

Schematic of a simulated moving bed process with four sections and two columns per section. The schematic on the left and right show the positions of the column before and after the switch.

Reproduced with permission from Journal of Chromatography A, *Elsevier.*

Sections 10.1 and 10.2, while the raffinate containing a product enriched in the weekly adsorbing component is collected between Sections 10.3 and 10.4. The countercurrent movement of the solid phase is simulated by periodically switching the inlet and the outlet ports in the direction of the fluid flow. The time period after which the switch is performed is called switch time, t^*.

As the unit is started and the feed is introduced, the unit goes through a transient period and ultimately reaching a cyclic steady state (CSS). This is in contrast to continuous non-cyclic processes that achieve a true steady state. At cyclic steady state the internal concentration profiles move in the direction of the fluid flow as shown in Fig. 10.8. The nature of the internal concentration profiles also result in a variation of the product concentrations within a switch time as shown in Fig. 10.8. As it can be seen during the transient period the cumulative concentration of both the extract and raffinate products vary and ultimately reach a constant value at CSS. From the picture is also clear that in most of the unit the two components overlap and it is sufficient if pure products can be obtained only near the product ports. This is a unique feature of the SMB where complete resolution of the two components is not required, thereby enabling the separation of low selectivity systems often using low efficiency columns.

10.6.1 Design of simulated moving bed chromatography

Each section of the SMB has a specific purpose. Sections 10.2 and 10.3 perform the separation, while Sections 10.1 and 10.4 are responsible for the regeneration of the solid and the fluid phases, respectively. In order to achieve the assigned duties, four inequalities need to be satisfied [27,28], namely:

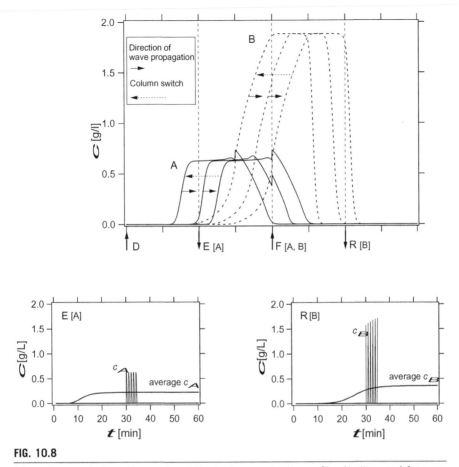

FIG. 10.8

Internal concentrations (*top row*) and Product concentration profiles (*bottom row*) for a an SMB process.

Reproduced with permission from Journal of Chromatography A, *Elsevier.*

$$t_{A,1}^{R} \leq t^{*}$$
$$t_{B,2}^{R} \leq t^{*} \leq t_{A,2}^{R}$$
$$t_{B,3}^{R} \leq t^{*} \leq t_{A,3}^{R} \tag{10.28}$$
$$t^{*} \leq t_{B,4}^{R}$$

The first condition ensures that Section 10.1 is operated in such a way that the strongly adsorbed component is eluted completely from Section 10.1 prior to the switch. Conditions (10.2) and (10.3) ensure that the strongly adsorbed component has a net movement towards the extract port while the weakly adsorbed component has a net movement towards the raffinate port. The final condition ensures that weakly adsorbed component is retained within Section 10.4. At this point, it is worth defining a new parameter, the dimensionless flow rate ratio m_j as [29,30]:

$$m_j = \frac{Q_j^{SMB} t^* - V\varepsilon}{V(1-\varepsilon)} = \frac{\text{net fluid flow rate}}{\text{net solid flow rate}} \qquad (10.29)$$

Combining the above equation with Eq. (10.28), for the case of linear isotherms and for a set of columns with 100% efficiency, it is possible to express Eq. (10.30) as

$$
\begin{aligned}
H_A &\leq m_1 \\
H_B &< m_2 \leq H_A \\
H_B &\leq m_3 \leq H_A \\
m_4 &\leq H_B
\end{aligned}
\qquad (10.30)
$$

The above equations provide explicit conditions that will guarantee complete separation for a mixture of two components, while achieving 100% recovery. The most important feature of the above relationships is that they are scale-independent. This provides an ideal methodology for scale-up of the processes. They also show that in order to design the separation for a system that follows a linear isotherm, the Henry constants of the two components are sufficient.

These conditions can be conveniently represented in a graphical form as shown in Fig. 10.9. Fig. 10.9A. shows the constraints on Sections 10.2 and 10.3, while Fig. 10.9B represents constraints on Sections 10.1 and 10.4. Let us consider Fig. 10.9A, which shows a variety of separation regions that are spanned on the (m_2, m_3) plane. It is clear that operating the unit within the triangle 'abw' while satisfying the conditions on Sections 10.1 and 10.4 (as given by Eq. 10.30) guarantees complete separation of the two components. It can be easily shown that the feed flow is proportional to the difference of m_2 and m_3 and hence the vertex of the triangle represents the operating condition that guarantees maximum productivity. A few

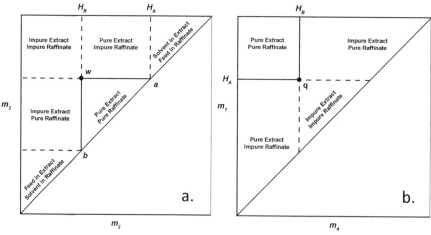

FIG. 10.9

Region of complete separation on (A) (m_2, m_3) plane and (B) (m_1, m_4) plane.

Reproduced with permission from Journal of Chromatography A, *Elsevier.*

features are worth noting. Fig. 10.9A. denotes that there is a range of operating conditions that guarantees complete separation, i.e., it is possible to operate anywhere within the triangle 'abw'. This makes the process robust. Further, it should be noted that operating near the edges of the triangle makes the process less robust, making the products susceptible for contamination. Hence, although it would be desirable to operate near the vertex, from a practical perspective the units are typically operated, within the triangular region, away from the edges. It is worth noting that the constraints have been derived for the case of a 100% efficient column. In practice, columns tend to have a finite efficiency and under those conditions, while the general topology of the plots shown in Fig. 10.9 remain the same, the region of complete separation shrinks [31].

The design of an SMB as illustrated above for the case of linear isotherms is straightforward. However, under non-linear conditions, the mathematical treatment is significantly complicated and is outside the scope of this chapter. However, it is very interesting that the analysis leads to results that are qualitatively similar to the plots shown in Fig. 10.10 [28,29,32,33]. The solution of the corresponding design problem on the (m_2, m_3) plane for the case of Langmuir isotherms is shown in Fig. 10.10A, with the equations of the boundaries provided in Table 10.1. It is worth noting that while the region of complete separation for the case of linear isotherms is a perfect right triangle, for the case of Langmuir isotherms, is a triangle-like shape with the boundaries intersecting the diagonal still at H_A and H_B. It is quite straightforward to notice that the boundaries are now dependent on the feed concentration of the two components. Fig. 10.10B shows the impact of varying the feed concentration on the region of complete separation. It can be seen that as the feed concentration decreases, the region of complete separation broadens eventually coinciding, as expected, with the boundary for the linear isotherm. This property is quite helpful to start-up SMB units as the operator can obtain an approximate idea of the complete separate region simply by measuring the Henry constants of the two solutes. For instance, the qualitative effect of increasing the Feed flow rate (F), Extract flow rate (E) and the switch time (t^*) are shown in Fig. 10.10C. Once the unit is started, the product purities can be measured and appropriate actions can be taken in order to move the operation to a desired region.

10.6.2 Practical design of a binary separation

In order to illustrate the key steps involved in design a practical separation, consider a binary mixture of two components that obey a Langmuir adsorption isotherm. The first step is to measure the Henry constants of the two solutes. This can be achieved by injecting a small dilute pulse of the two components in the stationary-phase–solvent combination that will be eventually used for the preparative separation. If the purification campaign is a significant one, then it is also desirable to measure non-linear isotherm parameters. Several methods are available and are described elsewhere. Once these parameters are available, and the feed concentrations are known, the separation regions on the (m_1, m_4) plane and (m_2, m_3) plane

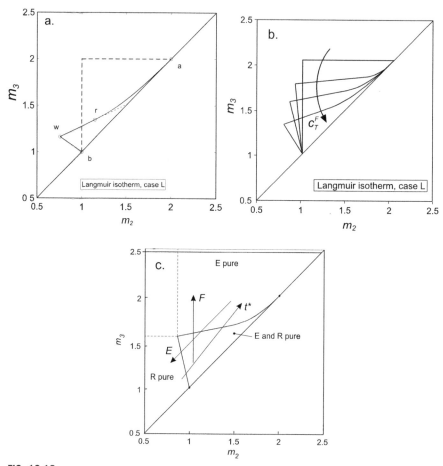

FIG. 10.10

Region of complete separation on (A) (m_2, m_3) plane for a system following a Langmuir isotherm; (B) effect of feed concentration on the region of complete separation region; (C) effect of varying key operating conditions: feed flow rate (F), extract flow rate (E) and switch time (t*) on (m_2, m_3) plane. For the region in (A) $H_A = 2$; $H_B = 1$; $K_A = K_B = 0.1$ L/g and $c_A = c_B = 1.5$ g/L.

Reproduced with permission from Journal of Chromatography A, *Elsevier.*

can be explicitly calculated. From these diagrams, the values of (m_1, m_2, m_3, m_4) can be selected based on the desired separation outcome. Note that in order to operate the SMB, the four internal flow rates (Q_1, Q_2, Q_3, Q_4) and the switch time, t^*, are required. Selecting the values of (m_1, m_2, m_3, m_4) satisfies four of the requirements. The fifth requirement typically arises from the maximum pressure drop that can be tolerated in the system (often limited by the maximum pressure achievable by the desorbent pump) or the velocity at which the maximum efficiency can be achieved.

Table 10.1 Mathematical expressions to calculate the boundaries of the region of complete separation region for a system following Langmuir isotherm as shown in Fig. 10.10.

Line or quantity	Expression
ab	$m_2 = m_3$
ar	$m_3 = m_2 + \left(\sqrt{m_2} - \sqrt{H_A}\right)^2 / (K_A c_A^F)$
rw	$m_2 H_A(\omega_1^F - H_B) + m_3 H_B(H_A - \omega_1^F) = \omega_1^F \omega_2^F(H_A - H_B)$
bw	$K_A c_A^F H_B m_3 + m_2[H_A - H_B(1 + K_A c_A^F)] = H_B(H_A - H_B)$
$m_{1,min}$	H_A
$m_{4,max}$	$\frac{1}{2}\left\{ m_3 + H_B + K_B c_B^F(m_3 - m_2) - \sqrt{[m_3 + H_B + K_B c_B^F(m_3 - m_2)]^2 - 4m_3 H_B} \right\}$

The maximum pressure drop constraint often provides the 'minimum switch time'. Solving Eq. (10.30) along with the suitable expression for the column pressure drop yields the values of m_j and t^*. Once the internal flow rates are determined, the controllable external flow rates, namely desorbent flow rate, D, extract flow rate, E, feed flow rate, F and raffinate flow rate, R, can be obtained using the following mass balance relationships:

$$D = Q_1 \tag{10.31}$$

$$E = Q_1 - Q_2 \tag{10.32}$$

$$F = Q_2 - Q_3 \tag{10.33}$$

$$R = Q_3 - Q_4 \tag{10.34}$$

Based on these values, the SMB can be started and the product qualities monitored. Measuring the purities of the extract and raffinate products, the operating parameters can be manually adjusted to reach the desired operating condition, using Fig. 10.10C.

10.6.3 Advanced configurations

In this chapter, the basic principles and the design of a 'classical' SMB was described. The cyclic, intermittent nature of the process that makes the process appear complex, in fact provides also many additional degrees of freedom that can be exploited to improve the performance. A number of such variations have been introduced such as modification of feed concentration (ModiCon) [34], variation in flow rates (PowerFeed) [35], independent switching of ports (VariCol) [36], selective collection of product fractions (Partial Discard) [37], re-introducing part of the product to the feed (Backfill) [38], addition of a column upstream of the feed port (FeedCol)

[39], over-purification followed by dilution (Bypass) [40], intermittent operations (I-SMB) [41,42], enriching the extract stream before re-introduction (enriched extract) [43], solvent [44] and temperature gradient [45] operation have been introduced. Extensive research has also been conducted into process optimisation [46,47] and control [48–51] of multi-column chromatographic processes. Simulated moving bed configurations for ternary separations have also been explored [52–54]. Another line of inquiry has also been into reducing the number of columns in the processes [55–57].

A significant advancement of the SMB concept has been seen in the separation of biomolecules where gradients in pH or solvent strength is required in order to elute components with a wide variety of adsorption affinities [58,59]. Further, very often these represent multi-component separations, where the target product elutes in between others. The multi-column solvent gradient processes (MCSGP) achieves this by a clever way of switching columns and introducing gradients [60,61]. This process is commercialised and has been used for the separation of a wide variety of bio molecules. Downstream separations, such as chromatography, have traditionally been a bottleneck in transitioning from batch operations to continuous operations. Continuous chromatography has played a significant role in enabling this transition which is expected to improve productivity of pharmaceutical processes and make them environmentally friendly [59].

10.7 Performance evaluation of chromatographic processes

In this chapter a variety of process configurations have been discussed starting from a simple elution chromatography to complex multi-column processes. Key metrics that are used to assess these processes are:

$$\text{Purity} = \frac{\text{Moles/Mass of the desired species in the product}}{\text{Moles/Mass of all species in the product}} \qquad (10.35)$$

$$\text{Recovery} = \frac{\text{Moles/Mass of the desired species in the product}}{\text{Moles/Mass of desired species in the feed}} \qquad (10.36)$$

$$\text{Productivity} = \frac{\text{Moles/Mass of the desired species collected in the product per unit time}}{\text{Mass/Volume of stationary phase}}$$
$$(10.37)$$

$$\text{Solvent consumption} = \frac{\text{Mass/Volume of solvent used}}{\text{Moles/Mass of desired species in the product}} \qquad (10.38)$$

Most chromatographic processes are designed to provide high-purity products. For the separation of high-value products, high recovery is desired. High productivity results in smaller footprint and a low solvent consumption leads to reduced operating

Table 10.2 Qualitative comparison of various modes of operation of chromatography processes.

Process	Adv/ disadv	kilo-Lab (few kgs)	Multi-purpose plant (100 kgs)	Large-scale dedicated plant (tonnes)
HPLC	+	• Versatility • Easy development • Multi-component separations possible	• Versatility • Easy development • Good if associated with solvent-recycling • Low productivity	• Easy to scale-up • Robust operation
	−	• Low productivity • Handling large solvent volumes		• Low productivity • High solvent consumption
SFC	+	• High productivity • Low organic solvent consumption • Easy development • Multi-component separations possible	• High productivity • Low organic solvent consumption	• High productivity • Low organic solvent consumption
	−	• Limited versatility (normal-phase separations) • Limited solubility	• Limited application • CO_2 consumption can be high owing to loss in collected fractions	• Very high investment costs linked to high-pressure operations • Solvent management requires additional infrastructure
SSR	+	• High productivity • Low solvent consumption	• Good trade-off between performance and investment costs	• High productivity • Low solvent consumption
	−	• Complex to design and implement • Typically limited to binary separations	• Complex to design and implement • Typically limited to binary separations	• Less robust compared to HPLC and SMB

Continued

Table 10.2 Qualitative comparison of various modes of operation of chromatography processes—*cont'd*

Process	Adv/ disadv	kilo-Lab (few kgs)	Multi-purpose plant (100 kgs)	Large-scale dedicated plant (tonnes)
MCC	+	• Very high productivity • Very low solvent consumption	• Very high productivity • Very low solvent consumption • High robustness	• Low production costs • High robustness at this scale • High investment costs
	−	• Complex equipment • Multi-component separations are complicated • Method development and optimisation is time consuming	• Complex equipment • Multi-component separations are complicated	

Abbreviations: HPLC, high-performance liquid chromatography; SFC, supercritical fluid chromatography; SSR, steady-state recycling; MCC, multi-column chromatography.
Reproduced with permission from Current Opinions in Chemical Engineering, Elsevier.

costs. Table 10.2 provides a qualitative guide that compares the various processes and how to select them for appropriate scales.

10.8 Conclusions

Chromatography that was considered primarily as an analytical separation process has made major progress in the past two decades. Now, it is considered as a well-established technique that is routinely used in the separation of high-value added products. Developments in the synthesis of novel stationary phases have made it possible to use more benign solvents and reduce their usage. Robust equipment and instrumentation have also made it possible to achieve a high level of optimised operation. The move from batch to continuous operation has been the most important development in the recent years. Chromatography has benefitted from culmination of appropriate equipment, design methods, and the availability of a wide variety of stationary phases. In the coming years, chromatography is expected to make major strides in the purification of biomolecules and contribute to the transition from batch to continuous manufacturing in areas such as bioprocessing.

References

[1] G. Guiochon, A. Fellinger, S. Golshan-Shirazi, Fundamentals of Preparative and Non-linear Chromatography, Academic Press, Boston, 2006, https://doi.org/10.1016/0003-2670(95)90109-4.

[2] R.-M. Nicoud, Chromatographic Processes, Cambridge University Press, 2015, https://doi.org/10.1017/cbo9781139998284.

[3] D.M. Ruthven, Principles of Adsorption and Adsorption Processes, John Wiley & Sons, 1984.

[4] P.C. Wankat, Large-Scale Adsorption and Chromatography, vols. 1 and 2, CRC Press, Boca Raton, FL, 1986.

[5] L. Miller, J.D. Pinkston, L.T. Taylor, Modern Supercritical Fluid Chromatography: Carbon Dioxide Containing Mobile Phases, John Wiley & Sons, 2019.

[6] G. Guiochon, Preparative liquid chromatography, J. Chromatogr. A 965 (1–2) (2002) 129–161.

[7] H. Schmidt-Traub, M. Schulte, A. Seidel-Morgenstern, Preparative Chromatography, second ed., 2013, https://doi.org/10.1002/9783527649280.

[8] A. Seidel-Morgenstern, Experimental determination of single solute and competitive adsorption isotherms, J. Chromatogr. A 1037 (1–2) (2004) 255–272, https://doi.org/10.1016/j.chroma.2003.11.108.

[9] D. DeVault, On the theory of chromatography, J. Am. Chem. Soc. 65 (4) (1943) 532–540.

[10] H.K. Rhee, R. Aris, N.R. Amundson, First-Order Partial Differential Equations, Vol. 2, Dover Publications, Mineola, NY, 2001.

[11] E. Glueckauf, Theory of chromatography: VII. The general theory of two solutes following non-linear isotherms, Discuss. Faraday Soc. 7 (1949) 12–25, https://doi.org/10.1039/DF9490700012.

[12] H.K. Rhee, R. Aris, N.R. Amundson, First-Order Partial Differential Equations, Vol. 1, Dover Publications, Mineola, NY, 2001.

[13] M. Mazzotti, A. Rajendran, Equilibrium theory-based analysis of nonlinear waves in separation processes, Annu. Rev. Chem. Biomol. Eng. 4 (1) (2013) 119–141, https://doi.org/10.1146/annurev-chembioeng-061312-103318.

[14] D. Basmadjian, The Little Adsorption Book: A Practical Guide for Engineers and Scientists, CRC Press, 1996.

[15] P.D. Lax, Hyperbolic Systems of Conservation Laws and the Mathematical Theory of Shock Waves, SIAM, 1973.

[16] E. Glueckauf, Theory of chromatography: part 10.—Formula for diffusion into spheres and their application to chromatography, Trans. Faraday Soc. 51 (1955) 1540–1551, https://doi.org/10.1039/TF9555101540.

[17] A.J. Martin, R.L. Synge, A new form of chromatogram employing two liquid phases: a theory of chromatography. 2. Application to the Micro-determination of the higher monoamino-acids in proteins, Biochem. J. 35 (12) (1941) 1358.

[18] J. van Deemter, F. Zuiderweg, A. Klinkenberg, Longitudinal diffusion and resistance to mass transfer as causes of nonideality in chromatography, Chem. Eng. Sci. 5 (6) (1956) 271–289.

[19] F. Gritti, G. Guiochon, The van Deemter equation: assumptions, limits, and adjustment to modern high performance liquid chromatography, J. Chromatogr. A (2013), https://doi.org/10.1016/j.chroma.2013.06.032.

[20] R. Hayes, A. Ahmed, T. Edge, H. Zhang, Core-shell particles: preparation, fundamentals and applications in high performance liquid chromatography, J. Chromatogr. A (2014), https://doi.org/10.1016/j.chroma.2014.05.010.

[21] C.J. Welch, N. Wu, M. Biba, R. Hartman, T. Brkovic, X. Gong, R. Helmy, W. Schafer, J. Cuff, Z. Pirzada, et al., Greening analytical chromatography, Trends Anal. Chem. (2010), https://doi.org/10.1016/j.trac.2010.03.008.

[22] C.M. Grill, L. Miller, Separation of a racemic pharmaceutical intermediate using closed-loop steady state recycling, J. Chromatogr. A 827 (2) (1998) 359–371, https://doi.org/10.1016/S0021-9673(98)00772-9.

[23] T. Sainio, M. Kaspereit, Analysis of steady state recycling chromatography using equilibrium theory, Sep. Purif. Technol. 66 (1) (2009) 9–18, https://doi.org/10.1016/j.seppur.2008.12.005.

[24] J. Siitonen, T. Sainio, M. Kaspereit, Theoretical analysis of steady state recycling chromatography with solvent removal, Sep. Purif. Technol. 78 (1) (2011) 21–32, https://doi.org/10.1016/j.seppur.2011.01.013.

[25] A. Rajendran, M. Mazzotti, Local equilibrium theory for the binary chromatography of species subject to a generalized Langmuir isotherm. 2. Wave interactions and chromatographic cycle, Ind. Eng. Chem. Res. 50 (1) (2011) 352–377, https://doi.org/10.1021/ie1015798.

[26] D.B. Broughton, C. Gerhold, Continuous Sorption Process Employing Fixed Bed of Sorbent and Moving Inlets and Outlets, 2985589, 1961.

[27] D.M. Ruthven, C.B. Ching, Counter-current and simulated counter-current adsorption separation processes, Chem. Eng. Sci. 44 (5) (1989) 1011–1038, https://doi.org/10.1016/0009-2509(89)87002-2.

[28] A. Rajendran, G. Paredes, M. Mazzotti, Simulated moving bed chromatography for the separation of enantiomers, J. Chromatogr. A 1216 (4) (2009) 709–738, https://doi.org/10.1016/j.chroma.2008.10.075.

[29] G. Storti, M. Mazzotti, M. Morbidelli, S. Carrà, Robust design of binary countercurrent adsorption separation processes, AIChE J. 39 (3) (1993) 471–492, https://doi.org/10.1002/aic.690390310.

[30] K. Kaczmarski, D.P. Poe, G. Guiochon, Numerical modeling of the elution peak profiles of retained solutes in supercritical fluid chromatography, J. Chromatogr. A 1218 (37) (2011) 6531–6539, https://doi.org/10.1016/j.chroma.2011.07.022.

[31] D.C.S. Azevedo, A.E. Rodrigues, Design of a simulated moving bed in the presence of mass-transfer resistances, AICHE J. 45 (5) (1999) 956–966, https://doi.org/10.1002/aic.690450506.

[32] M. Mazzotti, G. Storti, M. Morbidelli, Robust design of countercurrent adsorption separation processes: 4. Desorbent in the feed, AIChE J. 43 (1) (1997) 64–72, https://doi.org/10.1002/aic.690430109.

[33] M. Mazzotti, Equilibrium theory based design of simulated moving bed processes for a generalized Langmuir isotherm, J. Chromatogr. A 1126 (1–2) (2006) 311–322, https://doi.org/10.1016/j.chroma.2006.06.022.

[34] H. Schramm, A. Kienle, M. Kaspereit, A. Seidel-Morgenstern, Improved operation of simulated moving bed processes through cyclic modulation of feed flow and feed concentration, Chem. Eng. Sci. 58 (23–24) (2003) 5217–5227, https://doi.org/10.1016/j.ces.2003.08.015.

[35] Z. Zhang, M. Mazzotti, M. Morbidelli, PowerFeed operation of simulated moving bed units: changing flow-rates during the switching interval, J. Chromatogr. A 1006 (1–2) (2003) 87–99, https://doi.org/10.1016/S0021-9673(03)00781-7.

[36] O. Ludemann-Hombourger, R.M. Nicoud, M. Bailly, The "VARICOL" process: a new multicolumn continuous chromatographic process, Sep. Sci. Technol. 35 (12) (2000) 1829–1862, https://doi.org/10.1081/SS-100100622.

[37] Y.S. Bae, C.H. Lee, Partial-discard strategy for obtaining high purity products using simulated moving bed chromatography, J. Chromatogr. A 1122 (1–2) (2006) 161–173, https://doi.org/10.1016/j.chroma.2006.04.040.

[38] K.M. Kim, C.H. Lee, Backfill-simulated moving bed operation for improving the separation performance of simulated moving bed chromatography, J. Chromatogr. A (2013), https://doi.org/10.1016/j.chroma.2013.08.058.

[39] H.H. Lee, K.M. Kim, C.H. Lee, Improved performance of simulated moving bed process using column-modified feed, AIChE J. (2011), https://doi.org/10.1002/aic.12428.

[40] R.T. Maruyama, P. Karnal, T. Sainio, A. Rajendran, Design of bypass-simulated moving bed chromatography for reduced purity requirements, Chem. Eng. Sci. (2019), https://doi.org/10.1016/j.ces.2019.05.003.

[41] S. Katsuo, M. Mazzotti, Intermittent simulated moving bed chromatography: 1. Design criteria and cyclic steady-state, J. Chromatogr. A 1217 (8) (2010) 1354–1361.

[42] S. Jermann, S. Katsuo, M. Mazzotti, Intermittent simulated moving bed processes for chromatographic three-fraction separation, Org. Process Res. Dev. 16 (2) (2012) 311–322, https://doi.org/10.1021/op200239e.

[43] G. Paredes, H.K. Rhee, M. Mazzotti, Design of simulated-moving-bed chromatography with enriched extract operation (EE-SMB): Langmuir isotherms, Ind. Eng. Chem. Res. (2006), https://doi.org/10.1021/ie060256z.

[44] S. Abel, M. Mazzotti, M. Morbidelli, Solvent gradient operation of simulated moving beds—2. Langmuir isotherms, J. Chromatogr. A 1026 (1–2) (2004) 47–55, https://doi.org/10.1016/j.chroma.2003.11.054.

[45] C. Migliorini, M. Wendlinger, M. Mazzotti, M. Morbidelli, Temperature gradient operation of a simulated moving bed unit, Ind. Eng. Chem. Res. (2001), https://doi.org/10.1021/ie000825h.

[46] A. Toumi, S. Engell, Optimization-based control of a reactive simulated moving bed process for glucose isomerization, Chem. Eng. Sci. 59 (18) (2004) 3777–3792, https://doi.org/10.1016/j.ces.2004.04.009.

[47] Z. Zhang, K. Hidajat, A.K. Ray, M. Morbidelli, Multiobjective optimization of SMB and varicol process for chiral separation, AIChE J. 48 (12) (2002) 2800–2816, https://doi.org/10.1002/aic.690481209.

[48] K.U. Klatt, F. Hanisch, G. Dünnebier, Model-based control of a simulated moving bed chromatographic process for the separation of fructose and glucose, J. Process Control 12 (2) (2002) 203–219, https://doi.org/10.1016/S0959-1524(01)00005-1.

[49] G. Erdem, S. Abel, M. Morari, M. Mazzotti, M. Morbidelli, Automatic control of simulated moving beds II: nonlinear isotherm, Ind. Eng. Chem. Res. 43 (14) (2004) 3895–3907, https://doi.org/10.1021/ie0342154.

[50] C. Grossmann, M. Amanullah, M. Mazzotti, M. Morbidelli, M. Morari, Optimizing control of variable cycle time simulated moving beds, IFAC Proc. Vol. (2007) 201–206, https://doi.org/10.3182/20070606-3-MX-2915.00081.

[51] C. Langel, C. Grossmann, M. Morbidelli, M. Morari, M. Mazzotti, Implementation of an automated on-line high-performance liquid chromatography monitoring system for "cycle to cycle" control of simulated moving beds, J. Chromatogr. A 1216 (50) (2009) 8806–8815, https://doi.org/10.1016/j.chroma.2009.02.005.

[52] J. Nowak, D. Antos, A. Seidel-Morgenstern, Theoretical study of using simulated moving bed chromatography to separate intermediately eluting target compounds, J. Chromatogr. A 1253 (2012) 58–70, https://doi.org/10.1016/j.chroma.2012.06.096.

[53] J.S. Hur, P.C. Wankat, Two-zone SMB/chromatography for center-cut separation from ternary mixtures: linear isotherm systems, Ind. Eng. Chem. Res. 45 (4) (2006) 1426–1433, https://doi.org/10.1021/ie058046u.

[54] S.H. Jin, P.C. Wankat, New design of simulated moving bed (SMB) for ternary separations, Ind. Eng. Chem. Res. 44 (6) (2005) 1906–1913, https://doi.org/10.1021/ie040164e.

[55] J.M.M. Araújo, R.C.R. Rodrigues, M.F.J. Eusébio, J.P.B. Mota, Chiral separation by two-column, semi-continuous, open-loop simulated moving-bed chromatography, J. Chromatogr. A 1217 (33) (2010) 5407–5419.

[56] R.C.R. Rodrigues, R.J.S. Silva, J.P.B. Mota, Streamlined, two-column, simulated countercurrent chromatography for binary separation, J. Chromatogr. A 1217 (20) (2010) 3382–3391, https://doi.org/10.1016/j.chroma.2010.03.009.

[57] R.J.S. Silva, R.C.R. Rodrigues, J.P.B. Mota, Relay simulated moving bed chromatography: concept and design criteria, J. Chromatogr. A 1260 (2012) 132–142, https://doi.org/10.1016/j.chroma.2012.08.076.

[58] G. Carta, A. Jungbauer, Protein Chromatography: Process Development and Scale-Up, 2010, https://doi.org/10.1002/9783527630158.

[59] D. Pfister, L. Nicoud, M. Morbidelli, Continuous Biopharmaceutical Processes: Chromatography, Bioconjugation, and Protein Stability, Cambridge University Press, 2018.

[60] L. Aumann, M. Morbidelli, Method and Device for Chromatographic Purification, ETH Zurich, 2012, pp. 1–42.

[61] L. Aumann, M. Morbidelli, A continuous multicolumn countercurrent solvent gradient purification (MCSGP) process, Biotechnol. Bioeng. 98 (5) (2007) 1043–1055, https://doi.org/10.1002/bit.21527.

Further reading

H.K. Rhee, R. Aris, N.R. Amundson, First-Order Partial Differential Equations, vol. 1, Dover Publications, Mineola, NY, 2001.

D. Basmadjian, The Little Adsorption Book: A Practical Guide for Engineers and Scientists, CRC Press, 1996.

G. Guiochon, A. Fellinger, S. Golshan-Shirazi, Fundamentals of Preparative and Nonlinear Chromatography, Academic Press, Boston, 2006, https://doi.org/10.1016/0003-2670(95)90109-4.

G. Carta, A. Jungbauer, Protein Chromatography: Process Development and Scale-Up, 2010, https://doi.org/10.1002/9783527630158.

D. Pfister, L. Nicoud, M. Morbidelli, Continuous Biopharmaceutical Processes: Chromatography, Bioconjugation, and Protein Stability, 2018.

R.-M. Nicoud, Chromatographic Processes, Cambridge University Press, 2015, https://doi.org/10.1017/cbo9781139998284.

D.M. Ruthven, Principles of Adsorption and Adsorption Processes, John Wiley & Sons, 1984.

P.C. Wankat, Large-Scale Adsorption and Chromatography, vols. 1 and 2, CRC Press, Boca Raton, Florida, 1986.

H. Schmidt-Traub, M. Schulte, A. Seidel-Morgenstern, Preparative Chromatography, second ed., 2013, https://doi.org/10.1002/9783527649280.

Membrane separation processes

11

Ajay Kumar Ray

Department of Chemical and Biochemical Engineering, University of Western Ontario, London, ON, Canada

Nomenclature

		Units in SI system	Dimensions in M, N, L, T, θ, A
A_m	membrane area	m^2	$\mathbf{L}^2$
$a \ldots a_n$	osmotic coefficients	N/m^2	$\mathbf{ML}^{-1}\mathbf{T}^{-2}$
b	channel height	m	$\mathbf{L}$
C	solute concentration (mass fraction)	–	–
C_b	bulk concentration of particles	m^3/m^3	–
C_i	concentration of ions	$kmol/m^3$	$\mathbf{NL}^{-3}$
C_G	solute concentration in gel layer	–	–
C_f	solute concentration in bulk of feed	–	–
C_p	solute concentration in permeate	–	–
C_w	solute concentration at membrane	–	–
D_L	diffusion coefficient for liquid	m^2/s	$\mathbf{L}^2\mathbf{T}^{-1}$
d_e	channel hydraulic diameter	m	$\mathbf{L}$
e	void fraction (porosity)	–	–
$\mathbf{F}$	Faraday's constant	9.649×10^7 C/kmol	$\mathbf{N}^{-1}\mathbf{TA}$

Continued

Coulson and Richardson's Chemical Engineering. https://doi.org/10.1016/B978-0-08-101097-6.00011-0

		Units in SI system	Dimensions in M, N, L, T, θ, A
h_D	mass transfer coefficient	m/s	$\mathbf{LT^{-1}}$
i_{lim}	limiting current density	A/m^2	$\mathbf{L^{-2}A}$
J	membrane permeation rate (flux)	m/s	$\mathbf{LT^{-1}}$
J_{ss}	steady state membrane permeation rate	m/s	$\mathbf{LT^{-1}}$
l	thickness of boundary film	m	$\mathbf{L}$
ΔP	hydraulic pressure difference	N/m^2	$\mathbf{ML^{-1}T^{-2}}$
$\mathbf{R}$	universal gas constant	8314 J/kmolK	$\mathbf{MN^{-1}L^2T^{-2}\theta^{-1}}$
R	rejection coefficient	–	–
R_c	resistance of deposited layers	m^{-1}	$\mathbf{L^{-1}}$
R_f	resistance of film layer	m^{-1}	$\mathbf{L^{-1}}$
R_m	resistance of membrane	m^{-1}	$\mathbf{L^{-1}}$
$\mathbf{r}$	specific resistance of deposit	m^{-2}	$\mathbf{L^{-2}}$
T	absolute temperature	K	θ
t	time	s	$\mathbf{T}$
$\mathbf{t}_m$	transference number in membrane	–	–
$\mathbf{t}_s$	transference number in solution	–	–
u	cross-flow velocity	m/s	$\mathbf{LT^{-1}}$
v	partial molar volume	m^3/kmol	$\mathbf{N^{-1}L^3}$
V	volume of filtrate	m^3	$\mathbf{L^3}$
V_s	skeletal volume of particles deposited	m^3	$\mathbf{L^3}$
y	distance from membrane	m	$\mathbf{L}$
z	valence of ion	–	–
β	efficiency factor	–	–
ϕ	electrical potential	V	$\mathbf{ML^2T^{-3}A^{-1}}$
{\skew0 \dot {γ}}	shear rate at membrane surface	s^{-1}	$\mathbf{T^{-1}}$
μ	viscosity of permeate	Ns/m^2	$\mathbf{ML^{-1}T^{-1}}$
μ_i	chemical potential	J/kmol	$\mathbf{MN^{-1}L^2T^{-2}}$
Π	osmotic pressure	N/m^2	$\mathbf{ML^{-1}T^{-2}}$
ρ	fluid density	kg/m^3	$\mathbf{ML^{-3}}$

11.1 **Introduction**

Whilst effective product separation is crucial to economic operation in the process industries, certain types of materials are inherently difficult and expensive to separate. Important examples include:

(a) Finely dispersed solids, especially those which are compressible, and which have a density close to that of the liquid phase, have high viscosity, or are gelatinous.

(b) Low molecular weight, non-volatile organics, or pharmaceuticals and dissolved salts.

(c) Biological materials which are very sensitive to their physical and chemical environment.

The processing of these categories of materials has become increasingly important in recent years, especially with the growth of the newer biotechnological industries and with the increasingly sophisticated nature of processing in the food industries. When difficulties arise in the processing of materials of biological origin, it is worth asking, how does nature solve the problem? The solution which nature has developed is likely to be both highly effective and energy efficient, though it may be slow in process terms. Nature separates biologically active materials by means of membranes. As Strathmann [1] has pointed out, a membrane may be defined as 'an interphase separating two phases and selectively controlling the transport of materials between those phases'. A membrane is an interphase rather than an interface because it occupies a finite, though normally small, element of space. Human beings are all surrounded by a membrane, the skin, and membranes control the separation of materials at all levels of life, down to the outer layers of bacteria and subcellular components.

As discussed by Lonsdale [2], since the 1960s a new technology using synthetic membranes for process separations has been rapidly developed by materials scientists, physical chemists and chemical engineers. Such membrane separations have been widely applied to a range of conventionally difficult separations. They potentially offer the advantages of ambient temperature operation, relatively low capital and running costs, and modular construction. In this chapter, the nature and scope of membrane separation processes are outlined, and then those processes most frequently used industrially are described more fully.

11.2 **Classification of membrane processes**

Industrial membrane processes may be classified according to the size range of materials which they are to separate and the driving force used in separation. There is always a degree of arbitrariness about such classifications, and the distinctions which

Table 11.1 Classification of membrane separation processes for liquid systems.

Name of process	Driving force	Separation size range	Examples of materials separated
Microfiltration	Pressure gradient	10–0.1 μm	Small particles, large colloids, microbial cells
Ultrafiltration	Pressure gradient	<0.1 μm– 5 nm	Emulsions, colloids, macromolecules, proteins
Nanofiltration	Pressure gradient	~1 nm	Dissolved salts, organics
Reverse osmosis (hyperfiltration)	Pressure gradient	<1 nm	Dissolved salts, small organics
Electrodialysis	Electric field gradient	<5 nm	Dissolved salts
Dialysis	Concentration gradient	<5 nm	Treatment of renal failure

are typically drawn are shown in Table 11.1. This chapter is primarily concerned with the pressure driven processes, microfiltration (MF), ultrafiltration (UF), nanofiltration (NF) and reverse osmosis (RO). These are already well-established large-scale industrial processes. For example, reverse osmosis is used world-wide for the desalination of brackish water, with more than 1000 units in operation. Plants capable of producing up to 10^5 m^3/day of drinking water are in operation. As a further example, it is now standard practice to include an ultrafiltration unit in paint plants in the car industry. The resulting recovery of paint from wash waters can produce savings of 10% to 30% in paint usage, and allows recycling of the wash waters. The use of reverse osmosis and ultrafiltration in the dairy industry has led to substantial changes in production techniques and the development of new types of cheeses and related products. Nanofiltration is a process, with characteristics between those of ultrafiltration and reverse osmosis, which is finding increasing application in pharmaceutical processing and water treatment. Electrodialysis is a purely electrically driven separation process used extensively for the desalination or concentration of brackish water. There are about 300 such plants in operation. Economics presently favour reverse osmosis, however, rather than electrodialysis for such separations. The major use of dialysis is in haemodialysis of patients with renal failure, where it is most appropriate to use such a gentle technique. Haemodialysis poses many interesting problems of a chemical engineering nature, although dialysis is a relatively slow process not really suited to large-scale industrial separations.

11.3 The nature of synthetic membranes

Membranes used for the pressure-driven separation processes, microfiltration, ultrafiltration and reverse osmosis, as well as those used for dialysis, are most commonly made of polymeric materials [1]. Initially, most such membranes were cellulosic in nature. These are now being replaced by polyamide, polysulphone, polycarbonate, and a number of other advanced polymers. These synthetic polymers have improved chemical stability and better resistance to microbial degradation. Membranes have most commonly been produced by a form of phase inversion known as immersion precipitation. This process has four main steps: (a) the polymer is dissolved in a solvent to 10% to 30% by mass, (b) the resulting solution is cast on a suitable support as a film of thickness, approximately 100 μm, (c) the film is quenched by immersion in a non-solvent bath, typically water or an aqueous solution, (d) the resulting membrane is annealed by heating. The third step gives a polymer-rich phase forming the membrane, and a polymer-depleted phase forming the pores. The ultimate membrane structure results as a combination of phase separation and mass transfer, variation of the production conditions giving membranes with different separation characteristics. Most microfiltration membranes have a symmetric pore structure, and they can have a porosity as high as 80%. Ultrafiltration and reverse osmosis membranes have an asymmetric structure comprising a 1- to 2-μm-thick top layer of finest pore size supported by a ~100 μm thick more openly porous matrix, as shown in Fig. 11.1.

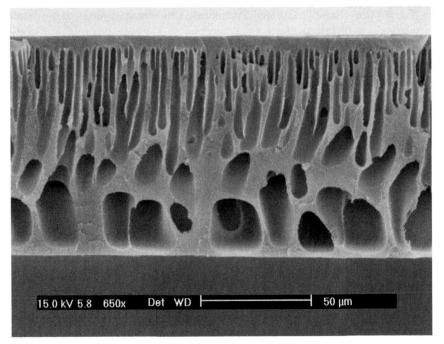

FIG. 11.1

Electron micrograph of a section of an asymmetric ultrafiltration membrane showing finely porous 'skin' layer on more openly porous supporting matrix.

Courtesy of Dr. Huabing Yin.

Such an asymmetric structure is essential if reasonable membrane permeation rates are to be obtained. Another important type of polymeric membrane is the thin-film composite membrane. This consists of an extremely thin layer, typically ~1 μm, of finest pore structure deposited on a more openly porous matrix. The thin layer is formed by phase inversion or interfacial polymerisation on to an existing microporous structure. Polymeric membranes are most commonly produced in the form of flat sheets, but they are also widely produced as tubes of diameter 10 to 25 mm and in the form of hollow fibres of diameter 0.1 to 2.0 mm.

A significant recent advance has been the development of microfiltration and ultrafiltration membranes composed of inorganic oxide materials. These are presently produced by two main techniques: (a) deposition of colloidal metal oxide on to a supporting material such as carbon, and (b) as purely ceramic materials by high-temperature sintering of spray-dried oxide microspheres. Other innovative production techniques lead to the formation of membranes with very regular pore structures. Zirconia, alumina, and titania are the materials most commonly used. The main advantages of inorganic membranes compared with the polymeric types are their higher temperature stability, allowing steam sterilisation in biotechnological and food applications, increased resistance to fouling, and narrower pore size distribution.

The physical characterisation of a membrane structure is important if the correct membrane is to be selected for a given application. The pore structure of microfiltration membranes is relatively easy to characterise, atomic force microscopy and electron microscopy being the most convenient methods and allowing the three-dimensional structure of the membrane to be determined. The limit of resolution of a simple electron microscope is about 10 nm, and that of an atomic force microscope is <1 nm, as shown in Fig. 11.2. Additional characterisation techniques, such as the bubble point, mercury intrusion or permeability methods, use measurements of the permeability of membranes to fluids. Both the maximum pore size and the pore size distribution may be determined. A parameter often quoted in manufacturer's literature is the nominal molecular weight cut-off (MWCO) of a membrane. This is based on studies of how solute molecules are rejected by membranes. A solute will pass through a membrane if it is sufficiently small to pass through a pore, if it does not significantly interact with the membrane and if it does not interact with other, larger solutes. It is possible to define a solute rejection coefficient R by:

$$R = 1 - \left(C_p/C_f\right) \tag{11.1}$$

where C_f is the concentration of solute in the feed stream and C_p is the concentration of solute in the permeate. For a given ultrafiltration membrane with a distribution of pore sizes, there is a relationship between R and the solute molecular weight, as shown in Fig. 11.3. The nominal molecular weight cut-off is normally defined as the molecular weight of a solute for which $R = 0.95$. Values of MWCO typically lie in the range of 2000 to 100,000 kg/kmol with values of the order of 10,000 being most common. High resolution electron microscopy does not allow the resolution of an extensive pore structure in the separating layer of reverse-osmosis membranes.

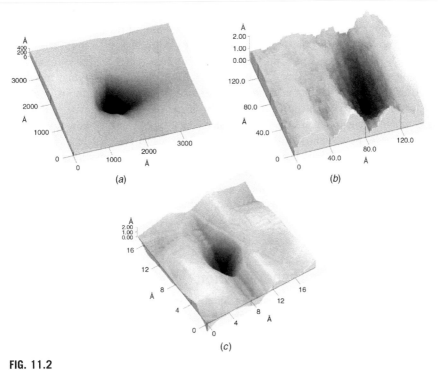

FIG. 11.2

AFM images of single pores in (A) microfiltration, (B) ultrafiltration and (C) nanofiltration membranes.

Courtesy Dr. Nichal Nidal.

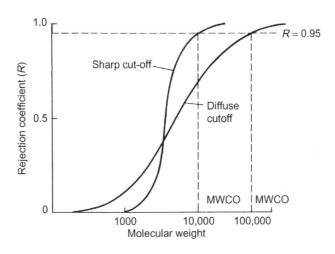

FIG. 11.3

Dependence of rejection coefficient on molecular weight for ultrafiltration membranes.

As discussed later, it is generally considered that reverse osmosis membranes do not contain pores and that they operate mainly by a 'solution-diffusion' mechanism.

Ion-exchange membranes, which are used for electrodialysis, usually consist of highly swollen charged gels prepared either by dispersing a conventional ion-exchange material in a polymer matrix, or from a homogenous polymer in which electrically charged groups such as sulphonic, carboxylic, or quarternised amine groups have been introduced as discussed by Lacey [3]. The first type is referred to as a heterogeneous membrane, whilst the second type is termed a homogeneous membrane. A membrane with fixed positive charges is referred to as an anion exchange membrane since it may bind and hence selectively transport anions from the surrounding solution. Similarly, a membrane containing fixed negative charges is termed a cation exchange membrane. Ion-exchange membranes exclude, that is, do not bind and do not allow the transport of, ions which bear charges of the same sign as the membrane.

11.4 General membrane equation

It is not possible at present to provide an equation, or set of equations, that allows the prediction from first principles of the membrane permeation rate and solute rejection for a given real separation. Research aimed at providing such a prediction for model systems is under way, although the physical properties of real systems, both the membrane and the solute, are complex. An analogous situation exists for conventional filtration processes. The *general membrane equation* is an attempt to state the factors which may be important in determining the membrane permeation rate for pressure driven processes. This takes the form:

$$J = \frac{|\Delta P| - |\Delta \prod|}{(R_m + R_c)\mu} \tag{11.2}$$

where J is the membrane flux,[a] expressed as volumetric rate per unit area, $|\Delta P|$ is the pressure difference applied across the membrane, the transmembrane pressure, $\Delta \Pi$ is the difference in osmotic pressure across the membrane, R_m is the resistance of the membrane, and R_c is the resistance of layers deposited on the membrane, the filter cake and gel foulants. If the membrane is only exposed to pure solvent, say water, then Eq. (11.2) reduces to $J = |\Delta P|/R_m\mu$. For microfiltration and ultrafiltration membranes where solvent flow is most often essentially laminar through an arrangement of tortuous channels, this is analogous to the Carman–Kozeny equation discussed in Chapter 4. Knowledge of such water fluxes is useful for characterising new membranes and also for assessing the effectiveness of membrane cleaning procedures. In the processing of solutes, Eq. (11.2) shows that the transmembrane pressure must exceed the osmotic pressure for flow to occur. It is generally assumed that the

---

[a]Membrane flux is denoted by J, the usual symbol in the literature on membranes. It corresponds with u_c, as used in Chapters 4 and 7 for flow in packed beds and filtration.

osmotic pressure of most retained solutes is likely to be negligible in the cases of microfiltration. The resistance R_c is due to the formation of a filter cake, the formation of a gel when the concentration of macromolecules at the membrane surface exceeds their solubility giving rise to a precipitation, or due to materials in the process feed that adsorb on the membrane surface producing an additional barrier to solvent flow. The separation of a solute by a membrane gives rise to an increased concentration of that solute at the membrane surface, an effect known as concentration polarisation. This may be described in terms of an increase in $\Delta\Pi$. It is within the framework of this equation that the factors influencing membrane permeation rate will be discussed in the following sections.

11.5 Cross-flow microfiltration

The solids–liquid separation of slurries containing particles below 10 µm is difficult by conventional filtration techniques. A conventional approach would be to use a slurry thickener in which the formation of a filter cake is restricted and the product is discharged continuously as a concentrated slurry. Such filters use filter cloths as the filtration medium and are limited to concentrating particles above 5 µm in size. *Dead end* or *frontal* membrane microfiltration, in which the particle containing fluid is pumped directly through a polymeric membrane, is used for the industrial clarification and sterilisation of liquids. Such a process allows the removal of particles down to 0.1 µm or less, but is only suitable for feeds containing very low concentrations of particles as otherwise the membrane becomes too rapidly clogged.

The concept of *cross-flow* microfiltration, described by Bertera et al. [4], is shown in Fig. 11.4 which represents a cross-section through a rectangular or tubular membrane module. The particle-containing fluid to be filtered is pumped at a velocity in the range of 1 to 8 m/s parallel to the face of the membrane and with a pressure difference of 0.1 to 0.5 MN/m^2 (MPa) across the membrane. The liquid permeates through the membrane and the feed emerges in a more concentrated form at the exit of the module.

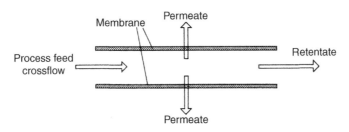

FIG. 11.4

The concept of cross-flow filtration [4].

All of the membrane processes listed in Table 11.1 are operated with such a cross-flow of the process feed. The advantages of cross-flow filtration over conventional filtration are:

(a) A higher overall liquid removal rate is achieved by prevention of the formation of an extensive filter cake.
(b) The process feed remains in the form of a mobile slurry suitable for further processing.
(c) The solids content of the product slurry may be varied over a wide range.
(d) It may be possible to fractionate particles of different sizes.

A flow diagram of a simple cross-flow system [4] is shown in Fig. 11.5. This is the system likely to be used for batch processing or development rigs and is, in essence, a basic pump recirculation loop. The process feed is concentrated by pumping it from the tank and across the membrane in the module at an appropriate velocity. The partially concentrated *retentate* is recycled into the tank for further processing whilst the *permeate* is stored or discarded as required. In cross-flow filtration applications, product washing is frequently necessary and is achieved by a process known as *diafiltration* in which wash water is added to the tank at a rate equal to the permeation rate.

In practice, the membrane permeation rate falls with time due to membrane fouling; that is blocking of the membrane surface and pores by the particulate materials, as shown in Fig. 11.6. The rate of fouling depends on the nature of the materials being processed, the nature of the membrane, the cross-flow velocity and the applied pressure. For example, increasing the cross-flow velocity results in a decreased rate of fouling. Backflushing the membrane using permeate is often used to control fouling as shown in Fig. 11.6C. Further means of controlling membrane fouling are discussed in Section 11.9.

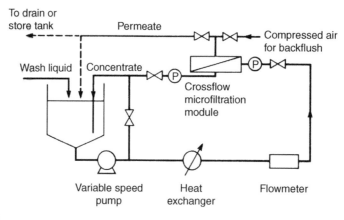

FIG. 11.5

Flow diagram for a simple cross-flow system [4].

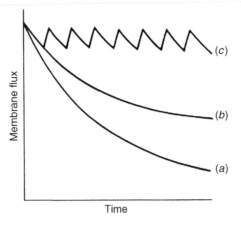

FIG. 11.6

The time-dependence of membrane permeation rate during cross-flow filtration: (a) Low cross-flow velocity, (b) Increased cross-flow velocity, (c) Backflushing at the bottom of each 'saw-tooth'.

Ideally, cross-flow microfiltration would be the pressure-driven removal of the process liquid through a porous medium without the deposition of particulate material. The flux decrease occurring during cross-flow microfiltration shows that this is not the case. If the decrease is due to particle deposition resulting from incomplete removal by the cross-flow liquid, then a description analogous to that of generalised cake filtration theory, discussed in Chapter 7, should apply. Eq. (11.2) may then be written as:

$$J = \frac{|\Delta P|}{(R_m + R_c)\mu} \tag{11.3}$$

where R_c now represents the resistance of the cake, which if all filtered particles remain in the cake, may be written as:

$$R_c = \frac{rVC_b}{A_m} = \frac{rV_s}{A_m} \tag{11.4}$$

where r is the specific resistance of the deposit, V the total volume filtered, V_s the volume of *particles* deposited, C_b the bulk concentration of particles in the feed (particle volume/feed volume) and A_m the membrane area. The specific resistance may theoretically be related to the particle properties for spherical particles by the Carman relationship, discussed in Chapter 4, as:

$$r = 180\left(\frac{1-e}{e^3}\right)\left(\frac{1}{d_s^2}\right) \tag{11.5}$$

where e is the void volume of the cake and d_s the mean particle diameter.

Combining Eqs (11.3), (11.4) gives:

$$J = \frac{1}{A_m}\frac{dV}{dt} = \frac{|\Delta P|}{(R_m + rVC_b/A_m)\mu} \tag{11.6}$$

Solution of Eq. (11.6) for V at constant pressure gives:

$$\frac{t}{V} = \frac{R_m\mu}{|\Delta P|A_m} + \frac{C_b r\mu V}{2|\Delta P|A_m^2} \tag{11.7}$$

yielding a straight line on plotting t/V against V.

Schneider and Klein [5] have pointed out that the early stages of cross-flow microfiltration often follow such a pattern although the growth of the cake is limited by the cross-flow of the process liquid. There are a number of ways of accounting for the control of cake growth. A useful method is to rewrite the resistance model to allow for the dynamics of polarisation in the film layer as discussed by Fane [6]. Eq. (11.3) is then written as:

$$J = \frac{1}{A_m}\frac{dV}{dt} = \frac{|\Delta P|}{(R_m + R_{sd} - R_{sr})\mu} \tag{11.8}$$

where R_{sd} is the resistance that would be caused by deposition of all filtered particles and R_{sr} is the resistance removed by cross-flow. Assuming the removal of solute by cross-flow to be constant and equal to the convective particle transport at steady state $(=J_{ss}C_b)$, then:

$$\frac{1}{A_m}\frac{dV}{dt} = \frac{|\Delta P|}{(R_m + (V/A_m - J_{ss}t)rC_b)\mu} \tag{11.9}$$

where J_{ss} can be obtained experimentally or from the film-model given in Eq. (11.15).

In a number of cases, a steady rate of filtration is never achieved and it is then possible to describe the time dependence of filtration by introducing an efficiency factor β representing the fraction of filtered particles remaining in the filter cake rather than being swept along by the bulk flow. Eq. (11.4) then becomes:

$$R_c = \frac{\beta rVC_b}{A_m} \tag{11.10}$$

where $0 < \beta < 1$. This is analogous to a *scour model* describing shear erosion at a surface. The layers deposited on the membrane during cross-flow microfiltration are sometimes thought to constitute dynamically formed membranes with their own rejection and permeation characteristics.

In the following section, film and gel-polarisation models are developed for ultrafiltration. These models are also widely applied to cross-flow microfiltration, although even these cannot be simply applied, and there is at present no generally accepted mathematical description of the process.

11.6 **Ultrafiltration**

Ultrafiltration is one of the most widely used of the pressure-driven membrane separation processes. The solutes retained or rejected by ultrafiltration membranes are those with molecular weights of 10^3 or greater, depending mostly on the MWCO of the membrane chosen. The process liquid, dissolved salts and low molecular weight organic molecules (500–1000 kg/kmol) generally pass through the membrane. The pressure difference applied across the membrane is usually in the range of 0.1 to 0.7 MN/m² and membrane permeation rates are typically 0.01 to 0.2 m³/m² h. In industry, ultrafiltration is always operated in the cross-flow mode.

The separation of process liquid and solute that takes place at the membrane during ultrafiltration gives rise to an increase in solute concentration close to the membrane surface, as shown in Fig. 11.7. This is termed concentration polarisation and takes place within the boundary film generated by the applied cross-flow. With a greater concentration at the membrane, there will be a tendency for solute to diffuse back into the bulk feed according to Fick's Law, discussed in Volume 1, Chapter 10. At steady state, the rate of back-diffusion will be equal to the rate of removal of solute at the membrane, minus the rate of solute leakage through the membrane:

$$J(C - C_p) = -D\frac{dC}{dy} \tag{11.11}$$

Here solute concentrations C and C_p in the permeate are expressed as mass fractions, D is the diffusion coefficient of the solute and y is the distance from the membrane. Rearranging and integrating from $C = C_f$ when $y = l$ the thickness of the film, to $C = C_w$, the concentration of solute at the membrane wall, when $y = 0$, gives:

$$-\int_{C_w}^{C_f} \frac{dC}{C - C_p} = \frac{J}{D}\int_0^1 dy \tag{11.12}$$

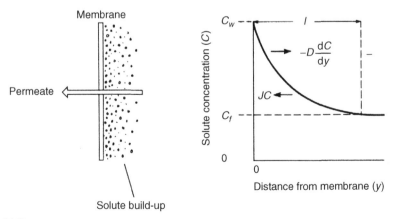

Membrane

Permeate

Solute build-up

FIG. 11.7

Concentration polarisation at a membrane surface.

or :
$$\frac{C_w - C_p}{C_f - C_p} = \exp\left(\frac{Jl}{D}\right) \tag{11.13}$$

If it is further assumed that the membrane completely rejects the solute, that is, $R=1$ and $C_p=0$, then:

$$\frac{C_w}{C_f} = \exp\left(\frac{Jl}{D}\right) \tag{11.14}$$

where the ratio C_w/C_f is known as the polarisation modulus. It may be noted that it has been assumed that l is independent of J and that D is constant over the whole range of C at the interface. The film thickness is usually incorporated in an overall mass transfer coefficient h_D, where $h_D = D/l$, giving:

$$J = h_D \ln\left(\frac{C_w}{C_f}\right) \tag{11.15}$$

The mass transfer coefficient is usually obtained from correlations for flow in non-porous ducts. One case is that of laminar flow in channels of circular cross-section where the parabolic velocity profile is assumed to be developed at the channel entrance. Here the solution of Lévêque [7], discussed by Blatt et al. [8], is most widely used. This takes the form:

$$Sh = 1.62\left(Re\ Sc\frac{d_m}{L}\right)^{1/3} \tag{11.16}$$

where Sh is the Sherwood number ($h_D d_m/D$), d_m is the hydraulic diameter, L is the channel length, Re is the Reynolds number ($u d_m \rho/\mu$), Sc the Schmidt number ($\mu/\rho D$), with u being the cross-flow velocity, ρ the fluid density and μ the fluid viscosity. This gives:

$$h_D = 1.62\left(\frac{uD^2}{d_m L}\right)^{1/3} \tag{11.17}$$

or for tubular systems:

$$h_D = 0.81\left(\frac{\gamma}{L}D^2\right)^{1/3} \tag{11.18}$$

where $\dot{\gamma}$, the shear rate at the membrane surface equals $8u/d_m$, as shown in Volume 1, Chapter 3.

For the case of turbulent flow the Dittus–Boelter [9] correlation given in Volume 1, Chapters 9 and 10, is used:

$$Sh = 0.023\ Re^{0.8} Sc^{0.33} \tag{11.19}$$

which for tubular systems gives:

$$h_D = 0.023\frac{u^{0.8}D^{0.67}}{d_m^{0.2}}\left(\frac{\rho}{\mu}\right)^{0.47} \tag{11.20}$$

and for thin rectangular flow channels, with channel height b:

$$h_D = 0.02\frac{u^{0.8}D^{0.67}}{b^{0.2}}\left(\frac{\rho}{\mu}\right)^{0.47} \tag{11.21}$$

For both laminar and turbulent flow it is clear that the mass transfer coefficient and hence the membrane permeation rate may be increased, where these equations are valid, by increasing the cross-flow velocity or decreasing the channel height. The effects are greatest for turbulent flow. For laminar flow the mass transfer coefficient is decreased if the channel length is increased. This is due to the boundary layer increasing along the membrane module. The mass transfer coefficient is, therefore, averaged along the membrane length.

This boundary-layer theory applies to mass-transfer controlled systems where the membrane permeation rate is independent of pressure, for there is no pressure term in the model. In such cases, it has been proposed that, as the concentration at the membrane increases, the solute eventually precipitates on the membrane surface. This layer of precipitated solute is known as the *gel-layer*, and the theory has thus become known as the *gel-polarisation* model proposed by Michaels [10]. Under such conditions C_w in Eq. (11.15) becomes replaced by a constant C_G the concentration of solute in the gel-layer, and:

$$J = h_D \ln \left(\frac{C_G}{C_f} \right)$$
(11.22)

If an increase in pressure occurs under these conditions, this produces a temporary increase in flux which brings more solute to the gel-layer and increases its thickness, subsequently reducing the flux to the initial level.

The agreement between theoretical and experimental ultrafiltration rates for macromolecular solutions can be said to be within 15% to 30%, as discussed by Porter [11]. Process patterns diagnostic of gel-polarisation type behaviour are shown in Fig. 11.8. The dependence of the membrane permeation rate on the applied pressure is shown in Fig. 11.8A. There is an initial pressure-dependent region followed by a pressure-independent region. The convergence of plots of the membrane permeation rate against ln C_f, as shown in Fig. 11.8B, is a test of Eq. (11.15). Finally, the slope of plots of the membrane permeation rate against the average cross-flow velocity confirms the usefulness of the correlations for laminar and turbulent flow. The gel-polarisation model also suggests, however, that in the pressure-independent region the gel concentration should be independent of membrane permeability and membrane type. As pointed out by Le and Howell [12], neither of these is observed in practice. This shows the need for a more detailed understanding of the nature of membrane-solute interactions. Further, for colloidal suspensions, experimental membrane permeation rates are often one to two orders of magnitude higher than those indicated by the Lévêque and Dittus–Boelter correlations [8]. This has been termed, *the flux paradox for colloidal suspensions* by Green and Belfort [13] and by Porter [14]. The *paradox* is most convincingly explained in terms of the *tubular-pinch effect* described by Sergre and Silberberg [15]. There is clear visual evidence that particles flowing through a tube migrate away from the tube wall and axis, reaching equilibrium at some eccentric radial position. The difficulty has been to produce a quantitative model incorporating lift forces to describe the effect, as pointed out by Belfort [16]. This is an area of both considerable mathematical complexity and controversy, though new models such as that proposed by Altena and

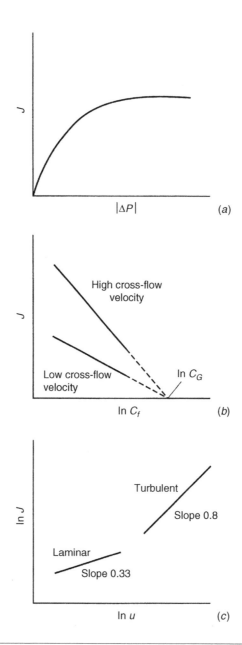

FIG. 11.8

Dependence of membrane flux J on (A) Applied pressure difference $|\Delta P|$, (B) Feed solute concentration C_f, (C) Cross-flow velocity (u) for ultrafiltration.

Belfort [17] appear to allow the prediction of membrane permeation rates for both macromolecular solutions and colloidal suspensions.

The explanation of the pressure-independent region during the ultrafiltration of macromolecules requires the arbitrary introduction of the concept of a gel-layer in the film model. A more complete description of the dependence of the membrane permeation rate on the applied pressure may be given by considering the effect of the osmotic pressure of the macromolecules as described by Wijmans et al. [18] Eq. (11.2) may then be written as:

$$J = \frac{(|\Delta P| - |\Delta \prod|)}{R_m \mu} \tag{11.23}$$

where $|\Delta\Pi|$ is difference in osmotic pressure across the membrane. The osmotic pressure of concentrated solutions is best represented in terms of a polynomial as:

$$\prod = a_1 C + a_2 C^2 + a_3 C^3 \tag{11.24}$$

where $a_1 a_2$ and a_3 are coefficients and C is the solute concentration expressed as mass fractions. In the present case, the difference in osmotic pressure across the membrane can be approximated as:

$$\left|\Delta \prod\right| = \prod = a C_w^n \tag{11.25}$$

where C_w is the concentration at the membrane surface and $n > 1$. Then, from Eqs (11.15), (11.23):

$$J = \frac{\left(|\Delta P| - a C_f^n \exp\left(nJ/h_D\right)\right)}{R_m \mu} \tag{11.26}$$

Taking derivatives of this equation provides valuable insights into the ultrafiltration process. This gives:

$$\frac{\partial J}{\partial |\Delta P|} = \left(R_m \mu + a C_f^n \frac{n}{h_D} \exp\left(\frac{nJ}{h_D}\right)\right)^{-1}$$

$$= \left(R_m \mu + \frac{n}{h_D}\left|\Delta \prod\right|\right)^{-1} \tag{11.27}$$

$$= \left(R_m \mu + \frac{n}{h_D}(|\Delta P| - J R_m \mu)\right)^{-1}$$

which gives the asymptotes:

$$\frac{\partial J}{\partial |\Delta P|} \to = (R_m \mu)^{-1} \text{ for} |\Delta P| \to 0 \text{ or} \left|\Delta \prod\right| \to 0$$

$$\text{and}: \frac{\partial J}{\partial |\Delta P|} \to 0 \text{ for} |\Delta P| \to \infty \text{ or} |\Delta P| \gg J R_m \mu$$

Thus, the basic features of the flux-pressure profiles (Fig. 11.8A) are accounted for without further assumptions:

(a) at low $|\Delta P|$ the slope is similar to that for pure solvent flow,
(b) as $|\Delta P|$ increases, the slope declines and approaches zero at high pressure.

The relationship between flux and solute concentration can be examined by rearranging Eq. (11.26), taking logarithms and differentiating to give:

$$\frac{\partial J}{\partial \ln C_f} = -\left(\frac{1}{h_D} + \frac{1}{n\left(\frac{|\Delta P|}{R_m\mu}\right) - J}\right)^{-1}$$

$$= -h_D\left(1 + \frac{R_m\mu h_D}{|\Delta \prod n|}\right)^{-1}$$

(11.28)

which shows that when polarisation is significant, that is $|\Delta P| >> J R_m \mu$ or $|\Delta \Pi| n / R_m \mu k >> 1$:

$$\frac{\partial J}{\partial \ln C_f} \to = -h_D$$

This is the same prediction for the limiting slope of a plot of J against $\ln C_f$ as for the gel-polarisation model. The value of the slope of such plots at all other conditions is less in magnitude than h_D.

Finally, from Eq. (11.26), when $J \to 0$:

$$|\Delta P| = aC_f^n = \prod \qquad (C_f \to C_{f\,lim})$$

(11.29)

that is, the limiting concentration is that giving an osmotic pressure equal to the applied pressure. This also implies that $C_{f,lim}=f|\Delta P|$, an important difference from the gel-polarisation model which predicts that $C_{f,lim}=C_g \neq f|\Delta P|$.

Osmotic pressure models can be developed from a very fundamental basis. For example, it is becoming possible to predict the rate of ultrafiltration of proteins starting from a knowledge of the sequence and three-dimensional structure of the molecule [19].

Example 11.1
Obtain expressions for the optimum concentration for minimum process time in the diafiltration of a solution of protein content S in an initial volume V_0.
(a) If the gel-polarisation model applies.
(b) If the osmotic pressure model applies.
It may be assumed that the extent of diafiltration is given by:

$$V_d = \frac{\text{Volume of liquid permeated}}{\text{Initial feed volume}} = \frac{V_p}{V_0}$$

Solution

(a) *Assuming the gel–polarisation model applies*

The membrane permeation rate, $J = h_D \ln(C_G/C_f)$ (Eq. 11.22)

where C_G and C_f are the gel and the bulk concentrations respectively.

In this case: $C_f = S/V_0$

and the volume V_d liquid permeated, $V_p = V_d S/C_f$

The process time per unit area, $t = V_p/J$

$$= V_d S/(C_f h_D \ln(C_G/C_f))$$

Assuming C_f and h_D are constant, then:

$$dt/dC_f = -V_d S/\left[h_D C_f^2 \ln\left(C_G/C_f\right)\right] + V_d S/\left\{h_D C_f^2 \left[\ln\left(C_G/C_f\right)\right]^2\right\}$$

If, at the optimum concentration C_f^* and $dt/dC_f = 0$, then:

$$1 = \ln\left(C_G/C_f^*\right)$$

$$\text{and} : \underline{\underline{C_f^* = C_G/e}}$$

(b) *Assuming the osmotic pressure model applies*

$$J = h_D \ln(C_G/C_f) \text{ (Eq. 11.22)}$$

Substituting for $|\Delta\Pi|$ at $C = C_w$, then:

$$J = [|\Delta P| - aC_f^n \exp(nJ/h_D)]/R_m\mu \text{ (Eq. 11.26)}$$

If $|\Delta\Pi|$ is very much greater than $JR_m\mu$, then:

$$J = (h_D/n) \ln\left(|\Delta P|/\left(aC_f^n\right)\right)$$

As before: $V_d = V_p/V_0$ and $C_f = S/V_0$.

Thus: $t = V_p/J$

$$= \left(V_d S/C_f\right)/\left[(h_D/n) \ln|\Delta P|/\left(aC_f^n\right)\right]$$

$$\text{and} : dt/dC_f = (V_d nS/h_D)\left[- \ln|\Delta P|/aC_f^n/C_f^2 + n/C_f^2\right]/ \ln\left(|\Delta P|/aC_f^n\right)$$

The process time, t, is a minimum when $dt/dC_f = 0$, that is when:

$$C_f^n = |\Delta P|/ae^n$$

11.7 Reverse osmosis

A classical demonstration of osmosis is to stretch a parchment membrane over the mouth of a tube, fill the tube with a sugar solution, and then hold it in a beaker of water. The level of solution in the tube rises gradually until it reaches a steady level. The static head developed would be equivalent to the osmotic pressure of the solution if the parchment were a perfect semipermeable membrane, such a membrane having the property of allowing the solvent to pass through but preventing the solute from passing through. The pure solvent has a higher chemical potential than the solvent in the solution and so diffuses through until the difference is cancelled out by the pressure head. If an additional pressure is applied to the liquid column on the solution side of the membrane then it is possible to force water back through the membrane. This pressure-driven transport of water from a solution through a membrane is known as *reverse osmosis*. It may be noted that it is not quite the reverse of osmosis because, for all real membranes, there is always a certain transport of the solute along its chemical potential gradient, and this is not reversed. The phenomenon of reverse osmosis has been extensively developed as an industrial process for the concentration of low molecular weight solutes and especially for the desalination, or more generally demineralisation, of water.

Many models have been developed to explain the semi-permeability of reverse osmosis membranes and to rationalise the observed behaviour of separation equipment. These have included the postulation of preferential adsorption of the solute at the solution–membrane interface, hydrogen bonding of water in the membrane structure, and the exclusion of ions by the membrane due to dielectric effects. They are all useful in explaining aspects of membrane behaviour, although the most common approach has been to make use of the theories of the thermodynamics of irreversible processes proposed by Spiegler and Kedem [20]. This gives a phenomenological description of the relative motion of solution components within the membrane, and does not allow for a microscopic explanation of the flow and rejection properties of the membrane. In the case of reverse osmosis, however, the thermodynamic approach is combined with a macroscopic *solution–diffusion* description of membrane transport as discussed by Soltanieh and Gill [21]. This implies that the membrane is non-porous and that solvent and solutes can only be transported across the membrane by first dissolving in, and subsequently diffusing through, the membrane.

For any change to occur a chemical potential gradient must exist. For a membrane system, such as the one under consideration, Haase [22] and Belfort [23] have derived the following simplified equation for constant temperature:

$$d\mu_i = v_i dP + \left(\frac{\partial \mu_i}{\partial C_i}\right)_T dC_i + z_i \mathbf{F} \, d\varphi \qquad (11.30)$$

where μ_i is the chemical potential of component i, v_i is the partial molar volume of component i, z_i is the valence of component i, ϕ is the electrical potential and $\mathbf{F}$ is Faraday's constant. This equation may be applied to any membrane process. For ultrafiltration, only the pressure forces are usually considered. For electrodialysis, the electrical and concentration forces are more important, whereas, in the present case of reverse osmosis both pressure and concentration forces need to be considered.

Integrating across the thickness of the membrane for a two-component system with subscript 1 used to designate the solvent (water) and subscript 2 used to designate the solute, for the solvent:

$$\Delta\mu_1 = \int \left(\frac{\partial\mu_1}{\partial C_1}\right)_{P,T} dC_1 + \int v_1 \, dP$$
$$= \int \left(\frac{\partial\mu_1}{\partial C_2}\right)_{P,T} dC_2 + \int v_1 \, dP$$

(11.31)

When $\Delta\mu_1$ becomes small, only the osmotic pressure difference $\Delta\Pi$ remains. Thus, for constant v_1:

$$\Delta\mu_1 = v_1\left(|\Delta P| - \left|\Delta\prod\right|\right)$$

(11.32)

For the solute:

$$\Delta\mu_2 = \int \left(\frac{\partial\mu_2}{\partial C_2}\right)_{P,T} dC_2 + \int v_2 \, dP$$

(11.33)

and for dilute solutions, $\mu_2 = \mu_2^0 + RT \ln C_2$, and for constant v_2:

$$\Delta\mu_2 = RT\Delta\ln C_2 + v_2|\Delta P|$$

(11.34)

where the second term on the right-hand side is negligible compared with the first term. In the present case:

$$\Delta\mu_2 = RT\Delta\ln C_2 \approx \left(\frac{RT}{C_2}\right)\Delta C_2 \text{(for low values of } C_2\text{)}$$

(11.35)

Incorporating the model of diffusion across the membrane, and writing Fick's law in the generalised form [24], (using $\mu = \mu^0 + RT \ln C$)[b]:

$$J = \frac{DC}{RT}\frac{d\mu}{dy}$$

(11.36)

where y is distance in the direction of transfer.

It is found for the solvent that:

$$J_1 = K_1\left(|\Delta P| - \left|\Delta\prod\right|\right)$$

(11.37)

where the permeability coefficient is described in terms of a diffusion coefficient, water concentration, partial molar volume of water, absolute temperature and effective membrane thickness. For the solute it is found that:

$$J_2 = K_2|\Delta C_2|$$

(11.38)

where K_2 is described in terms of a diffusion coefficient, distribution coefficient and effective membrane thickness. It is clear from these equations that solvent (water) flow only occurs if $|\Delta P| > |\Delta\Pi|$, though solute flow is independent of $|\Delta P|$. Thus, increasing the operating pressure increases the effective separation. This explains

---

[b]Eq. (11.36) represents the general form of Fick's law. The form used in previous chapters where driving force is expressed as a concentration gradient is a simplification of this equation.

why reverse osmosis plants operate at relatively high pressure. For example, the osmotic pressure of brackish water containing 1.5 to 12kg/m^3 salts is 0.1 to 0.7 MN/m^2 (MPa) and the osmotic pressure of sea water containing 30 to 50kg/m^3 salts is 2.3 to 3.7 MN/m^2 (MPa). In practice, desalination plants operate at 3 to 8 MN/m^2 (MPa).

The rejection of dissolved ions at reverse osmosis membranes depends on valence. Typically, a membrane which rejects 93% of Na^+ or Cl^- will reject 98% of Ca^{2+} or SO_4^{2-} when rejections are measured on solutions of a single salt. With mixtures of salts in solution, the rejection of a single ion is influenced by its relative proportion in the mixture. Thus, for 0.1kg/m^3 Cl^- in the presence of 1kg/m^3 SO_4^{2-} there would be only 50% to 70% rejection compared with 93% for solutions of a single salt. The rejection of organic molecules depends on molecular weight. Those with molecular weights less than 100kg/kmol may not be rejected, those with molecular weights of about 150kg/kmol may have about the same rejection as NaCl, and those with molecular weights greater than some 300kg/kmol are effectively entirely rejected.

The thermodynamic approach does not make explicit the effects of concentration at the membrane. A good deal of the analysis of concentration polarisation given for ultrafiltration also applies to reverse osmosis. The control of the boundary layer is just as important. The main effects of concentration polarisation in this case are, however, a reduced value of solvent permeation rate as a result of an increased osmotic pressure at the membrane surface given in Eq. (11.37), and a decrease in solute rejection given in Eq. (11.38). In many applications, it is usual to pretreat feeds in order to remove colloidal material before reverse osmosis. The components which must then be retained by reverse osmosis have higher diffusion coefficients than those encountered in ultrafiltration. Hence, the polarisation modulus given in Eq. (11.14) is lower, and the concentration of solutes at the membrane seldom results in the formation of a gel. For the case of turbulent flow the Dittus–Boelter correlation [9] may be used, as was the case for ultrafiltration giving a polarisation modulus of:

$$\frac{C_w}{C_f} = \exp\left(\frac{Jd^{0.2}}{0.023u^{0.8}D_L^{0.67}}\frac{\mu^{0.47}}{\rho^{0.47}}\right) \tag{11.39}$$

where solvent permeation rate, cross-flow velocity and solute diffusion coefficient have the greatest importance. The results are more complex for the case of laminar flow as $|\Delta\Pi|$ increases down the channel as the boundary layer is developed and water is removed. As a consequence, results do not appear in a closed form and finite-difference methods have been applied [10].

11.8 Membrane modules and plant configuration

Membrane equipment for industrial scale operation of microfiltration, ultrafiltration and reverse osmosis is supplied in the form of modules. The area of membrane contained in these basic modules is in the range of 1 to 20 m^2. The modules may be connected together in series or in parallel to form a plant of the required performance. The four most common types of membrane modules are tubular, flat sheet, spiral wound and hollow fibre, as shown in Figs 11.9–11.12.

(a) *Tubular modules* are widely used where it is advantageous to have a turbulent flow regime, for example, in the concentration of high solids content feeds. The

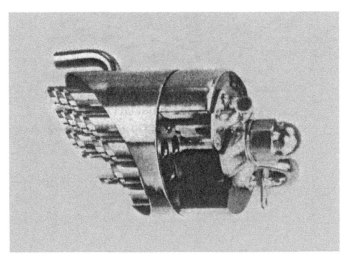

FIG. 11.9

Tubular module.

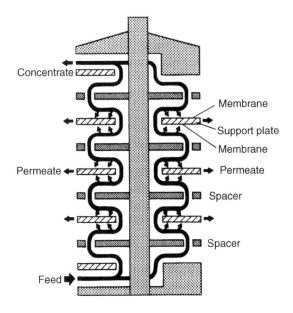

FIG. 11.10

Schematic diagram of flat-sheet module.

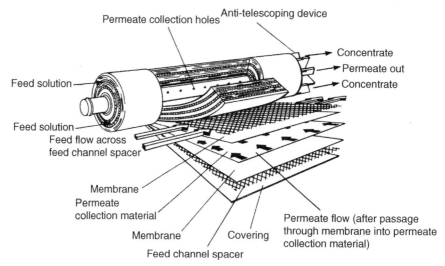

FIG. 11.11

Schematic diagram of spiral-wound module.

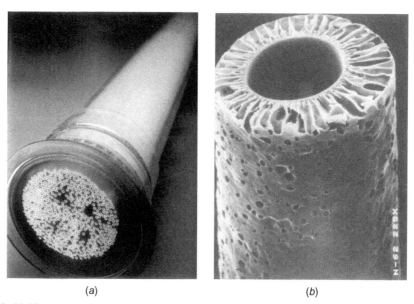

(a)   (b)

FIG. 11.12

(A) Hollow-fibre module and, (B) a single fibre.

membrane is cast on the inside of a porous support tube which is often housed in a perforated stainless steel pipe as shown in Fig. 11.10. Individual modules contain a cluster of tubes in series held within a stainless steel permeate shroud. The tubes are generally 10 to 25 mm in diameter and 1 to 6 m in length. The feed is pumped through the tubes at Reynolds numbers greater than 10,000. Tubular modules are easily cleaned and a good deal of operating data exist for them. Their main disadvantages are the relatively low membrane surface area contained in a module of given overall dimensions and their high volumetric hold-up.

(b) *Flat-sheet modules* are similar in some ways to conventional filter presses. An example is shown in Fig. 11.10. This consists of a series of annular membrane discs of outer diameter 0.3 m placed on either side of polysulphone support plates which also provide channels through which permeate can be withdrawn. The sandwiches of membrane and support plate are separated from one another by spacer plates which have central and peripheral holes, through which the feed liquor is directed over the surface of the membranes, The flow is laminar. A single module contains 19 m^2 of membrane area. Permeate is collected from each membrane pair so that damaged membranes can be easily identified, though replacement of membranes requires dismantling of the whole stack.

(c) *Spiral-wound modules* consist of several flat membranes separated by turbulence-promoting mesh separators and formed into a *Swiss roll*, as shown in Fig. 11.11. The edges of the membranes are sealed to each other and to a central perforated tube. This produces a cylindrical module which can be installed within a pressure tube. The process feed enters at one end of the pressure tube and encounters a number of narrow, parallel feed channels formed between adjacent sheets of membrane. Permeate spirals towards the perforated central tube for collection. A standard size spiral-wound module has a diameter of some 0.1 m, a length of about 0.9 m and contains about 5 m^2 of membrane area. Up to six such modules may be installed in series in a single pressure tube. These modules make better use of space than tubular or flat-sheet types, although they are rather prone to fouling and difficult to clean.

(d) *Hollow-fibre modules*, shown in Fig. 11.12, consist of bundles of fine fibres, 0.1 to 2.0 mm in diameter, sealed in a tube. For reverse-osmosis desalination applications, the feed flow is usually around the outside of the unsupported fibres with permeation radially inward, as the fibres cannot withstand high pressures differences in the opposite direction. This gives very compact units capable of high pressure operation, although the flow channels are less than 0.1 mm wide and are therefore readily fouled and difficult to clean. The flow is usually reversed for biotechnological applications so that the feed passes down the centre of the fibres giving better controlled laminar flow and easier cleaning. This limits the operating pressure to less than 0.2 MN/m^2, however, that is, to microfiltration and ultrafiltration applications. A single ultrafiltration module typically contains up to 3000 fibres and be 1 m long. Reverse osmosis modules contain larger numbers of finer fibres. This is a very effective means of incorporating a large membrane surface area in a small volume.

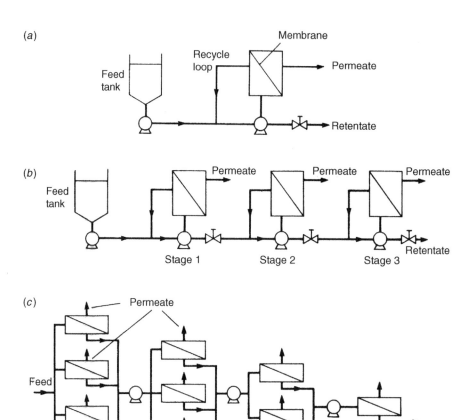

FIG. 11.13

Schematic flow-diagrams of (A) Single-stage 'feed and bleed', (B) Multiple-stage 'feed and bleed', (C) Continuous single-pass membrane plants.

Membrane modules can be configured in various ways to produce a plant of the required separation capability. A simple batch recirculation system has already been described in Section 11.5. Such an arrangement is most suitable for small-scale batch operation, but larger scale plants will operate either as *feed and bleed* or *continuous single-pass* operations, as shown in Fig. 11.13.

(a) *Feed and bleed.* Such a system is shown in Fig. 11.13A. The start-up is similar to that in a batch system in that the retentate is initially totally recycled. When the required solute concentration is reached within the loop, a fraction of the loop is continuously bled off. Feed into the loop is controlled at a rate equal to the permeate plus concentrate flow rates. The main advantage is that the final concentration is then continuously available as feed is pumped into the loop. The main disadvantage is that the loop is operating continuously at a concentration equivalent to the final concentration in the batch system and the flux is therefore lower than the average flux in the batch mode, with a correspondingly higher membrane area requirement.

Large-scale plants usually use multiple stages operated in series to overcome the low-flux disadvantage of the feed and bleed operation and yet to maintain its continuous nature as shown in Fig. 11.13B. Only the final stage is operating at the highest concentration and lowest flux, whilst the other stages are operating at lower concentrations with higher flux. Thus, the total membrane area is less than that required for a single-stage operation. Usually a minimum of three stages is required. The residence time, volume hold-up and tankage required are much less than for the same duty in batch operation. Feed and bleed systems also require less frequent sterilisation than batch processes in biotechnological applications, allowing longer effective operating times.

(b) *Continuous single pass.* In such a system, the concentration of the feed stream increases gradually along the length of several stages of membrane modules arranged in series as shown in Fig. 11.13C. The feed only reaches its final concentration at the last stage. There is no recycle and the system has a low residence time. Such systems must, however, either be applied on a very large scale or have only a low overall concentration factor, due to the need to maintain high cross-flow velocities to control concentration polarisation. The smallest possible single-pass system will have a single module in the final stage, with a typical feed flow rate of $0.1 \, \text{m}^3/\text{min}$. Such systems are used in large-scale reverse osmosis desalination plants but are unlikely to be used in biotechnological applications.

Example 11.2

As part of a downstream processing sequence, $10 \, \text{m}^3$ of a process fluid containing $20 \, \text{kg m}^{-3}$ of an enzyme is to be concentrated to $200 \, \text{kg/m}^3$ by means of ultrafiltration. Tests have shown that the enzyme is completely retained by a 10,000 MWCO surface-modified polysulphone membrane with a filtration flux given by:

$$J = 0.04 \, \ln(250/C_f)$$

where J is the flux in m/h and C_b is the enzyme concentration in kg m^{-3}. Four hours is available for carrying out the process.

(a) Calculate the area of membrane needed to carry out the concentration as a simple batch process, (b) Use the following approximation for estimating the average flux during a simple batch process:

$$J_{av} = J_f + 0.27(J_i - J_f)$$

where J_{av} is the average flux, J_i is the initial flux and J_f is the final flux. Is this approximation suitable for design purposes in the present case?

Solution

(a) From the gel–polarisation model:

$$J = \frac{1}{A}\frac{dV}{dt} = h_D \ln\left(\frac{C_G}{C_f}\right)$$

$$\text{Also}: C_f = C_0\left(\frac{V_0}{V}\right)$$

where C_0 and V_0 are the initial concentration and volume, respectively and C_f and V are the values at subsequent times.

Combining these equations gives:

$$\frac{dV}{dt} = A\left(h_D \ln\left(\frac{C_G}{C_f}\right) - h_D \ln\left(\frac{V_0}{V}\right)\right)$$

$$\int_{V_0}^{V_t} \frac{dV}{\left(J_0 - h_D \ln\left(\frac{V_0}{V}\right)\right)} = A\int_0^t dt$$

V	$(J_0 - h_D \ln(V_0/V))^{-1}$
10	9.90
5	13.64
3	18.92
2	27.30
1	112.40

The data are plotted in Fig. 11.14.
The area under the curve $= 184.4$.
Operation for four hours gives <u>46.1 m^2 membrane area</u>

(b) $J_0 = 0.04 \ln(250/20) = 0.101\,\text{m/h}$

$J_f = 0.04 \ln(250/200) = 0.008\,\text{m/h}$

$J_{av} = 0.008 + 0.27(0.101 - 0.008) = 0.033\,\text{m/h}$

For the removal of 9 m^3 filtrate in 4 h:

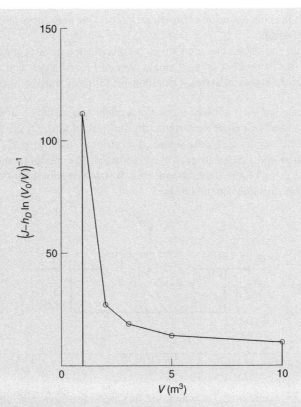

FIG. 11.14

Graphical integration for Example 11.2.

$$\text{Area} = (9/4)/0.033 = 68.2\,\text{m}^2 \text{ membrane}$$

The approximation is not suitable for design purposes.

Example 11.3

An ultrafiltration plant is required to treat 50 m³/day of a protein-containing waste stream. The waste contains 0.5 kg/m³ of protein which has to be concentrated to 20 kg/m³ so as to allow recycling to the main process stream. The tubular membranes to be used are available as 30 m² modules. Pilot plant studies show that the flux J through these membranes is given by:

$$J = 0.02 \ln\left(\frac{30}{C_f}\right) \text{m/h}$$

where C_f is the concentration of protein in kg/m³. Due to fouling, the flux never exceeds 0.04 m/h.

Estimate the minimum number of membrane modules required for the operation of this process (a) as a single *feed and bleed* stage, and (b) as two *feed and bleed* stages in series. Operation for 20 h/day may be assumed.

Solution

(a) with a single *feed and bleed* stage, the arrangement is shown in Fig. 11.15:

It is assumed that Q_0 is the volumetric flow rate of feed, Q_2 the volumetric flow rate of concentrate, C_0 the solute concentration in the feed, C_2 the solute concentration in the concentrate, F the volumetric flow rate of membrane permeate, and A the required membrane area. It is also assumed that there is no loss of solute through the membrane.

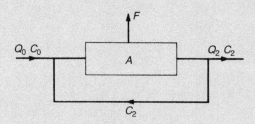

FIG. 11.15

Single 'feed and bleed' stage.

The concentration (C_l) at which the flux becomes fouling-limited is:

$$0.04 = 0.02 \ln\left(\frac{30}{C_l}\right)$$

or: $C_l \approx 4 \, \text{kg/m}^3$

That is, below this concentration the membrane flux is 0.04 m/h.

This does not pose a constraint for the single stage as the concentration of solute C_2 will be that of the final concentrate, 20 kg/m³.

Conservation of solute gives:

$$Q_0 C_0 = Q_2 C_2 \qquad (11.\text{i})$$

A fluid balance gives:

$$Q_0 = F + Q_2 \qquad (11.\text{ii})$$

Combining these equations and: substituting known values:

$$2.438 = A0.02 \ln\left(\frac{30}{20}\right)$$

and : $\underline{A = 301 \, \text{m}^2}$

Thus, 10 modules will almost meet the specification for the single-stage process.

(b) with two *feed and bleed* stages in series, the arrangement is shown in Fig. 11.16:

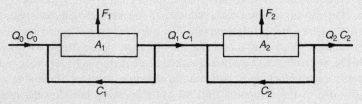

FIG. 11.16

Two 'feed and bleed' stages in series.

In addition to the symbols previously defined, Q_1 will be taken as the volumetric flow rate of retenate at the intermediate point, C_1 the concentration of solute in the retentate at this point, F_1 and F_2 the volumetric flow rates of membrane permeate in the first and second stages respectively, and A_1 and A_2 the required membrane areas in these respective stages.

Conservation of solute gives:

$$Q_0 C_0 = Q_1 C_1 = Q_2 C_2 \tag{11.iii}$$

A fluid balance on stage 1 gives:

$$Q_0 = Q_1 + F_1 \tag{11.iv}$$

A fluid balance on stage 2 gives:

$$Q_1 = Q_2 + F_2 \tag{11.v}$$

Substituting given values in Eqs (11.iv), (11.v) gives:

$$2.5 = \frac{1.25}{C_1} + 0.02 A_1 \ln\left(\frac{30}{C_1}\right) \tag{11.vi}$$

$$\text{or}: \frac{1.25}{C_1} = 0.0625 + 0.00811 A_1 \tag{11.vii}$$

The procedure is to use trial and error to estimate the value of C_1 that gives the optimum values of A_1 and A_2. Thus:

If $C_1 = 5\,\text{kg/m}^3$, then, $A_1 = 63\,\text{m}^2$ and $A_2 = 23\,\text{m}^2$.
That is, an arrangement of 3 modules −1 module is required.
If $C_1 = 4\,\text{kg/m}^3$, then $A_1 = 55\,\text{m}^2$ and $A_2 = 31\,\text{m}^2$.
That is, an arrangement of 2 modules −1 module is almost sufficient.
If $C_1 = 4.5\,\text{kg/m}^3$, then $A_1 = 59\,\text{m}^2$ and $A_2 = 27\,\text{m}^2$.

That is, an arrangement of 2 modules −1 module which meets the requirement.

This arrangement requires the minimum number of modules.

11.9 Membrane fouling

A limitation to the more widespread use of membrane separation processes is membrane fouling, as would be expected in the industrial application of very finely porous materials. Fouling results in a continuous decline in membrane permeation rate, an increased rejection of low molecular weight solutes and eventually blocking of flow channels. On start-up of a process, a reduction in membrane permeation rate to 30% to 10% of the pure water permeation rate after a few minutes of operation is common for ultrafiltration. Such a rapid decrease may be even more extreme for microfiltration. This is often followed by a more gradual decrease throughout processing. Fouling is partly due to blocking or reduction in effective diameter of membrane pores, and partly due to the formation of a slowly thickening layer on the membrane surface.

The extent of membrane fouling depends on the nature of the membrane used and on the properties of the process feed. The first means of control is therefore careful choice of membrane type. Secondly, a module design which provides suitable hydrodynamic conditions for the particular application should be chosen. Process feed pretreatment is also important. The type of pretreatment used in reverse osmosis for desalination applications is outlined in Section 11.11. In biotechnological applications, pretreatment might include prefiltration, pasteurisation to destroy bacteria, or adjustment of pH or ionic strength to prevent protein precipitation. When membrane fouling has occurred, backflushing of the membrane may substantially restore the permeation rate. This is seldom totally effective, however, so that chemical cleaning is eventually required. This involves interruption of the separation process, and consequently time losses due to the extensive nature of cleaning required. Thus, a typical cleaning procedure might involve: flushing with filtered water at 35°C to 50°C to displace residual retentate; recirculation or back-flushing with a cleaning agent, possibly at elevated temperature; rinsing with water to remove the cleaning agent; sterilisation by recirculation of a solution of 50 to 100 ppm of chlorine for 10 to 30 min (600–1800 s) at (293–303 K) (20–30°C); and flushing with water to remove sterilising solution. More recent approaches to the control of membrane fouling include the use of more sophisticated hydrodynamic control effected by pulsated feed flows or non-planar membrane surfaces, and the application of further perturbations at the membrane surface, such as continuous or pulsated electric fields.

11.10 Electrodialysis

As discussed by Pletcher [24], electrodialysis is an electrically driven membrane separation process. The main use of electrodialysis is in the production of drinking water by the desalination of sea-water or brackish water. Another large-scale application is in the production of sodium chloride for table salt, the principal method in Japan, with production exceeding 10^6 t per annum.

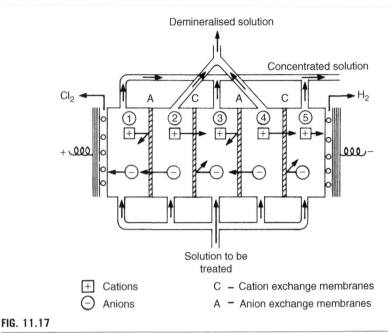

Demineralised solution

Concentrated solution

Solution to be treated

+ Cations
− Anions

C − Cation exchange membranes
A − Anion exchange membranes

FIG. 11.17

Schematic flow-diagram for an electrodialysis stack.

In electrodialysis, cation exchange membranes are alternated with anion exchange membranes in a parallel array to form solution compartments [3] of thickness $\sim$1 mm, as shown in Fig. 11.17. A single membrane stack will typically contain 100 to 400 membranes each of area 0.5 to 2.0 m^2. The process feed is pumped through the solution compartments. Cations are transported towards the cathode and anions towards the anode when an electrical potential is applied to the electrodes. However, to a first approximation the cations can be transported across the cation exchange membranes but not across the anion exchange membranes, and conversely for the anions. The net result is ion depletion and ion concentration in alternate compartments throughout the stack. The power requirements of a stack are typically 100 A at 150 V.

The membranes in electrodialysis stacks are kept apart by spacers which define the flow channels for the process feed. There are two basic types [3], (a) tortuous path, causing the solution to flow in long narrow channels making several 180° bends between entrance and exit, and typically operating with a channel length-to-width ratio of 100:1 with a cross-flow velocity of 0.3 to 1.0 m/s (b) sheet flow, with a straight path from entrance to exit ports and a cross-flow velocity of 0.05 to 0.15 m/s. In both cases, the spacer screens are also designed as turbulence promoters. These help reduce the concentration polarisation which occurs at the membranes, though with a rather different effect from that in the pressure driven membrane processes.

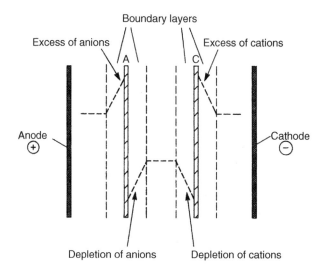

FIG. 11.18

Schematic representation of concentration polarisation with demineralisation of the central compartment.

In order to understand the occurrence of concentration polarisation in electrodialysis, it is necessary to investigate how ion transport occurs. The electric current is carried through the system of membranes and solution channels by the anions and cations. The fraction of the current carried by a given ion is termed its transference number, designated $\mathbf{t}^+$ for cations and $\mathbf{t}^-$ for anions. If the transference number of anions through the solution ($\mathbf{t}_s^-$) is 0.5 and the transference number of anions through the anion exchange membrane ($\mathbf{t}_m^-$) is 1.0; then only half as many ions will be transferred electrically through the solution to the static side of the boundary film on the side of the membrane that anions enter, as will be transferred through the membrane. The solution at the membrane interface will be depleted of anions, and for similar reasons the boundary film on the other side of the membrane will accumulate anions as shown in Fig. 11.18. The same effects will occur for cations at the cation exchange membranes. In practice, it is usually found that $\mathbf{t}_m \geq 0.90$. With any given thickness of boundary film a current density (and hence ion flux) can be reached at which the concentrations of electrolytes at the membrane interfaces on the depleting sides will approach zero. At such a current density, known as the *limiting current density*, H^+ and OH^- ions from ionisation of water will begin to be transferred through the membranes. This produces a loss of efficiency and often causes fouling of the membranes by precipitation of solute components due to pH changes. With the idealisation of completely static and completely mixed zones, a function known as the *polarisation parameter* may be defined as:

$$\frac{i_{\lim}}{zC_i} = \frac{D\mathbf{F}}{l(\mathbf{t}_m - \mathbf{t}_s)} \tag{11.40}$$

where i_{lim} is the limiting current density, D the diffusion coefficient, $\mathbf{F}$ is Faraday's constant, l the equivalent film thickness, C_i is ion concentration (kmol/m^3) and $\mathbf{t}_m$ and $\mathbf{t}_s$ the transference numbers in the membrane and solution respectively.

The value of i_{lim} is determined by the discontinuity in the dependence of cell current on applied cell voltage which occurs when the interfacial concentration approaches zero. The polarisation parameter is convenient in the design and scale-up of electrodialysis equipment. It can be easily measured in small-scale stacks at a given value of bulk concentration and then used to predict limiting current densities in larger stacks at other concentrations. Most stacks use operating values of the polarisation parameter that are 50% to 70% of the limiting values.

11.11 Reverse osmosis water treatment plant

The largest scale applications of membrane separation processes are those which form the key step in the desalination, or more generally demineralisation, of brackish water in the production of drinking water. In this section, an outline is given of such a plant capable of producing 70,000 m^3/day of drinking water for a large city in the Middle East, as described by Finlay and Ferguson [25]. The water to be processed is obtained from a deep well with a total dissolved solids (TDS) content of 1.4 kg/m^3; that is, it is of moderate salinity and hardness. The plant specification required that the product water should have a maximum TDS of 0.5 kg/m^3. A flow diagram of the overall process is shown in Fig. 11.19. Part of the flow bypasses the main

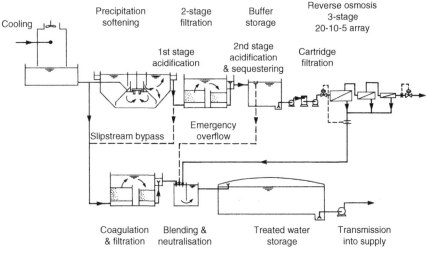

FIG. 11.19

Outline flow-diagram of a large-scale reverse-osmosis plant for the demineralisation of brackish waters [26].

pretreatment and demineralisation stages. This is possible because demineralisation by reverse osmosis produces a permeate with a TDS of about $0.22 \, kg/m^3$, significantly better than the product specification. The minimum plant output is $59,000 \, m^3/day$, requiring reverse osmosis demineralisation of $51,000 \, m^3/day$ of pretreated water, of which $6000 \, m^3/day$ is rejected, leaving $45,000 \, m^3/day$ to be blended with $14,000 \, m^3/day$ of slipstream. Wash-water and other losses are less than $1300 \, m^3/day$. Hence, a total flow of $66,500 \, m^3/day$ is required from the cooling towers, with an overall water loss of less than 11%. The main process steps are as follows:

11.11.1 Pretreatment

(a) Evaporative cooling to reduce the feed water temperature from 50–55°C to 30–35°C which is more compatible with satisfactory operation of the reverse osmosis unit and more suitable for final use.

(b) Precipitation softening by addition of slaked lime ($Ca(OH)_2$) and sodium aluminate or ferric chloride. The net result is part-removal of calcium, silica and especially colloids. The clarifiers used ensure completion of these processes within the tank.

(c) Acidification to optimise removal of residual coagulant,

(d) Prechlorination to ensure a disinfected supply to the reverse osmosis plant.

(e) Rapid gravity filtration to reduce further the content of particulate material. The first stage is upflow through a 1.5 m deep gravel bed, and the second stage downflow through a 1.1 m deep sand bed of effective particle size 0.9 mm. Identical units are used to filter the cooled and coagulated water in the slipstream.

(f) Acidification to reduce the pH to 5.0 for optimum life of the reverse osmosis membrane.

(g) Sequestering, addition of sodium hexametaphosphate to retard the precipitation of calcium sulphate which otherwise will exceed its solubility limit in the reject stream.

(h) Cartridge filtration with elements rated at 25 μm to protect the high pressure pumps and reverse osmosis membranes in the event of a break-through of particulate material.

11.11.2 Demineralisation by reverse osmosis

Reverse osmosis was chosen for the demineralisation step as it gave an economic solution in terms of both capital and running costs, allowed a high water recovery rate, was modular in construction and so could be easily extended, could cope with reasonable variations in feed salinity and had a proven track record in relatively large-scale installations. The prebooster pumps, cartridge filters and high pressure pumps are arranged in seven parallel streams, one of which is on standby. A total of thirteen reverse osmosis stacks is installed, any twelve of which will meet the required throughput. Each stack contains 210 reverse osmosis modules accommodated in 35 pressure vessels arranged in a series–parallel array of 20–10–5 to achieve the desired water recovery. These modules are of the spirally-wound type. The inlet to the reverse osmosis unit is instrumented for flow rate, pH, residual chlorine, turbidity, temperature and conductivity. The permeate from the unit is blended with the slipstream flow, with pH adjustment if necessary, to maintain a final water TDS $< 0.5\,kg/m^3$. The reject is discharged to evaporation ponds.

In temperate climate zones it may be more appropriate to instal a nanofiltration process rather than reverse osmosis. Nanofiltration allows the production of drinking water from polluted rivers. As for reverse osmosis, pretreatment is important to control fouling of the membranes. One of the largest such plants produces 140,000 m^3/day of water for the North Paris region [26].

11.12 **Pervaporation**

Two industrially important categories of separation problems are the separation of liquid mixtures which form an azeotrope and/or where there are only small differences in boiling characteristics. Pervaporation is a membrane process which shows promise for both of such separations as described by Rautenbach and Albrecht [27]. The process differs from other membrane processes in that there is a phase change from liquid to vapour in the permeate. The feed mixture is a liquid. The driving force in the membrane is achieved by lowering the activity of the permeating components at the permeate side. Components in the mixture permeate through the membrane and evaporate as a result of the partial pressure on the permeate side being held lower than the saturation vapour pressures. The driving force is controlled by applying a vacuum on the permeate side, as shown in Fig. 11.20.

Dense membranes are used for pervaporation, as for reverse osmosis, and the process can be described by a *solution–diffusion* model. That is, in an ideal case there is equilibrium at the membrane interfaces and diffusional transport of components through the bulk of the membrane. The activity of a component on the feed side of the membrane is proportional to the composition of that component in the feed solution. The composition at the permeate-phase interface depends on the partial pressure and saturation vapour pressure of the component. Solvent composition within the membrane may vary considerably between the feed and permeate sides interface in pervaporation. By lowering the pressure at the permeate side, very low concentrations can be achieved whilst the solvent concentration on the feed-side can be up to 90% by mass. Thus, in contrast to reverse osmosis, where such differences are not observed in practice, the modelling of material transport in pervaporation must take into account the concentration dependence of the diffusion coefficients.

Polyethylene (PE) is a standard material for separating organic mixtures, although selectivity and fluxes are too low for commercial application. A further development is the use of composite membranes in which a substantially insoluble support material is combined with an additive in which only one of the components of the mixture is highly soluble. An example is a membrane for separating benzene

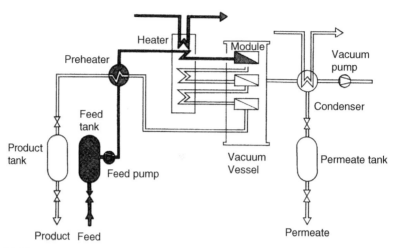

FIG. 11.20

Schematic diagram of a pervaporation unit.

from cyclohexane consisting of a cellulose acetate support matrix and incorporating polyphosphonates to improve the preferential permeability of benzene (CA-PPN). Such mixtures are usually separated by extractive distillation as the equilibrium curve is very shallow and shows an azeotropic point. As shown in Fig. 11.21 the separation characteristics are much more favourable in the case of pervaporation [28].

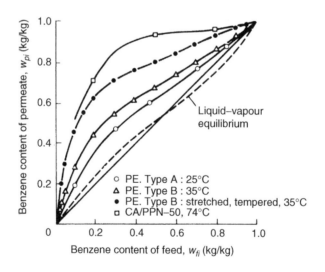

FIG. 11.21

Comparison of selectivity of pervaporation membranes and liquid–vapour equilibrium for benzene/cyclohexane mixtures.

   At present, there is one main commercial application of pervaporation, the production of high purity alcohol by a hybrid process which also incorporates distillation. Such separations use cellulose-acetate-based composite-membranes, with an active layer of polyvinyl alcohol, for example. Membrane fluxes are in the range of 0.45 to 2.2 kg/m^2 h. Pervaporation may have special potential if used following a conventional separation processes, such as distillation, when high product purities are required.

11.13 Liquid membranes

As discussed by Frankemfeld and Li [28] and del Cerro and Boey [29], liquid membrane extraction [28,29] involves the transport of solutes across thin layers of liquid interposed between two otherwise miscible liquid phases. There are two types of liquid membranes, emulsion liquid membranes (ELM) and supported liquid membranes (SLM). They are conceptually similar, but substantially different in their engineering.

   ELM are multiple emulsions of the water/oil/water or oil/water/oil types shown in Fig. 11.22. The membrane phase is that which is interposed between the continuous external phase and the encapsulated internal phase of the emulsion. Preparation of an ELM first involves emulsification of the inner phase in an immiscible solvent which may contain a surfactant and, depending on the application, other additives. The resulting emulsion is then dispersed in the continuous phase, either in mixing vessels or in column extractors. Globule diameters are in the range of 0.1 to 2.0 mm, whilst the internal emulsion droplets are 1 to 10 μm in diameter. The interfacial areas for mass transfer can be as high as 3000 m^2/m^3. Mass transport can be from the continuous phase to the inner encapsulated droplets, or the reverse. The position of equilibrium can be enhanced by additives which cause a chemical or enzymatic reaction in the receiving phase. After extraction of the required product, the emulsion is broken into its component parts, usually by thermal or electrical treatment. In SLM, the liquid separating layer is held within a solid microporous inert support by capillary forces. The support is typically a microporous membrane of the type used for the

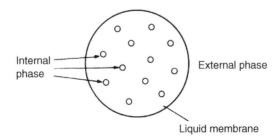

FIG. 11.22

Schematic diagram of emulsion liquid-membrane.

more conventional pressure-driven membrane process. Large areas for mass transfer, up to about 1000 m^2/m^3, can be achieved using spiral-wound or hollow fibre modules.

It is valuable to compare liquid membrane extraction with conventional solvent extraction in which the required product first partitions from an aqueous feed into an immiscible organic solvent. After separation of the phases, the product is extracted from the enriched solvent phase by contact with a stripping solution. Extraction and stripping occur simultaneously on either side of the membrane in liquid membrane extraction. As the liquid membrane is thin and large interfacial areas can be created, only short contact times are required to achieve a separation. Advantages include reductions both of solvent inventory and of equipment capacity.

The development of liquid-membrane extraction has been mainly in the fields of hydrometallurgy and waste-water treatment. There are also potential advantages for their use in biotechnology, such as extraction from fermentation broths, and biomedical engineering, such as blood oxygenation.

11.14 Gas separations

The main emphasis in this chapter is on the use of membranes for separations in liquid systems. As discussed by Koros and Chern [30] and Kesting and Fritzsche [31], gas mixtures may also be separated by membranes and both porous and non-porous membranes may be used. In the former case, Knudsen flow can result in separation, though the effect is relatively small. Much better separation is achieved with non-porous polymer membranes where the transport mechanism is based on sorption and diffusion. As for reverse osmosis and pervaporation, the transport equations for gas permeation through dense polymer membranes are based on Fick's Law, material transport being a function of the partial pressure difference across the membrane.

A number of polymers are suitable for gas permeation. The elimination of pores is crucial to the successful operation of membranes. Composite membranes have proved to be most suitable, for example silicone-coated polysulphone. The relative permeability of a number of gases in such a membrane is shown in Fig. 11.23. Systems with high packing density, such as hollow fibre or spiral wound modules, are used for gas permeation. Using very fine fibres, membrane-area packing-densities of up to 50,000 m^2/m^3 are achieved. Modules with high, up to 14.8 MN/m^2 or

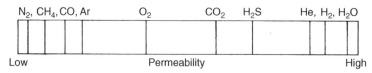

FIG. 11.23

Permeability of polysulphone–silicone membranes.

low, 0.8–0.9 MN/m^2 feed-side operating pressures are available. The most important application of membrane gas separation is the generation of N_2 from air. The production of oxygen from air is also significant. Other substantial applications are the recovery of hydrogen from refinery off-gases (CO, N_2, C_1, C_2) and vapours (C_{3+}, CO_2), and the removal of carbon dioxide from natural gas.

References

[1] H. Strathmann, Membrane separation processes (review), J. Membr. Sci. 9 (1981) 121.

[2] H.K. Lonsdale, The growth of membrane technology (review), J. Membr. Sci. 10 (1982) 81.

[3] R.E. Lacey, Dialysis and electrodialysis, in: P.A. Schweitzer (Ed.), Handbook of Separation Techniques for Chemical Engineers, second ed., McGraw-Hill, New York, 1988.

[4] R. Bertera, H. Steven, M. Metcalfe, Development studies of crossflow microfiltration, Chem. Eng. (401) (1984) 10.

[5] K. Schneider, W. Klein, The concentration of suspensions by means of cross flow microfiltration, Desalination 41 (1983) 271.

[6] A.G. Fane, Ultrafiltration: factors influencing flux and rejection, in: R.J. Wakeman (Ed.), Progress in Filtration and Separation, vol. 4, Elsevier, Amsterdam, 1986.

[7] M.D. Lévêque, Les lois de la transmission de chaleur pour convection, Ann. Mines 13 (1928) 201.

[8] W.F. Blatt, A. Dravid, A.S. Michaels, L. Nelson, Solute polarisation and cake formation in membrane ultrafiltration: causes, consequences and control techniques, in: J.E. Flinn (Ed.), Membrane Science and Technology, Plenum, New York, 1970.

[9] F.W. Dittus, L.M.K. Boelter, Heat Transfer in Automobile Radiators of the Tubular Type, vol. 2, University of California Publlications in Engineering, Berkeley, 1930, p. 443.

[10] A.S. Michaels, New separation technique for the CPI, Chem. Eng. Prog. 64 (1968) 31.

[11] M.C. Porter, Membrane filtration, in: P.A. Schweitzer (Ed.), Handbook of Separation Techniques for Chemical Engineers, second ed., McGraw-Hill, New York, 1988.

[12] M.S. Le, J.A. Howell, Ultrafiltration, in: M. Moo-Young (Ed.), Comprehensive Biotechnology, Pergamon, Oxford, 1985, p. 2.

[13] G. Green, G. Belfort, Fouling of ultrafiltration membranes: lateral migration and particle trajectory model, Desalination 35 (1980) 129.

[14] M.C. Porter, Concentration polarisation with membrane ultrafiltration, Ind. Eng. Chem. Prod. Res. Dev. 11 (1972) 234.

[15] G. Sergre, A. Silberberg, Radial particle displacements in Poiseuille flow of suspensions, Nature 189 (1961) 209.

[16] G. Belfort, Membrane separation technology: an overview, in: H.R. Bungay, G. Belfort (Eds.), Advanced Biochemical Engineering, John Wiley, New York, 1987.

[17] F.W. Altena, G. Belfort, Lateral migration of spherical particles in porous flow channels: application to membrane filtration, Chem. Eng. Sci. 39 (1984) 343.

[18] J.G. Wijmans, S. Nakao, C.A. Smolders, Flux limitation in ultrafiltration; osmotic pressure model and gel layer model, J. Membr. Sci. 20 (1984) 115.

[19] W.R. Bowen, P.M. Williams, Prediction of the rate of cross-flow ultrafiltration of colloids with concentration–dependent diffusion coefficient and viscosity–theory and experiment, Chem. Eng. Sci. 56 (2001) 3083.

[20] K.S. Spiegler, O. Kedem, Thermodynamics of hyperfiltration (reverse osmosis): criteria for efficient membranes, Desalination 1 (1966) 311.

[21] M. Soltanieh, W.N. Gill, Review of reverse osmosis membranes and transport models, Chem. Eng. Commun. 12 (1981) 279.

[22] R. Haase, Thermodynamics of Irreversible Processes, Addison Wesley, Massachusetts, 1969.

[23] G. Belfort, Desalting experience by hyperfiltration (reverse osmosis) in the United States, in: G. Belfort (Ed.), Synthetic Membrane Processes, Academic Press, Orlando, 1984.

[24] D. Pletcher, Industrial Electrochemistry, Chapman and Hall, London, 1982.

[25] W.S. Finlay, P.V. Ferguson, Design and operation of a turnkey reverse osmosis water treatment plant, in: Presented at the International Congress on Desalination and Water Re-use (Tokyo), 1977.

[26] C. Ventresque, V. Gisclon, G. Bablon, G. Chagneau, An outstanding feat of modern technology: the Mery-sur-Oise nanofiltration treatment plant, Desalination 131 (2001).

[27] R. Rautenbach, R. Albrecht, Membrane Processes, Wiley, Chichester, 1989.

[28] J.W. Frankenfeld, N.N. Li, Recent advances in membrane technology, in: R.W. Rousseau (Ed.), Handbook of Separation Process Technology, Wiley, New York, 1987, pp. 840–861.

[29] C. del Cerro, D. Boey, Liquid membrane extraction, Chem. Ind. (1988) 681–687.

[30] W.J. Koros, R.T. Chern, Separation of gaseous mixtures using polymer membranes, in: R.W. Rousseau (Ed.), Handbook of Separation Process Technology, Wiley, New York, 1987, pp. 862–953.

[31] R.E. Kesting, A.K. Fritzsche, Polymeric Gas Separation Membranes, Wiley, New York, 1993.

Further reading

G. Belfort (Ed.), Synthetic Membrane Processes, Academic Press, Orlando, 1984.

M. Cheryan, Ultrafiltration Handbook, Technomic Publishing Company, Pennsylvania, 1998.

R. Rautenbach, R. Albrecht, Membrane Processes, Wiley, 1989.

P.A. Schweitzer (Ed.), Handbook of Separation Techniques for Chemical Engineers, second ed., McGraw Hill, New York, 1988.

L.J. Zeman, A.L. Zydney, Microfiltration and Ultrafiltration. Principles and Applications, Marcel Dekker, New York, 1996.

Leaching

Ajay Kumar Ray

Department of Chemical and Biochemical Engineering, University of Western Ontario, London, ON, Canada

Nomenclature

		Units in SI system	Dimensions in M, L, T, θ
A	area of solid–liquid interface	m^2	$\mathbf{L}^2$
a	mass of solute per unit mass of solvent in final overflow	kg/kg	–
b	thickness of liquid film	m	$\mathbf{L}$
C_p	specific heat of solution	J/kg K	$\mathbf{L}^2\mathbf{T}^{-2}\theta^{-1}$
c	concentration of solute in solvent	kg/m^3	$\mathbf{ML}^{-3}$
c_o	initial concentration of solute in solvent	kg/m^3	$\mathbf{ML}^{-3}$
c_s	concentration of solute in solvent in contact with solid	kg/m^3	$\mathbf{ML}^{-3}$
D_L	liquid phase diffusivity	m^2/s	$\mathbf{L}^2\mathbf{T}^{-1}$
d	diameter of vessel	m	$\mathbf{L}$
F	total mass of material fed to thickener	kg	$\mathbf{M}$
h	heat transfer coefficient	W/m^2 K	$\mathbf{MT}^{-3}\theta^{-1}$
K_L	mass transfer coefficient	m/s	$\mathbf{LT}^{-1}$
k	thermal conductivity	W/m K	$\mathbf{MLT}^{-3}\,\theta^{-1}$
k'	a diffusion constant	m^2/s	$\mathbf{L}^2\mathbf{T}^{-1}$
L	mass flow of liquid retained by underflow	kg/s	$\mathbf{MT}^{-1}$
M	mass of solute transferred in time t	kg	$\mathbf{M}$
N	agitation speed (Section 12.2)	rpm	$\mathbf{T}^{-1}$
N	mass flow ratio of solid to liquid in underflow	kg/kg	–

Continued

Coulson and Richardson's Chemical Engineering. https://doi.org/10.1016/B978-0-08-101097-6.00012-2

		Units in SI system	Dimensions in M, L, T, θ
n	number of countercurrent washing thickeners	–	–
N_v	mass flow ratio of solid to liquid in overflow	kg/kg	–
N_Δ	mass fraction difference of insoluble component B	kg/kg	–
S	mass flow of entering solvent	kg/s	MT^{-1}
t	time	s	T
V	volume of solvent used for extraction (Section 12.2)	m^3	L^3
V	mass flow of overflow liquid	kg/s	MT^{-1}
x	mass fraction of solute A in underflow liquid	kg/kg	–
x_Δ	mass fraction difference of solute A	kg/kg	–
y	mass fraction of solute A in overflow liquid	kg/kg	–
Δ	difference between underflow and overflow	kg/s	MT^{-1}
μ	viscosity of solution or liquid	Ns/m^2	$ML^{-1}T^{-1}$
ρ	density of solution or liquid	kg/m^3	ML^{-3}

Suffixes

A, B, S, M refer to solute, insoluble solid, solvent, and mixed solution, respectively

1, ..., n, ..., N refer to liquid overflow or underflow from stages 1, ..., n, ..., N

0 refers to the liquid underflow feed to stage 1.

12.1 Introduction

12.1.1 Leaching operations

Leaching, sometimes called solid–liquid extraction, is concerned with the extraction of a soluble constituent (the solute) from a solid by means of a liquid solvent. The process can be used either for the production of a concentrated solution of a valuable solid material, or for the removal of an undesirable solute from a solid phase. Equipment is available for leaching under batch, semicontinuous, or continuous operating modes. Effluents from a leaching stage are essentially solid-free liquid, called *overflow*, and wet solids, the *underflow*. The final overflow from a multi-stage process contains some of the solvent and most of the solute, and the final underflow mainly contains the insoluble residues and the solvent.

The method used for leaching is determined by the proportion of soluble constituent present, its distribution throughout the solid, the nature of the solid, and the particle size. If the solute is uniformly dispersed in the solid, the material near the surface will be dissolved first, leaving a porous structure in the solid residue. The solvent has to penetrate this outer layer before it can reach further solute. The process will become progressively more difficult and the extraction rate will decrease. If the solute forms a very high proportion of the solid, the porous structure may break down almost immediately to give a fine deposit of insoluble residue, and access of solvent to the solute will not be impeded. Generally, the process can be considered in three steps: first the change of phase of the solute as it dissolves in the solvent, secondly its diffusion through the solvent in the pores of the solid to the outer surface of the particle, and thirdly the transfer of the solute from the solution in contact with the particles to the main bulk of the solution. Any one of these three processes may be responsible for limiting the leaching rate, though the first process usually occurs so rapidly that it has a negligible effect on the overall rate.

In some cases the soluble material is distributed in small isolated pockets in a material which is impermeable to the solvent such as gold dispersed in rock, for example. In such cases, the material is crushed so that all the soluble material is exposed to the solvent. If the solid has a cellular structure, the leaching rate will generally be low because the cell walls provide an additional resistance. In the extraction of sugar from beet, the cell walls perform the important function of impeding the extraction of undesirable constituents of relatively high molecular weight, and the beet should therefore be prepared in long strips so that a relatively small proportion of the cells are ruptured. In the extraction of oil from seeds, the solute is itself liquid.

12.1.2 **Industrial examples**

Leaching is widely used in mineral-processing industry for valuable metals. Metals normally exist in mixture with large amounts of undesirable constituents and leaching is used to extract the metals as soluble salts. For example, gold is recovered from low-grade ores with aqueous sodium cyanide. Copper is recovered by leaching the ore with acid and then extracting the dissolved copper with kerosene solutions of oximes. Nickel can be leached with combinations of sulphuric acid and ammonia.

In the biological and food industries, many products are separated from their original natural structures by leaching. Vegetable oils are currently extracted by leaching from a number of crops, including coconuts, flax seeds, palms, peanuts, soybeans, grape seeds, and sunflower seeds. Table sugar is extracted from sugarcane or sweet beets using hot water. In the pharmaceutical industry, different pharmaceutical products are obtained by leaching plant roots, leaves and stems. Moreover, proteins and other natural products are also recovered from bacterial cells or animal organs by leaching processes.

12.1.3 **Factors influencing the rate of leaching**

Rate of leaching is an important measure of the process efficiency. Because solute dissolution and mass transfer of the solute in porous solid and bulk solution are involved in the process, rate of leaching is therefore mainly influenced by the factors which are responsible for limiting the leaching rate. Four factors listed below are considered as key factors influencing the rate of leaching.

Particle size. Particle size influences the rate of leaching in a number of ways. The smaller the size, the greater is the interfacial area between the solid and liquid, and therefore the higher is the rate of transfer of material and the smaller is the distance the solute must diffuse within the solid as already discussed. On the other hand, the surface may not be so effectively used with a very fine material if circulation of the liquid is impeded, and separation of the particles from the liquid and drainage of the solid residue are made more difficult. It is generally desirable that the range of particle size should be small so that each particle requires approximately the same time for extraction and, in particular, the production of a large amount of fine material should be avoided as this may wedge in the interstices of the larger particles and impede the flow of the solvent.

Solvent. A good selective solvent with sufficiently low viscosity should be used for a leaching process. The ideal solvent for commercial leaching processes should have high solubility of leachant. In addition, the solvent should be unreactive and with lower boiling point for recycle.

Temperature. In most cases, the solubility of solutes increases with temperature to give a higher rate of leaching. Furthermore, the diffusion coefficient will be expected to increase with rise in temperature and this will also improve the rate of leaching. In some cases, the upper limit of temperature is determined by secondary considerations, such as, the necessity to avoid enzyme action during the extraction of sugar.

Agitation of the fluid. Agitation of the solvent is important because this increases the eddy diffusion and therefore the transfer of solute from the surface of the particles to the bulk of the solution is accelerated. In addition, agitation of suspensions of fine particles prevents sedimentation and the interfacial surface can be used more efficiently.

12.2 **Mass transfer in leaching operations**

Mass transfer rates within the porous residue are difficult to assess because it is impossible to define the shape of the channels through which transfer takes place. It is possible, however, to obtain an approximate indication of the rate of transfer from the particles to the bulk of the liquid. Using the concept of a thin film as providing the resistance to transfer, the equation for mass transfer is expressed as:

$$\frac{dM}{dt} = \frac{k' A(c_S - c)}{b} \tag{12.1}$$

where:

A is the area of the solid–liquid interface,
b is the effective thickness of the liquid film surrounding the particles,
c is the concentration of the solute in the bulk of the solution at time t,
c_s is the concentration of the saturated solution in contact with the particles,
M is the mass of solute transferred in time t, and
k' is the diffusion coefficient. (This is approximately equal to the liquid phase diffusivity D_L)

For a batch process in which V, the total volume of solution, is assumed to remain constant, then:

$$dM = V dc$$

$$\text{and :} \quad \frac{dc}{dt} = \frac{k' A(c_S - c)}{bV} \tag{12.2}$$

The time t taken for the concentration of the solution to rise from its initial value c_0 to a value c is found by integration, on the assumption that both b and A remain constant. Rearranging:

$$\int_{c_0}^{c} \frac{dc}{c_S - c} = \int_{0}^{t} \frac{k' A}{Vb} dt \tag{12.3}$$

Integrating from $t=0$ and $c=c_0$, we have

$$-\ln (c_S - c)\big|_{c_0}^{c} = \frac{k' A}{Vb} t$$

$$\text{and :} \quad \frac{c_S - c}{c_S - c_0} = e^{-(k'A/bV)t} \tag{12.4}$$

If pure solvent is used initially, $c_0=0$, thus,

$$c = c_S \left(1 - e^{-(k' A/bV)t}\right) \tag{12.5}$$

which shows that the solution approaches a saturated condition exponentially.

Example 12.1

In a pilot scale test using a vessel 1 m^3 in volume, a solute was leached from an inert solid and the water was 75 wt% saturated in 100 s. If, in a full-scale unit, 500 kg of the inert solid containing, as before, 28 wt% of the water-soluble component, is agitated with 100 m^3 of water, how long will it take for all the solute to dissolve, assuming conditions are equivalent to those in the pilot scale vessel? Water is saturated with the solute at a concentration of 2.5 kg/m^3.

Solution

For the *pilot-scale* vessel:

$$c = (2.5 \times 75/100) = 1.875 \text{ kg/m}^3$$
$$c_S = 2.5 \text{ kg/m}^3, \quad V = 1.0 \text{ m}^3 \text{ and } t = 10 \text{ s} \tag{12.6}$$

Thus, in Eq. (12.5):

$$1.875 = 2.5\left(1 - e^{-(k'A/1.0b)100}\right) \tag{12.6}$$

$$\text{and}: k'A/b = 0.139 \text{ m}^3/\text{s} \tag{12.6}$$

For the *full-scale* vessel:

$$
\begin{aligned}
c &= (500 \times 28/100) = 1.40 \text{ kg/m}^3 \\
c_S &= 2.5 \text{ kg/m}^3, \quad V = 100 \text{ m}^3
\end{aligned} \tag{12.6}
$$

Thus:

$$1.40 = 2.5\left(1 - e^{-0.139t/100}\right) \tag{12.6}$$

$$\text{and}: t = \underline{591 \text{ s}} \; (9.9 \text{ min}) \tag{12.6}$$

In most cases the interfacial area will tend to increase during the extraction and, when the soluble material forms a very high proportion of the total solid, complete disintegration of the particles may occur. Although this results in an increase in the interfacial area, the rate of extraction will probably be reduced because the free flow of the solvent will be impeded and the effective value of b will be increased.

Work on the rate of dissolution of regular shaped solids in liquids has been carried out by Linton and Sherwood [1], to which reference is made in Volume 1. Benzoic acid, cinnamic acid, and β-naphthol were used as solutes, and water as the solvent. For streamline flow, the results were satisfactorily correlated on the assumption that transfer took place as a result of molecular diffusion alone. For turbulent flow through small tubes cast from each of the materials, the rate of mass transfer could be predicted from the pressure drop by using the 'j-factor' for mass transfer. In experiments with benzoic acid, unduly high rates of transfer were obtained because the area of the solids was increased as a result of pitting.

The effect of agitation, as produced by a rotary stirrer, for example, on mass transfer rates has been investigated by Hixson and Baum [2] who measured the rate of dissolution of pure salts in water. The degree of agitation is expressed by means of a dimensionless group ($Nd^2\rho/\mu$) in which:

N is the number of revolutions of the stirrer per unit time,

d is the diameter of the vessel,

ρ is the density of the liquid, and

μ is its viscosity.

This group is referred to in Volume 1, Chapter 7 in the discussion of the power requirements for agitators.

For values of ($Nd^2\rho/\mu$) less than 67,000, the results are correlated by:

$$\frac{K_L d}{D_L} = 2.7 \times 10^{-5} \left(\frac{N d^2 \rho}{\mu}\right)^{1.4} \left(\frac{\mu}{\rho D_L}\right)^{0.5} \qquad (12.6)$$

and for higher values of $(N d^2 \rho / \mu)$ by:

$$\frac{K_L d}{D_L} = 0.16 \left(\frac{N d^2 \rho}{\mu}\right)^{0.62} \left(\frac{\mu}{\rho D_L}\right)^{0.5} \qquad (12.6)$$

where K_L is the mass transfer coefficient, equal to k'/b in Eq. (12.1).

Further experimental work has been carried out on the rates of melting of a solid in a liquid, using a single component system, and Hixson and Baum express their results for the heat transfer coefficient as:

$$\frac{hd}{k} = 0.207 \left(\frac{N d^2 \rho}{\mu}\right)^{0.63} \left(\frac{C_p \mu}{k}\right)^{0.5} \qquad (12.6)$$

for values of $(N d^2 \rho / \mu)$ greater than 67,000.

In Eq. (12.8), h is the heat transfer coefficient, k is the thermal conductivity of the liquid, and C_p is the specific heat of the liquid.

It may be seen from Eqs (12.7), (12.8) that at high degrees of agitation the ratio of the heat and mass transfer coefficients is almost independent of the speed of the agitator and:

$$\frac{K_L}{h} = 0.77 \left(\frac{D_L}{\rho C_p k}\right)^{0.5} \qquad (12.6)$$

Piret et al. [3] attempted to reproduce the conditions in a porous solid using banks of capillary tubes, beds of glass beads and porous spheres, and measured the rate of transfer of a salt as solute through water to the outside of the system. It was shown that the rate of mass transfer is that predicted for an unsteady transfer process and that the shape of the pores could be satisfactorily taken into account.

In a theoretical study, Chorny and Krasuk [4] analysed the diffusion process in extraction from simple regular solids, assuming constant diffusivity.

12.3 Equipment for leaching

Three distinct processes are typically involved in leaching operations: (a) dissolving the soluble constituent; (b) separating the solution, so formed, from the insoluble solid residue; and (c) washing the solid residue in order to remove unwanted soluble matter or to obtain as much of the soluble material as possible as the product. Leaching can be operated under batch or unsteady-state conditions as well as continuous or steady-state conditions. Both differential contactors and stage-wise contactors are used in batch or continuous operations. The type of equipment employed depends on the nature of the solid, i.e. whether it is granular or cellular and whether it is coarse or fine. The normal distinction between coarse and fine solids is that the former have sufficiently large settling velocities for them to be readily separable from the liquid, whereas the latter can be maintained in suspension with the aid of only a small amount of agitation. Generally, the solvent is allowed to percolate through beds of coarse materials, whereas a high resistance to solvent flow is exerted by fine solids.

As already pointed out, the rate of leaching, in general, is a function of the relative velocity between the liquid and the solid. In some plants, the solid is stationary and the liquid flows through the bed of particles, whilst in some continuous plants the solid and liquid move countercurrently.

12.3.1 Extraction from cellular materials

With seeds such as soya beans, containing only about 15% of oil, solvent extraction is often used because mechanical methods are not very efficient. Light petroleum fractions are generally used as solvents. Trichlorethylene has been used where fire risks are serious, and acetone or ether where the material is very wet. A batch plant for the extraction of oil from seeds is illustrated in Fig. 12.1. This consists of a vertical cylindrical vessel divided into two sections by a slanting partition. The upper section is filled with the charge of seeds which is sprayed with fresh solvent via a distributor. The solvent percolates through the bed of solids and drains into the lower compartment where, together with any water extracted from the seeds, it is continuously boiled off by means of a steam coil. The vapours are passed to an external condenser, and the mixed liquid is passed to a separating box from which the solvent

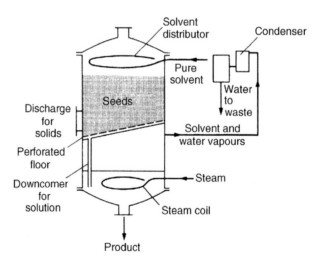

FIG. 12.1

Batch plant for extraction of oil from seeds.

is continuously fed back to the plant and the water is run to waste. By this means a concentrated solution of the oil is produced by the continued application of pure solvent to the seeds.

The Bollmann continuous moving bed extractor, as shown in Figs 12.2 and 12.3, which is described by Goss [5], consists of a series of perforated baskets, arranged as in a bucket elevator and contained in a vapour-tight vessel, is widely used with seeds which do not disintegrate on extraction. Solid is fed into the top basket on the downward side and is discharged from the top basket on the upward side, as shown in Fig. 12.3. The solvent is sprayed on to the solid which is about to be discarded, and passes downwards through the baskets so that countercurrent flow is achieved. The solvent is finally allowed to flow down through the remaining baskets in co-current flow. A typical extractor moves at about 0.3 mHz (1 revolution per hour), each basket containing some 350 kg of seeds. Generally, about equal masses of seeds and solvent are used and the final solution, known as miscella, contains about 25% of oil by mass.

The Bonotto extractor [5] consists of a tall cylindrical vessel with a series of slowly rotating horizontal trays. The solid is fed continuously on to the top tray near its outside edge and a stationary scraper, attached to the shell of the plant, causes it to move towards the centre of the plate. It then falls through an opening on to the plate beneath, and another scraper moves the solids outwards on this plate which has a similar opening near its periphery. By this means, the solid is moved across each plate, in opposite directions on alternate plates, until it reaches the bottom of the tower from which it is removed by means of a screw conveyor. The extracting liquid is introduced at the bottom and flows upwards so that continuous countercurrent flow is obtained, though a certain amount of mixing of solvent and solution takes place when the density of the solution rises as the concentration increases.

A more recently developed continuous extractor is the horizontal perforated belt extractor, probably the simplest percolation extractor from a mechanical view point. Here the basic principle is the extraction of an intermediate bed depth on a continuous belt without partitions. The extractor is fitted with a slow moving 'perforated belt' running on sprockets at each end of the extractor. A series of specially designed screens form a flat surface attached to the chains, made of wedge-bar type grids when non-powdery products are processed, or stainless steel mesh cloths for fine particle products. The flakes are fed into the hopper and flow on to the belt of the extractor, and the level is controlled by a damper at the outlet of the feeding hopper in order to maintain a constant flake bed height. This height can be adjusted when oil-bearing materials with lower percolation rates are to be processed. The two side walls of the extractor body provide support for the bed on the moving belt and, with no dividers in the belt, the bed of material becomes a continuous mass. This means that, under stationary conditions, miscella concentration in each and every point of the bed of material is constant as this concentration is not related to the position of a given compartment over the miscella collecting hopper. The belt speed is automatically controlled by the level of flakes in the inlet hopper, and is measured by a nuclear sensor which controls the infinitely variable speed drive.

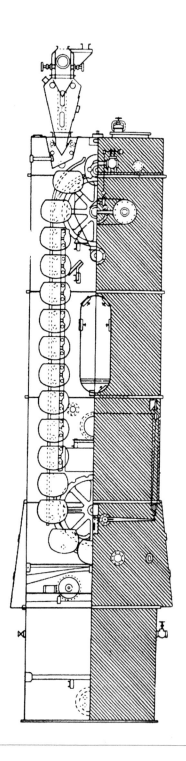

FIG. 12.2

Bollmann extractor.

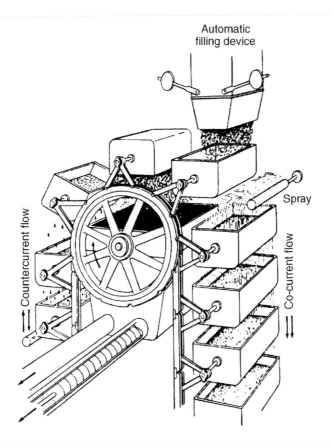

FIG. 12.3

Bollmann extractor – filling and emptying of baskets.

The raw material is sprayed with miscella during its entire passage through the extractor and fresh solvent, introduced at the discharge end of the extractor, circulates against the flow of flakes, under the action of a series of stage pumps. Each miscella wash has a draining section after which the top of the bed is scraped by a hinged rake which has two functions. Firstly, it prevents the thin layer of fines settled on the upper part of the bed from reducing the permeability of the bed of material, and secondly, it forms a flake pile at each draining section to prevent intermingling of miscella. At each wash section, a spray distributor ensures a uniform distribution of the liquid over the width of the bed and liquid flow rate is adjustable by individual valves. A manifold is fitted to permit miscella circulation through the same stage, or progression to the previous one. Before it is discharged, liquid in the material bed drains into the final collecting hoppers. Discharge of material into the outlet hopper is regulated by a rotary scraper which ensures an even feed of the extracted meal to the drainage section. The belt is effectively cleaned twice, first by fresh solvent just after material discharge and then at the other end of the return span, by means of miscella.

12.3.2 **Leaching of coarse solids**

Coarse solids are often leached by percolation through stationary solid beds in a cylindrical vessel with a perforated bottom to permit drainage of the solution. Sometimes, a number of tanks are used in series for a countercurrent system, called a extraction battery, in which fresh solvent is fed to the solid that is nearest to complete extraction, flowing through the tanks in series and being withdrawn from the one charged with fresh solid. The solvent may flow by gravity or be fed by positive pressure, and is generally heated before it enters each tank. Multiple piping is used so that tanks do not have to be moved for countercurrent operation. The system is called the *Shanks system*, which is widely used in the mineral and sugar industries for various products.

A continuous unit in which countercurrent flow is obtained is the tray classifier, such as the Dorr classifier described earlier. Solid is introduced near the bottom of a sloping tank and is gradually moved up by means of a rake. The solvent enters at the top and flows in the opposite direction to the solid, and passes under a baffle before finally being discharged over a weir (Fig. 12.4). The classifier operates satisfactorily provided the solid does not disintegrate, and the solids are given ample time to drain before they are discharged. A number of these units may be connected in series to give countercurrent flow.

A plant has been successfully developed in Australia for the extraction of potassium sulphate from alums containing about 25% soluble constituents. After roasting, the material which is then soft and porous with a size range from 12 mm to very fine particles, with 95% greater than 100-mesh (0.15 mm), is leached at 373 K with a solution that is saturated at 303 K, the flow being as shown in Fig. 12.5. The make-up water, which is used for washing the extracted solid, is required to replace that removed in the residue of spent solid, in association with the crystals, and by evaporation in the leaching tank and the crystalliser.

The leaching plant, shown in Fig. 12.6, consists of an open tank, 3 m in diameter, into the outer portion of which the solid is continuously introduced from an annular hopper. Inside the tank, a 1.8 m diameter vertical pipe rotates very slowly at the rate of about one revolution every 2400 s (0.0042 Hz). It carries three ploughs stretching to the circumference of the tank, and these gradually take the solid through holes into

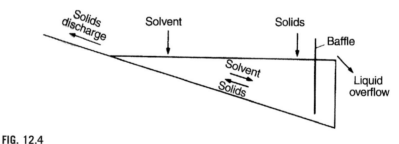

FIG. 12.4

Flow of solids and liquids in Dorr classifier.

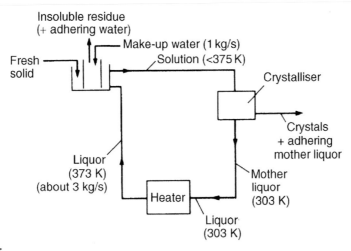

FIG. 12.5

Flow diagram for continuous leaching plant.

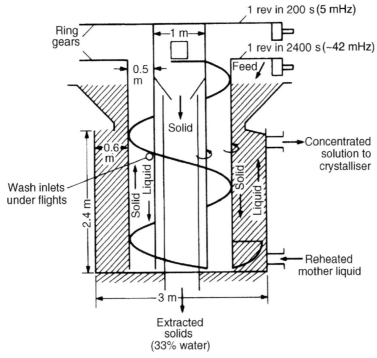

FIG. 12.6

Continuous leaching tank.

the inside of the pipe. A hollow shaft, about 1 m in diameter, rotates in the centre of the tank at about one revolution in 200 s (0.005 Hz) and carries a screw conveyor which lifts the solid and finally discharges it through an opening, so that it falls down the shaft and is deflected into a waste pipe passing through the bottom of the tank. Leaching takes place in the outer portion of the tank where the reheated mother liquor rises through the descending solid. The make-up water is introduced under the flutes of the screw elevator, flows down over the solid and then joins the reheated mother liquor. Thus countercurrent extraction takes place in the outer part of the tank and countercurrent washing in the central portion. The plant described achieves between 85 and 90% extraction, as compared with only 50% in the batch plant which it replaced.

12.3.3 Leaching of fine solids

Whereas coarse solids may be leached by causing the solvent to pass through a bed of the material, fine solids offer too high a resistance to flow. Particles of less than about 200-mesh (0.075 mm) may be maintained in suspension with only a small amount of agitation, and as the total surface area is large, an adequate extraction can be effected in a reasonable time. Because of the low settling velocity of the particles and their large surface, the subsequent separation and washing operations are more difficult for fine materials than with coarse solids.

Agitation may be achieved either by the use of a mechanical stirrer or by means of compressed air. If a paddle stirrer is used, precautions must be taken to prevent the whole of the liquid being swirled, with very little relative motion occurring between solids and liquid. The stirrer is often placed inside a central tube, as shown in Fig. 12.7, and the shape of the blades arranged so that the liquid is lifted upwards through the tube. The liquid then discharges at the top and flows downwards outside the tube, thus giving continuous circulation. Other types of stirrers are discussed in Volume 1, Chapter 7, in the context of liquid–liquid mixing.

An example of an agitated vessel in which compressed air is used is the Pachuca tank, shown in Fig. 12.8. This is a cylindrical tank with a conical bottom, fitted with a central pipe connected to an air supply. Continuous circulation is obtained with the central pipe acting as an air lift. Additional air jets are provided in the conical portion of the base and are used for dislodging any material which settles out.

The Dorr agitator which is illustrated in Fig. 12.9, also uses compressed air for stirring, and consists of a cylindrical flat-bottomed tank fitted with a central air lift inside a hollow shaft which slowly rotates. To the bottom of the shaft are fitted rakes which drag the solid material to the centre as it settles, so that it is picked up by the air lift. At the upper end of the shaft the air lift discharges into a perforated launder which distributes the suspension evenly over the surface of the liquid in the vessel. When the shaft is not rotating the rakes automatically fold up so as to prevent the plant from seizing up if it is shut down full of slurry. This type of agitator can be used for batch or continuous operation. In the latter case, the entry and delivery points are situated at opposite sides of the tank. The discharge pipe often takes

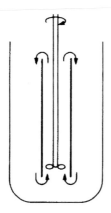

FIG. 12.7

Simple stirred tank.

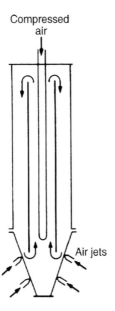

FIG. 12.8

Pachuca tank.

the form of a flexible connection which can be arranged to take off the product from any desired depth. Many of these agitators are heated by steam coils. If the soluble material dissolves very rapidly, extraction can be carried out in a thickener, such as the Dorr thickener described in Chapter 5. Thickeners are also extensively used for separating the discharge from an agitator, and are frequently connected in series to give countercurrent washing of the residue.

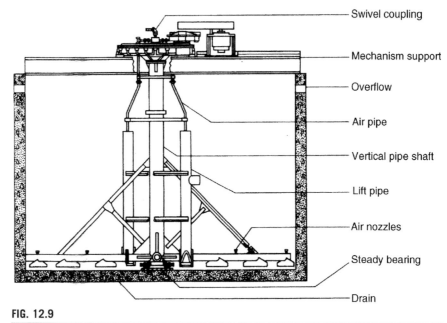

FIG. 12.9

Dorr agitator.

12.3.4 Batch leaching in stirred tanks

The batch dissolution of solids in liquids is very often carried out in tanks agitated by co-axial impellers including turbines, paddles and propellers, a system which may also be used for the leaching of fine solids. In this case, the controlling rate in the mass transfer process is the rate of transfer of material into or from the interior of the solid particles, rather than the rate of transfer to or from the surface of the particles, and therefore the main function of the agitator is to supply unexhausted solvent to the particles which remain in the tank long enough for the diffusive process to be completed. This is achieved most efficiently if the agitator is used to circulate solids across the bottom of the tank, or barely to suspend them above the bottom of the tank. After the operation is completed, the leached solids must be separated from the extract and this may be achieved by settling followed by decantation, or externally by filters, centrifuges or thickeners. The difficulties involved in separating the solids and the extract is one of the main disadvantages of batch operation coupled with the fact that batch stirred tanks provide only one equilibrium stage. The design of agitators in order to produce a batch suspension of closely sized particles has been discussed by Bohnet and Niesmark [6] and the general approach is to select the type and geometry of impeller and tank, to specify the rotational speed required for acceptable performance, and then to determine the shaft power required to drive the impeller at that speed.

Ores of gold, uranium and other metals are often batch-leached in *Pachua tanks* which are described in Section 12.3.3.

12.4 Equilibrium relations and: Single-stage leaching
12.4.1 Equilibrium relations in leaching

To analyse leaching process, the equilibrium relations between the two streams (overflow and underflow) are needed. In an ideal case, it is assumed that sufficient solvent is present to dissolve all the solute in the entering solid and intimate mixing of solid and solvent can be achieved in each unit (or stage), equilibrium is attained when a uniform solution is formed. Under this condition, the concentration of the liquid retained by the solid leaving any stage is identical to the concentration of the overflow liquid leaving from the same stage and the overflow is free of solids. Each unit then represents an ideal leaching stage and the equilibrium relationship is simply $y = x$. where y represents the mass fraction of solute in the overflow liquid; and x is the mass fraction of solute in the liquid retained by the underflow.

In many cases intimate mixing may not be achieved and contact time is no longer enough for a uniform solution to form. The equilibrium relationship between the overflow solution and the underflow solution needs to be determined by experiments. The mass ratio of solid to liquid in the underflow depends on the properties of phases (such as viscosity and density of the liquid) and the type of equipment. Hence, experimental data must be obtained to determine the mass ratio of solid to liquid in the underflow as a function of solute concentration. Sometimes, small amounts of fine solid particles may remain in the overflow due to the incomplete phase separation of solid particles from the liquid, mass ratio of solid to liquid in the overflow needs to be determined from tests with prototype equipment.

12.4.2 Equilibrium diagrams for leaching

The equilibrium data can be plotted on the rectangular diagram as mass fractions for the three components: solute **A**, insoluble constituent **B** and solvent **C**. The two phases are the overflow phase and underflow (slurry) phase. The mass ratios of insoluble solid **B** to solution mixtures of the two phases can be expressed as:

$$N = \frac{\text{kg solid}}{\text{kg solution}} = \frac{\text{kg } B}{\text{kg } A + \text{kg } C} \ (underflow) \tag{12.10}$$

$$N_V = \frac{\text{kg solid}}{\text{kg solution}} = \frac{\text{kg } B}{\text{kg } A + \text{kg } C} (overflow) \tag{12.11}$$

Both N and N_v may vary with the solute concentration, if the overflow is solid-free, then $N_v = 0$. The compositions of solute A in the underflow and overflow liquids are expressed as mass fractions:

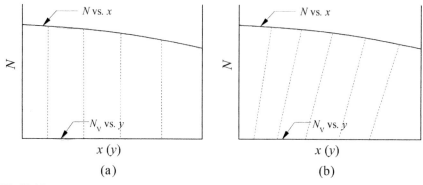

FIG. 12.10

Typical phase equilibrium diagrams: (A) case for vertical tie lines, (B) case where $y_A \neq x_A$ for tie lines.

$$x = \frac{kg \; solute}{kg \; solution} = \frac{kg \; A}{kg \; A + kg \; C} \; (underflow) \tag{12.12}$$

$$y = \frac{kg \; solute}{kg \; solution} = \frac{kg \; A}{kg \; A + kg \; C} \; (overflow) \tag{12.13}$$

Fig. 12.10A depicts a typical equilibrium diagram, where solute **A** is infinitely soluble in solvent **C**. The upper curve of N vs. x for the slurry underflow represents the separated solid under experimental conditions of actual stage process. The bottom line of N_V vs. y, where $N_v = 0$ on the axis, indicates that all the solid has been removed from the overflow liquid. Tie lines in Fig. 12.10A are vertical, which means uniform solution is obtained for the operation. In Fig. 12.10B, the tie lines are not vertical which usually results from insufficient contact time.

If the underflow curve of N vs. x is straight and horizontal, the amount of liquid associated with the solid in the slurry is constant for all concentrations. This would mean that flow rates of underflow and overflow solutions are constant throughout the various stages.

12.4.3 Single-stage leaching

The simplest leaching operation is the single-stage process as shown in Fig. 12.11A, where feed to be leached with a mass flow rate F (the sum of solution L_0 and inert solid B) and composition of N_0 and x_0 is in intimate contact of entering solvent with a mass flow rate of S and a solute concentration of y_S. The contacting time of the two streams is longer enough so that phase equilibrium is reached when the underflow and overflow leave from the unit. The material balance equations for single-stage operation are as follows for a total solution balance, component balances on solute **A** and insoluble component **B**, respectively.

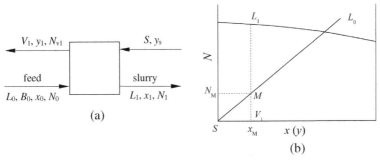

FIG. 12.11

Flow diagram and mass balance for single-stage leaching process.

$$L_0 + S = L_i + V_1 = M \tag{12.14}$$

$$L_0 x_0 + S y_s = L_1 x_1 + V_1 y_1 = M x_M \tag{12.15}$$

$$B = L_0 N_0 = L_1 N_1 + V_1 N_{v1} = M N_M \tag{12.16}$$

where M is the total flow rate of mixing solution; x_M and N_M are coordinates of this point M on the phase diagram; L and V represent the flow rates of underflow liquid and overflow liquid. Since phase equilibrium is reached, L_1 and V_1 must lie on the tie line passing through point M. Once the compositions of L_1 and V_1 are obtained, mass balance equations can be applied to determine the flow rates of L_1 and V_1 as shown in Fig. 12.11B.

Example 12.2

Oil is to be leached from 100 kg/min of soybean flakes, containing 20 wt% oil in a single equilibrium stage by 100 kg/min of hexane solvent. The value of N for the slurry underflow varies with solution concentration and the measured contents of underflows are listed in the table below. Assuming the overflow is solid-free and a uniform solution is reached, calculate compositions and flow rates of the underflow and overflow.

N	1.5	1.48	1.45	1.42	1.38	1.33
x	0	0.2	0.4	0.6	0.8	1.0

Solution

    Known parameters are as follows:

    For feed: $L_0 = 0.2F = 20\,\text{kg/min}$. $B = F\text{-}L_0 = 80\,\text{kg/min}$, $x_0 = 1.0$ and $N_0 = 4.0$

    For entering solvent: $S = 100\,\text{kg/min}$, $y_S = 0$

Mass balance equations on total solution and component A are:

$$L_0 + S = L_1 + V_1 = 120\,\text{kg/min}$$

$$L_0 x_0 + S y_s = L_1 x_1 + V_1 y_1 = 20\,\text{kg/min}$$

Because uniform solution is formed during the operation, thus $x_1 = y_1$, and:

$$x_1 = y_1 = x_M = 20/120 = \underline{0.167}$$

As $x_1 = 0.167$, based on phase equilibrium data as shown in Fig. 12.12, $N_1 = \underline{1.483}$, the flow rate of L_1 is:

$$L_1 = B/N_1 = 80/1.483 = 53.94\,\text{kg/min}$$

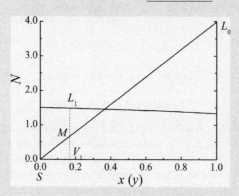

FIG. 12.12

Graphical solution to Example 12.2.

The flow rate of V_1 is

$$V_1 = L_0 + S - L_1 = 120 - 53.94 = \underline{66.06\,\text{kg/min}}$$

12.5 Countercurrent multistage leaching

12.5.1 Countercurrent washing

The most important method of leaching is the continuous countercurrent multistage leaching. Fig. 12.13 shows a process flow diagram for a continuous countercurrent cascade. The stages are numbered in the direction of solid flow. The V phase is the liquid that overflows from stage N to stage 1 in a direction counter to that of the solid phase. The L phase is the liquid carried by the solid as it moves from stage 1 to stage N. The leached solid leaves stage N and the final extract overflows from stage 1. The composition of the V phase is denoted by y and the composition of L phase by x as discussed in Section 12.4.2. Similar to the single-stage operation, the amounts of insoluble component **B** present in the overflow and underflow are denoted by N_V and N, respectively.

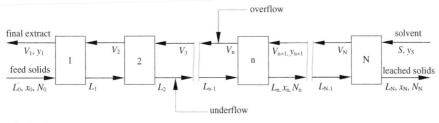

FIG. 12.13

Countercurrent multistage leaching system.

In order to derive the operating line equation, the mass balances on the total solution and solute **A** are made over the first n stages as shown in Fig. 12.13:

$$L_0 + V_{n+1} = L_n + V_1 \tag{12.17}$$

$$L_0 x_0 + V_{n+1} y_{n+1} = L_n x_n + V_1 y_1 \tag{12.18}$$

solving for y_{n+1},

$$y_{n+1} = \frac{L_n}{V_{n+1}} x_n + \frac{V_1 y_1 - L_0 x_0}{V_{n+1}} \tag{12.19}$$

The operating line Eq. (12.19), when plotted on a y–x diagram, passes through the terminal points (x_0, y_1) and (x_N, y_S). Usually, L_n and V_{n+1} vary from stage to stage, therefore, the slope of the operating line also vary from stage to stage, leading to a non-linear curve on the y–x plot.

Typically, solute and total-liquid material balances can be used to solve problems, including: (1) determination of number of stages required to achieve a specified degree of leaching; and (2) determination of the effect of leaching with a certain number of stages. Depending on the problem, either an analytical or a graphical solution method can be used.

12.5.2 Constant underflow

If the overflow is solid-free and the liquid retained by the underflow solids is independent of the concentration of the solute, then the constant underflow occurs. If the underflow is constant, so is the overflow. Under constant underflow condition, the operating line Eq. (12.19) is a straight line when plotted on the y–x plot. For constant underflow leaching process, special attention needs to be given to the first stage as L_1 is generally not equal to L_0 since the latter of the two contains little or no solvent. For a process shown in Fig. 12.13, we have

$$L_1 = L_2 = L_3 = L_n = \ldots\ldots\ldots = L_N \neq L_0 \tag{12.20}$$

$$S(V_{N+1}) = V_N = \ldots\ldots V_{n+1} = V_3 = V_2 \neq V_1 \tag{12.21}$$

A separate mass balance has to be made on stage 1 to obtain L_1 and V_1. Then, the straight operating line can be used as well as McCabe-Thiele method to step off the number of stages when the equilibrium line is plotted on the same diagram.

Example 12.3

120 kg/h of wax paper containing 25 wt% soluble wax and 75 wt% insoluble pulp are to be dewaxed by leaching with kerosene in a continuous countercurrent system. It is desired to reduce the wax content in the liquid adhering to the pulp leaving the last stage to 0.2 kg wax per 100 kg pulp. Kerosene entering the system is recycled from a solvent recovery unit and contains 0.05 wt% of wax. The final extract is to contain 5.0 wt% of wax. Experiments show that underflow from each stage contains 2.0 kg solution/kg insoluble pulp. Assuming contacting time of the overflow and underflow is longer enough for a uniform solution to be formed, determine the washing stages required by both analytical and graphical solution methods.

Solution

Let A = wax, B = inert pulp, C = kerosene, as feed contains 25 wt% of wax and 75 wt% of pulp, thus,

$$L_0 = A = 120(0.25) = 30 \text{ kg/h} \text{ and } x_0 = 1.0$$

$$B = 120(0.75) = 90 \text{ kg/h} \text{ and } N_0 = B/L_0 = 3.0$$

Since underflow from each stage contains 2.0 kg solution/kg insoluble pulp, therefore

$$N = 1/2.0 = 0.5$$

The constant underflow gives us:

$$L_n = B/N = 180.0 \text{ kg/h}$$

Because wax content in the liquid adhering to the pulp leaving the last stage is required to reduce to 0.2 kg wax per 100 kg pulp, the amount of wax in the final underflow is:

$$A_N = 0.2/100(B) = 0.002(90) = 0.18 \text{ kg/h}$$

The mass fraction of wax in the liquid of final underflow is:

$$x_N = A_N/L_N = 0.18/180 = 0.001$$

The overall material balance on liquid solution gives us:

$$L_0 + S = L_N + V_1$$

$$\text{Thus } V_1 = L_0 + S - L_N$$

The mass balance on wax over the entire process gives us

$$L_0x_0 + Sy_S = L_Nx_N + V_1y_1$$

Therefore

$$S = \frac{L_0x_0 - L_Nx_N - (L_0 - L_N)y_1}{y_1 - y_s} = \frac{30 - 0.001(180) - 0.05(30 - 180)}{0.05 - 0.0005}$$

$$= 753.94 \text{ kg/h}$$

$$\text{and,} \ V_1 = 30 + 753.94 - 180 = 603.94 \text{ kg/h}$$

Because of uniform liquid solution formed on each stage, i.e. $x_n = y_n$, and the operating line equation for each stage is

$$y_{n+1} = \frac{L_n}{V_{n+1}}x_n + \frac{V_1y_1 - L_0x_0}{V_{n+1}} = 0.239x_n + 2.613 \times 10^{-4}$$

We can find the number of stages required from the calculation results shown in the table below.

Stages	L_n, kg/h	V_{n+1}, kg/h	x_n	$y_{n+1} = x_{n+1}$
1	180.0	753.94	0.05	1.22×10^{-2}
2	180.0	753.94	1.22×10^{-2}	3.17×10^{-3}
3	180.0	753.94	3.17×10^{-3}	1.02×10^{-3}
4	180.0	753.94	1.02×10^{-3}	$5.05 \times 10^{-4} < 0.001$

Therefore, 4 stages are required for this process.

The operating line equation and equilibrium line can be plotted on y-x diagram shown in Fig. 12.14 for graphical solution. McCabe-Thiele method can be applied to determine the number of stages required.

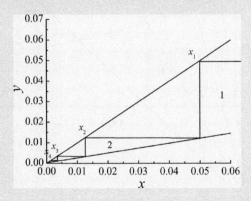

FIG. 12.14

Graphical solution to Example 12.3.

12.5.3 **Variable underflow**

In the systems considered so far, the quantity of solvent, or of solution, removed in association with the insoluble solids has been assumed to be constant and independent of the concentration of solution. A similar countercurrent system is now considered in which the amount of solvent or solution in the underflow is a function of the concentration of the solution and the underflow and overflow vary from stage to stage. For variable underflow system, the same notation will be employed as that used previously in Section 12.5.2.

The solution of the problem depends on the application of mass balances with respect to total solution and individual components, first over the system as a whole and then over the first n stages.

Mass balance on the system as a whole
For Solution

$$L_0 + S = L_N + V_1 = M \tag{12.22}$$

Solute A

$$L_0 x_0 + S y_S = L_N x_N + V_1 y_1 = M x_M \tag{12.23}$$

Insoluble B

$$L_0 N_0 = L_N N_N + V_1 N_{V1} = M N_M \tag{12.24}$$

where M is the flow rate of solution mixture, x_M and N_M are the coordinates of point M in Fig. 12.15. Usually the flows and compositions of L_0 and S are known

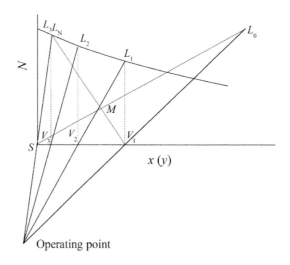

FIG. 12.15

Graphical construction for determining the number of stages.

and the desired exit concentration x_N (or y_1) is set. The coordinates x_M and N_M can then be calculated from Eqs (12.23), (12.24). Points L_0, S, M, L_N (or V_1) are plotted. As demonstrated in Section 12.4.2, L_0MS must lie on a straight line and V_1ML_N must lie on a straight line either.

Mass balance on stage 1 to stage n
For Solution

$$L_0 + V_2 = L_1 + V_1$$

$$L_1 + V_3 = L_2 + V_2, \text{and}$$

$$L_{n-1} + V_{n+1} = L_n + V_n$$

Rearranging the above equations for the difference flow Δ, which is also referred as operating point

$$\Delta = L_0 - V_1 = L_1 - V_2 = L_n - V_{n+1} = L_N - S \tag{12.25}$$

Solute A

$$L_0x_0 + V_2y_2 = L_1x_1 + V_1y_1$$

$$L_1x_1 + V_3y_3 = L_2x_2 + V_2y_2, \text{and}$$

$$L_{n-1}x_{n-1} + V_{n+1}y_{n+1} = L_nx_n + V_ny_n$$

Rearranging the above equation equations, we get,

$$\Delta x_\Delta = L_0x_0 - V_1y_1 = L_nx_n - V_{n+1}y_{n+1} = L_Nx_N - Sy_S \tag{12.26}$$

Then the x coordinate of the operating point Δ can be expressed as,

$$x_\Delta = \frac{L_0x_0 - V_1y_1}{L_0 - V_1} = \frac{L_Nx_N - Sy_S}{L_N - S} \tag{12.27}$$

Insoluble B

A balance made on insoluble B gives

$$N_\Delta = \frac{L_0N_0 - V_1N_{V1}}{L_0 - V_1} = \frac{L_NN_N}{L_N - S} \tag{12.28}$$

where N_Δ is the N coordinate of the operating point.

The operating point Δ is located graphically in Fig. 12.15 as the intersection of lines L_0V_1 and L_NS. To graphically determine the number of stages, we start from point V_1. A tie line (equilibrium line) through V_1 locates L_1. Line $L_1\Delta$ (operating line) is drawn to locate V_2, A tie line passing V_2 gives L_2. By alternatively applying the operating line and tie line, the number of stages can be determined when the desired L_N is reached.

Example 12.4

40,000 kg/h of flaked soybeans, containing 20 wt% oil, is to be leached of the oil with the same flow rate of pure n-hexane in a countercurrent flow system. It is desired to extract 96% of the oil from the soybean flakes and the final solid-free overflow solution is to contain 45 wt% oil. The retention of the solution by the inert solids of the soybeans varies as flows, where N is kg inert solid/ kg solution retained and x is the kg oil/ kg solution in the underflow. At equilibrium, weight fraction of oil in liquid of overflow is 80% of that in liquid of underflow at each stage ($y_n = 0.8 \times_n$). If the overflow is solid-free, calculate the amounts and composition of the exit streams and the total number of stages required.

x	0	0.2	0.40	0.60	0.80
N	3.57	2.94	2.5	2.13	1.82

Solution

Let A = oil, B = inert solids, C = n-hexane, as feed contains 20 wt% of oil and 80 wt% of inert solid, thus,

$$L_0 = A = 0.2F = 8,000 \text{ kg/h and } x_0 = 1.0$$

$$B = 0.8F = 32,000 \text{ kg/h and } N_0 = B/L_0 = 4.0$$

Pure solvent is used for the process and final solid-free overflow solution is to contain 45 wt% oil, thus: $y_S = 0$, $y_1 = \underline{0.45}$

As 96% of the oil is to be extracted by n-hexane, then 4 wt% of oil remains in the final exhausted solid, thus

$$L_N x_N = 0.04 L_0 x_0 \text{ and } V_1 y_1 = 0.96 L_0 x_0$$

The flow rate of final overflow solution can be calculated by

$$V_1 = 0.96 L_0 x_0 / y_1 = 0.96(8000)/0.45 = \underline{17,066.67 \text{ kg/h}}$$

The x-coordinate of the operating point Δ can be determined by

$$x_\Delta = \frac{L_N x_N - S y_s}{L_0 - V_1} = \frac{0.04 L_0 x_0}{L_0 - V_1} = \frac{0.04(8,000)}{8,000 - 17,066.67} = -0.0353$$

The intersection of the line $L_0 V_1$ and line $x_\Delta = -0.0353$ is the operating point Δ, Connecting Δ and S point and extending the line to intersect the N-x curve shown in Fig. 12.16 at point L_N, from which the composition of the final underflow can be determined.

$$x_N = \underline{0.037}$$

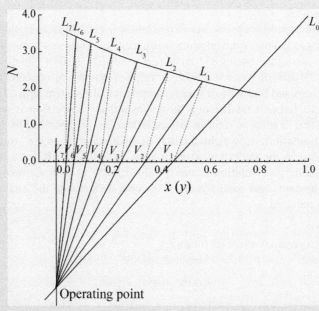

FIG. 12.16

Graphical solution for Example 12.4.

and

$$L_N = 0.04 L_0 x_0/x_N = 0.04(8,000)/0.037 = \underline{\underline{8648.65 \text{ kg/h}}}$$

From graphical solution, 7 stages are required for the operation.

For a design problem, the graphical method can be applied directly to find the number of stages required as the purity of the washed solid (or the purity of the final extract) is set. If in a given problem it is necessary to calculate the degree of washing obtained by the use of a certain number of stages, an initial assumption of the value of x_N (or y_1) must be made and a trial and error method has to be used to find the solution. The graphical method can then be applied for the number of thickeners specified in the problem, and the obtained and assumed values of x_N (or y_1) compared. If the obtained value is higher than the assumed value, the latter is too low. A new assumption of x_N (or y_1) needs to be made and the same procedures need to be repeated until the obtained and assumed values of x_N (or y_1) agree.

In case explicit phase equilibrium relationships are given, the problem can also be solved analytically as shown in the following example.

Example 12.5

A continuous multi-stage countercurrent leaching process is designed to exact hydrocarbons (solute) from oil sands using recycled water as solvent. The feed stream of 300 kg/h contains 85.0 wt% inert solids (B), 10.0 wt% hydrocarbons (A) and 5.0 wt% of water (C). The recycled water stream contains 0.5 wt% hydrocarbons and 99.5 wt% of water, entering into the system with a flow rate of 120 kg/h. Experimental studies have shown that the overflow is solid-free and the solid to solution ratio in the underflow can be described by $N = 4.0 - 0.8 \times$. At equilibrium, weight fraction of hydrocarbons in the overflow is 80% of that in the underflow at each stage ($y_n = 0.8 \times_n$). If we would like the final underflow solution contains only 2.0 wt% of hydrocarbons, please find the amounts and composition of the exit streams and the total number of stages required.

Solution

Known parameters are as follows:

For feed: $L_0 = 0.15F = 45$ kg/h; $B = 0.85F = 255$ kg/h

$$x_0 = 0.10F/0.15F = 0.667$$

For entering solvent: $S = 120$ kg/h, $y_S = 0.005$

Overflow is solid-free, therefore $N_v = 0$

Final underflow solution contains 2.0 wt% hydrocarbon, $x_N = \underline{0.02}$

From phase equilibrium, the solid to liquid ratio from the last stage, $N_N = 4 - 0.8x_N = \underline{3.984}$

Because there is no solid in the overflow, therefore $B = L_N x_N$, thus

$$L_N = B/x_N = 255/3.984 = \underline{64.0 \text{ kg/h}}$$

The flow rate of V_1 can be calculated from the mass balance of total solution,

$$V_1 = L_0 + S - L_N = 45 + 120 - 64 = \underline{101.0 \text{ kg/h}}$$

The mass balance on solute A from the inlet and outlet streams gives,

$$y_1 = \frac{L_0 x_0 + S y_S - L_N x_N}{V_1} = \frac{45(0.667) + 120(0.005) - 64(0.02)}{101}$$

$$= \underline{0.29}$$

The operating line equation for the countercurrent multistage operation is:

$$y_{n+1} = \frac{L_n}{V_{n+1}} x_n + \frac{V_1 y_1 - L_0 x_0}{V_{n+1}}$$

And the equilibrium relationship between the liquids of overflow and underflow follows:

$$y_n = 0.8x_n, \text{ thus } x_n = y_n/0.8$$

The solid to liquid ratio in the underflow follows the relationship of $N = 4 - 0.8x$ for each stage, As B is a constant in underflow, the underflow liquid flow rate for each stage can be calculated by $L_n = B/N$; the flow rate of $V_{n+1} = L_n + V_1 - L_0$. Therefore the number of required stages can be determined by applying mass balance on solute A for each stage until the final calculated x_n value is less than 2.0 wt%. The calculation results are demonstrated in the table below.

Stages	x_n	N_n	L_n, kg/h	V_{n+1}, kg/h	y_{n+1}
1	0.29/0.8	3.710	68.73	124.73	0.194
2	0.243	3.806	67.0	123.0	0.126
3	0.158	3.874	65.83	121.83	7.95×10^{-2}
4	9.94×10^{-2}	3.920	65.05	121.05	4.76×10^{-2}
5	5.95×10^{-2}	3.952	64.52	120.52	2.60×10^{-2}
6	3.24×10^{-2}	3.974	64.17	120.17	1.14×10^{-2}
7	$1.42 \times 10^{-2} < 0.02$				

Therefore, 7 stages are required for the operation.

References

[1] W.H. Linton, T.K. Sherwood, Mass transfer from solid shapes to water in streamline and turbulent flow, Chem. Eng. Prog. 46 (1950) 258.
[2] A.W. Hixson, S.J. Baum, Agitation: mass transfer coefficients in liquid–solid agitated systems. Agitation: heat and mass transfer coefficients in liquid–solid systems, Ind. Eng. Chem. 33 (1941) 478. 1433.
[3] E.L. Piret, R.A. Ebel, C.T. Kiang, W.P. Armstrong, Diffusion rates in extraction of porous solids–1. Single phase extractions; 2. Two-phase extractions, Chem. Eng. Prog. 47 (1951). 405 and 628.
[4] R.C. Chorny, J.H. Krasuk, Extraction for different geometries. Constant diffusivity, Ind. Eng. Chem. Process. Des. Dev. 5 (2) (1966) 206–208.
[5] W.H. Goss, Solvent extraction of oilseeds, J. Am. Oil Chem. Soc. 23 (1946) 348.
[6] M. Bohnet, G. Niesmark, Distribution of solids in stirred suspensions, German Chem. Eng. 3 (1980) 57.

Further reading

C.O. Bennett, J.E. Myers, Momentum, Heat and Mass Transfer, third ed., McGraw-Hill, New York, 1982.
C.J. Geankoplis, A.A. Hersel, D.H. Lepek, Transport Processes and Separation Process Principles, fifth ed., Prentice Hall, New York, 2018.

E.J. Henley, H.K. Staffin, Stagewise Process Design, John Wiley, New York, 1963.

A.L. Hines, R.N. Maddox, Mass Transfer Fundamentals and Applications, Prentice-Hall, Englewood Cliffs, 1985.

G. Karnofsky, The Rotocel extractor, Chem. Eng. 57 (1950) 109.

W.L. McCabe, J.C. Smith, P. Harriott, Unit Operations in Chemical Engineering, seventh ed., McGraw-Hill, New York, 2005.

F. Molyneux, Prediction of "A" factor and efficiency in leaching calculations, Ind. Chem. 37 (440) (1961) 485–492.

R.H. Perry, D.W. Green, J.O. Maloney (Eds.), Perry's Chemical Engineers' Handbook, seventh ed., McGraw-Hill, New York, 1997.

J.D. Seader, E.J. Henley, D.K. Roper, Separation Process Principles: Chemical and Biochemical Operations, third ed., John Wiley & Sons, Hoboken, 2011.

Engineering principles of bioseparations

Amarjeet S. Bassi

Chemical and Biochemical Engineering, Western University, London, ON, Canada

13.1 Introduction

The biopharmaceutical, biofuels, health-foods, nutraceuticals, agriculture, and environmental biotechnology industry are major economic contributors to society with sales in the billions of dollars [1]. The biopharmaceutical industry alone had annual revenues exceeding US$ 200 billion, with many biopharmaceuticals products in the healthcare field including proteins, peptides, and nucleic acids and is now a significant sector in the global market place [8]. Most biotechnological products are obtained from natural or genetically engineered microorganisms. Many important biopharmaceuticals such as monoclonal antibodies, hormones, and blood factors are manufactured using recombinant DNA technologies.

Downstream processing is currently a bottleneck and makes up roughly 70% to 85% of the total production costs in biopharmaceutical manufacturing industry [3]. Bioseparation is an important component of downstream processing operations in the biotechnology industry. In addition, many advances in bioprocessing have led to several fold improvements in average production A typical biomanufacturing process may begin with the cultivation of microbes in bioreactors under controlled conditions. Alternatively, the product of interest may be produced in enzyme reactors or in today's environment using cell-free systems. In all cases, the objectives are to recover the final product. In many cases, the product of interest is located intra-cellularly and in other situations, it is in the fermentation broth. Overall, the recovery of bioproducts from microbial systems which are cultivated in bioreactors is a difficult and challenging task. Hence several unit operations are required for the recovery and purification of the bioproducts. Table 13.1 lists the type of unit operations required for downstream processing in biotechnology.

Coulson and Richardson's Chemical Engineering. https://doi.org/10.1016/B978-0-08-101097-6.00013-4

Table 13.1 Downstream unit operations in biotechnological manufacturing.

Type of unit operation	Objectives
Cell disruption - High-pressure homogeniser - Osmotic shock - Ball mills - Bead mill - French press - Ultrasonic treatment	Breakup microbial cells to release intracellular products
Micro-filtration	Removal of cell debris or whole cells
Centrifugation	Removal of cell debris or whole cells
Ultra-filtration/Dia-filtration	Capture and concentration of soluble products such as proteins
Liquid–liquid extraction	Removal of inhibitory compounds from the fermentation broth
Chromatography: gel-filtration, preparatory HPLC	Concentration and purification of specific products of interest
Crystallisation	Final recovery of pure products

13.2 Importance of bioseparation in biotechnology

The products of biotechnology may include antibiotics enzymes vitamins, amino acids and biopharmaceutical drugs. Due to the wide range of industries and a large variety of products which are manufactured, a bioseparation unit operation train will vary significantly from one process to another. Bioseparation unit operations are a necessary sequence of steps to economically manufacture and purify the product of interest from the fermentation broth. Both the separation and purification types of unit operations will contribute towards large capital investment and operating costs. Thus a proper selection of the downstream processing train is crucial to maximise product yields and minimise cost. There are also constraints imposed by the manufacturing area as well. For example, the biopharmaceutical industry has very stringent requirements due to the need for GMP and usually requires stringent aseptic processes. On the other hand, downstream processing for other types of biomanufacturing such as biofuels or ethanol may be more flexible due to different process throughput needs.

13.3 Types of bioseparation unit operations

A typical downstream process will consist of the following specific sequence of steps.

1. The first section of downstream processing involves physical separation of suspended and insoluble materials. Cell harvesting requires processes such as filtration and centrifugation. Cell disruption is carried out where intra-cellular components need to be accessed and the cell debris is separated again through centrifugation.
2. Once the products have been removed from the cells, the next step is product isolation and concentration. A range off unit operations such as adsorption, ion exchange or solvent extraction could be utilised.
3. The next sequence of operations involves product purification using selective processes such as chromatographic approaches including preparatory HPLC.
4. The final sequence of operations involves the further purification of the specific product of interest, and crystallisation is usually an important unit operation for this purpose.

The principles of the various unit operations listed in Table 13.1 are now discussed in more detail below.

13.4 **Cell disruption**

Cell disruption by definition is the destruction of the microbial cell wall to gain access to the intra-cellular products [9]. This is an important unit operation as many genetically engineered microbes are being utilised and in many cases the products appear as inclusion bodies within the cell. An important engineering consideration required for selecting suitable unit operations for cell disruption requires an understanding of the microbial cell wall structure.

Both prokaryotes (bacteria) and eukaryotes (yeast, mould, microalgae, plant, and animal cells and protozoa) can be cultivated in bioreactors. The structural differences in the cell-walls of these microbial systems are significant. Thus, for instance, eukaryotes can have thick cell walls of varying and complex chemistry (e.g. microalgae, plant cells, yeast, moulds) to no cell walls (animal cells). The cell wall in eukaryotes may be 15% to 25% of the dry cell weight. Prokaryotes (bacteria) are either Gram positive or Gram negative hence have two different types of cell wall structures, see Fig. 13.1 below. For Gram negative bacterial cells, the cell wall consists of an outer membrane and a relatively thin peptidoglycan (murein) layer. The peptidoglycan chemistry provides the mechanical strength to the cell wall due to extensive cross-linking. Gram negative bacteria have also an inner cytoplasmic membrane. On the other hand, Gram positive bacteria have a thicker peptidoglycan layer and an inner cytoplasmic membrane. The removal of the peptidoglycan using enzymes such as lysozyme or using mechanical means can cause cell disruption. The cytoplasmic membrane is made up of lipids, proteins and may include other components such as lipo-polysaccharides.

Several polysaccharides including cellulose (amorphous and crystalline forms) and/or chitin are found in the cell walls of microalgae and others such as glucan in yeast. Due to the complex nature of the cell wall structures, as mentioned above,

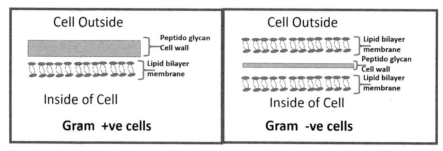

FIG. 13.1

Simplified schematic representation of bacterial cell walls.

several approaches are utilised including both mechanical and non-mechanical for cell disruption as discussed briefly here.

13.4.1 Mechanical approaches for cell disruption

Microbial cells can be disrupted using compression and shear forces [4]. Ball and bead mills, ultra-sonication and high-pressure homogenisation are examples of different techniques which are utilised.

High-pressure homogenisation

High-pressure homogenisation (HPH) is quite popular in large scale cell disruption operations. The principle of the technique is to first pressurise the microbial cell suspension using a positive displacement pump (50 to 500 MPa) and then force the suspension through a nozzle or small orifice gap to a pressure release zone. The rapid change in pressure causes the cell disruption. This process is highly efficient for the cell disruption of most microbial cells. Fig. 13.2 is a schematic diagram explaining the operation of the HPH. For the biopharmaceutical industry, the materials of construction are stainless steel due to the requirements of ease of cleaning, corrosion and wear resistance. The valves are usually made from tungsten carbide.

Ball mills and bead mills

Ball mills and bead mills consist of rotating chambers filled with balls or beads. In some cases, a mechanical agitator is used for the mixing of the beads. The microbial suspension is subjected to high shear due to repetitive collisions with the moving beads. This is a simplified approach for cell disruption as shown schematically in Fig. 13.3 below.

In a bead or ball mill, cell disruption occurs by shearing action with a destruction of the cell wall. Several variables which can affect the efficiency of cell disruption include bead loading, agitator speed, bead diameter, suspension viscosity and cell concentration. One significant disadvantage of bead mills is the rise in temperature which may be destructive to the product of interest including leading to the denaturation of proteins. Good cooling and temperature control is therefore important for bead milling operations.

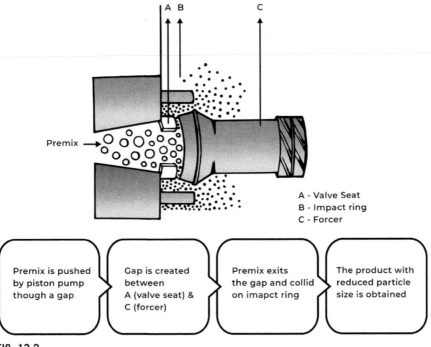

A - Valve Seat
B - Impact ring
C - Forcer

| Premix is pushed by piston pump though a gap | Gap is created between A (valve seat) & C (forcer) | Premix exits the gap and collid on imapct ring | The product with reduced particle size is obtained |

FIG. 13.2

High-pressure homogenisation [4].

From S.T.L. Harrison, Bacterial cell disruption: a key unit operation in the recovery of intracellular products, Biotechnol. Adv. 9 (1991) 217–240, with permission.

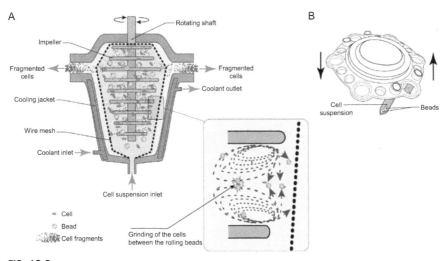

FIG. 13.3

Schematic diagram of a bead milling system. (A) Continuous. (B) Batch equipment [5].

From M. Koubaa, N. Imatoukene, L. Drévillon, E. Vorobiev, Current insights in yeast cell disruption technologies for oil recovery: a review, Chem. Eng. Process. Process Intensif. 150 (2020) 107868, with permission.

13.4.2 **Non-mechanical approaches for cell disruption**
Osmotic shock

Osmotic shock utilises the concept of a rapid change in osmotic pressure around the microbial cell's environment. Cells may be first equilibrated in a high ionic strength media such as 1 M sucrose solution and then exposed to water. This causes water to enter the cells and leads to cells swelling and bursting. The technique is, however, applicable to cells with weakened cell walls and may be utilised to release specific cell proteins whilst cells are metabolically viable (membranes are intact).

13.4.3 **Chemical treatment approaches**

Several chemical treatments can be applied for cell disruption. These include using chelating agents, chaotropes, detergents, solvents, and antibiotics. These are briefly discussed below.

Ethylenediamine tetra-acetic acid (EDTA)

Ethylenediamine tetra-acetic acid (EDTA) is a chelator which is applied to disrupt the lipopolysaccharide (LPS) layers and the outer membrane which surround Gram negative bacteria creating a permeable cell wall which can then be subject to further treatment such as by osmotic shock.

Chaotropes

Urea, guanidine or ethanol are chaotropes which can solubilise membrane proteins. In particular chaotropes may be used to solubilise inclusion bodies. Inclusion bodies are insoluble, denatured proteins created during the cultivation of genetically engineered bacteria.

Fig. 13.4 illustrates a section of a downstream processing scheme involving the use of chaotropes to dissolve inclusion bodies. Inclusion bodies contain denatured proteins in genetically engineered cells. These appear as insoluble particles within the bacterial cell and are released after lysis of the cell. The inclusion bodies need to be dissolved to extract and solubilise the denatured proteins. The example above illustrates the step where a denatured protein in an inclusion body is converted to a soluble form. As shown above, first urea and 2-mercaptoethnol (chaotropes are added to a reactor containing inclusion bodies. After a 6 to 8 h mixing in the reactor, the chaotropes are removed by flushing with water using the process of diafiltration to recover solubilised protein from the inclusion body.

Detergents

Detergents bind to the lipid membrane and cause cell wall disruption and membrane solubilisation by micelle formation. Anionic (sodium dodecyl sulphate (SDS) and other salts of fatty acids), cationic (tetra alkyl ammonium salts), and non-ionic (Triton X) are examples of the types of detergent molecules which are generally used in these applications.

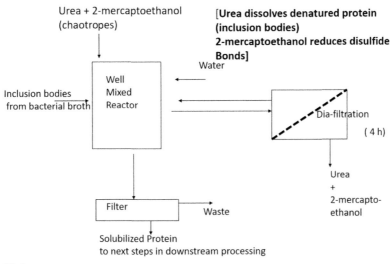

FIG. 13.4

Application of chaotropic agents to dissolve inclusion bodies.

Solvents

Many non-polar solvents such as toluene and chloroform may be applied for cell disruption. A severe disadvantage of using chemicals such as solvents and detergents is the added complexity of any further downstream processing and further purification of extracted products.

Antibiotics

Antibiotics that can be applied for cell lysis include β-lactam antibiotics which inhibit penicillin binding proteins (PBPs) involved in bacterial cell wall peptidoglycan synthesis. Despite their ability to cause rapid lysis, the use of antibiotics for large scale bioprocessing is not common due to its high cost and limited efficacy. The disposal of waste streams containing antibiotics can also become an issue.

13.4.4 Other approaches for cell disruption

There are also other approaches which can be considered when designing suitable cell disruption protocols. Heat assisted lysis is carried out by heating to 50–95°C. Enzymes can also be applied to attack the cell wall, degrading it from without, eventually leading to cell lysis and intracellular protein release. Lysozyme is an example of an enzyme found in saliva which attacks the peptidoglycan in bacterial cell walls. Enzymatic treatment may be combined with detergents for increasing the process efficiency.

13.5 Micro and ultrafiltration

Filtration is the unit operation for the physical separation of suspended and dissolved particulate materials in a slurry or solution. Macro and microfiltration are applied for removing large (coarse) solids and microorganisms whilst ultrafiltration and reverse osmosis are used to separate dissolved proteins and salts respectively. Filtration is a challenging operation for biological products due to the formation of cakes and decreasing efficiency with operational time. Surface filtration applies to the cases where the particles are retained on the external surface of the filter. In contrast, in deep filtration, particles penetrate the pores of the filter and are retained within the filter. The particle capture in depth filtration may be due to direct inertial impaction, interception, or even diffusion into the filter pores. Depth filtration is the most common mechanism in most filtration systems.

Conventional or 'dead-end' filtration occurs when the flow of solids suspension is normal to the filter area. Plate and frame filters and rotary vacuum filters are commonly used for microfiltration. The horizontal plate and frame filter is usually the choice for small-scale systems. It consists of several filter plates in parallel as shown in Fig. 13.5. In 'dead-end' plate and frame systems, the feed flow direction is normal to the membrane. Alternatively, 'cross flow' configurations have the feed moving tangentially across the surface of the membrane. Fouling is an issue and the plate and frame filter requires periodic cleaning to remove cake deposits.

The rotary vacuum filter is more commonly employed for large-scale operations for $>10\,m^3$ bioreactor volumes. It consists of a rotating drum filter pre-coated with a filter aid (diatomaceous earth) and immersed in a tank containing the slurry or suspension. A vacuum is applied inside the drum to draw the liquid into the drum. A thin coat of the filter cake forms on the surface of the drum and this is continuously removed using a knife shaped arrangement as the drum rotates. This leads

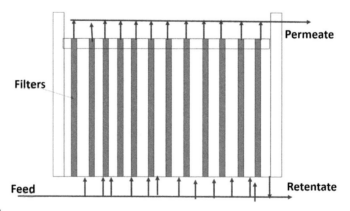

FIG. 13.5

Schematic of a plate and frame filter design. In the above configuration, feed flows in cross-flow configuration across the membrane surface.

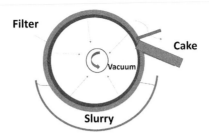

Filter

Cake

Vacuum

Slurry

FIG. 13.6

Schematic representation of the rotary vacuum filter commonly applied in large scale bioprocessing.

to a minimum cake build up and a high efficiency of operation. Fig. 13.6 shows a schematic representation of the principle of the rotary vacuum filter.

The flow rate through the filter is proportional to the pressure drop across the filter and inversely proportional to the filter thickness and fluid viscosity, in line with as Darcy's law, and given by:

$$\frac{F}{A} = \frac{k\Delta p}{\mu L} \tag{13.1}$$

where F is the volumetric feed flow rate, A is the cross sectional area of the filter, Δp is the pressure drop across the filter bed, μ is the fluid viscosity and L is the filter bed length or thickness. F/A is therefore also the velocity of the liquid. If the pressure drop is constant we can write:

$$\frac{1}{A}\frac{dF}{dt} = \frac{\Delta p}{\mu(R_m + R_c)} \tag{13.2}$$

where R_M is the flow medium resistance and R_C is the microbial cake resistance. These design equations can be used to estimate the required area of the filter and the flow velocity. The cake resistance is usually variable and further correlations are needed to take into account the compressibility of the filter cake.

The pressure drop across the filter medium can be a 'positive' pressure due the hydrostatic head of a liquid or pump flow or from the use of compressed air. In some cases in the biotechnology industry, a 'negative' pressure or vacuum can be applied using a vacuum pump. A second consideration is the change of media resistance as a cake may form on the filter surface due to the deposition of the particulates. The filtration efficiency drops progressively as the cake accumulates hence the cake thickness should be minimised. This can be achieved by different means. Cross-flow versus dead end filtration leads to a smaller cake thickness and a belter filtration efficiency. Mechanical means can used to scrape the cake such as in rotary filters or periodic washing. In addition, filter aids can be used to increase the filtration rate and decrease the tendency for cake formation. Diatomaceous earth is an example of a commonly used filter aid.

13.6 Equipment for filtration

Table 13.2 provides a summary of the filter media utilised in commercial scale filtration systems. The role of the filter media is to provide the impermeable barrier for filtration of the particulates. The filter media should have sufficient mechanical strength to withstand pressures, be inert and non-corrosive and offer low resistance to fluid flow.

13.6.1 Ultrafiltration

Ultrafiltration (UF) is distinguished from microfiltration (MF) based on the pore size of the membrane. Microfiltration is carried out using membranes with a pore size greater than $0.1\,\mu m$. Ultrafiltration (UF) membranes have pore sizes ranging from 0.001 to $0.1\,\mu m$. However UF membranes are described using molecular weight cut-off (MWCO). MWCO is defined as the molecular weight for 90% rejection of a solute such as a protein. Ultrafiltration membrane cut-off can range from $1000\,Da$ to $10^6\,Da$. Many different structures have been considered including charged membranes However, Fig. 13.7, a popular option for biotechnology applications is the anisotropic structure or asymmetric membrane Ultrafiltration is widely used to concentrate and recover proteins. For example, in the food industry, the recovery of cheese whey proteins is carried out using UF.

The membranes provide a microporous barrier made of polymers or ceramic. Many different structures have been considered (Fig. 13.7) including charged membranes. However a popular option for biotechnology applications is the anisotropic structure or asymmetric membrane. The asymmetric membrane structure is the most optimum for reducing membrane fouling. The asymmetric membrane structure shown below consists of two layers of different pore sizes. Typical industrial scale

Table 13.2 Filter media for commercial scale filtration.

Type of filtration and particle size	Filter media
Conventional ($\geq 100\,\mu m$)	Woven fabrics Metal screens Ceramic media
Sterile Air Filtration ($0.1\,\mu m$)	Polymer Membrane filters or depth filters of glass fibre, Cartridge filters
Liquid filtration in biotechnology (0.1 to $1.5\,\mu m$)	Autoclavable membrane filters
Ultrafiltration (0.001 to $0.1\,\mu m$)	cellulose acetate, polyamide, polycarbonate, polypropylene, polyethylene, regenerated cellulose, poly(vinyl chloride), poly (vinylidene fluoride) (PVDF), poly(tetrafluoroethylene) (PTFE), acrylonitrile copolymers, ceramics, zirconium oxide, borosilicate glass, stainless steel

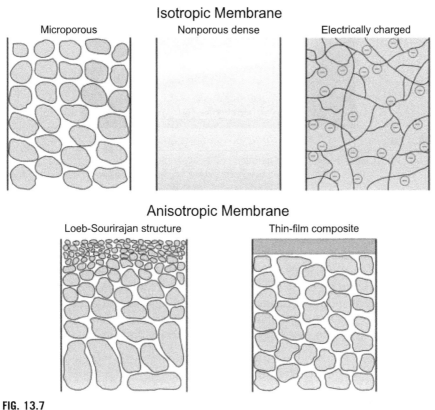

FIG. 13.7

Different configurations for process filtration.

From Lee and Darling, 2016. With permission from The Royal Society of Chemistry.

UF processes use both hollow fibre and tubular (spiral wound) membrane configurations (Fig. 13.8).

Three types of UF membranes are applied for industrial processes. These are: flat sheet membranes, tubular membranes, and hollow fibre membranes. The tubular membrane is the most common type and this is a hollow cylindrical tube where the walls of the tube act as the membrane. Such tubular membranes have typical dimensions of 0.5 to 2 cm. Hollow fibres mentioned earlier have diameters less than 1 mm. Fig. 13.8 depicts a hollow fibre configuration. The hollow fibre configuration provides the highest efficiency with the smallest footprint. However, the main limitations are low throughput compared to spiral wound tubular membranes. The throughput is measured as flux (flow per unit area of membrane) which is a function of the applied driving force and the concentration gradient. Prolonged use can lead to fouling of the membrane, which in turn can lead to rapid decrease in the permeate flux. For this reason, cross flow is the preferred configuration for UF processes.

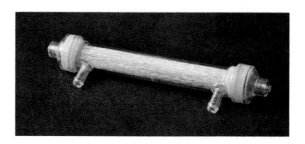

FIG. 13.8

A hollowfiber UF membrane configuration.

From https://www.con-vergence.com/product/hollow-fiber-membrane-potting/.

Concentration polarisation is the retention or accumulation of macro-molecules on or near the membrane surface. Fig. 13.9 illustrates this process. As the molecules accumulate, backward diffusion of the molecules will occur away from the membrane due to a higher concentration of the solute at the surface versus the bulk. Eventually a steady state is established with the molecules moving towards the membrane surface balanced by molecules moving away from the surface by diffusion. Concentration polarisation will therefore results in an extra resistance at the membrane surface and can lead to reduction in the transmembrane flux.

The flux in UF processes is given by the following design equation.

$$J_m = \frac{\Delta P - \Delta \pi}{R_m + R_c + R_i} \tag{13.3}$$

Here J_m is the membrane permeate flux; ΔP is the applied pressure drop, $\Delta \pi$ is the osmotic pressure and the denominator denotes the three resistances in series due to the membrane (R_m), concentration polarisation (R_c), and additional resistances due to impurities (R_i).

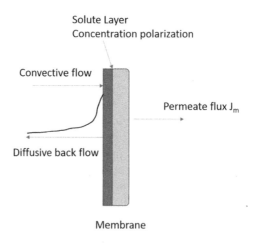

FIG. 13.9

Schematic description of concentration polarisation in UF process.

Table 13.3 Typical centrifugation force needed to separate cells or their components.

Cell or component of a cell	Centrifugal G force (× g)
Microorganisms (bacteria, yeast, protozoa)	1000–10,000
Animal cells	200–2000
Mitochondria, chloroplasts, lysosomes	15,000
Ribosomes	100,000–300,000

13.6.2 Centrifugation

Centrifugation is a unit operation that takes advantage of density differences between the solid and liquid phases present in the suspension. Centrifugation is an attractive option in large scale bioprocessing due to the ease of separation and the possibility of continuous operations. In bioprocessing, centrifugation operations are applied for the clarification of fermentation broths and cell lysates, recovery of inclusion bodies and density gradient separations. Typical densities of cells and components are given in Table 13.3.

The G-force or relative centrifugal force in Table 13.3 is defined as:

$$G = \frac{\varpi^2 r}{g} \tag{13.4}$$

where ω is the angular velocity, r is the radius of the centrifuge and g is the acceleration due to gravity.

The settling velocity in a centrifugal field is given by:

$$v_w = \frac{d^2}{18\mu}(\rho_s - \rho)w^2 r \tag{13.5}$$

where v_w is the terminal settling velocity, d is the diameter of the spherical particle, ρ_s is the solid density and ρ and μ are the density and viscosity of the fluid respectively. This equation is valid for $Re < 1$ a situation frequently encountered in bioprocessing systems.

Examples of process centrifuges include tubular centrifuges, disc centrifuges. For large scale bioprocessing and for continuous operations, the disc centrifuge is commonly used, shown schematically in Fig. 13.10.

A disc centrifuge (Fig. 13.10) consists of a series of inverted cones stacked with minimal spacing between them. As the feed stream moves upwards, rotation of the discs causes solid particles to move along radial direction towards the walls. The accumulated solids then move along the wall and can be continuously separated from the liquid. Depending on the centrifugal rotational speed, microbial cells and debris can be easily separated from the fermentation broth. The clear supernatant is withdrawn as the effluent stream as shown in the figure above.

13.6.3 Chromatograph purification

Chromatographic purifications are quite common in bioprocessing applications to separate and concentrate the product of interest from a mixture. Chromatographic methods take advantage of the fractionation of solutes based on their dynamic

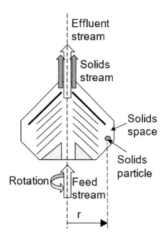

FIG. 13.10

The disc centrifuge.

From Chemical Engineering & Technology, Volume 41, Issue 12 p. 2289–2297.

distribution characteristics between the liquid (or gas) phase, also known as the mobile phase, and the solid phase known as the stationary phase. The stationary phase is usually packed in a column and the mobile phase flows through the column. The samples can be injected as pulses into the mobile phase. The time it takes for the solute molecules to travel through the column will depend on their respective interactions with the stationary phase. The greater is the interaction, the slower the solute will exit from the column. The interactions can be physical or chemical (ionic, hydrophobic etc.),

Gel filtration is based on size separation by running the mixture through a gel of certain pore size. Larger molecules will elute slower than smaller molecules. Ion exchange and other types of chromatography take advantage of differences in binding affinity between molecules in the liquid to the solid phase (stationary phase) based on surface charge or hydrophobicity or other characteristics. The solid phase is usually in a packed column and the liquid phase moves through the column and is termed the mobile phase. Chromatography allows the selective separation of the target molecules in a mixture of solutes. The mixture is introduced at one end of the packed column and differential separation causes an elution profiles or 'peaks' indicating different molecules as they emerge from the column. This is the typical operation in elution chromatography. Another example of such operations 'frontal chromatography' is used to separate a non-adsorbing solute from impurities which are retained in the column. The solid phase or adsorbents used in chromatography applications vary depending on the intended use. Hydrophobic and reverse-phase systems utilise silica-based resins coated with hydrophobic films. Ion-exchange and affinity ligand chromatographic systems employ polymer-based resins with pre-defined surface chemistries. The resins are porous with high specific surface

Table 13.4 Resins used in bioprocessing chromatography applications.

Type of resin	Applications
Silica	Size exclusion chromatography
Silica with hydrophobic coating	Reverse phase chromatography
Polymer-based resins	
styrene divinylbenzene	Ion exchange
polyacrylamide	Size exclusion chromatography
Cross linked agarose, dextran	Low-pressure chromatography

areas up to 1500 square meters per g. Table 13.4 summarises the properties of different types of resins used in chromatography.

Several variables influence the efficiency of the chromatographic separation. These include the resin size, shape and chemistry, and liquid distribution in the packed column and the packing technique. Good fluid distribution depends on several factors including: how well the resin particles are packed (consistency, uniformity); and the design of the column and limitation to any dead volume, or inaccessible areas for the analyte.

13.6.4 Chromatography column packing

Cylindrical, vertical in the column direction with frits made from stainless steel are typically used. The fluid flow is downwards to prevent air pockets. The resin packing is quite important for its efficient operation. For HPLC operations, columns are packed using compression along the axial direction through external, hydraulic piston pumps. Column packing usually begins by introducing a slurry containing the resin into the column. The packing density at this stage is low hence variable length columns may be used or secondary vessels connected to the column. The slurry is next. Replaced with a solvent. The resin may be degassed using vacuum to fully saturate the pores with the liquid solvent. Alternatively, ethanol can be used as a solvent to wet the particles. Isopropanol and/or salts in buffer may be added to reduce surface tension and prevent particle clumping and aggregation. To ensure the maximum efficiency of column packing, resin particle swelling should be minimised. This can be achieved by increasing the ionic strength. Mechanical pressure can then be applied in a controlled fashion to achieve the appropriate packing density,

13.6.5 Detectors in chromatography

Several detectors can be applied downstream of the column to collect the elution profile of products from the column. Examples of detectors can be electrical conductivity, UV (280 nm, 254 nm etc.), refractive index (RI), light scattering or electrochemical. Detection is dependent on effective resolution of the elution peaks which is a function of type of columns, affinity binding parameters, flow rate and the application of linear or exponential gradients.

13.6.6 **New trends in chromatography**

Reversed-phase chromatography

Reversed-phase chromatographic systems apply a hydrophobically coated silica-based resin to capture hydrophobic solutes from an aqueous mobile phase containing some organic solvents. Gradient flow is carried out to decrease the polarity of the mobile phase to allow partitioning of the solutes. Typically C8 (octyl) or C18 (Octyl decul), methyl or phenyl groups are attached to the silica resin solid phase. Aceto-nitrile, methanol, or isopropanol are commonly used as the mobile phase.

Hydrophobic interaction chromatography

Hydrophobic interaction chromatography is a technique where proteins can be cap-tured by 'salting out'. Hydrophobic resins used in the solid phase are polymers deri-vatised with groups such as butyl, phenyl or octyl. Under a high ionic strength or salt concentrations protein can aggregate on the solid phase. An aqueous mobile phase is used and protein will bind to the solid phase due to the 'salting out' effect.

Affinity chromatography

In the biotechnology industry, affinity chromatography plays a predominant role in downstream processing. Affinity chromatographic systems allow for the specific capture of biomolecules by ligands. The types of binding could include antibody–antigen, enzyme-substrate, lectin polysaccharide, or DNA, RNA strands with complementary sequences. Thus affinity chromatography can serve as a final step in a downstream processing train, to selectively capture a biomolecule of inter-est. As an example, the amino acid histidine has a specific affinity for metals. Metal affinity chromatography therefore utilises the binding of metals to poly-histidine. Specific proteins of value can be genetically engineered with histidine tags or as poly-histidine fusion proteins. Affinity chromatography can then be used to capture these proteins using metal ligands. Another example is the application of *Staphylo-coccus* Protein A for purifying and recovering monoclonal Antibodies (mAbs),

Currently highly specific Isotype G immunoglobulins (IgGs) are applied in the biopharmaceutical industry downstream operations as they can be engineered with highly specific complementarity for antigen binding regions. The Fc or invariant regions on these antibody proteins can be applied for binding to specific labelled sec-ondary antibodies for detection. Bozovičar et al. [2] have discussed recent approaches for the development of highly specific affinity ligands for affinity chro-matography using fragmented linear peptide libraries. Other ligands which have been utilised include 'Scaffold proteins' produced as single chain using recombinant methods, synthetic oligonucleotides, aptamers and polysaccharides. Lacki and Riske [6] have reviewed the recent advances in affinity chromatography for industrial scale bioprocesses. Li et al. [7] have also reviewed recent advances in the development and optimisation of biomaterials for applications in downstream processing specifically for therapeutic proteins. The authors have discussed key design features for the selec-tion of affinity ligands. In addition to the specific binding ability, other factors which need to be considered include stability in solutions and ease of immobilisation to the stationary phase,

Size exclusion chromatography or gel-filtration
Size exclusion chromatography (SEC) or gel filtration is a physical separation of various solutes which discriminates on the basis of their molecular size. As the mobile phase flows through the SEC resin, the specific pore size of the gel allows small molecules to 'filter' through faster whereas larger size molecules will move at a slower rate. The pore size of the gel is controlled by different degrees of polymerisation during their manufacture.

Two phase extraction
Two phase extraction is an unit operation which takes advantage of the partitioning of solutes between two immiscible or partially miscible phases. This principle is applied in cases where the product of interest may be preferentially soluble in one or other phase due to its hydrophobicity or hydrophilicity. Examples of products include from plant or microbial sources such as alkaloids and antibiotics, proteins, peptides or lipids. From an engineering point of view, the interfacial mass transfer is a critical factor therefore the design of contactors will determine the efficiency of the process. A mixer settler design with intensive mixing followed by settling to separate the extract and the raffinate is needed.

Precipitation
Precipitation is a unit operation which is commonly used for separating biomolecules such as proteins and nucleic acids. This method is also applied for fractionating mixtures of proteins. These biological macromolecules can be precipitated using a number of methods. The specific approaches include solvent addition, pH adjustment to the isoelectric point (iEP), temperature reduction, salts such as ammonium sulphate, urea, guanidine hydrochloride etc. As proteins are chains of amino acids, they acquire a certain charge based on the primary structure (sequence of amino acids). Thus, the solubility of proteins is a function of the solution pH. The isoelectric point is the pH where the protein has zero net charge, Proteins will precipitate at their isoelectric point and their differences in iEP can be exploited to fractionate proteins. Organic solvents will also lower the solubility of proteins and cause precipitation. However, the protein may be denatured and loose its functionality. Ethanol, methanol, acetone are examples of solvents which can be used for this purpose. Salts such as ammonium sulphate at high concentrations will also precipitate proteins (salting out).

Crystallisation
This is the last step in most downstream processing operations. The goals are to recover the product with a high degree of purity. In most bioprocessing settings, low temperatures are applied to minimise product biodegradation. Crystallisation is defined as the formation of solid particles of defined geometrically consistent

shapes (lattice) from the homogeneous solution (mother liquor). The mother liquor in the crystalliser is at high concentration. Most substances when crystallised will form specific shapes which are unique. The conditions used for crystallisation can however affect the overall size and shape of the product. These include any impurities and the solvent or solvents. Crystallisation is an important unit operation as it can produce products with the highest purity. An understanding of the crystallisation is important especially in the biopharmaceuticals industry.

Two important concepts in understanding crystallisation are the solubility limit and phase diagram (Fig. 13.11).

Both the soluble region (solute remains soluble) and the precipitation region (where amorphous structures or aggregates appear) are zones where no crystallisation occurs. Crystallisation is carried out by reducing the solubility by either cooling, evaporation or using anti-solvents. Once a super-saturated state is reached, crystallisation begins by nucleation. Nucleation occurs when a stable cluster of particles forms due to inter molecular forces of the particles.

In the bioseparation industry, continuous crystallisation is preferred and both tubular and well mixed (suspension) crystallisers are used. The former have shorter residence time and higher yields. Other approaches such as membrane crystallisers are also reported in the literature.

An important application in bioprocessing is the crystallisation of proteins. Usually this is the final step in the overall purification train. Several approaches can be used including isoelectric precipitation where the pH is adjusted to the isoelectric point of the protein, organic solvent addition; using dialysis or diafiltration to concentrate proteins and reduce the ionic strength. The main objective is to create a supersaturated solution. The addition of seed crystals is applied to trigger the process of nucleation. Proteins form macromolecular crystals and will still contain solvent. Thus they have low strength and can disintegrate easily. They are also heat and radiation sensitive. They are also polymorphic in shape.

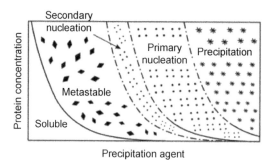

FIG. 13.11

Phase diagram for protein crystallisation, Solid line: solubility curve.

Permission from 'Bulk Protein Crystallization – Principles and Methods, in: Process Scale Bioseparations for the Biopharmaceutical Industry', CRC Press.

13.7 New directions in biotechnology: Cell-free protein synthesis

Modern biotechnology today involves the application of cell-free protein synthesis (CFPS) platforms to design and manufacture proteins and m-RNA for a myriad of applications including vaccine production. A cell free system is one where cellular processes are carried out in the absence of living cells. The cell is lysed and the internal cytoplasmic content is utilised. The key principle is the activation of transcriptional and translational machinery from crude cell extracts outside the living cell environment. The crude extracts containing the DNA template are supplemented with additional nucleotides, amino acids in buffers. CFPS provides advantages over traditional production and cultivation of cells (Fig. 13.12).

In general, downstream processing of CFPS should be simpler since there is a lack of cell debris. Cell-free reactions are usually small scale and automated. Whilst downstream operations are more simplified, CFPS offer new challenges for bioseparations. The first step is the selection of cell lysates. These are prepared by cultivating cells followed by lysis and removal of cell debris and genomic DNA. These extracts will contain the appropriate machinery for transcription and translation of genetic information including aminoacyl-tRNA synthetase (AAS), ribosomes, and factors necessary for elongation, initiation, and transcription. Protein synthesis is then carried out by adding the appropriate DNA template, amino acids, and other

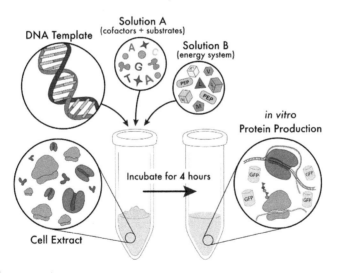

FIG. 13.12

A cartoon schematic of cell-free protein synthesis, a biotechnology that harnesses the cell's genetic code into a test tube.

From Genetic and Engineering and Biotechnology News, Feb 26, 2019 'Cell-Free Protein Synthesis Made Flexible and Accessible'.

components. CFPS offers several advantages: speed of synthesis in min to hours compared to in vivo cell cultivation, and also great flexibility for example to produce m-RNA vaccines, membrane proteins, or proteins toxic to cells, antibodies, hormones or other clinical biopharmaceuticals. Protein synthesis using recombinant elements (PURE) is a modified approach where the DNA template is mixed with purified components added individually to the mixture. Cell lysates from prokaryotic or eukaryotic systems are available. Wheat germ lysate (WGL) and cultured insect cells are two examples of eukaryotic systems.

13.8 New directions in biotechnology: Artificial intelligence in bioseparations

Matos et al. [10] have reviewed how the principles of artificial intelligence (AI) and big data analytics can be applied for the design and selection of affinity ligands for affinity chromatography applications. Fig. 13.13 shows the growth from the time of ligand discoveries in the early 20th century to the current and future trends.

In the case of affinity ligand design and selection, there is a plethora of data currently available including X-ray crystallography, NMR, nucleotide, proteomics data bases, and information on the chemistry of the molecular recognition and binding sites.

Several steps which can be implemented include data integration, new knowledge extraction using machine learning algorithms and optimisation of the bioseparation process based on the machine learning models. Data integration is a process which includes the collection of data from many sources, organisation of the data into a central integrated database. These are usually accomplished using machine readable interfaces or web services using XML or JavaScript languages. Several standards are being proposed in the biopharmaceutical industry to organise data and integrate information from different sources. Several examples applied to organise the data include relational databases, data warehouses based on star schemas, and resource description framework (RDF).

AI approaches take advantage of statistical techniques and machine learning or pattern recognition models. Algorithms are the basis of machine learning strategies for the manipulation and transform data typically in programming languages such as Python. Several types of algorithms include: (i) Bayesian algorithms: models based on specific assumptions; (ii) Clustering: Grouping data into classes or clusters; (iii) Decision tree: a branching structure to map specific outcomes; and (iv) Neural networks and deep learning.

Machine learning models are used to develop rules based on the set of available data. In the 'training' phase, where classification rules are developed based on a certain output or pattern. This modelling approach is called supervised machine learning. Machine learning algorithms can take advantage of multi-media sources of data to infer and develop rules. The model is then used to predict the 'labelling' of new information. Several software tools are available for developing these supervised

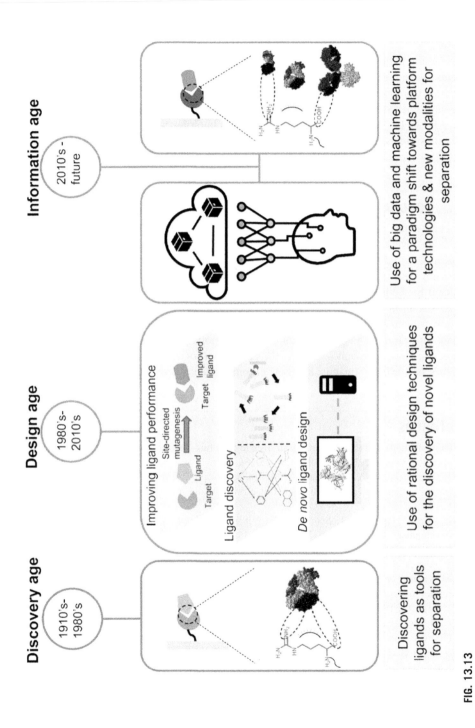

FIG. 13.13

Use of AI in bioseparation-ligand design.

machine learning models or deep learning (Keras, Weka etc.). Unsupervised machine learning is another approach where data sets are 'clustered' to discover patterns.

It is quite likely that the use of AI will become more common in bioseparation systems due to the increasing use of sensors to gather real time data during a process operation. Combining real time data with historical data can then be easily incorporated into supervised machine learning algorithms for process optimisation. In addition, AI can be very useful for the new biotechnology with Cell Free Protein synthesis systems.

13.9 Conclusions and future directions

This chapter has provided an overview of different unit operations in downstream processing. Over the last 40 years, biopharmaceuticals have played a key role as therapeutics. Due to their importance in health care, an intensive focus has also been on improving their manufacturing. Recombinant technologies have emerged as the dominant methods for many of these products. Improvements in upstream side of biomanufacturing have been on bioreactor design such as perfusion type systems. The downstream areas for bioseparation of these molecules have not undergone significant innovations and still account for a large portion of the overall cost of the bioprocess. New approaches such as the application of AI and machine learning, design of new affinity ligands, and new chromatography techniques are needed.

References

[1] Biotechnology Industry Organization, Healing, fueling, feeding: how biotechnology is enriching your life, Biotechnology 86 (2010).

[2] K. Bozovičar, B. Jenko Bizjan, A. Meden, J. Kovač, T. Bratkovič, Focused peptide library screening as a route to a superior affinity ligand for antibody purification, Sci. Rep. 11 (2021) 1–12.

[3] L. Gary, J. Hubbuch, T. Schroeder, E. Willimann, Shrinking the costs of bioprocess development, Bioprocess Int. 7 (2009).

[4] S.T.L. Harrison, Bacterial cell disruption: a key unit operation in the recovery of intracellular products, Biotechnol. Adv. 9 (1991) 217–240.

[5] M. Koubaa, N. Imatoukene, L. Drévillon, E. Vorobiev, Current insights in yeast cell disruption technologies for oil recovery: a review, Chem. Eng. Process. Process Intensif. 150 (2020), 107868.

[6] K.M. Łącki, F.J. Riske, Affinity chromatography: an enabling technology for large-scale bioprocessing, Biotechnol. J. 15 (2020) 1–11.

[7] Y. Li, et al., Emerging biomaterials for downstream manufacturing of therapeutic proteins, Acta Biomater. 95 (2019) 73–90.

[8] G. Walsh, Biopharmaceutical benchmarks 2018, Nat. Biotechnol. 36 (2018) 1136–1145.

[9] A.P.J. Middelberg, Process-scale disruption of microorganisms, Biotechnol. Adv. 13 (1995) 491–551.

[10] M.J.B. Matos, A.S. Pina, A.C.A. Roque, Rational design of affinity ligands for bioseparation, J. Chromatogr. A 1619 (2020).

Further reading

S. Ahuja (Ed.), Handbook of Bio-separations, first ed., vol. 2, Elsevier, 2021.

P.A. Belter, E.L. Cussler, W.-S. Hu, Bio-separations: Downstream Processing for Biotechnology, Wiley, NY, 1988.

G. Subramanian (Ed.), Bio separation and Bioprocessing: Bio-chromatography, Membrane Separations, Modeling, Validation, WILEY-VCH Verlag GmbH, 1998.

Problems

1.1. A liquid containing four components, **A, B, C**, and **D**, with 0.3 mol fraction each of **A, B**, and **C**, is to be continuously fractionated to give a top product of 0.9 mol fraction **A** and 0.1 mol fraction **B**. The bottoms are to contain not more than 0.5 mol fraction **A**. Estimate the minimum reflux ratio required for this separation, if the relative volatility of **A** to **B** is 2.0.

1.2. During the batch distillation of a binary mixture in a packed column, the product contained 0.60 mol fraction of the more volatile component when the concentration in the still was 0.40 mol fraction. If the reflux ratio used was 20:1, and the vapour composition y is related to the liquor composition x by the equation $y = 1.035 \times$ over the range of concentration concerned, determine the number of ideal plates represented by the column. x and y are in mole fractions.

1.3. A mixture of water and ethyl alcohol containing 0.16 mol fraction alcohol is continuously distilled in a plate fractionating column to give a product containing 0.77 mol fraction alcohol and a waste of 0.02 mol fraction alcohol. It is proposed to withdraw 25% of the alcohol in the entering steam as a side stream containing 0.50 mol fraction of alcohol.

Determine the number of theoretical plates required and the plate from which the side stream should be withdrawn if the feed is liquor at its boiling point and a reflux ratio of 2 is used.

1.4. In a mixture to be fed to a continuous distillation column, the mole fraction of phenol is 0.35, o-cresol is 0.15, m-cresol is 0.30, and xylenols is 0.20. A product is required with a mole fraction of phenol of 0.952, o-cresol 0.0474, and m-cresol 0.0006. If the volatility to o-cresol of phenol is 1.26 and of m-cresol is 0.70, estimate how many theoretical plates would be required at total reflux.

1.5. A continuous fractionating column, operating at atmospheric pressure, is to be designed to separate a mixture containing 15.67% CS_2 and 84.33% CCl_4 into an overhead product containing 91% CS_2 and a waste of 97.3% CCl_4, all by mass. A plate efficiency of 70% and a reflux of 3.16 kmol/kmol of product may be assumed. Determine the number of plates required. The feed enters at 290 K with a specific heat capacity of 1.7 kJ/kg and has a boiling point of 336 K. The latent heats of CS_2 and of CCl_4 are 25.9 MJ/kmol. The latent heat of CS_2 and CCl_4 is 25,900 kJ/kmol.

Mol% CS_2 in the vapour:	0	8.23	15.55	26.6	33.2	49.5	63.4	74.7	82.9	87.8	93.2
Mol% CS_2 in the liquor:	0	2.96	6.15	11.06	14.35	25.85	39.0	53.18	66.30	75.75	86.04

1.6. A batch fractionation is carried out in a small column which has the separating power of 6 theoretical plates. The mixture consists of benzene and toluene containing 0.60 mol fraction of benzene. A distillate is required, of constant composition, of 0.98 mol fraction benzene, and the operation is discontinued when 83% of the benzene charged has been removed as distillate. Estimate the reflux ratio needed at the start and finish of the distillation, if the relative volatility of benzene to toluene is 2.46.

1.7. A continuous fractionating column is required to separate a mixture containing 0.695 mol fraction n-heptane (C_7H_{16}) and 0.305 mol fraction n-octane (C_8H_{18}) into products of 99 mol% purity. The column is to operate at 101.3 kN/m^2 with a vapour velocity of 0.6 m/s. The feed is all liquid at its boiling point, and this is supplied to the column at 1.25 kg/s. The boiling point at the top of the column may be taken as 372 K, and the equilibrium data are:

Mole fraction of heptane in vapour	0.96	0.91	0.83	0.74	0.65	0.50	0.37	0.24
Mole fraction of heptane in liquid	0.92	0.82	0.69	0.57	0.46	0.32	0.22	0.13

Determine the minimum reflux ratio required. What diameter column would be required if the reflux used were twice the minimum possible?

1.8. The vapour pressures of chlorobenzene and water are:

Vapour pressure (kN/m^2)	13.3	6.7	4.0	2.7
Temperatures, (K)				
Chlorobenzene	343.6	326.9	315.9	307.7
Water	324.9	311.7	303.1	295.7

A still is operated at 18 kN/m^2, and steam is blown continuously into it. Estimate the temperature of the boiling liquid and the composition of the distillate if liquid water is present in the still.

1.9. The following values represent the equilibrium conditions in terms of mole fraction of benzene in benzene–toluene mixtures at their boiling point:

Liquid	0.51	0.38	0.26	0.15
Vapour	0.72	0.60	0.45	0.30

If the liquid compositions on four adjacent plates in a column are 0.18, 0.28, 0.41, and 0.57 under conditions of total reflux, determine the plate efficiencies.

1.10. A continuous rectifying column handles a mixture consisting of 40% of benzene by mass and 60% of toluene at the rate of 4 kg/s, and separates it into a product containing 97% of benzene and a liquid containing 98% toluene. The feed is liquid at its boiling point.

(a) Calculate the masses of distillate and waste liquor produced per unit time.
(b) If a reflux ratio of 3.5 is employed, how many plates are required in the rectifying part of the column?
(c) What is the actual number of plates if the plate efficiency is 60%?

Mole fraction of benzene in liquid	0.1	0.2	0.3	0.4	0.5	0.6	0.7	0.8	0.9
Mole fraction of benzene in vapour	0.22	0.38	0.51	0.63	0.7	0.78	0.85	0.91	0.96

1.11. A distillation column is fed with a mixture of benzene and toluene, in which the mole fraction of benzene is 0.35. The column is to yield a product in which the mole fraction of benzene is 0.95, when working with a reflux ratio of 3.2, and the waste from the column is not to exceed 0.05 mol fraction of benzene.

If the plate efficiency is 60%, estimate the number of plates required and the position of the feed point. The relation between the mole fraction of benzene in liquid and in vapour is given by:

Mole fraction of benzene in liquid	0.1	0.2	0.3	0.4	0.5	0.6	0.7	0.8	0.9
Mole fraction of benzene in vapour	0.20	0.38	0.51	0.63	0.71	0.78	0.85	0.91	0.96

1.12. The relationship between the mole fraction of carbon disulphide in the liquid and in the vapour evolved from the mixture during the distillation of a carbon disulphide–carbon tetrachloride mixture is:

x	0	0.20	0.40	0.60	0.80	1.00
y	0	0.445	0.65	0.795	0.91	1.00

Determine graphically the theoretical number of plates required for the rectifying and stripping portions of the column, if the reflux ratio = 3, the slope of the fractionating line = 1.4, the purity of the product = 99%, and the percentage of carbon disulphide in the waste liquors = 1%.

What is the minimum slope of the rectifying line in this case?

1.13. A fractionating column is required to distil a liquid containing 25% benzene and 75% toluene by mass, to give a product of 90% benzene. A reflux ratio of 3.5 is to be used, and the feed will enter at its boiling point. If the plates used are 100% efficient, calculate by the Lewis–Sorel method the composition of liquid on the third plate, and estimate the number of plates required using the McCabe–Thiele method.

1.14. A 50 mol% mixture of benzene and toluene is fractionated in a batch still which has the separating power of 8 theoretical plates. It is proposed to obtain a constant quality product with a mol% benzene of 95, and to continue the distillation until the still has a content of 10 mol% of benzene. What will be the range of reflux ratios used in the process? Show graphically the relation between the required reflux ratio and the amount of distillate removed.

1.15. The vapour composition on a plate of a distillation column is:

	C_1	C_2	$i-C_3$	$n-C_3$	$i-C_4$	$n-C_4$
Mole fraction	0.025	0.205	0.210	0.465	0.045	0.050
Relative volatility	36.5	7.4	3.0	2.7	1.3	1.0

What will be the composition of the liquid on the plate if it is in equilibrium with the vapour?

1.16. A liquor of 0.30 mol fraction of benzene and the rest toluene is fed to a continuous still to give a top product of 0.90 mol fraction benzene and a bottom product of 0.95 mol fraction toluene.

If the reflux ratio is 5.0, how many plates are required:

(a) if the feed is saturated vapour?
(b) if the feed is liquid at 283 K?

1.17. A mixture of alcohol and water containing 0.45 mol fraction of alcohol is to be continuously distilled in a column to give a top product of 0.825 mol fraction alcohol and a liquor at the bottom containing 0.05 mol fraction alcohol.

How many theoretical plates are required if the reflux ratio used is 3? Indicate on a diagram what is meant by the Murphree plate efficiency.

1.18. It is desired to separate 1 kg/s of an ammonia solution containing 30% NH_3 by mass into 99.5% liquid NH_3 and a residual weak solution containing 10% NH_3. Assuming the feed to be at its boiling point, a column pressure of 1013 kN/m², a plate efficiency of 60%, and that 8% excess over minimum reflux requirements is used, how many plates must be used in the column and how much heat is removed in the condenser and added in the boiler?

1.19. A mixture of 60 mol% benzene, 30% of toluene, and 10% xylene is handled in a batch still. If the top product is to be 99% benzene, determine:

(a) the liquid composition on each plate at total reflux,

(b) the composition on the 2nd and 4th plates for a reflux ratio $R=1.5$,
(c) as for (b) but $R=3$,
(d) as for (c) but $R=5$,
(e) as for (d) but $R=8$ and for the condition when the mole per cent benzene in the still is 10,
(f) as for (e) but with $R=5$.

The relative volatility of benzene to toluene may be taken as 2.4, and of xylene to toluene as 0.43.

1.20. A continuous still is fed with a mixture of 0.5 mol fraction of the more volatile component, and gives a top product of 0.9 mol fraction of the more volatile component and a bottom product containing 0.10 mol fraction.

If the still operates with an L_n/D ratio of 3.5:1, calculate by Sorel's method the composition of the liquid on the third theoretical plate from the top:

(a) for benzene–toluene, and
(b) for n-heptane–toluene.

1.21. A mixture of 40 mol% benzene with toluene is distilled in a column to give a product of 95 mol% benzene and a waste of 5 mol% benzene, using a reflux ratio of 4.

(a) Calculate by Sorel's method the composition on the second plate from the top.
(b) Using the McCabe and Thiele method, determine the number of plates required and the position of the feed if supplied to the column as liquid at the boiling point.
(c) Determine the minimum reflux ratio possible.
(d) Determine the minimum number of plates.
(e) If the feed is passed in at 288 K, determine the number of plates required using the same reflux ratio.

1.22. Determine the minimum reflux ratio using Fenske's equation and Colburn's rigorous method for the following three systems:

(a)			0.60 mol fraction C_6, 0.30 mol fraction C_7, and 0.10 mol fraction C_8 to give a product of 0.99 mol fraction C_6		
			Mole fraction	Relative volatility α	x_d
(b)	Components	**A**	0.3	2	1.0
		B	0.3	1	–
		C	0.4	0.5	–
(c)	Components	**A**	0.25	2	1.0
		B	0.25	1	–
		C	0.25	0.5	–
		D	0.25	0.25	–

1.23. A liquor consisting of phenol and cresols with some xylenols is fractionated to give a top product of 95.3 mol% phenol. The compositions of the top product and of the phenol in the bottoms are as follows:

	Compositions (mol%)		
	Feed	**Top**	**Bottom**
Phenol	35	95.3	5.24
o-Cresol	15	4.55	–
m-Cresol	30	0.15	–
Xylenols	20	–	–
	100	100	–

If a reflux ratio of 10 is used,

(a) Complete the material balance over the still for a feed of 100 kmol.
(b) Calculate the composition on the second plate from the top.
(c) Calculate the composition on the second plate from the bottom.
(d) Calculate the minimum reflux ratio by Underwood's equation and by Colburn's approximation.

The heavy key is m-cresol, and the light key is phenol.

1.24. A continuous fractionating column is to be designed to separate 2.5 kg/s of a mixture of 60% toluene and 40% benzene, so as to give an overhead of 97% benzene and a bottom product containing 98% toluene by mass. A reflux ratio of 3.5 kmol of reflux/kmol of product is to be used, and the molar latent heat of benzene and toluene may be taken as 30 MJ/kmol.

Calculate:

(a) The mass of top and bottom products per unit time.
(b) The number of theoretical plates and position of feed if the feed is liquid at 295 K, of specific heat capacity 1.84 kJ/kg K.
(c) How much steam at 240 kN/m^2 is required in the still.
(d) What will be the required diameter of the column if it operates at atmospheric pressure and a vapour velocity of 1 m/s.
(e) The necessary diameter of the column if the vapour velocity is to be 0.75 m/s, based on free area of column.
(f) The minimum possible reflux ratio, and the minimum number of plates for a feed entering at its boiling point.

1.25. A system that obeys Raoult's law shows that the relative volatility α_{AB} is P_A^0 / P_B^0, where P_A^0 and P_B^0 are the vapour pressures of the components **A** and **B** at the given temperature. From vapour pressure curves of benzene, toluene, ethyl benzene, and o-, m-, and p-xylenes, obtain a plot of the volatilities of each of the materials relative to m-xylene in the range 340–430 K.

1.26. A still has a liquor composition of *o*-xylene 10%, *m*-xylene 65%, *p*-xylene 17%, benzene 4%, and ethyl benzene 4%. How many plates at total reflux are required to give a product of 80% *m*-xylene, and 14% *p*-xylene? The data are given as mass per cent.

1.27. The vapour pressures of *n*-pentane and of *n*-hexane are:

Pressure								
(kN/m^2)	1.3	2.6	5.3	8.0	13.3	26.6	53.2	101.3
(mm Hg)	10	20	40	60	100	200	400	760

Temperature (K)								
C_5H_{12}	223.1	233.0	244.0	257.0	260.6	275.1	291.7	309.3
C_6H_{14}	248.2	259.1	270.9	278.6	289.0	304.8	322.8	341.9

The equilibrium data at atmospheric pressure are:

$x=0.1$	0.2	0.3	0.4	0.5	0.6	0.7	0.8	0.9
$y=0.21$	0.41	0.54	0.66	0.745	0.82	0.875	0.925	0.975

(a) Determine the relative volatility of pentane to hexane at 273, 293, and 313 K.

(b) A mixture containing 0.52 mol fraction pentane is to be distilled continuously to give a top product of 0.95 mol fraction pentane and a bottom of 0.1 mol fraction pentane. Determine the minimum number of plates that is the number of plates at total reflux by the graphical McCabe–Thiele method, and analytically by using the relative volatility method.

(c) Using the conditions in (b), determine the liquid composition on the second plate from the top by Lewis's method, if a reflux ratio of 2 is used.

(d) Using the conditions in (b), determine by the McCabe–Thiele method, the total number of plates required, and the position of the feed.

It may be assumed that the feed is all liquid at its boiling point.

1.28. The vapour pressures of *n*-pentane and *n*-hexane are given in Problem 1.27. Assuming that both Raoult's and Dalton's Laws are obeyed,

(a) Plot the equilibrium curve for a total pressure of 13.3 kN/m^2.

(b) Determine the relative volatility of pentane to hexane as a function of liquid composition for a total pressure of 13.3 kN/m^2.

(c) Estimate the error caused by assuming the relative volatility to be constant at its mean value.

(d) Would it be more advantageous to distil this mixture at a higher pressure?

1.29. It is desired to separate a binary mixture by simple distillation. If the feed mixture has a composition of 0.5 mol fraction, calculate the fraction it is necessary to vapourise in order to obtain:

(a) a product of composition 0.75 mol fraction, when using a continuous process, and
(b) a product whose composition is not less than 0.75 mol fraction at any instant, when using a batch process.

If the product of batch distillation is all collected in a single receiver, what is its mean composition?

It may be assumed that the equilibrium curve is given by:

$$y = 1.2x + 0.3$$

for liquid compositions in the range 0.3–0.8.

1.30. A liquor, consisting of phenol and cresols with some xylenol, is separated in a plate column. Given the following data complete the material balance:

Component	Mol%		
	Feed	Top	Bottom
C_6H_5OH	35	95.3	5.24
$o\text{-}C_7H_7OH$	15	4.55	–
$m\text{-}C_7H_7OH$	30	0.15	–
C_8H_9OH	20	–	–
Total	100	100	–

Calculate:

(a) the composition on the second plate from the top,
(b) the composition on the second plate from the bottom.

A reflux ratio of 4 is used.

1.31. A mixture of 60, 30, and 10 mol% benzene, toluene, and xylene, respectively, is separated by a plate column to give a top product containing at least 90 mol% benzene, a negligible amount of xylene, and a waste containing not more than 60 mol% toluene.

Using a reflux ratio of 4, and assuming that the feed is boiling liquid, determine the number of plates required in the column and the approximate position of the feed plate.

The relative volatility of benzene to toluene is 2.4 and of xylene to toluene is 0.45, and it may be assumed that values are constant throughout the column.

1.32. It is desired to concentrate a mixture of ethyl alcohol and water from 40 to 70 mol% alcohol. A continuous fractionating column, 1.2 m in diameter and having 10 plates, is available. It is known that the optimum superficial vapour velocity in the column at atmosphere pressure is 1 m/s, giving an overall plate efficiency of 50%.

Assuming that the mixture is fed to the column as a boiling liquid and using a reflux ratio of twice the minimum value possible, determine the feed plate and the rate at which the mixture can be separated.

Equilibria data:

Mole fraction alcohol in liquid	0.1	0.2	0.3	0.4	0.5	0.6	0.7	0.8	0.89
Mole fraction alcohol in vapour	0.43	0.526	0.577	0.615	0.655	0.70	0.754	0.82	0.89

1.33 Consider equilibrium in the ternary system of benzene (1), toluene (2), and o-xylene (3).

Given: $T = 393\,K$, $P = 1.2\,bar$, and $x_3 = 0.2$. Is the system completely defined?

1.34 Determine the vapour phase composition of a mixture in equilibrium with a liquid mixture of 0.5 mol fraction benzene and 0.5 mol fraction of toluene at 338 K. Will the liquid vapourise at a pressure of $101.325\,kN/m^2$?

1.35. What is the boiling point of an equimolar mixture of benzene and toluene at $101.325\,kN/m^2$?

1.36. What is the dew point of an equimolar mixture of benzene and toluene at $101.325\,kN/m^2$?

1.37. The overall mole fraction of benzene in a mixture of benzene and toluene is $z_F = 0.3$ at 1 atm total pressure and 95°C. State whether the mixture is a liquid, a vapour, or a two-phase mixture assuming the mixture behaves as an ideal solution.

(a) What is the minimum temperature at which a vapour of this composition can exist at 1 atm total pressure?

1.38. Mixtures of n-hexane and n-octane form essentially ideal solutions

(a) Prepare the bubble point, dew point and equilibrium curves (T–x–y^* and x–y^* plots) at 1 atm total pressure.

(b) Calculate the values of the relative volatility of n-hexane and n-octane in mixtures containing 10 mol% and 90 mol% hexane. Do these values suggest that the solutions may be ideal?

(c) A solution of concentration $x = 0.2$ is slowly heated. What is the composition of the initial vapour formed if the total pressure is (i) 1 atm, (ii) 2 atm?

(d) If a vapour of composition as in part (c)(ii) above is cooled to liquid state at 2 atm total pressure and a differential amount of vapour is then formed from the condensate, calculate its composition. How can the quantities be calculated using the T–x–y and x–y data?

1.39. The relative volatility of A in a mixture of A and B is $\alpha_{AB} = 1.5$. What is the mole fraction of B in the first droplet of liquid condensed from an equimolar saturated vapour mixture of A and B?

1.40. One mole of a solution of A and B (enthalpy $= 900$ kcal/kmol; $x_A = 0.4$) is mixed with two moles of another solution of A and B (enthalpy $= 1200$ kcal/kmol, $x_A = 0.8$). Calculate the enthalpy and composition of the mixture.

1.41. Calculate the bubble point of a mixture of 35 mol% methanol, 30 mol% ethanol, and 40 mole% n-propanol at 1 atm total pressure using K-value method.

(a) Calculate the dew point of a vapour of the above composition at 1.5 atm total pressure using K-value method.

Assume ideal behaviour of the solution.

1.42. A ternary solution of n-hexane (10 mol%), n-heptane (45 mol%), and n-octane may be considered to be ideal. The equilibrium vapourisation ratios are: $K_1 = 2.25, K_2 = 1.02$, and $K_3 = 0.6$ at 1 atm total pressure. State whether the solution is below its bubble point, saturated, or above its bubble point.

1.43. Ethanol forms a nearly ideal solution with iso-butanol and has a relative volatility of 2.2. A heated feed containing 40 mol% ethanol and 60 mol% iso-butanol enters a flash drum at a rate of 50 kmol/h.

(a) Prepare a table of the fractional yield of the distillate vs its composition.
(b) What fraction of the feed should be vapourised to have a bottom product containing not more than 10% ethanol?
(c) Consider now that there is a second flash drum that receives the bottom product from the first drum. If 60% of the feed is vapourised in each drum, find the vapour and the liquid flow rates from each chamber as well as their composition.

1.44. The molar enthalpies of saturated liquid and vapour mixtures of A and B at 1 atm total pressure (reference temp. $= 5°$C) may be expressed as: $H_L = 9000 + 1000$ x kJ/kmol; $H_v = 35,000 + 8000$ y kJ/kmol. The relative volatility is $\alpha_{AB} = 1.8$.

(a) Calculate the molar heats of condensation of saturated pure vapours of compounds A and B at 1 atm.
(b) We have one kmol of a liquid ($x = 0.3$, $H_L = 8500$ kJ/kmol), and one kmol of a vapour ($y = 0.5$, $H_v = 40,000$ kJ/kmol). What is the condition of the liquid (subcooled, saturated, or superheated) and the condition of the vapour (saturated or superheated)?

1.45. A mixture of 38 mol% propane, 22.5 mol% iso-butane, and 39.5 mol% n-butane is flashed in a drum. If 50 mol% of the feed vapourises, estimate the compositions of the liquid and the vapour phases. The K-values at the given conditions are as follows: Propane: 1.42; iso-butane: 0.86 and n-butane: 0.72.

1.46. Liquids A and B are miscible in all proportions and $p_A^v = 1.75$ p_B^v at all temperatures. The vapour pressure of B is given by.

$$\ln p_B^v \ (\text{mm Hg}) = 14.243 - 2570/(T + 232.5)$$

(a) At what temperature does an equimolar mixture of A and B boil if the total pressure is 1.5 bar?

(b) Neither A or B is miscible with water. If some water is added to the mixture, at what temperature should it boil if the total pressure remains unchanged?

The vapour pressure of water is given by.

$$\ln p_w^v \ (\text{mm Hg}) = 18.5882 - 3984.92/(T + 233.43).$$

1.47. Ten kmol of a feed having 65 mol% benzene and 35 mol% toluene is batch distilled at 1 atm pressure. Given $\alpha = 2.51$. Calculate the moles of distillate produced and the composition of the bottom product if distillation is done until.

(a) 75 mol% of the feed benzene leaves with the vapour.
(b) The still pot contains 35 mol% benzene.
(c) 50 mol% of the feed is vapourised.
(d) The accumulated distillate contains 75 mol% of benzene.

1.48. A student was asked to do flash calculation of an ideal mixture of four components having an overall composition (in mole fraction) of $w_1 = 0.2, w_2 = 0.15$, and $w_3 = 0.40$. At the condition of the flash drum, the equilibrium vapourisation ratios were: $K_1 = 2.1$, $K_2 = 1.02$, $K_3 = 0.6$, and $K_4 = 0.2$. The student reported that 38 mol% of the feed vapourised on flashing. Was the students' calculation correct?

1.49. An essential oil, virtually immiscible with water, is steam-distilled by passing live steam at 107°C through a mixture of the oil and water. The vapour pressure of water at 107°C is 1.3 bar and that of the essential oil is 6.55 mm Hg. How much steam is necessary to recover 80% of the essential oil? Neglect condensation of steam.

1.50. A mixture of benzene and p-xylene is batch distilled at atmospheric pressure. The rate of heat input to the still is 4000 kcal/h. Individual heats of vapourisation are: benzene = 100 kcal/kg, p-xylene = 85 kcal/kg, and the relative volatility of benzene in the mixture is 5.6. Calculate the instantaneous rate of vapourisation of p-xylene when the liquid in the still has 40 mol% benzene in it. Molecular weight of benzene = 78. Molecular weight of p-xylene = 106.

1.51. One thousand kilograms of an equimolar mixture of benzene and nitrobenzene is being separated by batch distillation. After an hour of operation, 500 kg of the mixture remains in the still. The operator takes a sample of the accumulated condensate and reports that it has 70 mol% benzene in it. The relative volatility of benzene in the mixture can be taken as 7. Is the reported concentration of the distillate reasonably accurate? Molecular weight of benzene = 78. Molecular weight of nitrobenzene = 123.

1.52. A quaternary solution is being distilled in a batch still. Over a period, 30 mol% of component 1 is distilled out. The following relative volatility values are given: $\alpha_{31} = 0.2$; $\alpha_{23} = 1.8$; $\alpha_{43} = 0.7$. What fraction of component 3 is distilled out over this period?

1.53. A saturated equimolar mixture of vapours of A and B enters a partial condenser at a rate of 1 kmol/h. The vapour leaves the condenser at a rate of 0.6 kmol/h.

If the relative volatility of B with respect to A is 0.3, calculate the composition of the vapour and the liquid leaving the partial condenser.

1.54. A mixture (40 mol% vapour, the rest liquid) of aniline and nitrobenzene (80 mol% aniline) is separated into a distillate having 98 mol% aniline and a bottom product with 3 mol% aniline. The reflux ratio used is 2.2. Determine the equation of the rectifying section operating line and of the feed line.

1.55. For distillation of an equimolar binary mixture of A and B, the equations of the operating lines are:

$$\text{Rectifying section}: y = 0.663x + 0.32$$
$$\text{Stripping section}: y = 1.329x - 0.01317$$

What is the condition of the feed?

1.56. A feed mixture of A and B (45 mol% A and 55 mol% B) is to be separated into a top product containing 96 mol% A and a bottom product having 95 mol% B. The feed is 50% vapour, and the reflux ratio is 1.5 times the minimum. What is the equation of the feed line? Determine the number of ideal trays required and the location of the feed tray. Given: $\alpha_{AB} = 2.8$.

1.57. A mixture of di- and tri-ethyl-amines containing 55 mol% of the former is fed to a distillation column at a rate of 40 kmol/h. The feed is at its bubble point. The column is to operate at atmospheric pressure and the top product should not have more than 2.5 mol% of the less volatile. Also, not more than 2% of the di-ethylamine in the feed should be allowed to leave at the bottom. The reflux to the column is a saturated liquid. Determine

(a) the minimum reflux ratio,
(b) the number of theoretical plates if the actual reflux ratio is 1.4 times the minimum,
(c) the slopes of the two operating lines, and
(d) the number of theoretical plates if the reflux is a subcooled liquid at such a temperature that one mole of vapour is condensed for twenty moles of reflux.

The equilibrium data in terms of the more volatile are:

x	0.02	0.039	0.052	0.065	0.09	0.092	0.14	0.215	0.43	0.601	0.782	0.853	0.932	0.985
y^*	0.042	0.085	0.124	0.153	0.225	0.243	0.316	0.449	0.678	0.802	0.910	0.948	0.970	0.993

1.58. A saturated vapour feed containing 30 mol% benzene and chlorobenzene is to be separated into a top product having 98 mol% benzene and a bottom product having 99 mol% chlorobenzene. Calculate the minimum reflux ratio. What is the corresponding boil-up ratio? The relative volatility of benzene in the mixture is 4.12.

1.59. A distillation column separates saturated liquid feed containing 25 mol% A and 75 mol% B. The relative volatility is $\alpha_{AB} = 2.51$. The liquid concentration on the 5th tray is $x_5 = 0.54$. The distillate has 98 mol% A and the reflux ratio is 3.0. Assume that the trays are ideal.

(a) Determine the concentration of the vapour
    (i) entering the 5th tray from the top, and
    (ii) leaving the 5th tray.
(b) Which section of the column (rectifying or stripping) does the 5th tray belong to?
(c) Calculate the enrichment of the vapour across the 4th tray.
(d) If 97% of A present in the feed goes to the top product, calculate the moles of liquid vapourised in the reboiler per mole of distillate and also the boil-up ratio.

2.1. Tests are made on the absorption of carbon dioxide from a carbon dioxide–air mixture in caustic soda solution of concentration 2.5 kmol/m^3, using a 250 mm diameter tower packed to a height of 3 m with 19 mm Raschig rings
    The results obtained at atmospheric pressure are:
    Gas rate, $G' = 0.34$ kg/m^{2}s. Liquid rate, $L' = 3.94$ kg/m^{2}s.
    The carbon dioxide in the inlet gas is 315 parts per million and in the exit gas 31 parts per million.
    What is the value of the overall gas transfer coefficient $K_G a$?

2.2. An acetone–air mixture containing 0.015 mole fraction of acetone has the mole fraction reduced to 1% of this value by countercurrent absorption with water in a packed tower. The gas flow rate G' is 1 kg/m^{2}s of air and the water flow rate entering is 1.6 kg/m^{2}s. For this system, Henry's law holds and $y_e = 1.75\times$, where y_e is the mole fraction of acetone in the vapour in equilibrium with a mole fraction x in the liquid. How many overall transfer units are required?

2.3. An oil containing 2.55 mol% of a hydrocarbon is stripped by running the oil down a column up which live steam is passed, so that 4 kmol of steam are used 100 kmol of oil stripped. Determine the number of theoretical plates required to reduce the hydrocarbon content to 0.05 mol%, assuming that the oil is non-volatile. The vapour–liquid relation of the hydrocarbon in the oil is given by $y_e = 33\times$, where y_e is the mole fraction in the vapour and x the mole fraction in the liquid. The temperature is maintained constant by internal heating, so that steam does not condense in the tower.

2.4. Gas, from a petroleum distillation column, has its concentration of H$_2$S reduced from 0.03 kmol H$_2$S/kmol of inert hydrocarbon gas to 1% of this value, by scrubbing with a triethanolamine–water solvent in a countercurrent tower, operating at 300 K and at atmospheric pressure.
    H$_2$S is soluble in such a solution and the equilibrium relation may be taken as $Y = 2\times$, where Y is kmol of H$_2$S kmol inert gas and X is kmol of H$_2$S/kmol of solvent.
    The solvent enters the tower free of H$_2$S and leaves containing 0.013 kmol of H$_2$S/kmol of solvent. If the flow of inert hydrocarbon gas is 0.015 kmol/m^{2}s of tower cross-section and the gas-phase resistance controls the process, calculate:

(a) the height of the absorber necessary, and
(b) the number of transfer units required.

    The overall coefficient for absorption $K_G'' a$ may be taken as 0.04 kmol/s m^3 of tower volume (unit driving force in Y).

2.5. It is known that the overall liquid transfer coefficient K_La for absorption of SO_2 in water in a column is $0.003\,kmol/s\,m^3$ $(kmol/m^3)$. Obtain an expression for the overall liquid film coefficient K_La for absorption of NH_3 in water in the same apparatus using the same water and gas rates. The diffusivities of SO_2 and NH_3 in air at $273\,K$ are 0.103 and $0.170\,cm^2/s$. SO_2 dissolves in water, so that Henry's constant $\mathcal{H}$ is equal to 50 $(kN/m^2)/(kmol/m^3)$. All the data refer to $273\,K$.

2.6. A packed tower is used for absorbing sulphur dioxide from air by means of a $0.5\,N$ caustic soda solution. At an air flow of $2\,kg/m^2s$, corresponding to a Reynolds number of 5160, the friction factor $R/\rho u^2$ is found to be 0.020.

Calculate the mass transfer coefficient in kg SO_2/s m^2 (kN/m^2) under these conditions if the tower is at atmospheric pressure. At the temperature of absorption, the diffusion coefficient of SO_2 is $0.116 \times 10^{-4}\,m^2/s$, the viscosity of the gas is $0.018\,mN\,s/m^2$, and the density of the gas stream is $1.154\,kg/m^3$.

2.7. In an absorption tower, ammonia is absorbed from air at atmospheric pressure by acetic acid. The flow rate of $2\,kg/m^2s$ in a test corresponds to a Reynolds number of 5100 and hence a friction factor $R/\rho u^2$ of 0.020. At the temperature of absorption, the viscosity of the gas stream is $0.018\,mN\,s/m^2$, the density is $1.154\,kg/m^3$, and the diffusion coefficient of ammonia in air is $1.96 \times 10^{-5}\,m^2/s$.

Determine the mass transfer coefficient through the gas film in kg/m^2 s (kN/m^2).

2.8. Acetone is to be recovered from a 5% acetone–air mixture by scrubbing with water in a packed tower using countercurrent flow. The liquid rate is $0.85\,kg/m^2s$ and the gas rate is $0.5\,kg/m^2s$.

The overall absorption coefficient K_Ga may be taken as $1.5 \times 10^{-4}\,kmol/[m^3s\,(kN/m^2)]$ partial pressure difference and the gas-film resistance controls the process.

What height of tower is required tower to remove 98% of the acetone? The equilibrium data for the mixture are:

Mole fraction acetone in gas	0.0099	0.0196	0.0361	0.0400
Mole fraction acetone in liquid	0.0076	0.0156	0.0306	0.0333

2.9. Ammonia is to be removed from 10% ammonia–air mixture by countercurrent scrubbing with water in a packed tower at $293\,K$ so that 99% of the ammonia is removed when working at a total pressure of $101.3\,kN/m^2$. If the gas rate is $0.95\,kg/m^2s$ of tower cross-section and the liquid rate is $0.65\,kg/m^2$ s, what is the necessary height of the tower if the absorption coefficient $K_Ga = 0.001\,kmol/m^3s$ (kN/m^2) partial pressure difference. The equilibrium data are:

Concentration							
(kmol NH$_3$/ kmol water)	0.021	0.031	0.042	0.053	0.079	0.106	0.150
Partial pressure NH$_3$							
(mm Hg)	12.0	18.2	24.9	31.7	50.0	69.6	114.0
(kN/m^2)	1.6	2.4	3.3	4.2	6.7	9.3	15.2

2.10. Sulphur dioxide is recovered from a smelter gas containing 3.5% by volume of SO_2, by scrubbing it with water in a countercurrent absorption tower. The gas is fed into the bottom of the tower, and in the exit gas from the top the SO_2 exerts a partial pressure of $1.14\,kN/m^2$. The water fed to the top of the tower is free from SO_2, and the exit liquor from the base contains $0.001145\,kmol\ SO_2/kmol$ water. The process takes place at $293\,K$, at which the vapour pressure of water is $2.3\,kN/m^2$. The water rate is $0.43\,kmol/s$.

If the area of the tower is $1.85\,m^2$ and the overall coefficient of absorption for these conditions $K_L''a$ is $0.19\,kmol\ SO_2/s\ m^3$ (kmol of SO_2 per kmol H_2O), what is the height of the column required?

The equilibrium data for SO_2 and water at $293\,K$ are:

kmol SO_2/1000 kmol H_2O	0.056	0.14	0.28	0.42	0.56	0.84	1.405
kmol SO_2/1000 kmol inert gas	0.7	1.6	4.3	7.9	11.6	19.4	35.3

2.11. Ammonia is removed from a 10% ammonia–air mixture by scrubbing with water in a packed tower, so that 99.9% of the ammonia is removed. What is the required height of tower? The gas enters at $1.2\,kg/m^2s$, the water rate is $0.94\,kg/m^2s$ and K_Ga is $0.0008\,kmol/s\,m^3\ (kN/m^2)$.

2.12. A soluble gas is absorbed from a dilute gas–air mixture by countercurrent scrubbing with a solvent in a packed tower. If the liquid fed to the top of the tower contains no solute, show that the number of transfer units required is given by:

$$\mathbf{N} = \frac{1}{\left[1 - \dfrac{mG_m}{L_m}\right]}\ \ln\left[\left(1 - \frac{mG_m}{L_m}\right)\frac{y_1}{y_2} + \frac{mG_m}{L_m}\right]$$

where G_m and L_m are the flow rates of the gas and liquid in $kmol/s\,m^2$ tower area, and y_1 and y_2 the mole fraction of the gas at the inlet and outlet of the column. The equilibrium relation between the gas and liquid is represented by a straight line with the equation $y_e = mx$, where y_e is the mole fraction in the gas in equilibrium with mole fraction x in the liquid.

In a given process, it is desired to recover 90% of the solute by using 50% more liquid than the minimum necessary. If the HTU of the proposed tower is $0.6\,m$, what height of packing will be required?

2.13. A paraffin hydrocarbon of molecular weight $114\,kg/kmol$ at $373\,K$, is to be separated from a mixture with a non-volatile organic compound of molecular weight $135\,kg/kmol$ by stripping with steam. The liquor contains 8% of the paraffin by mass and this is to be reduced to 0.08% using an upward flow of steam saturated at $373\,K$. If three times the minimum amount of steam is used, how many theoretical stages will be required? The vapour pressure of the paraffin at 373 K is 53 kN/m^2 and the process takes place at atmospheric pressure. It may be assumed that the system obeys Raoult's law.

2.14. Benzene is to be absorbed from coal gas by means of a wash oil. The inlet gas contains 3% by volume of benzene, and the exit gas should not contain more than 0.02% benzene by volume. The suggested oil circulation rate is 480 kg oil/100 m^3 of inlet gas measured at NTP. The wash-oil enters the tower solute-free. If the overall height of a transfer unit based on the gas phase is 1.4 m, determine the minimum height of the tower which is required to carry out the absorption. The equilibrium data are:

Benzene in oil (per cent by mass)	0.05	0.01	0.50	1.0	2.0	3.0
Equilibrium partial pressure of benzene in gas (kN/m^2)	0.013	0.033	0.20	0.53	1.33	3.33
(mm Hg)	0.1	0.25	1.5	4.0	10.0	25.0

2.15. Ammonia is to be recovered from a 5% by volume ammonia–air mixture by scrubbing with water in a packed tower. The gas rate is 1.25 m^3/sm^2 measured at NTP and the liquid rate is 1.95 kg/m^{2}s. The temperature of the inlet gas is 298 K and of the inlet water 293 K. The mass transfer coefficient is $K_Ga = 0.113$ kmol/m^{3}s (mole ratio difference) and the total pressure is 101.3 kN/m^2. What is the height of the tower to remove 95% of the ammonia. The equilibrium data and the heats of solutions are:

Mole fraction in liquid	0.005	0.01	0.015	0.02	0.03
Integral heat of solution					
(kJ/kmol of solution)	181	363	544	723	1084
Equilibrium partial pressures: (kN/m^2)					
At 293 K	0.4	0.77	1.16	1.55	2.33
At 298 K	0.48	0.97	1.43	1.92	2.93
At 303 K	0.61	1.28	1.83	2.47	3.86

Adiabatic conditions may be assumed and heat transfer between phases neglected.

2.16. A bubble-cap column with 30 plates is to be used to remove n-pentane from solvent oil by means of steam stripping. The inlet oil contains 6 kmol of n-pentane/ 100 kmol of pure oil and it is desired to reduce the solute content of

0.1 kmol/100 kmol of solvent. Assuming isothermal operation and an overall plate efficiency of 30%, what is the specific steam consumption, that is kmol of steam required/kmol of solvent oil treated, and the ratio of the specific and minimum steam consumptions. How many plates would be required if this ratio is 2.0?

The equilibrium relation for the system may be taken as $Y_e = 3.0X$, where Y_e and X are expressed in mole ratios of pentane in the gas and liquid phases respectively.

2.17. A mixture of ammonia and air is scrubbed in a plate column with fresh water. If the ammonia concentration is reduced from 5% to 0.01%, and the water and air rates are 0.65 and 0.40 kg/m²s, respectively, how many theoretical plates are required? The equilibrium relationship may be written as $Y = X$, where X is the mole ratio in the liquid phase.

2.18. It is required to absorb 95% of the acetone from a mixture with nitrogen containing 1.5 mol% of the compound in a countercurrent tray tower. The total gas input is 30 kmol/h, and water enters the tower at a rate of 90 kmol/h. The tower operates at 300 K and 1 atm. The equilibrium relation is $y = 2.53x$. Determine the number of ideal trays necessary for this separation. Use the Kremser equation.

2.19. Acetone from a mixture with air containing 2 mol% acetone is absorbed counter-currently in water in a plate tower containing four theoretical stages. The inlet gas rate is 40 kmol/h and acetone-free water is supplied to the column at a rate of 110 kmol/h. The equilibrium relation is $y = 2.5x$. Determine the concentration of the solute in the exit liquid analytically.

2.20. An aqueous waste stream containing a toxic volatile organic compound (VOC) must be air-stripped in a tray tower so that the air, loaded with the VOC may flow to the flare for incineration. The waste stream has 0.1 kg of the organic per 100 kg of water and the concentration must be reduced to 50 ppm. The equilibrium distribution of the solute between air and water is linear and can be expressed as.

$$Y = 4.35X \; (X, Y: \text{kg VOC per kg air or water})$$

A column of suitable diameter having 20 trays that are 50% efficient is available. Can the tower meet the requirement? The air rate is 1500 kg/h and the wastewater is to be treated at a rate of 4000 kg/h.

2.21. A feed having X_0 kg solute per kg carrier A is treated with a solvent B in an N-stage crosscurrent cascade. The feed rate is R_s kg/h (solute-free basis), and an equal amount of pure solvent, E_s kg/h, is supplied to each of the stages. If the equilibrium relation is linear ($Y = \alpha X$), determine the number of ideal stages required to reduce the solute concentration from X_0 to X_N.

2.22. A countercurrent absorption tower receives 100 kmol of a gas mixture per hour having 15% of a solute A. It is required to absorb 95% of the solute. 'Pure' solvent enters at a rate of 80 kmol/h at the top.

(a) What is the equation of the operating line if the concentrations are expressed in the mole ratio unit?

(b) If the mole fraction unit is used, what would be the slope of the operating line at a section where the bulk gas concentration is 10 mol%? What is the maximum slope of the operating line and where does it occur?

2.23. A gas mixture having 7 mol% of the solute A is to be scrubbed in a packed tower at a rate of 70 kmol/h. The solvent, water, is fed at a rate of 80 kmol/h. In the concentration range involved, the solubility of the gas is described by the equation: $y_A^* = 1.2\, x_A - 0.62\, x_A^2$. It is desired to absorb 98% of A present in the feed. Determine the equation of the operating line and the overall gas-phase driving force at a point in the column where the bulk liquid concentration is $x_A = 0.04$.

2.24. A solution of C in solvent A containing 25 mol% C is fed to a multi-stage separation device at a rate of 10 kmol/h in which it is to be treated with the extracting solvent B entering at equal molar rate. It is required to calculate and compare the performance of the separation device if it is (a) a co-current cascade, (b) a counter-current cascade, and (c) a cross-current cascade. The stages in a cascade are ideal in each case. Calculate the maximum fraction of the solute C possible to be removed from the solution (C in A) against the number of stages in each of the above three configurations. The equilibrium relation is $Y = X$.

2.25. A packed tower is to be designed for the absorption of 98% of the ammonia (A) from an air–ammonia mixture containing 4% ammonia at a rate of 187.4 kmol/h using water as the solvent. The tower operates at 105.1 kPa and 303 K. The equilibrium data for NH_3–water system at 303 K are given below:

Kg NH$_3$ per 100kg water	2	3	4	5	7.5	10
Partial pressure of NH$_3$ (mm Hg)	19.3	29.6	40.1	51.0	79.5	110

(a) Calculate the equilibrium data x_A vs p_A, x_A vs y_A, and X_A vs Y_A.
(b) Calculate the minimum water flow rate for the absorption. The inlet water is NH_3 free.

2.26. An absorption column is to be designed for reducing the concentration of a toxic vapour in an air emission from 1% to 0.02 mol%. The column will operate at 20°C and 0.5 bar gauge pressure. The scrubbing liquor flow rate is 1.3 times the minimum. The gas–liquid equilibrium relation ($p = H_x$) is linear and the Henry's law constant is 10 bar.

(a) How many trays are required to achieve this separation if the overall tray efficiency is 40%?
(b) After the column is designed, it becomes necessary to reduce the concentration of the toxic vapour in the vent air to 0.01% in order to meet a modified pollution control regulation. Is it possible to achieve this goal by operating the column at the same temperature but at an enhanced pressure? If so, at what pressure should the column be operated?

2.27. Formaldehyde present in an effluent air stream is to be absorbed into an effectively non-volatile solvent at 1 atm total pressure and 30°C. The inlet and outlet concentrations of the organic in the air stream are 4.25% and 0.08 mol%, respectively. The solvent entering the column has only traces of formaldehyde. The ratio of molar rates of liquid to gas input is 2.9 and the equilibrium relation is $y = 2.5x$. Determine:

(a) the equation of the operating line,
(b) the number of overall gas-phase transfer units, and.
(c) the maximum admissible ratio of the gas to liquid flow rates for this separation.

2.28. Hydrogen sulphide must be removed from a light refinery hydrocarbon stream before the gas is subjected to further processing. The feed gas contains 0.035 mole fraction of H_2S, 99% of which must be removed by scrubbing with an aqueous weakly basic solution. The 1.5 m diameter absorption column operates at 25°C and 101.3 kPa total pressure. The feed gas rate is 45 kmol/h/m^2 and the concentration of the solute (H_2S) in the absorbent solution leaving the tower is 0.015 mole H_2S per mole of the H_2S-free liquid. The equilibrium relation is linear and is given by $y = 1.95 \times$ and the overall gas-phase film coefficient is $K_y\bar{a} = 130$ kmol/m^3/h/Δy. Calculate

(a) the rate of solvent flow,
(b) the slope of the operating line in small concentration range if mole fraction units are used in the calculations,
(c) the overall driving forces at the top and at the bottom of the tower on (i) gas phase basis and (ii) liquid phase basis,
(d) the number of transfer units, and
(e) the packed height.

2.29. A gas mixture containing 10 mol% SO_2 and 90 mol% air at 1 atm total pressure and 30°C is to be scrubbed with water to remove 97% of the SO_2 in a tower packed with ceramic Raschig rings. The feed gas rate is 1500 kg/h. Tower cross-sectional area is 0.781 m^2.

Calculate (a) the minimum liquid rate, (b) the packed height if the liquid rate is 1.25 times the minimum.

Equilibrium data for SO_2–water system at 30°C.

$10^4\, x$	0	0.562	1.403	2.8	4.22	8.42	14.03	19.65	27.9
$10^3\, y$	0	0.790	2.23	6.19	10.65	25.9	47.3	68.5	104

The volumetric mass transfer coefficients at the given conditions are:

$$k_{x'}\,\bar{a} = 1.25\,\text{kmol/m}^3/\text{s}/\Delta x \text{ and } k_{y'}\,\bar{a} = 0.075\,\text{kmol/m}^3/\text{s}/\Delta y.$$

3.1. Tests are made on the extraction of acetic acid from a dilute aqueous solution by means of a ketone in a small spray tower of diameter 46 mm and effective height of 1090 mm with the aqueous phase run into the top of the tower. The ketone enters free from acid at the rate of $0.0014\,\mathrm{m}^3/\mathrm{s}\,\mathrm{m}^2$, and leaves with an acid concentration of $0.38\,\mathrm{kmol/m}^3$. The concentration in the aqueous phase falls from 1.19 to $0.82\,\mathrm{kmol/m}^3$.

Calculate the overall extraction coefficient based on the concentrations in the ketone phase, and the height of the corresponding overall transfer unit.

The equilibrium conditions are expressed by:

(Concentration of acid in ketone phase) $= 0.548$ (Concentration of acid in aqueous phase).

3.2. A laboratory test is carried out into the extraction of acetic acid from dilute aqueous solution, by means of methyl iso-butyl ketone, using a spray tower of 47 mm diameter and 1080 mm high. The aqueous liquor is run into the top of the tower and the ketone enters at the bottom.

The ketone enters at the rate of $0.0022\,\mathrm{m}^3/\mathrm{s}\,\mathrm{m}^2$ of tower cross-section. It contains no acetic acid, and leaves with a concentration of $0.21\,\mathrm{kmol/m}^3$. The aqueous phase flows at the rate of $0.0013\,\mathrm{m}^3/\mathrm{s}\,\mathrm{m}^2$ of tower cross-section, and enters containing $0.68\,\mathrm{kmol\ acid/m}^3$.

Calculate the overall extraction coefficient based on the driving force in the ketone phase. What is the corresponding value of the overall HTU, based on the ketone phase?

Using units of $\mathrm{kmol/m}^3$, the equilibrium relationship under these conditions may be taken as:

(Concentration of acid in the ketone phase) $= 0.548$ (concentration in the aqueous phase.)

3.3. Propionic acid is extracted with water from a dilute solution in benzene, by bubbling the benzene phase into the bottom of a tower to which water is fed at the top. The tower is 1.2 m high and $0.14\,\mathrm{m}^2$ in area, the drop volume is $0.12\,\mathrm{cm}^3$, and the velocity of rise is 12 cm/s. From laboratory tests, the value of K_w for forming drops is 7.6×10^{-5} kmol/s m^2 $(\mathrm{kmol/m}^3)$ and for rising drops $K_w = 4.2 \times 10^{-5}$ kmol/s m^2 $(\mathrm{kmol/m}^3)$.

What is the value of $K_w a$ for the tower in $\mathrm{kmol/sm}^3$ $(\mathrm{kmol/m}^3)$?

3.4. A 50% solution of solute **C** in solvent **A** is extracted with a second solvent **B** in a countercurrent multiple contact extraction unit. The mass of **B** is 25% that of the feed solution, and the equilibrium data are:

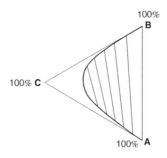

Determine the number of ideal stages required, and the mass and concentration of the first extract if the final raffinate contains 15% of solute **C**.

3.5. A solution of 5% acetaldehyde in toluene is to be extracted with water in a five stage co-current unit. If 25 kg water/100 kg feed is used, what is the mass of acetaldehyde extracted and the final concentration? The equilibrium relation is given by:

$$(\text{kg acetaldehyde/kg water}) = 2.20(\text{kg acetaldehyde/kg toluene})$$

3.6. If a drop is formed in an immiscible liquid, show that the average surface available during formation of the drop is $12\pi r^2/5$, where r is the radius of the drop, and that the average time of exposure of the interface is $3t_f/5$, where t_f is the time of formation of the drop.

3.7. In the extraction of acetic acid from an aqueous solution in benzene in a packed column of height 1.4 m and cross-sectional area $0.0045\,\text{m}^2$, the concentrations measured at the inlet and outlet of the column are:

Acid concentration in the inlet water phase, C_{w2}	$=0.69\,\text{kmol/m}^3$
Acid concentration in the outlet water phase, C_{w1}	$=0.685\,\text{kmol/m}^3$
Flow rate of benzene phase $=5.7 \times 10^{-6}\,\text{m}^3/\text{s}$	$=1.27 \times 10^{-3}\,\text{m}^3/\text{m}^2\text{s}$
Inlet benzene phase concentration, C_{B1}	$=0.0040\,\text{kmol/m}^3$
Outlet benzene phase concentration, C_{B2}	$=0.0115\,\text{kmol/m}^3$

The equilibrium relationship for this system is:

$$C_B^*/C_w^* = 0.247.$$

Determine the overall transfer coefficient and the height of the transfer unit.

3.8. It is required to design a spray tower for the extraction of benzoic acid from solution in benzene.

Tests have been carried out on the rate of extraction of benzoic acid from a dilute solution in benzene to water, in which the benzene phase was bubbled into the base of a 25 mm diameter column and the water fed to the top of the column. The rate of mass transfer was measured during the formation of the bubbles in the water phase and during the rise of the bubbles up the column. For conditions where the drop volume was $0.12\,\text{cm}^3$ and the velocity of rise 12.5 cm/s, the value of K_w for the period of drop formation was $0.000075\,\text{kmol/s}\,\text{m}^2$ (kmol/m^3), and for the period of rise $0.000046\,\text{kmol/s}\,\text{m}^2$ $(\text{kmol/s}\,\text{m}^3)$.

If these conditions of drop formation and rise are reproduced in a spray tower of 1.8 m in height and $0.04\,\text{m}^2$ cross-sectional area, what is the transfer coefficient, $K_w a$, $\text{kmol/s}\,\text{m}^3$ (kmol/m^3), where a represents the interfacial area in m^2/unit volume of the column? The benzene phase enters at the flow rate of $38\,\text{cm}^3/\text{s}$.

3.9. It is proposed to reduce the concentration of acetaldehyde in aqueous solution from 50% to 5% by mass, by extraction with solvent **S** at 293 K. If a countercurrent multiple contact process is adopted and 0.025 kg/s of the solution is treated with an equal quantity of solvent, determine the number of theoretical stages required and the mass and concentration of the extract from the first stage.

The equilibrium relationship for this system at 293 K is as follows:

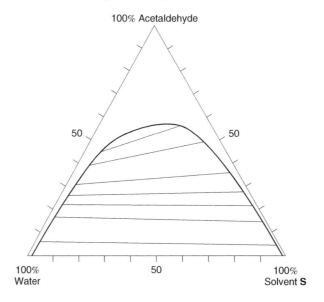

100% Acetaldehyde

50   50

100%   50   100%
Water   Solvent **S**

3.10. $160 \, \text{cm}^3/\text{s}$ of a solvent **S** is used to treat $400 \, \text{cm}^3/\text{s}$ of a 10% by mass solution of **A** in **B**, in a three-stage countercurrent multiple contact liquid–liquid extraction plant. What is the composition of the final raffinate?

Using the same total amount of solvent, evenly distributed between the three stages, what would be the composition of the final raffinate, if the equipment were used in a simple multiple contact arrangement?

Equilibrium data:

kg **A**/kg **B**	0.05	0.10	0.15
kg **A**/kg **S**	0.069	0.159	0.258
Densities (kg/m³):	$\rho_A = 1200$,	$\rho_B = 1000$,	$\rho_C = 800$

3.11. In order to extract acetic acid from a dilute aqueous solution with isopropyl ether, the two immiscible phases are passed countercurrently through a packed column 3 m in length and 75 mm in diameter. It is found that if $0.5 \, \text{kg/m}^2$ of the pure ether is used to extract $0.25 \, \text{kg/m}^2$ s of 4.0% acid by mass, then the ether phase leaves the column with a concentration of 1.0% acid by mass. Calculate:

(a) the number of overall transfer units, based on the raffinate phase, and
(b) the overall extraction coefficient, based on the raffinate phase.

The equilibrium relationship is given by: (kg acid/kg isopropyl ether)$=0.3$ (kg acid/kg water).

3.12. It is proposed to recover material **A** from an aqueous effluent by washing it with a solvent **S** and separating the resulting two phases. The light product phase will contain **A** and the solvent **S** and the heavy phase will contain **A** and water. Show that the most economical solvent rate, W (kg/s) is given by:

$$W = \left[(F^2 ax_0)/mb)\right]^{0.5} - F/m$$

where the feedrate is F kg/s water containing x_0 kg **A**/kg water, the value of **A** in the solvent product phase $=$ £a/kg **A**, the cost of solvent **S** $=$ £b/kg **S** and the equilibrium data are given by:

$$(\text{kg A/kg S})_{\text{product phase}} = m(\text{kg A/kg water})_{\text{water phase}}$$

where a, b and m are constants.

3.13. A solution of C in solvent A (L-phase) is contacted with solvent B (G-phase). The flow rates of the phases (on solute-free basis) are L_s and G_s *kmol*/h; feed concentration (C in A) is X_0 and the solvent B fed to the column is solute-free. The equilibrium relation is linear, $Y = \alpha X$. It is necessary to choose between two devices: a three-stage countercurrent device and a three-stage crosscurrent device.

(a) Which device between a three-stage countercurrent device and a three-stage crosscurrent device will offer better recovery of the solute? Compare by numerical calculation for $A = [L_s/G_s]/\alpha = 0.9$.

(b) If ideal cross-current cascade of infinite number of stages with a differential amount of G_s is added to each stage, what would be the maximum recovery of the solute?

3.14. The equilibrium distribution of a solute C between solvents A and B (up to 30% of C in solution in A) is given by.

$$Y = 3.75\ X.$$

where X and Y are the concentrations of C in A and B respectively, both in the mass ratio unit (i.e. mass of the solute per unit mass of solute-free solvent). The solvents A and B are practically immiscible.

It is required to calculate the amount of the solvent B required to separate 95% of C from 1000 kg of a 15% (by mass) solution of C in A for the following separation schemes:

(a) Ideal single-stage contact.
(b) An ideal three-stage crosscurrent contact, the amount of the solvent used in each stage being equal.
(c) A very large number of crosscurrent contacts using an infinitesimal amount of the solvent in each stage.

4.1. A single-effect evaporator is used to concentrate 7 kg/s of a solution from 10% to 50% of solids. Steam is available at 205 kN/m^2 and evaporation takes place at 13.5 kN/m^2. If the overall heat transfer coefficient is 3 kW/m^2 K, calculate the heating surface required and the amount of steam used if the feed to the evaporator is at 294 K and the condensate leaves the heating space at 352.7 K. The specific heat capacity of a 10% solution is 3.76 kJ/kg K, the specific heat capacity of a 50% solution is 3.14 kJ/kg K.

4.2. A solution containing 10% of caustic soda has to be concentrated to a 35% solution at the rate of 180,000 kg/day during a year of 300 working days. A suitable single-effect evaporator for this purpose, neglecting the condensing plant, costs £1600 and for a multiple-effect evaporator the cost may be taken as £1600 N, where N is the number of effects.

Boiler steam may be purchased at £0.2/1000 kg and the vapour produced may be assumed to be 0.85 N kg/kg of boiler steam. Assuming that interest on capital, depreciation, and other fixed charges amount to 45% of the capital involved per annum, and that the cost of labour is constant and independent of the number of effects employed, determine the number of effects which, based on the data given, will give the maximum economy.

4.3. Saturated steam leaves an evaporator at atmospheric pressure and is compressed by means of saturated steam at 1135 kN/m^2 in a steam jet to a pressure of 135 kN/m^2. If 1 kg of the high pressure steam compresses 1.6 kg of the evaporated atmospheric steam, what is the efficiency of the compressor?

4.4. A single-effect evaporator operates at 13 kN/m^2. What will be the heating surface necessary to concentrate 1.25 kg/s of 10% caustic soda to 41%, assuming a value of the overall heat transfer coefficient U of 1.25 kW/m^2 K, using steam at 390 K? The heating surface is 1.2 m below the liquid level.

The boiling-point rise of solution is 30 K, the feed temperature is 291 K, the specific heat capacity of the feed is 4.0 kJ/kg K, the specific heat capacity of the product is 3.26 kJ/kg K and the density of the boiling liquid is 1390 kg/m^3.

4.5. Distilled water is produced from sea-water by evaporation in a single-effect evaporator, working on the vapour compression system. The vapour produced is compressed by a mechanical compressor of 50% efficiency, and then returned to the calandria of the evaporator. Extra steam, dry and saturated at 650 kN/m^2, is bled into the steam space through a throttling valve. The distilled water is withdrawn as condensate from the steam space. 50% of the sea-water is evaporated in the plant. The energy supplied in addition to that necessary to compress the vapour may be assumed to appear as superheat in the vapour. Using the following data, calculate the quantity of extra steam required in kg/s.

The production rate of distillate is 0.125 kg/s, the pressure in the vapour space is 101.3 kN/m^2, the temperature difference from steam to liquor is 8 K, the boiling point rise of sea-water is 1.1 K and the specific heat capacity of sea-water is 4.18 kJ/kg K.

The sea water enters the evaporator at 344 K through an external heater.

4.6. It is claimed that a jet booster requires 0.06 kg/s of dry and saturated steam at 700 kN/m^2 to compress 0.125 kg/s of dry and saturated vapour from 3.5 to 14.0 kN/m^2. Is this claim reasonable?

4.7. A forward-feed double-effect evaporator, having 10 m^2 of heating surface in each effect, is used to concentrate 0.4 kg/s of caustic soda solution from 10% by mass. During a particular run, when the feed is at 328 K, the pressures in the two calandrias are 375 and 180 kN/m² respectively, while the condenser operates at 15 kN/m². For these conditions, calculate:

(a) the load on the condenser.

(b) the steam economy, and

(c) the overall heat transfer coefficient in each effect.

Would there have been any advantages in using backward feed in this case? Heat losses to the surroundings are negligible.

Physical properties of caustic soda solutions:

Solids concentration (per cent by mass)	Boiling point rise (K)	Specific heat capacity (kJ/kg K)	Heat of dilution (kJ/kg)
10	1.6	3.85	0
20	6.1	3.72	2.3
30	15.0	3.64	9.3
50	41.6	3.22	220

4.8. A 12% glycerol–water mixture is produced as a secondary product in a continuous process plant and flows from the reactor at 4.5 MN/m² and at 525 K. Suggest, with preliminary calculations, a method of concentration to 75% glycerol, in a plant which has no low-pressure steam available.

4.9. A forward feed double-effect vertical evaporator, with equal heating areas in each effect, is fed with 5 kg/s of a liquor of specific heat capacity of 4.18 kJ/kg K, and with no boiling point rise, so that 50% of the feed liquor is evaporated. The overall heat transfer coefficient in the second effect is 75% of that in the first effect. Steam is fed at 395 K and the boiling point in the second effect is 373 K. The feed is heated by an external heater to the boiling point in the first effect.

It is decided to bleed off 0.25 kg/s of vapour from the vapour line to the second effect for use in another process. If the feed is still heated to the boiling point of the first effect by external means, what will be the change in steam consumption of the evaporator unit? For the purpose of calculation, the latent heat of the vapours and of the steam may both be taken as 2230 kJ/kg.

4.10. A liquor containing 15% solids is concentrated to 55% solids in a double-effect evaporator, operating at a pressure in the second effect of 18 kN/m². No crystals are formed. The flow rate of feed is 2.5 kg/s at 375 K with a specific heat capacity of 3.75 kJ/kg K. The boiling-point rise of the concentrated liquor is 6 K and the steam fed to the first effect is at 240 kN/m². The overall heat transfer coefficients in the first and second effects are 1.8 and 0.63 kW/m² K, respectively. If the heat transfer area is to be the same in each effect, what areas should be specified?

4.11. Liquor containing 5% solids is fed at 340 K to a four-effect evaporator. Forward feed is used to give a product containing 28.5% solids. Do the following figures indicate normal operation? If not, why not?

Effect	1	2	3	4
Solids entering (per cent)	5.0	6.6	9.1	13.1
Temperature in steam chest (K)	382	374	367	357.5
Temperature of boiling solution (K)	369.5	364.5	359.6	336.6

4.12. 1.25 kg/s of a solution is concentrated from 10% to 50% solids in a triple-effect evaporator using steam at 393 K and a vacuum such that the boiling point in the last effect is 325 K. If the feed is initially at 297 K and backward feed is used, what is the steam consumption, the temperature distribution in the system and the heat transfer area in each effect, each effect being identical?

For the purpose of calculation, it may be assumed that the specific heat capacity is 4.18 kJ/kg K, that there is no boiling point rise, and that the latent heat of vapourisation is constant at 2330 kJ/kg over the temperature range in the system. The overall heat transfer coefficients may be taken as 2.5, 2.0 and 1.6 kW/m^2 K in the first, second and third effects, respectively.

4.13. A triple-effect evaporator concentrates a liquid with no appreciable elevation of boiling point. If the temperature of the steam to the first effect is 395 K, and vacuum is applied to the third effect so that the boiling point is 325 K, what are the approximate boiling points in the three effects? The overall heat transfer coefficients may be taken as 3.1, 2.3, 1.3 kW/m^2 K in the three effects, respectively.

4.14. A three-stage evaporator is fed with 1.25 kg/s of a liquor which is concentrated from 10% to 40% solids by mass. The heat transfer coefficients may be taken as 3.1, 2.5, and 1.7 kW/m^2 K, respectively, in each effect. Calculate the steam flow at 170 kN/m^2 and the temperature distribution in the three effects, if:

(a) the feed is at 294 K, and
(b) the feed is at 355 K.

Forward feed is used in each case and the values of U are the same for the two systems. The boiling point in the third effect is 325 K and the liquor has no boiling point rise.

4.15. An evaporator operating on the thermo-recompression principle employs a steam ejector to maintain atmospheric pressure over the boiling liquid. The ejector uses 0.14 kg/s of steam at 650 kN/m^2, and superheated by 100 K, and produces a pressure in the steam chest of 205 kN/m^2. A condenser removes surplus vapour from the atmospheric pressure line. What is the capacity and economy of the system, and how could the economy be improved?

The feed enters the evaporator at 293 K and the concentrated liquor is withdrawn at the rate of 0.025 kg/s. The concentrated liquor exhibits a boiling point rise of 10 K. Heat losses to the surroundings are negligible.

For the ejector, the nozzle efficiency is 0.95, the efficiency of momentum transfer is 0.80 and the efficiency of compression is 0.90.

4.16. A single-effect evaporator is used to concentrate 0.075 kg/s of a 10% caustic soda liquor to 30%. The unit employs forced circulation in which the liquor is pumped through the vertical tubes of the calandria which are 32 mm o.d. by 28 mm i.d., and 1.2 m long. Steam is supplied at 394 K, dry and saturated, and the boiling point rise of the 30% solution is 15 K. If the overall heat transfer coefficient is 1.75 kW/m^2 K, how many tubes should be used, and what material of construction would be specified for the evaporator? The latent heat of vapourisation under these conditions is 2270 kJ/kg.

4.17. A steam-jet booster compresses 0.1 kg/s of dry and saturated vapour from 3.4 kN/m^2 to 13.4 kN/m^2. High pressure steam consumption is 0.05 kg/s at 700 kN/m^2. (a) What must be the condition of the high pressure steam for the booster discharge to be superheated through 20 K? (b) What is the overall efficiency of the booster if the compression efficiency is 100%?

4.18. A triple-effect backward-feed evaporator concentrates 5 kg/s of liquor from 10% to 50% solids. Steam is available at 375 kN/m^2 and the condenser operates at 13.5 kN/m^2. What is the area required in each effect, assumed equal, and the economy of the unit?

The specific heat capacity is 4.18 kJ/kg K at all concentrations and there is no boiling-point rise. The overall heat transfer coefficients are 2.3, 2.0 and 1.7 kW/m^2 K respectively in the three effects, and the feed enters the third effect at 300 K.

4.19. A double-effect climbing film evaporator is connected so that the feed passes through two preheaters, one heated by vapour from the first effect and the other by vapour from the second effect. The condensate from the first effect is passed into the steam space of the second. The temperature of the feed is initially 289, 348 K after the first heater and 383 K and after the second heater. The vapour temperature in the first effect is 398 K and in the second 373 K. The flow rate of feed is 0.25 kg/s and the steam is dry and saturated at 413 K. What is the economy of the unit if the evaporation rate is 0.125 kg/s?

4.20. A triple-effect evaporator is fed with 5 kg/s of a liquor containing 15% solids. The concentration in the last effect, which operates at 13.5 kN/m^2, is 60% solids. If the overall heat transfer coefficients are 2.5, 2.0 and 1.1 kW/m^2 K, respectively, and the steam is fed at 388 K to the first effect, determine the temperature distribution and the area of heating surface required in each effect, assuming the calandrias are identical. What is the economy and what is the heat load on the condenser? The feed temperature is 294 K and the specific heat capacity of all liquors is 4.18 kJ/kg K.

If the unit is run as a backward-feed system, in which the coefficients are 2.3, 2.0 and 1.6 kW/m^2 K, determine the new temperatures, the heat economy and the heating surface required under these conditions.

4.21. A double-effect forward-feed evaporator is required to give a product which contains 50.0% by mass of solids. Each effect has 10 m^2 of heating surface and the heat transfer coefficients are 2.8 and 1.7 kW/m^2 K in the first and second effects respectively. Dry and saturated steam is available at $375\,kN/m^2$ and the condenser operates at $13.5\,kN/m^2$. The concentrated solution exhibits a boiling point rise of 3 K. What is the maximum permissible feed rate if the feed contains 10% solids and is at a temperature of 310 K? The latent heat of vapourisation is $2330\,kJ/kg$ and the specific heat capacity is $4.18\,kJ/kg\,K$ under all conditions.

4.22. For the concentration of fruit juice by evaporation, it is proposed to use a falling-film evaporator and to incorporate a heat-pump cycle with ammonia as the medium. The ammonia in vapour form enters the evaporator at 312 K and the water is evaporated from the juices at 287 K. The ammonia in the vapour–liquid mixture enters the condenser at 278 K and the vapour then passes to the compressor. It is estimated that the work in compressing the ammonia is $150\,kJ/kg$ of ammonia and that 2.28 kg of ammonia is cycled/kilogram of water evaporated. The following proposals are made for driving the compressor:

(a) To use a diesel engine drive taking 0.4 kg of fuel/MJ. The calorific value of the fuel is 42 MJ/kg, and the cost £0.02/kg.
(b) To pass steam, costing £0.01/10 kg, through a turbine which operates at 70% isentropic efficiency, between 700 and $101.3\,kN/m^2$.

Explain by means of a diagram how this plant will work, illustrating all necessary major items of equipment. Which method for driving the compressor is to be preferred?

4.23. A double-effect forward-feed evaporator is required to give a product consisting of 30% crystals and a mother liquor containing 40% by mass of dissolved solids. Heat transfer coefficients are 2.8 and $1.7\,kW/m^2\,K$ in the first and second effects, respectively. Dry saturated steam is supplied at $375\,kN/m^2$ and the condenser operates at $13.5\,kN/m^2$.

(a) What area of heating surface is required in each effect assuming the effects are identical, if the feed rate is 0.6 kg/s of liquor, containing 20% by mass of dissolved solids, and the feed temperature is 313 K?
(b) What is the pressure above the boiling liquid in the first effect?

The specific heat capacity may be taken as constant at $4.18\,kJ/kg\,K$, and the effects of boiling-point rise and of hydrostatic head may be neglected.

4.24. 1.9 kg/s of a liquid containing 10% by mass of dissolved solids is fed at 338 K to a forward-feed double-effect evaporator. The product consists of 25% by mass of solids and a mother liquor containing 25% by mass of dissolved solids. The steam fed to the first effect is dry and saturated at $240\,kN/m^2$ and the pressure in the second effect is $20\,kN/m^2$. The specific heat capacity of the solid may be taken as $2.5\,kJ/kg\,K$ both in solid form and in solution, and the heat of solution may be neglected. The mother liquor exhibits a boiling-point rise of 6 K. If the two effects are identical, what area is required if the heat transfer coefficients in the first and second effects are 1.7 and $1.1\,kW/m^2K$, respectively?

4.25. 2.5 kg/s of a solution at 288 K containing 10% of dissolved solids is fed to a forward-feed double-effect evaporator, operating at $14\,kN/m^2$ in the last effect. If the product is to consist of a liquid containing 50% by mass of dissolved solids and dry saturated steam is fed to the steam coils, what should be the pressure of the steam? The surface in each effect is $50\,m^2$ and the coefficients for heat transfer in the first and second effects are 2.8 and $1.7\,kW/m^2\,K$, respectively. It may be assumed that the concentrated solution exhibits a boiling-point rise of 5 K, that the latent heat has a constant value of 2260 kJ/kg and that the specific heat capacity of the liquid stream is constant at 3.75 kJ/kg K.

4.26. A salt solution at 293 K is fed at the rate of 6.3 kg/s to a forward-feed triple-effect evaporator and is concentrated from 2% to 10% of solids. Saturated steam at $170\,kN/m^2$ is introduced into the calandria of the first effect and a pressure of $34\,kN/m^2$ is maintained in the last effect. If the heat transfer coefficients in the three effects are 1.7, 1.4 and $1.1\,kW/m^2\,K$, respectively, and the specific heat capacity of the liquid is approximately 4 kJ/kg K, what area is required if each effect is identical? Condensate may be assumed to leave at the vapour temperature at each stage, and the effects of boiling point rise may be neglected. The latent heat of vapourisation may be taken as constant throughout.

4.27. A single-effect evaporator with a heating surface area of $10\,m^2$ is used to concentrate NaOH solution at 0.38 kg/s from 10% to 33.33% by mass. The feed enters at 338 K and its specific heat capacity is 3.2 kJ/kg K. The pressure in the vapour space is $13.5\,kN/m^2$ and 0.3 kg/s of steam is used from a supply at 375 K. Calculate:

(a) The apparent overall heat transfer coefficient.
(b) The coefficient corrected for boiling point rise of dissolved solids.
(c) The corrected coefficient if the depth of liquid is 1.5 m.

4.28. An evaporator, working at atmospheric pressure, is to concentrate a solution from 5% to 20% solids at the rate of 1.25 kg/s. The solution, which has a specific heat capacity of 4.18 kJ/kg K, is fed to the evaporator at 295 K and boils at 380 K. Dry saturated steam at $240\,kN/m^2$ is fed to the calandria, and the condensate leaves at the temperature of the condensing stream. If the heat transfer coefficient is $2.3\,kW/m^2\,K$, what is the required area of heat transfer surface and how much steam is required? The latent heat of vapourisation of the solution may be taken as being equal to that of water.

5.1. A saturated solution containing 1500 kg of potassium chloride at 360 K is cooled in an open tank to 290 K. If the density of the solution is $1200\,kg/m^3$, the solubility of potassium chloride is 53.55 kg/100 kg water at 360 K and 34.5 at 290 K calculate:

(a) the capacity of the tank required, and
(b) the mass of crystals obtained, neglecting loss of water by evaporation.

5.2. Explain how fractional crystallisation may be applied to a mixture of sodium chloride and sodium nitrate, given the following data. At 293 K, the solubility of sodium chloride is 36 kg/100 kg water and of sodium nitrate 88 kg/100 kg water. While at this temperature, a saturated solution comprising both salts will contain

25 kg sodium chloride and 59 kg sodium nitrate/100 kg of water. At 373 K, these values, again per 100 kg of water, are 40 and 176, and 17 and 160 kg, respectively.

5.3. 10 Mg (10 t) of a solution containing 0.3 kg Na_2CO_3/kg solution is cooled slowly to 293 K to form crystals of $Na_2CO_3.10H_2O$. What is the yield of crystals if the solubility of Na_2CO_3 at 293 K is 21.5 kg/100 kg water and during cooling 3% of the original solution is lost by evaporation?

5.4. The heat required when 1 kmol of $MgSo_4.7H_2O$ is absorbed isothermally at 291 K in a large mass of water is 13.3 MJ. What is the heat of crystallisation per unit mass of the salt?

5.5. A solution of 500 kg of Na_2SO_4 in 2500 kg water is cooled from 333 K to 283 K in an agitated mild steel vessel of mass 750 kg. At 283 K, the solubility of the anhydrous salt is 8.9 kg/100 kg water and the stable crystalline phase is $Na_2SO_4.10H_2O$. At 291 K, the heat of solution is -78.5 MJ/kmol and the heat capacities of the solution and mild steel are 3.6 and 0.5 kJ/kg K, respectively. If, during cooling, 2% of the water initially present is lost by evaporation, estimate the heat which must be removed.

5.6. A batch of 1500 kg of saturated potassium chloride solution is cooled from 360 K to 290 K in an unagitated tank. If the solubilities of KCl are 53 and 34 kg/100 kg water at 360 K and 290 K respectively and water losses due to evaporation may be neglected, what is the yield of crystals?

5.7. Glauber's salt, $Na_2SO_4 \cdot 10H_2O$, is to be produced in a Swenson–Walker crystalliser by cooling to 290 K a solution of anhydrous Na_2SO_4 which is saturated at 300 K. If cooling water enters and leaves the unit at 280 and 290 K, respectively, and evaporation is negligible, how many sections of crystalliser, each 3 m long, will be required to process 0.25 kg/s of the product? The solubilities of anhydrous Na_2SO_4 in water are 40 and 14 kg/100 kg water at 300 and 290 K respectively, the mean heat capacity of the liquor is 3.8 kJ/kg K and the heat of crystallisation is 230 kJ/kg. For the crystalliser, the available heat transfer area is $3 m^2$/m length, the overall coefficient of heat transfer is 0.15 kW/m^2 K, and the molecular weights are $Na_2SO_4 \cdot 10H_2O = 322$ kg/kmol and $Na_2SO_4 = 142$ kg/kmol.

5.8. What is the evaporation rate and yield of $CH_3COONa.3H_2O$ from a continuous evaporative-crystalliser operating at 1 kN/m^2 when it is fed with 1 kg/s of a 50% by mass aqueous solution of sodium acetate at 350 K? The boiling-point elevation of the solution is 10 K and the heat of crystallisation is 150 kJ/kg. The mean heat capacity of the solution is 3.5 kJ/kg K and, at 1 kN/m^2, water boils at 280 K at which the latent heat of vapourisation is 2.482 MJ/kg. Over the range 270–305 K, the solubility of sodium acetate in water s at T (K) is given approximately by:

$$s = 0.61T - 132.4 \text{ kg/100 kg water.}$$

Molecular weights : $CH_3COONa.3H_2O = 136$ kg/kmol, $CH_3COONa = 82$ kg/kmol.

6.1. A wet solid is dried from 25% to 10% moisture, under constant drying conditions in 15 ks (4.17b). If the equilibrium moisture content is 5% and the critical moisture content is 15%, how long will it take to dry to 8% moisture under the same conditions?

6.2. Strips of material 10 mm thick are dried under constant drying conditions from 28% to 13% moisture in 25 ks. If the equilibrium moisture content is 7%, what is the time taken to dry 60 mm planks from 22% to 10% moisture under the same conditions, assuming no loss from the edges? All moistures are expressed on the wet basis. The relation between E, the ratio of the average free moisture content at time t to the initial free moisture content, and the parameter J is given by:

E	1	0.64	0.49	0.38	0.295	0.22	0.14
J	0	0.1	0.2	0.3	0.5	0.6	0.7

It may be noted that $J = kt/l^2$, where k is a constant, t is the time in ks and $2L$ is the thickness of the sheet of material in mm.

6.3. A granular material containing 40% moisture is fed to a countercurrent rotary dryer at 295 K and is withdrawn at 305 K containing 5% moisture. The air supplied, which contains 0.006 kg water vapour/kg of dry air, enters at 385 K and leaves at 310 K. The dryer handles 0.125 kg/s wet stock.

Assuming that radiation losses amount to 20 kJ/kg of dry air used, determine the mass flow rate of dry air supplied to the dryer and the humidity of the outlet air.

The latent heat of water vapour at 295 K is 2449 kJ/kg, the specific heat capacity of dried material is 0.88 kJ/kg K, the specific heat capacity of dry air is 1.00 kJ/kg K and the specific heat capacity of water vapour is 2.01 kJ/kg K.

6.4. 1 Mg (1 t) of dry mass of a non-porous solid is dried under constant drying conditions with air at a velocity of 0.75 m/s. The area of surface drying is 55 m^2. If the initial rate of drying is 0.3 g/m^{2}s, how long will it take to dry the material from 0.15 to 0.025 kg water/kg dry solid? The critical moisture content of the material may be taken as 0.125 kg water/kg dry solid. If the air velocity were increased to 4.0 m/s, what would be the anticipated saving in time if the process is surface-evaporation controlled?

6.5. A 100 kg batch of granular solids containing 30% of moisture is to be dried in a tray dryer to 15.5% of moisture by passing a current of air at 350 K tangentially across its surface at the velocity of 1.8 m/s. If the constant rate of drying under these conditions is 0.7 g/s m^2 and the critical moisture content is 15%, calculate the approximate drying time. It may be assumed that the area of the drying surface is 0.03 m^2/kg dry mass.

6.6. A flow of 0.35 kg/s of a solid is to be dried from 15% to 0.5% moisture based on a dry basis. The mean heat capacity of the solids is 2.2 kJ/kg K. It is proposed that a co-current adiabatic dryer should be used with the solids entering at 300 K and, because of the heat sensitive nature of the solids, leaving at 325 K. Hot air is available at 400 K with a humidity of 0.01 kg/kg dry air, and the maximum allowable mass velocity of the air is 0.95 kg/m^{2}s. What diameter and length should be specified for the proposed dryer?

6.7. 0.126 kg/s of a product containing 4% water is produced in a dryer from a wet feed containing 42% water on a wet basis. Ambient air at 294 K and 40% relative humidity is heated to 366 K in a preheater before entering the dryer which it leaves at 60% relative humidity. Assuming that the dryer operates adiabatically, what is the amount of air supplied to the preheater and the heat required in the preheater?

How will these values be affected if the air enters the dryer at 340 K and sufficient heat is supplied within the dryer so that the air leaves also at 340 K and again with a relative humidity of 60%?

6.8. A wet solid is dried from 40% to 8% moisture in 20 ks. If the critical and the equilibrium moisture contents are 15% and 4%, respectively, how long will it take to dry the solid to 5% moisture under the same conditions? All moisture contents are on a dry basis.

6.9. A solid is to be dried from 1 kg water/kg dry solids to 0.01 kg water/kg dry solids in a tray dryer consisting of a single tier of 50 trays, each 0.02 m deep and 0.7 m square completely filled with wet material. The mean air temperature is 305 K and the relative humidity across the trays may be taken as constant at 10%. The mean air velocity is 2.0 m/s and the convective coefficient of heat transfer is given by:

$$h_c = 14.3 G'^{0.8} \qquad W/m^2 \ deg \ K$$

where G' is the mass velocity of the air in kg/m²s. The critical and equilibrium moisture contents of the solid are 0.3 and 0 kg water/kg dry solids respectively and the density of the solid is 6000 kg/m³. Assuming that the drying is by convection from the top surface of the trays only, what is the drying time?

6.10. Skeins of a synthetic fibre are dried from 46% to 8.5% moisture on a wet basis in a 10 m long tunnel dryer by a countercurrent flow of hot air. The air mass velocity, G, is 1.36 kg/m²s and the inlet conditions are 355 K and a humidity of 0.03 kg moisture/kg dry air. The air temperature is maintained at 355 K throughout the dryer by internal heating and, at the outlet, the humidity of the air is 0.08 kg moisture/kg dry air. The equilibrium moisture content is given by:

$$w_e = 0.25 \ (\text{per cent relative humidity})$$

and the drying rate by:

$$R = 1.34 \times 10^{-4} G'^{1.47} (w - w_c)(\mathcal{H}_w - \mathcal{H}) \quad kg/s \ kg \ dry \ fibres$$

where $\mathcal{H}$ is the humidity of the dry air and $\mathcal{H}_w$ the saturation humidity at the wet bulb temperature. Data relating w, $\mathcal{H}$ and $\mathcal{H}_w$ are as follows:

w (kg/kg dry fibre)	$\mathcal{H}$ (kg/kg dry air)	$\mathcal{H}_w$ (hg/hg dry air)	Relative humidity (per cent)
0.852	0.080	0.095	22.4
0.80	0.0767	0.092	21.5
0.60	0.0635	0.079	18.2
0.40	0.0503	0.068	14.6
0.20	0.0371	0.055	11.1
0.093	0.030	0.049	9.0

At what velocity should the skeins be passed through the dryer?

7.1. Spherical particles of 15 nm diameter and density of 2290 kg/m^3 are pressed together to form a pellet. The following equilibrium data were obtained for the sorption of nitrogen at 77 K. Obtain estimates of the surface area of the pellet from the adsorption isotherm and compare the estimates with the geometric surface. The density of liquid nitrogen at 77 K is 808 kg/m^3.

P/P^0	0.1	0.2	0.3	0.4	0.5	0.6	0.7	0.8	0.9
m^3 liquid N$_2 \times 10^6$/ kg solid	66.7	75.2	83.9	93.4	108.4	130.0	150.2	202.0	348.0

where P is the pressure of the sorbate in the gas and P^0 is its vapour pressure at 77 K.

7.2. A 1 m^3 volume of a mixture of air and acetone vapour is at a temperature of 303 K and a total pressure of 100 kN/m^2. If the relative saturation of the air by acetone is 40%, how much activated carbon must be added to the space in order to reduce the value to 5% at 303 K?

If 1.6 kg carbon is added, what is relative saturation of the equilibrium mixture assuming the temperature to be unchanged?

The vapour pressure of acetone at 303 K is 37.9 kN/m^2 and the adsorption equilibrium data for acetone on carbon at 303 K are:

Partial pressure acetone $\times 10^{-2}$ (N/m^2)	0	5	10	30	50	90
x_r (kg acetone/kg carbon)	0	0.14	0.19	0.27	0.31	0.35

7.3. A solvent, contaminated with 0.03 kmol/m^3 of a fatty acid, is to be purified by passing it through a fixed bed of activated carbon to adsorb the acid but not the solvent. If the operation is essentially isothermal and equilibrium is maintained between the liquid and the solid, calculate the length of a bed of 0.15 m diameter to give one hour's operation when the fluid is fed at 1×10^{-4} m^3/s. The bed is free of adsorbate initially and the intergranular voidage is 0.4. Use an equilibrium, fixed-bed theory to obtain the length for three types of isotherm:

(a) $C_s = 10\ C$.
(b) $C_s = 3.0\ C^{0.3}$ (use the mean slope).
(c) $C_s = 10^4\ C^2$ (the breakthrough concentration is 0.003 kmol/m^3).

C and C_s refer to concentrations in kmol/m^3 in the gas phase and the absorbent, respectively.

7.4 Laboratory test data on a newly developed adsorbent material show the material to be quite attractive for the removal of colour from wastewater from a textile dying plant. The following equilibrium relation was derived from the test data

$Y = 17.08\,X^{0.48}$. where $Y=$ kg colouring matter adsorbed per kg of the adsorbent, and $X=$ kg colour per kg water.

(a) What per cent of the colouring matter will be removed if 100 kg of the wastewater containing 1 part colouring matter per 100 part of water is treated with 1 kg of the adsorbent in the following ways?

    (i) The entire adsorbent is used in a single ideal batch.

    (ii) Half of the adsorbent is used to treat the solution in an ideal batch, the adsorbent is removed and the water is further treated with the rest of the adsorbent again.

(b) Calculate the quantity of adsorbent required to remove 99% of the colour from 100 kg solution containing 0.8% of colour by weight in

    (i) a single ideal batch.

    (ii) a very large number of 'ideal crosscurrent contact' in which a differential amount (infinitesimal amount) of the absorbent is added to each stage.

7.5 An adsorbent, which is a modified clay, is used to separate an organic compound A from an aqueous solution. One kilogram of the solution containing 10% organic was treated with varying amounts of clay in several laboratory tests. The following data was collected.

Gram clay used	15	40	60	100	135	210	325	450
% A in the solution in equilibrium	9.1	7.81	6.93	5.66	4.76	3.38	2.44	1.48

(a) Do the test data fit the Freundlich adsorption isotherm in the form
    $Y = \alpha X^{\beta}$, where

$X =$ gram solute adsorbed per gram clay, and $Y =$ gram solute in solution per gram solute-free solvent?

(b) How much of the adsorbent is required to recover 90% of the solute from 1000 kg of 10% solution?

(c) If 628 kg of the adsorbent is used per 1000 kg of the solution, how much of the solute is recovered if the treatment is done in two stages, using half of the clay in each stage?

8.1 A solution is passed over a single pellet of resin and the temperature is maintained constant. The take-up of exchanged ion is followed automatically and the following results are obtained.

t (min)	2	4	10	20	40	60	120
x_r (kg/kg)	0.091	0.097	0.105	0.113	0.125	0.128	0.132

On the assumption that the resistance to mass transfer in the external film is negligible, predict the values of x_r, the mass of sorbed phase per unit mass of resin, as a function of time t, for a pellet of a resin twice the radius.

9.1. Describe the principle of separation involved in elution chromatography and derive the retention equation

$$t_R = t_M \left[1 + \left(\frac{1 - \varepsilon}{\varepsilon} \right) K \right]$$

where t_R is the retention time of solute in the column, t_M is the mobile phase hold-up time in the column. ε is the packing voidage and K is the distribution coefficient.

9.2. In chemical analysis, chromatography may permit the separation of more than a hundred components from a mixture in a single run. Explain why chromatography offers such a large separating power. In production chromatography, the complete separation of a mixture containing more than a few components is likely to involve two or three columns for optimum economic performance. Why is this?

9.3. By using the chromatogram in Fig. 19.3, show that $k_1'=3.65$, $k_2'=4.83$, $\alpha=1.32$, $R_s=1.26$, and $N=500$. Show also that if $\varepsilon=0.8$ and $L=1.0$m, then $K_1=14.6$, $K_2=19.3$, and $H=2.0$mm, where R is the retention ratio, R_s is the resolution, d is the obstruction factor, H is the plate height, K_1 and K_2 are the distribution coefficients, k_1' and k_2' are the capacity factors, ε is the packing voidage, L is the length of the column, and N is the number of theoretical plates. Calculate the ratio of plate height to particle diameter to confirm that the column is inefficient, as might be anticipated from the wide bands in Fig. 19.3. It may be assumed that the particle size is that of a typical GC column as given in Table 19.3.

9.4. Suggest one or more types of chromatography to separate each of the following mixtures:

(a) α- and β-pinenes
(b) blood serum proteins
(c) hexane isomers
(d) purification of cefonicid, a synthetic β-lactam antibiotic.

Index

Note: Page numbers followed by *f* indicate figures, *t* indicate tables, and *b* indicate boxes.